Immunodiagnosis of Cancer

IMMUNOLOGY SERIES

Editor-in-Chief
NOEL R. ROSE
Professor and Chairman
Department of Immunology and
  Infectious Diseases
The Johns Hopkins University
School of Hygiene and Public Health
Baltimore, Maryland

European Editor
ZDENEK TRNKA
Basel Institute for
  Immunology
Basel, Switzerland

1. Mechanisms in Allergy: Reagin-Mediated Hypersensitivity
   Edited by Lawrence Goodfriend, Alec Sehon and Robert P. Orange
2. Immunopathology: Methods and Techniques
   Edited by Theodore P. Zacharia and Sidney S. Breese, Jr.
3. Immunity and Cancer in Man: An Introduction
   Edited by Arnold E. Reif
4. *Bordetella pertussis:* Immunological and Other Biological Activities
   J.J. Munoz and R.K. Bergman
5. The Lymphocyte: Structure and Function (in two parts)
   Edited by John J. Marchalonis
6. Immunology of Receptors
   Edited by B. Cinader
7. Immediate Hypersensitivity: Modern Concepts and Development
   Edited by Michael K. Bach
8. Theoretical Immunology
   Edited by George I. Bell, Alan S. Perelson, and George H. Pimbley, Jr.
9. Immunodiagnosis of Cancer (in two parts)
   Edited by Ronald B. Herberman and K. Robert McIntire
10. Immunologically Mediated Renal Diseases: Criteria for Diagnosis and Treatment
    Edited by Robert T. McCluskey and Giuseppe A. Andres
11. Clinical Immunotherapy
    Edited by Albert F. LoBuglio
12. Mechanisms of Immunity to Virus-Induced Tumors
    Edited by John W. Blasecki
13. Manual of Macrophage Methodology: Collection, Characterization, and Function
    Edited by Herbert B. Herscowitz, Howard T. Holden, Joseph A. Bellanti, and Abdul Ghaffar
14. Suppressor Cells in Human Disease
    Edited by James S. Goodwin
15. Immunological Aspects of Aging
    Edited by Diego Segre and Lester Smith
16. Cellular and Molecular Mechanisms of Immunologic Tolerance
    Edited by Tomáš Hraba and Milan Hašek

17. Immune Regulation: Evolution and Biological Significance
    Edited by Laurens N. Ruben and M. Eric Gershwin
18. Tumor Immunity in Prognosis: The Role of Mononuclear Cell Infiltration
    Edited by Stephen Haskill
19. Immunopharmacology and the Regulation of Leukocyte Function
    Edited by David R. Webb
20. Pathogenesis and Immunology of Treponemal Infection
    Edited by Ronald F. Schell and Daniel M. Musher
21. Macrophage-Mediated Antibody-Dependent Cellular Cytotoxicity
    Edited by Hillel S. Koren
22. Molecular Immunology: A Textbook
    Edited by M. Zouhair Atassi, Carel J. van Oss, and Darryl R. Absolom
23. Monoclonal Antibodies and Cancer
    Edited by George L. Wright, Jr.
24. Stress, Immunity, and Aging
    Edited by Edwin L. Cooper
25. Immune Modulation Agents and Their Mechanisms
    Edited by Richard L. Fenichel and Michael A. Chirigos
26. Mononuclear Phagocyte Biology
    Edited by Alvin Volkman
27. The Lactoperoxidase System: Chemistry and Biological Significance
    Edited by Kenneth M. Pruitt and Jorma O. Tenovuo
28. Introduction to Medical Immunology
    Edited by Gabriel Virella, Jean-Michel Goust, H. Hugh Fudenberg, and Christian C. Patrick
29. Handbook of Food Allergies
    Edited by James C. Breneman
30. Human Hybridomas: Diagnostic and Therapeutic Applications
    Edited by Anthony J. Strelkauskas
31. Aging and the Immune Response: Cellular and Humoral Aspects
    Edited by Edmond A. Goidl
32. Complications of Organ Transplantation
    Edited by Luis H. Toledo-Pereyra
33. Monoclonal Antibody Production Techniques and Applications
    Edited by Lawrence B. Schook
34. Fundamentals of Receptor Molecular Biology
    Donald F. H. Wallach
35. Recombinant Lymphokines and Their Receptors
    Edited by Steven Gillis
36. Immunology of the Male Reproductive Organs
    Edited by Pierluigi E. Bigazzi

37. The Lymphocyte: Structure and Function
    Edited by John J. Marchalonis
38. Differentiation Antigens in Lymphohemopoietic Tissues
    Edited by Masayuki Miyasaka and Zdenek Trnka
39. Cancer Diagnosis In Vitro Using Monoclonal Antibodies
    Edited by Herbert Z. Kupchik
40. Biological Response Modifiers and Cancer Therapy
    Edited by J. W. Chiao
41. Cellular Oncogene Activation
    Edited by George Klein
42. Interferon and Nonviral Pathogens
    Edited by Gerald I. Byrne and Jenifer Turco
43. Human Immunogenetics: Basic Principles and Clinical Relevance
    Edited by Stephen D. Litwin, with David W. Scott, Lorraine Flaherty, Ralph A. Reisfeld, and Donald M. Marcus
44. AIDS: Pathogenesis and Treatment
    Edited by Jay A. Levy
45. Cell Surface Antigen Thy-1: Immunology, Neurology, and Therapeutic Applications
    Edited by Arnold E. Reif and Michael Schlesinger
46. Immune Mechanisms in Cutaneous Disease
    Edited by David A. Norris
47. Immunology of Fungal Diseases
    Edited by Edouard Kurstak
48. Adoptive Cellular Immunotherapy of Cancer
    Edited by Henry C. Stevenson
49. Colony-Stimulating Factors: Molecular and Cellular Biology
    Edited by T. Michael Dexter, John M. Garland, and Nydia G. Testa
50. Introduction to Medical Immunology, Second Edition
    Edited by Gabriel Virella, Jean-Michel Goust, and H. Hugh Fudenberg
51. Tumor Suppressor Genes
    Edited by George Klein
52. Organ-Specific Autoimmunity
    Edited by Pierluigi E. Bigazzi, Georg Wick, and Konrad Wicher
53. Immunodiagnosis of Cancer, Second Edition
    Edited by Ronald B. Herberman and Donald W. Mercer

Additional Volumes in Preparation

Immunodiagnosis of Cancer

SECOND EDITION

edited by

RONALD B. HERBERMAN

University of Pittsburgh School of Medicine
and
Pittsburgh Cancer Institute
Pittsburgh, Pennsylvania

DONALD W. MERCER

Montefiore Hospital
and
Pittsburgh Cancer Institute
Pittsburgh, Pennsylvania

MARCEL DEKKER, INC.   New York • Basel • Hong Kong

Library of Congress Cataloging-in-Publication Data

Immunodiagnosis of cancer / edited by Ronald B. Herberman and Donald W. Mercer. — 2nd ed.
    p.   cm. — (Immunology series : v. 53)
    Includes bibliographical references.
    Includes index.
    ISBN 0-8247-8299-2
    1. Cancer––Immunodiagnosis. 2. Tumor antigens – Analysis.
3. Tumor proteins–Analysis.   I. Herberman, Ronald B.
II. Mercer, Donald W.   III. Series.
    [DNLM: 1. Immunologic Tests. 2. Neoplasms–diagnosis. W1 IM53K
v. 53 / QZ 241 I33]
RC 270.3.I45I46   1990
616.99'40756–dc20
DNLM/DLC
for Library of Congress   90-3889
                                                                                          CIP

This book is printed on acid-free paper

Copyright © 1990 by MARCEL DEKKER, INC. All Rights Reserved

Neither this book nor any part may be reproduced or transmitted in any form or by any means, electronic or mechanical, including photocopying, microfilming, and recording, or by any information storage and retrieval system, without permission in writing from the publisher.

MARCEL DEKKER, INC.
270 Madison Avenue, New York, New York 10016

Current printing (last digit):
10 9 8 7 6 5 4 3 2 1

PRINTED IN THE UNITED STATES OF AMERICA

Series Introduction

What has happened in the field of cancer immunodiagnosis during the past decade to warrant a second edition of *Immunodiagnosis of Cancer*? Quantitatively, the number of "tumor markers" has increased and their clinical interpretation has been more finely honed. By themselves, these two factors would amply justify an updating. In addition, however, a comparison of the second with the first edition makes it clear that a profound change in thinking has occurred.

Experience accumulated in the last 11 years has shown that few, if any, constituents are entirely unique for cancer cells and that cancer must be thought of as a shift in the antigenic composition of normal cells. Yet these changes are often highly characteristic and, for this reason, useful indicators diagnostically. More frequently, the main clinical value is the monitoring of tumor growth or recurrence. The value of immunological tools in many malignant diseases is now so well accepted that it no longer needs special justification.

Cancer immunology is still in its infancy as far as fundamental research is concerned. Empiricism, rigorously controlled and validated, has brought us this far. Now the time has come to look more deeply into the inner workings of the proliferating cells for the causes of unchecked growth with the aid of immunodiagnosis.

This second edition of *Immunodiagnosis of Cancer* is a treasure house of present-day knowledge and a promise of the new age of intervention to come.

Noel R. Rose
Series Editor

Preface to the Second Edition

It has been over ten years since the publication of the first edition of *Immunodiagnosis of Cancer*. At that time, very few tumor markers, mainly carcinoembryonic antigen, human chorionic gonadotropin, and alpha-fetoprotein, had found their way into large-scale clinical investigations and applications. Even for those initial tumor markers, there were only limited insights into their molecular characteristics. The field of immunodiagnosis has progressed in many important ways and this new edition reflects these advances. Many new markers for an increasingly wide array of tumor types have been discovered and characterized. The advent of monoclonal antibody technology has provided many reagents and assays with a high degree of specificity and reproducibility. The majority of the chapters in the second edition are focused on the current status of markers for most of the major types of human malignancy. In addition to having a much wider panel of immunodiagnostic assays, there also is much better understanding of the practical value of the assays and their limitations.

A fundamental biological problem has continued to limit the value of immunodiagnostic assays. It is now fully appreciated that there are very few truly tumor-specific antigens. Rather, careful characterization of almost all monoclonal antibodies and other reagents has indicated reactivity with some normal cells, and most putative tumor-associated antigens have eventually been found to be some type of differentiation antigen or oncogene product, expressed at least in small amounts in nonmalignant or even some normal cells.

Another fundamental biological limitation that has become increasingly appreciated is that circulating levels of almost all tumor markers become detectable only when there are substantial numbers of tumor cells, and often only when there is invasion of the tumor into normal tissues or metastatic spread. Some of the chapters describe various approaches that have been developed to circumvent this problem and procedures that are not subject to this limitation. These include testing for tumor markers in body fluids or in cells obtained directly from the region of tumor growth, assessing tumor specimens themselves for the presence of prognostically useful markers, and in vivo imaging with radiolabeled monoclonal antibodies. to detect occult primary or metastatic lesions.

In any event, for a variety of reasons it is quite rare for immunodiagnostic assays to be useful in the detection of cancer, even among individuals in high-risk groups. Immunodiagnostic approaches have also failed in most situations to help in the initial diagnosis of cancer. Despite these major limitations, immunodiagnostic procedures are being shown to be very useful tools in assessing prognosis, helping to determine the extent and type of therapy to be administered, and aiding in the management and follow-up of cancer patients. Most of the chapters in this volume provide practical information about the situations in which immunodiagnostic assays are useful, along with indications about the specific strategies to be utilized.

This second edition of *Immunodiagnosis of Cancer* also reflects the advances in our understanding of the methodologies that need to be applied to the evaluation of immunodiagnostic assays and the development of a much wider array of technical procedures for detecting tumor antigens. Many of the assays that are now available to clinicians have high levels of sensitivity, reproducibility, and ease of performance. This volume provides a practical guide to clinical chemists as well as oncologists for the evaluation and management of patients with cancer.

Ronald B. Herberman
Donald W. Mercer

Preface to the First Edition

This volume on the immunodiagnosis of cancer has been assembled in an attempt to provide tumor immunologists and clinicians with a reference describing the breadth and current status of this field. The different aspects are discussed not only with respect to the latest progress but also with respect to the theoretical basis for the approaches and the current problems and limitations. In many ways the direction of needed studies is suggested. In addition, articles have been included in the design of appropriate clinical studies and in the statistical evaluation of diagnostic assays.

In recent years there has been growing interest in the clinical application of immunologic assays for the evaluation of patients with cancer. This has been due in part to recent advances in radioimmunoassays and other highly sensitive immunologic techniques which allow measurement of picogram or nanogram quantities of antigens. In addition, immunologic assays can be exquisitely specific, and in some instances can discriminate between molecules with differences in a single amino acid or sugar. Therefore, one can anticipate that immunologic procedures will be developed to distinguish between various types of neoplastic cells and normal cells.

There are now a wide variety of studies in the broad area of immunodiagnosis which have given promising results, and thus the large number of articles which have been included in this book. We have attempted to present most of the major areas of active research in immunodiagnosis, and have included tests that are currently being used or evaluated for their potential clinical usefulness. However, it should be noted that this compilation cannot be complete, for a variety of reasons, one being the rapidity of progress in this field. We felt it was important to have the book published without delay so that the information would be reasonably current; therefore the contributors have had less than 6 months to complete their articles. Despite all efforts, there are certain to be promising new tests and interesting leads developing after the contents of the book are planned. We have also included only those tests or approaches that were familiar to us or brought to our attention. Since the work in the area of immunodiagnosis has truly been international, it is possible that we have been unaware of potentially important results that might have been suitable for inclusion. Perhaps a future edition will take care of these omissions.

Despite the large variety of tests which are being performed and the high level of effort being expended in this area of research, as yet there are very few immunodiagnostic tests which have been definitively shown to have a place in clinical medicine. This should not be a reason for discouragement or pessimism, but rather it is a reflection of the difficulties involved in satisfactory transfer of technology from the research laboratory to the bedside. Many of the problems are not unique to immunodiagnosis but are also true for other types of diagnostic tests, including some which have been available for many years and some which have been incorporated into widespread use without real validation or objective assessment of utility.

Although advances in immunodiagnosis are highly dependent on advances in basic research, there are a number of aspects of development and evaluation of potential tests which can be planned. A number of principles emphasized in this book transcend the problems associated with a particular technique. It is obvious that most investigators with clinical interests would appreciate the possible application of their laboratory findings. To facilitate the transfer of laboratory findings to the clinical situation it is usually necessary for the scientists to interact closely with clinicians, epidemiologists, and statisticians in order to design the appropriate studies for clinical evaluation. There is also frequent need for interaction among biochemists, cellular immunologists, and investigators with other types of expertise. Because of the variety of disciplines involved, it becomes necessary to take each into account in the design of experiments.

There are a number of issues involved in the practical development of a potential immunodiagnostic test which are mentioned in these chapters, but which need to be emphasized. There is a need for development, whenever possible, of standardized reagents (antigens, antisera, target cells, etc.) which come from reproducible sources, such as tissue culture cell lines. Many studies have indicated that results may vary substantially with different preparations. There is also a need for controls for interassay variation (e.g., inclusion of some internal standard in each experiment) and for data analysis which will be objective and sensitive. As discussed in Chapter 2, determining sensitivity in a test must be established on the basis of considerations of the competing desires for high specificity and high sensitivity. Beyond these technical issues is the very important need for a properly designed clinical study. There have been major defects in design in most studies performed in this area, and consequently it has usually taken several years to properly assess the clinical value of potentially useful tests. As indicated [in the first edition], much more thought and attention need to be given to proper design of studies, in order to rapidly evaluate a test for a specific clinical application. In initial evaluations, there is a tendency to compare young normal donors with older patients with advanced cancer, with little attention given to the fact that most forms of cancer arise out of the background of an older population with a variety of underlying benign chronic diseases. This is particularly true since there is an increased risk of cancer in individuals with various chronic inflammatory diseases. Therefore, for a test to be expected to be useful for the initial diagnosis of cancer, it should be able to discriminate between patients with cancer and patients with benign diseases of the same organ system, e.g., chronic ulcerative colitis, hepatitis, cirrhosis, polyps, atrophic gastritis, chronic cystic mastitis.

For the past few years, the Immunodiagnosis Program of the National Cancer Institute has made coded serum panels, of speciments from cancer patients and appropriate controls, available to investigators with potentially useful diagnostic tests. Our experience

Preface to the First Edition

has been that only a low proportion of tests, all of which were initially thought to be able to discriminate between cancer patients and controls, could identify the cancer specimens better than by chance. As already mentioned, a major reason for these disappointing results could be improper study design. Another explanation may be the natural enthusiasm of the investigators. To avoid unconscious bias, there is a need to incorporate early in the studies a test design with coded samples. However, it should be noted that it is important also to have some known positive and negative controls, or to uncode the results rapidly after data analysis has been completed, so that technical or other problems in the test can be rapidly identified.

A further major problem is the difficulty in making the transition from population studies to a study in which statements can be made regarding the individual patients, upon which therapy or other important clinical decisions may be based. All too often, one is given only anecdotal information which supports the value of a test for a particular application. This is a particular problem in evaluation of a test for its ability to predict survival or recurrence of disease. As pointed out by Mack [in the first edition], it is equally important to evaluate and follow up individuals who are test negative or who are not tested, as well as those who have positive tests. The close serial following of patients, especially those with good response to therapy, may give valuable information as to the predictive value of an assay.

We hope that this volume will provide the reader with some insight into the studies needed to evaluate potentially useful immunodiagnostic assays adequately and the immunologic approaches which are worthy of further investigation. The many articles in this book clearly illustrate the wide variety of promising opportunities. It is now the responsibility of investigators in this area of research to translate this promise into a number of practical tests which will be of value in the detection, diagnosis, and monitoring of cancer patients.

Ronald B. Herberman
K. Robert McIntire

Contents

Series Introduction (Noel R. Rose)   iii
Preface to the Second Edition   v
Preface to the First Edition   vii
Contributors   xv
Introduction (Garry I. Abelev)   xix

I. Experimental Design

1. Predictive Value of Immunodiagnostic Cancer Tests   3
   Robert S. Galen

2. Some Statistical Methods for Immunodiagnostic Cancer Tests   13
   Mitchell H. Gail

3. Multivariate Techniques to Assess Laboratory Tests in Cancer Patients   27
   Per Winkel and Bernard E. Statland

4. Use of Multiple Markers to Enhance Clinical Utility   39
   Donald W. Mercer

5. Serum and Tissue Banks for Biological Markers   55
   Theresa L. Whiteside and Ronald B. Herberman

II. Methodology

6. Development and Evaluation of Monoclonal Antibody-Based Immunoassays: Breast Carcinoma-Associated Mucins as Tumor Markers   69
   Joseph P. Brown, Peter S. Linsley, and Diane Horn

7. Strategies for Heterogeneous Enzyme Immunoassays for Tumor Markers   83
   William J. King, Joseph T. Tomita, Barry L. Dowell, and Dennis M. Delfert

8. Enhanced Detection Systems for Enzyme-Linked Heterogeneous Immunoassays: Luminescence 95
   John C. Edwards and Carl R. Moon

9. Enzyme-Linked Heterogeneous Immunoassays with Electrochemical Detection 107
   H. Brian Halsall, Sarah H. Jenkins, and William R. Heineman

10. Novel Detection System for Homogeneous Enzyme Immunoassays 125
    Yoshihiro Ashihara

11. Rapid Response Fiber Optic Evanescent Wave Immunosensors 145
    Barry I. Bluestein and Shin-Yih Chen

12. Recent Advances in Flow Cytometric Techniques for Cancer Detection and Prognosis 171
    J. Philip McCoy, Jr., and John L. Carey

13. Use of Monoclonal Antibodies as Probes for Oncogene Products 189
    Henry L. Niman

14. Use of Immunologic Techniques in Gene Analysis 205
    Emilia Turco, Reid Fritsch, and Massimo Trucco

III. Organ Site-Associated Tumor Markers

15. Breast Epithelial Antigens in the Circulation of Breast Cancer Patients 223
    Roberto L. Ceriani and Ernest H. Rosenbaum

16. Lymphocytic Leukemia and Lymphomas 243
    Hans G. Drexler and Jun Minowada

17. Immunodiagnosis of Acute Nonlymphocytic Leukemia 265
    Craig A. Hurwitz, Lewis C. Strauss, and Curt I. Civin

18. Serum Tumor Markers for Pancreatic Carcinoma 289
    Herbert A. Fritsche, Jr., and Frank B. Gelder

19. Melanoma: The Development of Immunoconjugates from a Preclinical Viewpoint 297
    Paul L. Beaumier, Alton C. Morgan, Jr., Clive S. Woodhouse, and Kevin S. Marchitto

20. Markers of Central Nervous System Tumors 317
    Georg D. Birkmayer

21. Immunodiagnosis of Ovarian Tumors 323
    Ian Jacobs and Robert C. Bast, Jr.

22. Prostate Cancer-Associated Markers 339
    T. Ming Chu

23. Tumor Markers of Head and Neck Carcinoma 357
    David E. Eibling, Robin L. Wagner, and Jonas T. Johnson

24. Uterine Tumors 385
    Hiroshi Kato

Contents

25. Testicular Tumors — 397
    Deborah J. Lightner and Paul H. Lange

26. Markers for Hepatocellular Carcinoma — 403
    Alan H. B. Wu and Stewart Sell

27. Bone Tumors — 423
    Morton K. Schwartz

28. Thyroid Tumors — 431
    Hinrich G. Seesko and Samuel A. Wells, Jr.

29. Biological, Molecular, and Clinical Markers for the Diagnosis and Typing of Lung Cancer — 453
    Adi F. Gazdar, James L. Mulshine, and Barnett S. Kramer

30. Immunodiagnosis of Renal Cell Carcinoma — 469
    Thomas Ebert and Neil H. Bander

31. Recent Diagnostic Modalities for Bladder Cancer — 485
    Joel Sheinfeld, Carlos Cordon-Cardo, and Neil H. Bander

IV. Broad or Non-Organ Site-Restricted Tumor Markers

32. Immunoglobulins in Diagnosis and Monitoring of Neoplasia — 501
    Daniel E. Bergsagel and Waldemar Pruzanski

33. Acute Phase Reactant Proteins — 521
    Edward H. Cooper

34. Fibronectin — 537
    Boleslaw H. Liwnicz and Raymond Sawaya

35. Immune Complexes and Malignancy — 555
    Daniel R. Vlock

36. Pancarcinoma T and Tn Epitopes: Autoimmunogens and Diagnostic Markers that Reveal Incipient Carcinomas and Help Establish Prognosis — 587
    Georg F. Springer, Wilfred Wise, Sheila C. Carlstedt, Parimal R. Desai, Herta Tegtmeyer, Rhona Stein, and Edward F. Scanlon

37. Serum Isoenzymes in Cancer Diagnosis and Management — 613
    Donald W. Mercer

38. Mucin Glycoproteins as Tumor Markers — 631
    William F. Feller, Jerry G. Henslee, Robert J. Kinders, George L. Manderino, Joseph T. Tomita, and Harry G. Rittenhouse

39. Placental Proteins as Tumor Markers — 673
    Glenn D. Braunstein

Index — *703*

Contributors

Yoshihiro Ashihara, Ph.D.   Fujirebio Inc., Tokyo, Japan

Neil H. Bander, M.D.   The New York Hospital–Cornell University Medical Center and Memorial Sloan-Kettering Cancer Center, New York, New York

Robert C. Bast, Jr., M.D.   Duke University Medical Center, Durham, North Carolina

Paul L. Beaumier, Ph.D.   NeoRx Corporation and University of Washington Schools of Medicine and Pharmacy, Seattle, Washington

Daniel E. Bergsagel, M.D., D.Phil.   Ontario Cancer Institute and Princess Margaret Hospital, Toronto, Ontario, Canada

Georg D. Birkmayer, M.D., Ph.D.   Laboratory for Bio-Analytics and Med-Info., Vienna, Austria

Barry I. Bluestein, Ph.D.   Ciba-Corning Diagnostics Corporation, Medfield, Massachusetts

Glenn D. Braunstein, M.D.   Cedars-Sinai Medical Center–UCLA School of Medicine, Los Angeles, California

Joseph P. Brown, Ph.D.* Genetic Systems, Seattle, Washington

John L. Carey, M.D.   Henry Ford Hospital, Detroit, Michigan

Sheila C. Carlstedt, B.A.[†]   Evanston Hospital, Evanston, Illinois

Roberto L. Ceriani, M.D., Ph.D.   John Muir Cancer and Aging Research Institute, Walnut Creek, California

Shin-Yih Chen, Ph.D.   Ciba-Corning Diagnostics Corporation, Medfield, Massachusetts

Present affiliation:
*Bristol-Meyers Squibb Co., Wallingford, Connecticut
[†] University of Health Sciences, Chicago Medical School, and Veterans Administration Medical Center, North Chicago, Illinois

T. Ming Chu, Ph.D. Roswell Park Memorial Institute, New York State Department of Health, Buffalo, New York

Curt I. Civin, M.D. The Johns Hopkins Oncology Center, Baltimore, Maryland

Edward H. Cooper, M.D., D.Sc. University of Leeds School of Medicine, Leeds, United Kingdom

Carlos Cordon-Cardo, M.D., Ph.D. Memorial Sloan-Kettering Cancer Center, New York, New York

Dennis M. Delfert, Ph.D. Abbott Laboratories, North Chicago, Illinois

Parimal R. Desai, Ph.D.* Evanston Hospital and Northwestern University, Evanston, Illinois

Barry L. Dowell, Ph.D. Abbott Laboratories, North Chicago, Illinois

Hans G. Drexler, M.D.† Royal Free Hospital School of Medicine, University of London, London, United Kingdom

Thomas Ebert, M.D. The New York Hospital-Cornell University Medical Center and Memorial Sloan-Kettering Cancer Center, New York, New York

David E. Eibling, M.D. U.S. Air Force Medical Center, Lackland Air Force Base, Texas

John C. Edwards, M.A., D.Phil. Amersham International, Little Chalfont, Buckinghamshire, United Kingdom

William F. Feller Georgetown University School of Medicine, Washington, D.C.

Reid Fritsch Pittsburgh Cancer Institute and University of Pittsburgh School of Medicine, Children's Hospital, Pittsburgh, Pennsylvania

Herbert A. Fritsche, Jr., Ph.D. University of Texas M. D. Anderson Cancer Center, Houston, Texas

Mitchell H. Gail, M.D., Ph.D. National Cancer Institute, Rockville, Maryland

Robert S. Galen, M.D., M.P.H. Case Western Reserve University School of Medicine, Cleveland, Ohio

Adi F. Gazar, M.D. NCI-Navy Medical Oncology Branch, National Cancer Institute and Naval Hospital, Bethesda, Maryland

Frank B. Gelder, Ph.D. Louisiana State University Medical Center, Shreveport, Louisiana

H. Brian Halsall, Ph.D. University of Cincinnati, Cincinnati, Ohio

William R. Heineman, Ph.D. University of Cincinnati, Cincinnati, Ohio

Jerry G. Henslee, Ph.D. Abbott Laboratories Diagnostic Division, North Chicago, Illinois

Ronald B. Herberman, M.D. University of Pittsburgh School of Medicine and Pittsburgh Cancer Institute, Pittsburgh, Pennsylvania

---

Present affiliations:

*University of Health Sciences, Chicago Medical School, and Veterans Administration Medical Center, North Chicago, Illinois

†DSM (German Collection of Microorganisms and Cell Cultures), Braunschweig, Federal Republic of Germany

Contributors

Diane Horn, Ph.D.   Oncogen, Seattle, Washington

Craig A. Hurwitz, Ph.D.*   The Johns Hopkins Oncology Center, Baltimore, Maryland

Ian Jacobs, M.A., M.B.B.S.   The London Hospital, Whitechapel, London, United Kingdom

Sarah H. Jenkins, Ph.D.   University of Cincinnati, Cincinnati, Ohio

Jonas T. Johnson, M.D.   Eye and Ear Institute, Pittsburgh, Pennsylvania

Hiroshi Kato, M.D.   Yamaguchi University School of Medicine, Ube, Japan

Robert J. Kinders, Ph.D.   Abbott Laboratories Diagnostic Division, North Chicago, Illinois

William J. King, Ph.D.   Abbott Laboratories, North Chicago, Illinois

Barnett S. Kramer   NCI-Navy Medical Oncology Branch, National Cancer Institute, and Uniformed Services University of the Health Sciences, Bethesda, Maryland

Paul H. Lange, M.D.   University of Washington, Seattle, Washington

Deborah J. Lightner, M.D.[†]   Veterans Administration Medical Center, Minneapolis, Minnesota

Peter S. Linsley, Ph.D.   Oncogen, Seattle, Washington

Boleslaw H. Liwnicz, M.D., Ph.D.   University of Cincinnati Medical Center, Cincinnati, Ohio

George L. Manderino, Ph.D.   Abbott Laboratories Diagnostic Division, North Chicago, Illinois

Kevin S. Marchitto, Ph.D.   NeoRx Corporation, Seattle, Washington

J. Philip McCoy, Jr., Ph.D.   Pittsburgh Cancer Institute, Pittsburgh, Pennsylvania

Donald W. Mercer, Ph.D.   Montefiore Hospital and Pittsburgh Cancer Institute, Pittsburgh, Pennsylvania

Jun Minowada, M.D.   Fujisaki Cell Center, Fujisaki, Okayama, Japan

Carl R. Moon, Ph.D.   Amersham International, Little Chalfont, Buckinghamshire, United Kingdom

Alton C. Morgan   NeoRx Corporation and University of Washington Schools of Pharmacy and Medicine, Seattle, Washington

James L. Mulshine, M.D.   NCI-Navy Medical Oncology Branch, National Cancer Institute and Naval Hospital, Bethesda, Maryland

Henry L. Niman, M.D., Ph.D.[‡]   Progenx, Inc., San Diego, California

Waldemar Pruzanski, M.D.   University of Toronto and Wellesley Hospital, Toronto, Ontario, Canada

---

Present affiliations:
*St. Jude's Children's Research Hospital, Memphis, Tennessee
†Abbott-Northwestern Hospital, Minneapolis, Minnesota
‡Graduate School of Public Health, University of Pittsburgh, Pittsburgh, Pennsylvania

Harry G. Rittenhouse, Ph.D.* Abbott Laboratories Diagnostic Division, North Chicago, Illinois

Ernest H. Rosenbaum, M.D. Mt. Zion Hospital and University of California, San Francisco School of Medicine, San Francisco, California

Raymond Sawaya, M.D. University of Cincinnati Medical Center, Cincinnati, Ohio

Edward F. Scanlon, M.D. Northwestern University, Evanston, Illinois

Morton K. Schwartz, Ph.D. Memorial Sloan-Kettering Cancer Center, New York, New York

Hinrich G. Seesko, M.D.[†] Washington University School of Medicine, St. Louis, Missouri

Stewart Sell, M.D. University of Texas Medical School, Houston, Texas

Joel Sheinfeld, M.D. Memorial Sloan-Kettering Cancer Center, New York, New York

Georg F. Springer, M.D.[‡] Evanston Hospital and Northwestern University, Evanston, Illinois

Bernard E. Statland, M.D., Ph.D. Methodist Hospital of Indiana, Indianapolis, Indiana

Rhona Stein Center for Molecular Medicine and Immunology, Newark, New Jersey

Lewis C. Strauss The Johns Hopkins Oncology Center, Baltimore, Maryland

Herta Tegtmeyer [§] Evanston Hospital, Evanston, Illinois

Joseph T. Tomita, Ph.D. Abbott Laboratories Diagnostic Division, North Chicago, Illinois

Massimo Trucco, M.D. Pittsburgh Cancer Institute and University of Pittsburgh School of Medicine, Children's Hosptial, Pittsburgh, Pennsylvania

Emilia Turco, M.D. Pittsburgh Cancer Institute and University of Pittsburgh School of Medicine, Children's Hospital, Pittsburgh, Pennsylvania

Daniel R. Vlock, M.D. Pittsburgh Cancer Institute, University of Pittsburgh School of Medicine, and Veterans Administration Medical Center, Pittsburgh, Pennsylvania

Robin L. Wagner, B.S. Eye and Ear Institute, Pittsburgh, Pennsylvania

Samuel A. Wells, Jr., M.D. Washington University School of Medicine, St. Louis, Missouri

Theresa L. Whiteside, Ph.D. University of Pittsburgh School of Medicine and Pittsburgh Cancer Institute, Pittsburgh, Pennsylvania

Per Winkel, M.D. University Hospital of Copenhagen, Copenhagen, Denmark

Wilfred Wise, M.D. [†] Evanston Hospital, Evanston, Illinois

Clive S. Woodhouse, Ph.D. NeoRx Corporation, Seattle, Washington

Alan H. B. Wu, Ph.D. University of Texas Medical School, Houston, Texas

---

Present affiliations:
*Specialty Laboratories, Inc., Santa Monica, California
†Phillips Universitaet, Marburg, West Germany
‡University of Health Sciences, Chicago Medical School, and Veterans Administration Medical Center, North Chicago, Illinois
§St. Francis Hospital and Loyola University Medical School, Evanston, Illinois

Introduction

Between the two editions of this book, there has been a decade of surprising moments and some disappointments. The first edition of *Immunodiagnosis of Cancer* appeared in 1979 when it became evident that this approach exists not only as a theoretical possibility but as a clinical reality. Alpha-fetoprotein (AFP), carcinoembryonic antigen (CEA), human chorionic gonadotropin (HCG) and the "old pair"–monoclonal serum Ig vis-à-vis Bence Jones protein (BJP)—were the first emissars of tumor immunology in clinics, creating cancer immunodiagnosis. The pioneer of leukemia immunophenotyping–CALLA (common antigen of acute lymphocytic leukemia)—made its appearance at that time, demonstrating its efficacy in ALL classification and prognosis. A wide scope of new antigenic markers holds new possibility in tumor serodiagnosis, but host antitumor reactions, which have been demonstrated both in vivo and in vitro, appear to hold the greatest potential for new cancer immunodiagnosis based on host-tumor relationships. The second part of the first edition is devoted completely to this subject.

But what has taken place in reality? As always, a few unpredicted discoveries have led to unexpected theoretical and clinical advances. First, monoclonal antibodies (MABs), absent in the Subject Index of the first edition, revolutionized the analytical tools used in the study of tumor antigens—both secreted and cellular. The very technology of hybridoma production and especially of their screening and selection permits one to detect the slightest components of the immune response and to enlarge them to an absolutely stable and easily reprodudible degree. The hybridoma technique created an "immunological microscope" with the magnification and resolving power of the best microscopes as compared with usual vision. The technique appears to be especially useful for detection of small antigenic differences, which had always been evaluated as the minutest, between the normal and malignant tissues. Indeed, several new markers, epitopes, were introduced in serological immunodiagnosis. They belong mainly to carbohydrate moieties of glycoproteins. These are the so-called CA 19.9- and CA 125-specific epitopes of glycoproteins defined by MABs and produced by different carcinomas, mainly gastrointestinal (CA 19.9) and ovarian (CA 125). As in the majority of other markers, the level of these epitopes is more informative for evaluation of prognosis and for postoperative monitoring than for primary immunodiagnosis. Several prospective serum tumor markers are under consideration at present. They include, first of all, markers of breast cancer and pancreatic carcinoma. In spite of numerous attempts, lung cancer remains without a serological marker.

The first edition of this book contained the initial steps of immuno-imaging, or immunodetection in situ of primary tumors and their metastases. This field of cancer immunodiagnosis has significantly progressed during the last decade due to the introduction of MABs. Minor amounts of the injected antibody protein; the increase in specificity, affinity, and specific radioactivity; and the use of hybrid "humanized" MABs consisting of mouse Fab and human Fc have considerably improved the final results. The majority of these investigations were made using CEA as a target for radioactively labeled MABs. Broad and active investigations on cell surface antigens of human melanoma have revealed several well-characterized markers that have the potential for use in immuno-imaging and even for immunotherapy of these malignant tumors.

However, the main areas of progress in immunodiagnosis, conditioned by MABs, involve two directions: immunophenotyping of leukemia based on cell surface antigens, and precise pathomorphological detection of tumor nature and the origin of metastases from an unknown primary source. For the first purpose—immunophenotyping of leukemia and lymphoma—a large collection of cell surface differentiation antigens defined by MABs is used. This allows the leukemia or lymphoma at issue to be related to a discrete line and stage of progenitor cell differentiation that is significant for the evaluation of prognosis and choice of therapy. This approach is especially important for identification of discrete forms in morphologically indistinguishable acute hemoblastoses. The "old" CALLA, coming from the previous decade, but detected with the modern MABs, has retained its critical role in such immunophenotyping. The detailed study of tumor cell surface antigens dictated by needs of immunodiagnosis have important applications in tumor therapy. The corresponding MABs are used for detection of single tumor cells in bone marrow samples and for purging of autologous patient's bone marrow. Such "cleaned" autologous bone marrow may be used for retransplantation into its host after drastic chemo- or radiotherapy. This approach is being widely investigated in clinical trials.

Another important new area opened by MABs during the last decade is fine morphological classification and identification of tumors and their metastases. This approach is based on the discovery of tissue specificity of intermediate filaments consisting of so-called cytokeratins, clearly defined by MABs. This allows precise determination of tissue origin of different tumors, preferably carcinomas, detection of their micrometastases in the surrounding connective tissue or distant lymph nodes, as well as single tumor cells in the smears of blood or bone marrow. The contemporary histological diagnosis is hardly possible without cytokeratin detection with MABs. Different companies prepare the corresponding sets of tissue-specific anti-cytokeratine MABs.

Thus, rapid and significant progress has been achieved in the field of immunohisto- and cytodiagnosis, mainly due to introduction of MABs.

Application of MABs to identification of new tumor-associated antigens permitted the establishment of standards for their determination. Such standard approaches are widely employed by different companies in an empirical search for new tumor markers used for diagnostic purposes. There is an area in oncological literature dealing mainly with esti-

Introduction

mation of specificity thresholds for different markers, the percentage of false-positive and negative results obtained with them, and evaluation of their clinical value, without many fundamental investigations. I have no objections to these approaches, but would like to note that they are quite different from those which led to the discovery of AFP, CEA, and other "old-fashioned" tumor markers. The basic study of "old" tumor markers has progressed tremendously. For instance, the complete primary and tertiary structure of AFP as well as the great majority of its functions have become known. Its gene was cloned and sequenced and its regulatory sites were identified. The cell interaction with extracellular matrix was shown to be crucial for AFP expression in mature hepatocytes. Thus, we have come very close to understanding the reason for AFP reexpression in hepatocellular carcinomas.

Important progress has also been achieved in CEA studies. Cloning of its and related genes disclosed a large superfamily of genes, related to immunoglobulin superfamily and also including genes for placenta-specific proteins. The genetic analysis suggests that CEA functions as a cell adhesion molecule and gives rise to interesting speculations about its role in malignant growth.

Fundamental structural studies of carbohydrates used as tumor-associated antigens demonstrated that here we often also deal with "oncodevelopmental" substances arising in most cases as a consequence of incomplete synthesis of carbohydrate moieties of glycolipid and glycoproteins.

And, finally, a few comments on host-tumor relations as a basis for tumor immunodiagnosis. At least until the present time they have not justified their hopes in clinical practice. Use of anti-EBV antibodies in diagnosis of Burkitt's lymphoma and differential diagnosis of nasopharyngeal carcinoma with other head and neck tumors has developed since the time of the first edition of this book. Antibodies to HTLV-1 and HPVs are used mainly in epidemiological investigations, but not in the clinical setting. Various tests based on cell-mediated immunological reactions have not yet found their place in cancer immunodiagnosis and most likely will not be employed there in the future.

What should be said of the future development of immunodiagnosis? Is it exhausted or in the stage of rapid growth? Any prognosis in this field is very dangerous. However, a few points are worthy of mention. First, oncogenes: It is impossible to believe that oncoproteins will not find their important place at least in characteristic and prognostic evaluation of different tumors. The first results in this direction are presented in this book.

We could expect that together with oncoproteins the new molecular technology of their detection (i.e., of their mRNA by means of in situ hybridization) will enter tumor immunodiagnosis. And the first hopeful steps of polymerase chain reaction—the most specific and unbelievably sensitive test for certain mRNA-species—created new possibilities in detection of tumor with characteristic chromosomal translocations or expression of unique genes. Its value in diagnosis of residual disease at least in certain types of leukemia is beyond doubt. Good luck!

Garry I. Abelev
Cancer Research Center
Moscow, USSR

Immunodiagnosis of Cancer

I
Experimental Design

1
Predictive Value of Immunodiagnostic Cancer Tests

ROBERT S. GALEN

*Case Western Reserve University School of Medicine,
Cleveland, Ohio*

I. INTRODUCTION

The predictive value model has been applied by a number of investigators for evaluating the clinical performance and effectiveness of clinical laboratory tests. In evaluating diagnostic procedures, the model facilitates selection of the best test. For a particular test, the model facilitates selection of the best methodology. Furthermore, the predictive value model can be expanded to deal with multiple tests. It has proven to be an expedient empirical tool for optimizing laboratory test selection, strategies, and interpretation.

II. PREDICTIVE VALUE OF LABORATORY DIAGNOSIS

Sensitivity, specificity, predictive value, and efficiency define a laboratory test's diagnostic accuracy [1]. Sensitivity indicates the frequency of positive test results in patients with a particular disease; specificity indicates the frequency of negative test results in patients without that disease. The predictive value of a positive test result indicates the frequency of diseased patients in all patients with positive test results. The predictive value of a negative test result indicates the frequency of nondiseased patients in all patients with negative test results. The efficiency of a test indicates the percentage of patients correctly classified (diseased and nondiseased) by the test (Table 1).

In screening for disease, the predictive value of the positive result is of the utmost importance. In the discussion that follows, predictive value will be used to refer the predictive value of the positive test result. A marked change in predictive value occurs when there is a change in the prevalence of the disease in the population under study. For example, a laboratory test that was positive in 95% of diseased patients and negative in 95% of nondiseased patients is evaluated in Table 2. Note the change in predictive value that occurs for this test with changing prevalence of disease. It can readily be seen that a particular test has a higher predictive value when the disease occurs with a higher prevalence. This explains why a good diagnostic test frequently fails as a screening test when the drop in prevalence is marked. When clinical judgment is used in ordering laboratory

Table 1 Predictive Value of Test Applied to Healthy and Diseased Populations

	No. with Positive Test Result	No. with Negative Test Result	Totals
No. with disease	TP	FN	TP + FN
No. without disease	FP	TN	FP + TN
Totals	TP + FP	FN + TN	TP + FP + TN + FN

TP, true positives: number of diseased patients correctly classified by the test; FP, false positives: number of nondiseases patients misclassified by the test; FN, false negatives: number of diseased patients misclassified by the test; TN, true negatives: number of nondiseased patients correctly classified by the test.

Sensitivity = positivity in disease, expressed as % = $\dfrac{TN}{TP + FN} \times 100$

Specificity = negativity in health, or absence of a particular disease, expressed as % = $\dfrac{TN}{FP + TN} \times 100$

Predictive value of positive test = % of patients with positive test results who are diseased =

$\dfrac{TP}{TP + FP} \times 100$

Predictive value of negative test = % of patients with negative test results who are nondiseased =

$\dfrac{TN}{TN + FN} \times 100$

Efficiency of test = % of patients correctly classified as diseased and nondiseased = $\dfrac{TP + TN}{TP + FP + FN + TN} \times 100$

Table 2 Predictive Value as a Function of Disease Prevalence[a]

Prevalence of Disease (%)	Predictive Value (%)
1	16.1
2	27.9
5	50.0
10	67.9
15	77.0
20	82.6
25	86.4
50	95.0

[a]For a laboratory test with 95% sensitivity and 95% specificity.

tests, the patient suspected of having a particular disease is placed in a new population with a higher prevalence or probability of disease. Therefore, the test performs much better.

Although predictive value theory has only recently found its way into the literature of laboratory medicine, the formula and concept are not at all new. The formula for calculating predictive value is frequently referred to as Bayes' formula and was published posthumously in 1763 [2]. It is clear from the current medical literature that Bayes' equation is becoming increasingly popular. The formula may be presented in several variations, and it is important to recognize these variations as the same equation presented here:

$$\text{predictive value} = \frac{(\text{prevalence})(\text{sensitivity})}{(\text{prevalence})(\text{sensitivity}) + (1 - \text{prevalence})(1 - \text{specificity})}$$

There has been considerable confusion in the literature concerning the meaning of the term *false-positive rate*. In evaluating the statistical skills of physicians, Dr. Berwick and associates expected their subjects to know that the "false-positive rate" was equal to the number of false-positives divided by the number of true-negatives plus false-positives, or 1 minus specificity. They considered incorrect an answer in which the number of false-positives was divided by the total number of subjects studied as well as an answer in which the number of false positives was divided by the total number of positive results, or 1 munus predictive value [3].

Unfortunately, theirs is an inappropriate and misleading definition. It is inappropriate because a rate, in scientific parlance, usually refers to a quantity per unit of time. It is also misleading because there are, in fact, four useful false-positive ratios or proportions, not just one.

One false-positive ratio is defined by the proportion of results that are falsely positive for a particular disease when the test is used in the study of a population in which that disease is totally absent, that is, (FP)/(TN + FP). This proportion is equal to 1 minus specificity

A second false-positive ratio is defined by the proportion of all results, both positive and negative, that are falsely positive for disease when the test is used in the study of a population in which a particular disease is both present and absent: (FP)/(TN + FP + TP + FN).

A third false-positive ratio is defined by the proportion of all positive results that are falsely positive for disease when the test is used in the study of a population in which a particular disease is both present and absent: (FP)/(TP + FP). This portion is equal to 1 minus the predictive value of a positive result.

A fourth and final false-positve ratio is defined by the proportion of all false results that are falsely positive when the test is used in the study of a population in which a particular disease is both present and absent: (FP)/(FP + FN).

Therefore, the term *false-positive rate* should be abandoned. There are at least four false-positive proportions and each must be clearly defined to avoid confusion [4].

III. APPLICATION OF THE MODEL

In 1977, Foti and co-workers reported on a radioimmunoassay (RIA) procedure for prostatic acid phosphatase [5]. Table 3 summarizes Foti's findings. Foti's study and the editorial by Gittes that followed resurrected acid phosphatase not only as a tumor marker, but also as a cancer screening test:

Table 3 Sensitivity and Specificity of RIA for PAP

Group	No. of Patients	Sensitivity (%)
Patients with prostatic cancer	113	70
Stage I	24	33
Stage II	33	79
Stage III	31	71
Stage IV	25	92
		Specificity (%)
Patients without prostatic cancer	217	94
Normal controls	50	100
Benign prostatic hyperplasia	36	94
Total prostatectomy	28	96
Other cancers	83	89
Gastrointestinal disorders	20	95

Source: Data summarized from [5].

The finding of an elevated prostatic acid phosphatase may soon no longer mean that cure is out of the question and that only palliative therapy may be used. A new excitement over this oldest of "tumor markers" is in the air [6].

The grim finding has been that overall, 90% of cases are first detected when they have already metastasized. The clear implication of the accompanying report is that mass screening on the basis of a blood test alone can reverse this gloomy experience [6].

Both Foti and Gittes committed the cardinal sin of ignoring disease prevalence and incidence. Is the test described by Foti suitable for mass screening? It is quite simple to evaluate the usefulness of this test using the predictive value model [7].

Foti and co-workers determined the sensitivity of radioimmunoassay (based on current clinical staging criteria) to be 33, 79, 71, and 92% in stages I-IV, respectively. The specificity of radioimmunoassay has been measured in specific subgroups, but is not available for the at-risk population as a whole. However, on the basis of specificity of 100% in normal subjects, 94% in benign prostatic hypertrophy, and 89% in patients with other carcinomas, a reasonable estimate of specificity would be 95%. The third factor, prevalence, is a function of age and the method of diagnosis. Estimates from autopsy series with step sections yield the highest rates and clinical series yield the lowest rates. An overall estimate of prevalence of 25% for men over age 60 seems more than reasonable.

Finally, the cases must be staged. Clinical data, which are probably biased toward later stages, combined with autopsy data, were used to estimate that 15, 25, 35, and 25% of the patients with carcinoma would be classified into stages I-IV, respectively [7]. Figure 1 demonstrates how this test can be evaluated using the predictive value model. On the basis of these estimates, the predictive value of RIA can be calculated to be 83%, with an overall sensitivity of 73%. Table 4 summarizes the cases from Figure 1 in the fourfold table format.

These calculations are representative of an unscreened population. If the population has been previously screened for prostatic carcinoma, the test would be detecting the incident, rather than the prevalent cases. An incidence rate of 5% per year (5-10 times

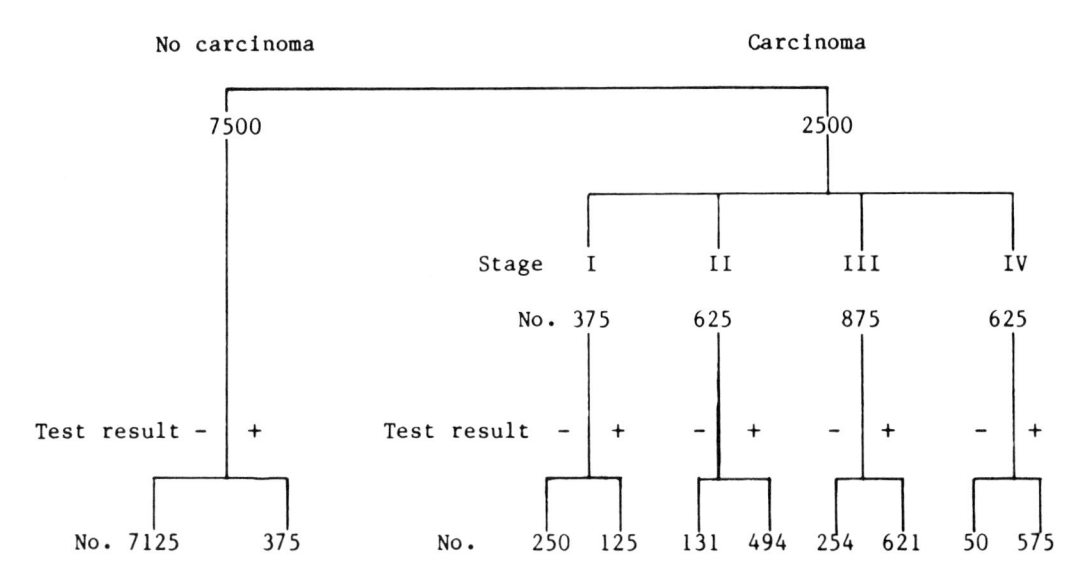

Figure 1 Testing for carcinoma of the prostate: distribution of results in a hypothetical unscreened population. Prevalence = 25%; specificity = 95%.

the maximal reported clinical incidence rate) is more than adequate to explain the rise in prevalence with age seen in autopsy specimens. This rate yields a predictive value of a positive test of 43%. Therefore, 57% of the follow-up work would be unnecessary. With this rate, it would be neither medically prudent nor cost-effective to use this test for screening purposes. With a lower incidence, the test would perform even worse. With a 1% incidence, the predictive value would be only 13%. Table 5 demonstrates the effect of prevalence on the predictive value of this test.

More recently, Watson and Tang, writing in *The New England Journal of Medicine* (the source of the first study and exuberant editorial), tried to answer some very fundamental and reasonable questions about the RIA test. They calculated that the predictive value of a positive screening test in a randomly selected man was 0.41%. Therefore, only 1 of every 244 subjects (100/0.41) with a positive test would have carcinoma of the prostate:

Table 4 Predictive Value of PAP

	RIA Results		
	Positive	Negative	Total
Prostatic carcinoma	1815	685	2500
Other	375	7125	7500
Total	2190	7810	10,000

Prevalence = 25%; sensitivity = 72.6%; Specificity = 95.0%.
PV + = 82.9%; PV − = 91.2%; EFF = 89.4%.

Table 5 Effect of Prevalence on Predictive Value (PV) of RIA-PAP Test

Prevalence (%)	PV (+)%	PV (−)%	EFF%
1	12.8	99.7	94.8
2	22.9	99.4	94.6
5	43.3	98.5	93.9
10	61.7	96.9	92.8
15	71.9	95.2	91.6
20	78.4	93.3	90.5
25	82.9	91.2	89.4
50	93.6	77.6	83.8

> These calculations by themselves are persuasive evidence against the utility of the RIA-PAP as a primary screening examination for prostatic carcinoma in the general male population and particularly in males whose prostate is normal according to rectal examination [8].

What about screening high-risk populations?

> Even in the older age groups in which the prevalence is greatest, the RIA-PAP would have limited value as a screening test. Considering the high prevalence of latent histologic carcinoma in this age group (over 75), we predict a far superior rate of detection from "blind" biopsies of the prostate than from the RIA-PAP [8].

Other authors, writing in the same issue of *The New England Journal of Medicine*, evaluated 10 tests or procedures in 300 symptomatic elderly men to determine which was most accurate in the detection of prostate cancer. The conclusion was "that a skillful rectal examination is still the most accurate office test for the detection of prostatic carcinoma" [9].

Of the 10 tests or procedures, the *least efficient* test was the acid phosphatage by RIA. In fact, both the counter immunoelectrophoresis [CIEP] and the Roy method for determining acid phosphatase performed better than RIA in their study. At the present time, we cannot help but agree with Watson and Tang, who say:

> With regard to the initial detection of prostatic carcinoma it seems that the most economical and most reliable probe for detection remains not a needle in the patient's arm, but the gloved finger of the physician ... [8].

The introduction of a radioimmunoassay procedure for prostatic acid phosphatase generated some controversy in the medical literature. More assays were developed, clinical trials were conducted, and Dr. Ruben Gittes wrote another editorial on the subject.

> Six years later, it is now clear that the radioimmunoassay did not and indeed could not live up to that hope...the specificity of only 94% undermined the potential of the test for screening purposes.
>
> Subsequent reports that the level of prostatic acid phosphatase was elevated in at least 6% of patients with benign prostatic hypertrophy strengthened the prediction that such an elevation was likely by overwhelming odds to be due to benign prostatic

hypertrophy rather than prostatic cancer. The intervening years confirmed the futility of measuring prostatic acid phosphatase as a screening test.

A marker specific for prostatic cancer may yet be found, which can overcome the mathematical advantage of the overwhelming incidence of benign prostatic hypertrophy. But for now the only reliable probe for detection of prostatic cancer is the examining finger of a physician performing a thorough examination [10].

The predictive value of an immunoassay test should be and must be a criterion of the assay's evaluation. The fiasco with the RIA-PAP test demonstrates what is likely to happen if the predictive value model is not used to evaluate new laboratory procedures prior to their introduction.

It is interesting to note that not only did Dr. Foti and co-workers in their original study and Dr. Gittes in his original editorial overlook the predictive value of the RIA test under evaluation, but the reviewers of these articles in the *New England Journal* did as well. This is particularly interesting in light of the "mixed reviews" received by *Beyond Normality* when it was published in 1975 [1]. The *New England Journal of Medicine* reported "This book is a brief, somewhat chatty exposition of a general mathematical relation developed nearly two centuries ago by the Reverend Thomas Bayes" [11].

In summary, three relationships are possible between sensitivity and specificity for any test: the sensitivity may be greater than the specificity; the sensitivity may be equal to the specificity; or the sensitivity may be less than the specificity. Let us review what happens to the predictive value and efficiency under each of these circumstances in the face of fixed, as well as increasing, disease prevalence.

At a given prevalance, an incremental increase in specificity results in a greater increase in the predictive value of a positive test than does the same incremental increase in sensitivity. As prevalence increases, the predictive value of a positive test increases, regardless of the relationship between sensitivity and specificity.

At a given prevalence, an incremental increase in sensitivity results in a greater increase in the predictive value of a negative test than does the same incremental increase in specificity. As prevalence increases, the predictive value of a negative test decreases, regardless of the relationship between sensitivity and specificity.

At a given prevalence (up to 50%), an incremental increase in specificity results in a greater increase in the efficiency of a test than does the same incremental increase in sensitivity. As prevalence increases, the efficiency of a test increases only if sensitivity is greater than specificity. When sensitivity is less than specificity, the efficiency of a test decreases with increasing prevalence. When sensitivity equals specificity, test efficiency is independent of prevalence and equal to sensitivity (or specificity). The value of test efficiency in all other cases falls between the value of test sensitivity and specificity, regardless of the relationship between sensitivity, specificity, and prevalence.

IV. COMBINATION TESTING

Although it is simple to apply the predictive value model to the evaluation of a single test, it can similarly be used to evaluate more than one test. In reality, it is rare to use the result of a single test as the final arbiter of a medical decision. But if we decide to use more than one test, a number of questions need to be answered: Which tests should be use? In what sequence should they be performed? How many tests are enough in a profile or battery of tests? How should the results be interpreted?

Table 6 Multiple Testing: Hypothetical Data for Two Tests (A and B)

	Test Results				Totals
	A + B −	A − B +	A + B +	A − B −	
Diseased	190	40	760	10	1,000
Nondiseased	9,800	4,850	100	84,250	99,000
Totals	9,990	4,890	860	84,260	100,000

To see how the predictive value model can be used to answer these questions, let us consider situations in which we use two tests. Examples includ the serological test for syphilis (VDRL) and fluorescent treponemal antibody test (FTA); metanephrines and vanillylmandelic acid (VMA); creatine phosphokinase (CPK) and lactate dehydrogenase (LDH) isoenzymes. Two independent tests in a screening or diagnostic situation can be used in three different ways:

1. Test A is applied first, and all those with a positive result are retested with Test B (series approach: [++] = +).
2. Test B is applied first, and all those with a positive result are retested with Test A (series approach: [++] = +).
3. Test A and B are used together, and all those with positive results for either or both tests are considered positives (parallel approach [+ −] = +; [++] = +; [− +] = +).

Which approach or sequence is best; This depends on the testing situation and the sensitivity and specificity of the individual tests and their combinations. For the sake of discussion, we will examine the hypothetical data presented in Table 6.

The sensitivity of test A is (190 + 760)/1,000 = 95%. The sensitivity of test B is (40 + 760)/1,000 = 80%. The sensitivity of the series combination (A and B positive) is 760/1,000 = 76%, but the sensitivity of the parallel combination (A or B positive) is (190 + 40 + 760)/1,000 = 99%. With parallel testing, the combined sensitivity is greater than the individual sensitivities of the contributing tests.

The specificity of test A is (4,850 + 84,250)/99,000 = 90%. The specificity of test B is (9,800 + 84,250)/99,000 = 95%. The specificity of the series combination, however, is (9,800 + 4,850 + 84,250)/99,000 = 99.9%, since A+,B−; A−,B+; and A−,B− are all interpreted as negative results in series testing. The specificity of the parallel combination is 84,250/99,000 = 85.1%, since only A−, B− is considered as a negative response.

Parallel testing results in the highest sensitivity but the lowest specificity, whereas series testing results in the lowest sensitivity but the highest specificity. Table 7 summa-

Table 7 Combination Testing for Hypothetical Data

Test	Sensitivity (%)	Specificity (%)
Single (test A)	95.0	90.0
Single (test B)	80.0	95.0
Series (test A and B)	76.0	99.9
Parallel (test A or B)	99.0	85.1

rizes these findings. For tests run in parallel (A and B determined simultaneously), but considered positive if either component is positive and negative only if both are negative, the sensitivity is higher and the specificity is lower than in comparable series testing. The sensitivity is increased because some diseased patients are positive on one test but not on the other. There are likewise more false-positive results in nondiseased patients. Of all the approaches, the parallel approach requires the most laboratory work since both tests are performed on all patients in the population.

In evaluating clinical laboratory tests, it is essential that the laboratory have at its disposal simple ways of analyzing data sets. The predictive value model has proven to be effective in designing test strategies and evaluating the usefulness of laboratory tests. The widespread use of computers in laboratory medicine should permit this approach to data analysis to become routine in the next few years. A number of computer programs have already been developed using this general approach [12-14].

REFERENCES

1. Galen RS, Gambino SR. The predictive value and efficiency of medical diagnoses. New York: John Wiley, 1975.
2. Bayes T. An essay toward solving a problem in the doctrine of chance. Philos Trans R Soc London 1763;53:370-418.
3. Berwick DM, Fineberg HV, Weinstein MC. When doctors meet numbers. Am J Med 1987;71:991-998.
4. Gambino SR, Galen RS. One man's rate is another man's ratio. Am J Clin Pathol 1983;80:127-128.
5. Foti AG, Cooper JF, Herschman H, Malvaez RR. Detection of prostatic cancer by solid-phase radioimmunoassay of serum prostatic acid phosphatase. N Engl J Med 1977;297:1357-1361.
6. Gittes R. Acid phosphatase reappraised. N Engl J Med 1977;297:1398-1399.
7. Fink DJ, Galen RS. Immunologic detection of prostatic acid phosphatase: critique II. Hum Pathol 1978;9:621-623.
8. Watson RA, Tang DB. The predictive value of prostatic acid phosphatase as a screening test for prostatic cancer. N Engl J Med 1980;303:497-499.
9. Guinan P, Bush I, Ray V et al. The accuracy of the rectal examination in the diagnosis of prostate carcinoma. N Engl J Med 1980;303:499-503.
10. Gittes RF. Serum acid phosphatase and screening for carcinoma of the prostate. N Engl J Med 1983;309:852-853.
11. McNeil BJ. Beyond normality: the predictive value and efficiency of medical diagnoses. Book review. N Engl J Med 1976;294:1016.
12. Galen RS, Weiss S. PVC-predictive value calculator. Beaumont, TX: Helena Laboratories, 1986.
13. MacNair DS. Predictive value, software review. JAMA 1987;285:548.
14. Krieg AF, Beck JR, Bongiovanni MB. Evaluating diagnostic performance of clinical tests by spreadsheet modeling. Arch Pathol Lab Med 1988;112:588-593.

2
Some Statistical Methods for Immunodiagnostic Cancer Tests

MITCHELL H. GAIL

National Cancer Institute, Rockville, Maryland

I. INTRODUCTION

In a previous paper [1] I described some statistical methods for determining whether a new immunodiagnostic test was potentially useful for diagnosis, prognostication, monitoring treated patients, or screening large populations. Although I touch on each of these applications in this chapter, new material centers on the use of multiple assays for diagnosis, the comparison of two diagnostic tests through their "receiver operating curves" (ROC), and the use of time-dependent covariates [2] and Markov models [3] for monitoring treated patients. In this volume, Statland and Winkel (Chap. 3) [4] discuss multivariate methods for survival data, methods for monitoring, and the establishment of reference ranges of assay values.

II. EVALUATION OF A SINGLE NEW IMMUNODIAGNOSTIC TEST FOR MEDICAL DIAGNOSIS

A. General Comments

For a new diagnostic test (DT) to have potential utility for diagnosis, it must be able to differentiate those with the disease of interest from normal individuals and from individuals who have other diseases likely to be found in a clinical practice. One would typically evaluate the DT in frozen sera from subjects with known diagnosis. As an example [5], we shall consider the differentiation of patients with advanced lung cancer from patients with a variety of benign lung diseases. We shall refer to this example often and shall rely on the reader to generalize to other diseases. If one can demonstrate good differentiation of those with advanced lung cancer from subjects with benign lung diseases, one might proceed to try to differentiate patients with localized lung cancer from those with benign disease. If the DT were successful in this more difficult task, one might undertake a prospective evaluation of the DT in a clinic to see how well it worked in practice.

A good experimental design is needed to establish that the DT differentiates those with advanced lung cancer from normal subjects with benign lung diseases. The extent of cancer should be carefully and objectively defined, because differentiation is usually more difficult for those with less extensive disease [6]. Likewise, it is important to describe the mix of diagnoses in the population of patients with benign diseases, because a given assay may distinguish some benign diseases from lung cancer more readily than others. The age and sex distributions in the study populations should also be examined. Often patients with lung cancer tend to be older than available normal controls and subjects with benign lung diseases. An assay that measures a variable that tends to increase with age, for example, could yield spuriously elevated values in those with lung cancer, compared to younger normal controls, even if the assay had no intrinsic diagnostic value for lung cancer. To deal with this problem, one can either design the samples to have nearly the same age and sex distributions among lung cancer cases as among control groups, or one can rely on statistical adjustment procedures, such as stratification according to age or sex or incorporation of age and sex into logistic discrimination models [5] to see whether the DT adds to age and sex in identifying lung cancer cases.

Whenever possible, the DT should be performed by a technician who has no access to other clinical data. Such "blinded" assessments are needed to ensure an objective evaluation. Special studies are needed to ensure that assay measurements on frozen sera are stable over the period of storage. It is important to collect information on concomitant therapies that might confuse the interpretation of assays performed on frozen sera. In the ideal situation, the blood sample will have been drawn before the initiation of therapy.

In addition to these precautions to protect the validity of the assay readings, it is important to ensure that diagnostic procedures are performed uniformly on all study subjects. One has little control over this aspect of design in retrospective studies on frozen sera, but in prospective studies one should plan that all study subjects undergo a well defined work-up to establish the diagnosis. To avoid "verification bias" [7], the clinician responsible for making the diagnoses should be blinded to the values of the DT being investigated. Begg [7] reviews other biases and necessary precautions needed for the assessment of DTs.

Evaluations of DTs can be strengthened by including both a "training sample" and a "validation sample" [5]. The training sample is used to determine the cutoff values for the DT required to achieve desired levels of sensitivity and specificity, and the validation sample is used to determine whether, in fact, the anticipated performance is achieved in an independent sample. The validation phase is even more important when multiple DTs are used, as described below, because discriminiation rules derived from the training data tend to perform unrealistically well when reapplied to the training data, especially with multivariate models.

B. Data Description

Immunodiagnostic test results are usually expressed as positive numbers. Often there is a lower threshold of detectability. For example, one might be told only that a given carcinoembryonic antigen (CEA) level is less than 1.0 ng/ml. Frequently, the distributions of the DT are skewed to the right, with quite a few extremely large values. These features are seen in the frequency plots of Figure 1, which compare the distributions of CEA in four clinical subgroups [1]. Note in particular that many CEA values are seen at 1.0 ng/ml, the lower limit of detectability, and many values of 20 ng/ml correspond to results of 20 ng/ml or more. From such a plot one gets the impression that CEA can usefully dif-

Statistical Evaluation of Immunodiagnostics 15

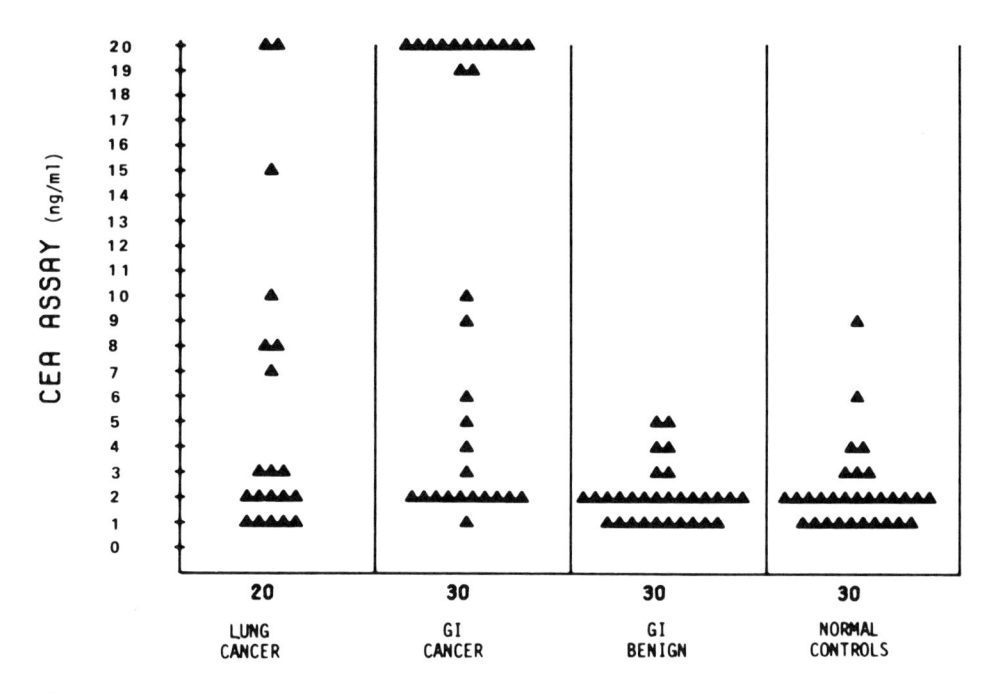

Figure 1 Frequency plot of CEA values (ng/ml) for each of four clinical subgroups. Each triangle corresponds to one subject [1].

ferentiate patients with gastrointestinal cancer and, possibly, patients with lung cancer, from normal subjects and from those with benign gastrointestinal diseases.

Such data can also be described by plots of the cumulative distributions (Fig. 2). Corresponding to an assay value X on the abscissa, the ordinate represents the cumulative proportion of individuals with assay levels ≥X. Figure 2 [5] shows the cumulative distributions of $X = \log_{10}(CEA + 1)$ for patients with advanced lung cancer (C), benign lung diseases (B), and normal controls (N). A CEA value of 9 ng/ml leads to an abscissal value

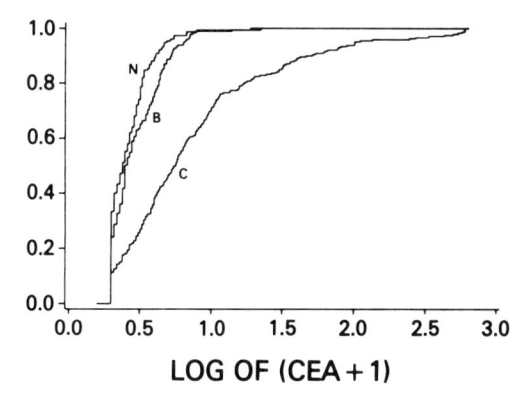

Figure 2 Cumulative distributions of $X = \log_{10}(CEA + 1)$ for samples of patients with advanced lung cancer (C), benign lung diseases (B), and normal controls (N). The ordinate shows the proportion of assay values less than or equal to X [5].

$\log_{10}(9 + 1) = 1$. Hence, from Figure 2 we read that about 67% of advanced lung cancer patients have CEA values ≤ g/ml, whereas 99.9% of those in groups N and B do. The cumulative proportion of individuals with CEA values below any given level may be read from Figure 2. Note that the cumulative distributions take a jump at $\log_{10}(1 + 1) = 0.301$, which is the threshold of detectability.

Data such as those in Figures 1 and 2 are not normally distributed and are not well summarized in terms of the mean and standard deviations. Instead, sample percentiles, which can be estimated even in the presence of a threshold of detectability should be used. For example, the median (50th percentile) of CEA (ng/ml) is 1.4 in normal men, 1.6 in men with benign lung disease, and 4.2 in men with advanced lung cancer [5]. The corresponding 90th percentiles are, respectively, 3.2, 3.9, and 37.2. These percentiles can be read from cumulative distributions like those in Figure 2. For example, the median for group C in Figure 2 corresponds to an abscissal value of about $0.8 = \log_{10}(\text{CEA} + 1)$, which yields a median CEA of 5.3 ng/ml for the combined male and female populations.

To see if apparent differences in distribution are statistically significant, one can apply two-sample tests, such as the Wilcoxon rank-sum test modified for ties [8]. Tests based on the assumption of normality should not be used ordinarily. Application of the Wilcoxon rank-sum test shows that the values of CEA do tend to be higher among the 30 patients with gastrointestinal cancer in Figure 1 than among the 30 patients with benign gastrointestinal disease [1].

A simple chi-square test may also be used to determine whether a higher proportion of those with cancer have positive assay results than of those without cancer. Suppose an investigator prespecifies an assay cutoff value such that the test is negative below the cutoff. Then the differentiation subjects with cancer from those with benign disease can be summarized as in Table 1. There $T_1 = n_{11} + n_{12}$, $T_2 = n_{21} + n_{22}$, $T_3 = n_{11} + n_{21}$, $T_4 = n_{12} + n_{22}$ and $T = n_{11} + n_{12} + n_{21} + n_{22}$. Discriminating those with gastrointestinal cancer from subjects with benign gastrointestinal disease (Fig. 1) with cutoff CEA ≥ 2.5 ng/ml, one finds $n_{11} = 19$ true positives, $n_{12} = 5$ false positives, $n_{21} = 11$ false negatives, and $n_{22} = 25$ true negatives. A simple chi-square test [9] is based on

$$\chi^2 = (|n_{11} n_{22} - n_{12} n_{21}| - T/2)^2 \, T/T_1 T_2 T_3 T_4. \tag{1}$$

In this example, $\chi^2 = 11.736$ corresponds to a one-sided significance level $\alpha/2 = 0.00031$, where α is the probability that a chi-square variate with one degree of fredom exceeds 11.736. For tables with smaller numbers, Fisher's exact test is recommended. The Fisher exact one-sided significance level is $\alpha/2 = 0.00024$ in this case [10]. Thus, the data demonstrate significantly higher CEA levels among those with gastrointestinal cancer than among those with benign gastrointestinal disease.

Table 1 2 X 2 Table Representing Outcomes of an Immunodiagnostic Test

	Cancer	Benign	Total
Test positive	n_{11}	n_{12}	T_1
Test negative	n_{21}	n_{22}	T_2
Total	T_3	T_4	T

One would often like to use a test based on Table 1 when no prespecified cutoff value is available. By examining the data, one can find the cutoff value that minimizes the number of misclassifications, $n_{12} + n_{21}$. For that selected cutoff value, the DT will appear to perform too well, when judged by the chi-square test in Eq. (1). Gail and Green [11] give tables for determining whether the smallest attainable $n_{12} + n_{21}$ is small enough to be judged statistically significant. In differentiating gastrointestinal cancer from benign disease, a CEA level ≤ 1.3 ng/ml leads to only $n_{12} + n_{21} = 16 + 2 = 18$ total misclassifications, significant at $p < 0.01$. Methods have recently been developed to test the significance level of the chi-square statistic when the cutoff level has been chosen to yield a maximal chi-square value [12,13].

C. Sensitivity, Specificity, ROC Curve, Positive Predictive Value, and Negative Predictive Value

Assume that a cutoff value has been chosen, and let p_{11}, p_{12}, p_{21} and p_{22} denote, respectively, the probability that a cancer patient has a positive test result (true positive), the probability that a subject without cancer has a positive test result (false positive), the probability that a cancer patient has a negative test result (false negative), and the probability that a subject without cancer has a negative test result (true negative). The quantity p_{11} is called the sensitivity of the test, and p_{22} is called the specificity. The probabilities p_{11}, p_{12}, p_{21}, and p_{22} may be estimated from the data in Table 1 by n_{11}/T_3, n_{12}/T_4, n_{21}/T_3, and n_{22}/T_4, respectively. For a given assay and cutoff value, the sensitivity depends on the mix of cancer patients and the specificity on the mix of noncancer patients. If it is reasonable to assume that the carefully staged cancer patients in one clinic are similar to carefully staged patients in another clinic, it is reasonable to assume that the DT sensitivity will be similar in the two clinics. Likewise the specificity measured in one clinic against well-defined groups of subjects without cancer may be applicable in other clinics. In this sense, sensitivity and specificity are regarded as intrinsic properties of the DT.

Vecchio [14] pointed out that sensitivity and specificity are not sufficient to characterize the performance of a DT; the prevalence of cancer in the target population must also be considered. If this prevalence is π_1, and if $\pi_2 = 1 - \pi_1$, the probability that an individual with a positive test indeed has cancer is $\pi_1 p_{11}/(\pi_1 p_{11} + \pi_2 p_{12})$, which Vecchio called the positive predictive value (PPV). Likewise, the probability that an individual with a negative test has no cancer is $\pi_2 p_{22}/(\pi_2 p_{22} + \pi_1 p_{21})$, the negative predictive value (NPV). In a target population with low cancer prevalence, PPV will tend to be low but NPV will tend to be high. In a population with high cancer prevalence, such as a tumor clinic, PPV will tend to be higher. For example, from the data in Figure 1, a screening test based on cutoff CEA ≥ 2.5 ng/ml yields estimates of sensitivity $\hat{p}_{11} = n_{11}/T_3 = 19/30 = 0.63$ for detecting gastrointestinal cancer and specificity $\hat{p}_{22} = n_{22}/T_4 = 25/30 = 0.83$ against benign gastrointestinal disease. If this test is used in a general medical clinic in which the prevalence of cancer among those with gastrointestinal complaints is only $\pi_1 = 0.02$, one estimates $\hat{P}PV = 0.02 \times 0.63/(0.02 \times 0.63 + 0.98 \times 0.17) = 0.07$ and $\hat{N}PV = 0.98 \times 0.83/(0.98 \times 0.83 + 0.02 \times 0.37) = 0.99$. Thus, in this setting with low cancer prevalence, negative test results will be reliable (1% false negatives) and positive tests will be unreliable (93% false positives). If the same test is applied instead in a tumor referral clinic in which the prevalence of cancer is 0.50, one finds $\hat{P}PV = 0.5 \times 0.63/(0.5 \times 0.63 + 0.5 \times 0.17) = 0.79$ and $\hat{N}PV = 0.5 \times 0.83/(0.5 \times 0.83 + 0.5 \times 0.37) = 0.69$.

Thus positive test results are much more reliable and negative test results are less reliable in this setting.

It is evident from Figure 1 that one can increase the specificity for differentiating gastrointestinal cancer from benign gastrointestinal diseases by raising the CEA cutoff. This change will result in a corresponding loss of sensitivity, however. For example, if the CEA cutoff is raised from 2.5 ng/ml to 5.0 ng/ml, specificity increases from 25/30 to 30/30 and sensitivity decreases from 19/30 to 16/30. More generally, as one varies the cutoff levels from very low values to very high values, the test sensitivity goes from 1.0 to 0.0 while the specificity goes from 0.0 to 1.0. These relationships are described by a plot of sensitivty on the ordinate versus 1 − specificity on the abscissa as all possible cutoff values are tried (Fig. 3). The curve in Figure 3 represents the CEA assay data in Figure 1 for gastrointestinal cancer and benign gastrointestinal disease. Such curves are called ROC curves. They emphasize that a DT should not be characterized by any single sensitivity or specificity but by a whole family of values. A test DT1 dominates another test DT2 if the sensitivity (ROC ordinate) of DT1 is everywhere higher than the sensitivity of DT2. In some cases one DT may have higher sensitivity over some portion of the ROC curve and lower sensitivity in other regions, so that neither test dominates.
In such circumstances, Hanley and McNeil [15] suggest using the area under the ROC curve as a figure of merit to rank two DTs, the test having the larger area being preferred.

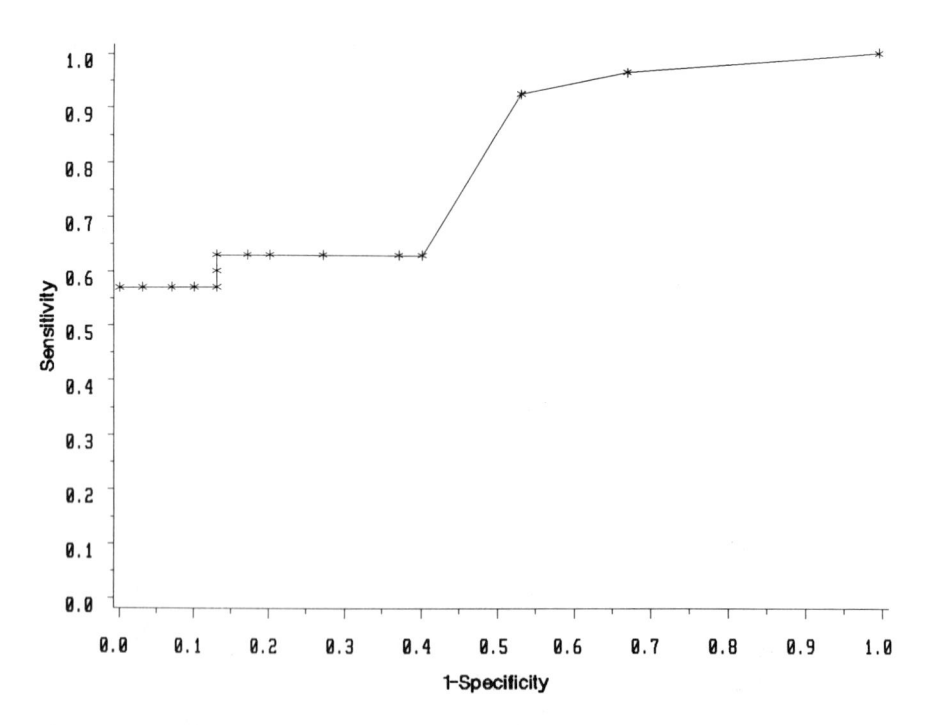

Figure 3 The ROC curve for CEA for differentiating gastrointestinal cancer from benign gastrointestinal disease (Fig. 1). Asterisks indicate where a change in cutoff values for the data in Figure 1 produces a change in the points (sensitivity, 1-specificity). These points are then connected by straight lines (original data from [1]).

This criterion gives equal weight to all possible specificities, however, and in many clinical and screening applications one demands that the specificity exceed a minimum level. For example, one may only be interested in tests with specificity of 0.8 or higher, corresponding to abscissal values near the origin in Figure 3. S. Wieand and co-workers [16] have developed nonparametric methods that allow one to compare two DTs in terms of the area under restricted portions of the ROC curve, such as a region of high specificity.

Sometimes it is sufficient to present several points from the ROC curve. For example, Gail et al. [1] tabulated three sensitivities for CEA for detecting advanced lung cancer: 0.82, 0.52, and 0.44, corresponding to specificities 0.50, 0.90, and 0.95, respectively, against benign lung disease. Any number of such points on the ROC curve can be constructed from the data in Figure 2. For example, the cutoff for $X = \log_{10}(CEA + 1)$ required to obtain specificity 0.90 is found from locus B to be about 0.75. Using this value, the sensitivity to detect lung cancer is estimated from Figure 2 as 0.52.

The optimal cutoff value to use for a given test depends on the application, and in particular on the "cost" c_{21} of a false-negative result, the "cost" c_{12} of a false-positive result, the disease prevalence π_1, and the probabilities p_{21} and p_{12} of false-negative and false-positive results, which are functions of the cutoff value. In principle, one seeks to determine the cutoff that minimizes the expected loss $\pi_1 c_{21} p_{21} + \pi_2 c_{12} p_{12}$. The optimal cutoff will be higher in a screening clinic for healthy individuals in which π_1 is small than in a clinic with higher cancer prevalence. Some of the practical difficulties in estimating an optimal cutoff have been described [1].

III. MULTIPLE MARKERS

A. Forming Discrimination Rules Based on Multiple Markers

Often several markers are available for distinguishing cancer from benign disease and one hopes to achieve improved discrimination by using the markers in combination. For example, Gail et al. [5] examined 11 markers in an effort to improve on the sensitivity and specificity of any single marker for differentiating patients with advanced lung cancer from those with benign lung diseases. The stragegy was to develop discrimination rules using multiple markers based on data in a "training" panel of sera from such patients and then to evaluate the performance of these rules on an independent "validation" panel. Although such independent validation is useful for evaluating single markers, it is crucial for evaluating complex discrimination rules based on multiple markers, because performance on the validation panel provides a much more realistic assessment of complex rules than would be obtained by simply reclassifying the training panel.

Although each of the 11 markers had been thought to be promising on the basis of earlier research, univariate assessments as outlined above suggested that only CEA, lipotropin, ferritin, lipid-bound sialic acid (LSA), and total sialic acid (TSA) would prove useful, alone or in combination.

A useful first step in assessing the potential joint utility of two assays is to plot the joint assay values for cancer cases and for those with benign disease (Fig. 4). The plotted data for TSA and CEA + 1 (on a log scale) indicate that cancer patients tend to have large values both of CEA and of TSA, whereas those with benign disease tend to have smaller values of both assays. The dashed straight line in Figure 4 is a "linear discriminant" because it divides the plane into the positive region of all points "northeast" of the line and

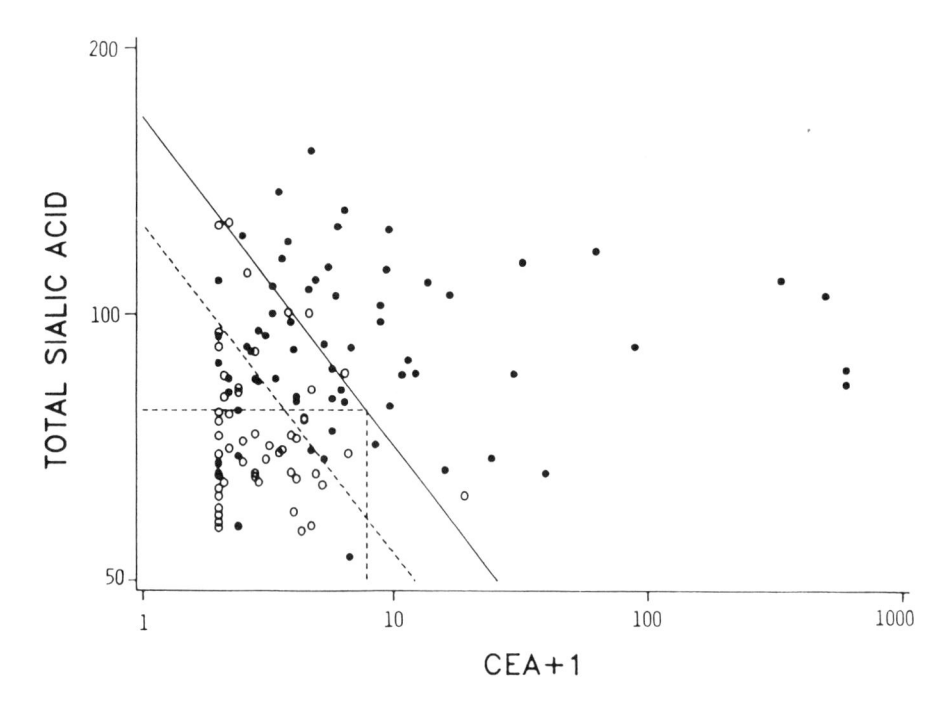

Figure 4 Differentiation of advanced lung cancer (●) from benign respiratory disease (○). Discrimination rules with moderate specificity are indicated by dashed lines; the solid line corresponds to a linear discriminant with high specificity. These solid and dashed lines were derived from "training data." Values northeast of the straight linear discriminate lines are positive, as are values outside the dashed rectangle obtained from recursive partitioning. The assay points shown are from independent "validation data" and are plotted on $\log_{10}$ scales [5].

the negative region of all points "southwest" of the line. The dashed rectangle might be called a "logical discrimination rule," since it partitions the plane into the positive region outside the rectangle (positive if either CEA exceeds 6.852 ng/ml or TSA > 77.98) and a complementary negative region inside the rectangle. Notice that both types of discrimination rules differentiate the cancer patients from those with benign disease well. In fact, the discrimination rules in Figure 4 were obtained from training data and the data points shown are from an independent validation panel, making performance of these rules even more impressive.

If one could use only TSA, one would try various horizontal lines to differentiate cancer patients from those with benign lung disease in Figure 4. However such discrimination would fail to identify those lung cancer patients with low TSA but high or moderately high CEA. Likewise, if one could use only CEA, one would try a series of vertical discrimination lines on Figure 4. Again, one would fail to detect those lung cancer patients with low CEA values but high or moderately high TSA. Graphic assessment thus suggests that the combination will perform better than either assay alone.

The straight line linear discriminants in Figure 4 were obtained from training data by the technique of logistic discrimination [17,18], which is more robust than linear discrimination based on assumptions of normality [19,20]. Before applying such techniques, one should transform the data to reduce extreme skewness. For example, had CEA rather than log(CEA + 1) been used, many of the extreme CEA values would not have even appeared on Figure 4 and would have affected adversely the resulting estimated linear discriminant. When three assays are combined, logistic discrimination yields a plane that differentiates cancers on one side from those with benign disease on the other, and for four or more markers, hyperplanes are produced.

Better separation may occasionally be obtained by curved loci derived from the method of quadratic discrimination [20,21] than by linear discriminant.

The space in Figure 4 could also have been divided into a set of mutually exclusive and exhaustive rectangles such that the DT is "positive" in some rectangles and "negative" in others. A particularly simple case is shown in Figure 4, in which the test is positive if *either* CEA *or* TSA is elevated and negative within the rectangle near the origin. The logical rule that says the DT is positive of *both* CEA *and* TSA are elevated would be positive only in a northeast quadrant (rectangle) if Figure 4. As one concatenates several markers with *fixed* cutoff levels by calling the combined test positive if any test is positive ("and/or rule"), one produces combined DTs with increasing sensitivity and decreasing specificity. If instead one calls the combined test positive only if all markers are positive ("both/and rule"), the combined test will tend to have lower sensitivity and higher specificity. Other more complicated rectangular partitionings correspond to positivity within jointly specified ranges of both assays. With more assays, these rectangles become rectangular parallelepipeds. Such rules can also be conveniently expressed by branching graphs. For example, the rectangle rule in Figure 4 is expressed as follows: At the first node one asserts the DT as positive if CEA > 2.75 ng/ml. If not, one proceeds to the descendent node and determines whether TSA > 81.53 mg/dl, in which case the combined DT is positive, or TSA ⩽ 81.53, in which case the DT is negative. Such branching decision structures are produced by an algorithm called recursive partitioning [22,23] that was used [5] to produce the rectangle in Figure 4.

The rules in Figure 4 were established on the basis of a training sample and were evaluated on an independent validation sample consisting of the points shown. The dashed linear discriminant and the recursive partitioning rules (dashed rectangle) had respective sensitivities of 85.3% and 87.7% compared to only 73.5% for CEA alone. The respective specificities against benign lung diseases were 67.3% and 50.0%, compared to 59.6% for CEA alone. Formal comparison based on the McNemar test for paired data indicate that both combined rules perform somewhat better than CEA alone [5].

The solid linear discriminant was obtained by increasing the intercept to the point at which the specificity in the training sample was 95%. On validation, the resulting sensitivity was 51.5% and the validated specificity was 92.3%. In a lung tumor clinic with cancer prevalence 0.30, this test would have a positive predictive value of 82% and a negative predictive value of 82%. When similar methods were used to differentiate local lung cancer from benign lung disease [6], positive predictive values of 76% and negative predictive values of 78% were obtained. This negative predictive value is not high enough to eliminate the need for other diagnostic procedures. However, positive marker results may indicate a need for further evaluation.

It was surprising that in an attempt to combine 11 assays the discrimination rules that performed well in the validation phase only required two markers, TSA and CEA. This was due in part to the fact that five of the assay performed poorly when evaluated individually. In part, the need for more assays was reduced because some assays, such as LSA and TSA, were highly correlated. Thus, once TSA information is taken into account, LSA provides little additional information for discrimination.

B. Screening and Prognostication

The same principles of discrimination discussed in connection with diagnosis also apply to screening and prognostication. However, in order to avoid an overwhelming number of false-positive results (low NPV) when screening large healthy populations, one must insist on extremely high levels of specificity to compensate for the low disease prevalence. This is especially true when the "costs" of a false-positive result are high, as in screening for the human immunodeficiency virus. In this context, concern about false-positive results has been expressed, even for combined assays with specificity 0.9999. Other aspects of screening are described very briefly elsewhere [1].

Evaluation of prognostic factors has become an important aspect of clinical trials research and epidemiological studies. Baseline marker values and other clinical features are used to differentiate patients with good future outcomes from those with poor future outcomes. The same discrimination methods used for diagnosis may be used to predict which patients will have a favorable prognosis and which will have an unfavorable prognosis [1]. Thus, baseline CEA values and Duke's histological classification may be combined through logistic regression or recursive partitioning to predict which patients with resected colon cancer will have a recurrence within 1 year. Multivariate methods are also available to predict entire survival curves, rather than simple dichotomous outcomes. Winkel and Statland [4] discuss this topic in Chapter 3. Methods and results for predicting survival following resection of lung cancer are provided elsewhere [24].

IV. MONITORING

Even when marker sensitivity and specificity are insufficient for screening large populations or for eliminating the need for other more invasive or costly diagnostic procedures, serial marker measurements may be useful in evaluating patients after treatment. For example, serial CEA measurements have been advocated to assist in the management of patients following resection for colorectal cancer; references favoring and opposing such use of CEA are cited in [2] and [25]. We shall concentrate on this example of patients at risk of recurrence following resection of colorectal cancer, although the same ideas apply more broadly. To establish the usefulness of a marker, one must answer a series of questions:

1. Is some feature of the serial marker profile associated with increased risk of recurrence?
2. What is the absolute chance of recurrence in a patient with a particular marker profile?
3. Can monitoring with the marker improve the survival of marker-positive patients who have no other evidence of recurrence?
4. Can the use of serial marker data in conjunction with second-look surgery improve overall 5 year survival rates or otherwise improve management for an entire population of patients with initially resected disease?

Questions 3 and 4 require information on the effectiveness of therapeutic options, such as second-look surgery. Question 4 also requires information on the proportion of

Statistical Evaluation of Immunodiagnostics

individuals who will develop abnormal marker profiles before other evidence of recurrence. These issues are treated elsewhere [25] and will not be considered here. Even to answer questions 1 and 2 requires a rigorous approach to clinical observation.

An ideal experiment to address question 1 would include the following features, most of which could be incorporated into protocols for clinical trials on therapeutic agents:

1. The patients would be carefully staged at the time of initial resection, and pertinent baseline data would be recorded.
2. The subsequent follow-up and procedures for diagnosing recurrence should not be allowed to depend on the patient's evolving marker values, and, indeed, the clinician diagnosing recurrence should not have access to marker data. Otherwise, a special effort might be made to establish recurrence whenever suggestive marker patterns were observed, thus introducing a potential bias.
3. The timing of the marker measurements should follow a predetermined schedule that, as far as possible, should not depend on the patient's clinical status. Otherwise, the investigator might take extra marker measurements whenever he or she thinks a recurrence is about to be diagnosed, thus introducing a potential bias in marker measurements.
4. One should plan to observe about 50 patients with recurrences [25] and take serial measurements with sufficient frequency that most of those with recurrences will have had enough marker data to be able to allow one to classify the marker profile pattern at the time of recurrence.

Although any feature of the serial marker profile may be studied to determine whether it is associated with risk of recurrence, several features have been emphasized in previous work [2,25], including:

1. The current CEA level just before the time the recurrence in question occurs ("current CEA")
2. The CEA level 6 months before the time the recurrence in question occurs (so called "lagged CEA"); lags of 12 months, 18 months, and so forth are also useful
3. The largest CEA observed up to the time of the recurrence in question
4. The rate of change of CEA in the 6 months preceding the recurrence in question.

Serial marker data usually have two features that pose special problems for analysis. The serial measurements are obtained at irregular intervals and data are often missing at particular times. Also, many patients are free of evidence of recurrence when their follow-up ends; their recurrence data are said to be "right censored." Gail [2] used the proportional hazards model of Cox [26] to analyze such data by treating the chosen aspect of the marker profile as a time-dependent covariate. In effect, the marker profile of a patient who develops recurrence at a particular follow-up time, t, is compared with the profiles of all the other individuals who have survived without recurrence to time t or beyond. The analysis is very similar to that for case-control studies, and, as in case-control studies, yields information on the relative risk of recurrence for persons in two marker states, not on the absolute risk. For example, Gail [2] estimated that an individual with CEA level >10 ng/ml had 6.8 times the risk of recurrence of another individual who had survived just as long and had a current CEA $\leqslant 2.5$ ng/ml. If, instead, one compared two individuals at risk at time t with respect to their CEA values measured 100 days earlier, the one with CEA > 10 ng/ml was estimated to have 5.0 times as great a risk as the other with CEA $\leqslant 2.5$ ng/ml.

Analyses like these can answer question 1, because they enable one to test whether certain aspects of the marker history are associated with increased risk. A shortcoming of the time-dependent covariates technique, however, is that the absolute risk of disease cannot be estimated (question 2).

If one contemplates second-look surgery, one wishes to know not only that the markers indicate a high relative risk but also, and more importantly, that the absolute risk of finding a recurrence is high. A rigorous assessment of absolute risk requires one to obtain both marker and recurrence data at regular intervals so that the entire marker profile may be related to absolute risk of disease. Markov models provide a useful structure for modeling absolute risk and for relating risk to marker levels. Myers et al. [3] applied a Markov model to calculate both relative and absolute risks of death from prostate cancer among patients with varying levels of alkaline and acid phosphatases. To accomplish this, they classified each patient at regular 90 day intervals into one of five living states (according to whether alkaline phosphatase levels were elevated or not, acid phosphatase levels were elevated or not, or marker data were missing) and the dead state. To use such models reliably, one must have relatively complete and regular marker data, and one must carefully test the stationarity and order assumptions of the Markov model.

Time series approaches to the analysis of marker data and additional details on the use of the Markov model are discussed in Chapter 3.

V. SUMMARY

This paper complements an article [1] in the previous edition of this book by emphasizing recent development in the use of ROC curves, the analysis of multiple markers, and methods of analysis for serial marker data. What has not changed is the need for statisticians and clinicians to work together closely to define clinical issues and their statistical formulations.

ACKNOWLEDGMENT

I would like to thank Mrs. Jennifer Donaldson for preparing Figure 3 and for typing the manuscript.

REFERENCES

1. Gail MH. Some statistical methods for evaluating immunodiagnostic tests. In: Herberman R, McIntire R, eds. Immunodiagnosis of cancer. New York: Marcel Dekker, 1979:20-37.
2. Gail MH. Evaluating serial cancer marker studies in patients at risk of recurrent disease. Biometrics 1981;37:67-78.
3. Myers LE, Paulson DF, Berry WR, Cox EB, Laszlo J, Stanley W. A time-dependent statistical model which relates current clinical status to prognosis: application to advanced prostatic cancer. J Chron Dis 1980;33:491-499.
4. Winkel P, Statland BE. Chapter 3, this volume.
5. Gail MH, Muenz L, McIntire KR, et al. Multiple markers for lung cancer diagnosis: validation of models for advanced lung cancer. J Natl Cancer Inst 1986;76:805-816.

6. Gail MH, Muenz L, McIntire KR, et al. Multiple markers for lung cancer diagnosis: validation of models for localized lung cancer. J Natl Cancer Inst 1988;80:97-101.
7. Begg CB. Biases in the assessment of diagnostic tests. Stat Med 1987;6:411-423.
8. Lehmann EL. Nonparametrics: statistical methods based on ranks. San Francisco: Holden-Day, 1975.
9. Snedecor GW, Cochran WG. Statistical methods. Ames, IA: Iowa State University Press, 1967.
10. Finney DJ, Latscha R, Bennett BM, Hsu P, Pearson ES. Tables for testing significance in a 2 X 2 contingency table. London: Cambridge University Press, 1963.
11. Gail MH, Green S. A generalization of the one-sided two-sample Kolmogorov-Smirnov test for evaluating diagnostic tests. Biometrics 1976;32:561-570.
12. Miller R, Siegmund D. Maximally selected chi-square statistics. Biometrics 1982; 38:1011-1016.
13. Halpern J. Maximally selected chi-square statistics for small samples. Biometrics 1982;38:1017-1023.
14. Vecchio TJ. Predictive value of a single diagnostic test in unselected populations. N Engl J Med 1966;274:1171-1173.
15. Hanley JA, McNeil BJ. The meaning and use of the area under a receiver operating characteristic (ROC) curve. Radiology 1982;143:29-36.
16. Wieand S, Gail MH, James BR, James KL. A family of nonparametric statistics for comparing diagnostic markers with paired or unpaired data. Biometrika 1989;76: 585-592.
17. Anderson JA. Logistic discrimination with medical applications. In: Cacoullos T, ed. Discrimination analysis and applications. New York: Academic Press, 1973: 1-15.
18. SAS Supplemental Library User's Guide. Cary, NC: SAS Institute, 1980:83-102.
19. Anderson TW. An introduction to multivariate statistical analysis. New York: John Wiley, 1958:126-153.
20. SAS User's Guide: Statistics. Cary, NC: SAS Institute, 1982:381-396.
21. Lachenbruch PA, Clarke WR. Discriminant analysis and its applications in epidemiology. Methods Inform Med 1980;19:220-226.
22. Friedman JH. A recursive partitioning decision rule for nonparametric classification. IEEE Trans Comput 1977;C-26:404-408.
23. Breiman L, Friedman JH, Olsen RA, et al. Classification and regression trees. Belmont, CA: Wadsworth, 1984.
24. Gail MH, Eagan RT, Field R et al. Prognostic factors in patients with resected stage I non-small cell lung cancer. Cancer 1984;54:1802-1813.
25. Gail MH. The evaluation of serial marker measurements for monitoring patients at risk of recurrent cancer: application to colorectal cancer. In: Mastromarino AJ, ed. The biology and treatment of colorectal cancer metastases. Boston: Martinius Nijhoff, 1986:235-251.
26. Cox DR. Regression models and life tables (with discussion). J R Stat Soc [B] 1972; 34:187-220.

3

Multivariate Techniques to Assess Laboratory Tests in Cancer Patients

PER WINKEL

University Hospital of Copenhagen, Copenhagen, Denmark

BERNARD E. STATLAND

Methodist Hospital of Indiana, Indianapolis, Indiana

I. INTRODUCTION

Laboratory measurements have three potential functions in patients known to have or expected to have cancer: screening for possible malignancy, classifying patients with documented malignancy, and monitoring the disease course and/or response to therapy in cancer patients. Often a plethora of laboratory test results (and clinical data) are available to the clinician and it may be difficult to extract the essential information and summarize the results. For this purpose various techniques of multivariate analysis are available that may prove helpful to clinicians in their efforts to produce a synthesis of the laboratory results.

Multivariate analysis may be defined as the collection of techniques appropriate for situations in which the random variation in several variates has to be studied simultaneously. It may be applied when each of several quantities is measured once in each of the patients studied (one observation per patient) or the same quantity (or set of quantities) is measured repeatedly in the same patient (multiple observations per patient).

This chapter discusses various multivariate strategies used to assign patients (classification) into clinically relevant groups. Initial data available at or near the time of first presentation with malignancy will be discussed briefly, but the emphasis will be on the analyses of sequential results (monitoring). The particular illustrative situation is laboratory testing in patients with documented breast cancer. Although the conditions, settings, and objectives of classification and monitoring may differ somewhat, the approaches should be applicable to patients with all types of cancers.

II. USE OF REFERENCE VALUES

Laboratory test values are often transformed before being used to denote whether the value is higher (or lower) than the upper (or lower) reference limit (usually determined as the mean ±2 standard deviations [SD] of the reference population). The value is then recorded as being normal or abnormal.

a)

O = CONTROL
● = RECURRENCE

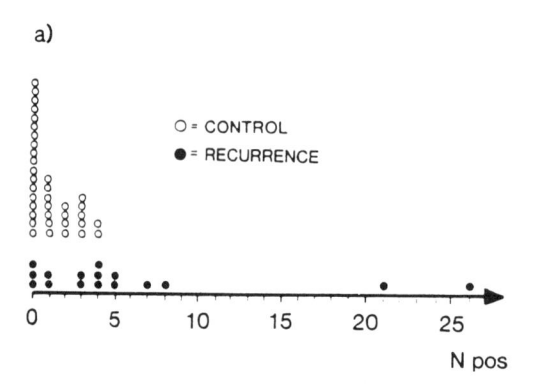

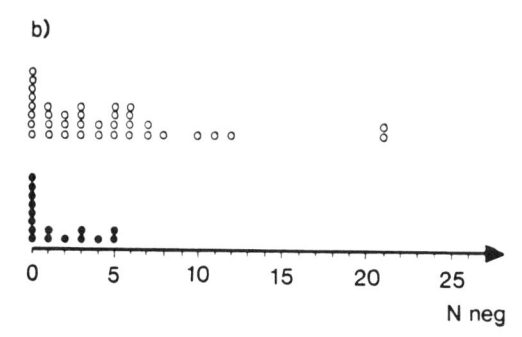

N pos

b)

N neg

Figure 1 (a) Distribution of the number of lymph nodes with tumor cells (N pos) found at surgery for primary breast cancer in 16 patients in whom recurrence developed within 25 months following operation and in 36 patients in whom the disease did not recur. (b) Distribution of the number of lymph nodes without tumor cells (N neg) found at surgery in the same two groups of patients [1].

Sometimes this approach may not be very wise. In relying on reference values as reported from the laboratory, there is the risk that a reference population is introduced that may be irrelevant in the given context because clinical chemistry values often are based on data from a group of *healthy* subjects.

Another disadvantage of categorizing the data into usual and unusual results as assessed from a reference or control group is that important relationships among the quantities may be missed completely. Figure 1a shows the distribution of the number of lymph nodes with metastases (positive nodes) found at surgery in two groups of patients with breast cancer. Members of one group had recurrence of disease within 25 months and members of the other did not. In retrospect, it is noted that the use of cutoff value of more than four positive lymph nodes would have resulted in the correct classification of 37.5% of the recurrences without any of the controls being misclassified as recurrences. Figure 1b shows the distribution of the number of lymph nodes without tumor involvement (negative nodes) that were found at surgery. Although, on the average, the controls have more negative nodes than those experiencing recurrence, it is not possible to define a lower limit that will safely differentiate the latter group from controls. The

Multivariate Assessment of Laboratory Tests

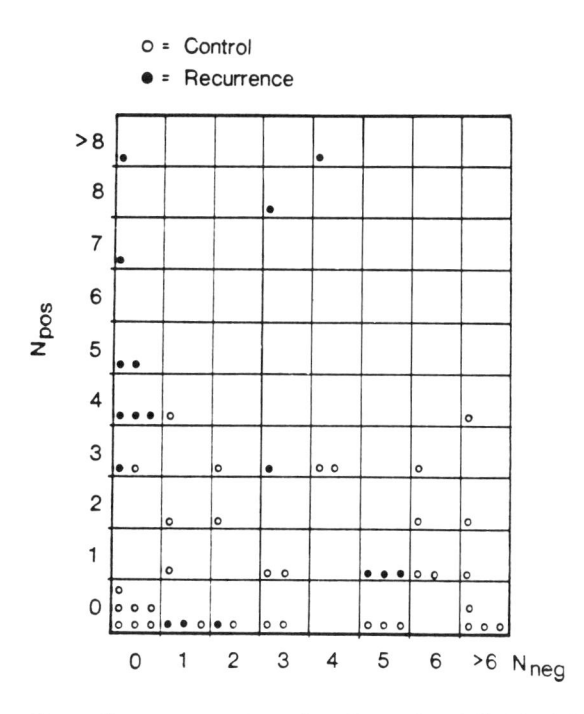

Figure 2 N pos vs. N neg (see Fig. 1 for definition) is depicted for each of 52 patients who underwent surgery for primary breast cancer [1].

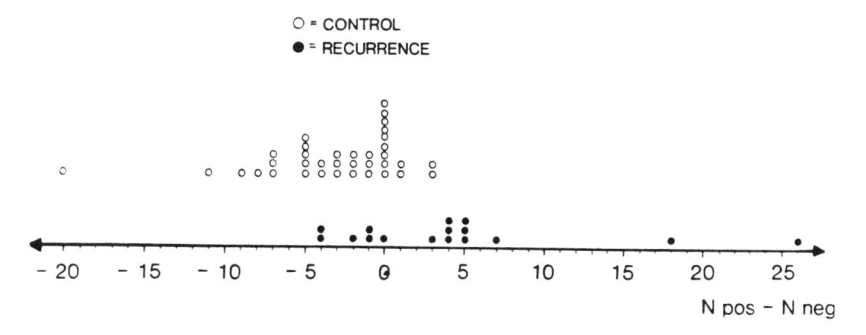

Figure 3 N pos − N neg (see Fig. 1 for definition) is depicted for each of 52 patients who underwent surgery for primary breast cancer [1].

SINGLE SAMPLE VALUES FROM 36 CONTROLS

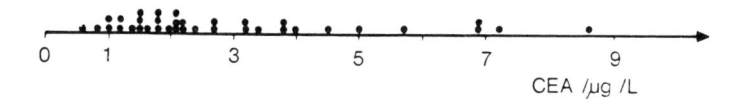

Figure 4 Serum CEA values measured 1.5-2 months following surgery for primary breast cancer in each of 36 patients in whom recurrence of disease did not develop within 25 months of surgery [1].

absence of negative nodes is actually the most common finding among the patients experiencing recurrence of disease as well as among the controls. On this basis, therefore, one might be tempted to discard the quantity since it does not appear to discriminate well.

The number of negative nodes is still a very useful quantity. This is apparent in Figure 2, which compares the number of positive nodes with the number of negative nodes for each member of the control group and each member of the recurrence group. As a rule, the negative nodes outnumber the positive nodes in the controls, while the opposite is true for the majority of those with recurrence. A simplistic way to quantitate this information would be to construct a new quantity that is the difference between the number of positive and negative nodes. The distribution of this quantity in each of the two groups is shown in Figure 3. There a cutoff value of 3 allows us to identify 56% of the cases of recurrence without misclassifying any of the controls.

In the analysis of cumulative data obtained from a particular patient whom we want to evaluate over time, the use of a group based reference limit obviously is inappropriate even if the reference group is the clinically relevant one and not just a group of healthy subjects usually used by most laboratories. A group-based reference interval or upper limit (and most, if not all, reference intervals are group based) is based on single values: one value obtained from each of a number of subjects from the reference population. The variability of these values therefore reflects the variability seen in the same subject over time as well as the variability among the subjects with regard to their mean levels. However, when we evaluate a particular subject we wish to determine if his or her values are within the range that is usual for him or her and not the range that is usual for a group of peers. This point is illustrated in Figure 4, which shows the serum carcinoembryonic antigen (CEA) levels documented shortly after operation in 36 subjects who underwent surgery for primary breast cancer. None of these subjects had recurrence of disease within 25 months of the operation. From these data, the upper 99% reference limit is calculated to be 7.4 µg/L. (If the calculations are based on a log gaussian distribution, the limit is 11.9). Figure 5 shows the cumulative serum CEA values obtained from three patients (subjects R1, R7, and R9) in each of whom metastatic disease developed. When each value is compared with those previously measured in the same subject (using time series analysis), the recurrence was predicted 6.5, 2.9, and 1.3 months in advance in subjects R1, R7, and R9, respectively. Had we used the group based reference limit, we would have missed two of the cases while the third (R7) would have been picked up at the same time that it was when the time series analysis was used.

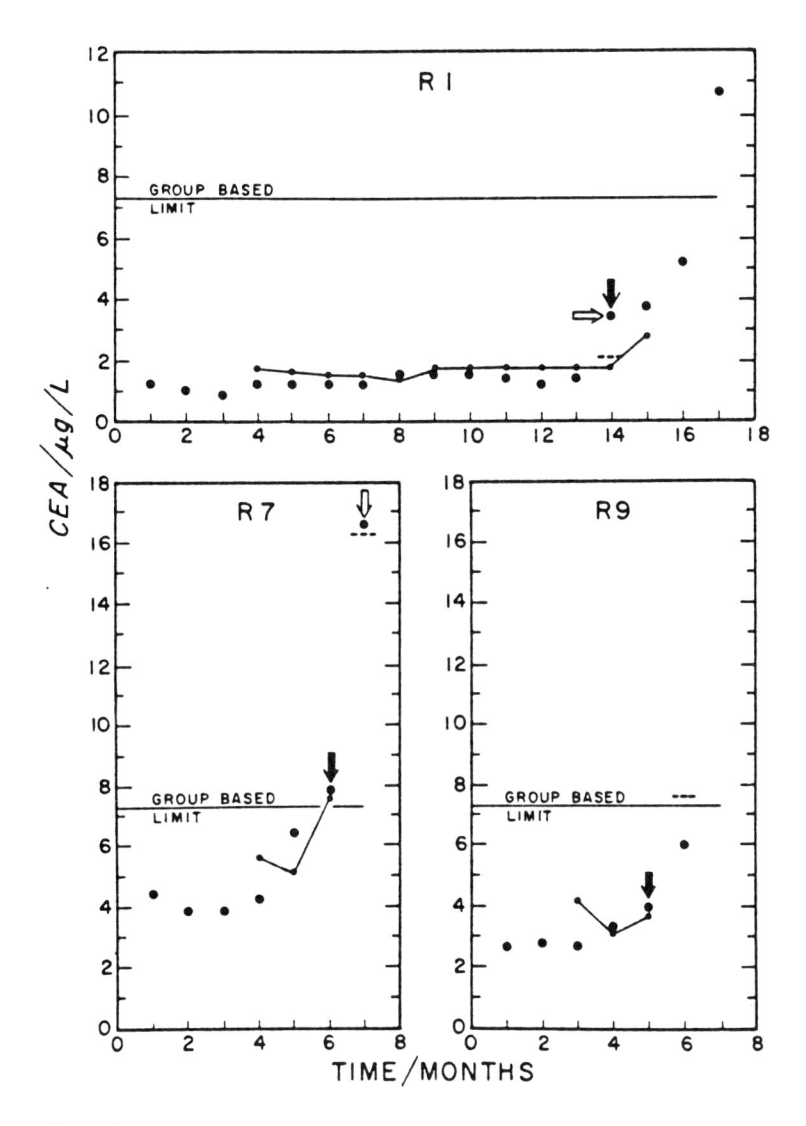

Figure 5 Postoperative serum CEA values observed in each of three patients who underwent surgery for primary breast cancer. Clinically overt disease recurred in all three after the last serum specimen was obtained. The black arrow indicates when the recurrence was predicted by time series analysis using two significance levels. The white arrow indicates when the recurrence was predicted using time series analysis based on one significance level (see text) [13].

III. GROUP-BASED MULTIVARIATE ANALYSIS

When only a single set of laboratory results measured in the same specimen is available for a given patient, we must compare these values to those obtained from other patients or healthy subjects to make inferences about the patient on the basis of the laboratory results. Thus, the analysis of the data has to be group based. By contrast, when previous results are available from the same patient, we may compare these present values to those previously obtained when we want to make inferences about the patient. In this case, the analysis of the data is subject based. In the first situation (group-based analysis), we may identify two situations. In the first one, our objective is to classify the patients into one of two or more groups of subjects (patients or healthy subjects). In a typical situation we want to determine whether the patient has a specified type of cancer. In the second one, our objective is to predict the time that will elapse until some specified event (death or recurrence of disease) occurs. In the former situation, a discriminant analysis or a logistic regression analysis is appropriate, whereas in the latter situation a survival analysis based on Cox's model will usually be the method of choice. For a review and discussion of discriminant function analysis and logistic regression analysis, the reader is referred to Chapter 2. We will briefly review here the Cox model.

A. Survival Analysis Based on the Cox Model

With the use of this technique [2], the risk of death (or recurrence) as a function of time is predicted by the background variates in a method that is very similar to regression analysis. The risk of recurrence is the dependent variate predicted by the independent variates (in the example described above the independent variates would be the number of positive lymph nodes, the grade of anaplasia, and so forth). The death rate or hazard rate $\lambda(t, Z_1, \ldots, Z_p)$ at time t after entrance for a patient with covariates or prognostic factors $Z_1, \ldots, Z_p$ has the form

$$\lambda(t; Z_1, \ldots, Z_p) = \lambda_0(t) e^{\beta_1 Z_1 + \cdots + \beta_p Z_p}, t > 0.$$

Here $\lambda_0(t)$, the so-called underlying death rate, is an unknown and unspecified function of time and $\beta_1, \ldots, \beta_p$ are unknown regression coefficients. If a regression coefficient is zero, the corresponding covariate has no influence on the death rate. Based on actual observations, the regression coefficients may be estimated and so may $\lambda_0(t)$, the cumulative underlying death rate. Estimates of survival probabilities ($\hat{S}[t]$) for patients with given values of the covariates may also be calculated together with an estimated standard error of S(t).

The hypothesis that certain, for example, k β_is are equal to zero (taking into account the remaining p-k covariates) may be tested. This fact may be exploited in various standard procedures in which prognostic factors that are redundant because they are correlated with other prognostic factors may be eliminated, leaving a minimal set of factors with influence on the prognosis. The reader is referred to other works [3-5] for details, computer programs, and illustrative examples. An application of the proportional hazards model of Cox with a time-dependent covariate is found in [6].

IV. MULTIPLE OBSERVATIONS PER PATIENT

Multivariate statistical analysis may involve an analysis of values of a certain quantity (or set of quantities) obtained at successive time points. In the following we will refer to such a series of values as a time series and the analysis of such a series of values as time series analysis.

Time series analysis can be said to be concerned with the monitoring of a system on the basis of repeated observation of some aspects of the system such as the diameter of screws produced by a machine. The purpose of such monitoring is to detect when the system is no longer in a state of control. To achieve this purpose, it is necessary to obtain observations from the system when it is known to be in a state of control. Once the general and specific behavior of the observations during such a period of control is known, the system can be monitored to see if the observations continue to occur as predicted.

We therefore, need initial observations during a state of control to characterize a specific system, in our case, the patient. This approach, however, represents a major problem because we cannot wait forever before we begin monitoring the patient. Therefore, specification of the parameters of the model must be based on a few observations, perhaps two or three. This immediately excludes the majority of statistical models available today because the assumption inherent in most models is that a considerable number of initial observations are available.

One way out of this dilemma might be to use the average values of the parameters obtained from a number of control patients. However, the intersubject variability of the parameters may prove to be prohibitively large. A systematic review of all the various types of statistical time series models and other models applicable for monitoring is beyond the scope of this chapter. For a brief introduction to the topic of time series analysis, the reader is referred to [7]. The standard textbook, however, is that of Box and Jenkins [8]. We will focus here on two simple models that have proven useful for monitoring of cancer patients: the Markov chain and the autoregressive model.

A. The Markov Chain

The Markov chain [9] (a so-called stochastic process model) is a simple model that is useful for a number of applications. A stochastic process represents the patient at any given time as being in one of a number of states. Time is a discrete variable in the model and is measured using a predefined unit. The study by Myers et al. [10] is a good illustration of the application of this model. They used it to represent changes in advanced prostate cancer patients' prognosis. Based on serum alkaline and acid phosphatase values, which were characterized as being normal or elevated, they defined six states including death. The time unit used was 90 days. The various states are defined in the legend to Figure 6, which documents the state of a given patient during four time units (i.e., during a period of 1 year). Given any history (i.e., sequence of states), a patient can either remain in his present state or make a transition to one of the other states. The likelihood of undergoing a particular change of state, given the preceding sequence of states, is called a transition probability. In the Markov chain model, the following simplifying assumptions are made:

1. At any point in a patient's history, the transition probability is independent of the patient's past history except for the state most recently occupied.
2. Transition probabilities depend only on the states of departure and arrival and not on the time of departure (i.e., the transition probabilities are stationary).

Based on these assumptions, the transition probability of going from one state (e.g., the state where both enzyme activities are elevated [state 2 in Fig. 6]) to another state (e.g., the state of death [state 1 in Fig. 6] is estimated as the number of times this particular transition is observed in the patient material over all observed transitions with the former state being the state of departure). In Figure 6 the transition from state 2 to state 1 is observed once and a transition with state 2 being the state of departure is observed three times.

19 March 1975   Patient No. 33   18 February 1976

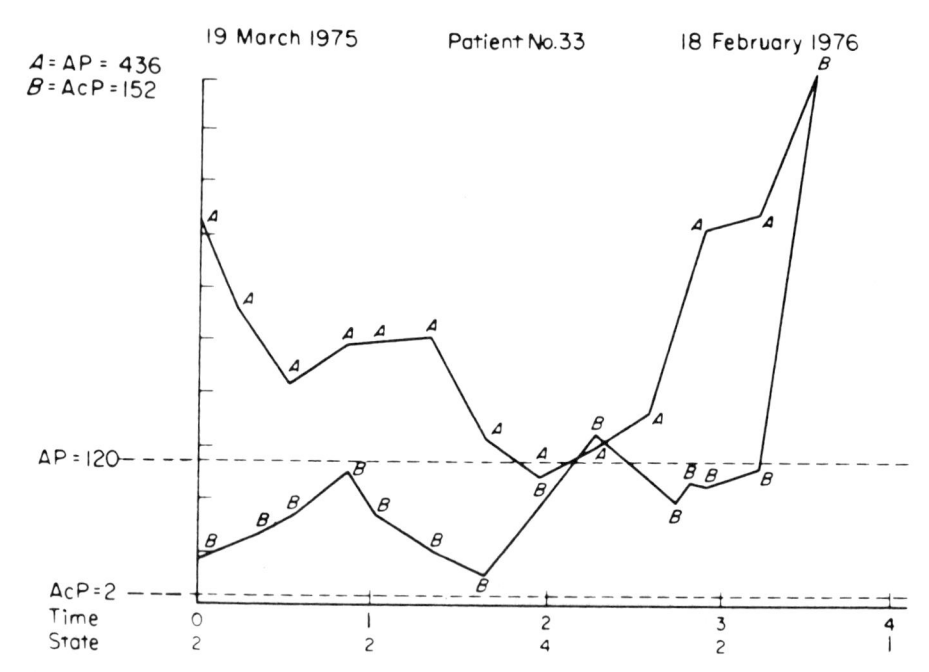

A = AP = 436
B = AcP = 152

AP = 120

AcP = 2

Time   0   1   2   3   4
State   2   2   4   2   1

Figure 6 Definition of the states for a patient during four time intervals (90 days intervals). The two curves depict the results of A (serum alkaline phosphatase) and B (serum acid phosphatase). The states are defined as follows; 1, dead; 2, A > 120 and B > 2 (both unfavorable); 3, A > 120 and B ⩽ 2; 4, A ⩽ 120 and B > 2; 5, A ⩽ 120 and B ⩽ 2 (both favorable). At time 0, both phosphatase measurements exceed their thresholds, so that the patient enters into state 2. He or she remains there throughout the first 90 day interval, but in the second interval the alkaline phosphatase level drops below 120 while the acid phosphatase level remains above 2, so that the patient moves (makes a transition) to state 4 at time 2. The patient then regresses to state 2 at time 3 and is dead (state 1) at time 4 [10].

These simplifying assumptions reduce considerably the number of parameters to be estimated but they may not always be realistic from a clinical point of view. However, they may be tested quite easily; even if they do not hold, this information may be valuable in itself.

B. The Homeostatic Autoregressive Model

A time series model (the homeostatic model) sufficiently simple for medical applications was brought to the attention of clinical chemists by Harris [11,12], who published an excellent review on the potential applications in clinical chemistry of this as well as other related models. We will review here the theory and application of this model.

The homeostatic model is of the so-called autoregressive type with an autocorrelation equal to zero. The model is based on the assumption that the quantity values observed in a patients over time vary at random relative to a mean value that characterizes the particular patient. When the patient is in a state of control, the mean and variance of the values are both stable. When one is applying this model to a patient, the mean and variance are estimated from the first two observations. The predicted value of the third observation

Multivariate Assessment of Laboratory Tests

is the mean of the first two observations. To test if the third observation is within the predicted limits, a simple t-test, based on the mean and variance of the first two observations, is performed. If the observation does not deviate significantly (at the significance level chosen) from the predicted value, it is added to the first two observations and an updated estimate of the mean and variance is calculated and used to test if the fourth observation is within the predicted limits, and so forth.

A more rigorous and mathematically precise definition of this model is given by the equation:

$$X_i = u + (X_{i-1} - u)\rho + b_i + a_i \quad (i = 2,n)$$

where u is the mean or set point characterizing the subject, ρ is the autoregressive correlation coefficient, b_i (i = 2,n) are independent random biological deviates following a Gaussian distribution with a mean zero and standard deviation σ_B, a_i (i = 2,n) are independent analytical deviates following a Gaussian distribution with mean zero and standard deviation σ_A, and a_i and b_i are statistically independent.

When $\rho = 0$ (in the homeostatic model), the equation reduces to $X_i = u + C_i$ and $C_i = b_i + a_i$ having a mean of zero and a standard deviation of

$$\sigma_B^2 + \sigma_A^2$$

Since the observations follow a Gaussian distribution with mean u and the standard deviation shown above, the predicted value of the ith observation is given by

$$\sum_{j=1}^{i-1} X_j/(i-1) = \overline{X}_{i-1}$$

the estimated variance is given by

$$\sum_{j=1}^{i-1} (X_j - \overline{X}_{i-1})^2/(i-2) = \sigma_{i-1}^2$$

and the upper alarm limit at the α level of significance is given by

$$\overline{X}_{i-1} + t_{i-2,1-\alpha} \cdot \sigma_{i-1}(i/(i-1))^{1/2}$$

where $t_{i-2,1-\alpha}$ is the $1 - \alpha$ fractile of the t-distribution with $i - 2$ degrees of freedom.

1. Adapting the Homeostatic Model to Specific Clinical Problems

The homeostatic time series analysis model seems reasonable from a biochemical point of view. The model also offers a number of practical advantages. Few observations are needed to obtain the initial estimates of the parameters for a patient. It is not necessary to use the same time interval between consecutive measurements and, as long as the analytical variation is stable, it is not necessary to know its actual value.

However, there is an unresolved statistical problem as well as problems related to specificity and the proper use of the data. The statistical problem involves the significance level to be chosen when testing if a single value deviates significantly from the predicted one. If n independent significance tests are performed at the 5% level of significance, the probability of obtaining at least one significant result is not 5% but $1 - (1 - 0.05)^n$ (i.e., one minus the probability of obtaining an insignificant result in each of the n trials).

The problem related to specificity is caused by an inherent weakness in this as well as a number of other statistical models. These models are not explicitly designed to be sensitive to the types of deviations expected to be found when a patient develops a pathologic lesion, such as, for example, the recurrence of disease.

The third problem relates to the variance components. The variance observed in a patient is the sum of the biological and analytical variances. Its estimated magnitude is initially based on very few observations. The analytical variance, however, may be estimated independently and quite reliably. Therefore, this added advantage should be incorporated in the application of this model.

To resolve these problems, Winkel et al. [13] proposed to use an independent estimate of the analytical variance instead of the total variance estimated on the basis of the patient's data whenever the latter was smaller than the former. The reason for this is that if the estimated total variance is smaller than a reliable estimate of the analytical variance, it must be caused by chance due to the low precision of the total variance estimate. Furthermore, they adjusted the significance test procedure to make the model less sensitive to unspecific aberrations and more sensitive to persistent deviations believed to be unique for the development of recurrence. This was done by using two significance levels: a high level that a single value must exceed in order to create an alarm condition (type 1) and a lower level that two consecutive or more than two values must exceed in order to create an alarm condition (type 2). They finally adjusted the two significance levels so that the overall probability of obtaining an alarm condition of either category is 1% when 10 independent trials are performed, that is, when 10 consecutive values are examined in a patient who is in a state of control. The two significance levels were chosen in such a way that the probability of a type 1 alarm condition is 0.2% and that of a type 2 alarm condition is 0.8%, provided that the patient is in a state of control.

Figure 5 shows the serum CEA levels in three patients, all of whom developed recurrence following operation for primary breast cancer. Had the homeostatic model been used to monitor these patients, one threshold value would have been used instead of two, so that the overall probability of obtaining at least one alarm condition in 10 consecutive trials was 1%. The white arrows in Figure 5 indicate when this procedure would have caused an alarm condition. It is noted that had the conventional procedure been used, recurrence in one patient (R9) would have been missed, the detection of recurrence in another patient (R7) would have been delayed for 1 month, while recurrence in the third patient would have been detected at the same time, which actually was the case when the modified method was used. Recurrence in a fourth patient, whose data are not shown in the figure, was detected at the same time as detection occurred clinically. Had the conventional approach been used, recurrence in this patient would have passed unnoticed.

This type of model evaluation based on relatively few clinical examples obviously is not very satisfactory. However, once approximate estimates of the important quantities have been obtained, a more extensive analysis may be carried out using simulation techniques.

A simulation is done by a computer typically programmed to generate 10,000 time series simulating 10,000 sets of patient data. The mean, autocorrelation, trend, and the analytical and biological coefficients of variation are first presented to the computer. Based on the mean, the autocorrelation, and the trend, the computer determines the expected value for each data point in a time series. To each value a random deviation is

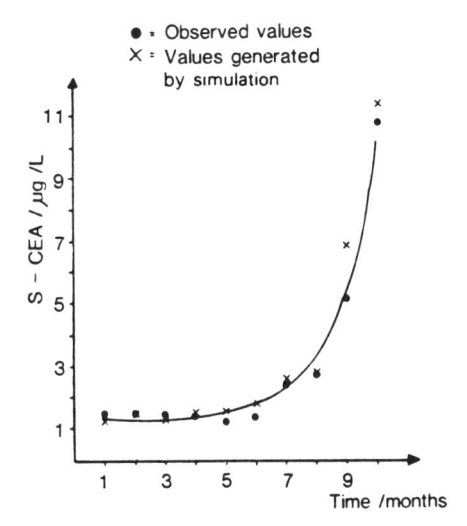

Figure 7 Postoperative serum CEA values (●) from a patient who underwent surgery for primary breast cancer. Clinically overt recurrence was noted shortly after the last serum specimen was obtained. The curve is fitted to the observations according to a mathematical model. Values (x) generated by computer simulation and based on this model are also shown. The model states that the serum CEA level is a sum of two components: one not related to the tumor and one that reflects the size of the tumor. It is furthermore assumed that the growth of the tumor is exponential. The following equation specifies the model:

$$Y(t) = \alpha + kT(t_0)e^{\beta(t-t_0)}$$

where $Y(t)$ is the CEA concentration at time t, α the constant component and $kT(t_0)e^{\beta(t-t_0)}$ the variable component (i.e., the trend). In the latter expression, k is the (unknown) proportionality constant relating tumor size to plasma CEA level, $T(t_0)$ is the tumor size at some arbitrary time (t_0), and β is the growth rate of the tumor [13].

then added that is generated on the basis of the specified analytical and biological standard deviations and a program for random number generation.

To generate the trend it is necessary to construct a mathematical model that is sensible from a biological point of view and also realistically describes the data obtained from patients in whom disease recurs. Figure 7 explains the model of a CEA-secreting tumor [13]. A simulation study, in which this model was used to generate the trend, showed that the modified homeostatic time series model was superior to the conventional homeostatic model and that both time series models were superior to a group-based decision level [13].

V. SUMMARY

In this chapter the application of multivariate techniques for the assessment of laboratory tests in cancer patients has been reviewed. We emphasize that the transformation of laboratory test values into just two categories (normal or abnormal) may entail a considerable loss of information. For instance, correlation between two laboratory tests that may be important for differentiating among various clinical categories of patients may disappear when this procedure is used.

When only a single set of laboratory results measured in the same specimen is available for a given patient, we must compare these values to those obtained from other patients or healthy subjects to make inferences about the patient on the basis of the laboratory results. Thus, the analysis of the data must be group based. Discriminant analysis, logistic regression analysis, and survival analysis based on Cox's regression model are the techniques most often used in this situation.

By contrast, when previous results are available from the same patient we may compare his or her present values to those previously obtained when we want to make inferences about the patient. Our objective is to make a prediction about the time that will elapse until some specified event (death or recurrence of disease) occurs. Two models that have been applied in this situation—the Markov chain and the autoregressive time series model—were reviewed and examples of specific medical applications presented.

REFERENCES

1. Winkel P, Statland BE, Bentzon MW. Mathematical models used to assess laboratory measurements in patients with cancer; biological time-series and multivariate analysis. In Statland BE, Winkel P, eds. Laboratory measurements in malignant disease. Saunders Co., Philadelphia: WB Saunders, 1982.
2. Cox DR. Regression models and life tables. J R Stat Soc [B] 1972;34:187-220.
3. Kay R. Proportional hazard regression models and the analysis of censored survival data. Appl Stat 1977;26:227-237.
4. Reinhardt PS (ed). SAS Supplemental Library User's Guide. Cary, NC: SAS Institute, 1980.
5. Drzewiecki KT, Kragh Andersen P. Survival with malignant melanoma: a regression analysis of prognostic factors. Cancer 1982;49:198-203.
6. Gail MH. Evaluating serial cancer marker studies in patients at risk of recurrent disease. Biometrics 1981;37:67-78.
7. Metzger BL, Schultz S. Time series analysis: an alternative for nursing. Nurs Res 1982;31:375-379.
8. Box GEP, Jenkins GM. Time series analysis, forecasting and control. San Francisco: Holden-Day, 1970.
9. Kemeny J, Snell JL. Finite Markov chains. New York: Springer Verlag, 1976.
10. Myers LE, Paulson DF, Berry WR, Cox EB, Laslo J, Stanley W. A time-dependent statistical model which relates current clinical status to prognosis: application to advanced prostatic cancer. J Chron Dis 1980;33:491-499.
11. Harris EK. Some theory of reference values. I. Stratified normal ranges and a method for following an individual's clinical laboratory values. Clin Chem 1975; 21:1457-1464.
12. Harris EK. Some theory of reference values. II. Comparison of some statistical models of intraindividual variation in blood constituents. Clin Chem 1976;22: 1343-1350.
13. Winkel P, Bentzon MW, Statland BE, et al. Predicting recurrence in patients with breast cancer based on cumulative laboratory results. A new technique for the application of time series analysis. Clin Chem 1982;28:2057-2067.

4
Use of Multiple Markers to Enhance Clinical Utility

DONALD W. MERCER

Montefiore Hospital and Tumor Marker Laboratory, Pittsburgh Cancer Institute, Pittsburgh, Pennsylvania

I. INTRODUCTION

The search for a unique biochemical "marker" of malignancy has been undertaken for decades by biochemists and immunologists. Although a wide variety of biochemical substances (oncofetal proteins, isoenzymes, mucin-type glycoproteins, and hormones) have been assayed in connection with the management of patients with late-stage neoplastic disease, no single serological marker has proven to be a sensitive and specific indicator of early-stage malignancy.

A promising approach to overcoming the nonspecificity and insensitivity of single tumor markers is the simultaneous assay of several markers, based on the premise that cancer cells are biochemically heterogeneous and may synthesize a broad spectrum of possible tumor markers [1]. To avoid the possibility of missing a potential cell marker, a battery of assays might offer the best opportunity to find one or more markers of malignant disease.

Previous reports describing the use of multiple markers were reviewed by McIntire [2] in the first edition of this book. Since that time, many new markers have been introduced. This chapter will examine the recent published work of investigators attempting to use simultaneous measurements of both traditional and new markers in the circulation as aids to the management of cancer. We will examine panels of markers with respect to particular types of cancer and only those types with currently promising panels will be discussed.

II. COLORECTAL CANCER

Since the introduction of carcinoembryonic antigen (CEA), the relationship between CEA and colonic carcinoma has been studied extensively. Persistently high or increasing concentrations of CEA in serum in the postoperative period have been found to be strongly suggestive of residual or metastatic disease. However, the hope that CEA or other markers would support early and specific diagnoses of malignancy has not been realized. At present, many colonic neoplasms are detected at a fairly late stage of disease, when curative therapy is not available.

Table 1 Serum Tumor Marker Panels in Colorectal Cancer

Investigator	Cancer Stage	Cutoff Level	Marker	Sensitivity No. Pts. (%) Single	Panel	Net Gain (%)
Mercer [3]	L	4.0 ng/ml	CEA	40/71 (78)	45/51 (88)	10
		35 U/ml	CA 19-9	33/51 (65)		
Novis [4]	E&L	5.0 ng/ml	CEA	163/220 (74)	170/220 (77)	3
			CA 19-9	96/220 (44)		
Guar [5]	E&L	NR	CEA	14/33 (42)	22/33 (67)	19
		10 U/ml	CA 195	16/33 (48)		
Chu [6]	E&L	5.0 ng/ml	CEA	42/72 (58)	57/72 (79)	21
		107 mg/dl	AAG	41/72 (57)		
Gupta [7]	Dukes' C	NR	CEA	12/35 (34)	17/29 (59)	18
		15 U/ml	CA 195	11/29 (37)		
Quentmeier [8]	NR	5.0 ng/ml	CEA	64/177 (36)	80/177 (45)	9
		25 U/ml	CA 19-9	35/177 (20)		
		27 U/ml	CA 125	18/177 (10)		
Kawahara [9]	E&L	5.0 ng/ml	CEA	32/74 (43)	42/74 (57)	14
		37 U/ml	CA 19-9	25/74 (34)		
		35 U/ml	CA 125	5/74 (7)		
		20 U/ml	SLEX	21/74 (28)		
Mercer[10]	E&L	3.6 ng/ml	CEA	12/23 (52)	22/23 (96)	22
		0.5 mg/dl	CRP	17/23 (74)		
		77 mg/dl	AAG	17/23 (74)		
		6.5 U/L	FHAP	10/23 (44)		
		1.5 U/L	CK-M2	13/56 (56)		

L, late; E, early; NR, not reported; CEA, carcinoembryonic antigen; AAG, alpha-1 acid glycoprotein, CRP, C-reactive protein; FHAP, fast homoarginine-sensitive alkaline phosphatase; CK-M2, macro-creatine kinase type 2; SLEX, sialylated Lewis x antigen; CA, cancer antigen.

In the late 1970s, several investigators began to perform simultaneous measurements of markers as an alternative approach to the use of CEA alone. As reviewed by McIntire [2], this approach was not successful due to the limited availability of colon-specific markers. Many new tumor-associated markers of the gastrointestinal tract such as the mucin-type glycoproteins (CA 19-9, CA 195, and CA 50) and creatine kinase isoenzymes (CK-BB and CK-M2) have been investigated and reports utilizing these potential colon markers, both singly and in combination with CEA, are now appearing in the literature.

Table 1 lists several recent reports describing the use of multiple markers in sera of patients with colorectal cancer. The detection rates for the single marker CEA are compared with combinations of various markers. Five panels of two markers, one panel of three markers, one panel of four markers, and one panel of five markers are reported. These studies demonstrate the value of the multiple-marker approach, with significant gains (10% or greater) in sensitivity observed for six of eight panels.

The panel of CEA/CA 19-9 was evaluated by two different investigators and slightly different results, most likely due to differences in patient groups, were observed. The report of Mercer [3] demonstrated sensitivity gains of 10% in patients with late-stage disease and Novis [4] found very little difference in sensitivity (3%) in a group of patients with both early- and late-stage disease. The panel CEA/CA 195 was also evaluated by two different investigators and almost identical gains in sensitivity (18 and 19%) were observed. The highest sensitivity score for a colon panel was reported by Mercer and Talamo [10]. A five-test combination of two acute-phase proteins, CEA, and two isoenzymes resulted in 96% sensitivity. Thus, the combination of CEA with either mucin-type glycoproteins markers (CA 195 and CA 19-9) or acute-phase proteins (AAG and CRP) and isoenzymes (CK-M2 and FHAP) resulted in panels that were considerably more sensitive for colorectal cancer than CEA alone.

However, results of panel testing in patients with benign disease have revealed problems of nonspecificity. For example, in four studies in which samples from patients with benign disease were tested, specificity scores for panels were as follows: Mercer CEA/CA 19-9 (83%), Chu AAG/CEA (67%), Quentmeier CEA/CA 19-9/CA 125 (89%) and Mercer CEA/CRP/AAG/FHAP/CK-M2 (70%). These scores in general were about 16% lower than those of CEA alone.

Although loss in specificity due to the use of colon panels was exhibited here, strategies such as serial testing and the use of intraindividual normal ranges would appear to assist in minimizing this problem. It is unfortunate that only isolated case reports and preliminary clinical studies with serial testing are presently available. Thus, more studies utilizing strategies described in this chapter are needed to evaluate properly the clinical value of these colon panels.

III. PANCREATIC CANCER

Early attempts to detect pancreatic cancer with markers focused on CEA, pancreatic oncofetal antigen, elastase, and ribonuclease. Unfortunately, these markers exhibited only limited sensitivity in early-stage cases and poor specificity in differentiatiing pancreatic cancer patients from those with benign disease.

Several pancreatic tumor-associated antigens such as PaA, PCAA, CA 19-9, CA 50, DU-PAN-2, and CA 195 have recently been reported to exhibit some clinical utility in the management of pancreatic cancer and they are now being extensively evaluated, both singly and in combination.

Table 2 lists the results of seven studies that illustrate the use of panels in the management of pancreatic cancer. Seven panels, consisting of two markers each, are presented. Six of seven panels contain the marker CA 19-9, which is considered by many to be the best single marker for pancreatic cancer. Improved sensitivity (greater than 10%) was exhibited in four of seven panels. Most impressive were those of Takasaki [15] CA 19-9/DU-PAN-2 and Sakamoto [16] CA 19-9/CA 125, with sensitivity of 95 and 97%, respectively. Conflicting results were found when sensitivity gains in three CA 19-9/CA 50 panels were compared; Harmenberg [14] observed a 10% gain in sensitivity but Haglund [12] and Paganuzzi [13] reported little or no gain.

Evaluations of specificity have been performed in four of these pancreatic panels and test results in patients with nonmalignant disease are encouraging. Minimal specificity loss was observed for the Loor [16] panel (91% single, 85% panel) and the Sakamota [15] panel (92% single, 89% panel). Slightly greater specificity loss was found for the

Table 2 Serum Tumor Marker Panels in Pancreatic Cancer

Investigator	Cancer Stage	Cutoff Level	Marker	Sensitivity No. Pts. (%) Single	Panel	Net Gain (%)
Loor [11]	Late	19 µg/ml	PCAA	40/76 (53)	68/76 (90)	24
		21.5 ng/ml	PaA	50/76 (66)		
Haglund [12]	Late	37 U/ml	CA 19-9	73/95 (77)	74/95 (78)	1
		17 U/ml	CA 50	67/95 (71)		
Paganuzzi [13]	Late	37 U/ml	CA 19-9	22/26 (85)	22/26 (85)	0
		17 U/ml	CA 50	21/26 (81)		
Harmenberg [14]	Late	37 U/ml	CA 19-9	59/72 (82)	66/72 (92)	10
		14 U/ml	CA 50	58/72 (81)		
Takasaki [15]	Late	37 U/ml	CA 19-9	16/21 (73)	20/21 (95)	22
		300 U/ml	DU-PAN-2	14/21 (67)		
Sakamoto [16]	Late	37 U/ml	CA 19-9	26/30 (87)	29/30 (97)	10
	Over 50 yr	32 U/ml	CA 125	20/30 (67)		
	Under 50 yr	143 U/ml				
Haglund [17]	Late	37 U/ml	CA 19-9	74/95 (78)	80/95 (84)	6
		35 U/ml	CA 125	43/95 (45)		

PCAA, pancreas cancer-association antigen; PaA, pancreas-specific antigen.

two panels of Haglund [12,17]; CA 19-9/CA 125 (78% single, 60% panel) and CA 19-9/CA 50 (76% single, 65% panel).

Although sensitivity and specificity scores with the above pancreatic panels are not yet perfect (100%), future studies involving strategies such as serial testing and the use of intraindividual normal range levels should help considerably to minimize losses in specificity and maximize gains in sensitivity.

IV. LUNG CANCER

In the past, several tumor-associated products such as adrenocorticotropic hormone (ACTH), calcitonin (CT), human chorionic gonadotropin (HCG), parathyroid hormone (PTH), carcinoembryonic antigen (CEA), and alpha fetoprotein (AFP) have been evaluated with regard to the diagnosis, staging, and treatment of lung cancer. Although numerous markers have been associated with small-cell lung cancer and to a lesser extent non-small-cell lung cancer, none of these markers has proven to be highly sensitive and specific. Various combinations of potential markers have been studied in an attempt to rectify

Table 3 Serum Tumor Marker Panels in Lung Cancer

Investigator	Cancer Stage	Cutoff Level	Marker	Sensitivity No. Pts. (%) Single	Panel	Net Gain (%)
Dnistrian [18]	All types Mets	5 ng/ml	CEA	27/42 (64)	36/42 (86)	12
		20 mg/dl	LASA	31/42 (74)		
Gail [19]	All types Advanced	3.2 ng/ml	CEA	104/171 (61)	148/171 (87)	27
	M	92.1 ng/dl	SA	70/171 (41)		
	F	78.5 ng/dl	SA			
Ganz [20]	SCLC E&L	5.0 ng/ml	CEA	23/46 (50)	43/46 (93)	4
		70.0 mg/ml	AAG	50/56 (89)		
Jacques [21]	SCLC E&L	5.0 ng/ml	CEA	84/193 (44)	97/116 (84)	18
		12.5 ng/ml	NSE	111/168 (66)		
		10.0 ng/ml	CK-BB	40/123 (32)		
Bork [22]	SCLC E&L	3.5 ng/ml	CK-BB	16/25 (64)	20/35 (80)	16
		15 ng/ml	NSE	10/25 (40)		
		75 pmol/l	ACTH	9/36 (28)		
		36 pmol/l	CT	8/29 (27)		
Buccheri [23]	All types E&L	100 U/l	TPA	51/98 (52)	83/98 (85)	33
		5 ng/ml	CEA	36/98 (37)		
		240 U/l	LDH	25/98 (25)		
		10 mIU/ml	β-HCG	18/98 (18)		
Kawahara [9]	All types E&L	5.0 ng/ml	CEA	41/89 (46)	57/89 (64)	18
		37 U/ml	CA 19-9	18/89 (20)		
		35 U/ml	CA 125	29/89 (33)		
		20 U/ml	SLEX	20/89 (22)		

SCLC, small-cell lung cancer; M, male; F, female; SA, sialic acid; CT, calcitonin, Mets, metastases; E&L, early and late.

this situation; these initial attempts were reviewed by McIntire [2] in the first edition of this book. Although past results with various marker panels have been encouraging, no panel was found to be effective in the detection of early-stage malignancy. There was also no ability to differentiate patients with lung cancer from those with benign disease.

To improve the utility of markers in the clinical setting, there has been an intensive search for new lung markers. A serum mucin-type glycoprotein for lung cancer, which is similar to the tumor-associated mucin markers of colon, breast, and pancreas, has not yet been discovered. However, several new isoenzyme-type markers such as neuron-specific enolase (NSE), and creatine kinase BB (CK-BB) along with lipid-associated sialic acid (LASA) and the neuropeptide bombesin have recently been reported to be lung markers of considerable potential and panels incorporating these markers are now being evaluated.

Table 3 lists several recent reports describing the use of multiple markers in sera of patients with lung cancer. Three panels of two markers, one panel of three markers, and three panels of four markers are reported. As observed with colorectal panels, CEA, which is present in six of the seven panels here, is frequently used as the reference marker against which to measure the value of a new lung marker. Dnistrian et al. [18] found that the combination of CEA and LASA enhanced sensitivity by 12% and Gail [19] found total sialic acid and CEA to be a good combination (27% sensitivity gain). Gail's study included seven other markers (ferritin, beta 2-microglobulin, lipotropin, beta-HCG, alpha-HCG, calcitonin, and PTH). All of these markers were tried in various combinations and the panel of CEA and total sialic acid was found to be best for differentiating patients with advanced cancer from those with benign lung disease and from healthy normal controls. A subsequent study by Gail [24] in patients with early lung cancer found CEA alone to be the best marker. Chapter 2 provides additional details concerning the statistical analysis involved here.

Use of the isoenzymes NSE and CK-BB by Jacques [21] and Bork [22] in combination with CEA resulted in sensitivity gains of 18 and 16%, respectively. However, the greatest gain in sensitivity (33%) was demonstrated by Buccheri [23] with a panel of four markers: CEA, tissue polypeptide antigen (TPA), lactate dehydrogenase (LDH), and β-HCG.

Evaluation of specificity has been limited to the panels of CEA/LASA and CEA/sialic acid in which specificity scores of 70% and 60% have been reported. The low specificity scores observed for these two panels indicate that CEA, LASA, and sialic acid are very nonspecific lung markers and the search for new lung markers must continue. Perhaps the other lung panels described here with markers such as NSE, CK-BB, and TPA will provide improved specificity, especially when studies involving serial testing are utilized.

V. BREAST CANCER

The use of traditional types of tumor markers such as CEA in the diagnosis and management of breast cancer has been attempted in the same manner as reported for gastrointestinal cancer. Although CEA levels are sometimes found elevated in patients with breast cancer, low incidence levels and insensitivity to early-stage disease have resulted in the limited use of CEA in treatment.

Initial attempts to measure other markers simultaneously with CEA have been previously reviewed by McIntire [2]. Results from these reports suggested that panels of mul-

Table 4 Serum Tumor Marker Panels in Breast Cancer

Investigator	Cancer Stage	Cutoff Level	Marker	Sensitivity No. Pts. (%) Single	Panel	Net Gain (%)
Tondini [27]	Mets	3.0 ng/ml	CEA	14/24 (58)	20/24 (83)	8
		22 U/ml	CA 15-3	18/24 (75)		
Stacker [28]	Mets	2.5 ng/ml	CEA	35/57 (61)	56/57 (98)	5
		300 IU	MSA	53/97 (93)		
Bray [29]	Active	5.0 ng/ml	CEA	16/51 (31)	30/51 (59)	6
		10 U/L	CA 549	27/51 (53)		
Mercer [30]	Active	60 U/ml	CA M26	43/72 (60)	62/66 (94)	19
		22 U/ml	CA M29	53/72 (74)		
		3.5 U/L	CK-BB	40/71 (56)		
Linsley [31]	Mets	48 U/ml	CA M26	43/72 (60)	69/72 (96)	3
		18 U/ml	CA M29	67/72 (93)		
		24 U/ml	CA 15-3	60/72 (83)		
Kawahara [9]		5.0 ng/ml	CEA	16/92 (17)	40/92 (43)	24
		37 u/ml	CA 19-9	13/92 (14)		
		35 U/ml	CA 125	17/92 (18)		
		20 U/ml	SLEX	17/92 (19)		

MSA, mammary serum antigen.

tiple markers are capable of providing clinically significant information. Although these initial reports are encouraging, subsequent studies by Coombes et al. [25,26] with a variety of markers (n = 21) demonstrated that most of the markers are nonspecific and few if any are abnormal in early breast cancer. Nevertheless, three markers from this group (alkaline phosphatase, CEA, and gamma-glutamyl transpeptidase) were found to provide a useful combination for detecting metastatic disease, in about half of the patients studied, before it was detected by clinical examination.

Recently, with the technological development of monoclonal antibodies directed toward mucin-type glycoproteins, several new immunoassays (CA 15-3, CA 549, CA M26, CA M29, and MSA) for breast cancer have been developed and reports of studies with these new assays, both singly and in combination with traditional markers such as CEA, are now appearing in the literature.

Table 4 lists several recent studies designed to assess the performance of multiple markers in breast cancer. Here the sensitivity scores for single markers are compared with marker panels. Three panels of two markers, two panels of three markers, and one panel of four markers are reported. As demonstrated in the first three studies, CEA is often used as a marker with which to compare new breast cancer markers. In all three studies the new breast markers, CA 15-3, CA 549, and MSA alone, performed much better than CEA alone. Moreover, the combinations of CEA with CA 15-3, CA 549, or

MSA offered no significant advantage over the use of CA 15-3, CA 549, or MSA alone. However, most impressive were the marker panels of CA M26/CA M29/CK-BB and CA M26/CA M29/CA 15-3, with sensitivity scores of 94 and 96%, respectively. Although losses in specificity were expected and reported for the Mercer [30] panel of CA M26/CA M29/CK-BB (100% single, 80% panel), serial testing and the establishment of intraindividual normal range levels for each panel marker should help to minimize this problem. Many panel combinations of the above markers have not yet been tried and more studies of this type will perhaps yield other panels of even greater clinical utility.

VI. OVARIAN CANCER

The search for a reliable and sensitive tumor marker to detect early ovarian carcinoma, especially of the epithelial type, has been continuing for many years. There is a real need for tumor markers here, since at the time of diagnosis about 70% of epithelial ovarian cancers have already spread beyond the pelvis and treatment by surgery or radiation therapy is no longer effective. Unfortunately, unlike the rare germ cell tumors of the ovary, the more common epithelial neoplasms do not produce large amounts of AFP or HCG and none of the traditional markers such as CEA have proven to be very useful since incidence levels are low (30-50%) and levels of elevation are modest.

Several mucin-type tumor-associated ovarian antigens such as CA 125, DF 3, CA 19-9, NG/70K, HMFG-1, and HMFG-2, along with the isoenzyme placentyl alkaline phosphatase (PLAP), have recently been introduced with encouraging results. Table 5 lists several recent studies designed to assess the performance of these potential ovarian markers, both singly and in various combinations.

Four panels of two markers, three panels of three markers, and two panels of four markers are presented. All panels contained CA 125, which is considered by many investigators to be the most sensitive ovarian marker available. In three of the seven panels, the combination of CA 125 with DF 3, CA 19-9, or CEA/CA 19-9 offered only minimal gains in sensitivity (average, 6%). However, in six panels 10% or greater increases in sensitivity were observed when panel results were compared to the use of CA 125 alone. Very impressive is the study of Negishi [39] in which a four-test combination of CA 125, TPA, ferritin, and IAP resulted in 100% sensitivity and 100% specificity. Also of note is the work of Ward [38], who found a marked increase in sensitivity (37%) in patients with early-stage disease when a three-test combination of CA 125, HMFG-2, and PLAP was used.

Unfortunately, in most of these reports, patients with benign disase were not studied. However, other studies have shown CA 125 levels to be elevated in patients with various nonmalignant liver and biliary diseases [40] as well as nonovarian malignancies [41]. Thus, a problem of specificity exists for CA 125, and several investigators have attempted to improve it by combining CA 125 with other markers such as CA 15-3 or CA 72-4 and restricting positive results to samples that express abnormal amounts of both markers. These studies are summarized in Table 6; specificity approaching 100% with gains ranging from 13 to 20% are observed.

Wu [45] approached the specificity problem in another manner. His group found that the use of the ratio of CA 125/CEA improved the specificity of CA 125 for ovarian cancer. The mean ratio for nonovarian disease was 0.94 (n = 94); for ovarian cancer patients 916 (n = 48). Another investigator [46] has suggested that PLAP might be an interesting marker to combine with CA 125 because of PLAP's high specificity and mod-

Table 5 Serum Tumor Marker Panels in Epithelial Ovarian Cancer

Investigator	Cancer Stage	Cutoff Level	Marker	Sensitivity No. Pts. (%) Single	Panel	Net Gain (%)
Dembo [32]	NR	NR	CA 125	NR (80)	NR (98)	18
			NB/70K	NR (80)		
Sekline [33]	2&3	35 U/ml	CA 125	33/40 (83)	36/40 (90)	7
		30 U/ml	DF 3	20/40 (50)		
Canney [34]	E&L	35 U/ml	CA 125	42/55 (76)	44/55 (80)	4
		35 U/ml	CA 19-9	16/55 (29)		
Fioretti [35]	E&L	65 U/ml	CA 125	15/21 (71)	21/21 (100)	29
		65 U/ml	CA 19-9	10/21 (48)		
Dhokia [36]	E&L	35 U/ml	CA 125	72/85 (85)	81/85 (95)	10
		0.13 OD	HMFG1	48/85 (56)		
		0.13 OD	HMFG2	55/85 (65)		
Bast [37]	E&L	2.5 ng/ml	CEA	29/105 (37)	95/105 (90)	7
		37 U/ml	CA 19-9	18/105 (17)		
		35 U/ml	CA 125	87/105 (83)		
Ward [38]	1&2	35 U/ml	CA 125	2/11 (27)		37
		1500 U	HMFG2	3/11 (27)	7/11 (64)	37
		NR	PLAP	3/11 (27)		
Negishi [39]	E&L	35 U/ml	CA 125	18/22 (80)		
		110 U/ml	TPA	14/22 (64)	22/22 (100)	20
		500 µg/ml	IAP	16/22 (73)		
		120 ml	Ferr	14/22 (64)		
Kawahara [9]	E&L	5.0 ng/ml	CEA	12/82 (15)	44/82 (54)	11
		37 U/ml	CA 19-9	14/82 (17)		
		35 U/ml	CA 125	35/82 (43)		
		20 U/ml	SLEX	8/82 (10)		

PLAP, placentyl alkaline phosphatase; IAP, immunosuppressive acidic protein; TPA, tissue polypeptide antigen; Ferr, ferritin; NR, not reported; OD, optical density.

Table 6 Serum Tumor Marker Panels in Benign Disease

Investigator	Cutoff Level	Marker	Specificity No. Pts. (%) Single	Panel[a]	Net[b] Gain(%)
Yedema [42]	35 U/ml	CA 125	24/32 (75)	30/32 (95)	20
	30 U/ml	CA 15-3	29/32 (91)		
Soper [43]	65 U/ml	CA 125	40/47 (85)	46/47 (98)	13
	10 U/ml	CA 72-4	46/47 (98)		
Einhorn [44]	65 U/ml	CA 125	134/166 (81)	163/166 (98)	17
	10 U/ml	CA 72-4	NR		

[a]Both test results elevated.
[b]Compared to CA 125 alone.

erate sensitivity for ovarian cancer but clinical studies demonstrating the effectiveness of this combination have not yet been presented.

VII. TESTICULAR CANCER

In the late 1970s, preliminary reports began to appear demonstrating the clinical usefulness of AFP and HCG as unique tumor markers for testicular germ cell cancer. These initial reports of simultaneous marker use were reviewed about 10 years ago by McIntire [2] and since that time AFP and HCG determinations have helped to increase the cure rate for testicular cancer from 60% to over 90% in all but the most advanced cases.

Since germ cell tumors are so often made up of diverse types of cells, some patients with nonseminomatous tumors have elevations of only one marker, whereas others have elevations of both. Furthermore, the proportion of the various type of cells may change over time, thus altering patterns in serum. Thus, continued measurement of both markers throughout patient's clinical course is required for effective clinical use.

Table 7 lists three panels consisting of two markers designed to monitor testicular cancer. As shown by Scardino [47] over 10 years ago, combined use of AFP and HCG as tumor markers for testicular cancer of the nonseminomatous type resulted in a 24% sensitivity gain (77% single, 91% panel). However, the AFP/HCG panel was ineffective in documenting germ cell testicular tumors of the seminoma type since this type of tumor does not produce AFP and HCG is only present in about 20% of patients. Thus, other markers are necessary here. As shown in Table 7, a panel consisting of HCG and PLAP has been examined by Lange [48]: sensitivity gains of 14% have been achieved (57% single, 71% panel). For additional information about the potential value of PLAP and LDH-LDH-1 as markers for seminoma-type testicular tumors, see Chapter 25 in this volume.

Table 7 Serum Tumor Marker Panels in Testicular Cancer

Investigator	Cancer Stage	Cutoff Level	Marker	Sensitivity No. Pts. (%) Single	Panel	Net Gain (%)
Scardino [47]	Seminoma	1 ng/ml	HCG	2/9 (22)	2/9 (22)	0
		40 ng/ml	AFP	0/9 (0)		
Lange [48]	Seminoma	1 ng/ml	HCG	14/28 (50)	20/28 (71)	14
		1.9 ng/ml	PLAP	16/28 (57)		
Scardino [47]	Nonseminoma	1 ng/ml	HCG	28/36 (77)	33/36 (91)	24
		40 ng/ml	AFP	21/36 (58)		

VIII. PROSTATE CANCER

The search for a reliable prostate-associated tumor marker has been underway for more than 50 years. From the mid-1950s to the early 1980s, interest focused on the isoenzyme prostatic acid phosphatase (PAP). Recently attention has turned to prostate-specific antigen (PSA), which appears to be a more sensitive indicator of early-stage disease (see Chapter 22 for a more detailed discussion of PSA). Currently the best marker is PSA, but it is still far from ideal.

Since PSA is biochemically and immunologically distinct from PAP, the possibility exists that these two prostate markers might perform better in combination as a panel. Studies by Kuriyama [49], as shown in Table 8, have demonstrated that combined testing of PAP and PSA results in an average sensitivity gain of 16% for all stages of prostatic cancer. They also found excellent specificity (90%) with only a 7% loss observed for the PAP/PSA panel.

Recently Siddall [50] tried the combination of gamma-seminoprotein (GSP) and PSA and found it no more effective than either GSP or PSA alone. Several new markers of prostate cancer have appeared (see Chapter 22) and these markers in combination with PAP and PSA also need to be evaluated.

Killian [51] studied the prognostic significance of five markers: acid phosphatase, total alkaline phosphatase (TAP), bone alkaline phosphatase (BAP), PAP, and PSA for progression of prostate cancer. Although various combinations of these markers were not evaluated, the apparent order of prognostic reliability for progression of disease as determined with Cox's regression model for each marker was shown to be PSA > PAP > BAP > ACP > TAP.

Although false-positive PAP and PSA values, especially from patients with benign prostatic hyperplasia (BPH), are a major problem, the use of serial testing as discussed by Cohen and Dix [52] would appear to be an excellent way to improve specificity. They recommended the establishment of more accurate intraindividual normal ranges for markers through serial testing. This approach to normal values would make it easier to interpret slight increases in the concentration of a marker in a given patient, because the patient's own normal value would form the basis of comparison.

Table 8 Serum Tumor Marker Panels in Prostate Cancer

Investigator	Cancer Stage	Cutoff Level	Marker	Sensitivity No. Pts. (%) Single	Panel	Net Gain (%)
Kuriyama [49]	A	15 ng/ml	PAP	4/12 (33)	7/12 (58)	25
		7.5 ng/ml	PSA	3/12 (25)		
	B	15 ng/ml	PAP	15/36 (42)	21/36 (58)	16
		7.5 ng/ml	PSA	10/36 (28)		
	C	15 ng/ml	PAP	18/28 (64)	19/28 (68)	4
		7.5 ng/ml	PSA	10/28 (36)		
	D	15 ng/ml	PAP	86/116 (74)	106/116 (92)	18
		7.5 ng/ml	PSA	74/116 (64)		

Investigator	Cancer Stage	Cutoff Level	Marker	Specificity No. Pts (%) Single	Panel	Net Loss (%)
Kuriyama [49]	BPH	15 ng/ml	PAP	27/29 (93)	26/29 (90)	7
		7.5 ng/ml	PSA	28/29 (97)		

IX. CONCLUSION

The field of tumor markers has progressed recently, primarily due to the advent of monoclonal antibody technology. Although current markers are considered by many to be excellent for monitoring the clinical status of a patient with established tumors, there is still a need for continued development of better markers since existing markers are frequently elevated in benign disease and absent in early-stage cancer.

In addition to searching for more sensitive and specific markers, another approach that may significantly enhance the clinical value of existing markers is the coordinated use of markers as tumor-specific panels. In this report the progress of multiple marker research is reviewed and results demonstrating the advantages and disadvantage of panel testing are reported. The major advantage observed here is undoubtedly that the use of panels provides a broader data base within which to find one or more markers capable of assisting in the management of a patient with an established tumor. This advantage has unfortunately been restricted to patients with extensive disease, since most of the panels described here were highly sensitive to late-stage disease but insensitive to early-stage disease.

The major disadvantage in the use of panels is the nonspecific nature of the markers in current use. Combining several such markers in a panel tends to magnify this problem, making it difficult to differentiate benign from malignant disease. To solve this problem, new markers are needed for panels that are highly specific (i.e., no elevations occurring in the absence of cancer). These markers do not need to be highly sensitive individually, but in combination as a panel they need to be extremely sensitive and capable of detecting most true-positive cases in early-stage cancer. However, based on past experience, the discovery of highly specific markers of this type is unlikely. Thus, until better markers are identified, strategies must be developed to exploit existing markers with due regard for their limitations.

One strategy described here for ovarian cancer [40-44] is the use of a highly sensitive marker (CA 125) in combination with a highly specific one (CA 72-4, CA 15-3, PLAP, or CEA). A similar approach was evaluated by Jacobs [53] in a prospective study of screening for ovarian cancer. In that study CA 125 was combined with ultrasound and specificity scores of 99.8% were achieved. This approach appears to be one way to improve specificity without experiencing a significant loss in sensitivity.

Another way to compensate for the lack of specificity, as suggested by Sulitzeanu [54], is to use only markers that are found in considerably higher amounts in sera of cancer patients compared to sera of healthy individuals. However, few existing markers are capable of providing quantitative differences of such magnitude to differentiate clearly between normal and early-stage marker levels.

A strategy frequently used to improve sensitivity involves panels composed of different types of markers. For example, a breast panel [30] containing two mucin-glycoproteins and one isoenzyme exhibited a sensitivity gain of 19%; a colon panel [10] containing one oncofetal antigen, two isoenzymes, and two acute-phase proteins gave a sensitivity gain of 22%; and an ovarian panel [39] consisting of a mucin-glycoprotein, a polypeptide, a glycoprotein, and an iron-storage protein demonstrated a 20% increase in sensitivity. In contrast, the use of panels with similar type markers resulted in little or no gain in sensitivity. For example, a pancreatic panel [11-13] containing mucin-glycoproteins CA 19-9 and CA 50 exhibited an average sensitivity gain of only 4%.

In general, other strategies that would appear to enhance the use of panel testing include use of serial measurements. For example, steadily rising marker levels over time would usually reflect increasing tumor burden while decreasing marker levels would indicate decreasing tumor burden. In contrast, a transient increase in marker or minor fluctuation would most likely be associated with benign or biological variation within the individual patient. Serial determinations of marker panels in this manner should help to minimize the loss of specificity usually associated with marker panels.

Another technique that might help to eliminate false-positive measurements and improve detection in early-stage disease would be the use of intraindividual normal ranges as suggested by Dix and Cohen [52]. This approach to normal values would make it easier to interpret slight increases in the concentration of a marker in a given patient because the patient's own normal value would form the basis of comparison. Winkel [55] has also recommended the use of statistical time-series analysis when evaluating cumulative data, in order to eliminate intersubject variability (see Chapter 3 in this volume).

There is not enough clinical information about most panels to justify their use in the management of cancer treatment at this time. Additional studies are needed that use the strategies outlined above to validate the clinical utility of promising panels.

REFERENCES

1. Heppner GH. Tumor heterogeneity. Cancer Res 1984;44:2259-2265.
2. McIntire RK. Use of multiple immunoassays for circulating tumor markers. In: Herberman RB, McIntire KR, eds. Immunodiagnosis of cancer, part 2. New York: Marcel Dekker, 1979: 521-539.
3. Mercer DW. Evaluation of CA 19-9 as a serum marker in colorectal cancer with samples from Mayo-NCI serum diagnostic bank. Unpublished observations.
4. Novis BH, Gluck E, Thomas GH, et al. Serial levels of CA 19-9 and CEA in colonic cancer. J Clin Oncol 1986;4:987-993.
5. Gaur PK, Lofaro LR, Bray KR. Relationship between cancer associated antigen 195 (CA-195) and CEA. Proc Am Assoc Cancer Res 1987;28:357.
6. Chu LYT, Lai LTY, Pokala HP. Value of plasma alpha-1-acid glycoprotein assay in the detection of human colorectal cancer: comparison with CEA. J Nat Cancer Inst 1982;68:75-79.
7. Gupta MK, Arciaga R. Measurement of a monoclonal antibody defined antigen CA-195 and CEA in patients with colon cancer. Clin Chem 1988;34:1295.
8. Quentmeier A, Schlag P, Geisen HP, et al. Evaluation of CA 125 as a tumor marker for gastric and colorectal cancer in comparison to CEA and CA 19-9. Eur J Surg Oncol 1987;13:197-201.
9. Kawahara MD, Terasaki PI, Chia D, et al. Use of four monoclonal antibodies to detect tumor markers. Cancer 1986;58:2008-2012.
10. Mercer DW, Talamo TS. Multiple markers of malignancy in sera of patients with colorectal carcinoma: preliminary clinical studies. Clin Chem 1985;31:1824-1828.
11. Loor R, Kuriyama M, Bodziak MLM, et al. Simultaneous evaluation of a pancreas-specific antigen and a pancreatic cancer-associated antigen in pancreatic carcinoma. Cancer Res 1984;44:3604-3607.
12. Haglund C, Kuusela P, Jalanko H, et al. Serum CA 50 as a tumor marker in pancreatic cancer; a comparison with CA 19-9. Int J Cancer 1987;39:477-481.
13. Paganuzzi M, Onetto M, Merroni P, et al. CA 19-9 and CA 50 in benign and malignant pancreatic and biliary diseases. Cancer 1988;61:2100-2108.
14. Harmenberg V, Wahren B, Wiechel KL. Tumor markers carbohydrate antigens CA 19-9 and CA 50 and CEA in pancreatic cancer and benign diseases of the pancreatobiliary tract. Cancer Res 1988;48:1985-1988.
15. Takasaki H, Vchida E, Tempero MA, et al. Correlative study on expression of CA 19-9 and DU-PAN-2 in tumor tissue and in serum of pancreatic cancer patients. Cancer Res 1988;48:1435-1438.
16. Sakamoto K, Haga Y, Yoshimura R, et al. Comparative effectiveness of the tumor diagnostics, CA 19-9, CA 125 and CEA in patients with diseases of the digestive system. Gut 1987;28:323-329.
17. Haglund C. Tumor marker antigen CA 125 in pancreatic cancer; a comparison with CA 19-9 and CEA. Br J Cancer 1986;54:897-901.
18. Dnistrian AM, Schwartz MK, Esakof DD. Clinical significance of plasma lipid bound sialic acid and CEA in lung cancer. Proc Am Assoc Cancer Res 1982;23:155.
19. Gail MH, Mueng L, McIntire KR, et al. Multiple markers for lung cancer diagnosis: validation of models for advanced lung cancer. J Natl Cancer Inst 1986;76:805-816.
20. Ganz PA, Ma PY, Wang HJ, et al. Evaluation of three biochemical markers for serially monitoring the therapy of small-cell lung cancer. J Clin Oncol 1987;5:472-479.
21. Jaques G, Bepler G, Holle R, et al. Prognostic value of pretreatment CEA, neuron specific enolase, and creatine kinase-BB levels in serum of patients with small-cell lung cancer. Cancer 1988;62:125-134.

22. Bork E, Hansen M, Urdal P, et al. Early detection of response in small cell bronchogenic carcinoma by changes in serum concentrations of creatine kinase, neuron specific enolase, calcitonin, ACTH, serotonin and gastrin releasing peptide. Eur J Cancer Oncol 1988;24:1033-1038.
23. Buccheri GF, Violante B, Sartoris AM, et al. Clinical value of a multiple biomarker assay in patients with bronchogenic carcinoma. Cancer 1985;57:2389-2396.
24. Gail MH, Muenz L, McIntire KR, et al. Multiple markers for lung cancer diagnosis: validation of models for localized lung cancer. J Natl Cancer Inst 1988;80:97-101.
25. Coombes RC, Gazer JC, Sloane JP, et al. Biochemical markers in human breast cancer. Lancet 1977;1:132-134.
26. Coombes RC, Dearnaley DP, Morag LE, et al. Markers in breast and lung cancer. Ann Clin Biochem 1982;19:263-268.
27. Tondini C, Hayes DF, Gelman R, et al. Comparison of CA 15-3 and CEA in monitoring the clinical course of patients with metastatic breast cancer. Cancer Res 1988;48:4107-4112.
28. Stacker SA, Sacks NPM, Golder J, et al. Evaluation of MSA as a serum marker in breast cancer: a comparison with CEA. Br J Cancer 1988;57:298-303.
29. Bray KR, Suchocki KM, Gaur PK. Incidence of elevated serum levels of CA-549 and CEA in breast cancer patients. Clin Chem 1988;34:1296.
30. Mercer DW, Barry G, Losos F, et al. Multiple markers of malignancy in sera of patients with breast carcinoma. Clin Chem 1988;34:1297.
31. Linsley PS, Brown JP, Magnani JL, et al. Monoclonal antibodies reactive with mucin glycoproteins found in sera from breast cancer patients. Cancer Res 1988;48:2138-2148.
32. Dembo AJ, Chang PL, Malkin A, et al. Plasma ovarian cancer antigen NG/70K: clinical correlations. Proc Soc Gynecol Oncol 1985;16:30.
33. Sekine H, Hayes DF, Ohnoi T, et al. Circulating DF3 and CA 125 antigen levels in serum from patients with epithelial ovarian carcinoma. J Clin Oncol 1985;10:1355-1362.
34. Canney PA, Wilkinson PM, James RD, et al. CA 19-9 as a marker for ovarian cancer: alone and in comparison with CA 125. Br J Cancer 1985;52:131-133.
35. Fioretti P, Gadducci A, Ferdeghini M, et al. Correlation of CA 125 and CA 19-9 serum levels with clinical course and second-look findings in patients with ovarian carcinoma. Gynecol Oncol 1987;28:278-283.
36. Dhokia B, Canney PA, Pectasides D, et al. A new immunoassay using monoclonal antibodies HMFG1 and HMFG2 together with an existing marker CA 125 for the serological detection and management of epithelial ovarian cancer. Br J Cancer 1986;54:891-895.
37. Bast RC, Klug TL, Schaetzl E, et al. Monitoring human ovarian carcinoma with a combination of CA 125, CA 19-9 and CEA. Am J Obstet Gynecol 1984;149:553-559.
38. Ward BG, Cruickshank DJ, Tucker DF, et al. Independent expression in serum of three tumor associated antigens. CA 125, placentyl alkaline phosphatase and HMFG2 in ovarian carcinoma. Br J Obstet Gynecol 1987;94:696-698.
39. Negishi Y, Furukawa T, Oka T, et al. Clinical use of CA 125 and its combination assay with other tumor marker in patients with ovarian carcinoma. Gynecol Obstet Invest 1987;23:200-207.
40. Touitou Y, Bogdan A. Tumor markers in non-malignant diseases. Eur J Cancer Clin Oncol 1988;24:1083-1091.
41. Finkler NJ, Knapp RC, Bast RC. Monoclonal antibodies in ovarian cancer. In Kupchik HZ, ed. Cancer diagnosis in vitro using monoclonal antibodies. New York: Marcel Dekker, 1988.

42. Yedema C, Massuger L, Hilgers J, et al. Pre-operative discrimination between benign and malignant ovarian tumors using a combination of CA 125 and CA 15-3 serum assays. Int J Cancer 1989;in press.
43. Soper JT, Hunter K, Tanner M, et al. Use of CA 125, CA 72 and CA 15-3 to discriminate malignant from benign pelvic masses. Proc Am Assoc Cancer Res 1987;28:25.
44. Einhorn N, Zurawski VR, Knapp RC, et al. Pre-operative elevations of CA 125, CA 72, and CA 15-3 in patients with nonmucinous epithelial ovarian cancer. Proc Am Assoc Cancer Res 1987;28:357.
45. Wu JT, Miya T, Knight JA, et al. Improved specificity of the CA 125 enzyme immunoassay for ovarian carcinomas by use of the ratio of CA 125 to CEA. Clin Chem 1988;34;1853-1857.
46. Eerdekens MW, Nouwen EJ, Pollet DE, et al. Placental alkaline phosphatase and CA 125 in sera of patients with benign and malignant diseases. Clin Chem 1988;31:687-690.
47. Scardino PT, Cox DH, Waldmann TA, et al. The value of serum tumor markers in staging and prognosis of germ cell tumors of the testis. J Urol 1977;118:994-999.
48. Lange PH, Millan JL, Stigbrand T, et al. Placental alkaline phosphatase as a tumor marker for seminoma. Cancer Res 1982;42:3244-3247.
49. Kuriyama M, Wang MG, Lee CL, et al. Multiple marker evaluation in human prostate cancer with the use of tissue-specific antigen. J Natl Cancer Inst 1982;68:99-105.
50. Siddall JK, Shetty SD, Cooper EH, et al. Measurements of serum gamma-seminoprotein and prostate specific antigen evaluated for monitoring carcinoma of the prostate. Clin Chem 1986;32:2040-2043.
51. Killian CS, Emrich LJ, Vargas FP, et al. Relative reliability of five serially measured markers for prognosis of progression in prostate cancer. J Natl Cancer Inst 1986;76:179-185.
52. Cohen P, Dix D. Routine screening for prostatic cancer by assay of serum acid phosphatase: a modest proposal. Clin Chem 1984;30:171.
53. Jacobs I, Bridges J, Reynolds C, et al. Multimodal approach to screening for ovarian cancer. Lancet 1988;1:268-271.
54. Sulitzeanu D. Human cancer-associated antigen: present status and implications for immunodiagnosis. Adv Cancer Res 1985;44:1-42.
55. Winkel P, Bentzon MW, Statland BE. Mathematical models used to assess laboratory and other measurements in patients with cancer. Biological time series and multivariate analysis. Clin Lab Med 1982;2:657-678.

5

Serum and Tissue Banks for Biological Markers

THERESA L. WHITESIDE and RONALD B. HERBERMAN
University of Pittsburgh School of Medicine and Pittsburgh Cancer Institute, Pittsburgh, Pennsylvania

The idea of banking sera and tissues of patients with cancer is not new. Following the introduction and improvements of freezing units after World War II, researchers and clinicians interested in monitoring changes detectable in sera and tissues of patients with tumors realized the importance of banking and proceeded to collect and freeze specimens from selected groups of patients. Contents of these "private" collections were and continue to be determined by individual interests and the needs of each collector. Pathology departments have taken a lead in accumulating cancer patients' specimens, partly because of a long-standing tradition and partly because of the unique position of the pathologist in cancer diagnosis and follow-up of cancer patients. However, the best and most extensive collections of cancer sera have been accumulated by oncologists with special, well-defined interests in one tumor type. Although such collections represent valuable sources of specimens from carefully staged patients with cancer, they are not generally available, and their usefulness depends on the generosity or commitment of clinicians who accumulated them. Furthermore, conditions for the preservation, processing, and freezing of such specimens may not be uniform, so that the user must assume that they are appropriate.

To provide an organized pool of appropriately preserved and retrievable serum specimens for use in cancer research, the idea of a serum bank as a central or core facility has developed. Several years ago, the National Cancer Institute established a serum bank at the Mayo Clinic, which can supply serum samples to investigators for cancer marker and other studies [1]. This bank became a prototype of a multiuser serum bank and an important resource for cancer researchers. Likewise, a central tissue bank for collecting fresh or frozen, but not fixed, tumor biopsies or cells can be visualized as an important, even essential, facility for research/clinical/teaching institutions.

We discuss here organizational and operational aspects of central serum and tissue banks at a large teaching/research institution. Since requirements for the successful operation of a serum bank are clearly different from those for a tissue bank, the two will be described separately. On the other hand, a number of practical and quality control details are applicable to collections of both sera and tissues. These will be emphasized, as will the option of operating both banks in a single location. This chapter discusses the rationale, highlights and organization, stresses the need for quality assurance, and gives selected examples of the uses of central serum and tissue banks.

I. RATIONALE

The main reason for operating a serum and tissue bank is to enhance the quality of current and future cancer diagnosis and research. Development and introduction into clinical practice of new tests for the diagnosis and/or monitoring of patients with cancer has been rapid, and this pace will probably continue in the future. To evaluate any new assay for a tumor marker in serum or tissue, it is clearly necessary to have available large panels of specimens from patients with cancer as well as from normal individuals and patients with benign diseases as matching controls. To evaluate a potentially useful assay expeditiously, banked specimens are invaluable, since the prospective collection of specimens, especially in rare diagnostic categories, may take many months. Accurate information about the type, stage, extent of the disease, and prior or current therapy is also essential for meaningful evaluations of cancer markers. Follow-up information on patients whose specimens are banked allows for the evaluation of possible prognostic value of a marker. Newer technologies in cancer diagnosis and research (e.g., immunological monitoring, DNA analysis of tumor cells, oncogene studies) all require the availability of fresh, cryopreserved tumor cells or fresh-frozen tissues. Many future technologies undoubtedly will also depend on the availability of unfixed tumor tissues and cells as well as properly stored sera or plasma. Institutions with clinical or research programs in cancer are thus increasingly dependent on banked specimens for implementing state-of-the-art assays and procedures as they become available. A well-stocked and organized serum/tissue bank is no longer a luxury but a necessity.

The second reason for a central bank is a need for large-scale interlaboratory comparisons of the performance of new and established tumor markers. Such comparisons are essential to validate diagnostic assays using sera from reliably diagnosed and staged patients. Interlaboratory comparisons should include sera from cancer patients at different stages of disease, from individuals with no disease, and from those suffering with diseases other than cancer. For example, banked sera from groups of individuals at risk of developing certain cancers are an important resource for evaluating new biological markers and for comparing their performance with that of established markers.

The other major reason for operating a serum or tissue bank as a central facility is economy. Although establishing and running such a bank appears expensive (see below), it is necessary to consider the savings generated by pooling together multiple and often redundant "private" banks into a central one. Savings in technician time, equipment, electricity, liquid nitrogen, and computer time add up to significant cost reductions over time. It is also considerably more economical to maintain a single bank for a variety of future studies rather than to organize separate, prospective studies for new applications as they arise. A well-operated tissue or serum bank is in a good position to charge users for specimens, and in this way defray its expenses, at least in part.

Biological Marker Banks

II. ADVANTAGES

The most obvious attraction of a tissue/serum bank is its convenience. A user is able to avoid making numerous telephone calls, rummaging in old freezers, and bargaining with colleagues, by placing a request for a computer listing of bank specimens from a central facility. It is also possible to make individual arrangements with the bank for storage of specimens of particular interest or usefulness. The bank, which maintains computerized files, can also provide information to accompany each specimen and flag serial specimens. There are other advantages as well: banked specimens are handled uniformly and according to an approved operational protocol; their storage is monitored, so that defrosting due to equipment and/or power failures is highly unlikely; flow of specimens, once organized, to one place encourages cooperation from clinicians, pathologists, and surgeons; and retrieval of specimens is easy and efficient, because computerized listings of locations for every vial as well as the number of availabile vials are maintained.

In providing fresh-cryopreserved tumor cells needed as targets for cytotoxicity assays [2], for drug-sensitivity studies [3], or for production of vaccines for patients with cancer [4], a tissue bank can provide a unique and invaluable service. Enzymatic digestion and separation of tumor cells from fresh tumor biopsy specimens under controlled and carefully monitored conditions to yield viable tumor cells requires trained and experienced personnel. These procedures are most often done in research laboratories without strict quality control measures. This results in tumor cell preparations that are variable in viability, sterility, and/or purity, which leads to unreliable results in assays involving these cells. It is obviously advantageous to have fresh tumor cells prepared, cryopreserved, and supplied by a bank that operates according to established standards and maintains a quality assurance program.

Availability of numerous specimens from patients with one type of cancer as well as specimens from those with different cancers is essential for multiple serum-marker analyses, described in Chapter 4. For that reason, a bank needs to establish a wide base of operations and collect control and patient specimens from newly diagnosed patients as well as those with advanced cancer, from untreated and treated patients, and those with primary and metastatic diseases. Only a central facility specifically devoted to the task of collecting sera and/or tissues will be able to provide in a timely manner and as needs arise the sufficient number of patient specimens, each accompanied by the required clinical and follow-up information.

One of the criticisms often made of banks is that it may not be possible to predict future needs adequately, and, therfore, collection and storage conditions may be neither adequate nor appropriate. While this may be partially valid, banks have proven their usefulness in development of new technologies. For example, when a new generation of serum marker immunoassays utilizing enzyme-labeled reagents (ELISA) was introduced in the late 1970s, banked serum specimens were essential in evaluating the initial performance of the method and establishing that its sensitivity equals or surpasses that of radioimmunoassay (RIA) or immunoradiometric assay (IRMA). The issue of fresh versus stored specimens has to be addressed through comparative testing for all new methods. Once the use of stored specimens is proven acceptable, serum banks can serve as a convenient source of appropriate samples. Adequate storage of samples in sufficient numbers, from patients with diverse diseases and from control individuals, can be ensured at the time a bank is being established through a carefully planned and thoughtful operational protocol. Series of specimens stored under controlled conditions and accompanied by accurate and complete clinical and follow-up information are likely to be an important resource now or in the future. Since centrally operated banks are more likely to fulfill the requirements for a high-quality service, their value for future cancer research and diagnosis will remain undiminished.

Another criticism of banks concerns the expense and effort expended in storing specimens that may never be utilized. This criticism may be valid; however, central banks that serve large hospital and research centers would be very unlikely to find themselves in such a situation. While some of the stored specimens may indeed never be utilized, others are likely to be in demand. A central bank is in a position to tailor its collections to the projected needs of an institution, and it should be organized and operated on the basis of well-defined priorities. A central bank should be flexible and sensitive to needs developing locally as well as nationally. This flexibility in storage priorities and technology represents the most advantageous aspect of a bank as a central facility. A bank that successfully satisfies local needs can also supply specimens to other institutions, including commercial and private requests.

A bank is also a useful corollary to clinical trials with biological response modifiers (BRM). Tissue specimens collected serially prior to, during, and after a trial serve to monitor the effects of a BRM on the immune or inflammatory cells, as measured through the presence of their activation products (e.g., cytokines, soluble IL2 receptors, neopterin) in serum or on tumor cells as determined in studies of tumor markers in serum or tumor tissues. A tissue bank may be in a position to obtain and cryopreserve tumor cells from pretrial surgery that can then be used for in vitro monitoring of antitumor responses during the trial. It can also prepare tumor-infiltrating lymphocytes for in vitro activation and generation of effector cells for adoptive immunotherapy [5].

Thus advantages of a centrally located and operated serum/tissue bank appear to be many and important for the successful performance of clinical trials, for patient monitoring, for establishing the validity of diagnostic procedures, and for the performance of adoptive immunotherapy in cancer patients.

III. ORGANIZATION OF A CENTRAL SERUM BANK

A bank needs to be established in a single, centrally convenient location. It may, for example, be practical to place a bank near a clinical laboratory, where routine delivery of blood specimens is already established. Timely specimen delivery is important, since prolonged storage (i.e., longer than 2 hr) at room temperature may adversely affect certain serum markers, possibly due to products released from platelets and other cells. On the other hand, it is not advisable to place specimens on ice, since this may interfere with complete clotting.

Although a bank generally collects and stores sera, plasma may also be acceptable. However, significant method-dependent differences may exist between serum and heparinized plasma. Heparin may interfere with some analyses [6] and tumor markers [7]. However, it has been demonstrated that most circulating markers can be reproducibly measured in plasma if heparin is removed. Methods are being developed for removing heparin and/or other anticoagulants through the use of an ion-exchange column (D. Mercer, personal communication) if they interfere with a tumor marker assay. In some cases, a chelating agent may be preferable to heparin, and thus a bank must be prepared to determine the best handling of specimens for a given assay. This can best be done through establishing collaborative arrangements with a local clinical chemistry or tumor marker laboratory. Patients who participate in clinical trials, especially those who are treated with BRMs, may be required to donate large volumes of blood for immunological monitoring of leukocyte numbers and functions. Such blood specimens are treated with anticoagulants, and the bank may be obliged to bank plasma rather than serum, since it may not be possible to obtain additional blood specimens for collection of sera. Such serial specimens from patients participating in clinical trials, and particularly those treated with BRMs, represent a valuable resource for studies of tumor markers and cell-activation markers.

Equipment needed by a serum bank includes large-capaicty $-80°C$ freezers with alarm systems and liquid nitrogen emergency supplies, a refrigerated centrifuge, a laminar flow hood, and a computer. The management of the bank should be entrusted to a professional, who is then responsible for hiring and training personnel, implementing quality control, and supervising daily operations. A bank technologist and a part-time nurse-clinical coordintor staff the bank. Before the bank can begin its operations, a protocol outlining aims, methods, procedures, amounts and frequencies of blood specimens to be drawn, and specimen retrieval procedures must be prepared and submitted for approval to the Internal Review Boards (IRBs) of all participating institutions. A procedure manual, which describes specimen intake, handling, storage, and retrieval, should be prepared and regularly updated by the bank director. Informed consent needs to be obtained prior to specimen collection. Once in operation, the bank needs to be prepared to accept all blood specimens that arrive in labeled Vaccutainer tubes and are accompanied by a referral card containing necessary information as specified in the procedure manual. Specimens received by the bank are clocked, given a number, logged in on the daily log sheet, and entered into a computer. Following centrifugation in the cold, all serum specimens are aliquoted (1 ml/vial) into Nunc vials under sterile conditions, labeled with a daily consecutive number, and placed in storage racks at $-80°C$. For each specimen, the information on the referral card is recorded in the computer, as well as time and date of specimen freezing, amount of serum received (ml), and storage location. Daily temperatures of all freezers are recorded on a chart, and a written record is maintained of all power failures and/or emergencies.

The bank is also responsible for maintaining diagnostic files on all patients whose specimens are in the bank. The file is opened at the time of specimen intake. The diagnostic file is identified by a unique number for every patient. Information in this file includes the patient's name, Social Security number, daily consecutive number, project/protocol number, birth date, death date, sex, primary diagnosis, primary site of diagnosis, date of diagnosis, date of onset, stage, date of metastasis, metastasis sites, date of recurrence, recurrent sites, surgical treatment type and dates, extent of disease, radiation field, amount, and dates, chemotherapy agents: amount and dates, and tumor response. It is the responsibility of the nurse-coordinator to seek and complete a diagnostic file within a week of banking specimens. Diagnostic files are reviewed each week by a bank technician who flags all files containing missing information. Equally important is providing follow-up information on all specimens in the bank. Provisions need to be made to update this information regularly, entering events such as recurrences, treatments, responses, and deaths. The follow-up data are necessary for adequate evaluations of prognostic and diagnostic significance of markers measured and for the correct interpretation of serial measurements.

Access to a central serum bank is provided to all participating physicians and those investigators who have an approved clinical/research protocol. It is important to establish and specify in the procedure manual how such approval can be obtained and who is entitled to use the bank. The participating physicians who provide specimens to the bank need to be given access to bank specimens, but they too need to have a valid reason and an approval for requesting these specimens. A committee of individuals interested in the bank activities may be created and given the responsibility of considering and approving all requests for serum samples. For the sake of fairness, individuals requesting serum specimens should do so in writing and should specify the reasons and numbers of specimens requested. To avoid conflict, the committee should consider each written request and make its decision on the basis of scientific merit. The bank director must authorize release of bank specimens in writing and should do so only after receiving the written recommendation of the committee.

One of the services the bank can provide is a computer listing of all specimens available to users. The diagnostic and follow-up file maintained on each specimen allows the bank to provide computer listings according to cancer type, primary site, disease stage or treatment (e.g., pretreatment versus postreatement). The number of specimens available in each category can be easily determined. In this way, the bank is able to facilitate planning of studies requiring multiple serum specimens for tumor marker determinations.

Determining how long it may be satisfactory to bank specimens is difficult. A bank specimen should be usable after many years of storage at $-80°C$, provided it is not defrosted and refrozen. However, a certain amount of desiccation may occur over the years of storage, and this could alter results of serum marker assays. Consideration may be given to storing serum vials under vacuum or to sealing glass vials to avoid evaporation. From a practical point of view, storing specimens for up to 10 years may be reasonable. On the other hand, this may depend on the bank intake and utilization. In an exceptionally busy and successful serum bank, it may become necessary to eliminate specimens older than 5 years.

Another issue is the amount of serum/plasma that is banked. This decision has to be based on institutional guidelines for blood drawing from patients, and on storage space. However, if the bank services a large health center, it must be prepared to accept multiple requests for certain specimens and have facilities to bank multiple serum/plasma vials for every specimen received. It may be more frugal to aliquot serum/plasma in a volume of 1 ml/vial, thus ensuring the availability of multiple vials. On the other hand, if prolonged storage is anticipated, it may be better to freeze aliquots of larger volumes to minimize desiccation.

IV. ORGANIZATION OF THE CENTRAL TISSUE BANK

A tissue bank is responsible for storing fresh, unfixed tissues (tumor, benign and normal) and cells for use in cancer research. The successful operation of the tissue bank depends on cooperative interactions with a local surgical pathology laboratory. While cooperation from surgeons is important, tumor tissues must be examined by a pathologist prior to any processing at a tissue bank. For that reason, arrangements that the bank can make with the pathologists are crucial to tissue bank activities. Since surgical pathology laboratories are primarily concerned with service, special handling and special attention that the tissue bank may require creates an additional and unwelcome burden. For example, a tissue bank may require that specimens processed for cells to be cultured or used in long-term cytotoxicity assays be handled under sterile conditions. Another requirement may be that no necrotic tissues be banked. Usually, only viable tumor cells are useful, so that surgical specimens received by the bank need to be fresh and placed in sterile tissue culture medium immediately after surgery. This means that the bank should receive a specimen within 1-2 hr after surgery. Special arrangements are, therefore, necessary between an operating room, a pathologist, and tissue bank personnel. These arrangements may, for example, involve a pathology assistant who is designated to handle all tumor specimens to be banked and is specially trained by a pathologist as well as by the tissue bank to meet the sterility and processing requirements of both laboratories. The tissue bank needs to provide an operating room with sterile containers of medium, organize access to a laminar flow hood, and make available sterile instruments for tissue dissections for a pathologist or pathology assistant. The bank also needs to be prepared to accept surgical specimens whenever they arrive, which is often after regular working hours. A successful tissue bank can be established only if all concerned, surgeons, pathologists, oncologists, and hospital administrators, are convinced and cognizant of the importance of bank activities for the future of cancer research.

A tissue bank should also be centrally placed, possibly close to the surgical pathology laboratories. Equipment required by the tissue bank is extensive and includes a laminar flow hood for sterile processing, a rate-control device (Cryomed) for freezing viable cells, $-80°C$ freezers with alarm and liquid nitrogen systems for storing of tissue blocks and tissues for DNA extraction, liquid nitrogen storage tanks for cryopreservation of cells in N_2 vapors, centrifuges, microscopes, water baths, magnetic stirrers, and a computer. Although some cells may be adequately cryopreserved at $-80°C$, rate-controlled freezing is usually preferable. Many tumor cells are fragile, and require rate-controlled freezing and special handling. A tissue bank needs to be directed by a professional who is trained in handling tumor tissues and cells and aware of special requirements imposed upon the bank. A technician experienced in the processing of tumor tissues, a clinical coordinator for maintaining diagnostic files, and a part-time messenger represent the minimum staff requirement. Since tumor biopsy specimens may arrive after working hours, there is a need for a back-up network to process these specimens.

In general, tumor specimens accepted by the bank should not be smaller than 1 g in wet weight and as large as possible. They should be sampled from the outer aspect of the tumor to avoid necrotic tissue. Large metastatic tumor masses are generally good sources of tumor tissues. However, each specimen must be handled individually, and the pathologist must determine how much and what part of the tumor specimen is appropriate for banking. Tumor tissues needed for therapeutic purposes (i.e., tumor vaccine) must be handled expeditiously by a pathologist, who needs to be informed by the surgeon or the clinician that a tumor is going to be used for therapy. A pathologist, who is familiar with the vaccination protocol, will be especially generous with the tissue and ensure its prompt and sterile handling for the benefit of a patient. Fluids from cancer patients represent an excellent source of viable tumor cells, and a tissue bank should be especially interested in collecting large volumes of such fluids. To recover a maximal number of tumor cells, fluids need to be anticoagulated immediately after being harvested.

Each specimen received by the bank is weighed and divided into two or three parts depending on its size. One part is cut into pieces (0.25-0.5 g), and the pieces wrapped in foil, snap frozen in isopentane, placed in a plastic vial, labeled, and cryopreserved at $-80°C$. The second is subjected to enzymatic digestion with collagenase and DNAase [8], the cells are placed on an appropriate gradient to separate tumor cells from tumor-infiltrating lymphocytes [9], extensively washed, checked for viability, and cryopreserved using a control-rate freezing device. A cytocentrifuge preparation of the tumor cells is submitted for cytologic examination to verify the diagnosis. The third part is also divided into chunks, which are wrapped in foil, placed in a plastic bag, labeled, and banked at −80°C. In this manner, each tumor specimen can be preserved for immunohistological examination (part 1); as fresh, viable tumor cells or tumor-infiltrating cells (part 2); and for molecular biology studies (part 3). If a tumor specimen is too small, it should be processed according to priorities defined by each bank and specified in the operation manual.

Banking of viable tumor cells is labor-intensive and requires personnel specially trained in handling enzymatic digestion and cell separation procedures, viability counts, and cryopreservation of live cells. These procedures must be carried under sterile conditions, because the recovered cells may be required for tumor cell vaccines, long-term cultures of mononuclear cells separated from tumors, or short- or long-term in vitro assays (e.g., cytotoxicity assays) that utilize viable tumor cells. When specimens are processed for the recovery of tumor cells, it is essential to prepare smears of separated tumor cells. Such smears, stained with the Papanicolaou stain, are submitted for cytological examination to confirm that they contain tumor cells. The percentage of tumor cells is determined by a cytologist, reported on a special form to the bank, and recorded in the bank computer.

When a tissue specimen arrives at the bank, the time of arrival is recorded in the log book, manually, and in a computer. Only specimens accompanied by a referral card containing information required for a diagnostic file are accepted. Each tissue specimen received is given a daily consecutive number and logged on the daily sheet and entered in the computer. An individual diagnostic file, identified by a unique number for each patient, is maintained on every patient whose specimens are in the tissue bank. Information contained in this file includes the patient's name, social security number, daily consecutive number, and the pathological diagnosis as well as additional clinical and treatment information as listed above (see section on operation of serum bank). The follow-up data are added to the file at specified intervals. The diagnostic file is computerized and reviewed weekly for identification of missing data. A clinical coordinator at the bank is responsible for seeking and providing information needed for a complete diagnostic and follow-up file on each banked specimen.

The disbursal of tumor tissues and/or cells from any disease category by the bank should be specified in the operation manual. It is helpful to establish a committee to handle requests for tissue disbursement. The committee should include a pathologist, surgeons, oncologists, and the bank director. Requests for tumor tissues may be submitted in writing to the committee and be accompanied by a brief justification and explanation of how the material requested is to be used. The committee should be charged with developing a detailed protocol for the disbursement of tissues and defining priorities of disbursement. The committee should recognize that physicians who provide the tissue deserve to be given special consideration. Furthermore, the committee may recognize that special handling of tissues may be necessary for individual projects; users requesting special processing may be asked by the bank to do so themselves. To facilitate the process of disbursement, the bank may prepare standard and simple request forms for tissue/cells from the bank and make them availabile throughout an institution.

V. QUALITY CONTROL IN TISSUE AND SERUM BANKS

Although requirements for long-term storage of sera or plasma are relatively simple, those for cryopreservation of tissues and cells are complex. Storage of tumor tissue and cells under conditions that prevent desiccation and other alterations of tissue and ensure viability of tumor cells is a very important function of the bank. A quality assurance program must therefore be an essential component of a well-managed serum/tissue bank. A quality control program may comprise three elements. First, education of personnel in special requirements for banking is essential. Training a technician in processing tumor tissues, for example, represents a considerable time investment on the part of a director or supervisor. A certain degree of judgment is required from a technologist to deal with a variety of heterogeneous tissue specimens that a bank receives. It is essential to impress upon a technologist that a constant flow of communication between all those involved in specimen banking is necessary. A technologist needs free access to a supervisor, bank director, and a pathologist at all times for consultations and decision making. The process of education also involves establishing relationships between operating room and surgical pathology personnel. It is important to provide feedback to the bank personnel about utilization of the specimens and provide examples and/or publications illustrating applications of specimens in cancer research or treatment. Highly motivated and well-educated staff provides the best assurance that high-quality standards will be maintained in the bank.

An equally important part of the quality assurance program is the supervision and implementation of control procedures to confirm that the process of banking yields appropriately preserved, usable specimens. These control checks need to be performed at regular intervals by a supervisor and/or bank director and should apply to all aspects of the banking operation, including specimen intake, labeling, log entry, equipment maintenance, recording of daily temperatures, keeping records of equipment failure, records of equipment service, and other bank practices as listed in the operation manual. Quality control procedures are also necessary. Since there are no established guidelines for such control procedures, a bank director assumes the responsibility for devising his or her own. Some examples of such procedures are as follows. In a serum bank, a specimen aliquoted in many vials may be obtained, banked in different locations, and then periodically submitted to a tumor marker laboratory for determinations of a given serum marker. This will ensure that the quality of a stored specimen does not deteriorate with time, at least with respect to the particular serum marker measured. In a tissue bank, a similar approach can be taken with respect to banked tumor cells aliquoted into numerous vials. A vial can be periodically defrosted and the viability of banked tumor cells can be checked. These tumor cells can also be placed in culture to determine sterility of the preparation.

The third component of a quality assurance program involves the users of the bank. It is important to encourage the users to communicate with the bank regarding the quality of the specimens received. The bank should make an attempt to survey users at regular intervals and seek their opinions on the bank's handling and quality of specimens.

VI. FISCAL ASPECTS OF BANKING

The operation of serum, and particularly tissue/cell, bank is expensive. In addition to the initial outlay of funds for equipment, there are expenditures related to personnel, specimen procurement, processing, and storage. A central serum/tissue bank used by a multidisciplinary roster of investigators represents a shared resource and as such qualifies for support from Cancer Center Grants. In addition, the bank can charge for its services. The operating and personnel expenses and fixed costs of the bank can serve as a basis for calculating a price for each type of specimen banked. A serum bank can institute a charge per vial of serum or plasma. A tissue bank should develop a price for a block of frozen tissue for cryosection, vial of cryopreserved viable tumor cells, or a gram of frozen tissue for other studies. Once a price schedule is developed, all investigators who expect to use the bank may be asked to include the estimated cost of specimens needed for their studies in research proposals submitted for external funding. Forms specifying bank prices could be created and distributed to all prospective users.

For financial and operational ease, it appears to make sense to combine serum and tissue banks into one unit. They can be operated at the same location, and the same personnel can be utilized for banking sera and tissues. The same professional and clinical coordinator can service both banks, and further cost savings can be generated by sharing a computer, equipment maintenance service, and quality assurance programs. An additional advantage in some studies may be the ability to match serum and tissues from the same patient.

Establishing and operating a serum/tissue bank represents an institutional commitment to the quality of cancer research and cancer therapy. It should not be expected that such banks become self-supporting. A considerable portion of the support should come from the institution itself. The size of a bank has to be determined by the current and projected needs of an institution. It is an important decision, because once a bank is established, its operation should complement programs and activities already underway without becoming a financial burden.

VII. SELECTED EXAMPLES OF UTILIZATION

Many different programs have a need for and use serum/tissue banks. For example, medical oncology and surgical oncology programs may wish to bank tumor tissues of patients with cancer for immunizations and in vitro sensitization studies [4]. An Eastern Cooperative Oncology Group (ECOG) protocol for vaccination of patients with colon cancer with their own tumor cells and bacillus Calmette-Guérin (BCG) [10] that has been underway for several years as a multi-institutional study is a good example of how a tissue bank can contribute to innovative and potentially more effective care of patients with cancer.

Adoptive immunotherapy with lymphokine-activated killer (LAK) cells has been used to treat patients with solid tumors refractory to conventional therapy [11,12] and is being evaluated for the treatment of patients with leukemias. The tissue bank can supply freshly cryopreserved tumor cells from autologous or allogeneic tumors of the same histological type as that of a patient to be treated which may be necessary for in vitro evaluation of effectiveness of LAK cells. For example, Figure 1 shows in vitro reactivities of LAK cells generated at our institution from the blood of patients with acute nonlymphocytic leukemias (ANLL) and tested in in vitro cytotoxicity assays for their ability to kill autologous or allogeneic leukemic blasts. These leukemic blasts were banked, defrosted when LAK cells were ready for in vitro testing, labeled with radioactive chromium, and used as targets in cytotoxicity assays. In these experiments, spontaneous release from banked leukemic targets was less than 30%. The results in Figure 1 indicate that banked leukemia cells performed well in ^{51}Cr release assays. The quality of banking of fresh tumor cells as targets for cytotoxicity assays is important, since it determines the percentage spontaneous release of a radioactive label. Cells with more than 30% spontaneous release are unsuitable for use as targets in cytotoxicity assays.

In the context of adoptive immunotherapy, serial monitoring of patients for immunological functions is essential to evaluate effects of therapy. Such monitoring of antitumor

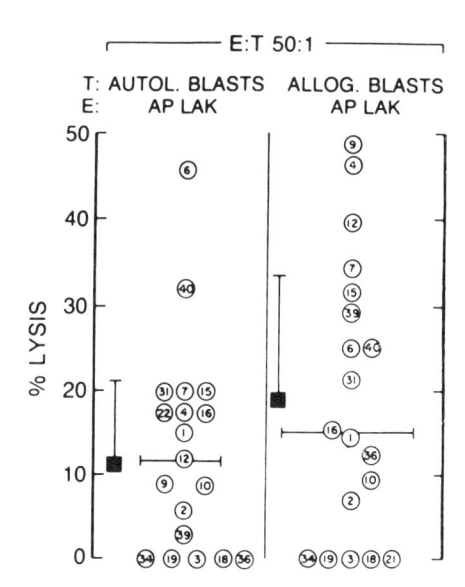

Figure 1 Cytotoxicity of lymphokine-activated killer (LAK) cells generated by culturing peripheral blood mononuclear cells of patients with acute nonlymphocytic leukemias in the presence of 1000 U/ml of rIL2 for 2-3 weeks. Fresh autologous or allogeneic leukemic blasts were used as targets. Cytotoxicity was determined in 4 hr ^{51}Cr release assays [20].

functions can only be performed if viable tumor cells obtained and cryopreserved from a variety of human tumors are available to the immunological monitoring laboratory for in vitro assays. Again, a tissue bank is needed to provide fresh tumor cells as targets or stimulating antigens.

Basic research programs in the immunology of human tumors have a continuous need for fresh tumor cell targets for functional studies of tumor-infiltrating lymphocytes [13], encapsulation experiments [14], or studies of tumor target sensitivity to immune effector cells [3]. For example, the nature of effector cells in interleukin 2 (IL2)-expanding populations of tumor-infiltrating lymphocytes (TIL) from human ovarian cell carcinomas was detected in the cell sorting and cytotoxicity assays with freshly banked human ovarian carcinoma cells, as shown in Figure 2. These studies indicated that the Leu19+ effectors were responsible for all of the cytotoxic activity in cultures of TIL isolated from human ovarian tumors and expanding in the presence of rIL2 [15]. This example illustrates the importance of the availability of fresh tumor cell targets for basic research investigators.

Two specific examples of the use of serum banks are cited. These emphasize the unique role that a serum bank can play in cancer research and diagnosis. Availability of banked sera makes it possible to conduct collaborative studies evaluating multiple simultaneous serum markers to determine the best combination for managing patients with cancer. In a study performed a few years ago with sera from lung cancer patients, identical aliquots of the same serum sample were obtained from the serum bank at the National Cancer Institute (NCI)/Mayo Clinic and distributed to different laboratories to perform quantitative assays for a number of markers [16]. The results obtained from such multicenter studies are very important for the selection of optimal groupings of markers for the different types of cancer and for the evaluation of different biostatistical techniques designed to maximize the benefits derived by multiple marker determinations.

FRESH AUTOLOGOUS TUMOR (OVARIAN CA)

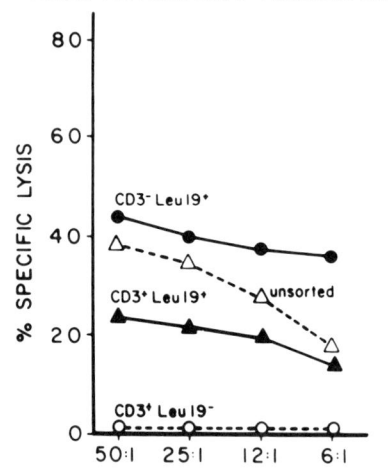

Figure 2 Cytotoxicity of Leu19+ effector cells against fresh autologous tumor cell targets. Tumor-infiltrating lymphocytes (TIL) were obtained from a human ovarian carcinoma, expanded in the presence of rIL2, and then sorted using a flow cytometer into CD3$^+$Leu19$^+$, CD3$^-$Leu19$^+$, and CD3$^+$Leu19$^-$ populations. These cells were used as effectors in a 4 hr ^{51}Cr release assay with fresh autologous ovarian carcinoma targets.

The second example demonstrates how the availability of banked sera from patients in clinical trials with interferon could be used to verify activation of immune cells and, at the same time, measure the result of treatment. It has been suggested that neopterin, which is measurable as a serum marker, may be a good correlate of macrophage activation by interferons [17], including interferon gamma, which is a product of activated lymphocytes. Interferons have also been reported to increase the expression of tumor-associated antigens such as TAG 72 and carcinoembryonic antigen (CEA) on the surface of tumor cells and increase shedding of these antigens into body fluids [18]. As a measure of immune cell activation, soluble receptor for IL2 (IL2R) has been shown to be a useful and reproducible marker [19]. These observations suggest that the use of serum markers such as neopterin, IL2R, CEA, TAG 72 antigen [18], and probably others could be alternatives to functional assays to evaluate the effects of interferon not only on the immune cells but also on the tumor cells. Before these hypotheses can be verified, serum levels of these markers in serially collected specimens from patients treated with interferons must be correlated with immunological data accumulated during interferon trials.

REFERENCES

1. Masnyk IJ, Sears M, Coele D, Go VLW, Dunsmore M. Serum bank for biological markers for breast cancer. J Natl Cancer Inst 1983;71:875-876.
2. Grimm EA, Mazumder A, Zhang HZ, Rosenberg SA. Lymphokine-activated killer cell phenomenon lysis of natural killer-resistant fresh solid tumor cells by Interleukin 2-activated autologous human peripheral blood lymphocytes. J Exp Med 1982;155:1823-1841.
3. Tanmigaioa N, Mizuno Y, Hashimura T, et al. Comparison of drug sensitivity among tumor cells within a tumor between primary tumor and metastases, and between differ-

ent metastases in the human tumor colony-forming assay. Cancer Res 1984;14:2309-2312.
4. Hoover HC Jr, Surdyke M, Dangel RB, Peters LC, Hanna MG Jr. Delayed cutaneous hypersensitivity to autologous tumor cells in colorectal cancer patients immunized with autologous tumor cells—bacillus Calmette-Guerin vaccine. Cancer Res 1984;44: 1671-1676.
5. Rosenberg SA, Packard BS, Aebersold PM, Solomon D, et al. Use of tumor-infiltrating lymphocytes and Interleukin 2 in the immunotherapy of patients with metastatic melanoma. A preliminary report. N Engl J Med 1988;00:1676-1680.
6. Harr R, Bond L, Trumbull D. A comparison of results for serum versus heparinized plasma for 30 common analyses. Lab Med 1987;18:449-455.
7. Wu JT. Interference of heparin in carcinoembryonic antigen radioimmunoassays. Clin Chim Acta 1983;130:47-54.
8. Whiteside TL, Miescher S, Hurlimann J, Von Fliedner V. Separation, phenotyping and limiting dilution analysis of T lymphocytes infiltrating human solid tumors. Int J Cancer 1986;37:803-811.
9. Whiteside TL, Miescher S, MacDonald HR, Von Fliedner V. Separation of tumor-infiltrating lymphocytes from human solid tumors. A comparison of velocity sedimentation and discontinuous density gradients. J Immunol Methods 1986;90:221-233.
10. Hoover HC, Harris JE, Dayal Y, Hanna MG, et al. Phase III protocol for the evaluation of combined modalities in the treatment of colonic carcinoma with positive nodes; Dukes' or with penetration through the muscularis propria or serosa (nodes negative) Dukes' B2, B3. Ecog Protocol, Est. 5283.
11. Rosenberg SA, Lotze MT, Muul LM, Chang AE, Avid FP, Leitman S, Linehan WM, Rovertson CN, Lee RE, Rubin JT, Seipp CA, Simpson CG, White DE. A progress report on the treatment of 157 patients with advanced cancer using lymphokine-activated killer cells and Interleukin-2 or high-dose Interleukin-2 alone. N Engl J Med 1987;316:889-897.
12. West WH, Tauer KW, Yannelli JR, Marshall GD, Orr DW, Thurman GB, Oldham RK. Constant-infusion recombinant Interleukin-2 in adoptive immunotherapy of advanced cancer. N Engl J Med 1987;316:898-905.
13. Whiteside TL, Heo DS, Takagi S, Johnson JT, Iwatsuki S, Herberman RB. Cytolytic anti-tumor effector cells in long-term cultures of human tumor-infiltrating lymphocytes in recombinant Interleukin 2. Cancer Immunol Immunother 1988;26:1-10.
14. Gorelik E, Ovegera A, Shoemaker K, Jarvis A, Alley M, Duff R, Mayo J, Herberman RB, Boyd M. Microencapusulated tumor assay—new short-term assay for in vitro evaluation of the effects of anticancer drugs on human tumor cell lines. Cancer Res 1987; 42:5739-5747.
15. Whiteside TL, Heo DS, Takagi S, Herberman RB. Characterization of novel anti-tumor effector cells in long-term cultures of human tumor-infiltrating lymphocytes. Transplant Proc 1988;20:347-350.
16. McIntire KR, Radovich BT, Gail MH, Go VL. Collaborative study for the evaluation of multiple simultaneous markers in lung cancer. Ann NY Acad Sci 1983;417:435-442.
17. Huber C, Batchelor JR, Fuchs D, Hausen A, Lang A, Niederwieser D, Reibnegger G, Swetly P, Troppmair J, Wachter H. Immune response-associated production of neopterin. J Exp Med 1984;160:310-316.
18. Borden EC. Augmented tumor-associated antigen expression by interferons. J Natl Cancer Inst 1988;80:148-149.
19. Pizzolo G. The soluble Interleukin-2 receptor as a new biological marker in diseases. Immunol Clin 1988;7:13-21.

20. Adler A, Chervenick P, Whiteside TL, Lotzova E, Herberman RB. Interleukin-2 induction of lymphokine activated killer (LAK) activity in the peripheral blood and bone marrow of acute leukemia patients. I. Feasibility of LAK generation in adult patients with active disease and in remission. Blood 1988;71:709-716.

II
Methodology

6

Development and Evaluation of Monoclonal Antibody-Based Immunoassays: Breast Carcinoma-Associated Mucins as Tumor Markers

JOSEPH P. BROWN*

Genetic Systems, Seattle, Washington

PETER S. LINSLEY and DIANE HORN

ONCOGEN, Seattle, Washington

I. INTRODUCTION

Carcinoma-associated mucins are becoming increasingly important as tumor markers [1]. Released into the circulation by most carcinomas, they are characterized by their high molecular weight and buoyant density, extensive O-linked glycosylation, and antigenic complexity. Monoclonal antibodies have been used to identify numerous epitopes of carcinoma-associated mucins, particularly in breast carcinoma [2-6].

In this chapter, we describe the generation and characterization of monoclonal antibodies to breast carcinoma-associated mucins. We address the problems of characterizing the antibodies on the basis of their patterns of immunoreactivity and the structure of their epitopes, with particular emphasis on the selection of antibodies for use in immunoassays. Finally, we describe two new immunoassays, CA M26 and CA M29, which offer significant advantages in clinical sensitivity and specificity for monitoring breast carcinoma patients.

II. MONOCLONAL ANTIBODIES

A. Antibody Generation

Mice were immunized with carcinoma cell lines; with milk fat globule membrane, which expresses mucin epitopes [7,8]; and with mucins purified from human milk and breast cancer effusions by density gradient fractionation and immunoaffinity chromatography, as described previously [9]. Hybridomas were generated by fusing spleen cells from the immunized mice with mouse myeloma cells [10] in nine separate fusions experiments, which are summarized in Table 1.

More than 6000 hybridomas were screened for production of antimucin antibodies by microtiter binding assays on purified mucins and on mucins captured from malignant effusions and cancer patients' sera by wheat germ agglutinin [9]. Monoclonal antibody binding was detected by using an IgG-specific goat antimouse immunoglobulin conjugate.

---

*Present affiliation: Bristol-Myers Squibb Co., Wallingford, Connecticut

Table 1 Antibodies to Breast Carcinoma-Associated Mucin

Antibody	Fusion Number	Mouse strain	Immunogen	Isotype	ATCC Number
M8	1	BALB/c	GM	IgG1	HB 9209
M10	2	BALB/c	CG	IgG1	HB 9244
M11	2	BALB/c	CG	IgG1	HB 9245
M12	2	BALB/c	CG	IgG1	HB 9246
M15	2	BALB/c	CG	IgG1	HB 9247
M16	3	C57/BALB[a]	CG	IgG1	HB 9216
M21	4	C57/BALB[a]	CG	IgG1	HB 9248
M22	5	BALB/c	CG	IgG1	HB 9249
M23	5	BALB/c	CG	IgG1	HB 9250
M25	3	C57/BALB[a]	CG	IgG1	HB 9217
M26	6	BALB/c	CE	IgM	HB 9212
M27	7	NZB	CG	IgG2a	HB 9229
M29	8	BALB/c	CM	IgG1	HB 9243
M38	9	BALB/c	E	IgG1	HB 9365

[a]C57BL/6 × BALB/c F1 hybrid.
C, carcinoma cells; E, effusion mucin; G, milk fat globule membrane; M, milk mucin.

Fourteen hybridomas were selected for investigation. All produced monoclonal antibodies of the IgG1 or IgG2a isotypes, except for antibody M26, which proved to be an IgM. Although mice immunized with milk fat globule membrane and carcinoma cells were used for most of the fusions, the three antibodies that proved to be of greatest value were obtained from mice immunized with purified mucins, either from milk (antibody M29) or effusions (antibodies M26 and M38).

Ascites fluids of mice were used as a source of purified antibody. IgG antibodies were purified by ion exhcange chromatography on diethylaminoethyl (DEAE)-Sephacel, and IgM antibodies were purified by size fractionation on Sephacryl S-300.

B. Antigen Identification

To confirm the antibodies were specific for mucins, they were tested on immunoblots of purified milk and effusion mucins, and in immunoprecipitation experiments with radiolabeled cell lysates. Although the antibodies differed from each other in their relative binding to different mucin preparations, they all bound high-molecular-weight molecules in at least one of the preparations [9]. Antibody M38 bound to both effusion and milk mucins, whereas antibodies M15, M22, M23, and M27 bound more strongly to milk mucin than to effusion mucin. Antibody M26 bound strongly to mucin purified from effusions, but not detectably to milk mucin. Antibodies M10, M11, and M12 bound strongly to mucins from carcinoma cell membranes, but poorly to milk and effusion mucins. Antibodies M21 and M29 did not bind detectably to immunoblots, but they did immunoprecipitate mucins from lung carcinoma cells labeled with tritiated glucosamine.

To obtain additional evidence that the antibodies bound mucins, they were tested in double determinant immunoassay (DDIA) [11]. The antibodies were adsorbed to microtiter wells and incubated with milk or effusion mucins. Bound mucin was quantitated by measuring binding of peroxidase-conjugated antibody W1, which had been previously shown to bind to a breast carcinoma-associated mucin epitope [9]. In this assay, all the antibodies bound to both milk mucin and effusion mucin, except antibody M26, which did not bind to milk mucin [9]. In this regard, antibody M26 might be considered to be more tumor-specific than the other antibodies.

Many of the antibodies that bound poorly to mucins on immunoblots gave clear positive results by DDIA. Antibody M8, for example, did not bind detectably to effusion mucin on immunoblots, but it worked well as a capture antibody in DDIA. These observations indicate that a lower epitope density is required for detection by DDIA than by immunoblotting.

These experiments provided evidence not only that all of the antibodies recognized mucin epitopes but also that these epitopes are present on the same population of mucin molecules, rather than antigenically distinct subpopulations. This conclusion was confirmed by further experiments, which showed that virtually every pair of antibodies tested functioned in DDIA.

C. Epitope Structure

To determine the biochemical nature of the epitopes recognized by the new antibodies, we tested chemically and enzymically treated mucins for antibody binding.

1. Periodate Oxidation

Many carbohydrate structures are destroyed by periodate oxidation. When periodate-treated mucins were tested, binding of antibody M26 was abolished, which suggested that carbohydrate is required for binding of this antibody. Binding of five other antibodies (M10, M11, M12, M21, and M29) was periodate-sensitive, but only to significantly higher periodate concentrations. Binding of the remaining eight antibodies (M8, M15, M16, M22, M23, M25, M27, M38) was not affected, even at the highest periodate concentration tested (0.1 M).

2. Neuroaminidase Digestion

Neuraminidase digestion removes terminal sialic acid structures. Antibody M26 did not bind to neuraminidase-treated mucin, which shows that sialic acid is required for binding of this antibody. However, binding of antibodies M8, M16, M21, M25, M27, and M29 to pleural effusion mucin was increased by neuraminidase treatment, which suggests that the epitopes recognized by these antibodies were unmasked by removal of sialic acid. Binding of the remaining seven antibodies (M10, M11, M12, M15, M22, M23, M38) was unaffected by neuraminidase treatment.

3. Hydrogen Fluoride Deglycosylation

Glycoproteins can be completely deglycosylated by treatment with anhydrous hydrogen fluoride [12]. When purified milk mucin was deglycosylated in this way, binding of antibodies M8, M16, M25, M29, and M38 was completely abolished. Antibodies M15, M22, M23, and M27, on the other hand, still bound, which suggested that these antibodies bind to epitopes on the core protein. Antibodies M10, M11, M12, M21, and M26 did not bind well to the starting milk mucin preparation and could not be tested.

4. Binding to Glycolipids

Many antibodies that recognize carbohydrate epitopes bind glycolipids, and we tested our antibodies on glycolipid preparations by immunostaining [13]. Antibody M26 stained several components in glycolipids from kidney and meconium and a single glycolipid band in a diasialylated ganglioside fraction, which showed that it binds with sialylated carbohydrate structures. Experiments are in progress to determine the carbohydrate structures of these glycolipids. None of the other antibodies bound glycolipids.

5. Binding to Synthetic Peptides

The observation that antibodies M15, M22, M23, and M27 bound to deglycosylated mucin suggested that they might bind polypeptide epitopes. A 20-residue peptide corresponding to the repeated structure identified in the mucin core protein [14] was synthesized and bound to microtiter wells. All four antibodies bound strongly to this peptide, but not to untreated wells (Table 2). The other antibodies did not bind. This provides strong evidence that antibodies M15, M22, M23, and M27 do indeed recognize peptide epitopes.

6. Epitope Mapping

To examine steric relationships between the epitopes, each antibody was tested for its ability to inhibit the binding of radiolabeled antibodies either to milk mucin or to cell

Table 2 Binding of Antibodies to Synthetic Peptide[a]

Antibody	Absorbance (A450 × 1000) Peptide	Control
None	33	26
M10	48	31
M11	40	31
M12	41	30
M15	2769	32
M16	45	36
M21	49	41
M22	384	35
M23	1677	31
M25	41	31
M26	41	33
M27	196	48
M29	46	43
M38	59	55

[a]Peptide SAPDTRPAPGSTAPPAHGVT [14] was adsorbed to polystyrene microtiter wells. The monoclonal antibodies were conjugated to peroxidase as described by Nakane and Kawoi [19] and tested for binding as described previously [9].

Table 3 Epitope Mapping

Labeled Antibody	M8	M10	M11	M12	M15	M16	M21	M22	M23	M25	M27	M29	M38
M8	++	–	–	–	–	++	–	–	–	+	–	–	+++
M10	+++	++	–	–	–	++	–	–	–	–	–	–	++
M11	–	–	+++	+++	–	–	–	–	–	–	–	–	–
M12	–	–	+++	+++	–	–	–	–	–	–	–	–	–
M15	–	–	–	–	+++	–	–	+++	++	–	+	–	–
M16	++	–	–	–	–	+++	–	–	–	+	–	–	+++
M21	–	–	–	–	–	–	+++	–	–	–	–	+++	–
M22	–	–	–	–	+	–	–	++	++	–	–	–	+++
M23	–	–	–	–	+++	++	–	+++	++	–	++	–	+++
M25	++	–	–	–	–	+++	–	–	–	++	–	–	+++
M27	–	–	–	–	+	–	–	++	++	–	+	–	–
M29	–	–	–	–	–	–	+++	–	–	–	–	+++	–
M38	–	–	–	–	–	–	–	–	–	–	–	–	+++

Antibodies were radioiodinated and tested for binding to purified milk mucin or carcinoma cells in microtiter wells. Unlabeled antibodies were tested for competition, and the antibody concentration required for half-maximal inhibition was determined: +++, <3.2 µg/ml; ++, 3.2-15.9 µg/ml; +, 16.0-50.0 µg/ml; –, > 50.0 µg/ml.

lines (Table 3). Antibody M26 was not included in this analysis, because its binding to immobilized mucin was impaired by radiolabeling.

Antibodies M21 and M29 gave the simplest result: they competed with one another but with none of the other antibodies. Antibodies M11 and M12 similarly competed with one another. At the other extreme, antibody M38 competed with 7 of the 13 antibodies tested, although none of the antibodies competed with it. Among the remaining antibodies, M8, M16, and M25 gave similar competition patterns, as did antibodies M15, M22, M23, and M27.

The competition patterns indicate the existence of several families of epitopes. Antibodies M15, M22, M23, and M27 form a distinct group, which is in good agreement with the biochemical data. Antibodies M21 and M29, antibodies M11 and M12, and antibodies M8, M16, and M25 likewise form distinct groups. This experiment also provides estimates of the avidities of the antibodies, and since most of the antibodies competed with themselves at concentrations below 15 µg/ml, their avidities must be at least 10^7 L/mole. It should also be pointed out that differences in avidity cannot account for the competition patterns observed.

7. Conclusions

The fourteen antibodies all recognize epitopes present on the same carcinoma-associated mucin molecules, as demonstrated by DDIA. The antibodies can be organized into three distinct groups on the basis of the structures of their epitopes. The first group (antibodies M15, M22, M23, M27) comprises antibodies that bind to the mucin core polypeptide. The second group contains only antibody M26, the only antibody demonstrated to bind a carbohydrate epitope. The third and largest group (antibodies M10, M11, M16, M21, M25, M29, M38) comprises antibodies whose epitopes have not been identified, but they may involve both carbohydrate and polypeptide structures. This group is heterogeneous, and it could be subclassified on the basis of epitope sensitivity or competition assays.

IV. IMMUNOASSAY DEVELOPMENT

A. Selection of Monoclonal Antibodies

The studies described above showed that monoclonal antibodies to many distinct mucin epitopes can be generated. Before attempting to identify the antibodies most suitable for use in immunoassays, it was important to identify the most desirable characteristics of an immunoassay for carcinoma-associated mucins. We decided that clinical sensitivity and specificity, the ability to differentiate between patients with breast cancer and controls were most important. Because of the presence of detectable amounts of carcinoma-associated mucins in the sera of healthy individuals, analytical sensitivity and specificity (the ability to detect small amounts of antigen) are of lesser importance.

Our experimental strategy was to set up and evaluate multiple prototype immunoassays. In early studies, we tested pools of 10-20 individual sera from breast cancer patients and controls. The results of these tests were informative; the assay values for the pools were the mean of the antigen levels in the individual sera forming the pools. However, the mean values of serum pools from cancer patients and in some cases from controls were often skewed by one or two sera with extremely high antigen levels. This made it difficult to extrapolate from the mean levels to levels in individual patients. We subsequently found it more satisfactory to evaluate the prototype tests on panels of approximately 40 sera from breast cancer patients and controls.

Antibody-coated microtiter wells were used as the solid phase, and peroxidase-conjugated antibodies were used as tracer (see below). Peroxidase conjugates were tested at a concentration of approximately 1 μg/ml. Some antibodies, such as M26 and M29, which gave low signals when used as tracers, gave best results when used as catchers with an antibody to an abundant epitope, such as antibody M38, as tracer. Tests are referred to in terms of the catcher and tracer. Thus "M26/M38" refers to a test with antibodies M26 and M38 as catcher and tracer, respectively.

1. Epitope Levels in Sera from Healthy Donors

As with other tumor-associated antigens, carcinoma-associated mucins are detectable in the sera of healthy donors, and for optimal assay performance it may be necessary to test diluted specimens. Thus for each of the assays to be evaluated, we tested serial dilutions of a pool of serum from healthy donors and determined the serum dilution that gave an absorbance value of approximately 0.1 in the enzyme immunoassay. Dilution factors ranged from 2 for M23/M23 to 1000 for M38/M38. Antibodies M26 and M29 were tested with antibody M38 as tracer since antibody M26 was not successfully conjugated to peroxidase and no signal was obtained with antibody M29 as catcher and tracer.

We believe that the simplest explanation for the large differences in epitope concentrations in normal serum is that the antibodies recognize different epitopes on the same mucin molecule and that the epitope densities vary greatly. In other words, there are many antibody M38 sites on each mucin molecule but very few sites for antibodies M26 and M29. An alternative explanation is that the differences observed are due to differences in antibody avididty, but we believe that the results of the competition experiments argue against this explanation. Antibody M23, for example, competed well, but its dilution factor was 2 compared with 1000 for antibody M38.

The assays were calibrated with standards prepared from breast carcinoma effusions. Units were chosen so that a pool of healthy donor sera contained approximately 20 units/ml.

2. Epitope Levels in Sera from Breast Cancer Patients

The assays were evaluated on a panel made up of approximately 40 sera from patients with metastatic breast carcinoma and 40 sera from healthy donors. The results for the first 10 of the breast cancer patients' sera are shown in Table 4. Although many of the assays differ in their patterns of immunoreactivity, similarities can be seen. For example, the assays using antibodies M15, M22, and M23 have similar patterns, as do the assays using antibodies M16 and M25.

The immunoreactivity patterns can be quantified in the form of a matrix of correlation coefficients. We found the statistical technique of cluster analysis to be very useful for this purpose, since it produces a readily interpretable tree structure from the correlation coefficients and provides an objective, statistical basis for classifying the tests into groups on the basis of their immunoreactivity patterns. The data for the panel of breast cancer and control sera were analyzed by cluster analysis on a personal computer, with software from BMDP Statistical Software (Los Angeles, CA). Four distinct clusters were identified. Within each cluster, correlation coefficents were as high as 0.90. Tests M15/

Table 4 Double Determinant Immunoassay of Sera from Patients with Metastatic Breast Cancer

Serum	M8 M8	M15 M15	M16 M16	M22 M22	M23 M23	M25 M25	M26 M38	M29 M38	M38 M38
1	328[b]	141	44	157	>212	52	14	>250	>375
2	26	92	16	131	99	12	141	201	106
3	51	374	45	>255	>212	56	>600	131	155
4	60	23	17	27	66	18	>600	101	126
5	75	56	7	80	129	6	11	>250	142
6	64	24	14	18	47	20	14	>250	253
7	17	15	8	26	56	8	12	188	138
8	25	11	3	12	24	4	66	20	22
9	·15	46	14	50	55	12	486	95	122
10	29	40	>170	55	54	>580	140	52	67

Antibodies Used in Double Determinant Immunoassay[a]

[a]The upper antibody was adsorbed to microtiter wells and used as catcher; the lower antibody was conjugated to peroxidase and used as tracer.
[b]Absorbance values for each tests were converted to arbitrary units/ml by comparison to a standard.

M15, M22/M22, and M23/M23, which use antibodies to polypeptide epitopes, form such a cluster, as do M8/M8, M29/M38, and M38/M38. Correlations between the clusters were much poorer; as low as 0.25 for the M26/M38 assay, which did not correlate well with any of the other tests.

3. Discriminant Analysis

For the same panel of sera, we evaluated test performance in terms of their ability to differentiate between the breast cancer patients and controls. Several methods of measuring and summarizing assay performance were considered. Graphic methods such as dot plots [15,16] and receiver-operator curves [17] were useful for obtaining an overview of the data, but we found that they were too cumbersome to use for comparing large numbers of tests. A popular approach has been to calculate sensitivity and specificity based on a selected cutoff value. Problems with this approach include the difficulty of objectively assigning the cutoff values and the influence of the relatively few data points close to the cutoff on the sensitivity and specificity.

Stepwise discriminant analysis was found to be the most useful method for comparing test performance. This statistical method compares the ability of tests to discriminate between two populations, with results being expressed as F values. The better a test discriminates between the populations, or the less overlap there is between test results for the two populations, the greater its F value. Stepwise discriminant analysis has several advantages over the other methods of comparing tests. First, it is efficient; more than 100 different tests can be compared in a few minutes using a personal computer. Second, the method does not require assignment of cutoffs. Third, it determines whether combinations of tests are more useful than single tests.

Stepwise discriminant analysis was performed with software from BMDP Statistical Software. The application of this method to the panel of sera from breast cancer patients and controls is summarized in Table 5. The highest F value in the first step of the analysis was 53.24 for M29/M38; this is thus the best single test. Note, however, that for all the tests the F values are statistically significant; all the tests differentiate between the two test populations to some extent. The second step of the analysis indicates that three tests, M22/M22, M23/M23, and M26/M38, contributed additional information, with M26/M38 providing the most. The third step of the analysis shows that none of the remaining tests adds more information.

What do these results mean? First, M29/M38 is the test that best differentiates between breast cancer patients and controls. This conclusion is confirmed by the observation that this test had the best performance in terms of sensitivity and specificity (Table 5). Second, M26/M38 contributes additional information. This implies that some breast carcinoma specimens with relatively low M29/M38 values should have elevated M26/M38 values. This was, in fact, observed. Sixteen of 39 sera from breast cancer patients had M29/M38 values no higher than the highest control value; of these, 10 had M26/M38 above the highest control value. These tests, referred to hereafter as CA M26 and CA M29, were chosen for further development and investigation.

B. CA M26 and CA M29 Enzyme Immunoassays

1. Assay Format

The CA M26 and CA M29 tests were configured as microtiter enzyme immunoassays. We adopted the microtiter format because of its speed and simplicity, and because such

Table 5 Stepwise Discriminant Analysis of Data from Immunoassays of Sera from Breast Cancer Patients and Controls

Test	F Value[a] Step 1	Step 2	Step 3	Sensitivity[b] (%)
M8/M8	17.83*	0.27	0.06	55
M15/M15	22.72*	2.29	0.51	65
M16/M16	15.71*	0.07	0.00	33
M22/M22	28.99*	4.30*	1.26	45
M23/M23	43.87*	7.28*	3.51	48
M25/M25	12.75*	0.04	0.07	72
M26/M38	29.59*	10.45*		67
M29/M38	53.24*			87
M38/M38	37.21*	0.02	0.52	67

[a]The F value in the first step of the stepwise discriminant analysis is a measure of the relative ability of the tests to differentiate between sera from breast cancer patients and sera from controls; the higher the F value, the better the discrimination. The second step of the analysis is carried out after taking into account the test that performed best in the first step. It is a measure of the ability of the remaining tests to provide additional information. The third step of the analysis indicates whether any of the remaining tests provide additional information.
[b]Specificity 95%.
*Significant at $p < 0.05$.

tests can be run either manually or with automation. If plates with removable strips are used, the microtiter format is suitable both for low- and high-volume testing.

Microtiter wells were coated with purified monoclonal antibodies M26 and M29 as described previously [18], blocked with a buffer solution containing 0.5% bovine serum albumin and 5% sucrose, and dried. Antibody M38 was conjugated with horseradish peroxidase as described previously by Nakane and Kawoi [19]. Test kits may be stored at 2-8°C for up to 1 year, with no detectable change in performance.

To perform the tests, standards, positive control, and specimens (diluted 1:11) are added to the antibody-coated wells. Antigen present in the sample binds to the adsorbed antibody during the initial 1 hr incubation period at room temperature, after which the wells are washed to remove unbound material. Peroxidase-conjugated antibody M38 is added, and during the second 1 hr incubation period it binds to the captured antigen. A final wash step removes unbound conjugate. Buffered substrate/chromogen reagent (hydrogen peroxide and 3,3′,5,5′ tetramethylbenzidine) is then added to each well. During the final incubation period, if antigen is present a blue color develops in a reaction with bound conjugate present in the well. The amount of color that develops is a measure of the amount of mucin antigen bound to the adsorbed antibody. The reaction is stopped by addition of 1 N sulfuric acid, which results in a color change from blue to yellow. The absorbance values for positive control, standards, and test samples are determined spectrophotometrically at 450 nm using a microplate reader.

Standard curves are constructed for each test by plotting absorbance value against concentration for each standard. The antigen concentrations of diluted samples and diluted positive control are then read from the standard curves. CA M26 and CA M29 test results are expressed in arbitrary units, based on standards maintained at Genetic Systems. For each test, standards are used at two concentrations: 30 and 75 units/ml for CA M26 units/ml, and 4 and 10 units/ml for CA M29. Sample diluent is used as a zero standard. The calibration curves are almost linear, and the two standards plus the zero standard are sufficient for constructing the standard curve by the linear interpolation or point-to-point method.

2. Analytical Peformance

Precision (within-run variability), reproducibility (between-run variability), and linearity were investigated and found to be acceptable for tests of this type, with precision better than 5% coefficient of variation (CV) and reproducibility better than 10% CV.

Potential interfering substances were tested. Bilirubin, hemoglobin, and a number of commonly used drugs were shown not to interfere with the CA M26 and CA M29 tests. A more serious potential source of interference is the presence of heterophile antibodies in the test samples [20]. We found that such antibodies occurred very infrequently, and that their effect could be essentially eliminated by the addition of an "irrelevant" IgG1 monoclonal antibody to the conjugate solution.

3. Normal Range

Serum samples from 600 healthy blood donors (sex and age unknown) were tested for CA M26 and CA M29. The mean CA M26 level was 33.6 U/ml and the mean CA M29 level was 7.3 U/ml. The effects of sex and age on CA M26 and CA M29 levels were investigated in 84 healthy donors. Mean CA M26 levels were between 28.8 and 36.6 U/ml, and mean CA M29 levels were between 8.8 and 10.9 U/ml. Analysis of variance showed no significant effect of either sex or age on CA M26 levels. CA M29 levels increased slightly with age in both males and females ($p = 0.016$).

Of the 684 specimens tested, 98.1% of the CA M26 values were less than or equal to 100 U/ml, and 99.7% of the CA M29 values were less than or equal to 20 U/ml. These antigen concentrations can be considered to be the upper limit of the normal range.

4. CA M26 and CA M29 Levels in Nonmalignant Disease

Serum samples from 134 patients (both sexes) with nonmalignant disease were tested for CA M26 and CA M29; 98.5% contained CA M26 levels less than 100 U/ml, and 89.6% contained CA M29 levels less than 20 U/ml. In a second study, 98.2% of 55 women with nonmalignant diseases had CA M26 levels lower than 100 U/ml and 98.2% had CA M29 levels lower than 20 U/ml.

5. CA M26 and CA M29 Levels in Breast Cancer Patients

In a preliminary study, 72 serum samples from patients with metastatic breast cancer and 92 samples from patients with benign breast disease were tested [9]. The assays were compared in terms of their ability to differentiate between the two groups of specimens, with cutoff values chosen to give approximately 90% clinical specificity (i.e., 90% of the control group tested negative). The CA M29 test gave the best performance; it detected elevated antigen levels in 93% of sera from patients with metastatic breast cancer. The sensitivity of the CA 15.3 test (83%) was lower than that of the CA M29 test (93%), but this difference was not statistically significant. Sensitivities of other assays were significantly lower, including CA M26 (60%, $p < 0.001$ by McNemar's test).

Table 6 Stepwise Discriminant Analysis of Data from CA M26, CA M29, and CA 15.3 Assays[a]

Test	F Value[a] Step 1	Step 2	Step 3	Sensitivity[b] (%)
CA M26	81.83*	4.14*		60
CA M29	259.44*			93
CA 15.3	176.87*	0.38	0.00	83

[a]See footnote to Table 5.
[b]Specificity 90%.
*Significant at $p < 0.05$.

Stepwise discriminant analysis of the same data was performed, and the results are shown in Table 6. The first step of the analysis showed that CA M26, CA M29, and CA 15.3 all differentiated significantly between breast cancer patients and controls, but that CA M29 differentiated better than the other tests. The second step of the analysis showed that CA M26 but not CA 15.3 contributed additional information. In the third step, CA 15.3 contributed no additional information. As in the previous analysis, the data indicate that the CA M26 and CA M29 should both be run for optimal performance.

This prediction was tested by evaluating the tests for their ability to complement the CA M29 assay. Using higher cutoffs that resulted in 100% specificity, 55 of 72 samples from breast cancer patients were elevated in either the CA M29 assay or the CA M26 test (76% sensitivity). This was significantly greater ($p < 0.001$ by McNemar's test) than the sensitivity (60%) obtained with the CA 15.3 test with the same criteria. No samples were elevated in the CA 15.3 test and negative for both CA M29 and CA M26, but samples were negative in the CA 15.3 test and positive in either CA M29 or CA M26. Thus when the CA M29 and CA M26 tests were used in combination, there was a statistically significant increase in test performance, which confirms the prediction of the stepwise discriminant analysis.

V. DISCUSSION

Monoclonal antibodies have been used to identify numerous epitopes of breast carcinoma-associated mucins [2-6] and these antibodies have been used to construct new immunoassays for the detection of circulating mucin antigens in patients with breast cancer [15,16,18,21-24].

The work that we describe here confirms and extends previous reports of the antigenic complexity of breast carcinoma-associated mucins [25]. These molecules express numerous epitopes, ranging from those formed by the mucin core polypeptide to those consisting solely of carbohydrate side chains, with the strong possibility that some epitopes contain both polypeptide and carbohydrate components. Because of this complexity, not all monoclonal antibodies are equally useful for constructing immunoassays, and it is important to determine their relative utility.

Our own experience in developing and evaluating tests for breast carcinoma-associated mucins is described in this chapter. We generated many antibodies to breast carcinoma-associated mucins and classified them on the basis of the chemical structure of their epitopes and their patterns of reactivity with mucins from different sources. We found that the antibodies recognized multiple epitopes, that epitope density varied greatly from antibody to another, and that mucins from different sources, such as milk and malignant effusions, differed significantly in their epitope compositions.

Enzyme immunoassays were constructed and used to test sera from breast cancer patients and controls. The assays gave different patterns of immunoreactivity with sera from breast cancer patients, and we found that the statistical technique of cluster analysis could be used to group the tests. The cluster analysis agreed well with the biochemical analysis; particularly noticeable was the clustering of tests constructed with antibodies to the mucin core polypeptide.

The most important observation was that the tests differed in the degree to which they differentiated between sera from cancer patients and controls, as measured by the statistical technique of stepwise discriminant analysis. Even tests constructed from antibodies from within the same cluster differed in performance.

It is clear, therefore that evaluation and comparison of immunoassays for breast carcinoma-associated mucins will not be straightforward. Different tests for the same analyte are commonly compared in terms of analytical performance (analytical sensitivity and specificity, precision, reproducibility, linearity). Fleisher et al. [26], for example, compared two tests for carcinoembryonic antigen (CEA) in these terms. The tests for breast carcinoma-associated mucins represent a new situation, since they use antibodies to different epitopes, and even though they may detect the same antigen molecule, they differ significantly in clinical sensitivity and specificity.

How should different tests for carcinoma-associated mucins be compared? To date they have been evaluated individually, or compared with other markers, such as CEA, and there are few reports of side-by-side comparisons of different tests for carcinoma-associated mucins. We believe that only by comparing test performance on the same panel of specimens will it be possible to determine the relative value of different carcinoma-associated mucin immunoassays.

The use of panels of tumor markers is gaining increasing acceptance [27-31], and our work shows that a combination of two tests for carcinoma-associated mucins (CA M26 and CA M29) gives better performance than either test alone. Although a full discussion of the use of panels of markers is beyond the scope of this chapter, the following points should be considered. First, there should be some clear statistical evidence that the combination of tests is advantageous. This question can be addressed by stepwise discriminant analysis. Second, there will generally be no advantage in combining tests that are highly correlated (correlation coefficient greater than 0.90). Third, one must specify how the results of several tests are to be interpreted; there must be an algorithm for combining the test results. As a final point, if the combination of tests is intended to increase clinical sensitivity, the effect on clinical specificity must also be considered, and vice versa.

In conclusion, carcinoma-associated mucins are a new and important family of tumor markers. The antigenic complexity of these molecules offers unprecedented opportunities and challenges for the development of new immunoassays. We describe here two new tests, CA M26 and CA M29, shown by objective statistical criteria to be superior to previously described tests. Further studies are underway to determine the clinical utility of these markers.

ACKNOWLEDGMENTS

We thank Drs. Gary Goodman, Donald Mercer, and Joyce Taylor-Papadimitriou for their advice and encouragement, and Anne Acheson, Sidney Hallam, Sharon Laska, and Pat Stewart for their excellent technical assistance.

REFERENCES

1. Rittenhouse HG, Manderino GL, Hass GM. Mucin-type glycoproteins as tumor markers. Lab Med 1985;16:556-560.
2. Arklie J, Taylor-Papadimitriou J, Bodmer W, et al. Differentiation antigens expressed by epithelial cells in lactating breast are also detectable in breast cancers. Int J Cancer 1981;28:23-29.
3. Ceriani RL, Peterson JA, Lee JY, et al. Characterization of cell surface antigens of human mammary epithelial cells with monoclonal antibodies prepared against human milk fat globule. Somat Cell Genet 1983;9:415-427.
4. Hilkens JF, Buijs J, Hilgers J, eta l. Monoclonal antibodies against human milk-fat globule membranes detecting differentiation antigens of the mammary gland and its tumors. Int J Cancer 1984;34:197-206.
5. Papsidero LD, Croghan GA, O'Connell MJ, et al. Monoclonal antibodies (F36/22 and M7/105) to human breast carcinoma. Cancer Res 1983;43:1741-1747.
6. Kufe D, Inghirami G, Abe M, et al. Differential reactivity of a novel monoclonal antibody (DF3) with human malignant versus benign breast tumors. Hybridoma 1984;3:223-232.
7. Shimizu M, Yamauchi K. Isolation and characterization of mucin glycoproteins in human milk fat globule membrane. J Biochem 1982;91:515-524.
8. Ormerod MG, McIlhinney J, Steele K, et al. Glycoprotein PAS-O from the milk fat globule membrane carries antigenic determinants for epithelial membrane antigen. Mol Immunol 1985;22:265-269.
9. Linsley PS, Brown JP, Magnani JL, et al. Monoclonal antibodies reactive with mucin glycoproteins found in sera from breast cancer patients. Cancer Res 1988;48: 2138-2148.
10. Köhler G, Milstein C. Continuous culture of fused cells secreting antibody of predefined specificity. Nature 1975;256:495-497.
11. Brown JP, Hellström KE, Hellström I. Use of monoclonal antibodies for quantitative analysis of antigens in normal and neoplastic tissues. Clin Chem 1981;27:1592-1596.
12. Mort AJ, Lamport DTA. Anhydrous hydrogen fluoride deglycosylates glycoproteins. Anal Biochem 1977;82:289-309.
13. Magnani JL, Nilsson B, Brockhaus M, et al. A monoclonal antibody defined-antigen associated with gastrointestinal cancer is a ganglioside containing lacto-N-fucopentose II. J Biol Chem 1982;257:14365-14369.
14. Gendler S, Taylor-Papadimitriou J, Duhig T et al. A highly immunogenic region of a human polymorphic epithelial mucin expressed by carcinomas is made up of tandem repeats. J Biol Chem 1988;263:12820-12823.
15. Burchell J, Wang D, Taylor-Papadimitriou J. Detection of the tumor-associated antigens recognized by the monoclonal antibodies HMFG-1 and 2 in serum from patients with breast cancer. Int J Cancer 1984;34:763-768.
16. Papsidero LD, Nemoto T, Croghan GA, et a.. Expression of ductal carcinoma antigen in breast cancer sera as defined using monoclonal antibody F36/22. Cancer Res 1984;44:4653-4657.

17. Krouwer JS. Cumulative distribution analysis grafhs—an alternative to ROC curves. Clin Chem 1987;33:2305-2306.
18. Linsley PS, Ochs V, Laska S, et al. Elevated levels of a high molecular weight antigen detected by antibody W1 in sera from breast cancer patients. Cancer Res 1986;46: 5444-5450.
19. Nakane PA, Kawoi A. Peroxidase-labeled antibody. A new method of conjugation. J Histochem Cytochem 1974;22:1084.
20. Boscato LM, Stuart MC. Heterophilic antibodies: a problem for all immunoassays. Clin Chem 1988;34:27-33.
21. Ceriani RL, Sasaki M, Sussman J, et al. Circulating human mammary epithelial antigens in breast cancer. Proc Natl Acad Sci USA 1982;79:5420-5424.
22. Hayes DF, Sekine H, Ohno T, et al. Use of a murine monoclonal antibody for detection of circulating plasma DF3 antigen levels in breast cancer patients. J Clin Invest 1985;75:1671-1678.
23. Hayes DF, Zurawski VR, Kufe DW. Comparison of circulating CA 15.3 and carcinoembryonic antigen levels in patients with breast cancer. J Clin Oncol 1986;4: 1542-1550.
24. Hilkens J, Kroezen V, Bonfrer JMG, et al. MAM-6 antigen, a new serum marker for breast cancer monitoring. Cancer Res 1986;26:2582-2587.
25. Burchell J, Durbin H, Taylor-Papadimitriou J. Complexity of antigenic expression of antigenic determinants recognized by monoclonal antibodies HMFG-1 and HMFG-2 in normal and malignant human mammary epithelial cells. J Immunol 1983;131: 508-513.
26. Fleisher M, Nisselbaum JS, Loftin L, et al. Roche RIA and Abbott EIA carcinoembryonic antigen assays compared. Clin Chem 1984;30:200-205.
27. Bast RC, Thomas LK, Schaetzl E, et al. Monitoring human ovarian carcinoma with a combination of CA 125, CA 19-9 and carcinoembryonic antigen. Am J Obstet Gynecol 1984;149:553-559.
28. Fuith L, Schrocksnadel H, Dworzak E, et al. Using a panel of tumor markers in the follow-up of breast carcinoma patients. J Tumor Marker Oncol 1987;2:177-180.
29. Steger G, Mader R, Dittrich C, et al. CA 15.3, CA 50, CEA, TPA, and a newly developed tumor marker score in breast cancer. J Tumor Marker Oncol 1987;2:169-176.
30. Aiginger P, Birkmayer GD. Advantages and pitfalls of a tumor marker screening (CEA, CA 125, CA 19-9) in patients with abdominal masses. J Tumor Marker Oncol 1988;3:137-140.
31. Montinari F, Beretta GD, Pancera G, et al. CEA, TPA and GICA during long-term follow-up and in therapy monitoring of gastrointestinal tumors. J Tumor Marker Oncol 1988;3:201-205.

7
Strategies for Heterogeneous Enzyme Immunoassays for Tumor Markers

WILLIAM J. KING, JOSEPH T. TOMITA, BARRY L. DOWELL, and DENNIS M. DELFERT

Abbott Laboratories, North Chicago, Illinois

I. BACKGROUND

Diagnostic assays have undergone significant transformations in the past few years with respect to their formats and technology. Previously the purview of specialized radiology or clinical chemistry laboratories, immunoassays have been simplified to permit routine testing of complex analytes in smaller hospital laboratories as well as in less traditional settings, such as doctors' offices. At the same time hospital and reference laboratories have faced increasing pressure to streamline operations to minimize overhead and assay turnaround time. These demands have fueled the development of new immunoassay systems that exploit a number of strategies for shortening assay incubation times and/or fully automating assay protocols.

From their beginnings in the 1970s, enzyme-linked immunosorbent assays (ELISA) utilized antibodies coupled to solid phase supports. These early configurations favored competitive inhibition assay formats in which antibody (ligand) was either coupled covalently to cellulose [1] or passively adsorbed to polystyrene tubes [2,3]. As the use of immunoassay technology expanded, microtiter plates [4], cellulose discs [5], Sepharose [6], acrylamide beads [7], glass rods [8], polystyrene beads [9,10], latex microparticles [11-13], and a variety of other support media also came into common use. Purified analytes labeled with enzymes such as alkaline phosphatase, horseradish peroxidase (HRPO), beta-galactosidase, or glucose oxidase provided a signal whose amplitude could be modulated through competition for binding to the immobilized ligand by unlabeled analyte in assay calibrators, controls, or unknowns. Alternative competitive inhibition configurations in which antigen is coupled to the solid phase support and antibody-enzyme conjugates provide the assay signal are, of course, also possible and began to appear shortly thereafter [14,15]. All of the assay configurations discussed in this chapter are "heterogeneous immunoassays" [16] in that the enzyme-labeled conjugate bound to the analyte is separated from unbound conjugate before enzyme activity is measured.

Although competitive assay formats continue to play a notable role, particularly with low-molecular-weight analytes in which the number of unique epitopes may be severely restricted, most tumor markers are not included. The generally larger size and antigenic complexity of these markers have permitted investigators to exploit noncompetitive or sandwich-type solid phase enzyme immunoassay (EIA) configurations. While these are potentially more sensitive than competition assays because of the opportunity for binding of several molecules of the conjugated probe to the analyte, their real value in tumor marker work may stem from the requirement that antibody binding occurs at two distinct sites. Some of these markers, carcinoembryonic antigen (CEA) being a notable example, are either incompletely characterized or are members of a family of closely related antigens. The obligation that antibody recognition occur at two distinct sites makes it possible in some cases to design reagent configurations with improved specificity for the analyte in question.

Two sandwich assay designs are commonly used. Direct sandwich assays use solid-phase-immobilized antibody (or other ligands, such as lectins) to capture antigen and an enzyme-labeled probe directed toward a second binding site to generate the assay signal. These secondary sites may be unique epitopes requiring a different ligand from that used for the solid phase, or redundant expressions of the primary binding sites, in which case the capture phase antibody may also serve as probe. In a typical CEA assay (Fig. 1) polystyrene beads (1/4 inch in diameter) coated with CEA antibody are incubated with specimens, standards, or controls. Within a fixed period of time a certain percentage of the analyte is bound to the solid phase. Unbound materials are then removed by aspiration and the beads are rinsed prior to a second incubation with HRPO-anti-CEA conjugate. Following this incubation the beads are washed again, a chromogenic HRPO substrate solution (o-phenylenediamine/H_2O_2) is added, and the reaction is quenched by the addition of acid after an appropriate period of color development. The intensity of this color is read spectrophotometrically and is proportional to the concentration of CEA in the sample [10,17,18]. With minor variations, these principles have also been applied to the development of heterogeneous EIAs for alpha fetoprotein (AFP) [19,20], prostatic acid phosphatase [21], prostate-specific antigen [22], the ovarian cancer marker CA 125 [23], estrogen and progesterone receptor proteins [24-26], and a variety of other putative tumor markers, hormones, and non-tumor-related analytes of diagnostic interest.

An alternative sandwich assay design, termed indirect (Fig. 2), is similar to the direct method except that the enzyme-labeled probe is directed against the ligand that binds to

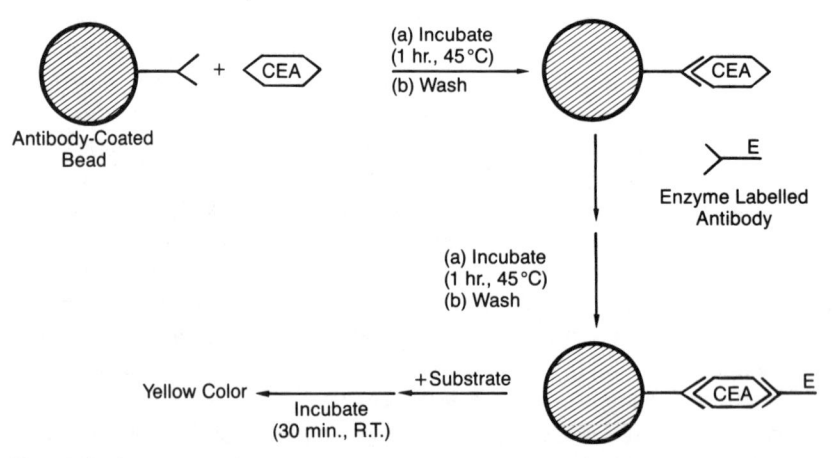

Figure 1 Sequence of events in a direct sandwich assay for CEA.

Heterogeneous Enzyme Immunoassays

A. Probe directed against secondary ligand

B. Probe directed against hapten on secondary ligand

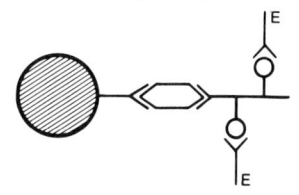

Figure 2 Indirect sandwich assay configurations. Assay steps are identical to those shown in Figure 1 except that antibody directed against the secondary epitope is either unlabeled (A) or labeled with a hapten (B) (e.g., biotin, fluorescein, etc.). In example A signal is generated by an enzyme-labeled probe directed against the second antibody. In B, the probe is directed against the hapten.

the secondary site on the analyte or, in some cases, to haptens attached to this ligand (HRPO-labeled antimouse, antigoat, antifluorescein, avidin, etc.). Indirect assay configurations may be especially useful in two situations: when functional groups on the secondary probe through which enzyme is conjugated are missing or are in areas of the probe required for analyte recognition, or when additional assay sensitivity is required.

As implied above, most currently available solid-phase immunoassays avoid the extended incubation times required for antibody-antigen (or more generally, "ligand-analyte") interactions to reach equilibrium by using a rate-endpoint methodology. This approach assumes that the rate of binding of the analyte to the solid capture phase is equivalent in specimens and standards provided with the assay kit. Because the capture phase concentration is kept constant, the binding reaction follows pseudo-first-order kinetics with respect to analyte concentration. If the binding reaction is terminated at any time before it reaches equilibrium, the concentration of analyte in the sample can be determined by quantifying the mass of analyte bound within this time and comparing it to the mass bound within the same time from a set of standards of known concentration. The actual length of time allowed for binding in such systems is usually driven by the sensitivity of the detection system used. Therefore, in order to shorten assay times significantly without sacrificing analytical sensitivity, one must either combine assay steps (the so-called "one-step" formats, in which the two antibody-binding steps are allowed to proceed simultaneously), improve the sensitivity of the detection system, or find some means of accelerating the rate of analyte capture.

In recent years it has become increasingly popular, especially for commercial immunoassays, to adopt a one-step format in the interest of shortening assay times [27]. Although these procedures have in general met or exceeded the standards set by two-step assays, such assays can pose special challenges for the assay developer. The simultaneous incubation of solid-phase antibody, antibody-enzyme conjugate, and specimen without the benefit of an intermediate wash step potentially increases the vulnerability of the system to interference by serum or plasma constituents. Interference may occur at the level of ligand binding or enzyme activity unless steps are taken in reagent development to circumvent these problems. Care must also be taken to ensure that "prozone" or "hook" effects [28], that is, decreases in assay signal in the presence of very high analyte levels, do not occur at analyte concentrations likely to be encountered in clinical specimens. Even when such problems have been successfully controlled, most assays still take 1-2 hr and are relatively labor-intensive.

II. LATEX MICROPARTICLE ASSAY TECHNOLOGY

Binding kinetics in solid-phase assay systems may be accelerated by increasing the number of interactions between the solid-phase ligand and the analyte. This can sometimes be accomplished by raising the temperature at which binding occurs, but this approach usually has limitations imposed by either the stability of one of the reactants or a concomitant increase in dissociation rate. Agitation of the reactants may also be beneficial. However, one of the most effective strategies for reducing the time needed for analyte binding is to increase the surface area of the solid phase. As an example, current technology using polystyrene beads (1/4 inch in diameter) presents approximately 1.25 cm^2 (if one ignores surface irregularities) of capture phase surface to the analyte. By coupling the capture phase reagent to latex microparticles (0.2-0.5 μm diameter) surface area may be increased by 5-10 times and analyte binding accelerated significantly (Fig. 3). As a result, immunoassay incubation times can be reduced from hours to a matter of minutes with no loss of analytical sensitivity. Microparticle suspensions may also be handled essentially as liquid reagents, which facilitates the automation of assay procedures. These particles are available from a number of vendors and may be obtained either as underivatized polystyrene for passive absorbtion of antibody or modified polystyrene to having functional groups for covalent antibody attachment.

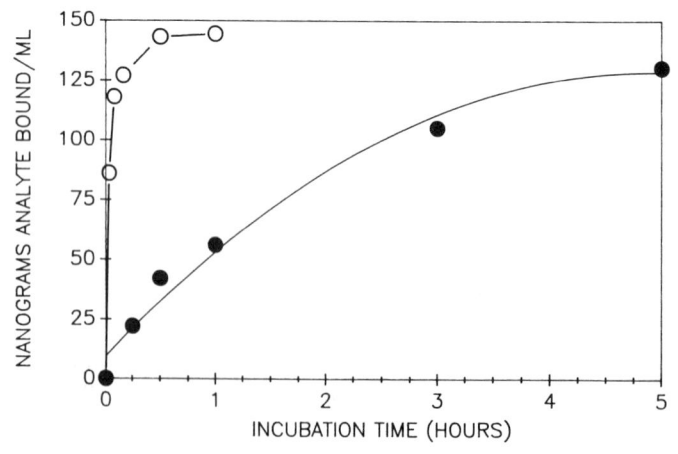

Figure 3 Comparison of time courses for association of analyte with solid-phase ligands. ●, ligand-coated 1/4-inch polystyrene bead; ○, similarly coated latex microparticles (0.125% solids, final concentration).

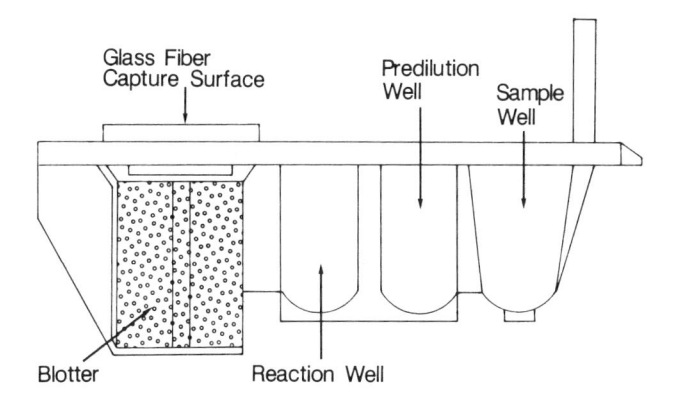

Figure 4 IMx reaction cell. This disposable accessory is loaded into a carousel that rotates beneath pipetting and optics stations within the IMx instrument.

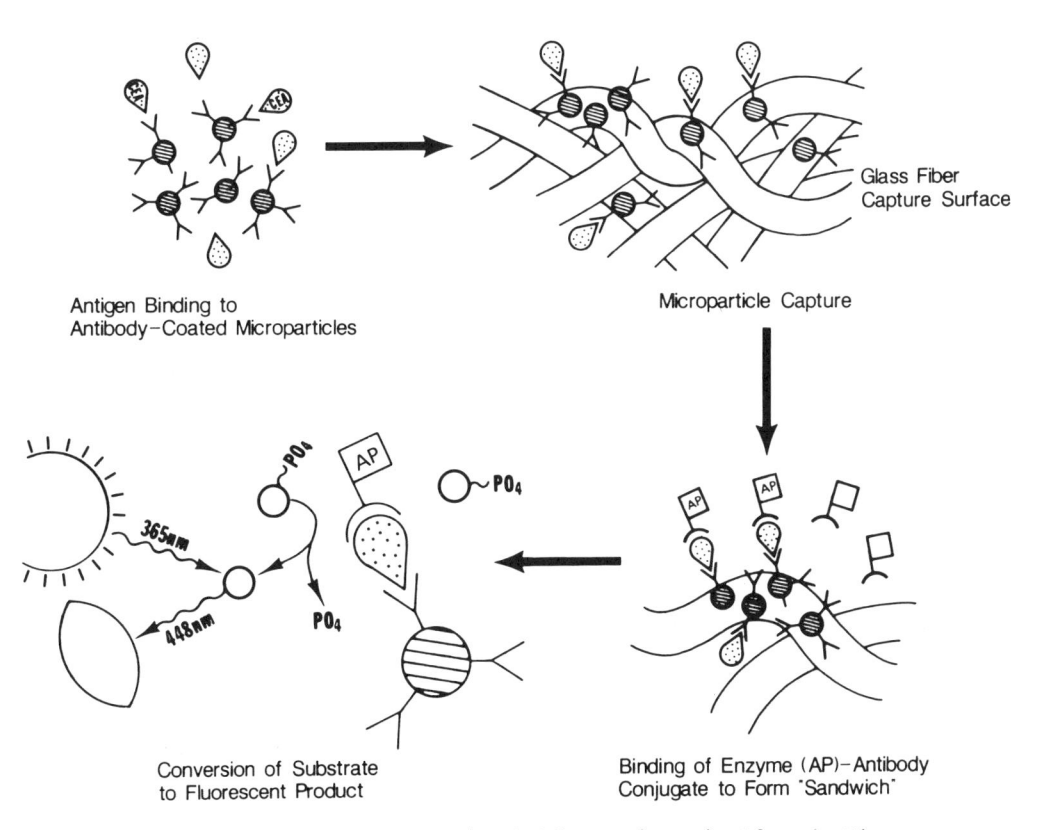

Figure 5 The MEIA reaction sequence (sandwich assay format). After pipetting approximately 100 µl sample into the specimen well of the reaction cell (Fig. 2), the operator loads the appropriate reagent kit into the instrument, selects the desired assay from the instrument menu, and starts the run. Twenty to forty minutes later, the concentrations of analyte in the samples are printed out.

IMx CEA Calibration Curve

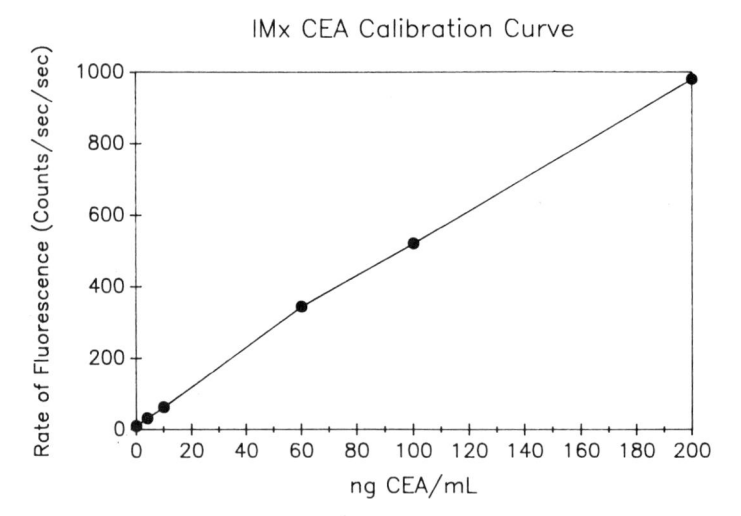

Figure 6 The MEIA (IMx CEA) calibration curve. Fluorescence rate is expressed in arbitrary instrument units of counts/sec/sec. After initial calibration, the curve is stored in instrument memory to be used for subsequent assay runs. The curve is replaced only when controls fall outside of specifications or critical instrument components are replaced.

Table 1 IMx CEA Precision

			Percentage CV		
Sample	n	ng CEA/ml	Within-Run	Between-Run	Between-Instrument
1	81	2.3	5.71	7.79	8.47
2	81	5.9	3.69	5.44	5.46
3	81	23.2	3.30	4.66	4.66
4	81	105.4	3.85	4.13	4.13

Three-day, three-instrument precision study for MEIA for CEA. Serum samples were assayed in replicates of three and between-run and between-instrument results were projected to reflect replicates of one, as is typically run on this instrument system. Coefficients of variation (% CV) were determined from components of variance from which a statistical estimate of the variation of individual values for multiple assay runs was derived [31].

Table 2 IMx AFP Precision

Sample	n	ng AFP/ml	Within-Run	Between-Run	Between-Instrument
1	90	9.7	4.26	5.95	8.00
2	90	41.1	3.80	4.81	5.24
3	90	69.4	3.71	4.92	5.55
4	90	160.1	2.72	4.85	5.75
5	90	266.2	2.86	5.91	7.50

Percentage CV

Five-day, three-instrument precision study for MEIA for AFP. Results were calculated as described in Table 1.

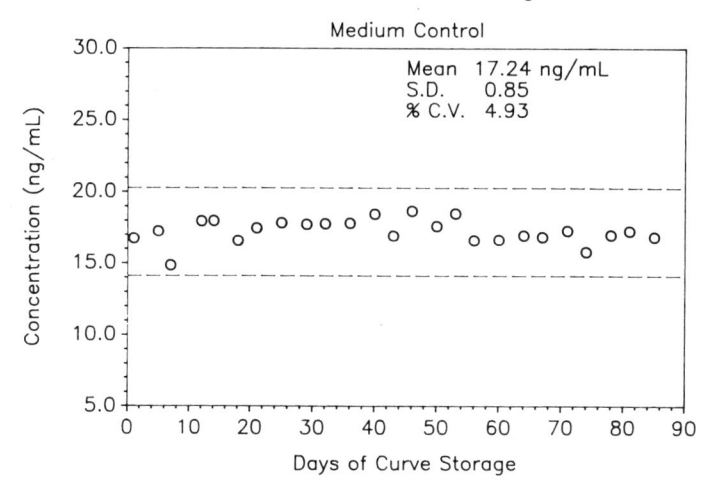

Figure 7 Extended curve storage in the IMx CEA assay. A single calibration curve was run on day 0 of the study. On subsequent days controls were run in replicates of one. The entire study was performed using a single reagent kit. For this experiment control ranges were set to approximate the acceptance ranges anticipated for final kit reagents. Data shown are for the medium level control, although similar results were seen for low-level (2.5 +/− 0.21 [SD] ng/ml) and high level (74.8 +/− 2.84 ng/ml) controls.

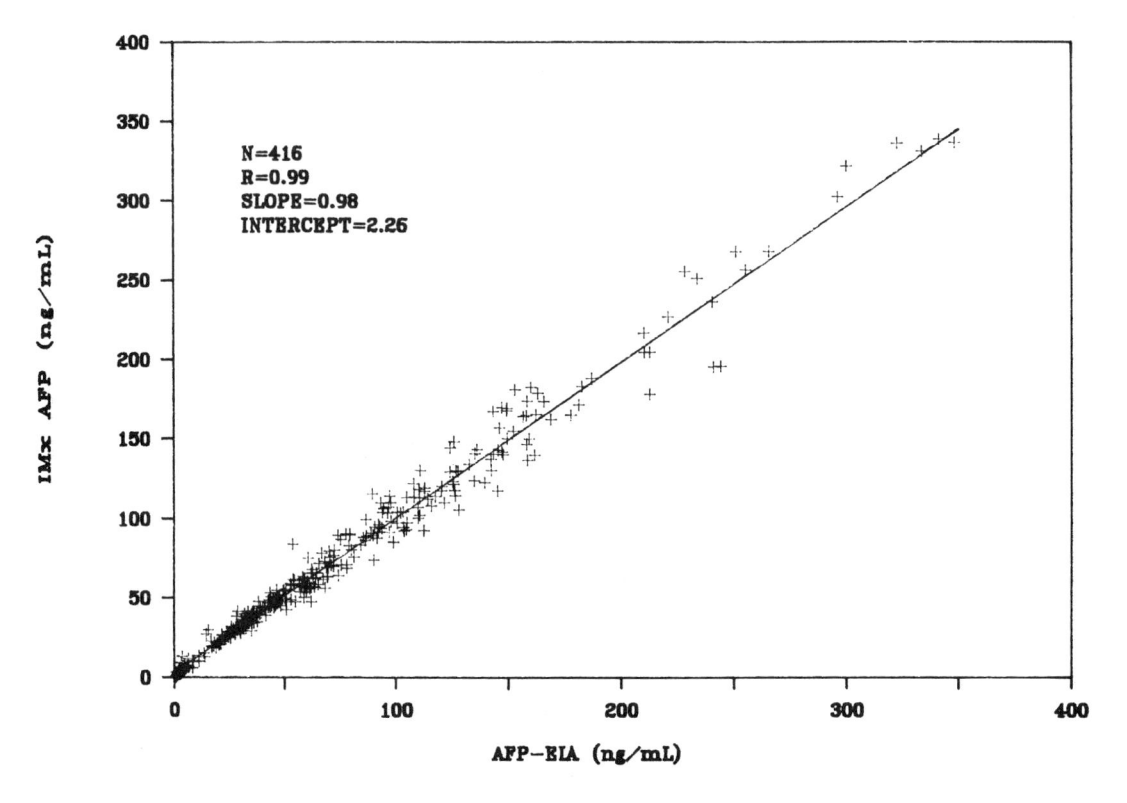

Figure 8 Correlation of AFP values obtained with matched specimens assayed by MEIA (IMx AFP) and conventional EIA (Abbott AFP EIA) procedures.

Examples of the application of this approach are illustrated by recent studies using immunological reagents for carcinoembryonic antigen (CEA) and alpha fetoprotein (AFP) to develop procedures for the Abbott IMx automated benchtop analyzer system [13]. The system combines instrument robotics and a controlled assay environment to perform both microparticle capture enzyme immunoassays (MEIAs) and fluorescence polarization immunoassays (FPIAs), which is a homogeneous assay format applied to low-molecular-weight analytes that will not be discussed here. In a typical MEIA batch procedure (Figs. 4,5), a sample that has been preloaded into the sample well of a disposable reaction cell is mixed by the instrument with the appropriate antibody-coated microparticle suspension. After this mixture is incubated for several minutes, it is transferred to a glass fiber capture surface of the reaction cell to which protein-coated microparticles tightly adhere. Unadsorbed materials are removed by capillary action driven by the underlying blotter; the efficiency of this separation is aided by subsequent buffer washes. An antibody-enzyme conjugate (e.g., anti-CEA/alkaline phosphatase) is then added to the washed microparticles and the second incubation takes place on the capture surface, completing the antibody "sandwich." Competition-type assay formats are also possible. Following the removal of unbound conjugate, enzyme substrate, 4-methyl umbelliferyl phosphate, is added and the matrix is moved beneath the instrument's optics. Light emitted at 448 nm from excitation of the reaction product following excitation at 365 nm is detected and quantified. The rate at which this fluorescence increases is compared to that occurring with calibrators of known analyte concentration (Fig. 6); the instrument performs the required interpolations to give the concentration of analyte in the sample. The procedure is complete within approximately 30 min from the time sample and assay reagents are loaded.

The rapidity with which these assays are completed does not compromise acceptable assay performance. For example, owing in part to the complete automation of the assay procedure, precise measurements of AFP or CEA are still obtained. Human sera enriched with analyte to several levels spanning the assay ranges (0-200 ng/ml for CEA, 0-350 ng/ml for AFP) were repeatedly assayed over a 3 day period on three different IMx instruments. The precision of these procedures (Tables 1 and 2) compared favorably to that seen with more conventional assay formats. The between-run reproducibility of reagents and the stability of the instrument environment eliminated the need for daily calibration, thus permitting extended storage of standard curves within the IMx instrument. In the study illustrated in Figure 7, for example, CEA controls were repeatedly assayed using the same reagent kit over a period of 88 days. Concentrations were determined from a single calibration curve run on the first day of the study. Analytical sensitivity, defined as the lowest concentration of CEA reliably differentiated from 0 ng/ml, was found to be better than 0.5 ng/ml. Similar results have been obtained for AFP. In a study comparing AFP values obtained with MEIA and conventional bead assay (Abbott AFP-EIA) procedures on 416 paired specimens, a (linear) correlation coefficient of 0.98 was observed (Fig. 8). Similarly, CEA values obtained by either technology are in excellent agreement (r = 0.988, n = 148).

Microparticle-based immunoassays therefore offer the opportunity to decrease greatly the turnaround time for routine testing without sacrificing the assay performance required to produce reliable clinical information. A version of this methodology, adapted for manual performance of qualitative or semiquantitative assays, is already in widespread use [29]. By wedding this technology to an instrument system capable of performing all assay steps and maintaining a consistent assay environment, the consistent and precise performance required for the clinical monitoring of tumor markers has been achieved.

REFERENCES

1. Engvall E, Perlmann P. Enzyme-linked immunosorbent assay (ELISA) quantitative assay of immunoglobulin G. Immunochemistry 1971;8:871-874.
2. Van Weemen BK, Schuurs AHWM. Immunoassay using antigen-enzyme conjugates. FEBS Lett 1971;15:232-236.
3. Engvall E, Perlamnn P. In: Peeters H, ed. Protides of the biological fluids, proceedings of the nineteenth colloquium, Brugge. Pergamon Press, Oxford; 1971:553-556.
4. Wolters G, Kuijpers L, Kacaki J, Schuurs A. Solid-phase enzyme-immunoassay of hepatitis B surface antigen. J Clin Pathol 1976;29:873-879.
5. Frackelton AR, Szaro RP, Weltman JK. A galactosidase immunosorbent test for carcinoembryonic antigen. Cancer Res 1976;36:2845-2849.
6. Kato K, Hamaguchi Y, Fukui H, Ishikawa E. Enzyme-linked immunoassay II. A simple method for the synthesis of the rabbit antibody-D-galactosidase complex and its general applicability. J Biochem 1975;78:423-425.

7. Guesdon JL, Avrameas S. Magnetic solid phase enzyme immunoassay. Immunochemistry 1977;14:443-447.
8. Hamaguchi Y, Kato K, Fukui H, Shirakawa I, et al. Enzyme linked sandwich immunoassay of macro molecular antigens using the rabbit antibody coupled glass rods as a solid phase. Eur J Biochem 1976;71:459-467.
9. Rose SP, Jolley ME, Waindle LM, et al. A sensitive solid-phase enzyme immunoassay for carcinoembryonic antigen. In: Lehman F, ed. Carcinoembryonic proteins, biology, chemistry, and clinical applications. Elsevier North-Holland, Amsterdam, 1978:206.
10. Maolini R, Bagrel A, Chavance C, et al. Study of an enzyme immunoassay kit of carcinoembryonic antigen. Clin Chem 1980;26:1718-1722.
11. Hechemy KE, Michaelson EE. Latex particle assay in laboratory medicine. Part I. Lab Manage 1984;22(6):27-40.
12. Hechemy KE, Michaelson EE. Latex particle assays in laboratory medicine. Part II. Lab Manage 1984;22(7):26-35.
13. Fiore M, Mitchell J. Doan T, et al. The Abbott IMx automated benchtop immunochemistry analyzer system. Clin Chem 1988;34:1726-1732.
14. Engvall E, Perlmann P. Enzyme-linked immunosorbent assay, ELISA. III. Quantitation of specific antibodies by enzyme-labeled anti-immunoglobulin in antigen-coated tubes. J Immunol 1972;109:129-135.
15. Belanger L, Sylvestre C, Dufour D. Enzyme-linked immunoassay for alpha-feto protein by competitive and sandwich procedures. Clin Chim Acta 1973;48:15-18.
16. Voller A. Heterogeneous enzyme-immunoassays and their applications. In: Maggio ET, ed. Enzyme-immunoassay. CRC Press, Boca Raton, FL, 1980:181-196.
17. Tomita JT, Kim YK, Schenck JR, et al. Ligand assay of CEA: solid-phase immunoassays for CEA. A decade of CEA testing. J Clin Immunoassay 1984;7:107-111.
18. Nisselbaum JS, Smith CA, Schwartz D, et al. Comparison of Roche RIA, Roche EIA, Hybritech EIA and Abbott EIA methods for measuring carcinoembryonic antigen. Clin Chem 1988;34:761-764.
19. Shaw NY, Guy JM, Borden KK, Bairstow CS. Measurement of alpha-fetoprotein (AFP) in maternal serum and amniotic fluid by solid-phase enzyme immunoassay. Clin Chem 1984;30:8.
20. Hibi N. Enzyme-immunoassay of human alpha-fetoprotein. Gann 1978;69:67-75.
21. Taylor EH, Gadsden RH. Serum prostatic acid phosphatase (PAP): monoclonal enzyme-linked immunoassay compared to polyclonal radioimmunoassay. Ann Clin Lab Sci 1984;14:21-26.
22. Ivor L, Schioppi L, Blase M, et al. Quantitation of serum prostate specific antigen (PSA) by simultaneous immunoenzymatic assay. Clin Chem 1987;33:928.
23. Markese JJ, Bryg K, Jackson CM, Shaw NY. Development of a solid-phase enzyme immunoassay for the measurement of CA-125. Tumor Biol 1987;8:320-321.
24. Nolan C, Pryzwara L, Miller L, et al. Characteristics of a solid phase enzyme immunoassay for human estrogen receptor. In: Ames F, Blumenschein G, Montague E, eds. Current controversies in breast cancer. University of Texas Press, Austin, 1984:433-441.
25. Weigand R, Cotter DL, Dunn RA, et al. Quantitation of progesterone receptor (PgR) in human breast tumors by double monoclonal enzyme immunoassay. Breast Cancer Res Treat 1986;8:87.
26. Smyth CM, Benn DE, Reeve TS. An enzyme immunoassay compared with a ligand-binding assay for measuring progesterone receptors in cytosols from breast cancers. Clin Chem 1988;34:1116-1118.
27. Halliday MI, Wisdom GB. A competitive enzyme-immunoassay using labeled antibody. FEBS Lett 1979;96:298-300.

28. Ryall RG, Story CV, Turner DR. Reappraisal of the causes of the "hook effect" in two-site immunoradiometric assays. Anal Biochem 1982;127:308-315.
29. Brown WE III. Microparticle-capture membranes: application of testpack hCG-urine. Clin Chem 1987;33:1567-1568.
30. Snedecor GW, Cochran WG. Statistical methods, 7th ed. The Iowa State University Press, Ames, IA, 1980;238-254.

8
Enhanced Detection Systems for Enzyme-Linked Heterogeneous Immunoassays: Luminescence

JOHN C. EDWARDS and CARL R. MOON

Amersham International, Little Chalfont, Buckinghamshire, United Kingdom

I. INTRODUCTION

The use of enzymes as tracer molecules in immunoassays has been gradually replacing the use of radioactive probes over the past decade. At the same time, in the field of high-sensitivity heterogeneous immunoassays, modifications to the established use of colorimetric enzymic reactions have been sought in order to improve assay performance. This chapter will review recent developments in the use of luminescence as the endpoint signal detection in enzyme immunoassays and their application to the immunodiagnosis of cancer. Other types of luminescence immunoassay involving direct labeling with substrates for luminescent reactions are not within the scope of this chapter. These methods are reviewed elsewhere [1,2].

II. LUMINESCENCE DETECTION IN ENZYME IMMUNOASSAYS

Luminescence is the emission of light from electronically excited molecules when they return to their ground state. If the light emission is observed from a chemical reaction, it is referred to as chemiluminescence; if it occurs from a living system or is derived from a living system, it is referred to as bioluminescence.

A. Enzyme Tracers

Various enzymes that have been used as tracer molecules for luminescence detection in immunoassays are listed in Table 1. Enzymes such as luciferase and horseradish peroxidase take part directly in the luminescent reaction, whereas the other enzymes shown in Table 1 are used to convert a substrate into a product that itself takes part in a luminescent reaction. For example, alkaline phosphatase has been used as a tracer to convert luciferin-O-phosphate to luciferin, which is then detected by a luminescent reaction catalyzed by luciferase [3]. These reactions may be allowed to take place simultaneously.

Table 1 Detection of Enzyme Tracers in Immunoassays Using Luminescence

Enzyme	Reference
Glucose-6-phosphate dehydrogenase	Wannlund et al. [4]
	Terouanne et al. [5]
Firefly luciferase	Wannlund et al. [4]
Bacterial luciferase	Jablonski [6]
Pyruvate kinase	Fricke et al. [7]
Horseradish peroxidase	Puget et al. [8]
	Velan and Halmann [9]
Alkaline phosphatase	Geiger and Miska [3]
Esterase	Schaap [10]
	Geiger and Miska [3]
Carboxypeptidase A and B	Geiger and Miska [3]
Arylsulfatase	Geiger and Miska [3]
Glucose oxidase	Botti et al. [11]
	Arakawa et al. [12]
Catalase	Lange et al. [13]
Beta-galactosidase	Takayasu et al. [14]
Invertase	Tsuji et al. [15]

B. Sensitivity

Luminescence offers a very sensitive method for the detection of enzymes. For example, the lowest amount of horseradish peroxidase that can be detected using luminescence has been estimated at 5×10^{-19} mol [8]. This compares very favorably with the sensitivity of other methods of detecting enzymes that are used in immunoassays [16].

C. Methods of Conjugation

Many different methods have been used to couple enzymes to haptens, high-molecular-weight antigens, and antibodies for use in immunoassays. The reader is referred to work by Ishikawa et al. [17] for a full review of these conjugation procedures.

D. Kinetics of Light Output

Production of light from the endpoint reaction in immunoassays can be in the form of a short-lived flash lasting less than a few seconds or of a continuous glow that can last up to several hours [18,19], depending on the particular system. Measurement of a short-duration flash of light requires more complex instrumentation since the sample must be placed in a light-tight enclosure while the initiating reagent is injected into the reaction vessel. Rapid mixing of reagents is essential if one is to achieve acceptable precision [20]. If the emission of light can be optimized to give a continuous output, the measurement system can be simplified since signal measurement will not be critically dependent on the time of addition of the initiating reagent. Luminescence reagents can then be conveniently added before the reaction vessel is introduced into the detection system.

E. Detection Systems

For systems giving a short-lived emission of light, signal measurement must be performed in a luminometer with the facility for injection of reagents into the reaction vessel in front of the detector. Many manual and automated instruments of this type are currently available. For sequential processing of large numbers of tests it is important to ensure that the reaction vessels do not dry out before the signal is measured [21]. The most commonly used detector for light emission is the photomultiplier tube, but other sensors, such as silicon photodiodes, have been used [22].

If the light emission is prolonged, it is also possible to use simple photographic detection systems to record the endpoints of immunological reactions from both immunoassay tests in a microtiter plate format [23], using cotton strands as the solid phase [19] and blots onto membranes [24,25]. With sensitive film rated at 20,000 ASA, exposure times of less than 1 min give a permanent record of immunoassay results on film. This simple system can be made semiquantitative and seems ideal for use in nonlaboratory testing conditions.

Automated luminometers have recently been developed specifically for use with microtiter plates. A number of these are now available commercially.

F. Immunoassay Solid Phase

All of the commonly used solid phases in immunoassays can be used for luminescent endpoint detection. Antibody-coated tubes or plastic beads contained in tubes are most often used for short-lived luminescence systems because they allow injection of reagents without spillage. For prolonged emission of light, other solid phases such as microtiter wells [26], particles [5], micropins for insertion into microtiter plates [22], nitrocellulose discs [22], or membranes [24] can be used.

III. ENHANCED LUMINESCENCE IMMUNOASSAYS

Horseradish peroxidase has been widely used as a tracer in immunoassays because of its ready availability, its relatively low molecular weight, and its high activity. Horseradish peroxidase can be detected by a number of reactions involving colorimetric [27], fluorimetric [28], and luminescent procedures [29-31].

The best known luminescent reaction involving horseradish peroxidase is the oxidation of luminol in the presence of hydrogen peroxide [8,9]. In 1983, a paper published by Whitehead et al. [18] described a marked increase in light emission when D-luciferin was added to this reaction. The report was followed by others [23,32] that identified other groups of compounds that had a similar effect on the light output. The phenomenon was called enhanced luminescence and, under optimum conditions, the emission of light compared to background signals could be increased up to 1000 times.

The mechanism of the enhancement phenomenon has been discussed in detail [33], and it appears that the enhancers may act by increasing the rate at which intermediate forms of horseradish peroxidase produced from the reaction with hydrogen peroxide and luminol are reconverted to active horseradish peroxidase, thus accelerating the overall catalytic cycle.

The technique of enhanced luminescence immunoassay has already been applied for the detection of a number of oncological markers. Initial publications by Thorpe et al. [18,34] demonstrated the application for assays of alpha fetoprotein (AFP) and carcinoembryonic antigen (CEA). Both of these assays used polystyrene beads as the solid phase but two different enhancers were used in the light emitting reaction: firefly luciferin and *p*-iodophenol.

Enhanced luminescence endpoints are also used in the Amerlite system (Amersham International plc, Amersham, Buckinghamshire, UK). A review of the development and performance of the system has been given by Edwards [26]. The enhanced luminescence reaction has been optimized to give a prolonged output of light. For endpoint determinations in the immunoassays, light emission can be initiated by the addition of a single reagent and the signal can then be detected during a period of at least 20 min after a 2 min stabilization period. The rapid processing of tests that can be achieved by using a microwell format means that the drift in light output during the measurement time for a complete 96-well assay is negligible.

The system can be used to measure both low-molecular-weight haptens and large proteins. Details of immunoassays specifically developed for the field of oncology are given in the following sections, with some of the clinical performance data obtained.

A. Alpha Fetoprotein

A two-site immunometric assay for measuring AFP using the Amerlite system has been described by White et al. [35]. The AFP in the specimen is incubated for 2 hr at 37°C with mouse monoclonal antibody to AFP coated onto microwells. Unbound material is removed by washing, and horseradish-peroxidase-labeled sheep anti-AFP is then incubated in the wells for 2 hr. The unbound conjugate is removed and a signal reagent containing the substrates for the luminescent reaction is added. The light emitted is then quantitated in a luminescence analyzer.

The assay has a working range of 0-750 kIU/L, sensitivity better than 1.0 kIU/L, and between-run precision of less than 8%. A typical standard curve is shown in Figure 1. Comparison with an established colorimetric enzyme immunoassay gave a correlation coefficient of 0.99 [35].

In clinical studies, AFP values from 195 healthy subjects indicated an upper limit of 5.0 kIU/L for the normal range (95% confidence limit), whereas 82% of patients with primary hepatocellular carcinoma had AFP values above 10 kIU/L. In 161 patients with nonseminomatous testicular cancer, 51% had AFP values above 10 kIU/L. None of the AFP values from 27 patients with seminomatous testicular cancer or 40 with ovarian cancer were above 10 kIU/L.

B. Carcinoembryonic Antigen

In this immunometric assay [36], a sheep antibody to CEA bound to microwells and a solution of mouse monoclonal antibody to CEA react simultaneously with CEA present in the specimen or standard. After incubation, unbound mouse antibody is removed by washing and is replaced by a second reagent containing a horseradish-peroxidase-labeled sheep antimouse immunoglobulin that reacts specifically with the bound mouse monoclonal antibody. After a second incubation, the unbound labeled antibody is removed by washing and the signal reagent is added. The enhanced luminescence signal is measured in a luminescence analyzer. The protocol for the CEA assay is summarized in Figure 2.

Luminescence in Immunoassays

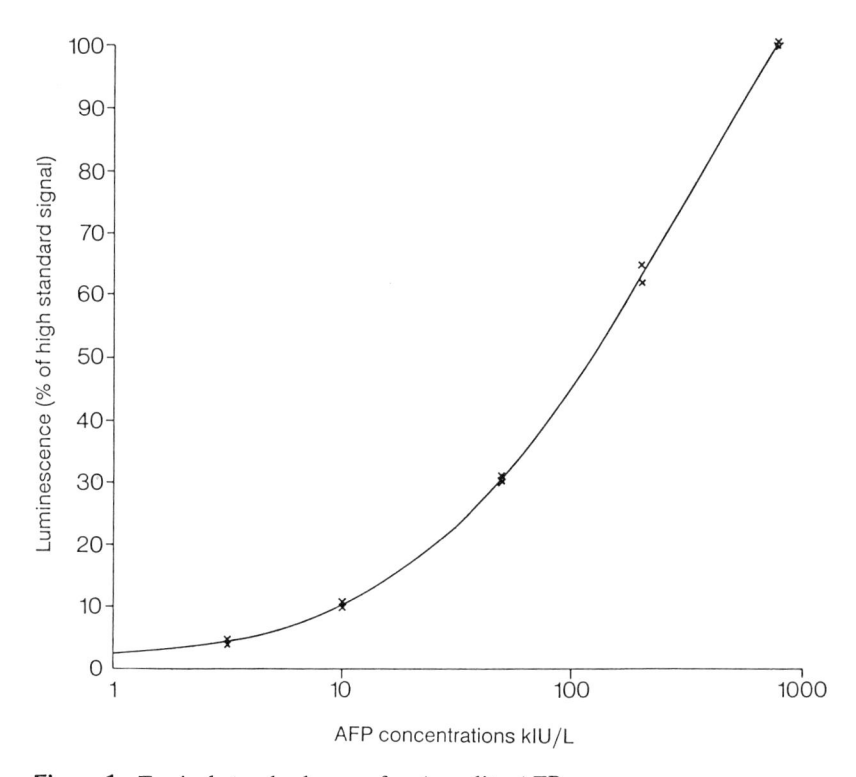

Figure 1 Typical standard curve for Amerlite AFP assay.

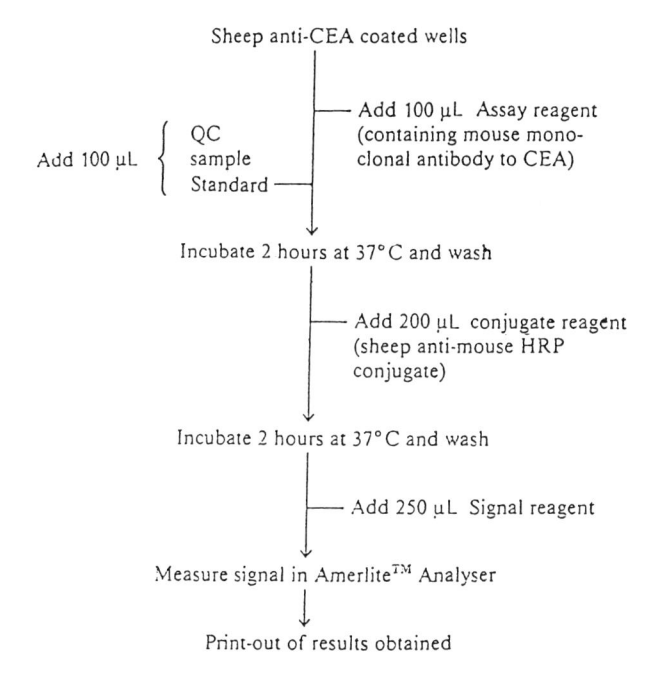

Figure 2 Protocol for Amerlite CEA assay.

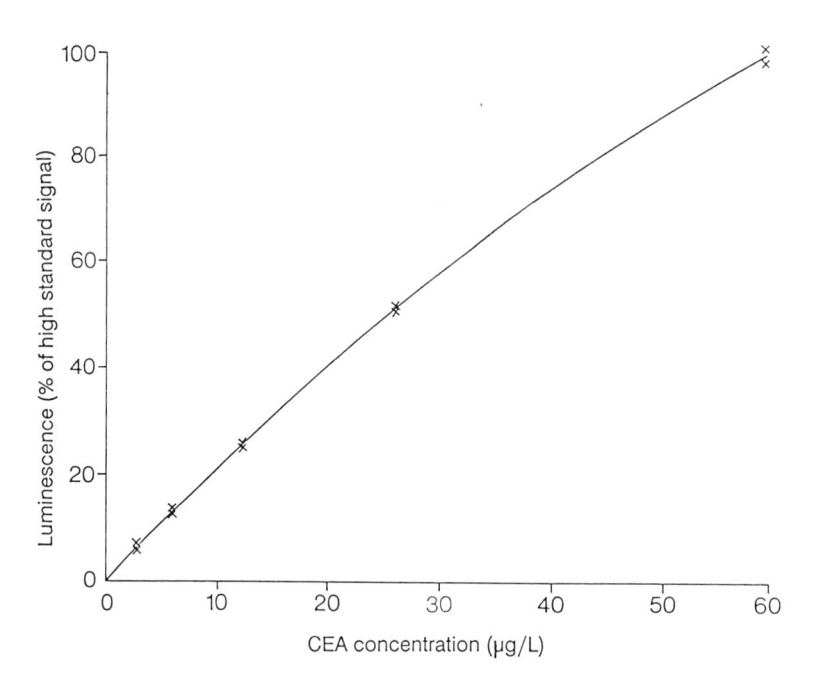

Figure 3 Typical standard curve for Amerlite CEA assay.

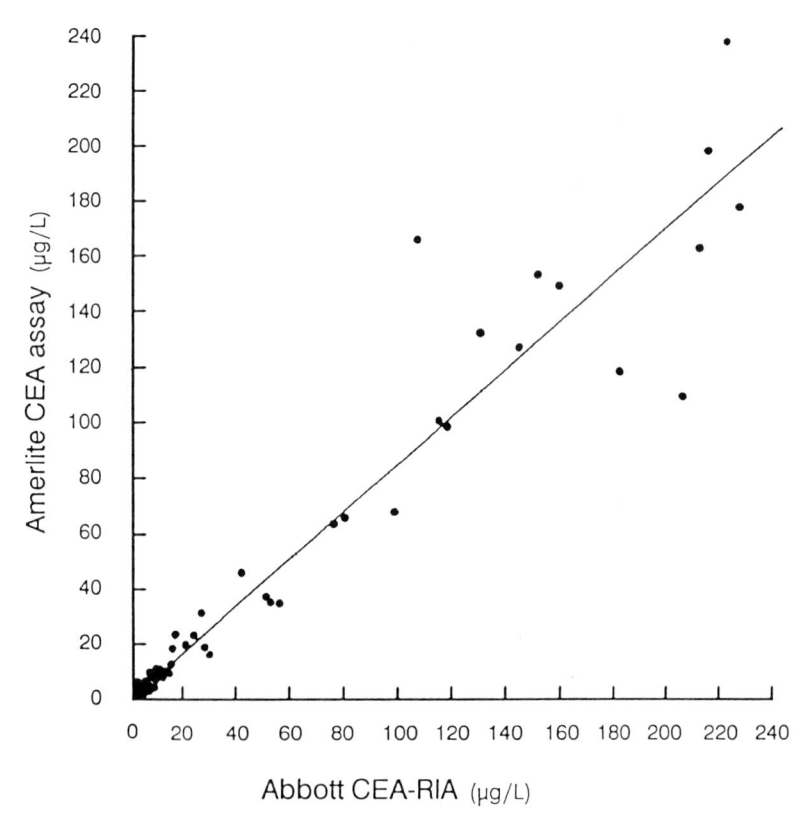

Figure 4 Linear regression analysis of CEA values from 67 patients with malignant disease (from ref. [37]).

Table 2 Amerlite CEA Assay: Between-Run Precision

Control Serum	Number of Determinations	Mean CEA Concentration (μg/L)	Coefficient of Variation (%)
1	16	2.78	10.3
2	16	5.80	9.1
3	16	9.92	6.2
4	16	28.18	8.3

The assay produces a working range of 0-60 μg/L (Fig. 3), and samples with higher concentrations can be diluted in a normal serum diluent before they are assayed. The sensitivity of the assay was found to be 0.2 μg/L. Between-run precision for control samples assayed on 16 occasions gave coefficients of variation ranging from 6.2 to 10.3% (Table 2).

The Amerilite CEA assay has been compared with an established radioactive immunoassay by Cooper and Forbes [37]. Linear regression analysis of results from 67 patients with malignant disease (Fig. 4) gave a correlation coefficient of 0.962 for samples with CEA values ranging from 0.2 to 232 μg/L. Samples with CEA values above the highest standard in the assay (60 μg/L) were diluted before analysis.

In 300 blood donors, the mean CEA level was 1.07 μg/L. Of the 300 samples tested, 29 samples (9.7%) had undetectable levels of CEA and 286 samples (95.3%) had values less than 3.0 μg/L. The highest CEA level encountered in a single normal subject was 8.4 μg/L.

The distribution of CEA levels in patients with lung cancer according to their histological type is shown in Table 3. These values contrast with those for benign lung lesions and also with controls without lung disease or cancer who are cigarette smokers. Of the patients with small cell lung carcinomas, 38% had CEA levels above 5.0 μg/L, whereas 99% of normal blood donors and 95% of patients with benign lung lesions had CEA values below 5.0 μg/L.

Table 3 Distribution of CEA Levels (μg/L) in Normals and in Patients with Lung Disease

Sample	Total No. of Patients	<3.0	3.1-5.0	5.1-10.0	10.1-75.0	>75
Normal blood donors[a]	300	286 (95.3)	10 (3.3)	4 (1.3)	0 (0)	0 (0)
Normal smokers	103	71 (69)	20 (19)	10 (10)	2 (2)[b]	0 (0)
Lung cancer:						
Small cell[c]	50	26 (52)	5 (10)	7 (14)	9 (18)	3 (6)
Nonsmall cell	44	26 (59)	7 (16)	6 (14)	4 (9)	1 (2)
Benign lung lesions[d]	20	18 (90)	1 (5)	1 (5)	0 (0)	0 (0)

[a]Includes smokers and nonsmokers.
[b]CEA values: 10.6, 12.7 μg/L
[c]At presentation.
[d]Lesions initially considered possible cancer.
Source: Ref. [37].

Table 4 Distribution of CEA Levels (μg/L) in Patients with Gastrointestinal Diseases

Sample	Total No. of Patients	<3.0	3.1–5.0	5.1-10.0	10.1-75.0	>75
Benign						
Crohn's disease	31	30 (97)	1 (3)	0 (0)	0 (0)	0 (0)
Ulcerative colitis	22	18 (82)	4 (18)	0 (0)	0 (0)	0 (0)
Liver disease	13	6 (46)	6 (46)	1 (8)	0 (0)	0 (0)
Other bowel diseases	58	45 (78)	7 (12)	5 (9)	1 (2)	0 (0)
Colorectal Cancer						
Duke's A and B[a]	26	14 (54)	2 (8)	4 (15)	4 (15)	2 (8)
C[a]	20	7 (35)	3 (15)	3 (15)	6 (30)	1 (5)
D[a]	15	4 (27)	0 (0)	2 (13)	4 (27)	5 (33)
Clinical remission	53	40 (75)	7 (13)	5 (9)	1 (2)	0 (0)
Recurrent disease	42	1 (2)	2 (5)	4 (10)	15 (36)	20 (47)

[a]Preoperative.
Source: Ref. [37].

Table 5 The HCG Levels (IU/L) Measured in Two Patients Receiving Treatment for Testicular Teratoma

Date	Details	HCG Level (1st IRP 75/537)
Jan 22	Preoperation	1439
Feb 2	Postoperation	1034
Feb 16	Starting chemotherapy	1126
Feb 24	Chemotherapy	1270
Mar 18	Chemotherapy	19.7
Apr 30	Chemotherapy	0.37
Feb 2	Preoperation	51290
Feb 10	Postoperation	35550
Feb 25	Chemotherapy	929
Mar 5	Chemotherapy	283
Mar 11	Chemotherapy	129

Source: Ref. [38].

The distribution of CEA levels in patients with colorectal cancer, prior to treatment, when in clinical remission, and in relapse are shown in Table 4. These are compared to the levels in benign disease of the gastrointestinal tract and liver. Ninety-five percent of patients with benign gastrointestinal disease had CEA levels below 5.0 µg/L, whereas 61% of patients with Dukes Stage C and D colorectal cancer had CEA values above this level.

C. Human Chorionic Gonadotropin

An enhanced chemiluminescence immunoassay that detects both whole human chorionic gonadotropin (HCG) and the beta-subunit of HCG has also been developed [38]. This is a one-step microwell assay using sheep antibody to beta-HCG immobilized on the solid phase and horseradish-peroxidase-labeled mouse monoclonal antibody to HCG. After a 1 hr incubation at 37°C the microwells are washed, and the enhanced luminescence signal is measured after addition of signal reagent.

The assay has a sensitivity of 1.0 IU/L, a working range up to 1000 IU/L, and between-run precision of less than 10%. Because of the specificity for both whole HCG and the beta-subunit, the assay has clinical utility in both pregnancy detection and oncology.

Use of the assay in monitoring treatment for testicular teratoma is shown in Table 5. In both patients, samples were taken just before and just after surgery, and further samples were taken during chemotherapy. The HCG levels dropped rapidly in response to treatment.

IV. OTHER LUMINESCENCE IMMUNOASSAYS IN ONCOLOGY

A direct bioluminescent immunoassay for AFP without a separation step has been developed by Terouanne et al. [39]. The assay uses anti-AFP monoclonal antibodies labeled with glucose-6-phosphate dehydrogenase and polyclonal antibodies against AFP coimmobilized on Sepharose with the bioluminescent enzymes luciferase and flavin mononucleotide (FMN) oxidoreductase. After 1 hr of incubation with the samples, luminescent reagent containing FMN, glucose-6-phosphate, nicotinamide-adenine dinucleotide (NAD), and a scavenging system is added and photon emission is measured. The assay has a sensitivity of 0.3 µg/L with a range up to 100 µg/L, and between-run precision of less than 7%.

Takayasu et al. [14] have described the chemiluminescent modification of a commercial enzyme immunoassay (MARKIT AFP, Dianippon Pharmaceutical Co., Osaka, Japan). The serum sample or standard and anti-AFP antiserum are incubated together for 1 hr at 37°C in a test tube. After incubation, AFP labeled with beta-D-galactosidase was added and preincubated for 10 min at 37°C. An insolubilized second antibody was then added and incubated for 1 hr at 37°C. The tubes were then centrifuged and the supernatant discarded. The chemiluminescent endpoint was developed in two steps. First a 2 hr incubation with lactose and glucose oxidase was carried out to generate hydrogen peroxide. This was then quantitated in a chemiluminescent reaction by the addition of *bis*(2,4,6-trichlorophenyl)oxalate. The assay gave similar performance to that of a commercially available enzyme immunoassay.

V. SUMMARY

The use of luminescence detection as the endpoint for enzyme immunoassays offers the sensitivity and performance that had previously been obtained only with radioactive immunoassays. For systems that are optimized to give a continuous output of light, simple and convenient processing of tests can be achieved without the need for accurate timing of the signal generation.

The applicability of these techniques for the analysis of a wide range of substances should lead to their establishment in cancer diagnosis as well as in other fields.

REFERENCES

1. Klingler W, Wiegand G, Knuppen R. Chemiluminescent labels for steroid immunoassays. J Steroid Biochem 1987;27:41-45.
2. Barnard GJR, Collins WP. The development of luminescence immunoassays. Med Lab Sci 1987;44:249-266.
3. Geiger R, Miska W. Bioluminescence enhanced enzyme immunoassay. J Clin Chem Clin Biochem 1987;25:31-38.
4. Wannlund J, De Luca M. A sensitive bioluminescent immunoassay for dinitrophenol and trinitrotoluene. Anal Biochem 1987;122:385-393.
5. Terouanne B, Carrie M-L, Nicolas J-C, Crastes de Paulet A. Bioluminescent immunosorbent for rapid immunoassays. Anal Biochem 1986;154:118-125.
6. Jablonski E. The preparation of bacterial luciferase conjugates for immunoassay and application to rubella antibody detection. Anal Biochem 1985;148:199-206.
7. Fricke H, Strasburger CJ, Wood WG. Development and optimisation of enzyme enhanced luminescence immunoassay (EE-LIA) for serum proteins and proteohormones. Frez Z Anal Chem 1982;311:373-379.
8. Puget K, Michelson AM, Avrameas S. Light emission techniques for the microestimation of femtogram levels of peroxidase. Anal Biochem 1976;79:447-456.
9. Velan B, Halmann M. Chemiluminescence immunoassay, new sensitive method for the determination of antigens. Immunochemistry 1978;15:331-333.
10. Schaap AP, Chen T-S, Handley RS, DeSilva R, Giri BP. Chemical and enzymatic triggering of 1,2-dioxetanes. 2: fluoride-induced chemiluminescence from tert-butyldimethylsilyloxy-substituted dioxetanes. Tetrahedron Lett 1987;28:1155-1158.
11. Botti R, Lenocini P, Neri P, Tarli P. Chemiluminescence enzyme immunoassay of human chorionic somatomammotropin. J Clin Chem Clin Biochem 1981;19:623-629.
12. Arakawa H, Maeda M, Tsuji A. Chemiluminescence enzyme immunoassay for thyroxine with use of glucose oxidase and a bis(2,4,6-trichlorophenyl)oxalate-fluorescent dye system. Clin Chem 1985;31:430-443.
13. Lange K, Simmet T, Peskar BM, Peskar BA. Determination of 15-keto-13,14-dihydro-prostaglandin E2 and prostaglandin D2 in human colonic tissue using a chemiluminescence enzyme immunoassay with catalase as the labelling enzyme. Adv Prostaglandin Thromboxane Leukotriene Res 1985;15:35-38.
14. Takayasu S, Maeda M, Tsuji A. Chemiluminescent enzyme immunoassay using beta-D-galactosidase as the label and the bis(2,4,6-trichlorophenyl)oxalate-fluorescent dye system. J Immunol Methods 1985;83:317-325.
15. Tsuji A, Maeda M, Arakawa H, Shimizu S, Tanabe K, Sudo Y. Chemiluminescence enzyme immunoassay using invertase, glucose-6-phosphate dehydrogenase and beta-D-galactosidase as label. In: Scholmerich J, Andreesen R, Kapp A, Ernst M, Woods WG, eds. Bioluminescence and chemiluminescence. Chichester: John Wiley, 1987:233-236.

16. Porstmann B, Porstmann T. Nugel E. Evers U. Which of the commonly used marker enzymes gives the best results in colorimetric and fluorimetric enzyme immunoassays: horseradish peroxidase, alkaline phosphatase or beta-galactosidase? J Immunol Methods 1985;79:27-37.
17. Ishikawa E, Imagawa M, Hashida S, Yoshitake S, Hamaguchi Y, Ueno T. Enzyme labelling of antibodies and their fragments for enzyme immunoassay and immunohistochemical staining. J Immunoassay 1983;4:209-327.
18. Whitehead TP, Thorpe GHG, Carter TJN, Groucutt C, Kricka LJ. Enhanced luminescence procedure for sensitive determination of peroxidase-labeled conjugates in immunoassay. Nature 1983;305:158-159.
19. Brown CR, Higgins KW, Frazer K, et al. Simultaneous determination of total IgE and allergen-specific IgE in serum by the MAST chemiluminescence assay system. Clin Chem 1985;31:1500-1505.
20. Taniguchi A, Hayashi Y, Yuki H. Study on some factors affecting the precision of luminescence analyses. Chem Pharm Bull 1986;34:3475-3480.
21. Wood WG. Luminescence immunoassays: problems and possibilities. J Clin Chem Clin Biochem 1984;22:905-918.
22. Thorpe GHG, Stott RAW, Sankolli GM, et al. Solid supports and photodetectors in enhanced chemiluminescent immunoassays. In: Scholmerich J, Andreesen R, Kapp A, Ernst M, Woods WG, eds. Bioluminescence and chemiluminescence. Chichester: John Wiley, 1987:209-213.
23. Thorpe GHG, Kricka LJ, Moseley SB, Whitehead TP. Phenols as enhancers of the chemiluminescent horseradish peroxidase-luminol-hydrogen peroxide reaction: application in luminescence-monitored enzyme immunoassays. Clin Chem 1985;31:1335-1341.
24. Laing P. Luminescent visualization of antigens on blots. J Immunol Methods 1986;92:161-165.
25. Schneppenheim R, Rautenberg P. A luminescence Western blot with enhanced sensitivity for antibodies to human immunodeficiency virus. Eur J Clin Microbiol 1987;6:49-50.
26. Edwards JC. Development of a nonradioactive immunoassay system. Int Clin Prod Rev 1987;Jan/Feb:30-37.
27. Avrameas S. Coupling of enzymes to proteins with glutardialdehyde. Use of the conjugates for the detection of antigens and antibodies. Immunochemistry 1969;6:43-52.
28. Guilbault GG, Brignac PJ, Juneau M. New substrates for the fluorometric determination of oxidative enzymes. Anal Chem 1968;40:1256-1263.
29. Whitehead TP, Kricka LJ, Carter TJN, Thorpe GHG. Analytical luminescence: its potential in the clinical laboratory. Clin Chem 1979;25:1531-1546.
30. Longeon A, Henry JP, Henry R. A solid phase luminescent immunoassay for human chorionic gonadotrophin using Pholas dactylus bioluminescence. J Immunol Methods 1986;89:103-109.
31. Ahnstrom G, Nilsson R. Activation of chemiluminescent oxidations catalysed by peroxidase and the differentiation of peroxidase reagents. Acta Chem Scand 1965;19:313-316.
32. Thorpe GHG, Kricka LJ, Gillespie E, et al. Enhancement of the horseradish peroxidase-catalysed chemiluminescent oxidation of cyclic diacyl hydrazides by 6-hydroxybenzothiazoles. Anal Biochem 1985;145:96-100.
33. Thorpe GHG, Kricka LJ. Enhanced chemiluminescence assays for horseradish peroxidase: characteristics and applications. In: Scholmerich J, Andreesen R, Kapp A, Ernst M, Woods WG, eds. Bioluminescence and chemiluminescence. Chichester: John Wiley, 1987:199-208.

34. Thorpe GHG, Kricka LJ, Whitehead TP. Applications of enhanced luminescent quantitation of horseradish peroxidase labels in immunoassays. Clin Chem 1985;31:913.
35. White GP, Jacoby B, Mabbett S, Mann J, Brockas AJ, Mashiter K. Amerlite immunoassays for CEA and AFP using enhanced luminescence. Clin Chem 1986;32:1124.
36. Wood WG. Initial results with a commercial luminescence enhanced enzyme immunoassay for the determination of carcinoembryonic antigen (CEA) in body fluids. J Clin Chem Clin Biochem 1987;25:699-703.
37. Cooper EH, Forbes MA. An evaluation of the Amerlite enhanced chemiluminescent immunoassay for carcinoembryonic antigen. Tumor Diagn Ther 1987;8:145-149.
38. Martin JK, Galloway T, Evans A, Lonsdale C, Valente E, White GP. Immunometric assay for human chorionic gonadotropin (HCG) based on enhanced luminescence. Clin Chem 1988;34:1220.
39. Terouanne B, Nicolas J-C, Crastes de Paulet A. Bioluminescent immunoassay for alpha-fetoprotein. Anal Biochem 1986;154:132-137.

9
Enzyme-Linked Heterogeneous Immunoassays with Electrochemical Detection

H. BRIAN HALSALL, SARAH H. JENKINS, and WILLIAM R. HEINEMAN
University of Cincinnati, Cincinnati, Ohio

I. GENERAL CONSIDERATIONS

The diagnosis of cancer, as discussed elsewhere in this volume, may involve the examination of bodily tissues and fluids using a wide variety of tools, both microscopic and macroscopic. One of the goals of molecular biology and biochemistry has been to relate the clinical state to reliable molecular "markers" present either in circulation or on the cell surface. In the ideal case, such markers could provide not only diagnostic information during treatment but also a very early warning system in the absence of symptoms. The search for more specific markers has been marked with frustration but is showing more promise with the discovery of the oncogene and its products [1]. The technology for quantitatively detecting such markers has made great advances in recent years, however, based principally on the immunoassay.

Quantitative immunoassays couple the principles of reaction equilibria to the specificity of the IgG molecule to provide protocols with detection limits typically at the femtomole level, but more recently at the attomole level and below [2,3]. This chapter discusses attainment of these lower detection limits.

Modern nonisotopic immunoassays equal or surpass the detection limits achieved by the radioimmunoassay, generally by making use of some form of amplification of the signal being detected [4]. Thus far, the most popular amplification method, and the one used in the work described here, is by use of an enzyme label. An enzyme is chosen for its ability to turn over a readily detected substrate-product system. (Other enzymatic properties are important, but ignored for the purposes of this brief discussion.) It is then covalently attached (becoming a "label" and forming a "conjugate") to a component of the immunoassay protocol chosen such that its concentration, directly or indirectly, reflects the concentration of the desired analyte. Amplification is achieved by permitting the enzyme conjugate to turn over large numbers of substrate molecules, after which the concentration of the product of the enzymatic reaction is measured. The most sensitive schemes currently use fluorescence [5,6], chemiluminescence [7,8], or electrochemical detection [9]; our work being in the latter area. Where the detection

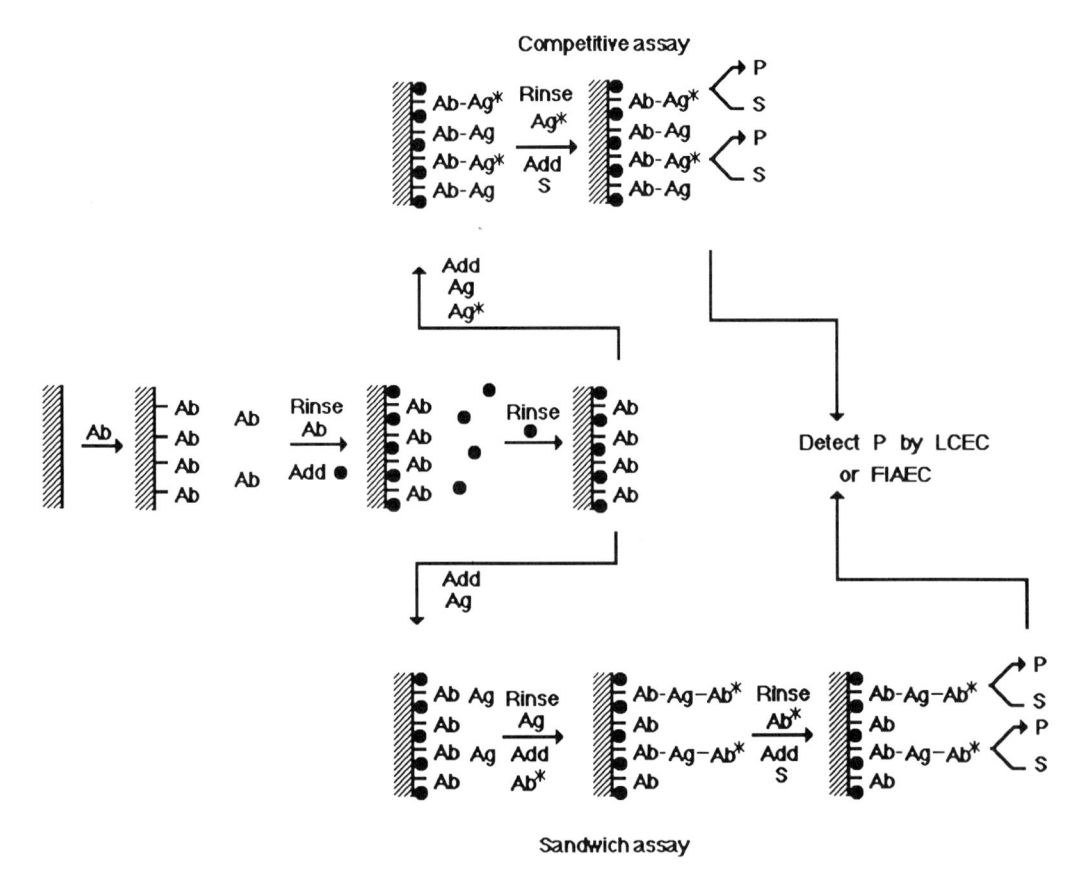

Figure 1 Heterogeneous assay protocol for both the competitive and noncompetitive or "sandwich" modes. ●; Tween 20; **Ab**, antibody; **Ab***, enzyme-labeled antibody; **Ag**, antigen; **Ag***, enzyme-labeled antigen; **S**, substrate, **P**, product.

limits are undemanding and amplification is unnecessary, labels other than enzymes may be used [10].

Enzyme immunoassays may be organized into two fundamental types; heterogeneous and homogeneous. Homogeneous assays rely on some change in the properties of the label when the enzyme conjugate is complexed with another component of the reaction mixture, as in the EMIT assays [11,12]. This is convenient because it eliminates the separation step that differentiates homogeneous from heterogeneous assays. Although we have developed both heterogeneous [2,13-15] and homogeneous [16-18] assays, the latter are not considered further since they have an inherently higher detection limit. Discussions of general strategies for electrochemically based immunoassays can be found elsewhere [19].

Figure 1 illustrates the principles of the heterogeneous assay for the competitive and noncompetitive or "sandwich" modes. The sandwich format is used for the very low detection limits because the capture reaction can be driven almost to completion under appropriate reaction conditions.

II. ELECTROCHEMICAL DETECTION

Electrochemical detection may use voltammetry or potentiometry. In voltammetry, the potential is controlled (either held constant or scanned) and the current due to the resulting redox processes is measured. This is a kinetic, or approach to equilibrium, method, with detection limits of 10^{-9}-10^{-10} M. Amperometry is voltammetry at constant potential. In potentiometry, the potential is measured at zero current. This is an equilibrium method with detection limits of 10^{-5}-10^{-6} M. Potentiometrically based devices are familiar as pH electrodes and ion-selective electrodes. Our work has been amperometrically based because of the lower detection limit, although others have used potentiometric approaches. Modern electrochemical methods are discussed by Lunte and Heineman [20].

The amperometric technique imposes constraints on the properties of the enzyme product being detected. It must be not only redox active but preferably also be so in an electrochemical matrix "window," that is, in a region of potential where interferences are minimal. To avoid the need to remove oxygen, the observed reaction should be an oxidation, and should occur between +200 and +900 mV, with the lower end being more desirable. More positive values incur the danger of solvent breakdown, while values approaching zero imply increasing instability. In addition, the substrate should be electroinactive at the measuring potential in order to minimize interferences. Although these requirements are moderately limiting, a number of enzyme/substrate/product combinations satisfy them, including some of the most widely used enzymes in the immunoassay field in general.

Alkaline phosphatase is probably the best example of such an enzyme and has applications in different types of spectrophotometrically based immunoassays [21] as well as DNA probe assays [22,23]. This popularity is not only due to obvious factors such as the availability at high specific activity, and a high turnover number, but also to the ability of the enzyme to cleave phosphate esters of enormous variety. This makes the

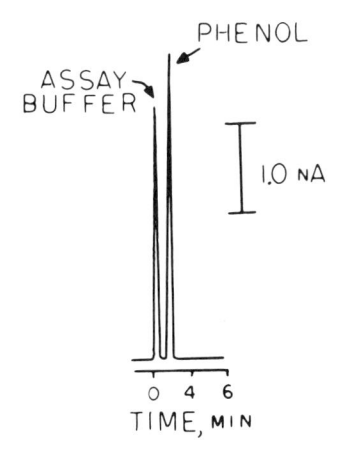

Figure 2 The LCEC chromatogram of a 20 µl injection of 5.3 × 10^{-7} M phenol in 0.05 M carbonate buffer. Flow rate, 1.2 ml/min; eluent, 0.1 M phosphate buffer (pH = 7.0) (from ref. [25]).

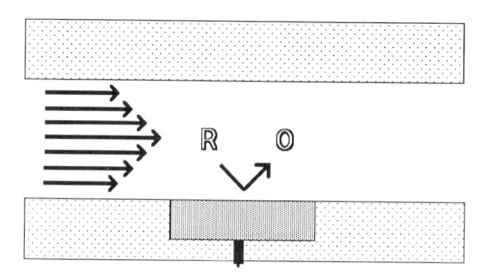

Figure 3 Schematic view of thin-layer flow cell. R and O are the reduced and oxidized species, respectively.

task of finding a product that fits into the electrochemical window much easier. The work described here used phenyl phosphate as substrate, which provides phenol for detection by oxidation. Recent work [24] has used p-amino phenyl phosphate as substrate, the product of which, p-amino phenol, has a lower oxidation potential. Both of these substrates are electroinactive at the detection potentials of the products.

How should the product-containing solution be presented to the electrode? Routine use demands that the system be reproducible, robust, simple to use, and rapid. Cost, very low detection limits, and the availability of clinical samples demand that the volume requirement be low. Many of these are met by using either flow injection analysis (FIAEC) or liquid chromatography (LCEC), with electrochemical detection (see Fig. 4). In FIAEC, the product containing sample is injected into a flowing stream that passes through a thin layer electrochemical cell, where the electrochemistry takes place. In LCEC, a chromatographic column is placed between the injection port and the electrode to perform desired chromatography. This is necessary at low current levels since the compositions of the injected solution and the mobile phase are not the same, and an apparent contribution to the current yield is obtained due only to the change in capacitance as the sample passes over the electrode. Thus, separation of the capacitance and redox peaks is essential when very low detection limits are desired. Figure 2 illustrates this for the injection of a standard phenol solution. An improvement of two orders of magnitude in detection limit is gained by the separation.

The heart of the LCEC system is the electrochemical detector itself, the basic principle of which is shown in Figure 3. The most popular of these are commercially available thin-layer devices with a minimum operating volume of about 1 μl. Thus the injected sample size need only be about 5 μl, to allow for chromatographic band broadening. This has important implications, as discussed later. The cells have excellent flow characteristics and a large ratio of electrode surface area to cell volume. Thus a high response is achieved with a low dead volume. The cells are easy to clean when necessary; changing to a different electrode material is also easy.

III. INTERFERENCES

Two forms of interference must be dealt with, apart from those due to the discriminatory abilities of the antibodies used. Electrochemical interferences due to compounds that oxidize at the same potential as the enzyme product are restricted to the components of the injected solution. Since the assay is heterogeneous, this is a simple mixture of known and predictable character. Other electrochemical interferences may occur due to either polymerization of the oxidized product on the electrode surface or adsorption of other components from the injected solution on to the electrode [25]. Solute concentrations, however, are very low in the applications under consideration, and the mobile phase serves to wash the electrode surface continuously so that passivation is not a factor.

The second, much more serious, interference is that of nonspecific adsorption of the conjugate to the reaction vessel where the immunoreactions take place, or to macromolecules contained within it (i.e., interactions with species other than the analytically specific target of the conjugate). This is a concern not only of electrochemically based immunoassay but of any immunoassay, as has been discussed by Ekins [24]. As he has also pointed out, where nonspecific adsorption exists, detection limits cannot be lowered merely by increasing the specific activity of the enzyme used in the conjugate.

The following section describes the results of experiments designed to lower the detection limit of the electrochemical enzyme immunoassay simply by reducing the level of nonspecific adsorption.

IV. ASSAY AT THE ATTOMOLE LEVEL

The model used in this study [2] was a sandwich assay (Fig. 1) for mouse IgG. In the assay, the primary IgG was antimouse IgG, and was passively adsorbed to polystyrene cuvettes. In the second step, the solution containing the mouse IgG analyte was presented to this; however, other proteins in the sample could, and did, also adsorb to the surface. Then the alkaline phosphatase-antimouse IgG conjugate was added. The desired reaction is the interaction of this with the antigen. Other interactions are nonspecific.

A. Materials and Methods

The LCEC apparatus is shown in Figure 4. The working electrode was paraffin-oil-based carbon paste, the reference electrode was Ag/AgCl, and the auxiliary electrode was stainless steel. The applied potential was 895mV vs. Ag/AgCl. Chromatography was provided by the 25 × 2 mm column packed with 10 μm C18 material. A 20 μl sample loop was used, and mobile phase was pumped at 1.2 ml/min.

Polystyrene cuvettes and trays (Gilford Instruments, Cleveland, OH) were used as the assay reaction vessel. Tween 20 (Fisher Scientific, Cincinnati, OH), mouse IgG and bo-

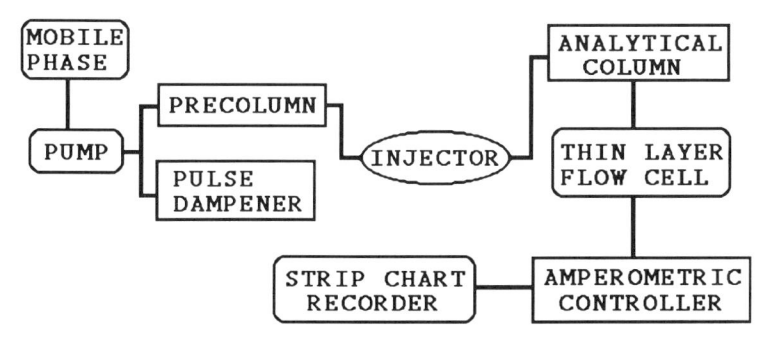

Figure 4 Schematic diagram of the LCEC apparatus (from ref. [2]).

vine serum albumin (BSA; Sigma, St. Louis, MO), rat and antimouse IgG, rat antimouse-alkaline phosphatase conjugate (Boehringer Mannheim, Indianapolis, IN), and phenyl phosphate (Calbiochem-Behring, La Jolla, CA) were the reagents used.

The principal buffer was 0.1M pH 7.0 phosphate (PB), from which were made PB + 0.05 v/v% Tween 20 (PBT), and PBT + 1 w/v% BSA (PBTBSA). The mobile phase was 0.05M Tris-HCL + 1 mM $MgCl_2$, pH 7.0.

Primary antibody coating solution was 1.8 µg/ml in PB. Standard antigen solutions were prepared in PBT from a 1 µl/ml stock solution. The conjugate was 2.5 U/ml in PBT. Substrate solution was 1.5 mM phenyl phosphate in 0.05 M Tris-HCl + 1 mM $MgCl_2$, pH 9, and was prepared immediately prior to use. The rinse solution between incubation steps was PBT.

For the assay, primary antibody was adsorbed to the polystyrene cuvettes for 12 hr, and rinsed three times for 15 min total. Antigen solution (750 µl) was added, left for 3 hr, and the cuvettes rinsed as before. Conjugate solution (750 µl) was added and treated similarly. The final steps were the incubation of substrate (750 µl) with the sandwich for 20 min, followed by the injection of 20 µl of the reaction mixture into the chromatograph. The concentration of phenol produced was determined by comparing the height of the peak current due to its oxidation to that given by a calibration curve.

The nonspecific adsorption of conjugate was measured by the amount of phenol produced by a layer consisting of primary antibody that had been incubated with substrate solution for 20 min in the absence of antigen.

Extraneous adsorption may come from not only the conjugate but also other solutions used for incubations. Therefore both the primary antibody and antigen preparations were assessed for adsorbable alkaline phosphatase by measuring the phenol produced by the primary antibody adsorbed layer alone, and by this layer plus the adsorbed antigen layer.

B. Results and Discussion

The calibration curve is shown in Figure 5. The current at zero antigen concentration has four possible components: the presence of phenol in the substrate solution; the presence of alkaline phosphatase in the primary antibody preparation, and its adsorption to the cuvette; the presence of alkaline phosphatase in the antigen preparation, and its adsorption to the cuvette; and adsorption of the conjugate to the cuvette. Table 1 lists the relative contribution of each of these to the background current; it is clear that by far the greatest is from the adsorption of the conjugate. (It is worth noting that the current due to the substrate is contaminant phenol in the commercial preparation of phenyl phosphate used, which was of the highest purity available. This does not create any difficulty, since it is easily subtracted as a blank; however, its ready detection illustrates the superior sensitivity of the electrochemical methods.)

When the primary antibody is adsorbed to the cuvette, the accessible surface is lessened and therefore conjugate adsorption should diminish. Figure 6 shows that this is indeed the case, and also removes the possibility that the conjugate is binding to the primary antibody layer. These results indicated that efforts should be made to modify the cuvette surface that remains uncovered after the adsorption of the primary antibody.

Adsorption to polystyrene surfaces is due principally to entropically driven hydrophobic interactions. These can be reduced using the nonionic detergent Tween 20 and BSA. The actual ability of Tween 20 to inhibit the adsorption of the enzyme conjugate

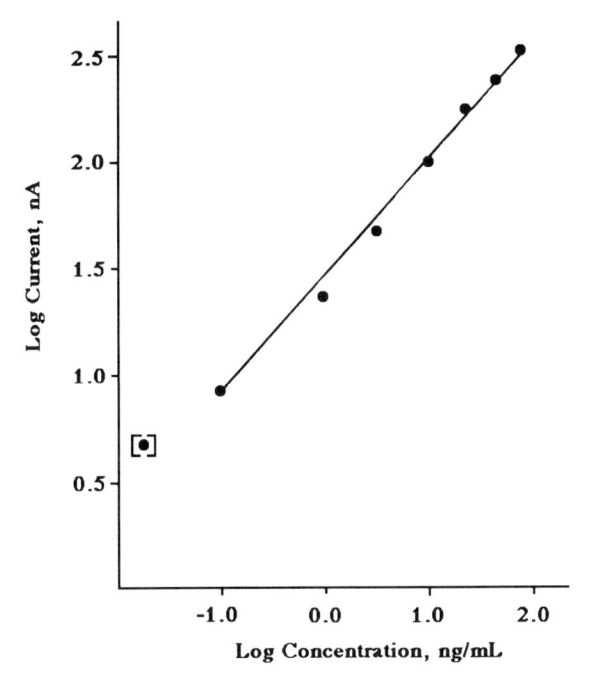

Figure 5 Standard calibration curve for the determination of mouse IgG. [●] corresponds to a 0.0 ng/ml mouse IgG concentration (from ref. [2]).

is shown in Figure 7. If Tween 20 is absent from all solutions, conjugate adsorption is very high. If, however, it is present in all solutions, the adsorption is drastically reduced but not eliminated. The additional benefits of using BSA as a site blocker are seen in Figure 8. If BSA is present in all solutions beginning with the first rinse, conjugate adsorption is reduced a further 96% from that observed when only Tween 20 is used. It can also be inferred from Figure 8 that the level of conjugate adsorption is independent of the presence of BSA in the primary antibody solution. However, adsorption of the primary antibody is critically dependent on whether BSA is present, as seen in Table 2,

Table 1 Origin of the Background Current

	Mean i_p(nA) ± SD	Percentage of Total Background
Substrate	0.80 ± 0.08	12.3
Adsorption of enzyme from primary antibody preparation	0.04 ± 0.08[a]	0.0
Adsorption of enzyme from antigen preparation	0.02 ± 0.08[a]	0.0
Adsorption of enzyme conjugate	5.48 ± 0.20[a]	84.3

[a]Data are corrected for the 0.80 nA contribution of endogenous phenol in the substrate.
Source: Ref. [2].

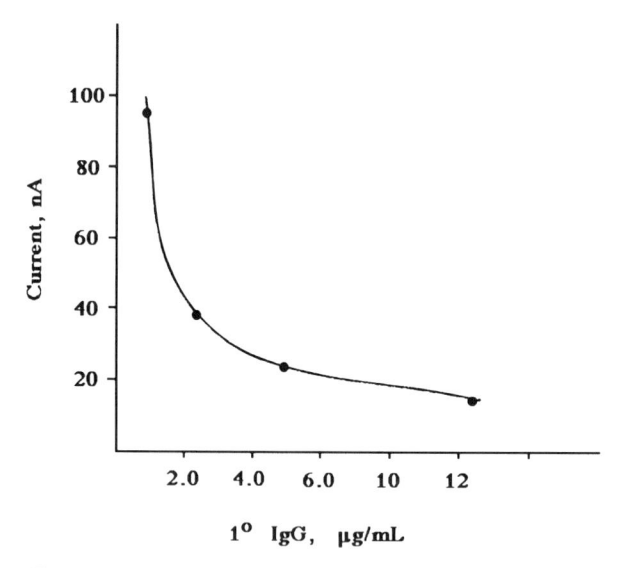

Figure 6 The nonspecific adsorption of the enzyme conjugate measured by the amount of current generated in the absence of antigen as a function of the primary antibody concentration (from ref. [2]).

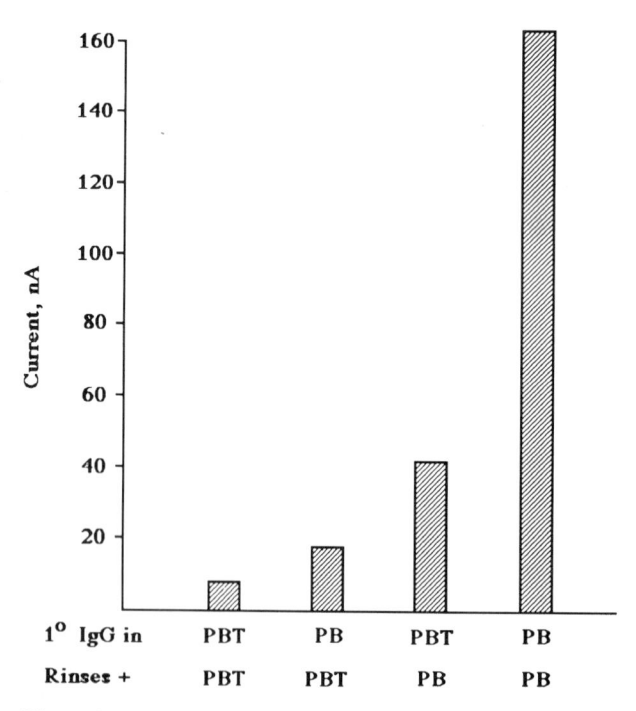

Figure 7 The nonspecific adsorption of the enzyme conjugate with and without the use of Tween 20. The primary antibody was present in either phosphate buffer (PB) or phosphate buffer with 0.05% Tween 20 (PBT). Rinses + refers to the buffer components of all solutions beginning with the first rinse solution (from ref. [2]).

Enzyme-Linked Heterogeneous Immunoassays

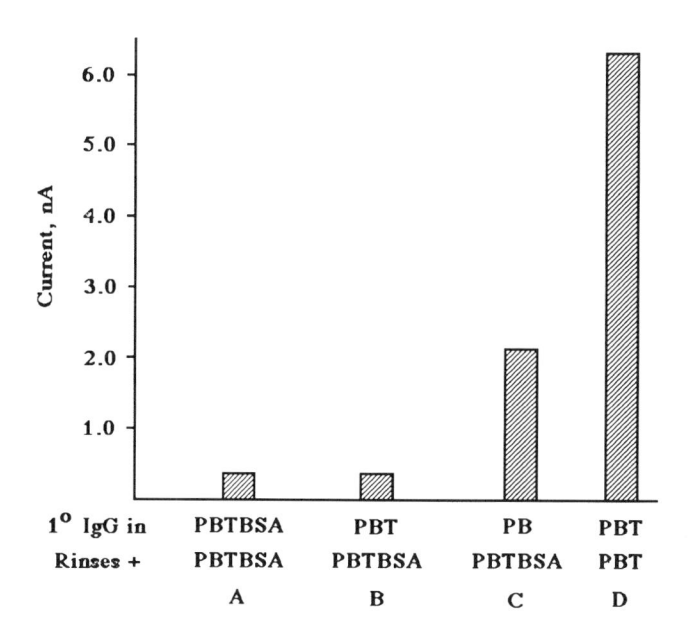

	A	B	C	D
1° IgG in	PBTBSA	PBT	PB	PBT
Rinses +	PBTBSA	PBTBSA	PBTBSA	PBT

Figure 8 The nonspecific adsorption of the enzyme conjugate with and without the use of BSA. The primary antibody was present in phosphate buffer (PB), phosphate buffer with 0.05% Tween 20 (PBT), or phosphate buffer with 0.05% Tween 20 and 1.0% BSA (PBTBSA). Rinses + refers to the buffer components of all solutions beginning with the first rinse solution (from ref. [2]).

Table 2 Amount of Primary Antibody Adsorbed Under Conditions of Competition with Either Tween 20 or BSA Using a Labeled Antibody Directed Against the Primary Antibody

Experiment	Mean i_p(nA) ± SD
1° IgG in PB	21.3 ± 1.05
1° IgG in PBT	17.4 ± 0.95
1° IgG in PBTBSA	0.0 ± 0.08

Data are corrected for the contribution of endogenous phenol in the substrate.
Source: Ref. [2].

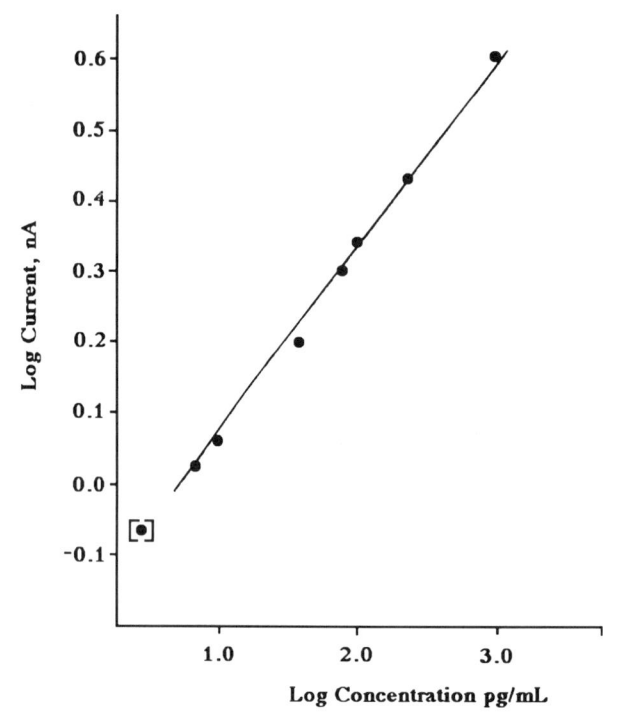

Figure 9 Standard calibration curve for mouse IgG using a combination of Tween 20 and BSA as surface-blocking agents. [●] represents the background current observed at 0.0 ng/ml mouse IgG (from ref. [2]).

because the BSA competes for those sites at which the primary antibody would otherwise adsorb.

A new calibration curve (Fig. 9) was now constructed that incorporated changes in the assay protocol to include Tween 20 and BSA. The primary antibody solution contained 0.05% Tween 20, and all other solutions (with the exception of the substrate solution) had 0.05% Tween 20 and 1% BSA. Since BSA was not present in the primary antibody solution, it was necessary to increase the length of the first set of rinse times to permit the BSA to adsorb (the first rinse was 20 min instead of 5 min). Comparison of Figures 9 and 5 shows that a considerable reduction in the detection limit has resulted because of lowering the background due to nonspecific adsorption. The detection limit has been lowered over 13-fold to 50 attomoles, while at the same time the high concentration range of the assay has been maintained, thus extending the working range by more than an order of magnitude.

V. ASSAY AT THE SUBATTOMOLE LEVEL

Certain clinical situations would benefit greatly from a reduction in the volume of sample necessary for assay of a particular analyte. Not only does this present less risk for the patients when sampling, for example, from restricted physiological regions such as the eye or brain, or in pediatric or geriatric cases, but it also reduces the time required for the cell culture of biopsy samples to produce testable quantities of shed or excreted cell products. Thus the aim here is not the quantitation of low concentrations of material but of low amounts. This section will describe our efforts to reduce detection limits by electrochemical enzyme immunoassay to the 10^{-21} moles level [27,28].

Improvements in assay sensitivity can be made in several ways, of which the most straightforward is to alter the geometry of the reaction vessel and reduce its volume. Since the concentration of product (phenol in this case) produced by a unit activity of enzyme depends upon the volume of the container, whereas the amount produced does not, increasing the surface area to volume ratio of the reaction vessel should drastically reduce the amount of enzyme conjugate needed to produce a detectable amount of product in a given period of time. Thus, since the amount of conjugate in a sandwich assay is proportional to the amount of antigen, a lower detection limit is obtained. In addition, the large surface area to volume ratio will reduce the diffusional path length and shorten the assay time.

Assay sensitivity can also be affected by the way in which the primary antibody is attached to the solid phase. This is a consequence of the requirement that for optimal interaction of the antibody and antigen, the antibody binding sites should be exposed and oriented distally from the solid surface. This means that in the ideal case, attachment of the antibody would be via the carboxy terminus of the heavy chains of the antibody. Passive adsorption methods do not provide such an orientation, with the result that binding capacity is reduced. Covalent attachment by means of coupling reagents such as glutaraldehyde [29] and water-soluble carbodiimides [30] may also be used. Although easy to do, glutaraldehyde coupling suffers from the disadvantage that it is mediated through the amino groups [31] on the antibody, which includes the amino terminus of the F_{ab}. Thus a significant number of sites are lost in this way. Water-soluble carbodiimides, however, couple through carboxyl groups [32]. Their use is particularly attractive if selectivity can be made for the carboxy terminus. This has been done by keeping the pH of the reaction mixture acidic, thus retarding the reaction with the side chains and gaining some selectivity for the carboxy terminus [33-35].

The combination of these (alteration of vessel geometry and oriented coupling of the primary antibody) should yield an assay protocol with subattomole potential. We have approached this by performing all assay incubations in a capillary tube whose surface has been modified by a polymeric covering. This polymer contains amino groups that can be coupled to carboxyl groups on the primary antibody. This section will demonstrate these concepts with a sandwich assay similar to that described above.

A. Materials and Methods

In addition to those described above, the following materials were used: 75 mm X 1.2 mm microhematocrit tubes (Thomas Scientific, Philadelphia, PA), 1-ethyl-3(3-dimethylaminopropyl)carbodiimide hydrochloride (EDC; Sigma), poly(vinyl benzyl chloride) (60:40 mixture of 3- and 4-isomers), and tris(2-aminoethyl)amine (TA) (Aldrich Chemical, Milwaukee, WI). The LCEC system used, and the conditions, were as described above, as were PBTBSA, mobile phase buffer, and enzymatic reaction buffer, except that the phosphate was 0.2 M.

Polymer coating solution (2.5 w/v%) was prepared in toluene. The TA and EDC solutions were prepared in 0.2 M phosphate buffers at pH 7 and 6, respectively. Primary

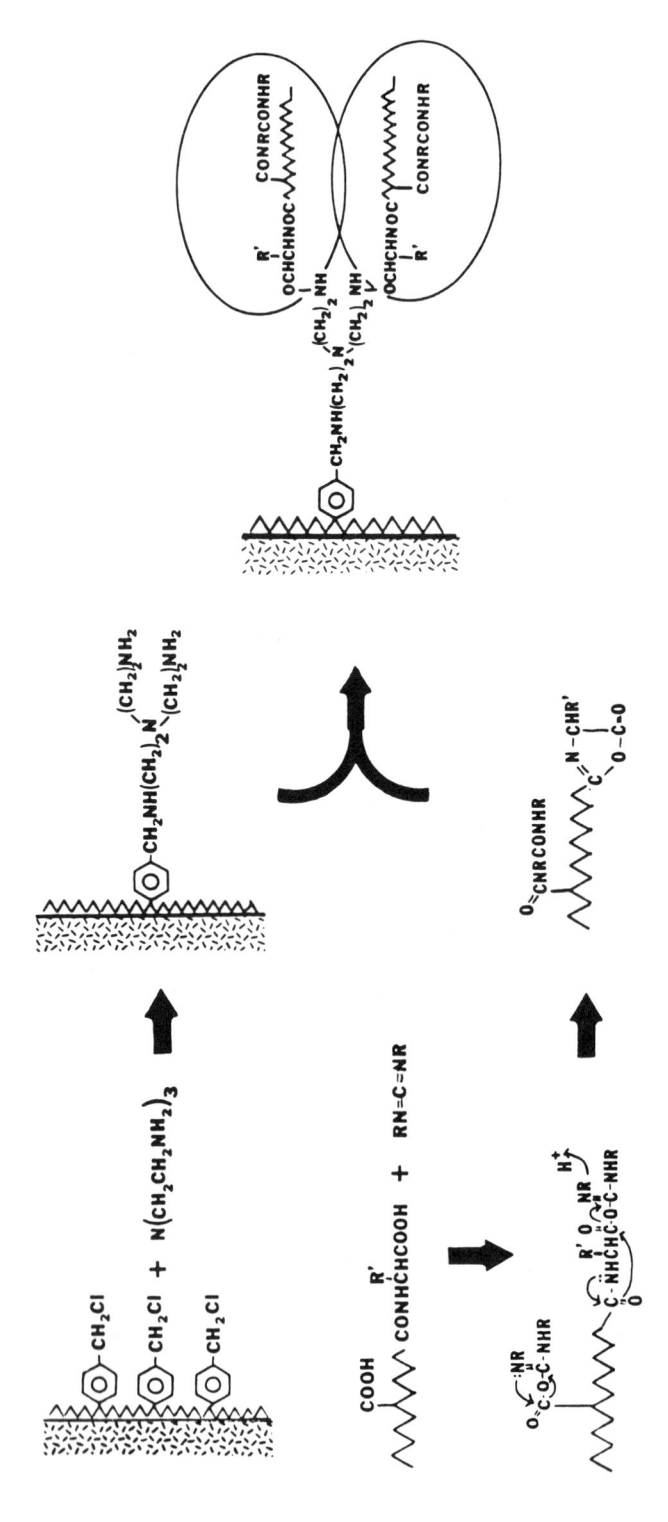

Figure 10 Reaction sequence for immobilization of the primary antibody onto the inner surface of the capillaries.

antibody was at 26 μg/ml in PB pH 6. Standard antigen solutions were in PBTBSA. Conjugate was 1 U/ml in unbuffered saline, and the substrate solution was at 2.75 mM.

The glass capillaries were first coated with the polymer by filling the tubes half full with the coating solution, inverting several times, removing the solution by aspiration, and drying the tubes overnight at 50°C. When the tubes were completely dry, TA (75 μl, 2%) was pipetted into them, allowed to incubate for 2 hr, removed by aspiration, and the capillaries rinsed once with PB.

Primary antibody (0.5 ml) was added to EDC (0.5 ml, 2 μM), both in PB pH 6, in a test tube. The reaction was timed from the point of mixing. The mixture (75 μl) was pipetted into the capillaries, and left for 40 min. The solution was removed and 0.1 M PBTBSA containing 0.1% NaN_3 added to each tube. Capillaries were capped and stored refrigerated until use. The reaction scheme for this immobilization is shown in Figure 10.

The assay procedure was essentially the same as used for the earlier work [2], except for the solution volumes, incubation times, and the mode of injection into the chromatograph. Thus, all reaction volumes were 70 μl (the size of the capillary), and the analytical sample was injected directly into the chromatograph from the capillary by means of a trivial modification of the injection port. Incubation times were 10 min for antigen, 10 min for conjugate followed by a 3 × rinse with PTBTSA, and 5 min for substrate.

B. Results and Discussion

The polymer coating strategy is particularly advantageous since it provides a robust substrate that can be applied to a wide variety of supports, and the coupling chemistry for the primary antibody layer to this is the same for all of these. Most protein-solid phase covalent attachment procedures, if not controlled, can be destructive to the protein structure. The coupling procedure was therefore optimized using a simplex procedure [36, 37]. The current generated when the assay was performed using a 40 pg/ml antigen standard was chosen as the response to be maximized. When this response was used, both the amount of primary antibody coupled and the amount retaining its antigen-binding capacity could be maximized simultaneously. Since the polymer coating and TA reactions were an adaptation of a successful application elsewhere [38], the conditions optimized were the concentration of EDC and the time of the surface/EDC/protein reaction. It was found that the blank response also varied depending on the conditions, so that each response was blank corrected. The two-dimensional simplex is shown in Figure 11. The optimum was found to be at 1 μM EDC and a 40 min reaction interval (inset). This was not a local optimum, as judged by the responses of additional points chosen outside the simplex that were not significantly above the blank. The optimum was therefore chosen for subsequent use.

After modification, however, the polymer coating produces a surface that is not efficiently blocked by the combination of Tween 20 and BSA, and a high zero antigen current was obtained that indicated significant nonspecific adsorption of the conjugate. This observation emphasizes the need to evaluate nonspecific adsorption carefully whenever the support matrix or attachment chemistry is modified. In this instance, the involvement of electrostatic interactions was indicated, since increasing the ionic strength or lowering the pH of the incubation buffers significantly reduced the background current. This source of nonspecific interaction is to be expected since the polymer surface had been modified to contain amino groups. Although other strategies for reducing this interaction are being investigated, it was found that including phosphate in the rinse solutions

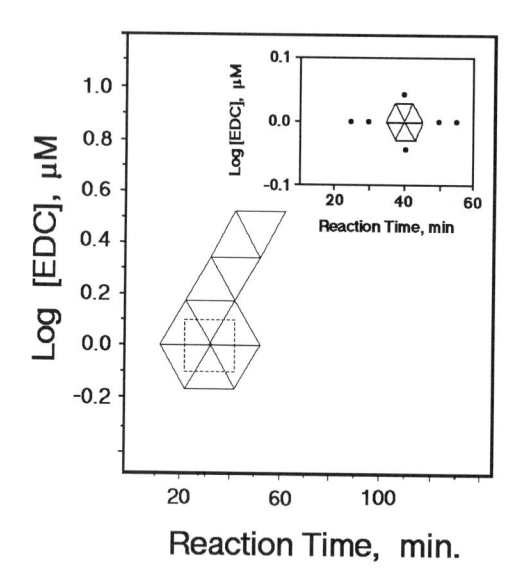

Figure 11 Simplex optimization of the primary antibody attachment procedure. The experimental conditions varied were the concentration of carbodiimide used and the protein/carbodiimide/amino support reaction time. The response to be maximized was the current signal recorded for the sandwich assay using a 40 pg/ml mouse IgG standard. Each vertex represents one set of experimental conditions; the responses were measured in triplicate. The ● represent other conditions where the response was measured. The initial optimum was located at a reaction time of 30 min and [EDC] = 1.0 μM. Inset: Simplex optimization was located with a smaller simplex size. The actual optimum was located at the same [EDC] as the initial optimum but at a slightly longer reaction interval of 40 min.

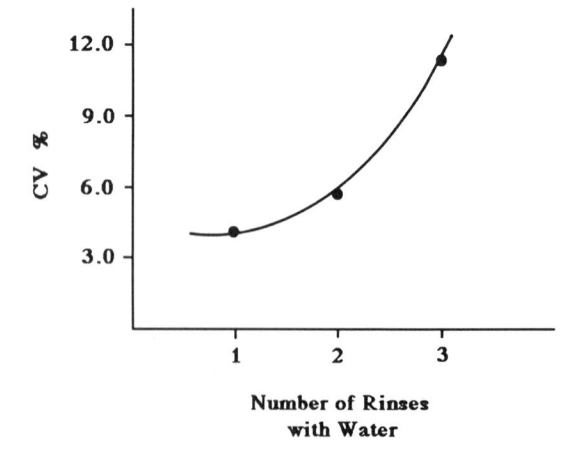

Figure 12 The CV% of the blank current response plotted against the number of rinses with water to remove phosphate before addition of substrate. The total number of rinses in each case was three; the remaining rinses were with PBTBSA.

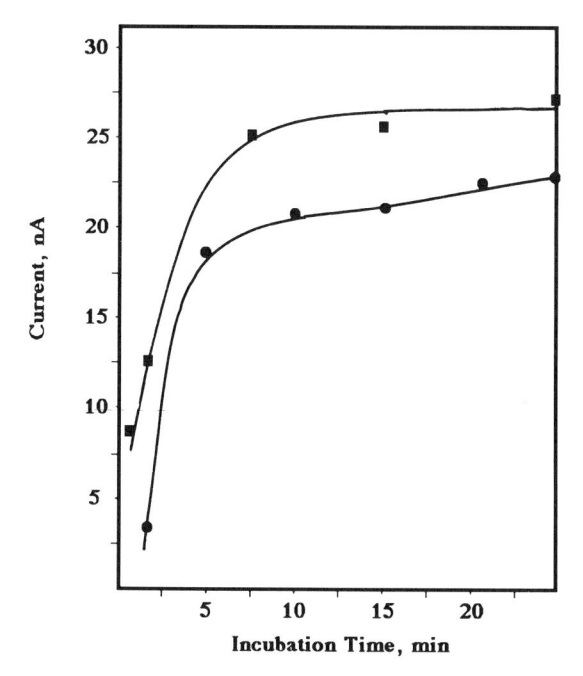

Figure 13 Equilibrium times for conjugate and antigen incubations. ■, Response for a 4.0 pg/ml mouse standard, incubated for different lengths of time, using a 15 min conjugate incubation; ●, response for a 4.0 pg/ml mouse standard using a 15 min Ag incubation and various conjugate incubation times.

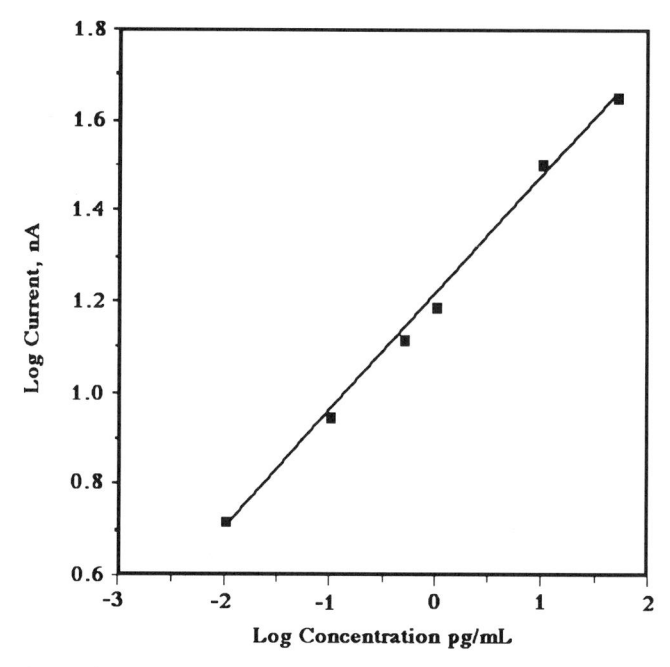

Figure 14 Calibration curve for the sandwich assay for mouse IgG using the capillary format.

after the conjugate incubation decreased the background signal. This appears to be due mostly to the inhibition of the enzyme rather than the reduction of the nonspecific interaction. As can be seen from Figure 12, reduction of the background current in this way brings it into a range where the coefficient of variation of its measurement is close to a minimum, and therefore the precision of the overall assay is improved.

Assay sensitivity is compromised if the antigen and conjugate reactions are not close to equilibrium. The times required for these are shown in Figure 13 and are both approximately 7 min. For convenience, 10 min incubations were used. Thus, from the injection of antigen solution to the final analytical result after LCEC took 27 min, with samples staggered at 2 min intervals.

Figure 14 shows the calibration curve for the assay in the optimized form. The assay has a linear range of over 4 orders of magnitude and a detection limit of a remarkable 4.6×10^{-21} moles, or 2800 molecules of antigen, with a standard deviation of 150 molecules, for the 70 μl sample used. This detection limit is approximately 3 orders of magnitude lower than the assay discussed above. Detailed comparison of the two assays is not possible at this point because of their radically different designs. However, it seems that the improvement is not due solely to the enhanced surface area to volume ratio, since this yields an improvement of about 100-fold. The assay described first was not optimized, and possibly the formation constants of the primary antibody and the antigen may not be the same in each assay due to the different modes of attachment of the antibody.

We have miniaturized the assay while attempting to suppress the nonspecific adsorption. It is to be expected that as the surface area/volume ratio of the reaction vessel increases, the deleterious effects of nonspecific adsorption will become more limiting, reflecting this feature of the geometry. This limitation is observed since the zero antigen background current in the capillaries is higher than in the cuvettes.

The sensitivity achieved here is meaningless if the assay efficiency is lost with real samples. We therefore performed recovery studies of the mouse IgG antigen that had been added to human serum as the sample matrix. Table 3 shows that sensitivity is not lost since over 90% of the antigen is recovered under the previously optimized assay conditions, and this can be raised by increasing the incubation time of the sample.

The second major advantage of this format is that the assay is extremely rapid, not only for these detection limits but also for immunoassays in general. We attribute this to the very short diffusional pathlengths in the capillaries that permit both short equilibration times and the rapid generation of measurable concentrations of phenol. Indeed, the phenol concentrations produced are far from the electrochemical detection limit for this substance by LCEC, which is about 10^{-9} M, and we are therefore in no way straining the capabilities of the electrochemical technique.

In principle, and with the nonspecific adsorption controlled, it should be possible using presently available commercial LCEC equipment and capillaries to sample and

Table 3 Recovery Studies of Mouse-IgG-Spiked Human Serum Controls

Standard Concentration (pg/ml)	Measured Concentration (pg/ml)	Recovery (%)
10.0	8.8	88
1.0	0.93	93

rapidly quantitate analytes at amounts of fewer than 250 molecules in sample of 5 μl. It is not unreasonable that much smaller volumes can be manipulated, and amounts quantitated, perhaps approaching the theoretical limit of the sandwich immunoassay, one molecule. This work is in progress.

REFERENCES

1. Sell S. Cancer markers: past, present and future. In: Reisfeld RA, Sell S, eds. Monoclonal antibodies and cancer therapy. New York: Alan R. Liss, 1985: 2-23.
2. Jenkins SH, Heineman WR, Halsall HB. Extending the detection limit of solid phase electrochemical enzyme immunoassay to the attomole level. Anal Biochem 1988;168:292-299.
3. Weeks I, Woodhead SJ. Chemiluminescence immunoassay. J Clin Immunoassay 1984;7:82-89.
4. Blaedel WJ, Bogulaski RC. Chemical amplification: a review. Anal Chem 1978;50: 1026-1032.
5. Gerson BJ. Fluorescence immunoassay. J Clin Immunoassay 1984;7:73-77.
6. O'Donnell CM, Suffin SC. Fluorescence immunoassay. Anal Chem 1979;51:33A-40A.
7. Wood WG. Luminescence immunoassay: problems and possibilities. J Clin Chem Biochem 1984;22:905-911.
8. Arakawa H, Maeda M, Tsuji A. Chemiluminescence enzyme immunoassay using glucose oxidase and bis(2,4,6-trichlorophenyl) oxalate fluorescent dye. In: Malvano R, ed. Immunoenzymatic techniques. Boston: Martinus Nijhoff, 1980;233-237.
9. Heineman WR, Halsall HB, Wehmeyer KR, Doyle MJ. Immunoassay by electrochemical techniques. Methods Enzymol 1983;92:432-444.
10. Collins WP. Alternative immunoassays. New York: John Wiley, 1985.
11. Ullman EF, Maggio ET. Principles of homogeneous enzyme immunoassay. In: Maggio ET, ed. Enzyme immunoassay. Boca Raton, FL: CRC Press, 1981:105-134.
12. Ngo TT. Linhoff HM. Recent advances in homogeneous and separation-free enzyme immunoassays. Appl Biochem Biotech 1981;6:53-61.
13. Wehmeyer KR, Halsall HB, Heineman WR, Volle CP, Chen I-W. Competitive heterogeneous enzyme immunoassay for digoxin with electrochemical detection. Anal Chem 1986;58:135-139.
14. Doyle MJ, Halsall HB, Heineman WR. Enzyme-linked immunoadsorbent assay with electrochemical detection for α_1-acid glycoprotein. Anal Chem 1984;56:2355-2360.
15. Wehmeyer KR, Halsall HB, Heineman WR. Heterogeneous enzyme immunoassay with electrochemical detection: competitive and "sandwich"-type immunoassays. Clin Chem 1985;31:1546-1549.
16. Heineman WR, Anderson CW, Halsall HB. Immunoassay by differential pulse polarography. Science 1979;204:865-866.
17. Wright DS, Halsall HB, Heineman WR. Digoxin homogeneous enzyme immunoassay using high performance liquid chromatography column switching with amperometric detection. Anal Chem 1986;58:2995-2998.
18. Eggers HM, Halsall HB, Heineman WR. Enzyme immunoassay with flow-amperometric of NADH. Clin Chem 1982;28:1848-1851.
19. Halsall HB, Heineman WR. Strategies for electrochemical immunoassay. Anal Chem 1985;57:1321A-1328A.

20. Lunte CE, Heineman WR. Electrochemical techniques in bioanalysis. Top Curr Chem 1988;143:1-31.
21. McComb RB, Bowers GN, Posen S. Alkaline phosphatase. New York: Plenum Press, 1979:
22. Marich JE, Wong DF, Chapman GD, Ruth JL. Development of a nonisotopic DNA probe test for *Salmonella*. 1987 AACC conference on gene probe technology. San Diego, CA, November 1987.
23. Risen LA, Freier SM, Menees LL, Ruth JL. Non-radioactive DNA probe-based technology compared to standard ELISA and bioassay methods for the detection of enterotoxigenic *Escherichia coli*. 1987 AACC conference on gene probe technology. San Diego, CA, November 1987.
24. Tang HT, Lunte CE, Halsall HB. Heineman WR. Competitive heterogeneous electrochemical enzyme immunoassay based on the detection of p-aminophenol. Anal Chim Acta 1988;214:187-195.
25. Wehmeyer KR, Doyle MJ, Wright DS, Eggers HM, Halsall HB, Heineman WR. Liquid chromatography with electrochemical detection of phenol and NADH for enzyme immunoassay. J Liquid Chromatogr 1983;6:2141-2146.
26. Ekins RP. Towards immunoassays of greater sensitivity, specificity and speed: an overview. In: Albertini A, Ekins RP, eds. Monoclonal antibodies and developments in immunoassay. New York: Elsevier, 1981:3-21.
27. Halsall HB, Heineman WR, Jenkins SH. Capillary immunoassay with electrochemical detection. Clin Chem 1988;34:1701-1702.
28. Halsall HB, Heineman WR, Jenkins SH, Wehmeyer KR, Doyle MJ, Wright DS. Electrochemical enzyme immunoassay. J Res Nat Bureau Stand 1988;93:491-493.
29. Ikariyama Y, Furnki M, Aizawa M. Sensitive bioaffinity sensor with metastable molecular complex receptor and enzyme amplifier. Anal Chem 1985;57:496-500.
30. Croft LS. Introduction to protein sequence analysis. New York: John Wiley, 1980; 66-83.
31. Weetal HH, Messing RA. Insolubilized enzymes on inorganic materials. In: Hair ML, ed. The chemistry of biosurfaces. New York: Marcel Dekker, 1972:563-592.
32. Weetal HH. Insolubilized antigens and antibodies. In: Hair ML, ed. The chemistry of biosurfaces. New York: Marcel Dekker, 1972:598-610.
33. Carray KL, Koshland DE. Carboiimide modification of proteins. Methods Enzymol 1979;25:616-623.
34. Previero A, Durancourt J, Coletti-Previero MA. Solid phase sequential analysis: specific linking of acidic peptides by their carboxyl ends to insoluble resins. FEBS Letts 1973;33:135-138.
35. Line WF, Kwong A, Weetal HH. Pepsin insolubilized by covalent attachment to glass: preparation and characterization. Biochim Biophys Acta 1971;242:194-202.
36. Deming SN, Morgan SL. Simplex optimization of variables in analytical chemistry. Anal Chem 1973;45:278A-282A.
37. Krause RD, Lott JA. Use of the simplex method to optimize analytical conditions in clinical chemistry. Clin Chem 1974;20:775-782.
38. Ikariyama Y, Suzuki S, Aizawa M. Luminescence immunoassay of human serum albumin with hemin as labeling catalyst. Anal Chem 1982;54:1126-1129.

10
Novel Detection System for Homogeneous Enzyme Immunoassays

YOSHIHIRO ASHIHARA

Fujirebio Inc., Tokyo, Japan

Many methods are available for the determination of substances in body fluid. These include high-pressure liquid chromatography (HPLC), bioassay, and receptor assay. Yalow and Berson [1] developed in 1956 a unique and highly sensitive immunoassay consisting of antibody and radiolabeled antigen for the determination of insulin.

Since then, radioimmunoassay (RIA) methods [2,3] have been routinely used for clinical diagnosis. However, RIA has several disadvantages, such as pollution of radiolabel, short half-life of the label, and hazards to human health. Hence enzyme immunoassays [4,5], which use enzyme as labels instead of radiolabels, have been developed to solve these problems. Rubenstein et al. [6] have improved the heterogeneous enzyme immunoassay into a simple and rapid system: homogeneous enzyme immunoassay (homogeneous EIA), which eliminates washing steps for the separation of bound and free labels. This homogeneous EIA has several advantages for use in clinical diagnosis, such as no need for a separation procedure and shorter assay times. In this chapter I describe homogeneous EIA for the determination of high-molecular-weight (HMW) antigens such as ferritin and alpha fetoprotein (AFP) since many tumor markers are macromolecular antigens consisting of protein, glycolipid, and/or carbohydrate [7-10].

I. CLASSIFICATION OF HOMOGENEOUS EIA

As indicated in Table 1, homogeneous EIAs can be mainly classified into two major types: competitive binding and noncompetitive binding.

Competitive assays in homogeneous EIA usually consist of antigen labeled with enzyme. In some cases, modulators such as enzyme inhibitor, enzyme activator, allosteric effector, or prosthetic group can be used. The assay is based on competitive reaction of antibody with antigen labeled with enzyme (conjugate) and antigen in sample. The enzyme activity is either inhibited or activated when the conjugate forms an immunocomplex with antibody. These assays are superior for the detection of low-molecular-weight (LMW) antigens rather than HMW antigens.

Table 1 Homogeneous Enzyme Immunoassays

Competitive binding assay
  Enzyme multiple immunoassay technique
  Substrate labeled fluorescence immunoassay
  Enzyme modulator-mediated immunoassay
  Apoenzyme reactivation immunoassay system
  Enzyme channeling immunoassay
  Biotin-enzyme avidin immunoassay
  Cloned enzyme donor immunoassay
Noncompetitive binding assay
  Hybrid antibody immunoassay
  Enzyme inhibitory immunoassay
  Proximal linkage immunoassay
  Enzyme enhancement immunoassay
  Associated enzyme sensitive technique

On the other hand, antibody labeled with enzyme is used in noncompetitive binding assays, which can be classified into two types. One is the proximal linkage assay system [11], based on substrate "channeling" [12-15] caused by close contact of two enzymes. The other is based on the modulation of enzyme activity due to binding of antigen to the conjugate. Both assays are excellent for the measurement of HMW antigens.

The sensitivity of homogeneous EIA depends not only on the specific activity of enzyme but also on the affinity of the antibody to the antigen. As reported by Ekins [16], theoretical sensitivity of the competitive assay by radioisotope is 1×10^{-14} mole/l. Since this value is less sensitive than those calculated for noncompetitive assays, noncompetitive homogeneous EIAs are more suited to the detection of HMW antigens.

Assay systems for different competitive homogeneous EIAs are listed in Table 2. These assay methods can be grouped into two types according to the modulation mode of enzyme activity or substrate. Both the enzyme multiple immunoassay technique

Table 2 Competitive Binding Homogeneous Enzyme Immunoassays

Assay Method	Enzyme	Modulator	Analyte
Direct modulation			
EMIT	Lysozyme		Morphine
	G6PDH		Digoxin, valproic acid, etc.
	MDH		Thyroxin
	β-galactosidase		IgG, ferritin
SLFIA	β-galactosidase		Theophylline, etc.
Indirect modulation			
EMMIA	HRPO	Anti-HRPO antibody	Dinitrobenzene
BAIA	Biotin enzyme	Avidin	Theophylline, IgG
ECIA	Hexokinase, G6PDH	–	Human IgG
ARIS	Glucose oxidase	FAD	Theophylline, TBG, etc.
CEDIA	β-galactosidase	Enzyme donor	Digoxin

Table 3 Noncompetitive Binding Enzyme Immunoassays

Assay Methods	Enzyme	Modulation manner	Antigen
Associated enzyme-sensitive technique	HRPO	Stabilization	AFP, IgE
Hybrid antibody immunoassay	G6PDH	Steric hindrance	Human IgM
Insoluble substrate immunoassay	Amylase	Steric hindrance	Ferritin, AFP
Proximal linkage immunoassay	G6PDH, hexokinase	Proximal linkage	hCG
Enzyme enhancement immunoassay	β-galactosidase	Charge effect	Human IgG, CRP

(EMIT) and substrate labeled fluorescein immunoassay (SLFIA) methods show that interaction between antibody and antigen attached to enzyme or substrate can directly change enzyme activity. On the other hand, in methods such as enzyme modulator mediated immunoassay (EMMIA), apoenzyme reactivation immunoassay system (ARIS), and cloned enzyme donor immunoassay (CEDIA) enzyme activities can be indirectly modulated. Most assays for LMW antigens listed in Table 2 are commercially available and easily adapted to automatic analyzers.

Various methods for the determination of HMW antigens have been investigated by noncompetitive binding assays. These assays, which are listed in Table 3, utilize antibody labeled with enzyme. There are no commercially assay methods available for HMW antigens by this method.

II. COMPETITIVE BINDING HOMOGENEOUS EIA

A. Enzyme Multiple Immunoassay Technique

The enzyme multiple immunometric technique (EMIT) assay systems [6,17,18] consist of an antibody and enzyme labeled with antigen (conjugate). The enzyme activity can be modulated by the interaction of antibody with the conjugate. There are two types of EMIT methods. One is based on a decrease in the enzyme activity by its modulation. It is due to steric hindrance or structural changes in the catalytic center of the enzyme caused by binding of the antibody to the conjugate (Fig. 1). The other [19] is based on an increase in enzyme activity observed by the combination of malate dehydrogenase (MDH) labeled with thyroxin and antibody. The MDH can bind to thyroxin, which is the competitive inhibitor [19,20]. If thyroxin combining chemically reactive groups is employed, the thyroxin is fixed covalently into the active center of the enzyme and therefore the enzyme activity is lost. However, antibody binds to antigen attached to the enzyme and this complex changes the structure, as illustrated in Figure 2. Its active center is exposed to express the catalytic activity. These methods are applied to therapeutic drug monitoring for digoxin, phenytoin, [17,21], and theophylline. In addition, these methods are also used in clinical laboratories for detection of abused drugs or hormones. The detection limit of an EMIT assay is dependent on the assay protocol. Highly

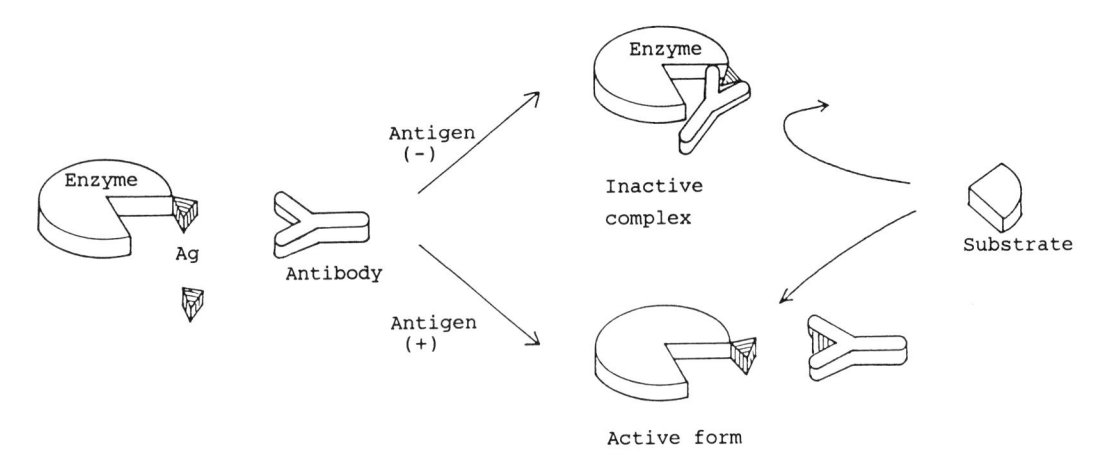

Figure 1 Enzyme multiple immunoassay technique (EMIT). In this system, the activity of enzyme as label is inhibited by binding of antibody to antigen attached with enzyme. Glucose-6-phosphate dehydrogenase and lysozime as enzymes are usually used in this method.

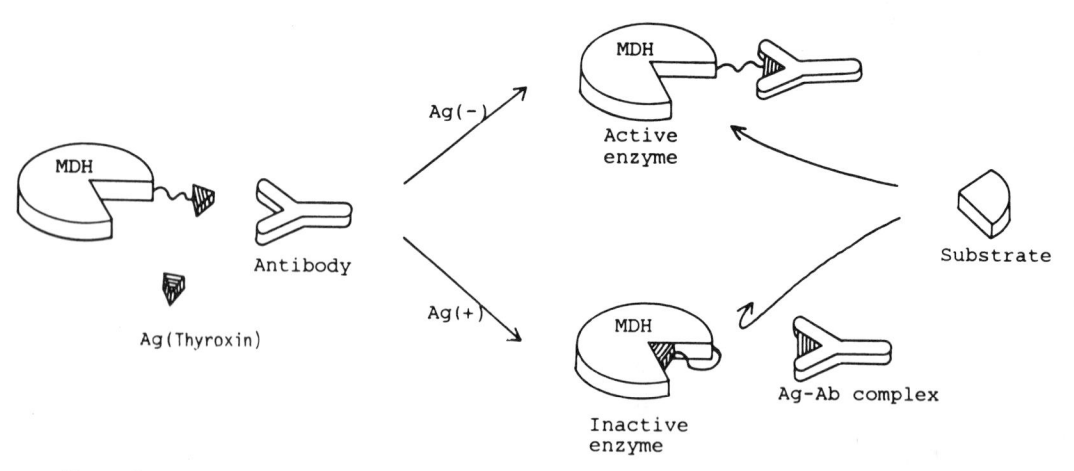

Figure 2 EMIT reactivation assay method. Malate dehydrogenase (MDH) conjugated with thyroxin as antigen is inactive enzyme. When antibody binds to the antigen in the conjugate, reactivation of the enzyme occurs.

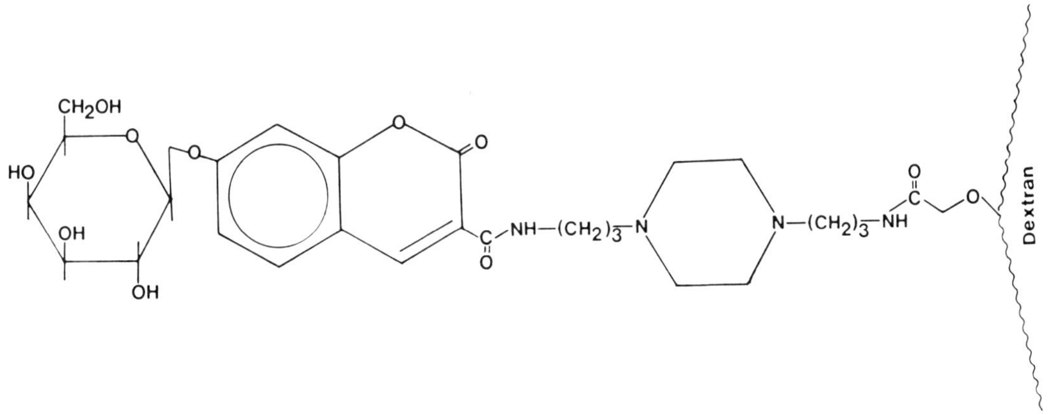

Figure 3 The fluorogenic macromolecular substrate for β-galactosidase. This substrate is employed in EMIT assay to determine HMW protein. Average molecular weight of the dextran is 40,000.

sensitive assays, such as measurement of 0.5-3 ng/ml of digoxin, require the pretreatment of sample to destroy endogenous enzyme activity.

The determination of HMW antigens by EMIT has been achieved by using synthetic macromolecular substrate [22,23] for β-galactosidase, as indicated in Figure 3. When an immune complex between enzyme-antigen conjugate and antibody is formed, the enzyme is not accessible to the substrate, and the enzyme cannot hydrolyze the substrate because of the bulky structure of the immune complex. The assay principle is similar to the EMIT method. By this method, IgG and ferritin were assayed and detected in the range of 25-500 ng/ml and 10-500 ng/ml, respectively.

B. Substrate-Labeled Fluorescein Immunoassay

Substrate-labeled fluorescein IA (SLFIA) methods [24-26] use characteristic fluorogenic substrate, umbelliferyl B-galactoside, attached to antigen. The substrate is hydrolyzed by B-galactosidase to release an umbelliferone as the fluorogen. When the substrate associates with antibody, it cannot be hydrolyzed by the enzyme due to its steric hindrance.

For formation of immune complex of antigen with antibody, free antigen in sample solution competes with antigen bound to the substrate. The substrate in excess of free antigen is released from the complex and is hydrolyzed. The fluorescence intensity of hydrolyzed fluorogen that shows maximum emission at 452 nm is therefore proportional to the antigen concentration in sample (Fig. 4).

The SLFIA method is used for the detection of antibiotics, anticonvulsant, and antiasthmatic drug such as amikacin, phenytoin, or theophylline. The detection range for theophylline is 5-40 μg/ml. However, this method may not be a sensitive enough assay because there is no amplification system. Although this system could also be applied to the determination of HMW antigens such as IgG [27], it was not useful for HMW antigens such as alpha fetoprotein (AFP) due to the high background fluorescence of the substrate at low concentrations.

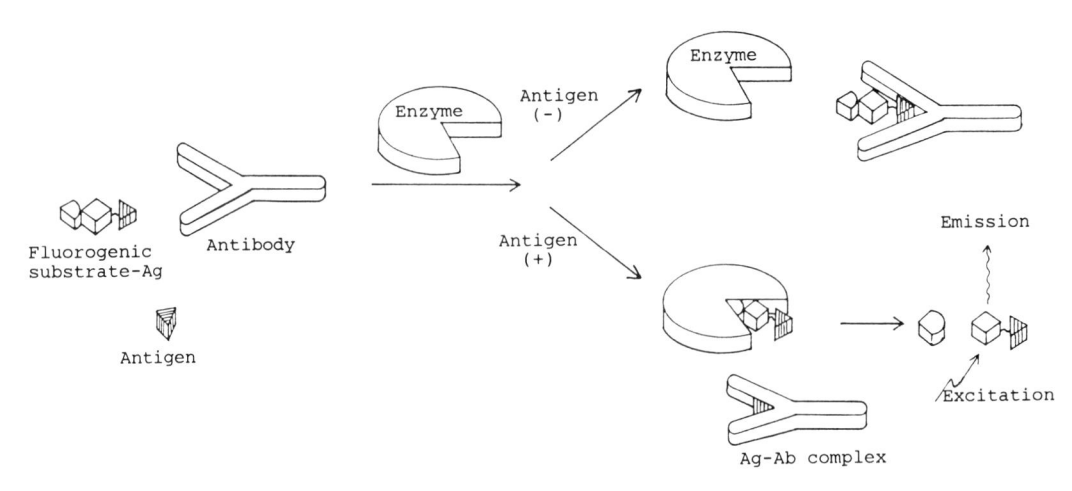

Figure 4 Assay principle of SLFIA. β-galactosylumbelliferone-3-theophylline is utilized as antigen (Ag), combined with fluorogenic substrate for β-galactosidase.

C. Enzyme Modulator-Mediated Immunoassay

In the enzyme modulator mediated IA (EMMIA) method, Ngo et al. [26,28,29] used enzyme modulators such as inhibitor, activator, receptor, or antibody (Ab) = specific to enzyme (E). The principle of the EMMIA is as follows. The modulator attached to antigen (Ag-M) and free antigen sample competitively bind to antibody. When Ag-M is complexed with antibody, the modulator cannot interact with the enzyme due to the bulky structure of the complex. In this assay, because the enzyme inhibitor is used as modulator, enzyme activity is directly proportional to the amount of free antigen in sample.

D. Apoenzyme Reactivation Immunoassay System

Morris et al. [30] have developed a unique homogeneous enzyme immunoassay using a prosthetic group combined with antigen and apoenzyme: flavin adenine dinucleotide (FAD)-antigen and apo-glucose oxidase, respectively. As shown in Figure 5, the assay system is based on the competitive binding between analyte and labeled antigen to antibody. Glucose oxidase from *Aspergillus niger* easily released FAD to form inactive apoenzyme under acidic pH at 1.4. FAD-antigen conjugate free from antibody can associate with apoenzyme and become an active holoenzyme, the activity of which can be measured by colorimetrically monitoring the generation of H_2O_2. On the contrary, when the conjugate binds to antibody, the association of the FAD moiety with apoenzyme is sterically inhibited so that enzyme activity directly relates to concentration of analyte. Morris and co-workers succeeded in obtaining a stable apoenzyme by using glycerol as stabilizer of the enzyme during the dissociation process. The measurable range for theophylline was 1-40 μg/ml for 5 min assay and the intra-assay precision was less than 7%.

Homogeneous Enzyme Immunoassay

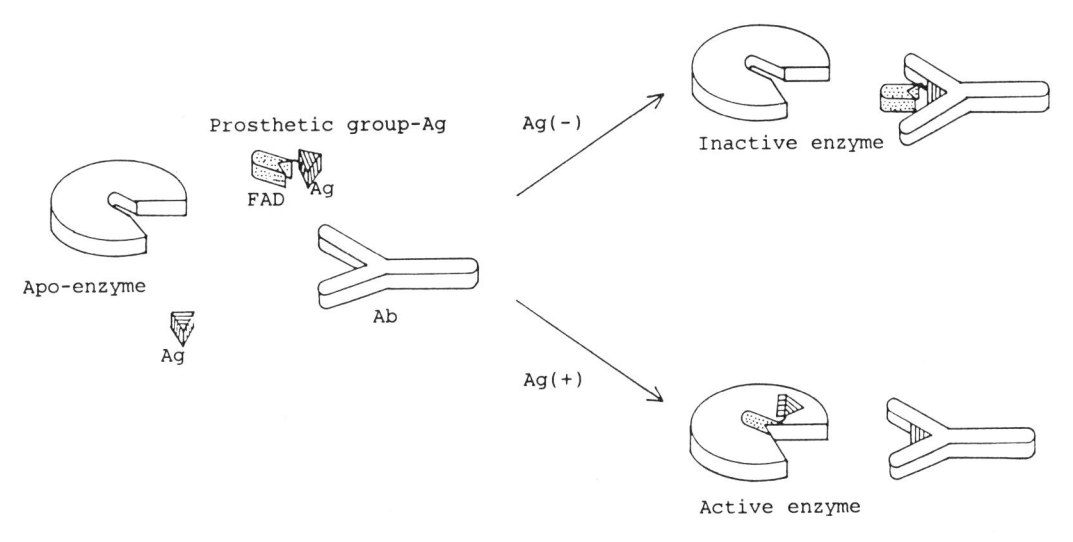

Figure 5 Apoenzyme reactivation immunoassay system (ARIS). Flavin adenine dinucleotide (FAD) attached to antigen (Ag) is used. Apoenzyme used is apoglucose oxidase.

E. Enzyme Channeling Immunoassay

Litman et al. [12] reported the homogeneous ECIA suitable for the determination of HMW antigen such as human IgG using two kinds of enzyme reaction systems. In this method, the substance produced by first enzyme can become a substrate for second enzyme. The assay principle is illustrated in Figure 6.

This method consists of a first enzyme labeled with antibody, second enzyme, and antigen. Among them, the second enzyme and the antigen are immobilized on a solid phase such as agarose. Hexokinase as first enzyme in the conjugate competitively bound to the solid phase produces a glucose-6-phosphate (G6P). This G6P can be used as the substrate for G6P dehydrogenase (G6PDH) as a second enzyme to form a gluconate-6-phosphate and NADH in the presence of nicotinamide-adenine dinucleotide (NAD). The G6P produced by free hexokinase disperses evenly into bulk solution and does not serve as the substrate for G6PDH. The activity of the second enzyme is inversely proportional to antigen concentration in the sample. The "channeling efficiency," namely substrate transfer from one to another enzyme, depends on physical factors such as viscosity of the solution and the size of the solid particles. Under optimum conditions, detection range for human IgG was 10-100 ng/ml. This method correlated well with the fluorescence excitation transfer immunoassay (r = 0.98, y = 0.98 X+ 0.98). This method can be improved further and adapted to noncompetitive assay, namely proximal linkage immunoassay, which will be described later.

F. Biotin Enzyme Avidin Immunoassay

Avidin possesses highest affinity with $K_a = 10^{-15}$ to biotin; therefore, it is a suitable modulator for the EMMIA method [26] as a strong and specific inhibitor of biotin enzyme. Avidin has been applied to EMMIA method for the determination of LMW antigen as well as HMW antigen such as IgG or IgM. The principle is illustrated in Figure

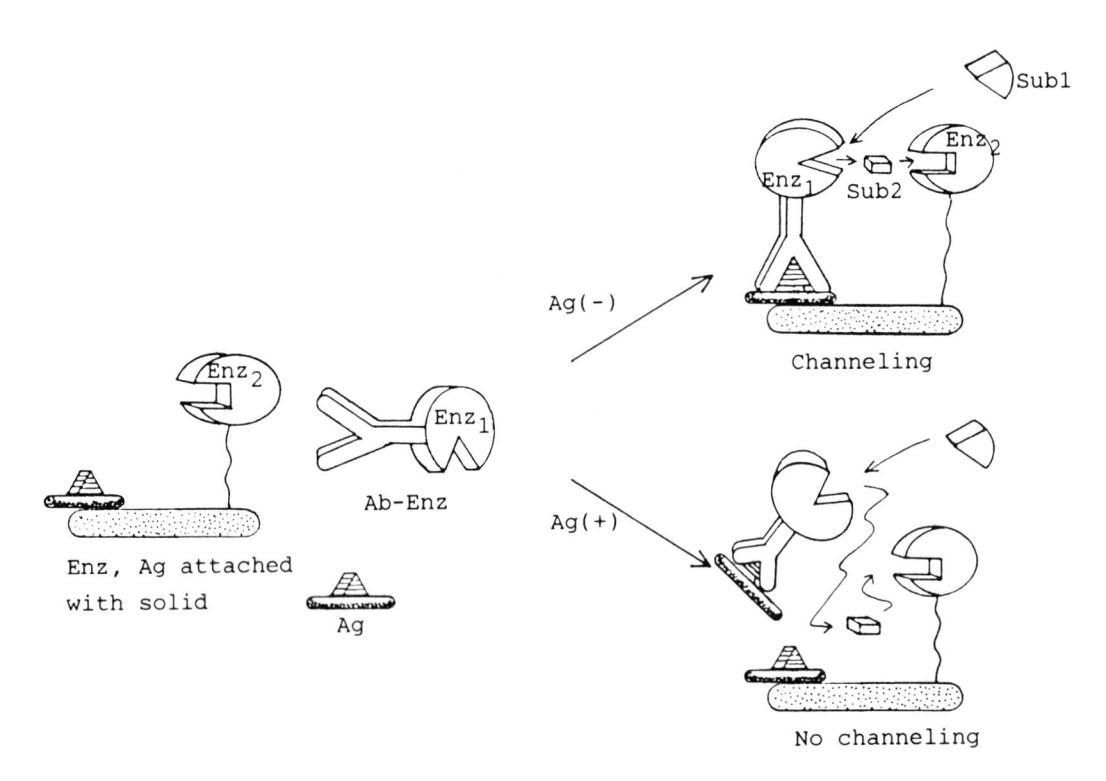

Figure 6 Principle of ECIA. Enz_1 and Enz_2 are hexokinase and G6PDH, respectively. In this assay, human IgG is determined as antigen (Ag). When antigen is absent in sample, antibody (Ab) labeled with Enz_1 binds to Ag on the solid phase. Enz_1 can catalyze substrate 1 (Sub_1) to produce substrate 2 (Sub_2) for enzyme 2 (Enz_2) attached to the solid phase.

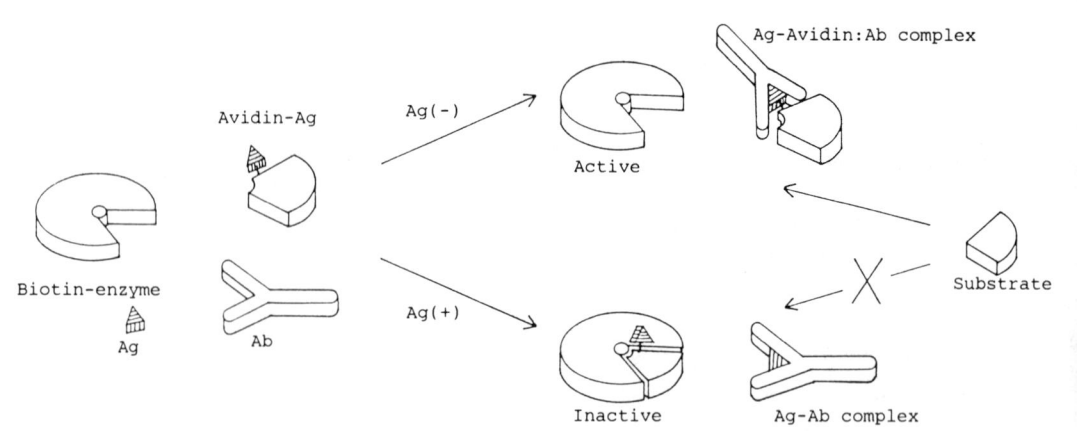

Figure 7 Biotin-enzyme avidin immunoassay (BAIA). Biotin-enzyme is propionyl Coa carboxylase. Goat IgG or theophylline can be used as antigen (Ag).

Table 4 Properties of Biotin Enzyme Obtained by Different Cultures

Enzyme	Specific Activity (U/mg)	Ki Value for Avidin (M)
Pyruvate carboxylase		
Bacillus stearothermophilus IFO 12550	120.0	2.0×10^{-5}
Pseudomonas citronellolis ATCC 13167	12.0	6.7×10^{-7}
Propionyl CoA carboxylase		
Mycobacterium smegmatis IFO 3082	43.0	3.5×10^{-8}
Mycobacterium smegmatis IFO 13167	44.0	2.0×10^{-8}
Mycobacterium smegmatis NIHJ 1628	44.0	3.0×10^{-10}
Mycobacterium smegmatis NIHJ 1629	41.0	3.0×10^{-9}
Mycobacterium rhodochrous IAM 12125	36.0	4.7×10^{-8}
Bovine liver	32.4	1.7×10^{-8}

7. The assay system consists of an avidin-analyte conjugate as immunomodulator and propionyl CoA carboxylase as enzyme. When antibody to the analyte reacts with the immunomodulator, the avidin in the complex is not available to inhibit the enzyme. In an excess of analyte, the antobody binds to the free analyte, whereas the biotin enzyme binds to the conjugate to form inactive complex regarding the enzymatic activity. The activity is inversely proportional to the analyte concentrations. We could apply this method to detect HMW antigen, such as goat IgG, as well as theophylline. An LMW analyte is the preferable antigen for labeling with modulator, because HMW antigens such as ferritin or AFP interfere sterically with the association of avidin-conjugate and biotin enzyme. However, avidin-Fab (goat IgG) conjugate inhibited 80% of the original activity of the enzyme. The detection range of goat IgG was 1-1000 μg/ml and 5-500 μg/ml for theophylline. The sensitivity of this method depends not only on the affinity of avidin to enzyme but also on the specific activity of enzyme used. We have prepared several enzymes from microbes. The properties of different biotin enzymes in regard to Ki and specific activity were studied. The results obtained are shown in Table 4. The enzyme from NIHJ 1628 strain obtained the strongest inhibition in all enzymes prepared and its the Ki value was 10^{-11}, but its specific activity was 40 μg/mg of protein, which is quite low. As a result, these two factors are canceled out. Thus, the detection limit for theophylline was not changed by the use of this system.

G. Cloned Enzyme Donor Immunoassay

Henderson et al. [31] have developed the CEDIA method by utilizing B-galactosidase obtained through genetic engineering. β-galactosidase consists of four subunits of polypeptide chain, and only tetramer of the subunit shows the enzyme activity. The deletion of NH2-terminal polypeptide of 10-60 amino acid residues of each subunit forms inactive β-galactosidase, called enzyme acceptor (EA). This EA is able to associate with the NH2-terminal small polypeptide which is a kind of enzyme (ED) corresponding to the deltion part of the β-galactosidase subunit.

The tetramer of EA bound to ED is enzymatically active. Since ED molecules can be bound covalently to antigen with cysteine and lysine residues of ED by use of bifunctional coupling reagent without any disturbance of the tetramer formation, whatever ED

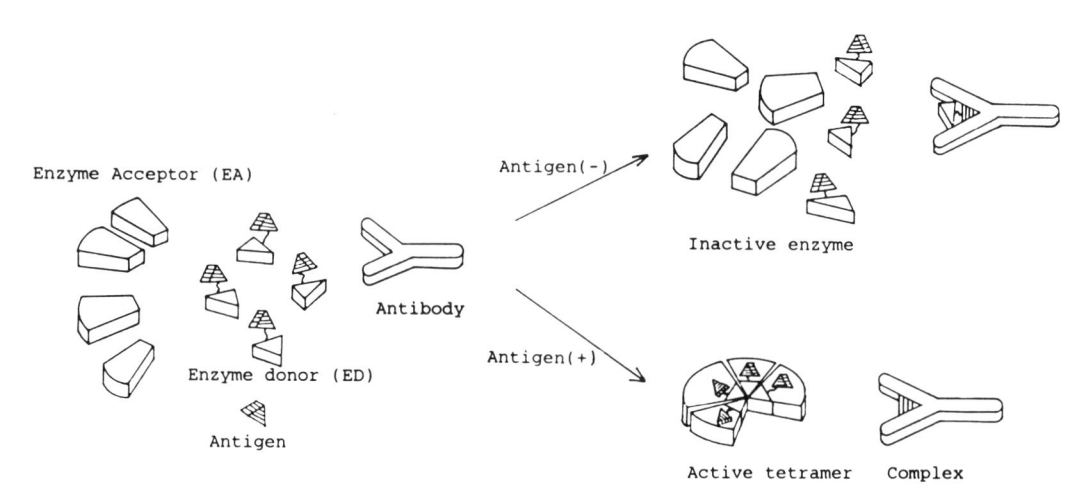

Figure 8 Principle of CEDIA. Enzyme acceptors (EAs) associate with enzyme donors (EDs) to form active β-galactosidase tetramer. Antibody inhibits the association of EA with ED-Ag.

is conjugated with antigen can associate with EA and form active tetramer. The complex of the ED-conjugate with antibody cannot associate with EA and results in no formation of the active tetramer. In a practical assay, as shown in Figure 8, free antigen in serum binds competitively to antibody with the antigen of ED. Thus, enzyme activity of the β-galactosidase is directly proportional to antigen concentration in serum. They have applied this method to the determination of digoxin in serum using a COBAS B10 automated analyzer. The standard curve was linear in the range of 0-4 ng/ml at 37°C for 15 min assay. The CEDIA for digoxin detection was well correlated with RIA: Y = 1.00 X 0.06. Coefficient variation of intra- and interassay values was less than 5%. The assay for the determination of HMW antigens by the CEDIA method may require improved preparation of conjugate because HMW antigen combined with ED interferes with the association between ED and EA due to its steric hindrance.

H. Improvement of Competitive Homogeneous EIA for HMW Antigens

Most cancer markers [8-10] are macromolecular antigens composed of glycolipid, carbohydrate, and/or protein moiety with a molecular weight greater than 10,000. As mentioned above, competitive assays of homogeneous EIA are best suited for the determination of LMW antigens, such as drugs and steroid hormones, but not for the sensitive detection of HMW antigens. When HMW antigens are labeled with enzyme, the antigen itself causes a change in activity of the enzyme by conformational modulation; therefore, the antigen used for labeling should be a small molecule. If synthetic peptides or sugar chains as epitopes of HMW antigens are used, HMW antigen could be determined by homogeneous EIA methods such as EMIT, SLFIA, ARIS, or CEDIA.

III. NONCOMPETITIVE BINDING ENZYME IMMUNOASSAY

A. Hybrid Antibody Immunoassay

Some antibodies to enzyme are known to modulate catalytic activity [32,33]. We have investigated the modulation of enzyme activity by antibodies to several enzymes such as G6PDH, hexokinase, β-galactosidase, and amylase. Monoclonal antibody specific to G6PDH inhibited 80% of the original activity of the enzyme and was therefore used as enzyme modulator in homogeneous EIA. The assay principle as indicated in Figure 9 is based on competitive enzyme inhibition by hybrid antibody.

The hybrid antibody is composed of two different antibodies attached covalently to each other. One antibody is specific to analyte and the other to G6PDH. The hybrid antibody inhibits the enzyme activity in the absence of analyte. However, when excess analyte is present, the enzyme is active. In this method, the size of analyte changes the degree of enzyme modulation and large molecules can inhibit the interaction between hybrid antibody and the enzyme.

The activity of the enzyme is proportional to the concentrations of analyte. A standard curve for human IgM shows a linear relationship in the range of 78-1000 μg/ml. Since the hybrid antibody was prepared by chemical treatment, the sensitivity of the present method is considerably affected by the method of preparation. Several investigators have recently developed hybrid monoclonal antibodies [34,35] that are able to bind to both analyte and enzyme. These techniques may be applicable to our method since this antibody can bind to not only antigen but also to enzyme. It is, therefore, necessary for the assay that once one of the binding sites of the antibody is occupied by either analyte or enzyme, the other binding site is not accessible. To attain this, the hybrid antibody's flexibility can be reduced by chemical or immunochemical fixation of the hinge region.

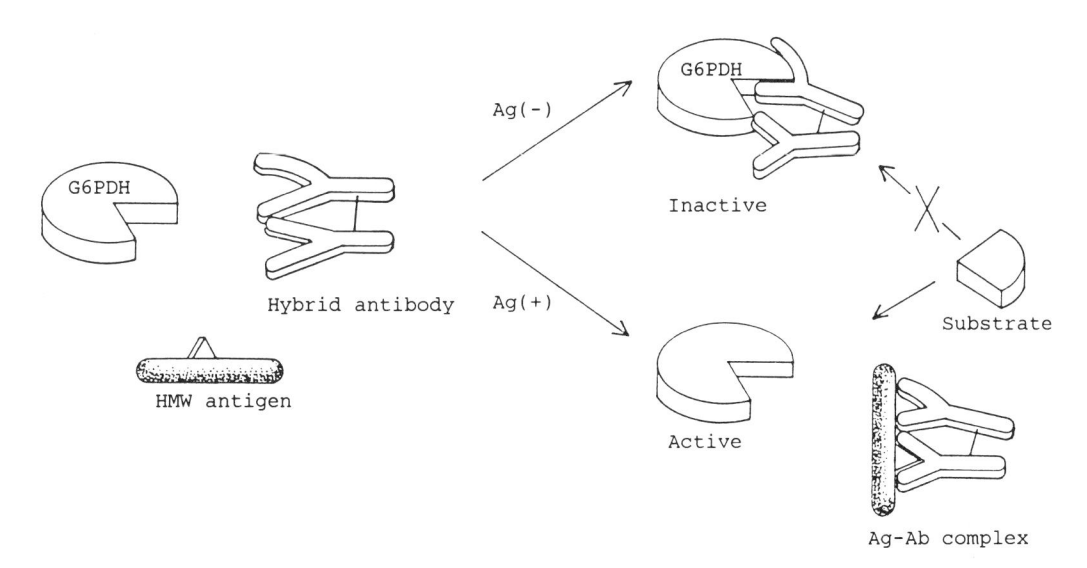

Figure 9 Hybrid antibody immunoassay (HAIA). Enzyme is G6PDH. Hybrid antibody is composed of anti-G6PDH antibody and antihuman IgG antibody.

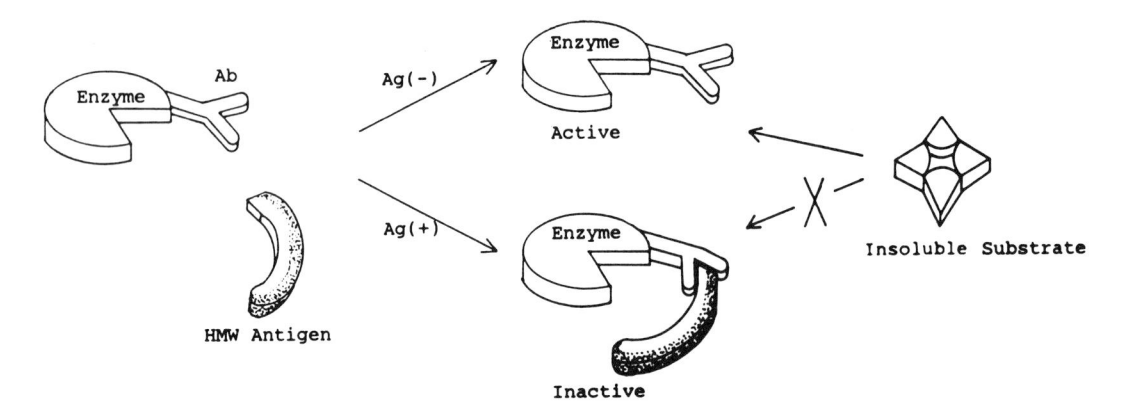

Figure 10 Enzyme inhibitory immunoassay (EIIA). Enzyme is α-amylase from *Bacillus subtilis* or dextranase from *Chaetomium gracile*. The HMW antigen (Ag) is ferritin or alpha fetoprotein.

B. Enzyme Inhibitory Homogeneous Immunoassay (EIHIA)

We developed a unique homogeneous EIA [36] that consists of antibody labeled with enzyme (conjugate) and insoluble substrate. The assay method is based on the reaction indicated in Figure 10. Insoluble substrate cannot be efficiently catalyzed by enzyme in the complex between antigen and the conjugate in excess of antigen. The activity of the enzyme is inversely proportional to amounts of antigen in sample. In the present method, α-amylase and insoluble dye starch were used as labeling enzyme and substrate, respectively. For the conjugation of antibody with enzyme, the thiol group of Fab of antibody was coupled to α-amylase by hetero bifunctional cross-linking reagent, N-(4-carboxycyclohexylmethy)maleimide. Even at a low concentration (200 ng/ml of ferritin), the immunological reaction between the conjugate and ferritin reached equilibrium rapidly after these materials were mixed. This method is sensitive enough to detect 10-800 ng/ml of ferritin and 5-200 ng/ml of alpha fetoprotein (AFP). The sensitivity of the method depends not only on the specific activity of enzyme but also on the affinity of antibody to conjugate. In the AFP assay, it was able to detect 1-50 ng/ml. This was achieved by either prolongation of incubation time for enzyme reaction or use of a higher dilution of the conjugate. Hence, highly sensitive assay will be obtained by improving the conditions for enzyme reaction. There was good correlation between the method and conventional RIA with 4 = 0.993, Y = 0.993X− 6.1 for ferritin. Since no washing step for separation of bound and free labels is required for the homogeneous system, these assays are sometimes interfered with by serum substances such as the mimic enzyme of the conjugate (i.e., serum amylase) or nonspecific adsorbent (i.e., rheumatoid arthritis factor). We have overcome these interferences by using specific inhibitor to serum amylase and Fab fragment instead of intact IgG. Results using the present method were consistent with these from RIA and highly reproducible, with a 4-5% CV for the within-run precision.

We also used dextranase [37] as labeling enzyme; this is superior to α-amylase because of the lack of mimic enzyme in serum. Ferritin and AFP were assayed and detected in the range of 8-200 ng/ml and 80-3000 ng/ml, respectively.

Table 5 Combination of Two Markers in Proximal Linkage Immunometric Assays

First Marker (M1)	Second Marker (M2)	Substance for M1	Substance for M2	Signal
Enzyme cycling				
Hexokinase	G6PDH	ATP, glucose	G6P, NAD$^+$	NADH
Glucose oxidase	Peroxidase	Glucose	H$_2$O$_2$, leuco dye	dye
FMN oxidoreductase	Luciferase	FMN, NADH	FMNH$_2$, aldehyde	hν
Enzyme/substrate				
Peroxidase	Luminol	H$_2$O$_2$	—	hν
Enzyme/allosteric effector				
phosphofructokinase	Phosphoenol pyruvate carboxylase	Fructose-6-phosphate	Phosphoenol pyruvate, NADH	NAD$^+$
aspartate aminotransferase	Phosphoenol pyruvate carboxylase	Oxalacetate, glutamate	phosphoenol pyruvate, NADH	NAD$^+$
Enzyme/prosthetic group				
Alkaline phosphatase	G6PDH	NADP$^+$	NAD$^+$, G6P	NADH

C. Proximal Linkage Immunoassay

Enzyme channeling immunoassay can be modified to noncompetitive assay by using two monoclonal antibodies specific to two different epitopes of one antigen, which are labeled with two different markers. Whole enzyme reactions can proceed by close interaction of these markers. Seviel et al. [11] have proposed several combinations of enzymes with markers for the assay systems and names this method proximal linkage immunoassay. Their examples and my consideration of the present method are listed in Table 5. The enzyme cycling system corresponds to the enzyme channeling assay. In addition, enzyme, modulator such as allosteric effector, substrate, prosthetic groups, and inhibitor can be also applicable to this method.

D. Enzyme Enhancement Immunoassay

In the enzyme enhancement IA (EEIA), presented here, β-galactosidase as the label attached to antibody, succinylated antibody [38] as enhancer, and cationic substrate are used. These materials have been also used in antigen inhibition assay methods. This method is based on the charge effect of highly anionic antibody on the enzyme activity, as illustrated in Figure 11. Here, the labeled enzyme is localized in an anionic environment followed by complex formation among anionic antibody, enzyme labeled with antibody, and antigen. The enzyme activity is also enhanced by this negative charge. A similar phenomenon to this has already been reported by studies [39-45] of immobilized enzyme. When the solid phase with the same charge as substrate was used, enzyme activity of the solid phase was obviously decreased by the force of electric repulsion. In contrast, if the solid phase with the opposite charge to the substrate is used, the enzyme activity was increased. In the EEIA method, the enzyme activity is linearly related to antigen concentrations. For human IgG and C-reactive protein, the dynamic range of

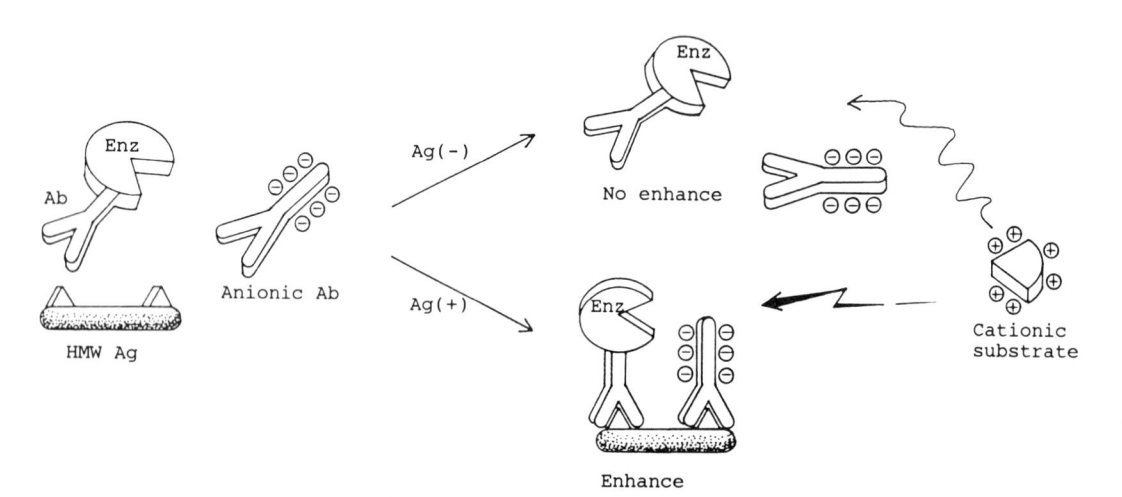

Figure 11 Enzyme enhancement immunoassay (EEIA). Substrate that carries positive charge at neutral pH is indicated in Figure 3. Two different antibodies are used: one is labeled with β-galactosidase (Enz) and the other carries a negative charge.

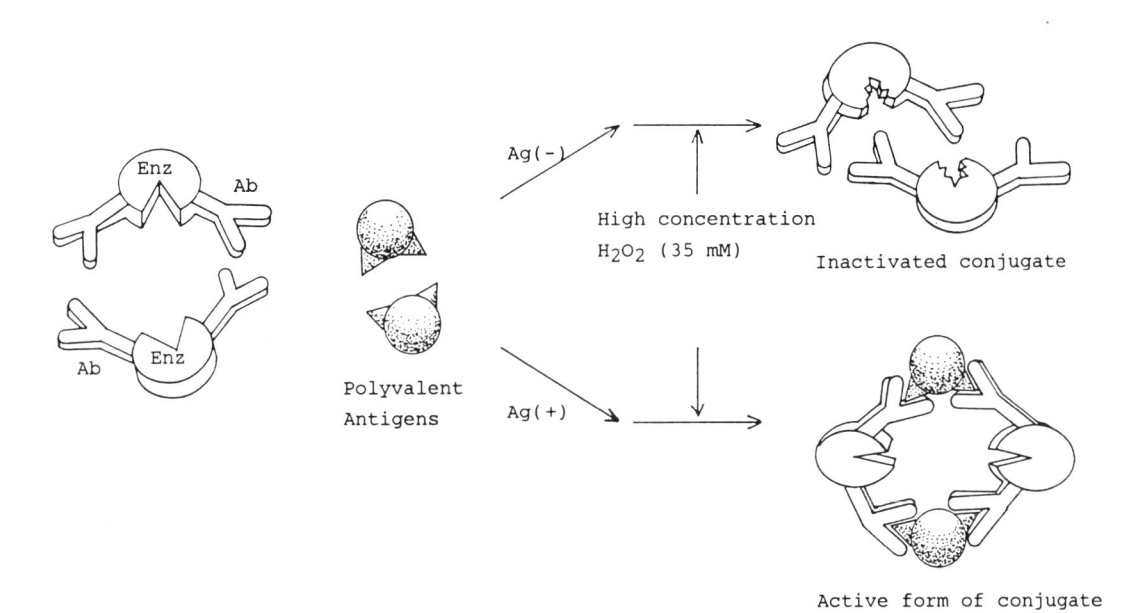

Figure 12 Associated enzyme sensitive technique (AIST). Alpha fetoprotein or IgE is used as polyvalent antigen. Monomeric peroxidase (Enz) bound to antibodies is inactivated at high concentrations of hydrogen peroxide (35 mM), whereas peroxidase in the complex of antigen with the labeling antibodies is stable at that concentration.

the assay for IgG was 10-10^6 ng/ml and for C-reactive protein it was 32-500 µg/ml. This assay may not be applicable in high concentrations of antigen because of a high-dose hook phenomena.

E. Associated Enzyme Sensitive Technique

Hoshino et al. [42] have observed that horseradish peroxidase (HRPO) is quite stable in aggregated form even at high concentration of H_2O_2, while monomer HRPO is irreversibly inactivated under the same condition when H_2O_2, phenol, and 4-aminoantipyrine were used. They applied this system to homogeneous EIA for the detection of HMW antigen such as alpha fetoprotein. The assay principle is based on stable aggregated HRPO conjugate, which is labeled with antibody (Fig. 12). A linear relationship exists between enzyme activity and increasing amounts of analyte. They compared the stability of monomeric HRPO with the aggregated form by using anti-HRPO antibody. Eight percent of the original activity of the aggregated enzyme was retained at more than 35 mM of H_2O_2, while activity of monomeric enzyme was 5% of the original. The polymerized HRPO with $NaIO_4$ showed activity similar to that of the aggregation of the HRPO molecule, which may be protected from the formation of inactive compound III, an intermediate form of HRPO. Detection range of this assay for AFP was 50-800 ng/ml. This method has also been applied to other analytes: digoxin and HMW protein such as ferritin, IgE, and beta 2-microglobulin. However, if the sample has a high concentration of antigen, one cannot measure accurately using this method because of the "high-dose hook" phenomenon, as mentioned in proximal linkage assay. Tandem assays

[47,48] using two types of monoclonal antibodies are useful for solving this problem. Hemoglobin in hemolytic sample acts as a mimic enzyme of the HRPO conjugate, and the assay therefore gave apparently higher than true values.

F. Interference with Homogeneous Enzyme Immunoassay

As mentioned in the section on EIHIA, there are several interfering substances in serum for the homogeneous EIA, such as mimic enzyme and antibodies to the labeling enzyme. In addition to the activity derived from the conjugate, mimic enzyme gives additional activity. For troubleshooting against such substances, a specific inhibitor to the mimic enzyme is selected to block the activity of the mimic enzyme but not the labeling enzyme. To diminish the interference by the antibodies, chemical or physical treatment is useful. Rheumatoid factors in serum also interfere with the assay. In such cases, the Fab' fragment instead of whole antibody can be used to avoid the interference because rheumatoid factors cannot bind to Fab' fragment [43].

IV. FUTURE TECHNOLOGY FOR HOMOGENEOUS EIA

Assays for tumor-associated antigens are very important in clinical laboratories, especially for monitoring therapy. Enzyme immunoassays that will become more commonly used than RIA should satisfy the following requirements: (1) no measurement of volume, (2) no control of temperature, (3) sandwich-type assay, (4) sensitivity less than 1 ng/ml of detection limit, (5) easily automated, and (6) highly reproducible.

To replace the labeling of antigen or antibody with enzyme by chemical conjugation, recombinant DNA technology will be useful for the preparation of conjugates. For example, a structural gene of enzyme is joined with that of antigen or antibody, and the constructed chimeric gene is inserted into target plasmid vector. Several investigators have already obtained chimeric antibodies or enzyme [44-47], but so far no homogeneous EIAs using these chimeric proteins are available.

In homogeneous EIA, as mentioned, an increase in sensitivity can be achieved only by decreasing the amounts of antigens or antibodies, both of which are labeled with enzyme in the reaction system. It is quite obvious that there is a limitation in lowering those concentrations to form immunocomplexes depending on the affinity of antibody. For one heterogeneous EIA [48,49], solid phase sandwich assay [50], excess amounts of antibodies can be used. According to theoretical calculation by this method, one can detect one molecule of analyte. However, homogeneous EIA of the sandwich type using excess of antibodies is not available. Recently, therefore, several heterogeneous EIAs [51-53] without any steps for the separation of bound and free label have been developed. These assays for HWM antigens are rapid, simple, and highly sensitive. There are already some commercially available kits for sandwich-type immunosorbent assay. These heterogeneous assay methods will be used in place of some homogeneous EIAs.

ACKNOWLEDGMENTS

I thank Dr. M Okada for helpful discussions and Mr. I. Nishizono for his valuable assistance.

REFERENCES

1. Yallow RS, Berson SA. Assay of plasma insulin in human subjects by immunological method. Nature 1959;184:1648-1649.
2. Wians FH Jr, Jacobson JM, Dev J, Heald JI, Ortiz G. Thyrotroph function assessed by sensitive measurement of thyrotropin with three immunoradiometric assay kits: analytical evaluation and comparison with the throliberin stimulation test. Clin Chem 1988;34:568-575.
3. John R, Evans PE, Scalon MF, Hall R. Clinical value of immunoradiometric assay of thyrotropin for patients with nonthyroidal illness and taking various drugs. Clin Chem 1987;33:566-569.
4. Engvall E, Perlman P. Enzyme-linked immuno assay (ELISA): quantitative assay of immunoglobulin G. Immunochemistry 1971;8:871-874.
5. Ollerich M. Enzyme-immunoassay: a review. J Clin Chem Clin Biochem 1984;22:895-904.
6. Rubenstein KE, Schneider RS, Ullman EF. "Homogeneous" enzyme immunoassay: a new immunochemical technique. Biochem Biophys Res Commun 1972;47:846-851.
7. Bara J, Decaens C. Blood-group-related antigens in colonic carcinogenesis. Ann Inst Pasteur Immunol 1987;138:891-899.
8. Buchegger F, Fournier K, Schreyer M, Carrel S, Mach JP. Swine monoclonal antibodies of high affinity and specificity to carcinoembryonic antigen. J Nat Cancer Inst 1987;79:337-342.
9. Dyatlovitskaya EV, Bergelson LD. Glycosphingolipids and antitumor immunity. Biochim Biophys Acta 1987;907:125-143.
10. Kjeldsberg CR, Marty J. Use of immunologic tumor markers in body fluid analysis. Clin Lab Med 1985;5:233-264.
11. Sevier ED, David GS, Martinis J, Desmond WJ, Bartholomew RM, Wang R. Monoclonal antibodies in clinical immunology. Clin Chem 1981;27:1797-1806.
12. Litman DJ, Hanlon TM, Ullman EF. Enzyme channeling immunoassay: a new homogeneous enzyme immunoassay technique. Anal Biochem 1980;106:223-229.
13. Terouanne B, Carrie ML, Nicolas JC, Paulet AC. Bioluminescent immunosorbent for rapid immunoassays. Anal Biochem 1986;154:118-125.
14. Carrie ML, Terouanne B, Brochu M, Nicolas JC, Paulet AC. Bioluminescent immunoassays of progesterone: a comparative study of three different procedures. Anal Biochem 1986;154:126-131.
15. Terouanne B, Nicolas JC, Paulet AC. Bioluminescent immunoassays of α-fetoprotein. Anal Biochem 1986;154:132-137.
16. Ekins R. Toward immunoassays of greater sensitivity, specificity and speed: an overview. In: Albertini A, Ekins R, eds. Monoclonal antibodies and developments in immunoassays. Amsterdam: Elsevier/North Holland Biochemical Press, 1981.
17. Schneider RS, Bastiani RJ, Rubenstein KE, Ullman EF, Jaklitsch K, Kameda N. Homogeneous enzyme immunoassay for diphenylhydantoin and phenobarbital in serum. Clin Chem 1974;20:869-872.
18. Wellington P, Rutkowski PB. Rapid estimation of plasma phenobarbital level by an enzyme immunoassay for barbiturates. Ther Drug Monit 1982;4:319-322.
19. Hoshino N, Hama M, Suzuki R, Kataoka Y, Soe G. A new homogeneous enzyme immunoassay. Its application to measurement of α-fetoprotein. J Biochem 1985;97:113-118.
20. Ullman EF, Yoshida R, Blakemore J, Maggio ET, Leute RK. Mechanism of inhibition of malate dehydrogenase by thyroxine derivative reactivation by antibodies. Biochem Biophys Acta 1979;576:66.

21. Varrone S, Consiglio E, Covelli I. The nature of inhibition of mitochondrial malate dehydrogenase by thyroxine, iodine cyanide and molecular iodine. Eur J Biochem 1970;13:305-312.
22. Ernst R, Williams L, Dalbey M, Collins C, Pankey S. Homogeneous enzyme immunoassay (EMIT) protocol for monitoring tricyclic antidepressants on COBAS-B10 centrifugal analyzer. Ther Drug Monit 1987;9:85-90.
23. Gibbons I, Skold C, Rowley GL, Ullman EF. Homogeneous enzyme immunoassay for proteins employing β-galactosidase. Anal Biochem 1980;102:167-170.
24. Burd JF, Wong RC, Feeney JE, Carrico RJ, Boguslaski RC. Homogeneous reactant-labeled fluorescent immunoassay for therapeutic drugs exemplified by gentamicin determination in human serum. Clin Chem 1977;23:1402-1408.
25. Li TM, Benovic JL, Buckler RT, Burd JF. Homogeneous substrate-labeled fluorescent immunoassay for theophylline in serum. Clin Chem 1981;27:22-26.
26. Ngo TT, Carrico RJ, Boguslaski RC, Burd JF. Homogeneous substrate-labeled fluorescent immunoassay for IgG in human serum. J Immunol Methods 1981;42:93-103.
27. Armenta R, Tarnowski T, Gibbons I, Ullman EF. Improved sensitivity in homogeneous enzyme immunoassays using fluorogenic macromolecular substrate: an assay for serum ferritin. Anal Biochem 1985;146:211-219.
28. Ngo TT, Lenhoff HM. Enzyme modulators as tools for development of homogeneous enzyme immunoassays. FEBS Lett 1980;116:285-288.
29. Ngo TT. Enzyme modulator mediated immunoassay (EMMIA). Int J Biochem 1983;15:583-590.
30. Morris DL, Ellis PB, Carrico RJ, Yeager FM, Schroeder HR, Albarella JP, Bogusraski RC. Flavin adenin dinucleotide as a label in homogeneous colorimetric immunoassays. Anal Chem 1981;53:658-665.
31. Henderson DR, Friedman SB, Harris JD, Manning WB, Zoccoli MA. CEDIA TM, a new homogeneous immunoassay system. Clin Chem 1986;32:1637-1641.
32. Ashihara Y, Nishizono I, Miyagawa E, Kasahara Y. Homogeneous enzyme immunoassay for macromolecular antigens using avidin biotin enzyme. J Clin Lab Anal 1987;1:80.
33. Rotman MB, Celada F. Antibody-mediated activation of a defective β-D-galactosidase extracted from an *Escherichia coli* mutant. Proc Natl Acad Sci USA 1968;60:660-667.
34. Ashihara Y, Nishizono I, Suzuki H, Jasahara Y. Homogeneous enzyme immunoassay for macromolecular antigens using hybrid antibody. J Clin Lab Anal 1987;1:77-79.
35. Karawajew L, Micheel B, Behrsing O, Gaestel M. Bispecific antibody-producing hybrid hybridomas selected by a fluorescence activated cell sorter. J Immunol Methods 1987;96:265-270.
36. Karawajew L, Behrsing O, Kaiser G, Micheel B. Production and ELISA application of bispecific monoclonal antibodies against fluorescein (FITC) and horseradish peroxidase (HRP). J Immunol Methods 1988;111:95-99.
37. Ashihara Y, Nishizono I, Tanimoto T, Tuchiya H, Yamamoto K. Kido Y, Miyagawa E, Kasahara Y. Enzyme inhibitory homogeneous immunoassay for high molecular antigen (I). J Clin Lab Anal 1988;2:138-142.
38. Nishizono I, Ashihara Y, Tanimoto T, Tuchiya H, Yamamoto K, Kasahara Y. Enzyme inhibitory homogeneous immunoassay for high molecular weight antigen (II). J Clin Lab Anal 1988;2:143-147.
39. Gibbons I, Hanlon TM, Skold CN, Russel ME, Ullman EF. Enzyme-enhancement immunoassay: A homogeneous assay for polyvalent ligands and antibodies. Clin Chem 1981;27:1602-1608.

40. Hornby WE, Lilly MD. The preparation and properties of ficin chemically attached to carboxymethylcellulose. Biochem J 1966;98:420-425.
41. Hornby WE, Lilly MD. Some changes in the reactivity of enzymes resulting from their chemical attachment to water-insoluble derivatives of cellulose. Biochem J 1968;107:669-674.
42. Chung ST, Hamano M, Aida K, Uemura T. Studies of ATP deaminase part III. Water insoluble ATP deaminase. Arg Biol Chem 1968;32:1287-1291.
43. Kato K, Umeda U, Suzuki F, Hayashi D, Kosaka A. Use of antibody Fab' fragments to remove interference by rheumatoid factors with the enzyme-linked immunoassay. FEBS Lett 1979;102:253-256.
44. Powell PP, Kyle JW, Miller RD, Pantano J, Grubb JH, Sly WS. Rat liver β-glucuronidase. cDNA cloning, sequence comparisons and expression of a chimeric protein in COS cells. Biochem J 1988;250:547-555.
45. Bruggemann M, Williams GT, Bindon CI, Clark MR, Walker MR, Jefferies R, Waldmann H, Neuberger MS. Comparison of the effector functions of human immunoglobulins using a matched set of chimeric antibodies. J Exp Med 1987;166:1351-1361.
46. Sun LK, Curtis P, Racowicz-Szulczynska R, Ghrayeb J, Chang N, Morrison SL, Koprowski H. Chimeric antibody with human constant regions and mouse variable regions directed against carcinoma-associated antigen 17-1A. Proc Natl Acad Sci USA 1987;84:214-218.
47. Pierard L, Jacobs P, Gheysen D, Hoylaerts M, Andre B, Topisirovic L, Cravador A, Foresta F, Herzog A, Collen D, Wilde MD, Bollen A. Mutation and chimeric recombinant plasminogen activator. Production in eukaryotic cells and preliminary characterization. J Biol Chem 1987;262:11771-11778.
48. Emancipator K, Cadoff EM, Burke MD. Analytical versus clinical sensitivity and specificity in pregnancy testing. Am J Obstet Gynecol 1988;158:613-616.
49. Wu AH, Wong SS, Waldron C, Chan DW. Automated quantification of choriogonadotropin: analytical correlation between serum and urine with creatinine correction. Clin Chem 1987;33:1424-1426.
50. Imagawa M, Yoshitake S, Ishikawa E, Niitu Y, Urushizaki I, Kanazawa R, Tachibana S, Nakazawa N, Ogawa H. Development of highly sensitive sandwich enzyme immunoassay for human ferritin using affinity-purified anti-ferritin labeled with β-galactosidase from *Escherichia coli*. Clin Chim Acta 1982;121:277-289.
51. Valkirs GE, Barton R. ImmunoConcentration^{TM}—a new format for solid-phase immunoassays. Clin Chem 1985;31:1427-1431.
52. Bardwin J, Hargreaves WR. A rapid, self-washing immunoassay system with a micro particular solid phase. Clin Chem 1987;33:1566-1567.
53. Brown WE. Microparticle-capture membranes: application to TestPack hCG-urine. Clin Chem 1987;33:1567-1568.

11
Rapid Response Fiber Optic Evanescent Wave Immunosensors

BARRY I. BLUESTEIN and SHIN-YIH CHEN

Ciba-Corning Diagnostics Corporation, Medfield, Massachusetts

I. INTRODUCTION

One of the prime objectives of today's analytical biochemist working in the field of medical technology is to develop methodology and instrumentation that is simple to use, technically insensitive, and provides rapid test results of medical value to the attending physician. When such technology is successful, it can result in the performance of quality "in vitro" diagnostic testing beyond the boundries of the central laboratory. Bringing the actual testing directly to the site(s) of primary patient care rather than transporting the sample to the lab can result in a number of medical and economic benefits. By having diagnostic test results in time frames approaching "real time," clinicians are in a position to institute acute interventional therapy based on more complete findings of clinical and biochemical data, monitor and adjust dosages of drugs with very restricted therapeutic and toxic windows, and reduce a patient's hospitalization and the psychological trauma of waiting for a diagnostic test result.

A. Immunometric Assay

Major advances have been made in recent years in the development of real-time, or close to real-time, sensors and devices to measure a variety of analytes of clinical relevance. These would include simple to use electrochemical sensors for the electrolytes Na^+ and K^+, and one-step protocol systems for the measurement of small metabolites such as glucose. Recently, a number of dry reagent chemistry systems such as the Seralyzer (Ames, Elkhart, IN), the Refletron (Boehringer-Mannheim, Indianapolis, IN), and the Ektachem DT-60 (Kodak, Rochester, NY) have appeared in medical facilities and they greatly simplify the measurement of routine chemical levels measured colorimetrically by spectrophotometric methods. Conspicuously absent from all these advances in simple, rapid measurement technology has been the ability to measure high-molecular-weight (>5000 Mr), ultralow-concentration (10^{-9}-10^{-12} mol/L) analytes. This class of molecules includes most protein hormones and tumor-specific antigens found in peripheral circulation. With the advent of genetic engineering and hybridoma technology, many of the therapeutic agents of the future such as interferons, growth factors, other biological response modifiers, and targeted organ-specific monoclonal antibody-chemotherapeutic drug conjugates for cancer therapy also will belong in this category of analytes.

The current methodology of choice to measure such analytes is based on the concept of heterogeneous solid-phase immunometric analysis [1]. This methodology uses at least two antibodies (Ab) directed against the analyte of choice, preferably against different recognition sites (epitopes) on the surface of the analyte (antigen, Ag) to be detected. One Ab is immobilized to a solid support such as glass beads, magnetic particles, or the surface of a microtiter plate; the second antibody is covalently coupled to a quantifiable reporter (tag) group. Early phases of this technology used radioactive isotopes as the reporter. However, due to the limited shelf-life of radioactive materials and the hazards and inconvenience of handling isotopes, modern refinements have substituted nonisotopic probes such as enzymes and chemiluminescent or fluorescent tracers.

Regardless of the labeled probe used, these technologies all suffer from the same inherent problem. Before the final readout, the solid phase must be washed rigorously to remove unbound Ab-probe. This means that the primary solid phase Ab-Ag-probe Ab reaction cannot be monitored directly and a number of postreaction preparation steps are required before the final signal quantitation of the solid phase. For nonisotopic immunoassays, there are further process steps such as substrate addition, secondary incubation with substrate to develop color or fluorescence, and reaction termination before the final readout. All these steps take time and processing by the user or complex instruments. Typical processing steps required to perform a standard enzyme immunoassay via enzyme-linked immunosorbent assay (ELISA) are shown in Table 1.

B. Immunobiosensors

The most suitable immunoassay system would embody the attributes of the extreme sensitivity afforded by heterogeneous (phase separation) immunometric procedures and the ease of use of today's ion-selective electrode sensors. In such a system, a solid phase antibody would act as a direct signal transducer capable of converting the Ag-Ab interaction directly into a detectable response. Such a system would require nothing more of the user than addition of the sample (possibly mixed with a single, stable reagent) without any further processing required. Numerous strategies for developing electrochemical,

Table 1 Process Steps in Enzyme Immunometric Assay

1. Aliquot sample
2. Aliquot enzyme-antibody conjugate
3. Incubate to endpoint
4. Separate solid-phase bound and unbound conjugate
5. Wash solid phase with buffer
6. Remove wash fluids
7. Add enzyme substrate: colorimetric or fluorescent
8. Incubate to allow product development
9. Stop enzyme reaction
10. Measure optical density or fluorescence of fluid

Fiber Optic Immunosensors

piezoelectric, or optical immunosensors and their current status and attendant problems have been reviewed by North [2]. Simple immunoassay test systems that require only the patient's sample and addition of tagged reagent and without phase separation (homogeneous) have been developed [3,4]. However, these systems function only in a competitive immunoassay format, can only measure analytes of very low molecular weight, and are generally restricted in their sensitivity to ranges of analytes around 10^{-6} mol/L.

C. Fiber Optic Sensors

1. Distal Sensing

Fiber optic probes or sensors can be organized into two distinct categories: distal sensing and evanescent wave. In the former, all chemistry reactions occur extrinsic to the fiber itself. A sensing element is attached to the distal tip of the fiber and light passing down the fiber is used to measure the attendant chemical reaction. Incident light modulated by the chemistry reaction is also conducted back into the fiber from the distal tip where light perturbations are detected with a photomultiplier tube or solid-state detector. Parameters monitored with the fiber include solution absorption of incident light (reflectance spectroscopy) and fluorescence spectroscopy, such as monitoring changes of pH-sensitive dyes [5-7]. In these approaches, the fiber is a passive agent, acting merely to conduct light via total internal reflection and to pick up and conduct either fluorescent emissions or reflected incident light. Fibers used in this fashion are complete fiber optic elements containing a core glass, a thin layer of lower refractive index glass cladding (required to satisfy optical requirements necessary for total internal reflection or conduction of light), and a plastic coating that gives mechanical flexibility and stability to the optical fiber.

2. Evanescent Wave Sensing

The second approach to fiber optic sensing is to use the fiber itself as an active transducer element. In this case, the core glass portion of the fiber is exposed by removal of the lower refractive index glass cladding and plastic coating, and the lateral surfaces are modified by coating with a biological recognition molecule (antibody). When such a process is performed, the fiber optic core glass (typically pure quartz) now becomes a solid phase antibody capable of being used in a heterogeneous immunoassay format. When used in conjunction with a fluorescent probe and a unique portion of electromagnetic radiation characteristic of total internal reflection elements (optical fibers and prisms) known as the "evanescent wave," a solid phase immunometric assay system can be developed that requires no user- or instrument-mediated separation of bound and free fluorescent probe before the detection of solid-phase, immunologically bound probe. This concept thus serves as the basis of a real-time, process-simple immunosensor that will be described here.

II. PRINCIPLES OF EVANESCENT WAVE IMMUNOSENSORS

A. Evanescent Theory

When a light ray propagating through an optically transparent medium of high refractive index (e.g., quartz) impinges at the interface of a medium of lower refractive index (e.g., water or serum) at angles exceeding the critical angle (θ_c), relative to the normal of the interface, total internal reflection occurs. This critical angle is given by the equation:

$$\theta_c = \sin^{-1}(n_2/n_1) \qquad (1)$$

where n_2 and n_1 are the refractive indices of the liquid and solid.

Although the incident light beam reflects totally internally and is conducted down the fiber, at the interface of incident and reflected light an electromagnetic field called the evanescent wave penetrates perpendicularly to the fiber axis a small distance into the lower index medium to a depth typically less than one wavelength. Since this field cannot go to zero instantaneously, it decays in amplitude exponentially with distance from the solid-liquid interface. The evanescent wave is capable of exciting fluorescent molecules that may be present near the surface of this interface. The electric field amplitude (E) at any distance from the interface is described as:

$$E = Eo \exp(-Z/dp) \qquad (2)$$

where Eo is the electric field amplitude at the surface, dp the distance for the electric field amplitude to fall to 1/e of its value at the interface, and Z the distance from the interface. The depth of penetration (dp) is determined by Eq. 3.

$$dp = \frac{\lambda}{4\pi}(n_1^2 \sin^2\theta - n_2^2)^{-1/2} \qquad (3)$$

The principles describing total internal reflection have been developed by Harrick [8] and most recently reviewed by Axelrod and colleagues [9]. Figure 1 depicts the exponential decay in amplitude of the evanescent wave. The word evanescent is derived from the Latin meaning "vanishing."

If a fluorescent molecule, either bound to the outer lateral surface of the fiber or free within the evanescent zone, is excited by radiation of the proper wavelength, it will emit a higher wavelength fluorescence that in turn can be picked up by the fiber and propagated in the fiber in both directions by total internal reflection. Hirschfeld [10] first demonstrated this effect for dilutions of fluorescein dye. Since excitation and capture of fluorescently emitted radiation is restricted to a small cylindrical zone surrounding the fiber, this measurement approach is relatively insensitive to quenching of surface localized fluorescence in light-scattering or highly absorbing solutions. This is because the light beam does not pass through the bulk solution on its way to the detector system and thus is not subjected to the extremes of spectral absorption of fluorescent emission present in conventional solution fluorimetry (inner filter effects) [11].

Although a number of optical configurations of the type of orientation of lamp source and detector and the geometry of the optical waveguide (flat plate vs. cylindrical fiber) have been used, we have chosen the system that allows for maximization of fluorescence detection. The optical configuration is based on the concept proposed by Hirschfeld and Block [12,13] and is shown in Figure 2. This orientation of lamp and detector optics at the same (proximal) end of the fiber minimizes the amount of excitation light that can reach the detector, since most of the excitation radiation transmitted into the fiber emerges from the distal end. Optical orientations have been described in which the detector is placed at the distal end of the fiber [14]. These approaches have the disadvantage of having to achieve extremely high levels of spectral filtration to separate excitation from fluorescent emission radiation. A complete and quantitative model of ray optics describing the nature of evanescent excitation and fluorescent light tunneling has recently been completed by our laboratories [15]. This model predicts that even under optimum conditions of high angle light launch into the fiber and spot size illumination

Fiber Optic Immunosensors

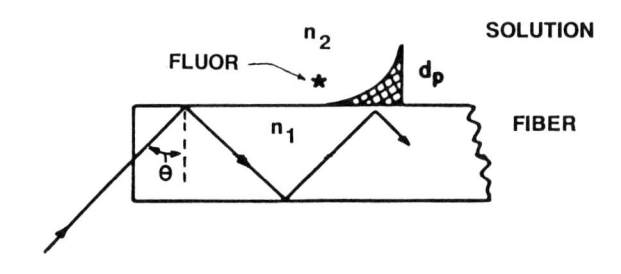

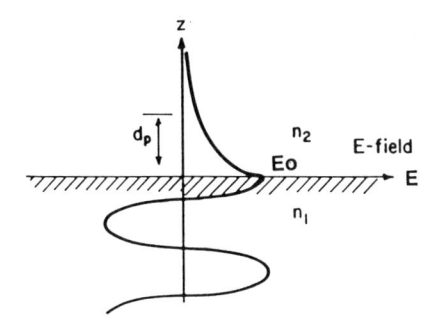

A.

B.

Figure 1 (A) Simplified model of a single excitation light ray launched into an evanescent wave sensor. When θ exceeds the critical angle, total internal reflection occurs. $n_1 > n_2$. Under these conditions, an evanescent wave is generated that extends into the lower index medium. (B) Electric field diagram of the evanescent wave. At the solid-liquid interface the field amplitude (Eo) is equivalent to that inside the fiber. As the wave extends into solution in the z direction, its amplitude decays exponentially with distance. dp is dependent on the launch angle of light into the fiber, the wavelength of light, and the refractive indices of the two media.

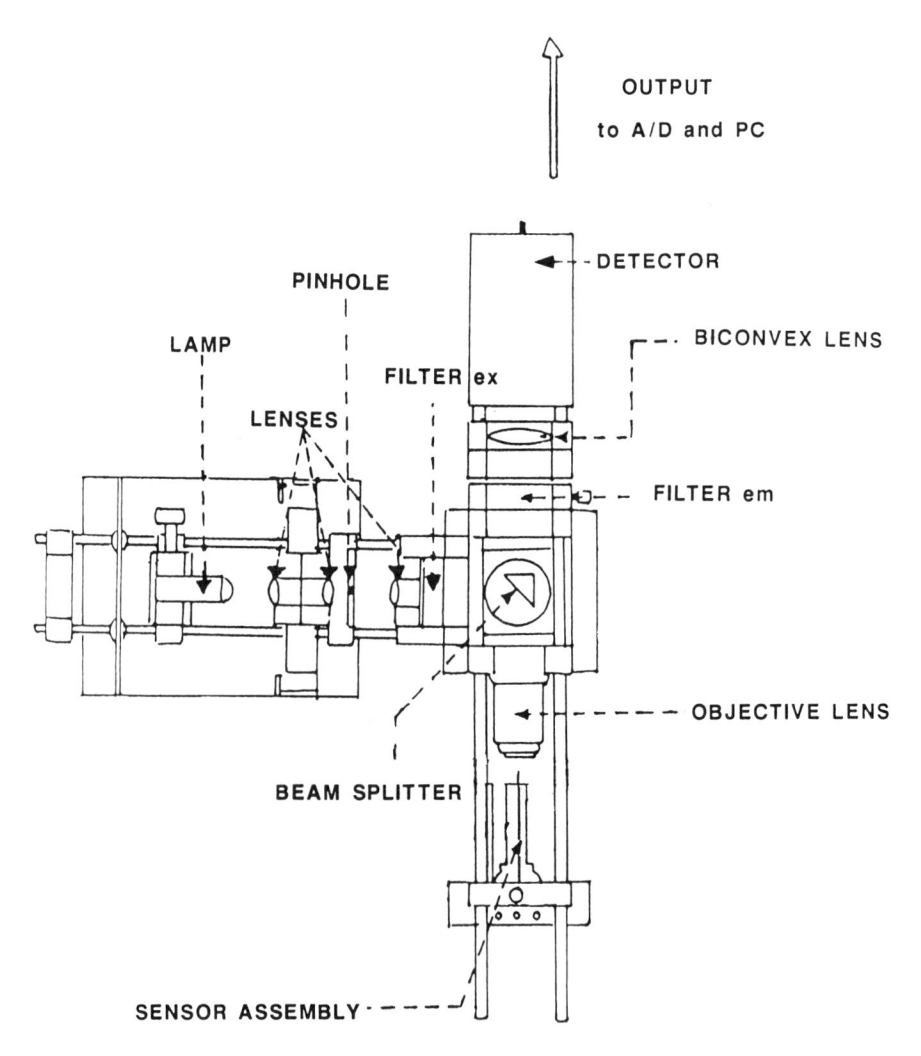

Figure 2 Optics diagram for the evanescent fluorimeter. Light from a tungsten-halogen lamp is imaged on the proximal surface of a fiber sensor contained in its sample chamber after passage through an excitation filter (485 nm), dichroic beam splitter, and focusing lenses. Higher-wavelength fluorescent light (525 nm) tunnels back into the fiber and is waveguided back to the proximal end of the fiber, through the beam splitter and emission filter, where it is imaged on the surface of a solid-state silicon diode detector.

Fiber Optic Immunosensors

on the proximal fiber face, only about 2% of the evanescent fluorescent light emitted by the fluorophore in solution is coupled back into the fiber. Since fluorescein dye (the primary fluorescent probe used in the conjugation of biochemical probes) has a Stokes shift of only about 30 nm (excitation 495 nm, emission 525 nm), the need to minimize excitation light at the detector face in addition to good spectral filtration cannot be overemphasized, if one expects to achieve high sensitivity.

The second aspect of this system that differs from other proposed total internal reflection fluorescence systems previously described is the choice of a cylindrical (fiber waveguide) rather than a flat plate total internal reflection element. In the cylindrical geometry, total internal reflections of fluorescent light that has tunneled back into the reflection element are confined to two of the three dimensions rather than one. Since spreading of light in a transverse dimension is essentially eliminated in a cylindrical fiber, once fluorescent light rays enter the optical element they are trapped and can be collected at the proximal surface by a small detector. Since at each point of intercession of an incident and reflected excitation ray an evanescent wave is generated, sensitivity is enhanced due to multiple internal reflections. In a cylindrical waveguide the total number of reflections in fiber of length L and radius a is given by:

$$N = \frac{L}{2a} \cot\theta \tag{4}$$

This implication of enhanced sensitivity is borne out by the fact that although a number of planar optical waveguide immunosensors have been described [16,17], the best sensitivity achieved in terms of analyte detection has been on the order of 10^{-8} mol/L, while in the system to be described analyte levels are as low as 10^{-11} mol/L have thus far been detected.

B. Evanescent Wave Immunosensors

From the previous discussion, it should be apparent that the evanescent wave sensor (EWS) is capable of measuring concentrations of fluorescent molecules localized on or near the outer surface of the fiber, even in the presence of turbid or highly absorbing solutions. If one could create an analyte-specific concentration of fluorescent probe within the evanescent zone, this would form the basis of a sensitive system to measure analytes even in the presence of nonspecific interfering components in solution.

All solid-phase immunoassays, both immunometric and competitive, are specific concentrators of mass. If the outer lateral surface of a fiber optic is coated with antigen-specific antibodies and is immersed in solution with unknown analyte and fluorescent probe (antigen or antibody conjugate), the basis of a unique solid-phase immunoassay is formed. Although this is a true heterogeneous (phase separation) system, as a natural consequence of the evanescent wave's ability to measure localized concentrations of fluorescent probe, as an immunoreaction occurs and antigen and labeled antibody are concentrated on the surface, rate changes in fluorescent signal due to interaction of the primary immunoreactants can be detected in "real time" as binding occurs. Any initial background interferences from the sample and the initial low concentration signal of added fluorescent tag can be blanked out at the initiation of the reaction. The immunoreaction can thus be monitored as it is occurring. With appropriate control of reaction timing, signals can be accumulated and analyzed in a rate or timed endpoint mode. This system is applicable to both immunometric assay architectures, where signal is directly

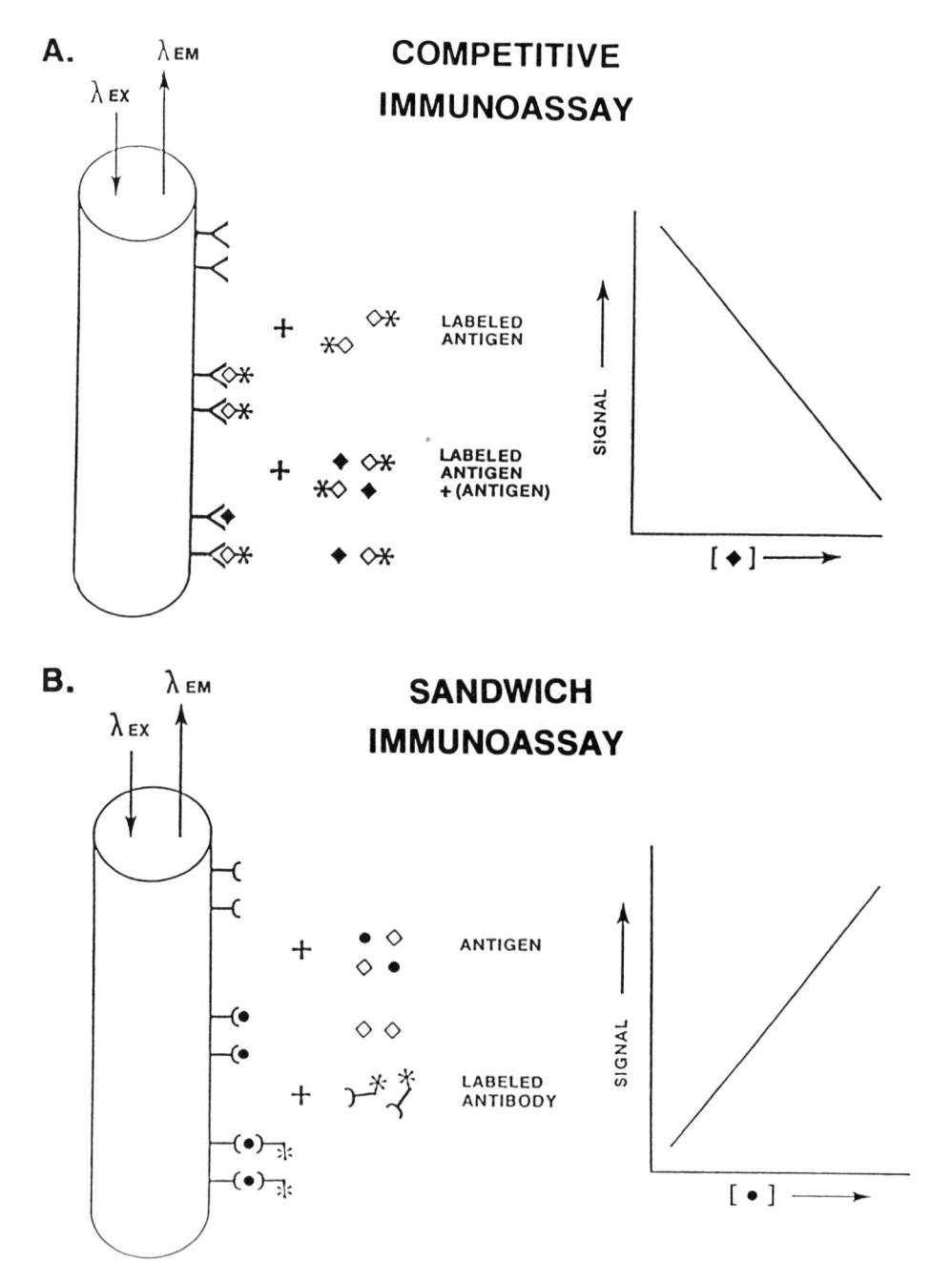

Figure 3 Schematic diagrams of solid phase EWS immunoassay architectures. (A) Competitive format for small molecules utilizing fluorescently labeled antigen. Signal is inversely proportional to analyte concentration. (B) Immunometric architecture. Multi-epitope antigens allow the use of dual antibodies: one bound to the solid phase (fiber) and one fluorescently labeled. Signal is directly proportional to concentration.

Table 2 ^{125}I[Anti ferritin-Ferritin] Complex Binding to Optical Fibers

Ferritin (ng/ml)	Reaction Conc. (ng/ml)	Initial Molar Ferritin Conc. ($\times 10^{-11}$)	Moles ^{125}I Ab Bound[a] ($\times 10^{-16}$)	Effective Molar Conc.[b] ($\times 10^{-6}$)	Apparent Fold Conc.	Total % Fiber Capacity Utilized[c]	Total Volume Serum Completely Extracted of Ferritin (μl)
6	3	0.67	0.56	2.37	3537	0.09	4.2
30	15	3.30	3.30	13.94	4224	0.54	5.0
64	32	7.10	8.50	36.52	5143	1.40	6.0
250	125	28.0	28.10	118.70	4239	4.70	5.0
640	320	71.0	49.30	208.30	2933	8.10	3.5
1100	550	120.0	58.60	248.0	2067	9.70	2.4

[a] Assay is for 10 min total reaction time.
[b] This assumed that the length of a "sandwich" of Ab-ferritin-tagged Ab is 30 nm. The differential volume of the fiber + "sandwich" cylinder and the fiber alone would be 2.36 $\times 10^{-9}$ l. This would be the volume into which all bound Ag would be concentrated during the reaction with the fiber.
[c] Based on an equilibrium binding capacity of 6.02 $\times 10^{-14}$ moles.
Source: From Ref. [18], with permission.

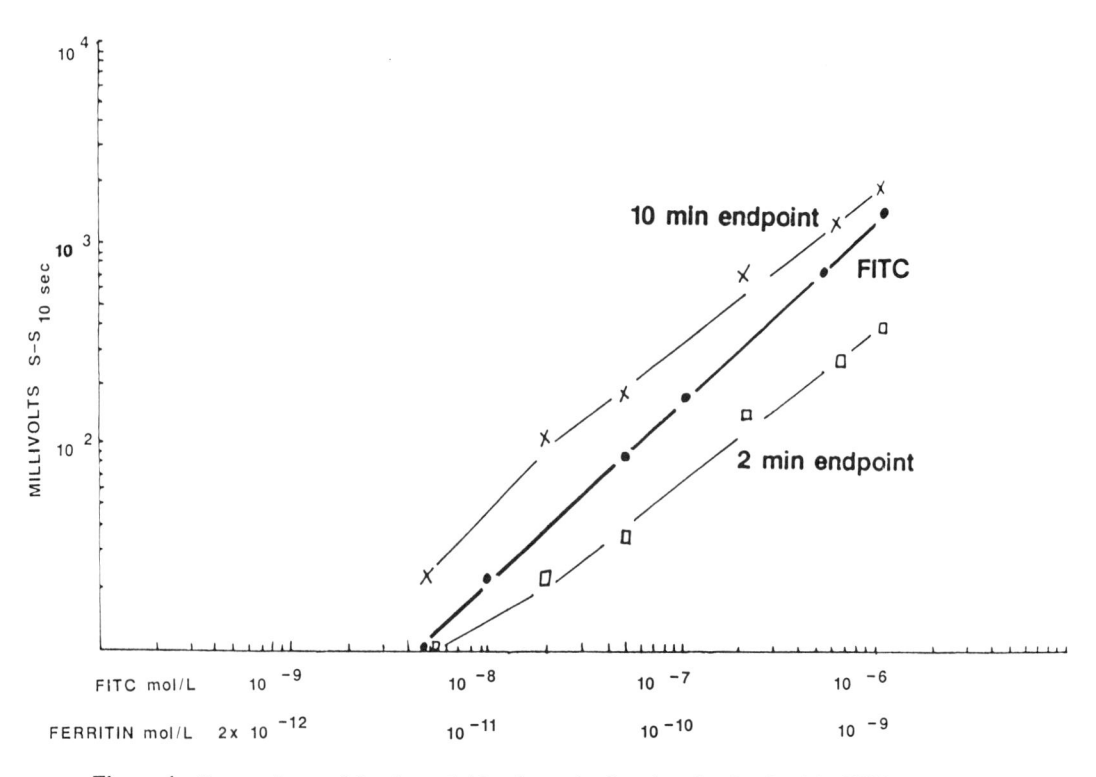

Figure 4 Comparison of 2 min and 10 min endpoint signals of a ferritin EWS to signals generated by free FITC dye assessed on a clean fiber surface without coupled antibody. For each ferritin concentration measured, a separate antibody-coated fiber was used. Reactions were run as described in text. Ferritin concentrations shown are those found in neat serum samples, which are diluted 1:2 with fluorescent antibody and buffer prior to testing (modified from ref. [18], with permission).

proportional to analyte concentration, and to competitive formats for small molecules, where signal is inversely proportional to analyte concentration and fluorescent small molecule conjugates replace fluorescently labeled antibodies. The comparative configuration of immunometric and competitive assays in EWS format is shown in Figure 3.

Recent studies in our laboratory fully developed the concept of the fiber optic as a solid-phase concentrator of mass [18]. In these studies, the high-molecular-weight iron storage protein ferritin (Mr 450,000) was used as a model to study the ability of EWS to measure a large protein molecule with a broad dynamic clinical range (5-1000 ng/mL, 1.2×10^{-11}-2.2×10^{-9} mol/L). In these studies, extensive isotopic modeling using ^{125}I-labeled monoclonal antiferritin Ab was used to assess the fiber as a solid phase support. The isotopic probe was then substituted with a fluorescein isothiocyanate probe (FITC-labeled antiferritin Ab) and studies repeated in a nonseparation EWS mode. Results (Table 2) show the marked mass concentrating capability of the solid phase. After a 10 min reaction, analyte was concentrated up to 2000-5000-fold compared to the initial solution concentration. This was also reflected in the fiber optic mode when signals from solutions of various concentrations of free FITC dye were compared to signals of ferritin concentrations determined at two different reaction time intervals (Fig. 4).

Fiber Optic Immunosensors

After correction for sample dilution and the fact that compared to free FITC, antibody coupled dye loses 50% of its quantum efficiency [18], extrapolation of the 10 min signal to the FITC dye curve shows signal concentration greater than 3 orders of magnitude when signal concentrations are compared to the actual analyte concentration detected.

Results in the following section will present some of our recent improvements in the EWS system designed to develop a durable, easy-to-use analytical system.

III. MATERIALS AND METHODS

A. Antibody Coating of Optical Fibers

Complete details of the preparation of optical quality quartz rod substrates, methods of coating antibody on the lateral surfaces, preparation of fluorescent antibody tag probes, and optical instrumentation have been described previously [15,18,19].

Optical quality quartz rods with diameters of 500, 1000, or 2000 μm were redrawn from General Electric Type 214 quartz (General Electric, Cleveland, OH) and polished to an optical quality surface on both ends. After being cleaned, fibers were dried and immersed in a solution of organic silane to modify the quartz surface prior to protein coupling.

Antibody coating solutions were prepared by diluting partially purified rabbit immunoglobulin with antibodies directed against either human spleen ferritin or the small molecule cardiac glycoside-digoxin to a final protein concentration of 50 μg/ml in phosphate-buffered saline (PBS), pH 7.4. For the immunometric system (ferritin), fibers were placed in a small tray and covered with the solution of antibody for several hours at room temperature. Following incubation, antibody solution was removed and fibers were washed with PBS, and PBS containing 1.5M NaCl to remove loosely adsorbed protein. Fibers were stored in bulk in 50 ml conical tubes containing 0.1% (w/v) bovine serum albumin (BSA)-PBS with 0.2% sodium azide as preservative. Immunosensors were stored in the refrigerator until use.

Preparations of fibers for the small-molecule competitive assay format (digoxin) were done in an identical fashion except that antidigoxin antibody was mixed with various amounts of an equal protein concentration of nonspecific rabbit immunoglobulin partially purified by identical techniques. This titering of antibody loaded onto the fiber surface was necessary to control the number of antigen-binding sites on the fiber. Antibody titration was optimized by evaluating Ab-coated fibers in a radioimmunoassay (RIA) procedure in which [^{125}I] digoxin (Ciba-Corning Diagnostics, Medfield, MA) and dilutions of unlabeled digoxin in human serum were allowed to compete with each other for Ag binding sites on Ab immobilized on the fiber. Optimum ratios of rabbit antidigoxin and nonspecific immunoglobulin used in coating solutions were determined based on this isotopic evaluation and on maximizing the displacement slope of a standard curve between 0 and 5 ng/ml of unlabeled digoxin.

B. Fluorescent Probes

Fluorescent probes were prepared using fluorescein conjugates of either monoclonal antiferritin antibody (Hybritech, La Jolla, CA) or fluoresceinated digoxin. Fluorescent antibody was prepared by conventional techniques using FITC (Isomer I, Sigma Chemical, St. Louis, MO) as described previously [18]. Following fluoresceination, FITC-Ab was assessed for immunological activity via a solid-phase immunometric titration assay utiliz-

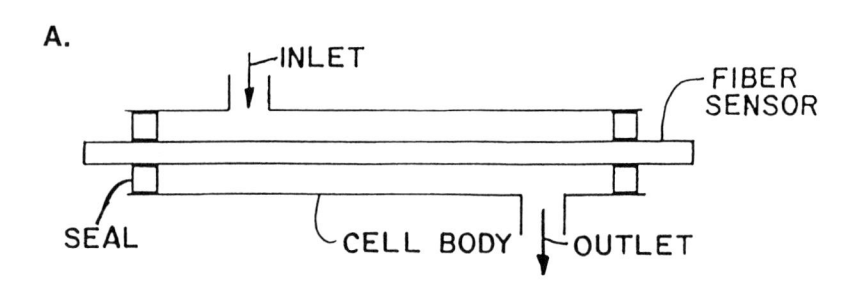

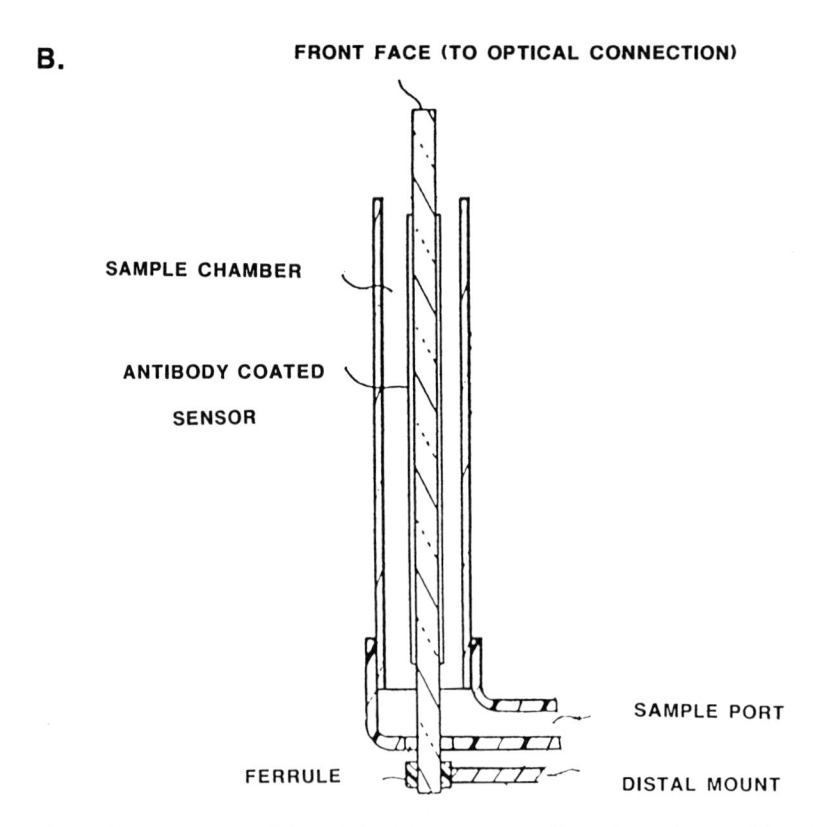

Figure 5 Sensor assemblies. (A) Thin contacting flow-through assembly. In this configuration, the sensor is held by two thin membrane caps that only slightly contact the fiber. Since fluorescent signal output is proportional to the eighth power of the system numerical aperture (NA), any interference or stripping of light by the proximal cap diminishes the sensitivity of the system. (B) Distal holding assembly. This modification eliminates any possibility of signal perturbation by the proximal seal, allowing the full system NA (serum/quartz refractive indices) to be utilized. The distal seal is increased in size markedly to secure the sensor rigidly in its assembly. Since the seal is beyond the area of antibody coating, it does not restrict transmission of immunological fluorescent signal back up the fiber to the proximal surface (modified from refs. [20 and 21]).

Fiber Optic Immunosensors

ing an antigen- (ferritin) specific support. Titration curves of fluoresceinated and parent monoclonal Ab were compared to each other and the FITC-Ab adjusted to a final concentration of 1000 ng/ml equivalent of unconjugated monoclonal antibody. The fluorescein content of this protein concentration as estimated in a conventional fluorimeter was approximately 8×10^{-9} mol/L when compared to a calibration curve of various concentrations of free fluorescein dye. This concentration of dye was at the low end of detection in the EWS fluorimeter.

Fluorescein-digoxin conjugate was purchased from Colony, Inc. (Richardson, TX). Fluorescent small-molecule probes for a variety of drugs are readily available since they are the basis of signal detection in commercially available fluorescence polarization immunoassay systems. Fluorescein-digoxin was adjusted for digoxin content via RIA analysis (MAGIC, Ciba Corning Diagnostics, Medfield, MA) and adjusted to a final working concentration of 2.5×10^{-9} mol/L. Analysis of fluorescein and immunoreactive digoxin content showed a fluorescein/digoxin ratio of 1:1.

C. Sensor Assemblies

In previous work in our laboratory, fiber/sample chamber assemblies were prepared by inserting the fiber into a tube (4 mm I.D. × 50 mm) with input and output ports positioned at opposite ends (Fig. 5A). The fiber was held in place with two specially fabricated thin silicone seals through which the fiber ends protruded. The proximal end of the fiber was secured in optical alignment to the input excitation beam while light emerging from the distal end was allowed to travel free. Only the lateral surfaces of the antibody coated fiber were exposed to solution.

The need for a very thin contacting with the fiber at this proximal seal was demonstrated by the fact that both input excitation and fluorescent emitted light making its way back to the detector was stripped out or absorbed by this seal if its contact with the fiber was too thick or refractive index greater than that of the fiber core material. In either case, the net effect is to reduce immunological signal detection. These effects and the optical requirements of this seal have been described [20]. The distal seal, on the other hand, since it is beyond the area of actual solution/surface immunochemistry reaction, can be of much greater thickness and durability since it will not interfere with signal detection.

To develop a more rigorous immunosensor/sample chamber assembly, we have redesigned this apparatus using a single, distal hold to secure the sensor in position. The principles of this device are described in a recently issued patent [22] and shown in Figure 5B. The key advantages of such a device are that it securely holds the fiber in position, thus allowing precise and easy optical alignment of the sensor in the optical beam, and that only sample fluid is contacting the sensor at the proximal (light input and output) end of the fiber. Based on optical principles of the importance to the system of numerical aperture (NA) on signal magnitude and the fact that NA is determined by the refractive indices of the fiber and any medium or material that contacts it, this approach maximizes the amount of excitation light that can be input into the sensor and the amount of return fluorescent light that can reach the detector. Data presented here were collected utilizing this latter device.

D. Assay Procedure

To perform a single analysis, an EWS contained within its sample chamber assembly is inserted into an assembly carrier and positioned in alignment with the optical beam. An aliquot of serum or plasma is then added to a premeasured aliquot of fluorescent Ab or hapten and the mix immediately injected via a syringe or pump into the sample chamber surrounding the fiber. Data are collected every 10 sec to monitor reaction rates for a period ranging from 2 to 5 min. Output from the solid-state detector is converted to digital output via an analogue to digital converter (Metrabye, Taunton, MA) and stored directly on a personal computer for data reduction. The initial data point at 10 sec is usually used as a background subtraction to differentiate immunological signal from nonspecific signal due to (1) back reflection from the fiber and surface, (2) endogeneous sample fluorescence or light scatter, and (3) fluorescence background emanating from the added fluorescent probe observed before immunoconcentration from Ag-Ab reaction. The assay is conducted in an identical fashion for both immunometric and competitive assay architectures. Each sensor is used only once and unknown sample values are calculated based on a stored standard curve derived on each batch of sensors prepared using standards of known value.

IV. IMMUNOMETRIC AND COMPETITIVE EVANESCENT WAVE IMMUNOASSAY RESULTS

A. Ferritin Immunometric Assay

1. Reaction Rate Monitoring

Using the distal hold configuration described in the previous section, concentrations of ferritin from 0 to 1500 ng/ml were assessed for real-time signal response due to fiber immobilized Ab-ferritin-FITC-Ab complex formation. As can be seen in Figures 6 and 7, reaction rates and signal magnitude monitored between 10 sec and 5 min (300s) were proportional to analyte concentration. Total serum sample volume used in this evaluation was 50 μl. By keeping the tag concentration relatively high, pseudo-first order rate kinetics could be maintained even at high analyte concentration for at least 3 min. Beyond 3 min the extraction of labeled Ab by the fiber results in a reduction of the solution concentration of tagged Ab to the point where it is no longer in great excess. As expected, this results in loss of a linear response first observed at high analyte concentrations.

2. Standard Calibration Curve: Rate Versus Endpoint

Since all immunoassays are based on referencing signals from unknown samples compared to a calibration curve developed utilizing samples at known concentrations, standard curves were constructed in two ways. In the first, timed points were taken from the reaction curves in Figure 6 at each concentration analyzed and plots were constructed of analyte concentration versus accumulated signal (log-log scales). Reaction times analyzed were 1, 2, and 5 min. Typical standard calibration curves at these timed intervals are shown in Figure 8. As expected, as the reaction is allowed to proceed for longer intervals, the total signal accumulated at each concentration increases. At the very high concentrations, there is curvature and ultimately plateauing of the signal typical of most immunometric systems regardless of the labeled probe used.

Fiber Optic Immunosensors

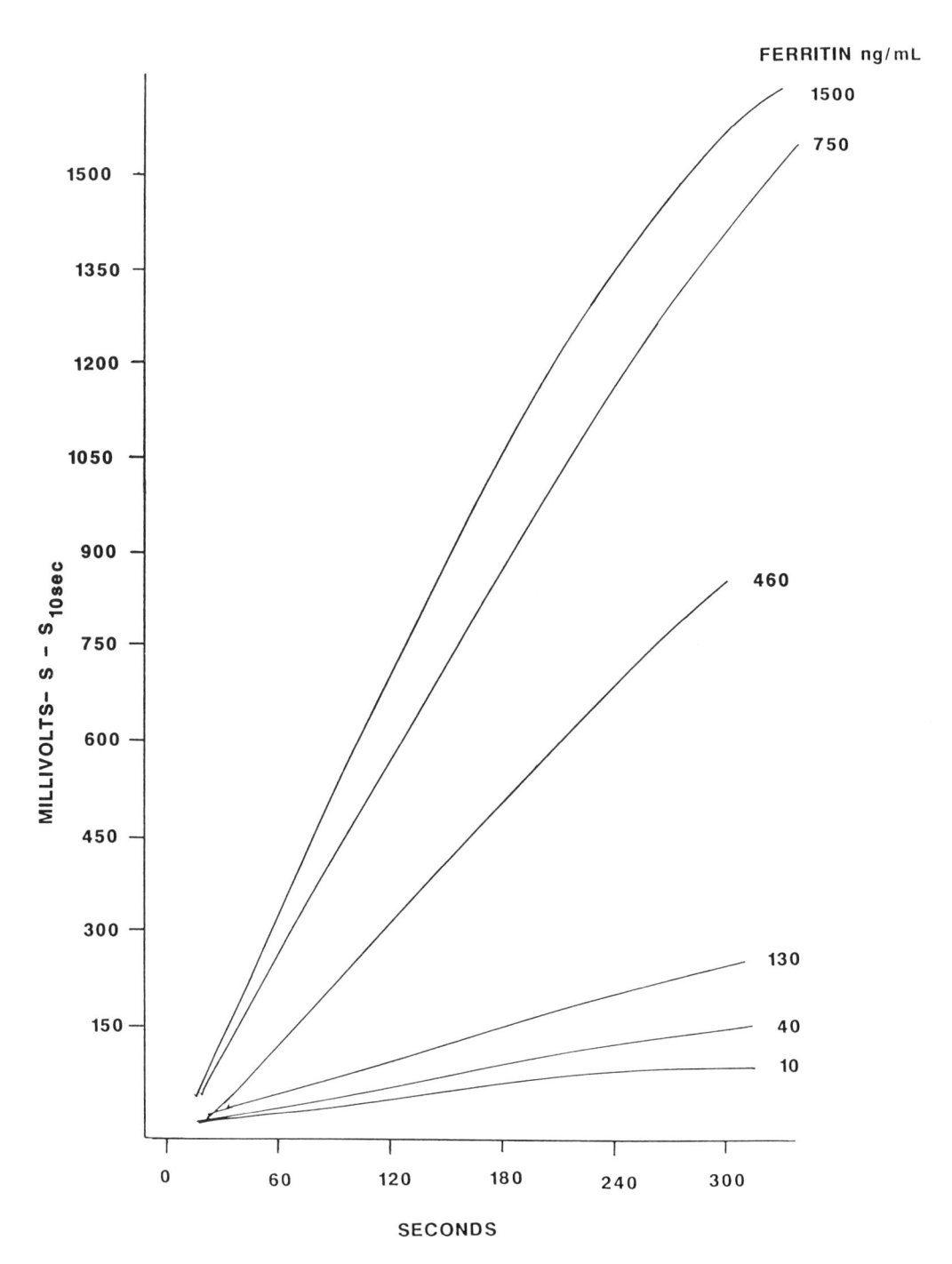

Figure 6 Ferritin EWS reaction curves. Each line represents a separate sensor. Serum (50 µl) was added to 50 µl of FITC-labeled antiferritin Ab and injected into the sensor chamber. After the initial 10 sec of reaction, measurements were taken at 10 sec intervals to monitor rates of immunocomplex formation. Concentrations from 10 to 1500 ng/ml were tested.

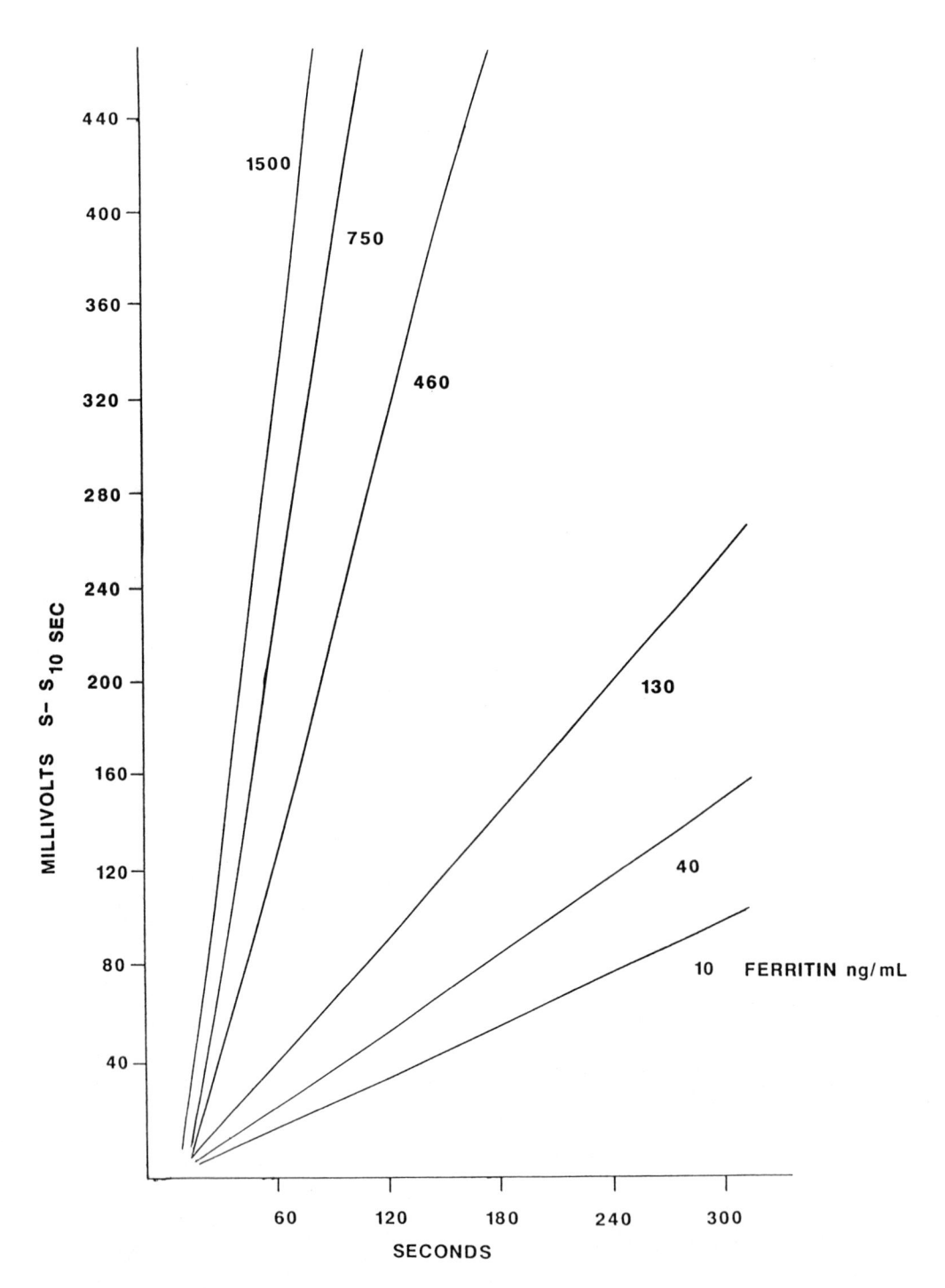

Figure 7 See Figure 6; an expanded Y axis here highlights the linear rate responses observed at the lower ferritin concentrations.

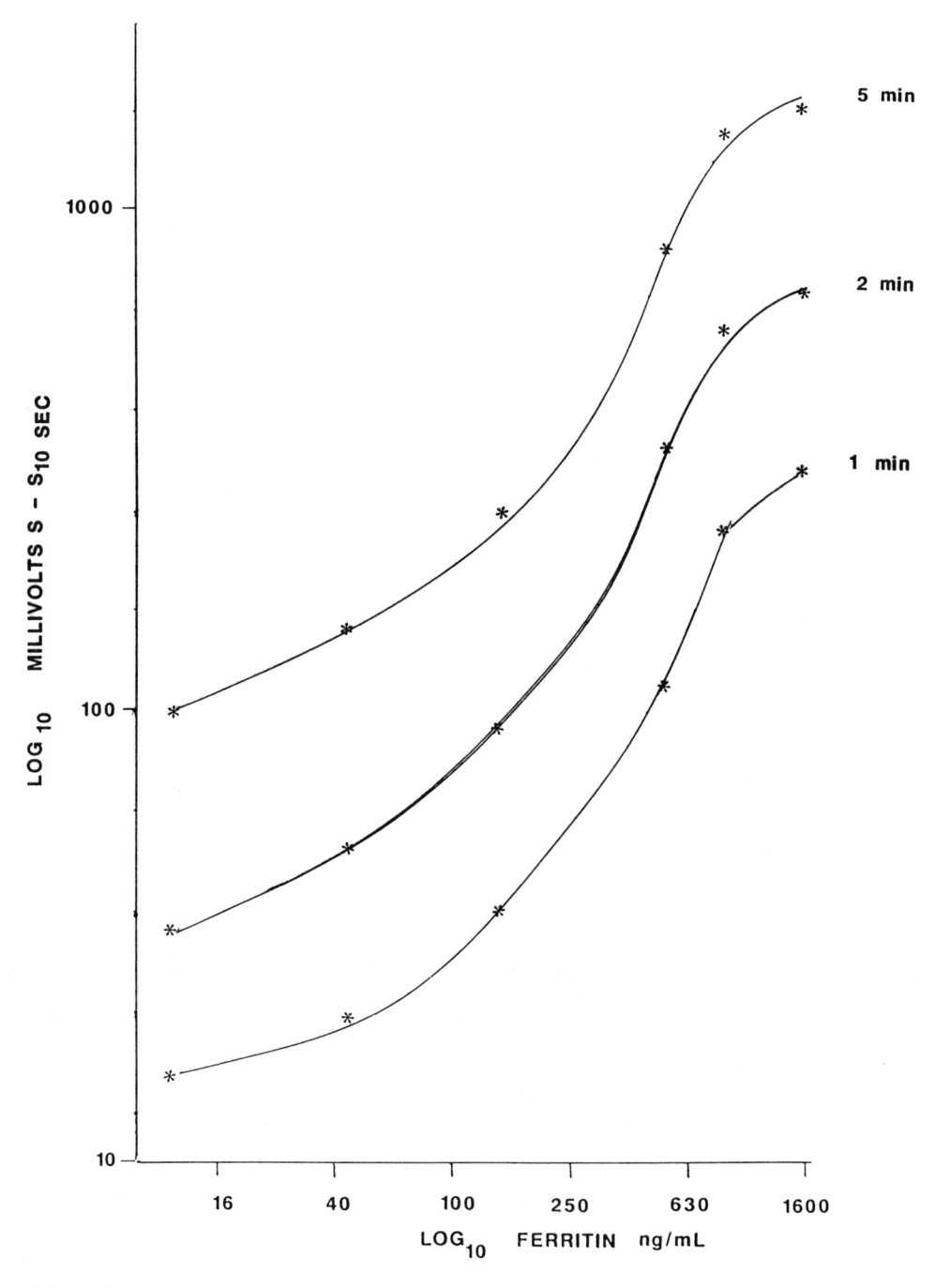

Figure 8 Ferritin EWS standard curves based on endpoints at 1, 2, and 5 min. Saturation of the system begins to occur between 750 and 1500 ng/ml. When the FITC-Ab concentration is kept high, high-dose hook effects were avoided.

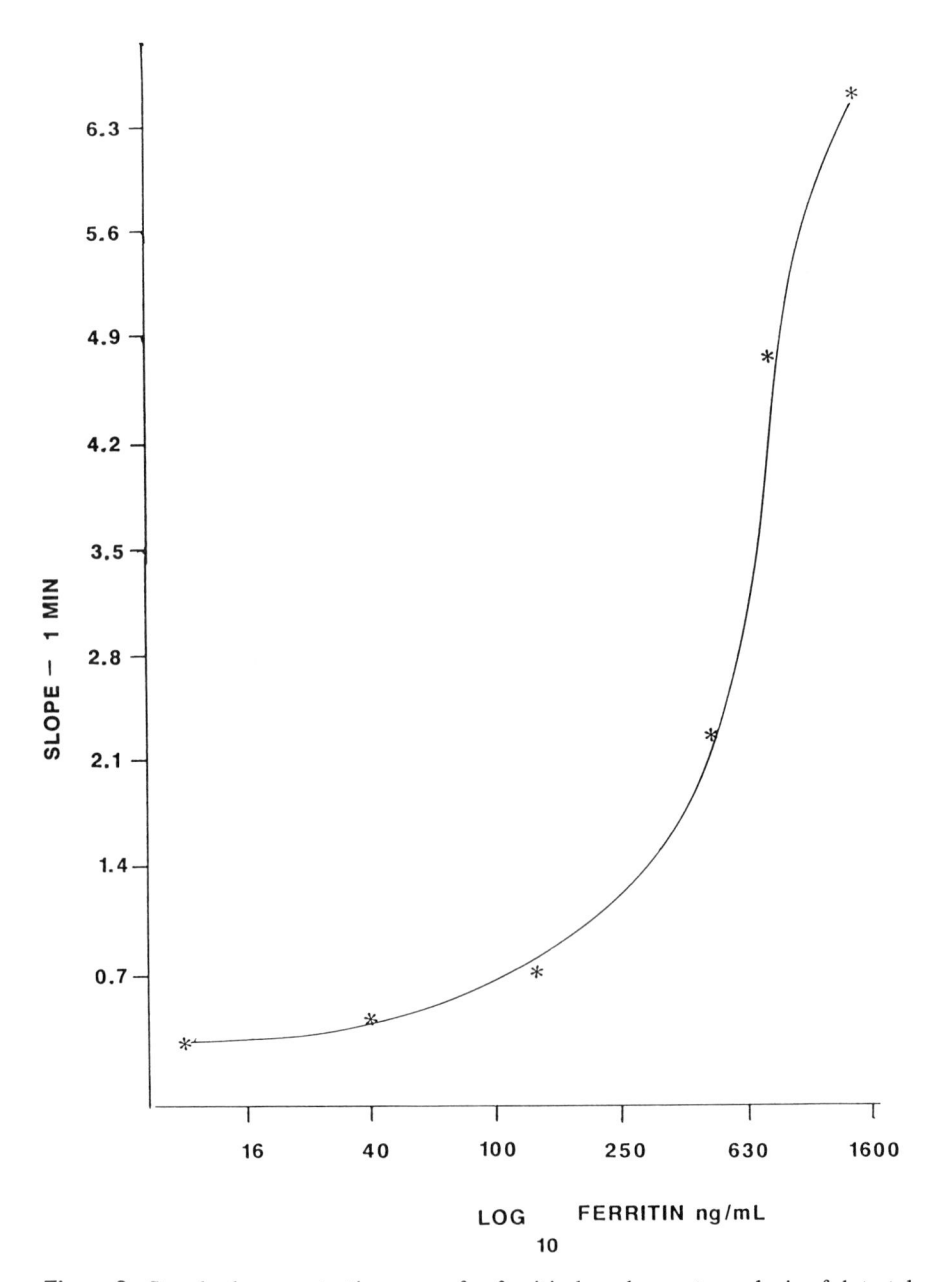

Figure 9 Standard concentration curve for ferritin based on rate analysis of data taken over the first minute of reaction. Data for slope determination were obtained from reaction rates measured between 10 sec and 1 min and were calculated by least-squares regression analysis.

A second and potentially simpler and faster method of determining sample concentration is based on construction of a calibration curve after analysis of reaction rates while the system is maintained in a first order condition. Analysis of the signal response slope of the linear regression line under such conditions for each concentration of analyte is shown in Figure 9.

Rate analysis showed that under the reaction conditions used, a first order reaction was maintained for up to 3 min at the highest ferritin concentration and that computation of the slope at each concentration displayed a uniform rate (Δ mV/min) in as little as 60 sec. Analysis of the kinetics of the primary Ag-Ab interaction not only frees the system from precise timing constraints, since rates can be monitored in any selected time interval once the convective forces due to sample/tag injection have dissipated, it also permits more data points to be used in the calculated response parameter at each concentration. Either timed endpoint or rate analysis may be used as an extremely rapid method of evaluation.

3. Precision

To develop an EWS immunoassay system, it is imperative that each sensor respond in an identical fashion since sensors are "unit use" in nature and can only perform a single analyte measurement and, unlike potentiometric sensors, which calculate electrolyte concentrations based on slope analysis via the Nernst equation, EWS require comparison of the unknown to calibration curves constructed with known analyte concentrations.

To achieve overall system precision, the EWS system requires control of a large number of variables in the following categories: fiber substrate material, solid phase chemistry, and instrument and sensor assembly. These variables have been described in detail [18]. Repetitive analysis at a single analyte concentration (250 ng/ml ferritin) is shown in Figure 10. Calculation of the coefficients of variation (CV) throughout the 5 min reaction show a uniformity in the less than 6% range in as little as 20 sec after reaction initiation. Greater imprecision in the very early time intervals (up to 11% CV) is most likely due to residual convective and turbulent effects as a result of injection of the sample/reagent mix into the chamber surrounding the fiber. These values are well within the expected CV range of today's conventional immunoassay procedures (2-12% CV).

B. Small-Molecule Competitive Immunoassay

Although the most notable application of EWS technology is in the analysis of large-molecule "sandwich" or immunometric assays, which up to now has been constrained by laborious multistep protocols, the EWS can also measure small molecules in a competitive binding architecture. One conducts an assay in identical fashion to the "sandwich" assay except that fluorescently labeled Ab is replaced by a fluorescein-hapten conjugate. Since the tagged species is competing for limited antibody-binding sites on the fiber, signal response curves are inversely proportional to the unknown concentration of the patient's analyte.

A typical set of reaction rate response curves for serum digoxin concentrations ranging from 0.5 to 5 ng/mL (6×10^{-10} to 6×10^{-9} mol/L) is shown in Figure 11. These values are the range assessed for this drug to differentiate between therapeutic and toxic levels. As expected, signal magnitude is inversely proportional to analyte concentration and the reaction appears overall to be second order in nature. Differentiation between signals at all concentrations is discernible in as little as 30 sec. However, it is probably preferable

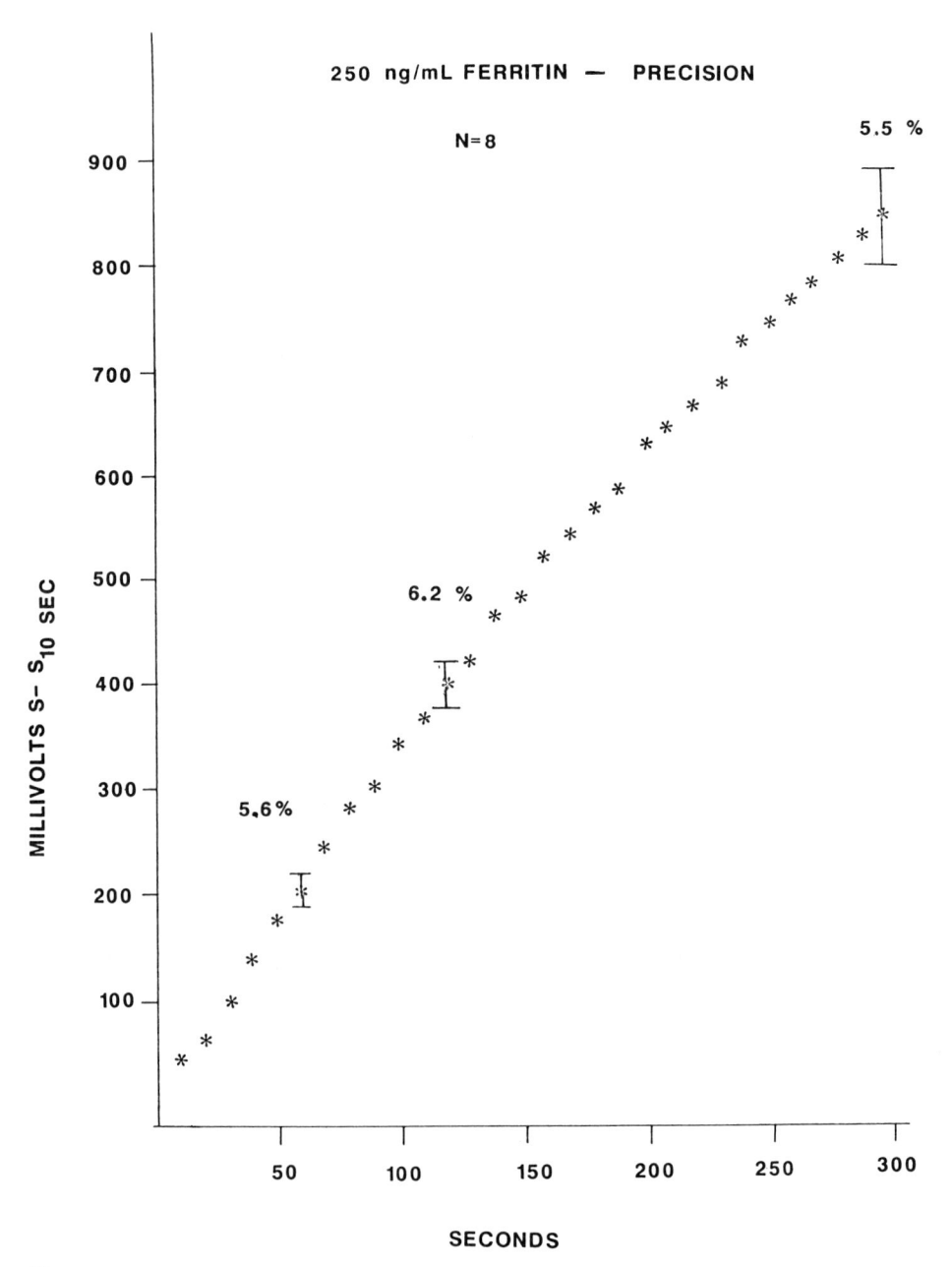

Figure 10 Multiple ferritin sensors were individually tested at a single concentration of ferritin (250 ng/ml; 5.5×10^{-10} mol/L) to evaluate the system's precision. Sensors and FITC-Ab were from the same batch preparations. Each sensor was measured in the same reusable distal hold assembly and assayed as described in text. Asterisks are the mean value of eight repetitions (+/− SD). Percentage values are coefficients of variation for the signal values (in millivolts) taken at 60, 120, and 300 sec.

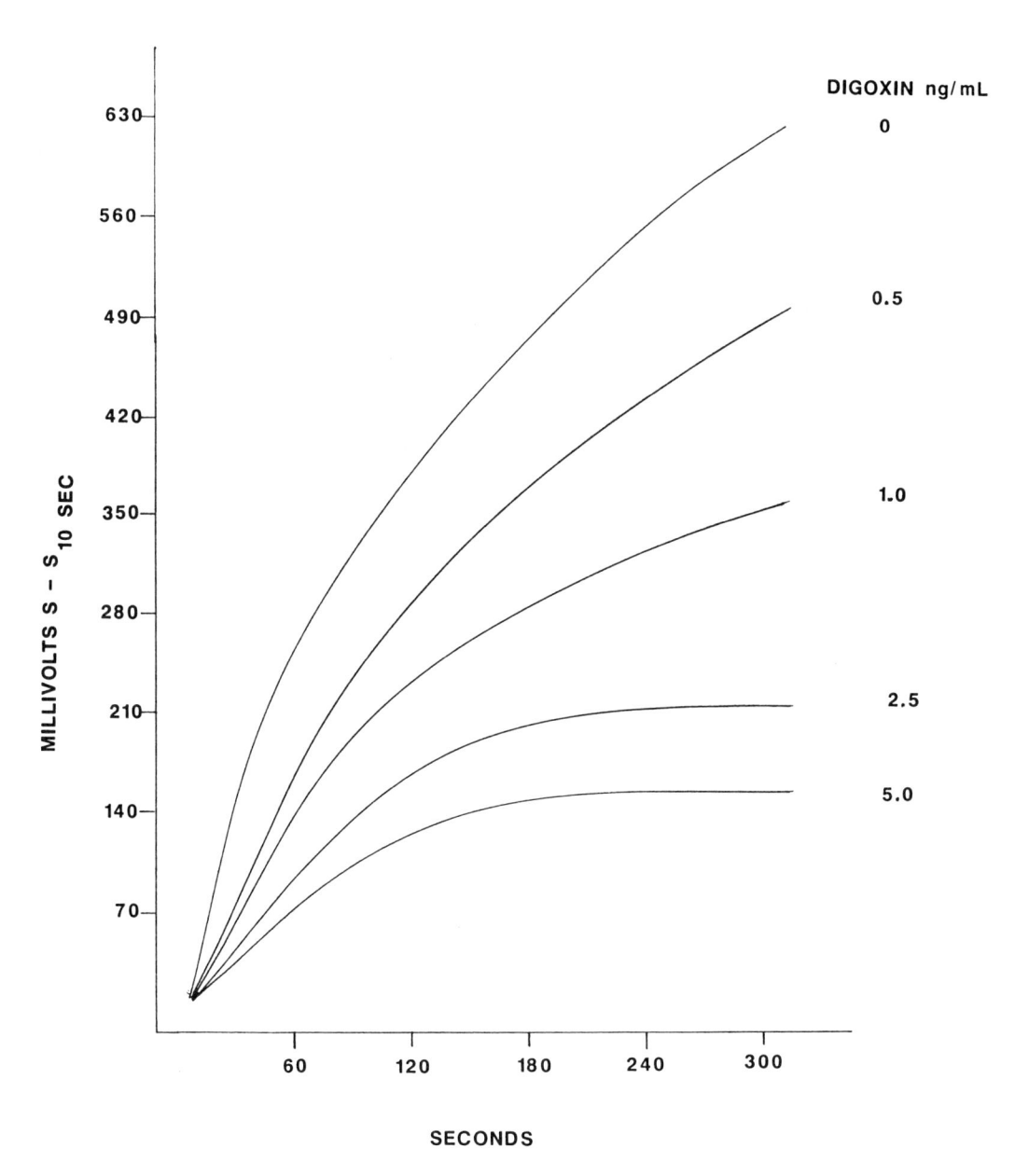

Figure 11 Digoxin competitive immunoassay EWS reaction rate curves at different concentrations in the clinically dynamic range. Results are displayed as millivolt signal minus signal at 10 sec to correct for fiber background and nonspecific signal.

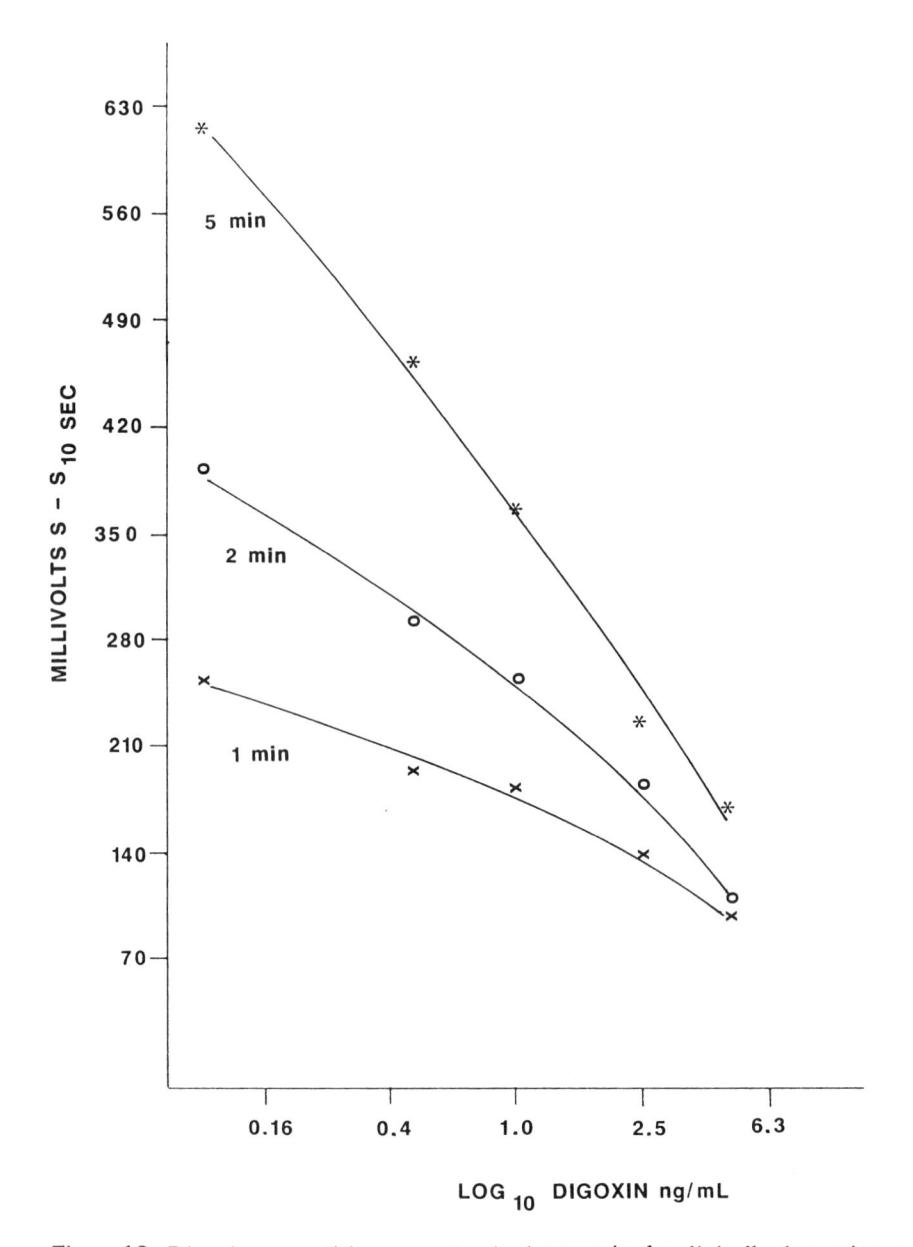

Figure 12 Digoxin competitive assay standard curves in the clinically dynamic range. Standard curves using the endpoint method are based on total immunological signal accumulated at 1, 2, and 5 min of reaction immediately after sample injection.

Fiber Optic Immunosensors

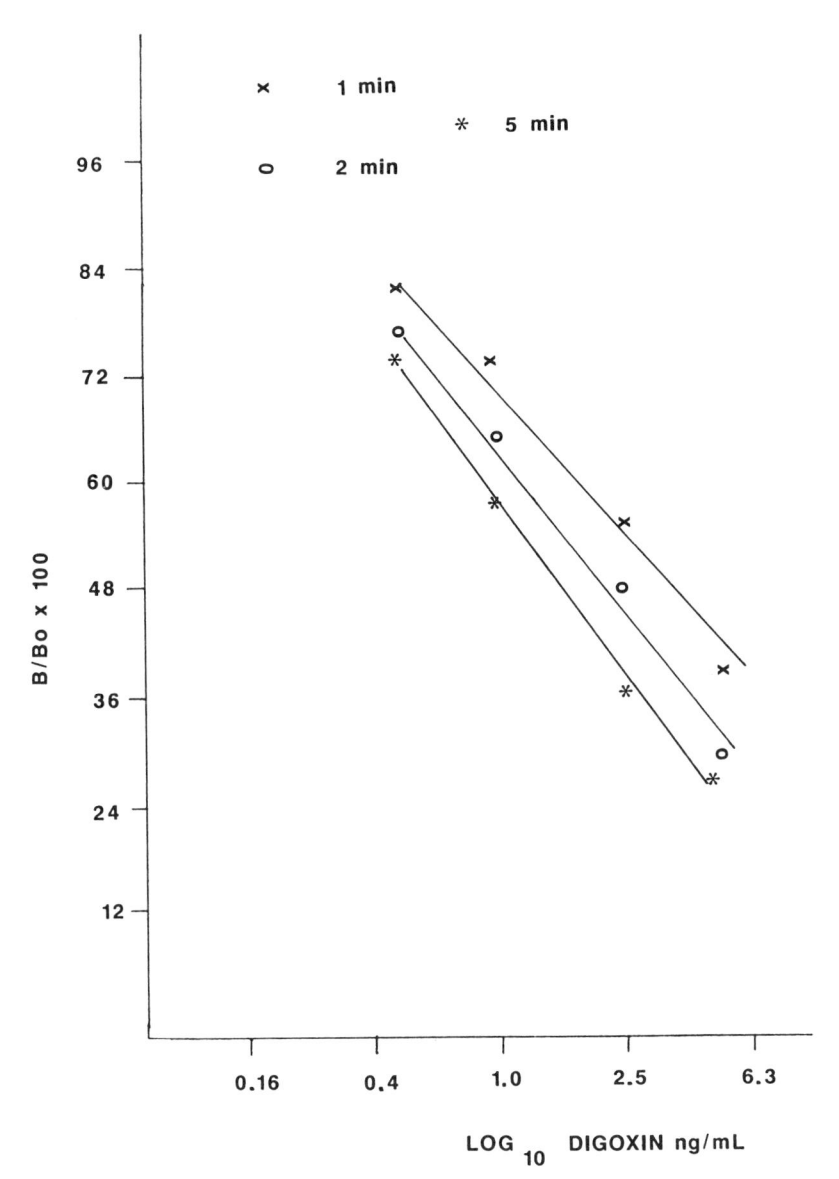

Figure 13 Data as in Figure 12 plotted as the ratio of signal at a given analyte concentration divided by signal at zero (B/Bo). Curve normalization by this technique shows that the curves at 2 and 5 min are nearly identical.

to continue to accumulate signal for slightly longer periods of time to maximize the differential slope of the standard curve and thus increase the point to point resolution resulting in both improved sensitivity and precision.

A series of endpoint standard curves for 1, 2, and 5 min timepoints are shown in Figures 12 and 13. As expected, the steepness of the slope increases with increased reaction time. When displayed as absolute signal voltage differentials (Fig. 12), a 5 min reaction time appears to give the largest differential between 0 and 0.5 ng/ml (an indicator of system sensitivity). However, when plotted as the ratio of signal analyte concentration to signal zero concentration (B/Bo), which along with the logit transformation is the preferred format for developing calibration curves for competitive immunoassay, these ratios show that by 2 min, 80-90% of the values found at 5 min have been achieved.

As with the EWS for large molecules, the digoxin sensor is extremely rapid in its output response and requires no user or system mechanics other than inserting the sensor/cartridge and injecting the sample/reagent mix.

V. CONCLUSIONS

Based on the findings presented in this chapter, the promise of a single, rapid, simple-to-use immunoassay capable of measuring large and small analytes is close to being realized. Because no separations are required before the final analyte determination, instrumentation need not be burdened with wash buffer reservoirs, vacuum pumps, or switching valves and can thus be made very compact. Given the nature of the technology, this opens up a wide realm of possibilities for the performance of quantitative immunoassay beyond the centralized specialty laboratory by personnel who need only minimal training in good laboratory practice. Such alternative sites may include satellite labs, chemotherapy units, hospital wards, doctors' offices, or even ambulances.

With the establishment of monoclonal antibody techniques and the consequent ability to develop highly specific, reagent-grade recognition molecules, immunodiagnostics have branched into medical specialty areas beyond their classic use in the evaluation of endorinological, and infectious diseases and anemia. Areas of expansion include monitoring of drugs (both therapeutic and illicit), cancer diagnosis and prognosis, myocardial infarction, and evaluation of thrombotic and fibrinolytic status. In many of these areas, since disorders result in acute, life-threatening situations and since new interventional therapies are now established in the fields or are on the immediate horizon, rapid methods of diagnosis and monitoring of potentially very toxic drugs are required.

Most immunoassay systems have not been able to be performed "stat" due to the centralization of the technology, which requires transport times and the extensive time required to conduct the assay. The EWS technology represents one way of bringing the technology to the patient and attending physician directly, resulting in a rapid turnaround time from sampling to answer. Such an approach enhances the quality and certainty of medical decision making and will result in more cost-effective delivery of medical services.

ACKNOWLEDGMENTS

We would like to thank the following people, whose efforts have made the development of fiber optic evanescent wave immunosensors possible: Ms. Irene Walczak, Ms. Cynthia Urciuoli, and Ms. Linda Stundtner for their work in the development of antibody coating technology; Dr. Walt Love and Mr. Tom Cook for efforts in the field of EWS optical theory; Dr. Rudolf Slovacek, Ms. Mary Craig, and Mr. Richard Schulkind for instrument development; Mr. Robert Green and Mr. Ray Souza for sensor assembly design and fabrication; and Mr. Herb Schulkind and Mr. Bill Stewart for software support. The development of a complete immunoassay system requires the skills of a number of professionals from diverse disciplines. We would like to thank Dr. Richard Falb, Vice-President, Research and Development, and Mr. John MacPhee, Manager, Sensor R & D, Ciba-Corning Diagnostics Corporation for their continued support in supplying timely resources to this project.

REFERENCES

1. Avrameas S. Heterogeneous enzyme immunoassays. In: Voller A, Bartlett A, Bidwell D, eds. Immunoassays for the eighties. Baltimore: University Park Press, 1981: 85.
2. North JR. Immunosensors: antibody based biosensors. Trends Biotechnol 1985;3: 180.
3. Dandliker WB, Kelly RJ, Dandliker J, Farquhar J, Levin J. Fluorescence polarization immunoassay, theory and experimental method. Immunochemistry 1973;7: 219.
4. Rupchock P, Sommer R, Greenquist A, Tybach R, Walter B, Zipp A. Dry reagent strips used for determination of theophylline in serum. Clin Chem 1985;31:737.
5. Sepaniak MJ. Fiberoptic sensors for bioanalytical measurements. Clin Chem 1986; 32:1041.
6. Seitz WR. Chemical sensors based on fiber optics. Anal Chem 1984;56:16A.
7. Angel SM. Optrodes: chemically selective fiber optic sensors. Spectroscopy 1987; 2:38.
8. Harrick NH. Internal reflection spectroscopy. New York: Interscience, 1967.
9. Axelrod D, Burghardt TP, Thompson NL. Total internal reflection fluorescence. Annu Rev Biophys Bioeng 1984;13:247.
10. Hirschfeld TB. Total reflection fluorescence. Can J Spectrosc 1965;10:128.
11. Wiechelman KJ. Empirical correction equation for the fluorescence inner filter effect. Am Lab 1986;18:49.
12. Hirschfeld TB, Block MJ. Fluorescent immunoassay employing optical fiber in capillary tube. U.S. Patent 4,447,546, 1984.
13. Block MJ, Hirschfeld TB. Assay apparatus and method. U.S. Patent 4,558,014, 1984.
14. Sutherland RM, Dahne C, Place JF, Ringrose AR. Optical detection of antibody-antigen reactions at a glass liquid interface. Clin Chem 1984;30:1533.
15. Love WF, and Slovacek RS. Optical characteristics of fiber optic evanescent wave sensors. In: Wingard, L. and Wise, DL, eds. Biosensors with Fiber Optics. Clifton, NJ: Humana Press, In press.
16. Sutherland RM, Dahne C, Place JF, Ringrose AR. Immunoassays at a quartz-liquid interface: theory, instrumentation and preliminary application to the fluorescent immunoassay of human immunoglobulin G. J Immunol Method 1984;74:253.
17. Badley RA, Drake RAL, Shanks IA, Smith, AM, Stephenson PR. Optical biosensors for immunoassays: the fluorescence capillary fill device. Phil Trans R Soc London 1987;316(1176):143.
18. Bluestein BI, Craig M, Slovacek R, Stundtner L, Urciuoli C, Walczak I, Luderer A. Fiber optic evanescent wave sensors for clinical diagnosis. In: Wingard, L and Wise DL, eds. Biosensors with Fiber Optics. Clifton, NJ: Humana Press, In press.

19. Lackie S, Glass TR, Block MJ. Theory and instrumentation for a cylindrical waveguide evanescent fluorosensor. In: Wingard, L and Wise, DL, eds. Biosensors with Fiber Optics. Clifton, NJ: Humana Press, In Press.
20. Cook TA, Slovacek RE, Love WF, Evanescent wave sensors. U.S. Patent 4,852,967, 1989.
21. Cook TA. Immunoassay apparatus. U.S. Patent 4,671,938, 1987.

12
Recent Advances in Flow Cytometric Techniques for Cancer Detection and Prognosis

J. PHILIP McCOY, Jr.

Pittsburgh Cancer Institute, Pittsburgh, Pennsylvania

JOHN L. CAREY

Henry Ford Hosptial, Detroit, Michigan

I. INTRODUCTION

There has been a dramatic increase over the past 10 years in the application of flow cytometry for the evaluation of neoplastic processes. While many of these techniques are still considered experimental, others have become integral components of the diagnosis or management of specific neoplasms. In certain instances, flow cytometry has led to a more precise delineation of the clonality or lineage of neoplastic lesions, resulting in new classifications or groupings. It should be emphasized, however, that flow cytometric assays alone are insufficient to establish proper diagnoses: these assays must be evaluated together with histomorphological examination of the specimen and a clinical evaluation of the patients to arrive at an appropriate diagnosis.

This chapter will focus on immunological techniques in flow cytometry that may be utilized for cancer diagnosis. It should be noted that certain nonimmunological techniques are also used in flow cytometry, particularly DNA content and cell cycle analyses. There is little doubt concerning the utility of these latter measurements in aiding in the diagnosis and determining the prognosis of various tumors. To focus this text better on immunodiagnostic assays, flow cytometric analyses of DNA content will not be addressed except in the content of multiple parameter analyses. The reader is referred to several excellent reviews for a comprehensive description of DNA content and cell cycle analytical methods and applications [1-3].

A recurrent theme in this chapter will be the detection of multiple features on individual cells by flow cytometry. It has become increasingly apparent that no single marker or parameter alone is definitive in diagnosing or predicting tumor behavior, but specificity may be improved by using combinations of markers. Flow cytometry is ideally suited to this task: individual cells may be examined extremely rapidly for six parameters (FLS, SSC, and four fluorescence detectors) or more on commercially available instrumentation. One is actually performing multiple parameter analyses when performing a one-color analysis using forward- and side-scatter patterns to set "gates," although multiple parameter analyses are generally regarded as involving several parameters other than light scatter.

We will discuss the technical aspects of flow cytometers (i.e., their design and principles of operation) only briefly as required in describing various assays. For a comprehensive review of this subject the reader is referred to a thorough and very readable text by Shapiro [4].

II. TUMOR PHENOTYPING

A. Hematological Neoplasms

The advent of immunophenotypic analysis has allowed new insights into both malignant and normal lymphohematopoiesis. These techniques have also proven to be valuable adjuvant tools in the diagnosis of leukemia/lymphoma, and in defining nonmorphological subsets of these neoplasms that have independent prognostic and therapeutic significance. The combined morphological and immunological approach to the diagnosis of lymphohematopoietic neoplasia centers on three primary questions: (1) Is the cellular proliferation of leukocyte origin? (2) Is the leukocyte proliferation malignant? (3) What is the lineage and stage of maturation of the leukemia/lymphoma?

The ability to answer these questions by flow cytometry has been greatly enhanced by the availability of well defined monoclonal antibodies (MABs). Three major international workshops [5-7] have carefully defined various differentiation- and/or function-associated antigens and the MABs that recognize them. These antigens and their associated MABs are given cluster designations (CD).

The leukocyte common antigen (LCA/CD45) is present on virtually all mature and immature leukocytes [8-12]. However, virtually all erythroid precursors, and perhaps the pluripotential stem cell and the very earliest of committed progenitor cells, are CD45 negative. The CD45 antigen is actually a family of four molecules of 180, 190, 200-205, and 220 kilodaltons [13]. The higher-molecular-weight LCAs (205 and 220; CD45R) are primarily found on B cells, B-cell precursors, natural killer (NK) cells, and a subset of T cells, while the lower-molecular-weight LCAs are present on T cells and myelomonocytic cells and their precursors.

This widespread expression of CD45 has been most useful in the differential diagnosis of large cell malignancies (i.e., lymphoma vs. carcinoma vs. melanoma) [8,9]. In fresh unfixed tissue, virtually all large cell lymphomas are CD45 positive. This is also true of most mature (nuclear TdT-negative) and a significant proportion of immature (nuclear TdT-positive) lymphomas and leukemias. Although CD45 retains its specificity in formalin-fixed, paraffin-embedded tissue, there is some loss of sensitivity, particularly with large cell lymphomas [9]. In the majority of plasma cell neoplasms, most plasma cells are CD45 negative.

If an atypical cellular proliferation is clearly of leukocyte lineage, the question arises as to whether it is neoplastic or reactive. In cases that are morphologically indeterminate, flow cytometric immunoanalysis may be helpful. This usually revolves around the ability to determine directly or indirectly whether this is polyclonal (reactive) or monoclonal (neoplastic) proliferation. Direct evidence would be a clonal rearrangement of the immunoglobulin or T-cell antigen receptor genes, and/or a monoclonal protein product of these genes [14]. Indirect evidence would be a dominant population of cells with an atypical antigen phenotype [14,16]. It must be remembered that a small neoplastic monoclone may not be detectable in a larger population of reactive cells, nor are all monoclonal proliferations necessarily neoplastic. The best way to avoid misinterpretation is the correlate immunophenotypic results carefully with a light microscopic and clinical features of the proliferation.

The definition of B-cell clonality is often determined by the proportional expression of surface or cytoplasmic immunoglobulin light chains (kappa; lambda) on a population of B cells. However, some malignant B-cell proliferations (i.e., large cell lymphomas) may lack detectable immunoglobulin [15,16], although this in itself is highly specific for certain malignant B-cell proliferations [16]. Indirect evidence of B-cell clonality would include coexpression of differentiation/activation antigens on a large proportion of the cells [15,16]. Specific examples include expression of CD5 on B-chronic lymphocytic leukemia/widely diffuse lymphocytic leukemia (CLL/WDLL), CD11c on hairy cell leukemia, and CD10 on nodular lymphomas or lymphoblastic leukemias/lymphomas [15-17].

The definition of T-cell clonality by immunohistochemistry is more difficult. There is no direct analogue of kappa/lambda light chain expression to define the clonality of the T-cell antigen receptor (Ti). The CD4/8 ratio does not reflect clonality, and may range from 0.5 to 10:1 in reactive states. A more useful, albeit indirect, indication of clonality is the atypical expression of differentiation/activation antigens on a large proportion of the cells. Specific examples would include loss of "pan" T-cell antigens (CD2, 3, 5, 7) [16]; (2) lack of Leu8/CD45R expression on most T cells in *Mycosis fungoides* [18]; (3) coexpression of CD4 and CD8 and/or CD1 expression on T lymphoblastic leukemia/lymphoma [19]; and (4) coexpression of CD16 on T gamma lymphoproliferative disease [20].

The stage of maturation of a neoplastic leukocyte proliferation may allow a more objective definition of prognostically and therapeutically significant groups. In some instances, these groups are defined solely by antigen phenotype. In acute leukemias, the delineation of acute myeloblastic from acute lymphoblastic leukemia (AML vs. ALL) is extremely important, both for determining the appropriate therapy and defining long-term prognosis. While the lineage of some acute leukemias may be readily apparent through light microscopic and cytochemical analysis, the lineage of others is not. In such instances, the differential expression with MABs to CD5, 7, 10, 13, 19, and 33, along with the determination of nuclear TdT expression, will allow confident differentiation between AML and ALL in greater than 95% of cases [21-23]. In addition, there are prognostically significant subsets of both AML and ALL that can be defined only by immunoanalysis. The childhood surface immunoglobulin (sIg) negative B-lineage ALLs (in particular, CD10+ cases) have the most favorable prognosis of ALLs [22-25]. In contrast, the B-cell (sIg+/CD10-) ALLs have the lowest survival. The T-lineage (CD7+/TdT) and pre-B (cytoplasmic Ig+/sIg-) ALLs have an intermediate prognosis.

The usefulness of immunoanalysis in defining prognostic subsets in mature (Tdt negative) lymphoid malignancies is controversial. It is uncertain whether diffuse large cell and high-grade lymphomas of T lineage are more aggressive than those of B lineage. Cossman et al. [26] reported no significant difference in mean survival of a group of T and B high-grade non-Hodgkin's lymphomas (NHL)s. However, Spier et al. [27] found a significant decrease in mean free survival of T large cell lymphomas as opposed to B large cell NHL. More large prospective studies will be required to ascertain whether T vs. B lineage is important in this subgroup of NHL.

While this conclusion is controversial, TdT+ AMLs appear to have a worse prognosis than those that are TdT negative [28,29]. Cases of ALL that coexpress a myeloid antigen ("mixed" lineage ALL) have been shown to have a more adverse prognosis in adults [30]. These immunophenotypic factors are not totally independent, and must be combined with the age, sex, white blood cell (WBC), and blast morphology to define most accurately the true prognosis, and hence the quality and quantity of chemotherapy given to the patient.

Myelodysplasias (preleukemic states) are often seen prior to the onset of AML. These may be low grade (refractory anemia with or without ringed sideroblasts) or high grade (refractory anemia with or without ringed sideroblasts) or high grade (refractory anemia with excess blasts and in transition). The percentage of CD13+ and/or CD33+ cells in the myelodysplastic marrows has been reported to correlate with risk of progression to AML, independently of morphological appearance [31]. If these results are confirmed in large studies, flow cytometric immunoanalysis may become a routine screening test for patients with myelodysplasia.

Mention should be made briefly of the analytical techniques involving multiple parameter analysis of hematological neoplasms. One example of how multiple parameter analyses have been exploited to address pertinent clinical questions is the method developed by Braylan and colleagues [32] for concomitant examination of DNA content and surface marker expression. Using staining of surface markers by fluorescein isothiocyanate (FITC)-conjugated antibodies followed by fixation in 50% ethanol and staining with propidium iodide, these investigators were able to assess the ploidy of neoplastic B lymphocytes, while excluding nonneoplastic non-B-cells from analysis [33]. This served not only to enhance the ability to detect aneuploidy in specimens with a low percentage of neoplastic cells, but also to demonstrate definitively the ploidy of cells bearing surface marker characteristics of the blasts. Cell cycle information may also be obtained for cells bearing the phenotype of the leukemic blasts without including the majority of nonneoplastic cells.

B. Solid Tumors

The application of flow cytometry to the detection of surface markers on solid tumors presents more technical obstacles than are encountered with hematological neoplasms. This is due to several factors: the need to dissociate solid tumors into single cell suspensions without severe perturbations of the cell membrane and surface antigens; intratumor heterogeneity in the expression of antigens; the ease and sensitivity of alternative (immunohistochemical) procedures, which allow visualization of tumor morphology as well as staining of the surface and cytoplasmic antigens (thus eliminating much of the motive for performing flow cytometry); and, in general, the lack of highly specific markers that will increase the accuracy of diagnosis or lineage determination, or act as prognostic indicators. Progress in the flow cytometric evaluation of solid tumor surface antigens is nevertheless being made, particularly with the improved dissociation procedures for solid tumor specimens and by the use of exfoliative specimens such as cervical scrapings and bladder washings.

1. Gynecological Neoplasms

A variety of markers have been used in the flow cytometric analysis of gynecological material in an attempt to delineate methods to prescreen patients for possible neoplastic processes or to predict tumor behavior. Sasaki et al. [34] and Suehiro and co-workers [35] have used antibodies to the TA-4 antigens to detect abnormal cells in solid tumor specimens and exfoliated cervical-vaginal cells, respectively. In Suehiro's study, the false-negative rate was determined to be 14.3% while the false-positive rate was 20.6%; the former was due primarily to inflammatory exudate and debris and the latter to cell aggregation. Further technical refinements or the use of this marker in combination with other cellular features may result in a useful prescreening technique using the TA-4 marker.

One approach in the flow cytometric analysis of gynecological specimens has been the use of surface marker/DNA analysis to detect and characterize abnormal cells. As an example, Flint et al. [36] used multiple parameter analysis to examine the expression of a cervical carcinoma antigen (CCA), DNA content, and light scatter of exfoliated cervical-vaginal specimens. By designating a specimen as abnormal if it was either aneuploid or expressed CCA, a high degree of correlation was found between flow cytometric detection of abnormal cells and manual evaluation of cytomorphology. Aurelian [37] has performed similar studies to detect abnormal cervical cells using herpes simplex virus antigen, DNA content, and cell size as screening parameters. Fowler et al. (38) used multiparameter flow cytometric analyses of dissociated ovarian tumors and ascites specimens to study to expression of CA125 and class I and class II MHC antigens on aneuploid tumor populations. While this latter study did not identify a prognostic indicator for ovarian tumors, it indicated the feasibility of this approach for subclassifying aneuploid tumors.

2. Urological Neoplasms

Huffman and colleagues [39] used flow cytometry to study 28 patients' bladder washings stained for the expression of a cytokeratin antigen and DNA content. While not specific for neoplastic tissue, the cytokeratin antibody (AE-1) demonstrated great specificity for uroepithelium, and thus allowed interrogation of the DNA content of bladder epithelium exclusively without interference with normal inflammatory or reactive cells. This resulted in a superior ability to detect aneuploid populations (indicative of a poor prognosis) in bladder irrigation specimens. Similar findings were reported by Fietz et al. [10] in a series of transitional cell carcinomas. Czerniak and Koss [41] examined surgically excised bladder tumors for the expression of the Ca-1 antigen and DNA content by flow cytometry, and found a high degree of correlation between invasiveness and aneuploidy, and invasiveness and high expression of the antigen. The loss of blood group antigens in concert with DNA analysis has been proposed by Orihuela and colleagues [42] and Devonec et al. [43] as a multiple parameter flow cytometric method for assessing the malignant potential of bladder washing specimens. The former group of investigators demonstrated the utility of this approach in monitoring the response of patients with superficial bladder cancer to intravesical bacillus Calmette-Guérin (BCG) therapy. In a study of transitional cell carcinomas, Orntoft and co-workers [44] examined the expression of cell surface carbohydrates in concert with DNA content by flow cytometry. Of particular note is the use of lectin binding for the detection of the cell surface carbohydrates in this study and the correlation between the binding of peanut agglutinin and the presence of tumor invasion.

3. Colorectal Neoplasms

In specimens of colorectal tumors, Durrant and co-workers [45] examined the expression of four antigens (carcinoembryonic antigen [CEA], Y haptenic blood group, 791T/36, and 791T-P72) in addition to DNA content in 50 colorectal tumors by flow cytometry.

Although this study did not identify a significant diagnostic or prognostic indicator, it demonstrated the feasibility of using multiparameter approaches to study tumor behavior and progression. Van den Ingh and co-workers [46] characterized colorectal adenomas with antibodies against CEA and gastric fucomucins and correlated the staining with DNA content. These data suggested that a multiparameter approach may be useful in subclassifying these neoplasms. The intensity of expression of HLA antigens in colorectal neoplasms has also been measured by flow cytometry in an unsuccessful attempt to delineate a prognostic indicator [47].

III. ASSESSMENT OF DRUG RESISTANCE AND TUMOR PROLIFERATION

A. Drug Resistance

The development of resistance to cytoreductive drug therapy in neoplastic cells is a major obstacle in effective cancer treatment [48]. The drug-resistant cell lines may exist de novo or occur after the first round of chemotherapy. Screening the cancer cells before beginning therapy has clear potential, particularly since well-defined probes to antigens associated with drug resistance have become available.

The p170 glycoprotein (GP170) is strongly associated with the presence of multiple drug resistance (MDR) [49-54], that is, the development of cross-resistance to a variety of natural chemotherapeutic agents including *Vinca* alkaloids, epipodophyllotoxins, anthracyclines, and colchicine. While the relationship between MDR and GP170 is clear, the biochemical mechanism of resistance has not been clearly elucidated. In Chinese hamster ovary (CHO) cells and human leukemic lymphoblasts, GP170/MDR expression is associated with decreased uptake and retention of the cytotoxic drug in most instances.

Polyvalent and monoclonal antisera have been produced that recognize cell surface and cytoplasmic domains of the GP170 molecule [49-51]. Utilizing flow cytometric analysis, Epstein et al. [53] demonstrated a direct correlation between GP170 expression in multiple myeloma cells and resistance to vincristine/adriamycin/daunarubicin therapy in a small series of patients. Ma et al. [54] found that, upon relapse, two patients with aggressive acute myelogenous leukemia expressed GP170. The number of positive cells continued to increase during the terminal stages of the disease.

Methotrexate (MTX) is a standard drug in the treatment of hematopoietic and nonhematopoietic neoplasia. The pharmacological effects of MTX stem from the binding of MTX to dihydrofolate reductase (DHFR), resulting in alterations of the intracellular pools of reduced folates. MTX resistance can be mediated by either decreased affinity of DHFR for MTX, or increased amounts of DHFR [55]. Since MTX binds stoichiometrically to DHFR, the intracellular levels of the latter can be determined by the steady-state amount of labeled MTX. Kaufman et al. [55], utilizing FITC-labeled MTX, demonstrated that increased MTX is seen in resistant murine sarcoma cell lines. The latter presumably reflects increased amounts of DHFR. One must be cautious in interpreting low amounts of FITC-MTX as indicating drug sensitivity, since MTX-resistance can be secondary to decreased uptake.

Glutathione (GSH) is an intracellular antioxidant that provides protection against free radicals, oxidants, and electrophiles. Thus, GSH can modulate cellular sensitivity to radiation and/or chemotherapy [36]. Utilizing a fluorescent thiol probe (monochlorobimane; MBCL), Rice et al. [56] reported that variations in fluorescence directly correlated with GSH concentrations, as assessed by high-performance liquid chromatography. In addition, CHO cell lines resistant to adriamycin or MTX had increased amounts of GSH. This may prove to be a useful probe in assessing the presence of GSH-mediated or GSH-associated drug resistance by flow cytometry.

B. Tumor Proliferation

In general, the vulnerability of tumors to cytoreductive chemotherapy or radiotherapy often correlates with the proliferation fraction (PF) of the neoplasm (i.e., G_1, S, G_2, M). In addition, the aggressiveness of untreated tumors is also proportional to the PF. Fluorescent dyes, such as propidium iodide (PI) and DAPI, bind stoichiometrically to DNA, and allow an objective determination the percentage of cells with S ($>2N$, $<4N$) or G_2 + M (4N) DNA content. However, these techniques cannot assess the fraction of cells in G_1, the critical phase when the cell commits itself to a cycle of proliferation. Several nuclear antigens (Ki-67, PCNA, p105) have recently been defined that are required and/or associated with the transition to G_1 [57-62] and also define some or all of the other proliferative phases. In addition, membrane/cytoplasmic antigens associated with leukocyte activation/proliferation (4F2, CD23, CD30, IL2/CD25, and transferrin/T9 receptors) may be useful in defining PFs and/or prognosis for non-Hodgkin's lymphomas [63,64].

1. Ki-67

Ki-67 is a nuclear antigen that is first expressed in the G_1 phase of the cell cycle and is variably expressed throughout the S, G_2, and M phases. It is essentially absent in G_0 [57,58,65,66]. Thus, Ki-67 expression is more sensitive than total DNA content measured by PI or DAPI in detecting the PF. In addition, Ki-67 is more specific than transferrin receptor (T9) for cells in the proliferative phases. This is because the T9 molecule is expressed in both proliferating and metabolically active, G_0 cells (i.e., macrophases) [67]. The Ki-67 proliferation index has been evaluated most widely on NHLs and breast carcinomas. Schwartz et al. [68] and Hall and co-workers [69] have both demonstrated good correlation between the grade of the NHL and PF determined by Ki-67. In addition, Schwartz et al. [68] found good correlation with nuclear DNA content (S, S + G_2 + M) and the Ki-67-determined PF. Hall et al. [69], reported that low-grade NHLs with a Ki-67 PF greater than 5% did worse than those with less than 5% and that high-grade NHLs with Ki-67 PFs greater than 80% did better than those with Ki-67 PFs less than 80%. In partial distinction, Spier et al. [31] found that diffuse large cell lymphomas (a subset of intermediate and high-grade NHLs) with a Ki-67 PF less than 60% had a better outcome than those with a Ki-67 PF >60%. Further large studies matched for NHL morphology and treatment will be required to assess the prognostic value of Ki-67 PFs.

The work in breast carcinomas is also somewhat contradictory. The primary goal is to define patients with either localized (surgically resectable) or metastatic disease at higher risk for relapse or aggressive clinical course. These latter patients will be more likely to be treated with adjuvant chemotherapy. Walther and Camplejohn [70] found good correlations between the Ki-67 PFs and S phase estimated by PI in 95 cases of breast carcinoma. Lelle et al. [71] reported a direct trend between Ki-67 PFs in 140 breast carcinomas and histological grade of tumor and node-positive carcinomas. However, Barnard et al. [72] found no correlation between Ki-67 PFs and tumor size, nodal metastasis, or estrogen receptor status. As above, further stage, prospective studies, balanced for tumor type, stage, and therapy will be required.

2. Proliferating Cell Nuclear Antigen

Proliferating cell nuclear antigens (PCNA) otherwise known as cyclin, is a 36 kilodalton polypeptide found in the nucleus [59,60]. It is apparently the delta subunit of DNA polymerase, and is not found during the G_0 phase. PCNA first appears in G_1, and reaches its maximum concentration in the S phase. During G_2 and M phases of the cell cycle, the PCNA level declines [60]. Preliminary studies have shown that PCNA can be detected by both immunohistochemical [73] and flow cytometric techniques [59,60]. While reports of PCNA expression in fresh neoplastic tissue are not readily available, evaluation of its presence in fixed tissue indicated that the PCNA-determined PFs directly correlate with mitotic activity of breast and colon carcinomas [73]. However, although there was a trend toward higher PCNA-determined PFs in high-grade breast and colon tumors, there were too few cases to reach significance. Takasaki et al. [74] reported increased numbers of PCNA+ cells at the onset of blast crises of chronic myeloid leukemia. As with Ki-67 and other proliferation-associated antigens, further work is required to elucidate more clearly any prognostic values.

3. p105 Antigen

The p105 antigen is a nuclear-matrix-associated protein, located primarily in the interchromatic regions [61,62]. It is present at low levels in G_0 cells, but increases in quantity progressively from early G_1 to M [61,62]. Its function is unknown. Bauer et al. [75] found that colonic carcinoma cells with aneuploid DNA contents had relatively high levels of p105 than did euploid ones. This may offer a possible insight into the association between aneuploidy and proliferation in tumors.

4. Lymphoid Proliferation Associated Antigens (4F2, IL2R/CD25):
   Transferrin Receptor/T9, T10/CD38, Ki-i/CD30

The 4F2, CD25, and T9 antigens are expressed in the early or late G_1 phase of cell proliferation [76]. Holte et al. [63] reported a significant correlation between the percentage of 4F2+ cells in 75 low-grade NHLs and poor patient survival. Although T9 expression correlated with DNA-PFs and treatment response, there was no correlation with overall survival in this group. There was also no correlation of CD25 expression with cell cycling (measured by fluorescent dyes), treatment responses, or survival of patients with low-grade NHLs [63]. Although CD23 has been reported to be associated with high-grade NHLs with high proliferative activity [64], there are too few cases in this report for this to be definitive. Also, CD23 and another activation marker, CD30, are not direct markers of proliferation, and antigens such as 4F2, Ki-67, and PCNA are preferable.

IV. IMMUNE RESPONSES TO TUMORS

In addition to phenotyping tumor cells, flow cytometry is being used to determine the phenotypes of lymphocytes and monocytes in the circulation and those infiltrating into tumor masses of patients with neoplastic processes. The hypothesis underlying these studies is that the immune response to a neoplasm is a critical determinant of the progression of the tumor. It may be possible to predict the clinical behavior of the neoplasm by monitoring alterations in the patients' immune status. Wolf et al. [77] examined lymphocyte subsets in the peripheral blood of patients with squamous cell carcinoma of the head and neck and found a significant correlation between low percentages of suppressor T lymphocytes and decreased disease-free survival.

In contrast, Dawson and colleagues [78] found a deficiency of helper-inducer T cells, but no alterations in the levels of other immune cells, in patients with squamous cell carcinomas of the head and neck. In a similar study of patients with gastrointestinal carcinoma (GIC), Boltze et al. [79] found a reduction in the number of suppressor T lymphocytes and an increase in the number of cells bearing monocyte markers in the peripheral blood of patients with GIC compared with normal controls or patients with non-GIC neoplasms. No correlation was made between tumor stage or progression and immune cell phenotyping.

Contrary data were reported by Eskinazi and co-workers [80] in a study of peripheral blood lymphocytes in patients with oral cancer. No alteration in the number of suppressor T cells was observed, although a significant increase was observed in the number of Leu 7+ (natural killer) cells in patients with oral cancer. Fracchia and colleagues [81] examined both the peripheral blood and bronchoalveolar lavage specimens from patients with lung cancer for alterations in T lymphocytes. No significant alterations were discovered in the peripheral blood of the cancer patients; however the lavages from both involved and uninvolved lungs of cancer patients demonstrated a significant increase in the number of total T cells as well as suppressor T cells, and a decrease in helper/suppressor ratio.

Correlations consistent from study to study have not been obtained and work in this area must be expanded for this method to be adequately assessed. As biological response modifiers (BRMs) become integral components of cancer therapy, flow cytometric analysis of the immune status of cancer patients will become increasingly important, not only as a general diagnostic or prognostic indicator but also for determining the efficacy and mode of action of the BRMs. For example, Phillips et al. [82] used dual-color flow cytometry in conjunction with functional assays to demonstrate the in vivo and in vitro activation of NK cells in patients receiving combined interleukin (IL)-2/lymphokine-activated iller (LAK) therapy.

V. ONCOGENES AND ONCOGENE PRODUCTS

There has recently been a great amount of interest in evaluating the use of oncogenes or their products as diagnostic or prognostic indicators. The vast majority of these studies have been conducted using various blotting techniques; the average expression of the oncogene has been detected on a large population of cells. Flow cytometric techniques for the detection and quantitation of oncogenes and oncogene products offer substantial advantages over blotting techniques: analysis is on a cell-by-cell basis rather than entire populations, multiple features of the cells can be examined concomitantly, and quantitation is relatively easy. With the ras p21 oncogene product, the majority of the work accomplished to date in this area has utilized cell lines rather than patient-derived specimens [83-85]. These studies have developed the methodology for detecting p21 by flow cytometry in conjunction with other parameters such as DNA content, thus allowing p21 expression to be correlated with cell cycle. Preliminary reports examining the expression of the ras p21 gene product by flow cytometry in clinical specimens of hematological neoplasms have indicated the feasibility of examining hematological neoplasms in this

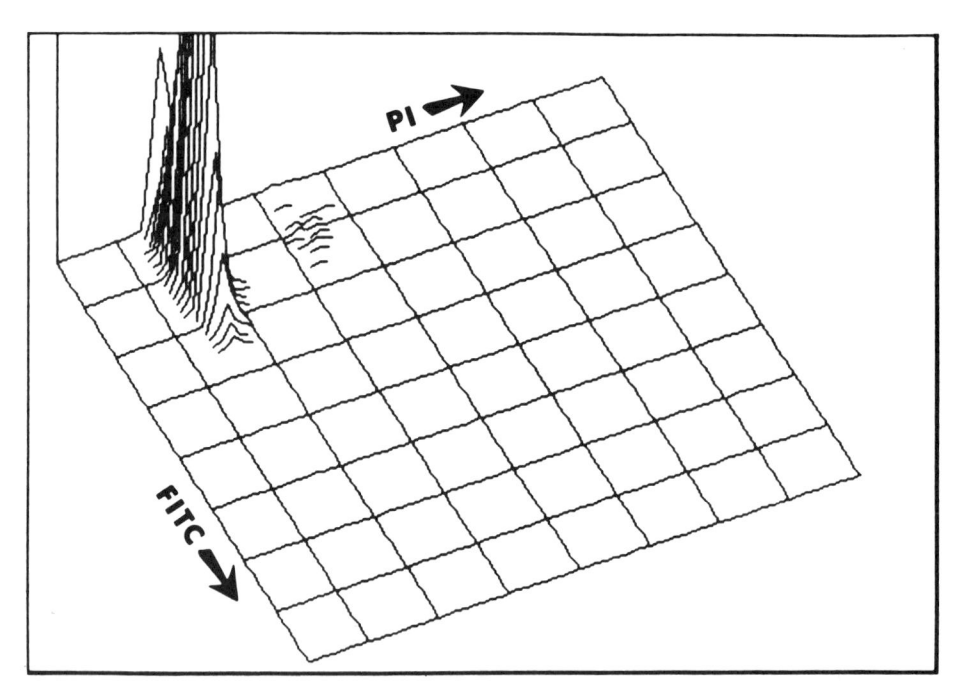

Figure 1 Blast cells from a cutaneous T-cell lymphoma stained with normal mouse IgG (control antibody), FITC-conjugated goat anti-mouse IgG, and propidium iodide (for DNA content).

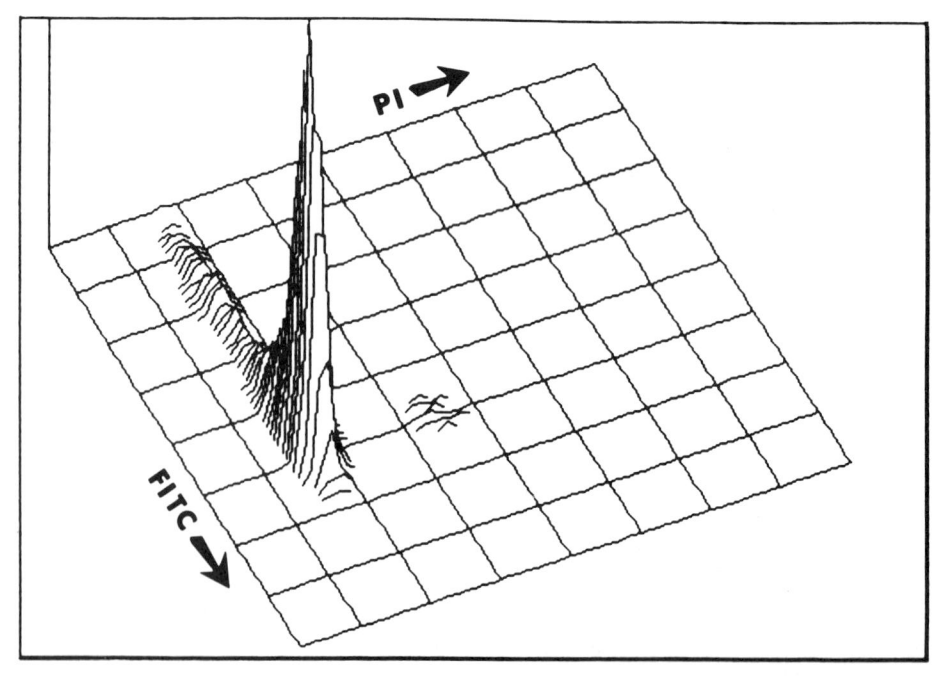

Figure 2 Blast cells from the same patient as in Figure 1 stained with a monoclonal anti-ras p21 antibody, FITC-goat anti-mouse IgG, and propidium iodide. Note the increase in the FITC staining compared to Figure 1 and that the staining is apparent in all phases of the cell cycle.

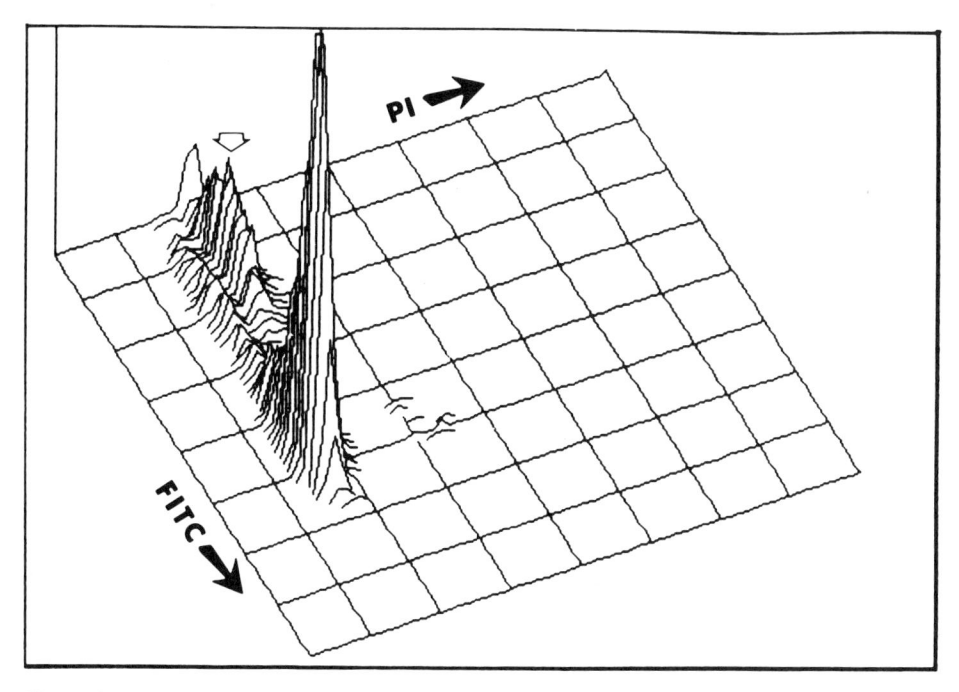

Figure 3 Blast cells from Figure 2 with the addition of propidium iodide-stained normal diploid lymphocytes (open arrow). Note that the normal lymphocytes stain more brightly with propidium iodide, indicating a hypodiploid blast population.

method for ras p21, although any definitive prognostic or diagnostic correlations have yet to be established. Studies from Barlogie's lab [86] have reported concomitant detection of DNA content and ras p21 in clinical specimens of multiple myeloma. Takeda et al. [87] have used single parameter analyses to examine ras p21 expression in an array of acute leukemias. They have also demonstrated the feasibility of two-color analyses of surface markers and oncogene products or DNA content and oncogene products (Figs. 1-3) and of three-parameter analyses in which leukemic blasts are stained with an antibody to a surface antigen followed by staining for ras p21, then by staining with propidium iodide for DNA content (unpublished data). Fan [88] has presented data suggesting that flow cytometric detection of ras p21 in prostatic adenocarcinoma may be of use as a clinical predictor of bone metastases.

A significant amount of work, led by the efforts of Watson and colleagues, has also examined the c-myc oncogene product by flow cytometry. Using concomitant staining for DNA content as well as for the expression of the c-myc p62 gene product, Watson and co-workers have examined the nuclei from a variety of paraffin-embedded tumors, including testicular [89], colonic [90,91], ovarian [92], and cervical [93] neoplasms for ploidy and elevated levels of c-myc p62. Their findings suggested that such flow cytometric investigations may prove to be of prognostic value in testicular neoplasms and of possible diagnostic value in the other neoplasms examined.

Interest in the area of flow cytometric evaluation of oncogenes and oncogene products is clearly growing. The ability to examine surface markers, oncogene products, and DNA content concomitantly on individual cells exists and almost certainly will be used to delineate better the subsets of neoplastic cells. One report has demonstrated the feasibility of in situ hybridization of DNA sequences in intact nuclei for flow cytometric analysis [94], thus possibly providing a technique for future examination of oncogenes directly by flow cytometry.

VI. SUMMARY

The application of flow cytometry to cancer diagnosis and the prediction of tumor behavior continues to expand. The ability of flow cytometry to examine multiple parameters of large numbers of individual cells is being exploited increasingly to characterize neoplastic processes better. This will allow closer examination of tumor heterogeneity and identification of subpopulations with different behavioral patterns.

Flow cytometry is being used with greater frequency in attempts to predict tumor behavior and response to therapy. Flow cytometry may have the most to offer in this area. Diagnosis of cancer by routine histopathological examination will not be replaced by flow cytometry; flow cytometry will be used in conjunction with morphological descriptions. The detection of proliferation antigens and oncogene products, as well as cell cycling, provides information on neoplastic progression that is not readily obtainable by other methods. In addition, flow cytometry will undoubtedly be used to measure other features of neoplastic cells, such as enzyme levels and ion fluxes, which may better characterize the behavior of the tumor.

Advances in flow cytometry instrumentation, light sources, fluorochrome development, and basic aspects of cellular and molecular biology will continue to permit this technology to define neoplastic cells and their behavior better, resulting in both improved patient care and a better understanding of tumor biology.

REFERENCES

1. Barlogie B, Raber MN, Schumann J et al. Flow cytometry in clinical cancer research. Cancer Res 1983;43:3982-3997.
2. Coon JS, Landay AL, Weinstein RS. Biology of disease: advances in flow cytometry for diagnostic pathology. Lab Invest 1987;57(5):453-479.
3. Herman CJ, McGraw TP, Marder RJ, Bauer KD. Recent progress in clinical quantitative cytology. Arch Pathol Lab Med 1987;111:505-512.
4. Shapiro HM. Practical flow cytometry, 2nd ed. New York: Alan R. Liss, 1988.
5. Bernard A, Boumsell L, Dausset J et al. (eds.). Leucocyte typing. Berlin: Springer-Verlag, 1984.
6. Reinherz E, Haynes B, Nadler L, Bernstein I (eds.). Leukocyte typing II. Berlin: Springer-Verlag, 1986.
7. McMichael A. Beverley P, Cobbold S et al. (eds.). Leucocyte typing III. Oxford: Oxford University Press, 1987.
8. Wood G, Link M, Warnke R et al. Pan-leukocyte monoclonal antibody L3B12. Am J Clin Pathol 1984;81:176-183.
9. Battifora H, Trowbridge I. A monoclonal antibody useful for the differential diagnosis between malignant lymphoma and nonhematopoietic neoplasms. Cancer 1983; 51:816-821.

10. Ritter M, Sauvage C, Pegram S. et al. The human leucocyte-common (LC) molecule: dissection of leukemias using monoclonal antibodies directed against framework and restricted antigenic determinants. Leukemia Res 1985;9:1249-1254.
11. Katz F, Lam G, Ritter M. The human leucocyte-common antigen: differential expression of framework and restricted antigenic determinants on early haemopoietic progenitors. Br J Haematol 1985;61:695-705.
12. Shah V, Civin C, Loken M. Flow cytometric analysis of human bone marrow: IV. Differential quantitative expression of T-200 common leukocyte antigen during normal hemopoiesis. J Immunol 1988;140:1861-1867.
13. Jing S, Ralph S, Thomas M, et al. Structural studies of the transferrin receptor and T200 glycoprotein (CD45). In McMichael A, Beverley P, Cobbold S, et al., eds. Leucocyte typing III. Oxford: Oxford University Press, 1987:899-905.
14. Wright J, Poplack D, Bakhshi A, et al. Gene rearrangements as markers of clonal variation and minimal residual disease in acute lymphoblastic leukemia. J Clin Oncol 1987;5:735-741.
15. Anderson K, Bates M, Slaughenhoupt B, et al. Expression of human B cell-associated antigens on leukemias and lymphomas: a model of human B cell differentiation. Blood 1984;63:1424-1433.
16. Picker L, Weiss L, Medeiros L, et al. Immunophenotypic criteria for the diagnosis of non-Hodgkin's lymphoma. Am J Pathol 1987;128:181-201.
17. Schwarting R, Stein H, Wang C-Y. The monoclonal antibodies S-HCL 1 (Leu-14) and S-HCL 3 (Leu M5) allow the diagnosis of hairy cell leukemia. Blood 1985; 65;974-984.
18. Weiss L, Wood G, Warnke R. Immunophenotypic differences between dermatopathic lymphadenopathy and lymph node involvement in mycosis fungoides. Am J Pathol 1985;120:179-185.
19. Link M, Stewart S, Warnke R, Levy R. Discordance between surface and cytoplasmic expression of the Leu-4 (T3) antigen in thymocytes and in blast cells from childhood T lymphoblastic malignancies. J Clin Invest 1985;76:248-253.
20. Bray R, Gottschalk L, Landay A, Gebel H. Differential surface marker expression in patients with CD-16+ lymphoproliferative disorders: in vivo model for NK differentiation. Hum Immunol 1987;19:105-115.
21. Drexler H. Classification of acute myeloid leukemias—A comparison of FAB and immunophenotyping (review). Leukemia 1987;1:697-705.
22. Kamps WA, Humphrey GB. Heterogeneity of childhood acute lymphoblastic leukemia—impact on prognosis and therapy (review). Semin Oncol 1985;12:268-280.
23. Nadler LM, Korsmeyer SJ, Anderson KC, et al. B cell origin of non-T cell acute lymphoblastic leukemia: a model for discrete stages of neoplastic and normal pre-B cell differentiation. J Clin Invest 1984;74:332-340.
24. Crist W, Pullen J, Boyett J, et al. Clinical and biological features predict a poor prognosis in acute lymphoid leukemias in infants: a Pediatric Oncology Group study. Blood 1986;67:135-140.
25. Crist W, Pullen J, Boyett J, et al. Acute lymphoid leukemia in adolescents: clinical and biologic features predict a poor prognosis—a Pediatric Oncology Group study. J Clin Oncol 1988;6:34-43.
26. Cossman J, Jaffe ES, Fisher RL. Immunologic phenotype of diffuse, aggressive non-Hodgkin's lymphomas—correlation with clinical features. Cancer 1984;54:1310-1317.
27. Spier C, et al. The aberrancy of immunophenotype as indicator of prognosis in diffuse large cell lymphoma. Mod Pathol 1988;1:87A (abstract).
28. Bradstock K, Hoffbrand A, Ganeshaguru K, et al. TdT expression in acute non-lymphoid leukemia. Br J Haematol 1981;47:133-143.

29. Jani P, Berby W, Greaves M, et al. TdT in acute myeloid leukemia. Leukemia Res 1983;7:17-29.
30. Sobol R, Mick R, Royton I, et al. Clinical importance of myeloid antigen expression in adult acute lymphoblastic leukemia. N Engl J Med 1987;316:1111-1117.
31. Kokland P, Kerndrup G, Griffin J, Ellegaard J. Analysis of leukocyte differentiation antigens in blood and bone marrow from preleukemia (refractory anemia) patients using monoclonal antibodies. Blood 1986;67:898-902.
32. Braylan RC, Benson NA, Nourse VA, Kruth HS. Correlated analysis of cellular DNA, membrane antigens and light scatter of human lymphoid cells. Cytometry 1982;2(5):337-343.
33. Braylan RC, Benson NA, Nourse VA. Cellular DNA of human neoplastic B-cells measured by flow cytometry. Cancer Res 1984;44:5010-5016.
34. Sasaki K, Nagai M, Kato H, et al. Flow cytometric analysis of tumor antigen TA-4 in cervical squamous cells. Gann 1984;75:703-706.
35. Suehiro Y, Kato H, Nagai M, Torigoe T. Flow cytometric analysis of tumor antigen TA-4 in cervical cytologic specimens. Cancer 1986;57:1380-1384.
36. Flint A, McCoy JP, Esch TR, et al. Simultaneous measurement of DNA content and detection of surface antigens of cervical vaginal cells by flow cytometry. Anal Quant Cytol Histol 1987;9(5):419-424.
37. Aurelian L. Herpes simplex virus diagnosis: antigen detection by ELISA and flow microfluorometry. Diagn Gynecol Obstet 1982;4(4):375-388.
38. Fowler WC, Maddock MB, Moore DH, Haskill S. Significance of multiparameter flow cytometric analysis of ovarian cancer. Am J Obstet Gynecol 1988;158:838-845.
39. Huffman JL, Garin-Chesa P, Gay H, et al. Flow cytometric identification of human bladder cells using a cytokeratin monoclonal antibody. Ann NY Acad Sci 1986;468:302-315.
40. Feitz WF, Beck HL, Smeets AW, et al. Tissue-specific markers in flow cytometry of urologic cancers: cytokeratins in bladder carcinoma. Int J Cancer 1985;36(3):349-356.
41. Czerniak B, Koss LG. Expression of Ca antigen on human urinary bladder tumors. Cancer 1985;55:2380-2383.
42. Orihuela E, Varadachay S, Herr HW, et al. The practical use of tumor marker determination in bladder washing specimens: assessing the urothelium of patients with superficial bladder cancer. Cancer 1987;60(5):1009-1016.
43. Devonec M, Fontaniere B, Blanc-Brunat N, et al. Revelation of AB cell surface antigens on bladder-washing specimens with the immunofluorescence technique: an approach to automated cytology. Eur Urol 1984;10:323-325.
44. Orntoft TF, Petersen SE, Wolf H. Dual-parameter flow cytometry of transitional cell carcinoma: quantitation of DNA content and binding of carbohydrate ligands in cellular subpopulations. Cancer 1988;61:963-970.
45. Durrant LG, Robins RA, Armitage NC, et al. Association of antigen expression and DNA ploidy in human colorectal tumors. Cancer Res 1986;46:3543-3549.
46. Van den Ingh HF, Bara J, Cornelisse CJ, Nap M. Aneuploidy and expression of gastric-associated mucus antigens M1 and CEA in colorectal adenomas. Am J Pathol 1987;87:174-179.
47. Durrant LG, Ballantyne KC, Armitage NC, et al. Quantitation of MHC antigen expression on colorectal tumours and its association with tumour progression. Br J Cancer 1987;56(4):425-432.
48. Dexter D, Leith J. Tumor heterogeneity and drug resistance (review). J Clin Oncol 1986;4:244-257.

49. Danks M, Metzger D, Ashmun R, et al. Monoclonal antibodies to glycoproteins of vinca alkaloid-resistant human leukemic cells. Cancer Res 1985;45:3220-3224.
50. Kartner N, Evernden-Porelle D, Bradley G, Ling V. Detection of P-glycoprotein in multidrug-resistant cell lines by monoclonal antibodies. Nature 1985;316:819-823.
51. Kartner N, Shales M, Riordan J, Ling V. Daunorubicin-resistant Chinese hamster ovary cells expressing multidrug resistance and a cell-surface P-glycoprotein. Cancer Res 1983;43:4413-4419.
52. Danks M, Yalowich J, Beck W. Atypical multiple drug resistance in a human leukemic cell line selected for resistance to teniposide (VM-26). Cancer Res 1987;47:1297-1301.
53. Epstein J, Barlogie B. VAD resistant cells in aneuploid multiple myeloma (MM) over-express mdr_1 associated p-glycoprotein. Fed Proc 1988;29:301.
54. Ma D, Davey R, Harman D, et al. Detection of a multidrug resistant phenotype in acute non-lymphoblastic leukemia. Lancet 1987;1:135-137.
55. Kaufman R, Bertino J, Schimke R. Quantitation of dihydrofolate reductase in individual parental and methotrexate-resistant murine cells. J Biol Chem 1978;253:5852-5860.
56. Rice G, Bump E, Shrieve D, et al. Quantitative analysis of cellular glutathione by flow cytometry utilizing monochlorobimane: some applications to radiation and drug resistance in vitro and in vivo. Cancer Res 1986;46:6105-6110.
57. Gerdes J, Schwab U, Lemke H, Stein H. Production of a mouse monoclonal antibody reactive with a human nuclear antigen associated with cell proliferation. Int J Cancer 1983;31:13-20.
58. Gerdes J, Lemke H, Baisch H, et al. Cell cycle analysis of a cell proliferation-associated human nuclear antigen defined by the monoclonal antibody Ki-67. J Immunol 1984;133:1710-1715.
59. Kurki P, Lotz M, Ogata K, Tan E. Proliferating cell nuclear antigen (PCNA)/cyclin in activated human T lymphocytes. J Immunol 1987;138:4114-4120.
60. Kurki P, Ogata K, Tan E. Monoclonal antibodies to proliferating cell nuclear antigen (PCNA)/cyclin as probes for proliferating cells by immunofluorescence microscopy and flow cytometry. J Immunol Methods 1988;109:49-59.
61. Bauer K, Clevenger C, Williams T, Epstein A. Assessment of cell cycle-associated antigen expression using multiparameter flow cytometry and antibody-acridine orange sequential staining. J Histochem Cytochem 1986;34:245-250.
62. Clevenger C, Epstein A, Bauer K. Modulation of nuclear antigen p105 as a function of cell-cycle progression. J Cell Physiol 1987;130:336-343.
63. Holte H, Davies C, Kvaloy S, et al. The activation-associated antigen 4F2 predicts patient survival in low-grade B-cell lymphomas. Int J Cancer 1987;39:590-594.
64. Williamson J, Grigor I, Smith M, et al. Ploidy; proliferative activity, cluster differentiation antigen expression and clinical remission in high grade non-Hodgkin's lymphoma. Histopathology 1987;11:1043-1054.
65. Palutke M, KuKuruga D, Tabaczke P. A flow cytometric method for measuring lymphocyte proliferation directly from tissue culture plates using Ki-67 and propidium iodide. J Immunol Methods 1987;105:97-105.
66. Schwarting R, Gerdes J. Niehus J, et al. Determination of the growth fraction in cell suspensions by flow cytometry using the monoclonal antibody Ki-67. J Immunol Methods 1986;90:65-70.
67. Pileri S, Gerdes J, Rivano M, et al. Immunohistochemical determination of growth fractions in human permanent cell lines and lymphoid tumors: a critical comparison of the monoclonal antibodies OKT9 and Ki-67. Br J Haematol 1987;65:271-276.
68. Schwartz B, Weinberg D, Pinkus G, Image analysis quantitation of proliferation in non-Hodgkin's lymphomas. Mod Pathol 1988;1:82A.

69. Hall P, Richards MA, Gregory WM, et al. The prognostic value of Ki-67 immunostaining in non-Hodgkin's lymphoma. J Pathol 1988;154:223-235.
70. Walker R, Camplejohn R. Comparison of monoclonal antibody Ki-67 reactivity with grade and DNA flow cytometry of breast carcinomas. Br J Cancer 1988;57:287-290.
71. Lelle R, Heidenreich W, Strauch G, Gerdes J. The correlation of growth fractions with histologic grading and lymph node status in human mammary carcinoma. Cancer 1987;59:83-88.
72. Barnard N, Hall P. Lemoine N, Kadar N. Proliferative index in breast carcinoma determined in situ by Ki-67 immunostaining and its relationship to clinical and pathological variables. J Pathol 1987;152:287-295.
73. Robbins B, de la Vega D, Ogata K, et al. Immunohistochemical detection of proliferating cell nuclear antigen in solid human malignancies. Arch Pathol Lab Med 1987;111:841-845.
74. Takasaki Y, Robinson W, Tan E. Proliferating cell nuclear antigen in blast crisis cells of patients with chronic myeloid leukemia. J Natl Cancer Inst 1984;73:655-661.
75. Bauer K, Clevenger C, Endow R, et al. Simultaneous nuclear antigen and DNA content quantitation using paraffin-embedded colonic tissue and multiparameter flow cytometry. Cancer Res 1986;46:2428-2434.
76. Cotner Y, Williams J, Christension L, et al. Simultaneous flow cytometric analysis of human T cell activation antigen expression and DNA content. J Exp Med 1983;157:461-472.
77. Wolf GT, Schmaltz S, Hudson JL, et al. Alterations in T-lymphocyte subpopulations in patients with head and neck cancer. Arch Otolaryngol Head Neck Surg 1987;113:1200-1206.
78. Dawson DE, Everts EC, Vetto RM, Burger DR. Assessment of immunocompetent cells in patients with head and neck squamous cell carcinoma. Ann Otol Rhinol Laryngol 1985;94(4Pt):342-345.
79. Boltze G, Penner E, Holzinger C, et al. Surface phenotypes of human peripheral blood mononuclear cells from patients with gastrointestinal carcinoma. J Cancer Res Clin Oncol 1987;113:291-297.
80. Eskinazi DP, Perna JJ, Mihail R. Mononuclear cell subsets in patients with oral cancer. Cancer 1987;60:376-381.
81. Fracchia A, Pacetti M, Barberis M, et al. Determination of T-lymphocyte subpopulations in patients with lung cancer: a comparison between lung lavage and peripheral blood by monoclonal antibodies and flow cytometry. Respiration 1987;51(3):161-159.
82. Phillips JH, Gemlo BT, Myers WW, et al. In vivo and in vitro activation of natural killer cells in advanced cancer patients undergoing combined recombinant interleukin-2 and LAK cell therapy. J Clin Oncol 1987;5:1933-1941.
83. Young HA, Klein RA, Shih TY, et al. Detection of the intracellular ras p21 oncogene product by flow cytometry. Anal Biochem 1986;156:67-71.
84. Czerniak B, Herz F, Wersto RP, Koss LG. Modification of Ha-ras oncogene p21 expression and cell cycle progression in the human colonic cancer cell line HT-29. Cancer Res 1987;47:2826-2830.
85. Freedman D, Auersperg N. Detection of an intracellular transforming protein (v-Ki-ras p21) using the flow activated cell sorter (FACS). In Vitro Cell Dev Biol 1986;22(10):621-624.
86. Tsuchiya H, Epstein J, Selvanayagam P, et al. Correlated flow cytometric analysis of H-ras p21 and nuclear DNA in multiple myeloma. Blood 1988;72:796-800.

87. Takeda T, Krause JR, Carey JL, McCoy JP. Detection of the ras p21 gene product in human acute leukemia by flow cytometry. J Clin Lab Analysis 1989;3:108-115.
88. Fan K. Heterogeneous subpopulations of human prostatic adenocarcinoma cells: potential usefulness of p21 protein as a predictor for bone metastasis. J Urol 1988;139(2):318-322.
89. Watson JV, Stewart J, Evan GI, et al. The clinical significance of flow cytometric c-myc oncoprotein quantitation in testicular cancer. Br J Cancer 1986;53(3):331-337.
90. Sundaresan V, Forgacs IC, Wight DG, et al. Abnormal distribution of c-myc oncogene product in familial adenomatous polyposis. J Clin Pathol 1987;40(11):1274-1281.
91. Watson JV, Sikora K, Evan GI. A simultaneous flow cytometric assay for c-myc oncoprotein and DNA in nuclei from paraffin embedded material. J Immunol Methods 1985;83:179-192.
92. Watson JV, Curling OM, Munn CF, Hudson CN. Oncogene expression in ovarian cancer: a pilot study of c-myc oncoprotein in serous papillary ovarian cancer. Gynecol Oncol 1987;28(2):137-150.
93. Hendy-Ibbs P, Cox H, Evan GI, Watson JV. Flow cytometric quantitation of DNA and c-myc oncoprotein in archival biopsies of uterine cervix neoplasia. Br J Cancer 1987;55(3):275-282.
94. Trask B, van den Engh G, Landegent J, in de Wal NJ, van der Ploeg M. Detection of DNA sequences in nuclei in suspension by in vitro hybridization and dual beam flow cytometry. Science 1985;230:1401-1403.

13
Use of Monoclonal Antibodies as Probes for Oncogene Products

HENRY L. NIMAN*

Progenx, Inc., San Diego, California

I. INTRODUCTION

The importance of oncogene-encoded and related proteins in the diagnosis and management of neoplasia is becoming increasingly apparent. The experimental manipulation of oncogenes began in the early 20th century [1]. Avian neoplasms were transmissible by filtrates of tumor extracts. In the 1950s, similar observations were made in the murine system. Viruses were implicated as the causative agents and these RNA tumor viruses were the center of the attention of virologists and then molecular biologists. Because of their simple genetic composition, these viruses provided a molecular system for the analysis of viral oncology. The genes responsible were isolated in the late 1970s and nucleic acid probes were generated. Related sequences were identified in a broad spectrum of organisms, which suggested that the gene products participated in important physiological functions such as cell division and differentiation. DNA technology also allowed more complex systems to be analyzed. In transfection assays (transfer of DNA to indicator cells that would transform when oncogenes were added) oncogenes in human tumors were identified [2]. Molecular analysis of these genes identified two *ras* genes (H-*ras* and K-*ras*) that had been transduced by retroviruses. Sequencing of the genes indicated that a single-point mutation produced an activated oncogene [3-8]. A third method of identifying oncogenes used preferred integration sites of retroviruses. In these cases, retroviruses that transduce oncogenes cause transformation by integration adjacent to normal genes, which become inappropriately expressed due to the transcriptional regulatory sequence of the retroviruses. A preferred integration site of avian leukosis virus near the *myc* oncogene identified this mechanism of transformation [9, 10]. Thus, viral transduction, DNA transfection, and preferred integration sites represent the three biological approaches used to identify most of the oncogenes.

Sequencing of these genes provided a powerful method for identifying structural and functional relationships among these genes. The most solid link between the neo-

---

**Present affiliation*: Graduate School of Public Health, University of Pittsburgh, Pittsburgh, Pennsylvania.

plastic transformation of oncogenes and the mitogenic action of growth factors was apparent when the predicted sequence of the *sis* oncogene was compared to the sequence of platelet-derived growth factor (PDGF) [11-16]. The homology of the oncogene with the β chain of PDGF provided a conceptual mechanism for transformation via altered signal transduction. Critical control points in this pathway were further identified by comparison of the sequence of the receptor for another growth factor (epidermal growth factor) with another oncogene (*erb* B) [17-19]. This relationship between oncogenes representing polypeptide growth factors or their receptors was strengthened by the similarity between the *fms* oncogene and the receptor for colony-stimulating factor 1 [20-22].

The clinical importance of these oncogenes was apparent from the point mutations in the various *ras* genes [23-27]. Additional lesions such as gene amplification or rearrangements were subsequently found. In the search for *myc* gene amplification, two related genes were identified. N-*myc* was found to be amplified in neuroblastoma [28] and L-*myc* was amplified in small cell lung carcinomas [29]. Moreover, amplification of N-*myc* correlated with a poor prognosis [30]. A gene (*HER*-2/*neu*) related to the Epidermal Growth Factor (EGF) receptor [*HER*-1] was identified via transfection experiments with a rat neuroblastoma cell [31] or cross-hybridization with a nucleic acid probe of *HER*-1 [32] or *erb* B [33]. This gene was found to be amplified in a subset of breast carcinomas and this lesion also predicted a poor prognosis [34]. Other clinical correlations arose from studies of the *abl* oncogene in patients with chronic myelocytic leukemia (CML) [35,36]. In this case, the *abl* oncogene was rearranged due to the Philadelphia chromosome translocation. Moreover, additional alterations of the *abl* oncogene were observed in patients with acute lymphoblastic leukemia (ALL) [37-39]. Thus use of sequences and probes of nucleic acid has identified and characterized a variety of oncogenes that have clinical utility.

II. COMBINING TECHNOLOGIES

Although much can be learned from the identification and sequencing of oncogenes, the transformation processes is mediated by the protein products. In the early 1980s it became apparent that synthetic peptides could provide the link between the sequence data accumulating for oncogenes and the ability to identify and characterize the protein products with antibody probes [40,41]. This approach allowed the investigator to select regions of the gene product, synthesize peptides that would represent the predicted sequence, and produce antibodies that would recognize the parent molecule. This approach was widely used [42] but was limited by the heterogeneity in the immune response. Indeed, this heterogeneity was thought to be the mechanism underlying the widespread success of the technique. Since it was known that peptides exist in a large number of conformations, the successful identification of the parent molecule was thought to be due to production of a heterogeneous array of antibodies that could represent each of the conformations the peptide could adopt. On rare occasions, one of these conformations would mimic the parent molecule. This stochastic model predicted difficulties in combining synthetic peptide and monoclonal antibody technologies because the number of conformations a peptide could adopt was thought to be as high as 1,000-10,000. Thus, the frequency for identifying monoclonal antibodies that would recognize both the peptide and the parent molecule would be low. It was fortunate that this stochastic model did not fit the observed frequency for isolating monoclonal antibodies that would recognize both the peptide and the parent molecule. A formal test of the model demonstrated that between 25 and 100% of the monoclonal antibodies selected solely on the basis of recognizing the synthetic peptide also recognized the parent molecule [43].

III. ONCOGENE FAMILIES

As the hybridoma/synthetic peptide technology was being optimized, sequencing of oncogene-encoded or related genes accelerated. In addition to the genes isolated by transduction, transfection, and integration analysis, related proteins were identified by cross-hybridization to various probes. The accumulating sequences allowed the oncogenes to be classified on the basis of a sequence homology. Although these techniques did not identify new growth factors that definitely bound to kinase receptors, the genes located near the two preferred integration sites of mouse mammary tumor virus (*int*-1 and *int*-2) encoded proteins that contain signal sequences suggesting the proteins are secreted [44, 45]. Indeed the *int*-2 oncogene was found to have homology to acidic and basic fibroblast growth factor [46-48] as well as a new oncogene isolated via transfection assays from a stomach tumor (*HST*) [49] or Kaposi's sarcoma (KS-3) [50]. The *int*-1 oncogene had extensive homology [51] with a gene controlling early developmental events in *Drosophila*. The number of growth factor receptors also grew markedly, as did the number of genes that encoded related proteins lacking the extracellular domain. The receptors formed subgroups on the basis of homologies in both the extracellular and kinase domains. *HER*-1 was closely related to *HER*-2. However, this subgroup also contained significant homology with the insulin receptor [52] as well as the insulin-like growth factor receptor [53]. The external domains of all four proteins had virtually identical spacings of their cysteine residues in addition to the relatedness in the kinase domain shared by many receptors and nonreceptor kinases. These homologies strongly suggest that the gene products occupy a similar position in the signal transduction pathway, although only *HER*-1 has been isolated via retroviral transduction and only *HER*-2/*neu* has been identified via the transfection assay. Similar observations can be made for other subgroups of oncogene families such as the oncogenes *fms*, *kit*, and the PDGF receptor [21,54,55]. These gene products also have extensive homology in the intra- and extracellular domains. Moreover, the three genes have a similarly sized exon inserted in the middle of the kinase domain as well as similar spacing of the cysteine residues in the extracellular domain.

In addition to the genes that contain hormone or growth-factor-binding domains, other genes encode only the kinase domain. These genes also form subgroups with extensive homology as well as similar intron/exon borders. The largest group contains at least eight members including *src*, *yes*, *fgr*, *lck*, *fyn*, *lyn*, *hck*, and *tkl* [56-66]. The region between the kinase domain and the variable sequences at the amino end of the molecule is also conserved. Homology with the *abl* oncogene is also found in this region [67]. Moreover, a recently sequenced gene *crk* [68] has homology with this region, as does phospholipase C [69].

Another group of receptors is related to the *erb* A oncogene. These receptors are located in the cytoplasm and are translocated to the nucleus after hormone binding. Sequencing of at least two *erb*-A-related genes have been published [70-72]. Both bind thyroid hormone but are located on different chromosomes and have different expression profiles. One may [71,72] be the human homologue of the avian *erb* A proto-oncogene [73] or may represent a third thyroid hormone receptor gene. This thyroid hormone receptor has extensive homology within the DNA-binding domain of a variety of steroid hormone receptors including glucocorticoid, estrogen progesten, mineralocorticoid, androgen, and vitamin D_3 [74-82]. There is also homology in the hormone-binding domain. Moreover, the thyroid hormone gene products are even more closely related to two genes encoding receptors for retinoic acid [83-86].

Other genes also act in the nucleus and are thought to be transcription regulatory genes. In addition to the c-*MYC*, N-*MYC,* and L-*MYC* genes mentioned above, which have distinct regions of homology [87-89], more distant homology is found with other nuclear oncogenes, particularly in the DNA-binding domain. These genes include *FOS* and *JUN* [90-83]. JUN may be identical to the transcription regulator gene *AP-1* [93] and has extensive homology with a second gene *JUN* B [94], while *FOS* is related to an additional gene *FRA*-1 [94]. Thus the additional sequences increase the number of potential peptides for the generation of monoclonal antibodies.

IV. CHOICE OF PEPTIDES

Peptides for monoclonal antibody production can be chosen from two major categories: sequences that are unique to a particular oncogene or sequences that are well conserved among different family or subgroup members. Most hydrophilic peptides 10-25 amino acids in length will usually elicit antibodies with binding activity with the intact protein. Although the utility of antibodies to unique regions is fairly obvious, antibodies to conserved regions offer several advantages, particularly if the reactive antigens are characterized beyond binding activity with the monoclonal antibody.

As the number of family members grow, it becomes increasingly easy to identify conserved regions. These regions are usually conserved because they are involved in essential functions of the proteins. Thus, antibodies directed against these regions have the potential of inihibiting the function of the protein. Antibodies to the conserved region also increase the likelihood that cross-reacting proteins will be encoded by additional family members. Because binding activity of multiple antibodies at distinct sites on a protein greatly increases the likelihood that the protein is an additional family member, assaying novel proteins with antibodies to conserved regions of the same family increases the likelihood of finding multiple antibodies against the new protein. The multiple antibodies are useful not only for the classification of the protein but also for designing assays that will capture and identify the unique ligand. Moreover, because each monoclonal antibody recognizes a stretch of six to eight amino acids, a panel of antibodies to a conserved region can be quickly characterized using binding activities with peptides representing homologous regions of other family members. Table 1 shows an example of the ATP-binding domain of a variety of protein kinases. This represents the human or mammalian oncogene sequences as well as additional kinases. As can be seen from the consensus sequence, many of the positions within the ATP-binding domain are conserved among the majority of the kinase. Moreover, many of the members of subgroups have identical sequences in this region (*PKCI/PKCII*; *HER*-1/*HER*-2; *FMS/KIT; ROS*/sev; *HIR/HILR; LYN/LCK*/tkl; *SRC/YES*). Thus antibodies to one peptide are directed against several family members.

Another conserved region is the DNA-binding domain of the steroid/thyroid/retinoic acid receptor family. This region contains cysteine residues that are held in position by a zinc molecule. The intervening residues are thought to form a finger structure that recognizes DNA sequences involved in transcriptional regulation [119]. In Table 2, con-

Table 1 ATP-Binding Domain

Gene	Sequence	Reference
pbk	L G R G V S S V V R R C	96
FES	I G R G N F G E V F S G	97
sea	I G R G H F G S V Y H G	98
MET	I G R G H F G C V Y H G	99
PKC I/PKC II	L G K G S F G K V M L A	100,101
PKC III	L G K G S F G K V M L S	100
capk	L G T G S F G R V M L V	102
CDC 28	I G E G T Y G V V Y K G	103
TRK	L G E G A F G K V F L A	104
RET	L G E G E F G E V Y R G	105
EPH	L G E G S F G I V Y K G	106
RAF-1	I G S G S F G T V Y K G	107
PKS	I G T G S F G T V F R G	108
pim-1	L G S F G F G S V Y S G	109
MOS	L G A G G F G S V Y K A	110
cgpk	L G V G G F G R V E L A	111
HER-1/HER-2	L G S G A F G T V Y K G	17,18,19
PDGF-R	L G S G A F G Q V V E A	55
FMS/KIT	L G A G A F G K V V E A	21,54
ROS/sev	L G S G A F G E V Y E G	112,113
HIR/HILR	L G Q G S F G M V Y E G	52,53
ABL	L G G G Q Y G E V Y E G	114
mlck	L G G G K F G A V C T C	115
LYN/lck/tk1	L G A G Q F G E V W M G	63,60,66
HCK	L G A G Q F G E V W M A	64,65
SRC/YES	L G Q G C F G E V W M G	56,57
FYN	L G N G Q F G E V W M G	61
FGR	L G T G C F G D V W L G	58

Table 2 Steroid Hormone Superfamily Receptors

Gene	DNA-Binding Domain	Reference
VDR	K G F F R R S M K R K A M P T C P F	82
ERB A-α	K G F F R R T I Q K N L H P T Y S C A R	71
c-erb A	K G F F R R T I Q K N L H P T Y S C A R	73
ERB A-β	K G F F R R T I Q K N L H P S Y S C A R	70
TR2	K G F F K R S I R K N L V Y S C R G	79
RAR2	K G F F R R S I Q K N M I Y T C H R	86
RAR1	K G F F K R S I Q K N M V Y I C H R	84,85
ER	K A F F K R S I Q G H̄ N D Y M C P A	75
ERR1	Y A F F K R T I Q G S I E Y S C P A	78
ERR2	Y A F F K R T I Q G N I E Y S C P A	80
GR	K V F F K R A V E G Q H N Y L C A G	74
MR	K V F F K R A V E G Q H N Y L C A G	77
AR	K V F F K R A A E G K Q K Y L C A S	79,80
PR	V V F F K R A M E G Q H N Y L C A G	76

Gene	Hormone-Binding Domain	Reference
VDR	H L A D L V S Y S I Q K V I G F A K	82
ERB A-β	H F T K I I T P A I T R V V D F A K	70
ERB A-α	E F T K I I T P A I T R V V D F A K	71
c-erb A	E F T K I I T P A I T R V V D F A K	73
RAR2	K F S EL A T K C I I K I V E F A K	86
RAR1	K F S EL S T K C I I K T V E F A K	84,85
ER	L L T N L A D R E L V H M I N W A K	75
ERR1	T L C D L F D R E I V V T I S W A K	78
ERR2	T L C D L A D R E L V F L I S W A K	78
GR	T L N M L G G R Q V I A A V K W A K	74
MR	T L N R L A G K Q M I Q V V K W A K	77
PR	S L N Q L G E R Q L L S V V K W S K	76

served regions in the DNA- and hormone-binding domains are shown. The sequences demonstrate how each family forms a subgroup. The table also displays residues that are well conserved among all family members, which suggests that structurally distinct ligands activate diverse genes through similar mechanisms.

V. CHARACTERIZATION OF ANTIBODIES

Antibodies to conserved domains can be characterized quickly using binding activities against related peptides. This analysis can be used to group a panel of antibodies raised against the same peptide. The binding data can also be used to predict binding to addi-

Table 3 Peptides Representing the ATP-Binding Domain

Peptide	Gene	Sequence
701	PKCI/II	L G K G S F G K V M L A (C)
711	PKCIII	L G K G S F G K V M L S (C)
620	v-sea	L G R G H F G S V Y H G (C)
121	v-fes/v-fps/FES	I G R G N F G E V F S G (C)
360	v-ros/ROS	L G S G A F G E V Y E G (C)
310	v-abl/ABL	L G G G Q Y G E V Y E G (C)
390	c-lsk/tck/LYN	L G A G Q F G E V W M G (C)
202	v-yes/v-src/SRC/YES	L G Q G C F G E V W M G
212	v-fgr/FGR	L G T G C F G D V W L G
371	HIR	L G Q G S F G M V Y E G (C)
294	v-kit	L G A G A F G K V V E A (C)
293	FMS/KIT	L G T G A F G L V V E A (C)
290	v-fms	L G A G A F G L V V E A (C)
179	HER 1/HER 2	L G S G A F G T V Y K G (C)
172	v-erb B	L G S G A F G T I Y K G (C)
169	MOS	L G A G G F G S V Y K A (C)
165	v-mos	I G S G G F G S V Y K A (C)
265	A-raf	I G T G S F G T V F R G (C)
251	v-raf/v-mil/RAF	I G S G S F G T V Y R G (C)
253	RAF	I G S G S F G T V Y K G (C)
254	PKS	I G T G S F G T V F R G (C)
600	TRK	L G E G A F G K V F L A (C)
340	cdc28	V G E G T F G V V Y K A (C)
320	cpk (C)	L G T G S F G R V M L V (C)
321	cpk (R)	D N H G S F G E L A L M (C)
681	PDGF-R	L G S G A F G Q V V E A (C)
901	pbk	L G R G V S S V V R R C (C)
911	cgk	L G V G G F G R V E L V (C)
771	PIM-1	L G S G G F G S V Y S G (C)

(C), a cysteine residue added for coupling to immunogenic KLH carrier protein.

tional family members. Moreover, cross-reactivity patterns help to localize the antibody-binding site and define the amino acids recognized by the antibody. Prediction of binding activities for newly sequences genes or newly synthesized peptides can then be made.

Table 3 lists some of the peptides used to characterize monoclonal antibodies directed against the ATP-binding domain of various kinases. In Table 4, the results of binding activities for a panel of *fes* monoclonal antibodies are listed. This demonstrates the wide range of binding activities for antibodies generated from a single fusion. The largest

Table 4 Cross-Reactivity of FES Antibodies Against ATP-Binding Domain Sequences

Peptide	121-19B10	121-09G10 04C07	121-03H05	121-01D05	121-03H04	121-05E05	121-18C05 14C09 03F07 03G03	121-01F09 04F06 04F08 15B10 16C07	121-08D08	121-09E05	121-06D10	121-11G08	121-06C09
121	2,4	2,3	2,3	2,3	2,3	3,3	2,3	2,3	2,3	3,3	0,2	0,3	0,2
165	2,0	2,0	2,0	1,0	1,0	3,1		1,0	2,0	2,1			1,0
169													
172										0,1		2,3	
179										2,2			
202	0,2	0,2		0,2									
212		0,1		0,1	1,1	0,1				0,1			
251												3,3	
253				2,2	0,1		1,1		2,3	2,2			
254		0,1		2,1	0,1	2,1			0,1	1,1			
290				0,2	0,1	0,2			0,1	0,3	2,2	0,3	

Antibody Probes for Oncogene Products 197

293			0,1		0,2				2,3
294	0,1								
310		0,1			1,2			0,1	
320					0,2			1,0	
321					1,2				
340			0,1	0,1	2,2				
360	3,3	3,3	1,1	1,1	0,1	0,2	1,0	2,2	3,3
371			3,2	0,1				1,1	
390	2,3		0,1						4,3
600									
620			1,2			0,2	4,4	2,3	
681	0,2								
701		1,1	1,1	0,2	0,1	2,3	2,2	3,3	3,1
711	0,1	1,3	1,2	1,2	0,2	3,3	3,3	3,4	0,1
771					0,1		1,3	3,3	
901			0,1			4,4	4,4	4,4	
911			0,1						

Numbers represent binding to free peptide or peptide coupled to keyhole-limpet hemocyanin.
The background (0.1 or 0.2 O.D. units) is subtracted from each reading.
The above background levels are represented as follows: A 0 or blank space represents 0–0,3; 1 is 0.3–0.7; 2 is 0.7–1.1; 3 is 1.1–1.5; 4 is > 1.5.

group contains five cell lines (121-1F9, 4F6, 4F8, 15B10, 16C7) that recognize peptides 620, 701, 711, and 901, in addition to the immunizing 121 peptide. Two other hybridomas (121-8D8, 9E5) also readily bind to these peptides as well as additional peptides. Thus, all 7 hybridomas recognize 5 of the 29 peptides tested. All 5 of these peptides contain a basic amino acid (lysine or arginine) as the third residue in contrast to the other 24 peptides tested. Thus, in the context of the consensus ATP-binding domain, a basic residue at a critical position defines binding activity for these seven antibodies. Moreover, the binding characteristics of these antibodies predicts that the antibodies would also recognize the MET sequence (see Table 1).

VI. CHARACTERIZATION OF ANTIGENS

Panels of monoclonal antibodies can also be used to characterize quickly antigens found in clinical samples [117-119]. Because antibodies to synthetic peptides frequently recognize an array of related proteins and the antibody-binding sites may be involved in protein interactions, the immunoblot technique offers a method for identifying and characterizing oncogene-encoded and oncogene-related proteins. Cross-reactivities with a panel of peptides can first be used to group the antibodies and these groupings can then be used to characterize related antigens. Thus, the seven antibodies described above as recognizing ATP-binding domain sequences that contain a basic residue at the third position also display similarities when tested in immunoblot assays of clinical samples. A protein of approximately 55 kD is recognized in urine samples from pregnant patients [119]. This protein does not appear to be a *fes* gene product because other antibodies to the same *fes* sequence as well as other regions of *fes* products fail to bind this protein. However, the 55 kD protein contains at least three *fes*-related epitopes based upon differential binding patterns of these antibodies with synthetic peptides as well as other proteins identified in tumor extracts. Thus, this group of *fes* antibodies appears to recognize a novel oncogene-related product.

VII. CLINICAL APPLICATIONS

The ability to recognize related proteins in clinical samples begins to demonstrate the power of large panels of antibodies for clinical analysis. Well over 350 oncogene-related proteins have been identified in clinical samples using immunoblot techniques [119]. This approach not only allows each of these gene products to be monitored independently, but also the size and approximate amount of the protein can be acquired automatically. Thus, a data base of expression patterns can be built rapidly. Correlations of these patterns with clinical data can then be used in detecting, classifying, grading, and monitoring neoplasia. These approaches should thus provide a wealth of data for improved management of cancer patients.

REFERENCES

1. Bishop JM. Cellular oncogenes and retroviruses. Annu Rev Biochem 1983;52:301-354.
2. Weinberg RA. Oncogenes of spontaneous and chemically induced tumors. Adv Cancer Res 1982;36:149-163.

3. Der CJ, Krontius TG, Cooper GM. Transforming genes of human bladder and lung carcinoma cell lines are homologous to the *ras* genes of Harvey and Kirsten sarcoma viruses. Proc Natl Acad Sci USA 1982;79:3637-3640.
4. Santos E, Tronick SR, Aaronson SA, et al. T24 human bladder carcinoma oncogene is an activated form of the normal human homologue of BALB- and Harvey-MSV transforming genes. Nature (London) 1982;298:343-347.
5. Taparowsky E, Suad Y, Fasano O, et al. Activation of the T24 bladder carcinoma transforming gene is linked to a single amino acid change. Nature (London) 1982; 300:762-765.
6. Tabin CJ, Bradley SM, Bargmann CI, et al. Mechanism of activation of a human oncogene. Nature (London) 1982;300:143-149.
7. Parada LF, Tabin CJ, Shih C, et al. Human EJ bladder carcinoma oncogene is the homologue of Harvey sarcoma virus *ras* gene. Nature (London) 1982;297:474-478.
8. Reddy EP, Reynolds RK, Santos E, et al. A point mutation is responsible for the acquisition of transforming properties by the T24 human bladder carcinoma oncogene. Nature (London) 1982;300:149-152.
9. Hayward WS, Neel BS, Astrin SM. Activation of a cellular *onc* gene by promoter insertion in ALV-induced lymphoid leukosis. Nature (London) 1981;290:475-480.
10. Neel BG, Haywood WS, Robinson HL, et al. Avian leukoses virus-induced tumors have common proviral integration sites and synthesize discrete new RNAs; oncogenesis by promoter insertion. Cell 1981;23:323-334.
11. Antoniades HN, Hunkapillar MW. Human platelet-derived growth factor (PDGF): amino-terminal amino acid sequence. Science 1985;220:963-965.
12. Devare SG, Reddy EP, Law JD, et al. Nucleotide sequence of the simian sarcoma virus genome: demonstration that its acquired cellular sequences encode the transforming gene product $p28^{sis}$. Proc Natl Acad Sci USA 1983;80:731-735.
13. Doolittle RF, Hunkapillar MW, Hood LE, et al. Simian sarcoma virus *onc* gene, v-*sis*, is derived from the gene (or genes) encoding platelet-derived growth factor. Science 1983;221;275-277.
14. Waterfield MD, Scrace GT, Whittle N, et al. Platelet-derived growth factor is structurally related to the putative transforming protein $p28^{sis}$ of simian sarcoma virus. Nature (London) 1983;304:35-39.
15. Robbins KC, Antoniades HN, Devare SG, et al. Structural and immunological similarities between simian sarcoma virus gene product(s) and human platelet-derived growth factor. Nature (London) 1983;305:605-608.
16. Niman HL. Antisera to a synthetic peptide of the *sis* viral oncogene product recognize human platelet-derived growth factor. Nature (London) 1983;307:180-183.
17. Downward J, Yarden Y, Mayes E, et al. Close similarity of epidermal growth factor receptor and v-*erb* B oncogene protein sequences. Nature (London) 1984;307:521-527.
18. Ullrich A, Coussens L, Hayflich JS, et al. Human epidermal growth factor receptor cDNA sequence and aberrant expression of the amplified gene in A431 epidermoid carcinoma cells. Nature (London) 1984;309:418-425.
19. Yamamoto Y, Nishida T, Miyajima M, et al. The *erb*-B gene of avian erythroblastosis virus is a member of the *src* gene family. Cell 1983;35:71-78.
20. Hampe A, Gobet M, Sherr CJ, et al. Nucleotide sequence of the feline retroviral oncogene v-*fms* shows unexpected homology with oncogenes encoding tyrosine-specific protein kinase. Proc Natl Acad Sci USA 1984;81:85-89.
21. Coussens L, Van Beveren C, Smith D, et al. Structural alteration of viral homologue of receptor proto-oncogene *fms* at carboxyl terminus. Nature (London) 1986;320: 277-280.
22. Sherr CJ, Rettenmeier CW, Sacca R, et al. The *fms* proto-oncogene product is re-

lated to the receptor for the mononuclear phagocyte growth factor CSF-1. Cell 1985;42:665-676.
23. Bos JL, Fearon ER, Hamilton SR, et al. Prevalence of *ras* gene mutations in human colorectal cancers. Nature (London) 1987;327:293-297.
24. Forrester K, Almoguera C, Han J, et al. Detection of high incidence of K-*ras* oncogenes during human colon tumorigenesis. Nature (London) 1987;327:298-303.
25. Hirai H, Kobayashi Y, Mano H, et al. A point mutation of codon 13 of the N-*ras* oncogene in myelodysplastic syndrome. Nature (London) 1987;327:430-432.
26. Liu E, Hjelle B, Morgan R, et al. Mutations of the Kirsten-*ras* proto-oncogene in human preleukaemia. Nature (London) 1987;330:186-188.
27. Liu E, Hjelle B, Bishop JM. Transforming genes in chronic myelogenous leukemia. Proc Natl Acad Sci USA 1988;85:1952-1956.
28. Schwab M, Alitalo K, Klempnauer KH, et al. Amplified DNA with limited homology to *myc* cellular oncogene is shared by human neuroblastoma cell lines and a neuroblastoma tumor. Nature (London) 1983;305:245-248.
29. Nau MM, Brooks BJ, Battey J, et al. L-*myc*, a new *myc*-related gene amplified and expressed in human small cell lung cancer. Nature (London) 1985;318:69-73.
30. Broeder GM, Seger RC, Schwab M, et al. Amplification of N-*myc* in untreated neuroblastoma correlates with advanced disease stages. Science 1984;224:1121-1124.
31. Bargmann CI, Hung M-C, Weinberg RA. The *neu* oncogene encodes an epidermal growth factor receptor-related protein. Nature (London) 1986;319:226-230.
32. Coussens L, Yang-Feng TL, Liao Y-C, et al. Tyrosine kinase receptor with extensive homology to EGF receptor shares chromosomal location with *neu* oncogene. Science 1985;230:1132-1139.
33. Yamamoto T, Ikawa S, Akiyama T, et al. Similarity of protein-encoded by the human c-erb-B-2 gene to epidermal growth factor receptor. Nature (London) 1986;319:230-234.
34. Slamon DJ, Clark GM, Wong SG, et al. Human breast cancer: correlation of relapse and survival with amplification of the HER-2/*neu* oncogene. Science 1987;235:177-182.
35. DeKlein A, van Kesse AG, Grosveld G, et al. A cellular oncogene is translocated to the Philadelphia chromosome in chronic myelocytic leukemia. Nature (London) 1982;300:765-768.
36. Konopka JB, Watanabe SM, Singer JW, et al. Cell lines and clinical isolates derived from Ph[1]-positive chronic myelogenous leukemia patients express c-*abl* proteins with a common structural alteration. Proc Natl Acad Sci USA 1985;82:1810-1814.
37. Kurzrick R, Shtalrid M, Romera P, et al. A novel c-*abl* protein product in Philadelphia-positive acute lymphoblastic leukaemia. Nature (London) 1987;325:631-635.
38. Chan LC, Karhi KK, Rayter SI, et al. A novel *abl* protein expressed in Philadelphia chromosome positive acute lymphoblastic leukaemia. Nature (London) 1987;325:635-637.
39. Clark SS, McLaughlin J, Timmons M, et al. Expression of a distinctive BCR-ABL oncogene in Ph[1]-positive acute lymphocytic leukemia (ALL). Science 1988;239:775-777.
40. Sutcliffe JG, Shinnick TM, Green N, et al. Chemical synthesis of a new polypeptide predicted from nucleotide sequence allows detection of a new retroviral gene product. Nature (London) 1980;287:801-805.
41. Walter G, Scheidtmann K-H, Carbone A, et al. Antibodies specific for the carboxy and amino terminal regions of simian virus to large tumor antigen. Proc Natl Acad Sci USA 1980;77:5197-5200.

42. Sutcliffe JG, Shinnick TM, Green N. Antibodies that react with predetermined sites on proteins. Science 1983;219:660-666.
43. Niman HL, Houghten RA, Walker LE, et al. Generation of protein-reactive antibodies by short peptides in an event of high frequency: implications for the structural basis of immune recognition. Proc Natl Acad Sci USA 1983;80:4949-4953.
44. van Ooyen A, Kwee V, Nurse R. The nucleotide sequence of the human *int*-1 mammary oncogene; evolutionary conservation of coding and noncoding sequences. EMBO J 1985;4:2905-2909.
45. Moore R, Case G, Brookes S, et al. Sequence, topography and protein coding potential of mouse *int*-2: a putative oncogene activated by mouse mammary tumour virus. EMBO J 1986;5;919-924.
46. Esch F, Baird A, Ling N, et al. Primary structure of bovine pituitary basic fibroblast growth factor (FGF) and comparison with the amino-terminal sequence of bovine brain acidic FGF. Proc Natl Acad Sci USA 1985;82:6507-6511.
47. Abraham JA, Whang JL, Tumolo A, et al. Human basic fibroblast growth factor: nucleotide sequence and genomic organization. EMBO J 1986;5:2523-2528.
48. Abraham JA, Mergia A, Whang JL, et al. Nucleotide sequence of a bovine clone encoding the angiogenic protein, basic fibroblast growth factor. Science 1986;233: 545-548.
49. Taira M. Yoshida T, Miyagawa K, et al. cDNA sequence of human transforming gene *hst* and identification of the coding sequence required for transforming activity. Proc Natl Acad Sci USA 1987;84:2980-2984.
50. Bovi PD, Curatloa AM, Kern FG, et al. An oncogene isolated by transfection of Kaposi's sarcoma DNA encodes a growth factor that is a member of the FGF family. Cell 1987;50:729-737.
51. Rijsewijk F, Schuermann M, Wagenaar E, et al. The drosophila homology of the mouse mammary oncogene *int*-1 is identical to the segmentation gene *wingless*. Cell 1987;50:649-657.
52. Ullrich A, Bell JR, Chen EY, et al. Human insulin receptor and its relationship to the tyrosine kinase family of oncogenes. Nature (London) 1985;313:756-761.
53. Ullrich A, Gray A, Tam AW, et al. Insulin-like growth factor I receptor primary structure: comparison with insulin receptor suggests structural determinants that define functional specificity. EMBO J 1986;5:2503-2512.
54. Yarden Y, Kuang W-J, Yang-Feng T, et al. Human proto-oncogene c-*kit*: a new cell surface receptor tyrosine kinase for an unidentified ligand. EMBO J 1987;6:3341-3351.
55. Gronwald RGK, Grant FJ, Halderman BA, et al. Cloning and expression of a cDNA coding for the human platelet-derived growth factor receptor: evidence for more than one receptor class. Proc Natl Acad Sci USA 1988;85:3435-3439.
56. Parker RC, Mardon G, Lebo RV, et al. Isolation of duplicated human c-*src* genes located on chromosomes 1 and 20. Mol Cell Biol 1985;5:831-838.
57. Sukegawa J, Semba K, Yamanashi Y, et al. Characterization of cDNA clones for the human c-*yes* gene. Mol Cell Biol 1987;7:41-47.
58. Nishizawa M, Semba K, Yoshida MC, et al. Structure expression and chromosomal location of the human c-*fgr* gene. Mol Cell Biol 1986;6:511-517.
59. Inoue K, Ikawa S, Semba K, et al. Isolation and sequencing of cDNA homolgous to the v-*fgr* oncogene from a human B lymphocyte cell line, IM-9. Oncogene 1987; 1:301-304.
60. Marth JD, Peet D, Krebs EG, et al. A lymphocyte-specific protein-tyrosine kinase gene is rearranged and overexpressed in the murine T cell lymphoma LSTRA. Cell 1985;43:353-364.

61. Semba K, Nishizawa M, Miyajima N, et al. *yes*-Related protooncogene, *syn*, belongs to the protein-tyrosine kinase family. Proc Natl Acad Sci USA 1986;83:5459-5463.
62. Kawakami T, Pennington CY, Robbins KC. Isolation and oncogenic potential of a novel human *src*-like gene. Mol Cell Biol 1986;6:4195-4201.
63. Yamanashi Y, Fukushige S, Semba K, Sukegawa J, et al. The *yes*-related cellular gene *lyn* encodes a possible tyrosine kinase similar to p56^{lck}. Mol Cell Biol 1987; 7:237-243.
64. Quintrell N, Lebo R, Varmus HE, et al. Identification of a human gene (HCK) that encodes a protein-tyrosine kinase and is expressed in hemopoietic cells. Mol Cell Biol 1987;7:2267-2275.
65. Ziegler SF, Marth JD, Lewis DB, et al. Novel protein-tyrosine kinase gene (hck) preferentially expressed in cells of hematopoietic origin. Mol Cell Biol 1987;7: 2276-2285.
66. Strebhardt K, Mullins JI, Bruck C, et al. Additional member of the protein-tyrosine kinase family: the *src* and *lck*-related protooncogene *c-tkl*. Proc Natl Acad Sci USA 1987;84:8778-8782.
67. Reddy EP, Smith MJ, Sunivasan A. Nucleotide sequence of ableson murine leukemia virus genome: structural similarity of its transforming gene product to other *onc* gene products with tyrosine-specific kinase activity. Proc Natl Acad Sci USA 1983; 80:3623-3627.
68. Mayer BJ, Hamaguchi M, Hanafusa H. A novel viral oncogene with structural similarity to phospholipase C. Nature (London) 1988;332:272-275.
69. Stahl ML, Ferenz CR, Kelleher KL, et al. Sequence similarity of phospholipase C with the non-catalytic region of *src*. Nature (London) 1988;332:269-272.
70. Weinberger C, Thompson CC, Ong ES, et al. The *c-erb*-A gene encodes a thyroid hormone receptor. Nature (London) 1986;324:641-646.
71. Benbrook D, Pfahl M. A novel thyroid hormone receptor encoded by a cDNA clone from a human testis library. Science 1987;238:788-791.
72. Thompson CC, Weinberger C, Lebo R, et al. Identification of a novel thyroid hormone receptor expressed in the mammalian central nervous system. Science 1987; 237:1610-1614.
73. Sap J, Munoz A, Damm K, et al. The *c-erb* A protein is a high-affinity receptor for thyroid hormone. Nature (London) 1986;324:635-640.
74. Hollenberg SM, Weinberger C, Ong ES, et al. Primary structure and expression of a functional human glucocorticoid receptor cDNA. Nature (London) 1985;318:641-646.
75. Greene GL, Gilna P, Waterfield M, et al. Sequence and expression of human estrogen receptor complementary DNA. Science 1986;231:1150-1154.
76. Misrahi M, Atger M, d'Auriol L, et al. Complete amino acid sequence of the human progesterone receptor deduced from cloned cDNA. Biochem Biophys Res Commun 1987;143:740-748.
77. Arriza JL, Weinberger C, Cerelli G, et al. Cloning of human mineralocorticoid receptor complementary DNA: structural and functional kinship with the glucocorticoid receptor. Science 1987;237:268-275.
78. Giguere V, Yang N, Segui P, et al. Identification of a new class of steroid hormone receptors. Nature (London) 1988;331:91-94.
79. Chang C, Kokontis J, Liao S. Molecular cloning of human and rat complementary DNA encoding androgen receptors. Science 1988;240:324-326.
80. Lubahn DB, Joseph DR, Sullivan PM, et al. Cloning of human androgen receptor complementary DNA and localization to the X chromosome. Science 1988;240:327-330.

81. Brumester JK, Maeda N, DeLuca HF. Isolation and expressi5n of rat 1-25-dihydroxyvitamin D_3 receptor cDNA. Proc Natl Acad Sci USA 1988;85:1005-1009.
82. Baker AR, McDonnel DP, Hughes M, et al. Cloning and expression of full-length cDNA encoding vitamin D receptor. Proc Natl Acad Sci USA 1988;85:3294-3298.
83. Giguere V, Ong ES, Segui P, et al. Identification of a receptor for the morphogen retinoic acid. Nature (London) 1987;330:624-629.
84. Petkovich M, Brand NJ, Krust A, et al. A human retinoic acid receptor which belongs to the family of nuclear receptors. Nature (London) 1987;330:444-450.
85. de The H, Marchio A, Tiollais P, et al. A novel steroid thyroid hormone receptor-related gene inappropriately expressed in human hepatocellular carcinoma. Nature (London) 1987;330:667-670.
86. Brand N, Petkovich M, Krust A, et al. Identification of a second human retinoic acid receptor. Nature (London) 1988;332:850-856.
87. Colleg WW, Chen EY, Smith DH, et al. Identification and nucleotide sequence of a human locus homologous to the v-*myc* oncogene of avian myelocytomatosis virus MC29. Nature (London) 1983;301:722-724.
88. Stanton LW, Schwab M, Bishop JM. Nucleotide sequence of the human N-*myc* gene. Proc Natl Acad Sci USA 1986;83:1772-1776.
89. Legouy E, DePinho R, Zimmerman K, et al. Structure and expression of the murine L-*myc* gene. EMBO J 1987;6:3359-3366.
90. van Straaten F, Muller R, Curran T, et al. Nucleotide sequence of a human c-*onc* gene: deduced amino acid sequence of human c-*fos* protein. Proc Natl Acad Sci USA 1983;80:3183-3187.
91. Maki Y, Bos TJ, Davis C, et al. Avian sarcoma virus 17 carries the *jun* oncogene. Proc Natl Acad Sci USA 1987;84:2848-2852.
92. Bohmann D, Bos TJ, Admon A, et al. Human proto-oncogene c-*jun* encodes a DNA binding protein with structural and functional properties of transcription factor AP1. Science 1987;238:1381-1392.
93. Angel P, Allegretto EA, Okino ST, et al. Oncogene *jun* encodes a sequence-specific trans-activator similar to AP-1. Nature (London) 1988;332:166-173.
94. Ryder K, Lau LF, Nathans D. A gene activated by growth factors is related to the oncogene v-*jun*. Proc Natl Acad Sci USA 1988;85:1487-1491.
95. Cohne DR, Curran T. Fra-1: A serum-inducible, cellular intermediate-early gene that encodes a *fos*-related antigen. Mol Cell Biol 1988;8:2063-2069.
96. Reimann EM, Titani K, Ericsson LH, et al. Homology of the subunit of phosphorylase b kinase with cAMP-dependent protein kinase. Am Chem Soc 1984;23:4185-4192.
97. Roebrock AJM, Schalken JA, Verbeek JS, et al. The structure of the human c-*fes/fps* proto-oncogene. EMBO J 1985;4:2897-2903.
98. Vogt PK. Personal communication 1987.
99. Park M, Dean M, Karl K, et al. Sequences of *MET* protooncogene c-DNA has features characteristic of the tyrosine kinase family of growth-factor receptors. Proc Natl Acad Sci USA 1987;84:6379-6383.
100. Parker PJ, Coussens L, Totty N, et al. The complete primary structure of protein kinase C—the major phorbol ester receptor. Science 1986;233:853-859.
101. Knopf JL, Lee M-H, Sultzman LA, et al. Cloning and expression of multiple protein kinase C cDNAs. Cell 1986;46:491-502.
102. Shoji S, Parmelle D, Wade R, et al. Complete amino acid sequence of the catalytic subunit of bovine cardiac muscle cyclic AMP-dependent protein kinase. Proc Natl Acad Sci USA 1981;78:848-852.
103. Lee MG, Nurse P. Complementation used to clone a human homologue of the fission yeast cell cycle control gene cdc2. Nature (London) 1987;327:31-35.

104. Martin-Zanca D, Hughes SH, Barbacid M. A human oncogene formed by the fusion of truncated tropomyosin and protein tyrosine kinase sequences. Nature (London) 1986;319:743-748.
105. Takahashi M, Cooper M. *ret* Transforming gene encodes a fusion protein homologous to tyrosine kinases. Mol Cell Biol 1987;7:1378-1385.
106. Hirai H, Maru Y, Hagiwara K, et al. A novel putative tyrosine kinase receptor encoded by the *eph* gene. Science 1987;238:1717-1720.
107. Bonner TI, Kerby SB, Sutrave P, et al. Structure and biological activity of human homologs of the *raf/mil* oncogene. Mol Cell Biol 1985;5:1400-1407.
108. Mark GE, Seeley TW, Shows TB, et al. *pks*, a *raf*-related sequence in humans. Proc Natl Acad Sci USA 1986;83:6312-6316.
109. Selten G, Cuypers HT, Boelens W, et al. The primary structure of the putative oncogene *pim*-1 shows extensive homology with protein kinases. Cell 1986;46:603-611.
110. Watson R, Oskarsson M, Vande Woude GF. Human DNA sequence homologous to the transforming gene (mos) of Moloney murine sarcoma virus. Proc Natl Acad Sci USA 1982;79:4078-4082.
111. Takio K, Wade RD, Smith SB, et al. Guanosine cyclic $3',5'$-phosphate dependent protein kinase, a chimeric protein homologous with two separate protein families. Biochemistry 1984;23:4207-4218.
112. Birchmeier C, Birnbaum D, Waitches G, et al. Characterization of an activated human *ros* gene. Mol Cell Biol 1986;6:3109-3116.
113. Hafen E, Basler K, Edstroem J-E, et al. Sevenless, a cell-specific homeotic gene of drosophila, encodes a putative transmembrane receptor with a tyrosine kinase domain. Science 1987;236:55-63.
114. Shivelman E, Lifshitz B, Gale RP, et al. Alternative splicing of RNAs transcribed from the human *abl* gene and from the *bcr-abl* fused gene. Cell 1986;47:277-284.
115. Takio K, Blumenthal DK, Edelman AM, et al. Amino acid sequence of an active fragment of rabbit skeletal muscle myosin light chain kinase. Biochemistry 1985;24:6028-6037.
116. Evans RM. The steroid hormone receptor superfamily. Science 1988;240:120-125.
117. Niman HL, Thompson AMH, Yu A, et al. Antipeptide antibodies detect oncogene-related proteins in urine. Proc Natl Acad Sci USA 1985;82:7924-7928.
118. Niman HL, Monoclonal antibodies to oncogene products. Prog Clin Biol Res 1989;288:35-52.
119. Niman HL. Unpublished findings.

14
Use of Immunologic Techniques in Gene Analysis

EMILIA TURCO, REID FRITSCH, and MASSIMO TRUCCO
Pittsburgh Cancer Institute and University of Pittsburgh School of Medicine, Children's Hospital, Pittsburgh, Pennsylvania

I. INTRODUCTION

The diagnosis of cancer at the DNA level is possible once we recognize that cancer is a genetic disorder. The genetic alterations responsible for the transformation of a certain cell into a neoplastic cell are generally assumed to be somatic, that is, limited to the affected tissue and acquired at a certain time. The somatic mutation hypothesis supports the idea that such a pathological switch may be due to numerous different, but generally not mutually exclusive, causes. Knudson [1] organized these causes into four major categories. The first category includes "spontaneous mutations," which are supposed to produce an irreducible background level of cancer. The second category includes "mutational agents," which favor the deletion or the addition of genetic material acting on this predisposed background. The third category includes the significant, genetically determined differences among individuals. These "genetic differences" may allow or impede the activation of the transformation process mediated or provoked by the previously mentioned mutagens. The last category considers the possible "inheritance" of a specific initiating mutation that predisposes an individual to "hereditary cancer." It must be noted, however, that what is inherited is the predisposition to cancer, and not the cancer itself. These somatic mutations will obviously be effective in the development of cancer when they involve critical genetic loci.

Oncogenes are those genes originally found to be responsible for the transforming capabilities of retroviruses. They seem to be highly conserved in evolution, and their nucleotide sequences are generally very similar even in unrelated species. Oncogenes in the human genome are called proto-oncogenes, and are considered to be the best candidates for mediating a neoplastic transformation after undergoing a mutational event [2]. It has been hypothesized that the proto-oncogenes play a normal role in early embryonic development or in normal cell growth, and it seems to be their mistimed or excessive activation that leads to an abnormal cellular proliferation. It has also been shown, at least in vitro, that the synergistic activity of two different, simultaneously activated oncogenes usully causes an uncontrolled cellular proliferation [3]. Among the possible mechanisms involved in this activation are point mutations, insertions and deletions, gene amplification, and chromosomal rearrangement. However, the proto-oncogenes are certainly not the only factors involved in the switch of a cell from normal to neoplastic [1]. Furthermore, probes specific for genes other than the oncogenes have been shown to be useful at least for making differential diagnoses of certain neoplasia. For example, hybridization of DNA from lymphoid neoplasia with immunoglobulin-specific or T-cell-specific gene probes can be used to determine whether a given tumor is derived from B cells or T cells [4,5].

The genetic changes caused by all these events can be monitored, at least theoretically, by using DNA hybridization techniques. The rationale of these techniques will be discussed in detail. Particular emphasis will be given to some immunologically oriented modifications to the standard hybridization techniques that may, in the near future, allow for a faster, less expensive, radioisotope-free, yet still sensitive molecular technique to diagnose cancer. These modifications will make these tests available to any clinical laboratory. The test will easily be more reliable than any of the diagnostic tests to date, and may even play a role in cancer prevention.

II. THEORETICAL BACKGROUND OF DNA HYBRIDIZATION METHODOLOGY

It is understood that the rationale for what molecular techniques may offer today to cancer diagnosis depends completely on the basic chemical structure of the DNA molecule itself. We will briefly discuss some aspects of this theoretical background to underscore which modifications are necessary to make some of the molecular procedures more suitable for our actual purposes.

Four bases are present in the DNA molecule: adenine (A), guanine (G), thymine (T), and cytosine (C). A and G are purines and T and C are pyrimidines. The bases are aligned in a specific sequence along a polymer of deoxyribose rings bonded to phosphate groups, with each base covalently linked to one deoxyribose ring. One base with its deoxyribose ring and the phosphate group constitutes a nucleotide (Fig. 1). The two complementary DNA strands that form the DNA alpha helix are kept together by hydrogen bonds present between complementary bases. Purines and pyrimidines of one strand oppose the corresponding pyrimidines and purines of the other stand; the purine A binds only with the pyrimidine T (A-T) and the pyrimidine C binds only with the purine G (C-G), forming base pairs (bp). High temperature or low ionic strength can destroy these hydrogen bonds and cause the separation of the two DNA strands, or denature the DNA. If one allows the DNA mixture to return gradually to physiological temperatures or osmolarity, the hydrogen bonds will reform between the homologous strands of DNA. This event is very specific and is known as DNA renaturation.

DNA can be blocked (fixed) in its denatured state, and an extra single-stranded stretch of DNA with known specificity, called a probe, can be introduced into the system. By restoring the renaturation conditions, the probe will eventually bind with its complementary base sequence on a strand of the original DNA molecule, generating a hybrid structure. If the DNA probe is somehow labeled, it will be possible to locate the site of hybridization.

In situ hybridization occurs when either DNA or messenger RNA (mRNA) is probed within a whole cell. In the case of DNA hybridization, a cell nucleus is fixed in metaphase, and a probe is introduced that will hybridize with its corresponding gene (or gene segment) on a chromosome. The mRNA present in the cytoplasm of a cell can also be

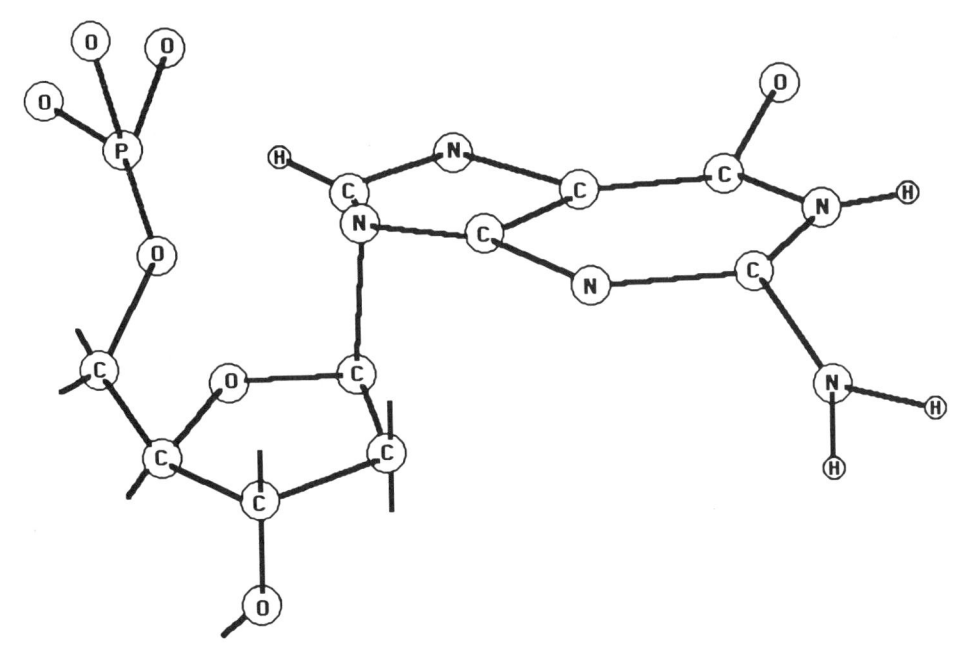

Figure 1 A DNA nucleotide. The base is attached to a deoxyribose ring that is in turn bonded to a phosphate group (modified from ref. [6]; reprinted with the permission of Scientific American Books, Inc.).

probed. If a chromosomal rearrangement, such as an interchromosomal translocation, has taken place, the results of the in situ hybridization, can be superimposed on a cell's karyotype (i.e., the banding patterns of its chromosomes) to reveal both the breaking point of the donor chromosome and the new position of the segment on the recipient chromosome. A classic example of translocation occurs in the majority of patients with Burkitt's lymphoma [7]. The c-*myc* oncogene in these patients' tumor cells has moved from chromosome 8 to chromosome 14, close to the genes for the constant region of the immunoglobulin heavy chain.

Another informative technique involves hybridization between denatured genomic DNA (i.e., all the DNA present in the nucleus of a human cell) fixed to a nitrocellulose filter (or other bonding matrix) and a specific probe. Genomic DNA (which, in the human, is composed of approximately 6×10^9 bp in 23 pairs of chromosomes) can be cut into smaller fragments by many different bacterial endonucleases, termed restriction enzymes. A restriction enzyme cuts or digests the DNA when it finds a particular sequence of 4-10 bases. These sequences or restriction sites are specific and unique for each endonuclease. The various DNA fragments, generated by endonuclease digestion, are electrically negative (as is all DNA), and are generally small enough to migrate in an agarose gel from the negative to the positive pole of an electric field. The smaller fragments move faster than the larger ones, which are delayed by the agarose acting as a molecular "sieve." An enzyme active on a six-base restriction site may generate an average of about 1 million different 2-5 kb fragments per human genome (i.e., 3×10^9 bp in the haploid human genome divided by an average fragment of 3,000 bp equals 1×10^6

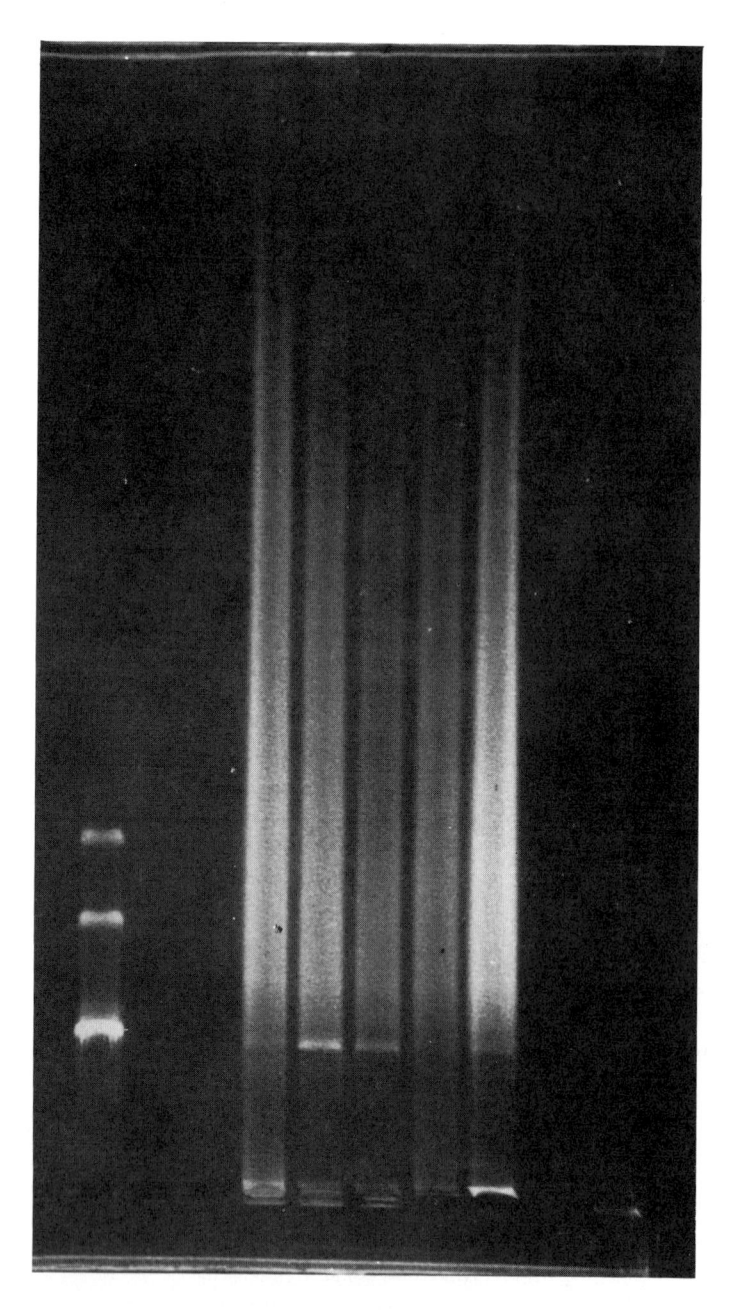

Figure 2 Agarose gel. Endonuclease (PstI) digested genomic DNA from five different donors. HindIII digested lambda phage DNA was loaded into a separate slot as fragment size marker.

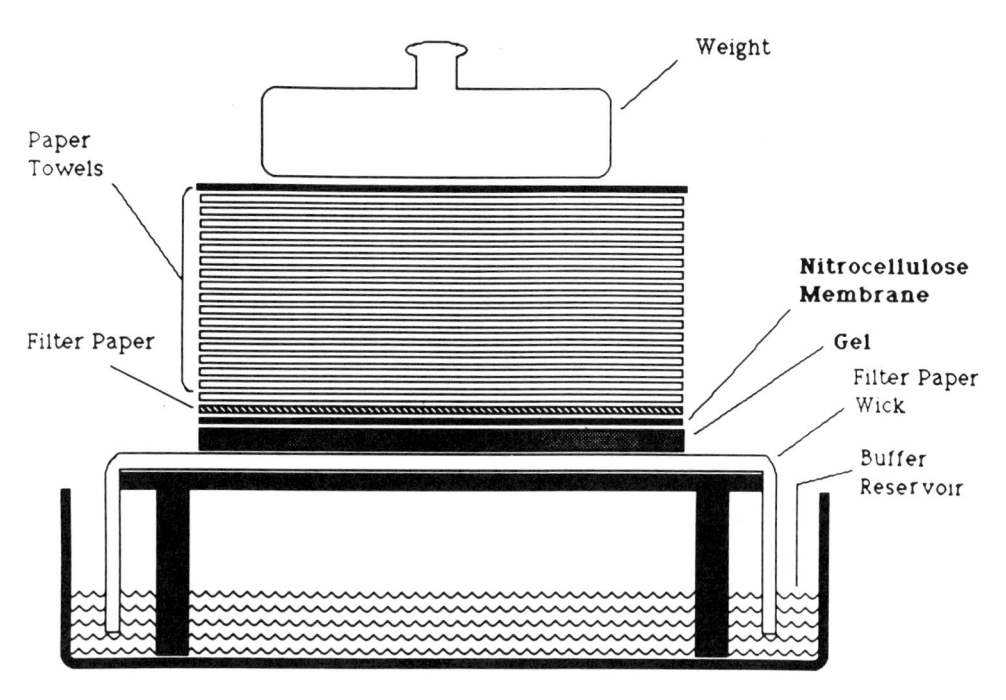

Figure 3 Southern blot. Double-stranded DNA fragments are separated by size, using agarose gel electrophoresis. The gel is soaked first in an acidic solution and then in an alkaline solution to break down further and then denature the DNA fragments, facilitating transfer. The gel is placed onto filter paper, the ends of which rest in a concentrated salt buffer reservoir. A sheet of nitrocellulose is placed on top of the gel and a large stack of absorbent paper towels is placed on top of the nitrocellulose. The salt solution is drawn up through the gel and the nitrocellulose by the capillary action of the paper towels. As the salt solution moves through the gel, it carries along the DNA fragments. The DNA fragments cannot pass through the nitrocellulose, because the nitrocellulose binds single-stranded DNA. The fragments are deposited and bound onto the nitrocellulose in the same pattern that they were in the gel. The nitrocellulose with the bound DNA is either baked or exposed to UV light to fix the DNA and can then be used for hybridizations [8]. Techniques to transfer the DNA to other bonding matrices are similar.

fragments). The agarose gel can be stained with a DNA-specific dye such as ethidium bromide so that the DNA appears, under ultraviolet (UV) light, as a smear of bands (Fig. 2). After alkaline denaturation, the various fragments can be transferred in order onto a bonding matrix.

The most commonly used technique to transfer DNA onto a matrix is the Southern blot (Fig. 3) [8]. This technique uses 0.5-1% agarose gels, which resolve fragments ranging in size from 50,000 to 5,000 bp. (Fragments that are larger do not transfer well, and fragments that are smaller do not bind efficiently to the matrix, so other techniques must be used.) The DNA is irreversibly bound to the matrix by exposing it to high temperatures or to UV light. Since the denatured DNA has been fixed, a labeled probe can be used to reveal the presence and the relative position(s) of fragment(s) carrying the corresponding gene without damaging the original DNA. The hybridized DNA can then be denatured and the probe washed off. This process can be repeated several times using different probes. Each new hybridization may take place under different renaturation or stringency conditions. These conditions may be varied to allow the hybridization of DNA sequences with many mismatched bases (low stringency) or a few mismatched bases (high stringency).

We certainly do not know the location of all the restriction sites specific for a certain restriction enzyme within the human genome. In situ hybridization techniques can tell us on which chromosome one gene is located and if that gene has been translocated. However, we cannot tell if a restriction site is present in a stretch of DNA that actually encodes for a certain protein product (exon), or if it is instead between two exons (intron), or even between two genes (spacer DNA). Fortunately, the stochastic distribution of the restriction sites can be used to our advantage since it can give us useful, albeit indirect, information about a given gene. We can also study a hypothetical gene for which we do not have a specific probe but that is suspected of being implicated in the final outcome of the disease, as long as it is physically close to another for which we have a probe.

Let us assume that the restriction enzyme we use first, A (Fig. 4, panel 1), cuts the gene for which we have a probe into a 5 and a 7 kb fragment. Each fragment carries part of the gene of interest, so that both fragments will hybridize with one part of the radioactively labeled probe, and both will be revealed as discrete bands on an x-ray film. The second enzyme, B, cuts the DNA on both sides of the gene, generating a single 15 kb fragment and revealing a single band when probed. Using this information and the results of a double digestion with both enzymes (Fig. 4, panel 1, A + B), we can organize the restriction sites in order, making a restriction map of the area where our gene is located. Once this restriction map is established using DNA from normal tissue, any mutations in the DNA sequence can be revealed using the same enzymes to digest DNA from potentially abnormal tissue from a patient. Among the mutations that can be identified are point mutations affecting a restriction site within the gene (Fig. 4, panel 2), the deletion of a stretch of DNA containing a restriction site within the gene (Fig. 4, panel 3), and a deletion that occurred outside the gene (Fig. 4, panel 4). If a certain disease or a certain type of cancer is found to be associated with the presence of, for example, a B fragment that is shorter than the B fragment found in normal tissue, and the A fragment sizes are the same, one can infer that the gene responsible for the disease is between restriction site A on the right and the adjacent restriction site B (Fig. 4, panel 5). Once this second gene is somehow affected by a mutational event, a pathological condition may result. In this example, it has been hypothesized that the deletion of the second gene generates the pathological condition.

Using this technique, correct diagnoses can be made of diseases such as alpha and beta thalassemia and sickle cell anemia, which are associated with the alpha and beta globin genes (reviewed in [9]). In some cases, bladder, lung, and colon carcinomas have been found to be associated with changes in the nucleotide sequence that codes for the amino acid in position 12 of the K-*ras* oncogene [10].

The presence of an increased number of copies of a particular proto-oncogene (the amplification of this gene) is usually caused by the duplication of the segments of chromosomal DNA containing the proto-oncogene [11], and is frequently found associated with the neoplastic transformation of a cell. This amplification of the gene can be revealed by quantitative hybridization in situ or on DNA fixed to a matrix. In this type of analysis it is critical to use the appropriate quantitative standards to estimate properly the number of copies of the gene in the DNA to be studied and of the same gene ex-

Immunological Gene Analysis 211

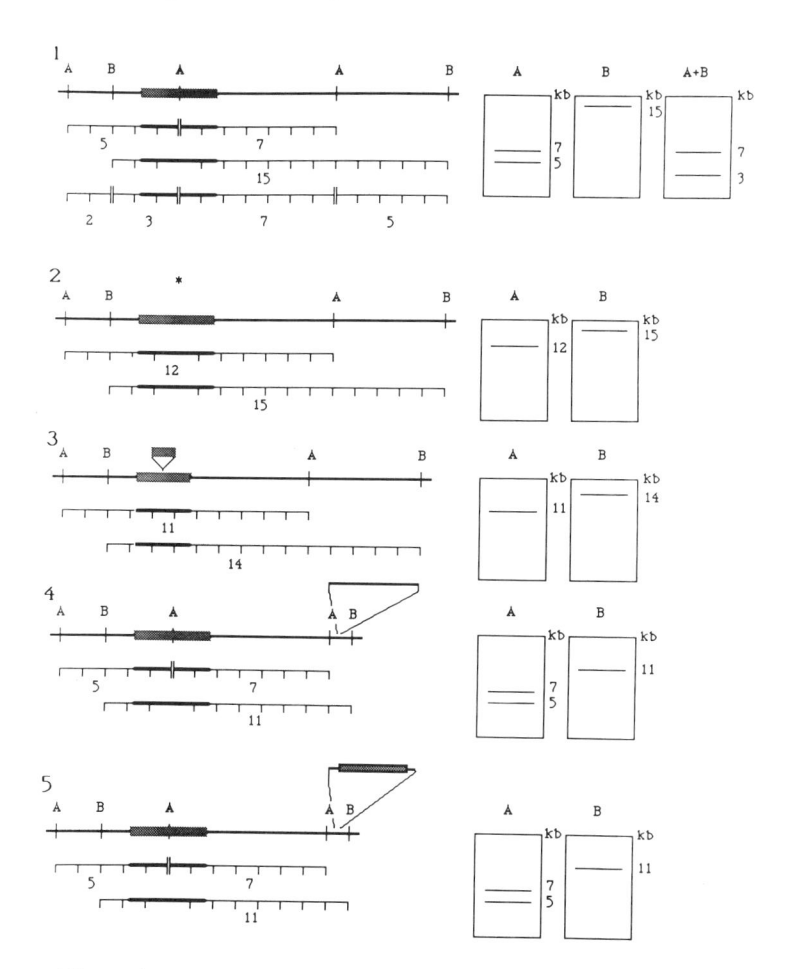

Figure 4 DNA modifications detectable by Southern blotting. Panel 1, the normal gene; panel 2, point mutation affecting a restriction site within the gene; panel 3, deletion of a DNA segment containing a restriction site within the gene; panel 4, deletion that occurred outside of the gene; panel 5, deletion of a second gene involved in the pathology (modified from [9]; reprinted with the permission of Academic Press).

pressed in normal DNA. The number of silver grains (in the in situ hybridization) or the intensity of the signal emitted by a given quantity of hybridizing radioactive probe (in hybridization with DNA fixed on a matrix) is assumed to be proportional to the number of gene copies present in equivalent volumes of normal and neoplastic tissues [12].

As previously mentioned, in situ hybridization can also be used to test the quantity of specific mRNA present in the cytoplasm of an active cell. One recent study that used this technique documented the expression of the N-*myc* gene in neuroblastoma [13]. The data obtained from in situ hybridization correlated well with the data obtained from blot analysis, and the increased N-*myc* expression was found to be correlated with a poor prognosis.

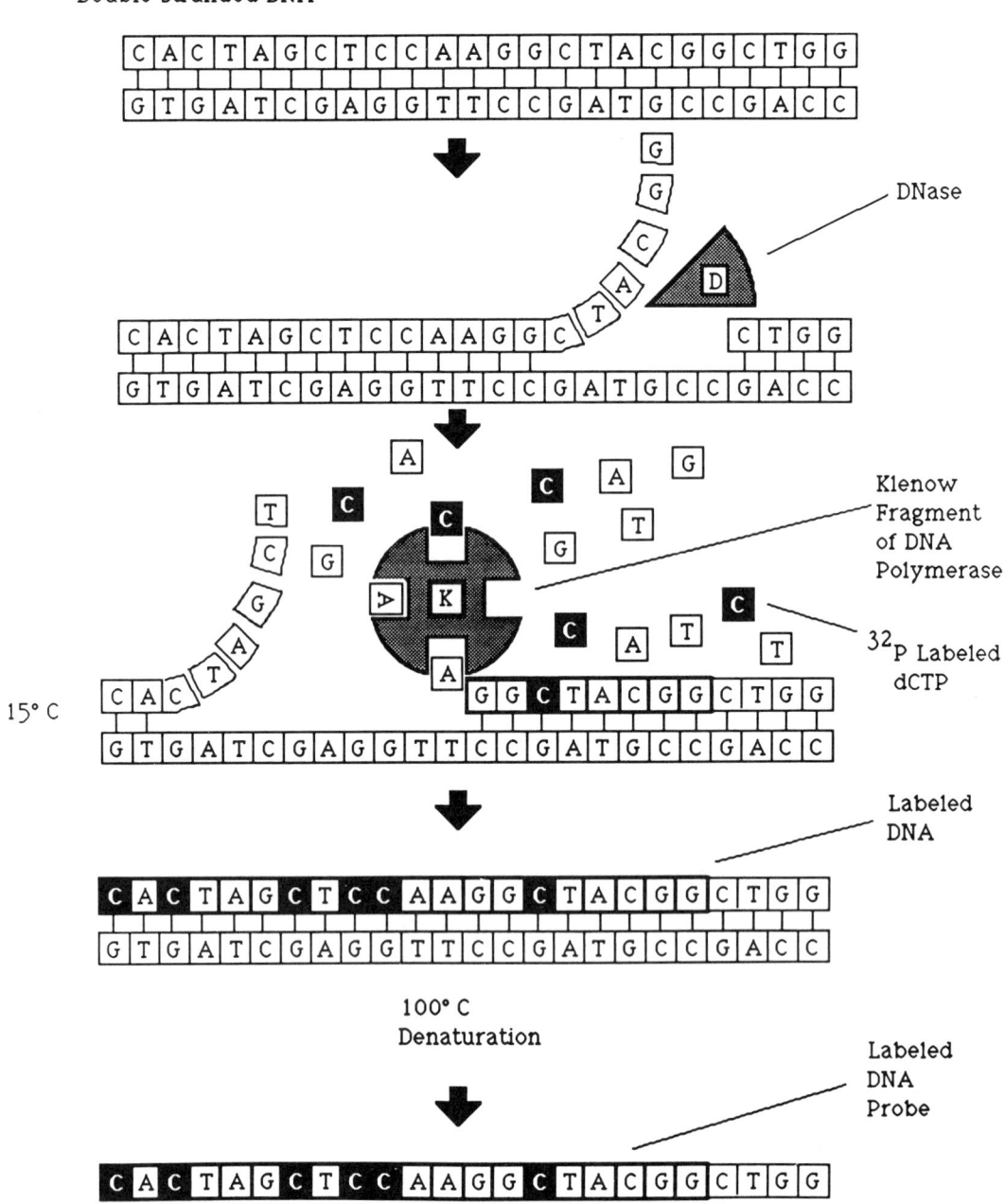

Figure 5 Nick translation. A nick is introduced into double-stranded DNA by DNase I and is used as the starting point for the subsequent DNA synthesis. The Klenow fragment of DNA polymerase I is added and the reaction is carried out at 15°C to prevent the enzyme from making more than one copy of DNA. As the DNA polymerase (represented as a cogged wheel) moves along the strand of DNA, it displaces the complementary strand and synthesizes a new complementary strand of DNA by adding the appropriate residues. This new strand will contain a number of radioactively labeled nucleotides, which replace the nonradioactive nucleotides. The DNA is heated to 100°C to separate the two strands so that the radioactively labeled strand can be used as a probe [14].

Immunological Gene Analysis

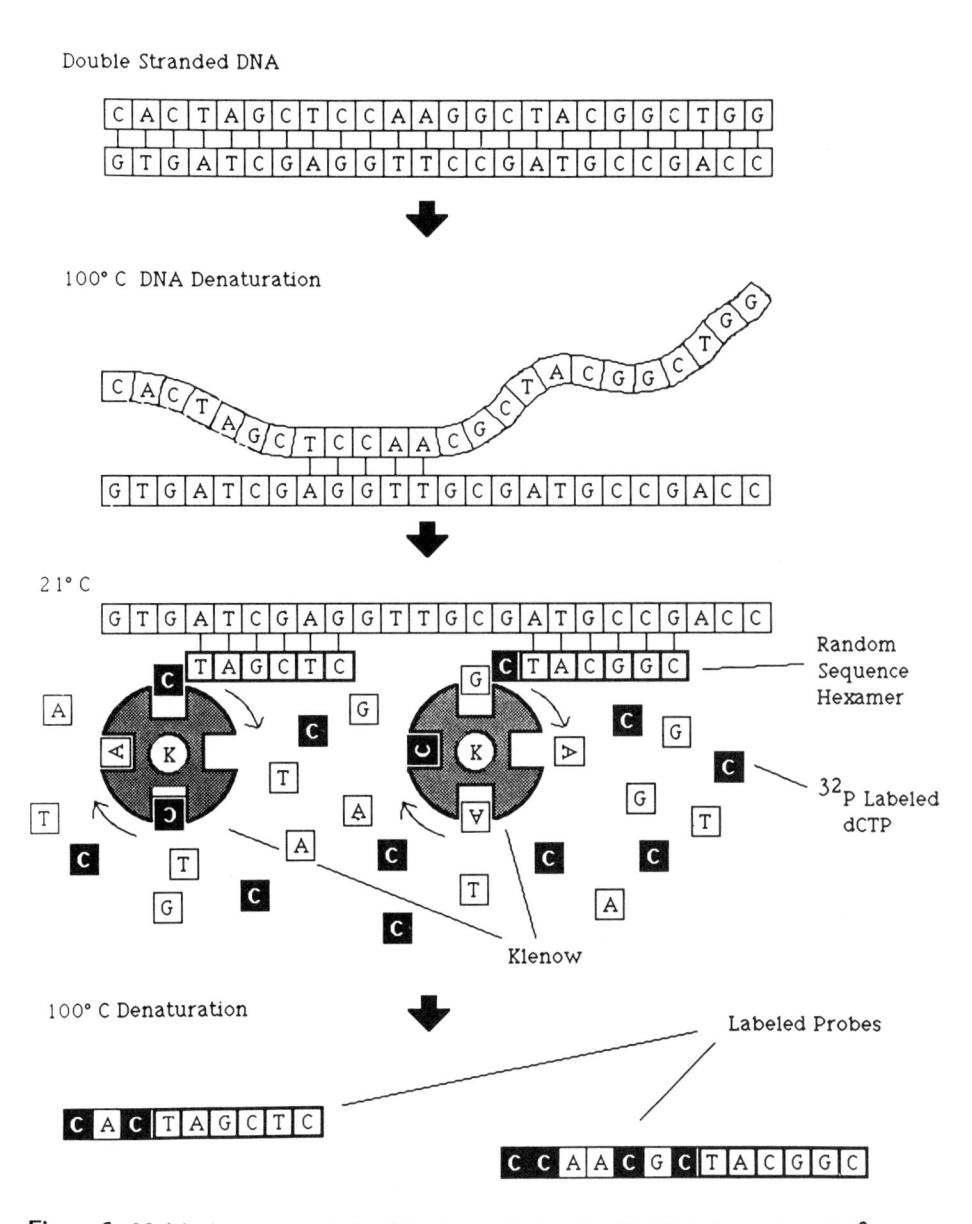

Figure 6 Multiprime assay. A double-stranded stretch of DNA is heated to 100°C to separate the strands. The denatured DNA is added to a reaction mixture containing both radioactive and nonradioactive nucleotides and small (six bp) random sequences of DNA, termed hexamers. These hexamers bind to these complementary sites on the separated strands of the original DNA. The Klenow fragment of DNA polymerase I, represented as a cogged wheel, is now added and will eventually locate the bound hexamers. Using them as starting points, it begins to synthesize new pieces of DNA. As the DNA polymerase proceeds, it uses the nucleotides present in the mix to generate a radioactively labeled piece of DNA. The resulting DNA is heated again to 100°C to separate the complementary strands of DNA, which can then be used as probes [15].

III. IMMUNOLOGICAL TECHNIQUES THAT MAY ALLOW THE ROUTINE UTILIZATION OF DNA HYBRIDIZATION METHODOLOGY IN CANCER DIAGNOSIS

Many theoretical questions about the use of DNA hybridization techniques in disease diagnosis must still be answered. However, practical aspects of hybridization technology are also hampering the popularization of such techniques. Indeed, DNA hybridization methodology needs to be further developed to allow its routine use in the diagnosis of diseases such as cancer. We refer in particular to the speed with which such an analysis must be completed.

To make the diagnosis feasible in a time short enough to make it useful, the limiting factor seems to be the use of probes that are radioactively labeled. The two most frequently used techniques for labeling both involve the substitution of one or more phosphate residues with the ^{32}P isotope. These techniques, "nick translation" (Fig. 5) and "multiprime labeling" (Fig. 6), present the advantage of generating very specific and sensitive probes; because of this, scientists have been reluctant to abandon them.

The sensitivity of a system is defined as the minimum amount of the specific DNA fragment necessary to detect a gene present in that particular fragment, and is a critical feature of any system. Sensitivity can be calculated as follows. There are 6.575×10^{-12} g of DNA per human cell.* If our fragment is, for example, $1/10^6$ of the entire genome,† or 6.575×10^{-18} g, and if 10^5 cells are necessary to detect a signal on the x-ray film to which the hybridized DNA has been exposed, the sensitivity of our system will be equal to 6.575×10^{-18} g/cell $\times 10^5$ cells, or 6.575×10^{-13} g. This is roughly 1 picogram (10^{-12} g), which is sufficient to detect a fragment of average size.

Radiolabeled probes generally offer this level of sensitivity. However, radioactivity cannot be used in many laboratories. Radioactivity is expensive, lasts for a limited amount of time, and can only be revealed using x-ray techniques that require time (between 3 and 30 days, depending on the technique) and specific equipment.

Furthermore, the number of units of radioactive isotope that can be incorporated into a certain stretch of DNA are limited by the size of the stretch itself, and their signals can only be amplified to a certain extent.

In this context the methodology developed for solving immunological problems may be used to eliminate most of the drawbacks of isotopes. Two examples will be given to indicate how DNA hybridization technology is changing to meet the need for methods of diagnosing cancer at the molecular level that can be used on a routine basis. One method uses biotinylated analogues of thymine triphosphate (TTP), for example, to prepare biotin-labeled DNA probes. A different system is based instead on the production of chemically modified nucleic acids for synthesizing probes that can be detected using immunological techniques. The characteristics of these two methods are presented below and are compared with the classic radiolabeling techniques.

---

*$6 \times 10^9 \times 660 = 396 \times 10^{10}$ daltons (i.e., number of base pairs in the human haploid genome multiplied by the molecular weight of a pair of bases) must be expressed in g for calculating the weight of 1 mole of genome. This will be obtained by dividing 396×10^{10} g by Avogadro's number (by 6.023×10^{23}).

†3×10^9 base pairs of the haploid human genome divided by 3×10^3 base pairs of an average-sized fragment.

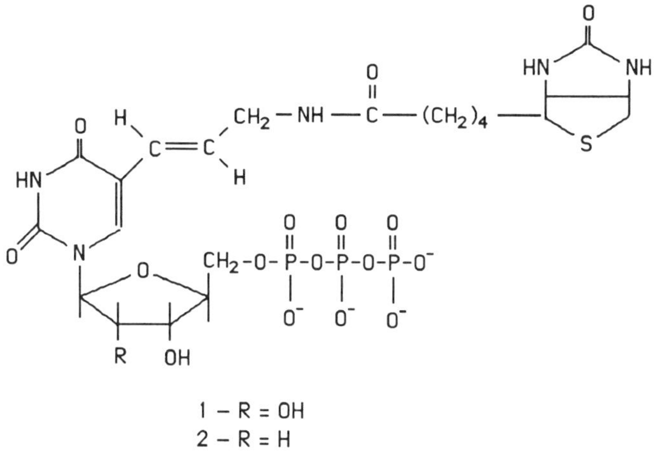

Figure 7 Deoxythymine triphosphate (dTTP) contains a biotin molecule covalently bound to the C-5 position of the pyrimidine ring through an allylamine linker arm.

A. Enzymatic Synthesis of Biotin-Labeled Polynucleotides

Ward's group [16] has synthesized analogues of deoxyuridine and uridine triphosphate (dUTP and UTP, respectively) and of deoxythymine and thymine triphosphate (dTTP and TTP, respectively) that contain a biotin (vitamin H) molecule covalently bound to the C-5 position of the pyrimidine ring through an allylamine linker arm (Fig. 7). Because polynucleotides can be generated that contain 50 molecules of biotin per kb, and because the biotinilated dUTP and dTTP analogues were shown to be excellent substrates for *E. coli* DNA polymerase I, both nick translation and multiprime labeling can be used successfully to prepare biotinylated polynucleotides (i.e., biotinylated DNA probes or bioprobes). These probes have denaturation, reassociation, and hybridization characteristics similar to those of controls that do no have biotin substitutions.

Many of biotin's features make these probes more suitable for our purposes than radioactive probes. In fact, the interaction between biotin and avidin, a 68,000 dalton glycoprotein from egg white, has one of the highest binding constants known (K_{dis} = 10^{-15}) [17]. Therefore, bioprobes can be recognized simply by adding an appropriate indicator molecule coupled with avidin to the hybridized DNA. This indicator can be fluorescent dyes, electron-dense proteins, enzymes, or antibodies. A notable feature of avidin molecules is that they can multiply the number of indicator molecules that can attach to the biotin residues, thus enhancing detection (Fig. 8).

As an alternative, antibiotin antibodies can be used without avidin to detect or locate specific sequences in chromosomes or in DNA fixed to a matrix. For example, protocols have been developed that use either rabbit antibiotin antibodies and a fluorescein-labeled goat antirabbit IgG or antibodies conjugated with alkaline phosphatase to turn a colorless substrate (5-bromo-4-chloro-3-indolyl phosphate and nitro blue tetrazolium salt [NTB]) into a purple/blue precipitate, thus identifying the positions of the hybridized biotinylated DNA probes.

These approaches offer many advantages over the use of radioactivity. Results are available in a short period of time, and the probes are stable so that they do not have to

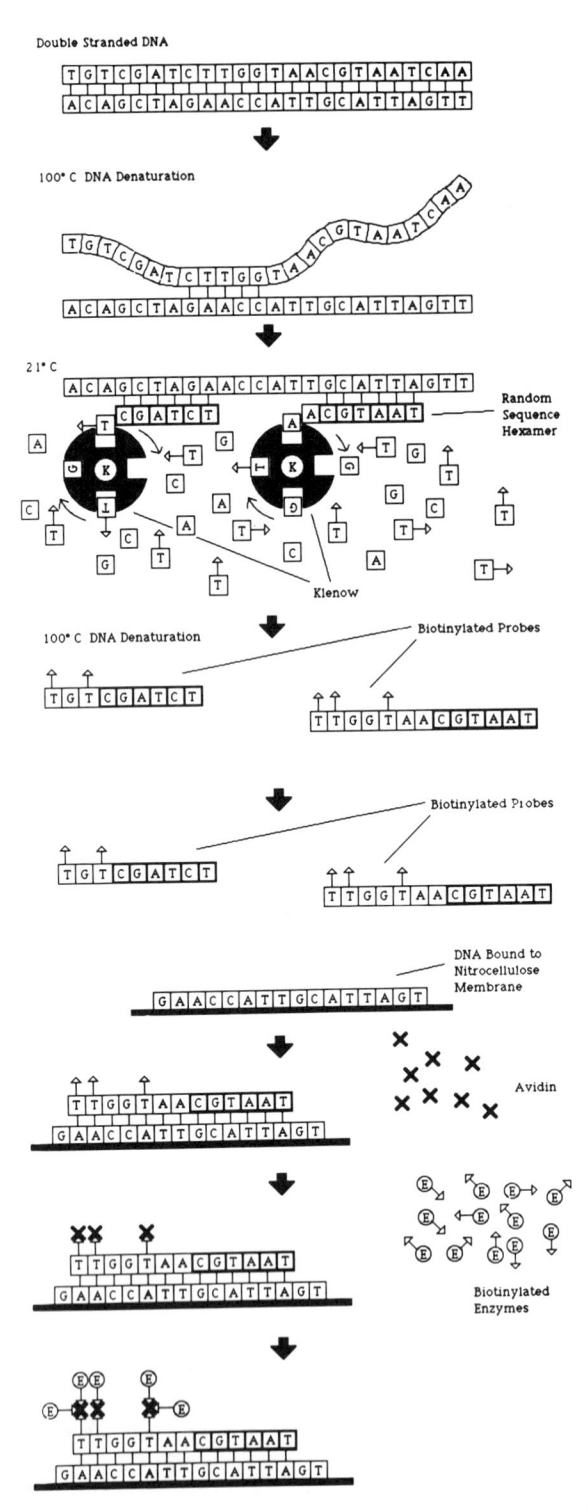

Figure 8 Multiprime assay incorporating into DNA probes a biotinylated (⇪) base: thymine (T). The revealing system proposed here consists of avidin (X) plus a biotinylated (⇪) enzyme (E).

be freshly prepared each time. In addition, any laboratory can perform these techniques because they use only nonhazardous material. If one starts with less than a million cells, biotin labeling permits the detection of target sequences in the 1-10 pg range after enzyme incubation periods of 1 hr or less. Since the biotin-labeled probe concentrations can be much higher (between 250 and 750 ng/ml) than the concentrations used with radiolabeled probes without necessarily increasing nonspecific binding to the matrix, hybridization times can be dramatically reduced to only 1-2 hr.

B. Generation of Immunodetectable Probes by Nucleic Acid Chemical Modificaton

Guanine residues in nucleic acids can be modified by treating them with N-acetoxy-N-2-acetylaminofluorene (AAF) or its 7-iodo derivative (AAIF) (Fig. 9) in an in vitro nonenzymatic reaction [18]. Such modified single- or double-stranded RNA or DNA molecules are suitable for detecting specific DNA sequences. They can be hybridized with denatured, fixed DNA using the previously described procedures, and recognized by specific antibodies [19]. Such probes are characterized by their high stability, retaining their properties for more than 2 years at +4°C. When used with alkaline-phosphatase-linked second antibodies, they also offer sensitivity in the picogram range (Fig. 10). Unlike radioactive probes, these immunonucleic probes also can be used at high concentration without a significant increase in background to allow short hybridization time. Furthermore immunochemical detection provides high resolution in a short reaction time.

Kits for nonradioactive DNA labeling and detection are already commercially available. One of them, from Boehringer Mannheim Biomedicals (Cat. # 1093657), uses the same technical approach: the dUTP is linked via a spacer arm to the steroid hapten digoxigenin (Dig-dUTP). Hybrid DNA is detected by enzyme-linked immunoassay using a specific antidigoxigenin antibody conjugated with alkaline phosphatase and an appropriate enzyme-catalyzed color reaction.

C. Restriction analysis of Amplified DNA Sequences

A novel, rapid, nonradioactive technique has recently been developed to identify certain genetic disorders. This technique does not utilize a DNA probe. Its specificity is based

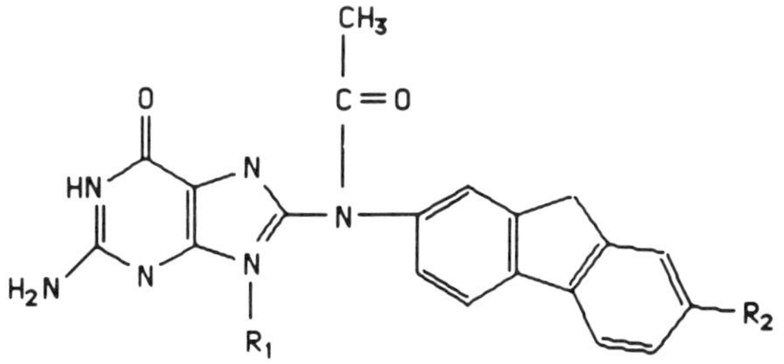

R_1 = 1'-ribosyl or 1'-deoxyribosyl;
R_2 = H(AAF) or I(AAIF)

Figure 9 N-acetoxy-N-2-acetylaminofluorene (AAF) or its 7-iodo derivative (AAIF).

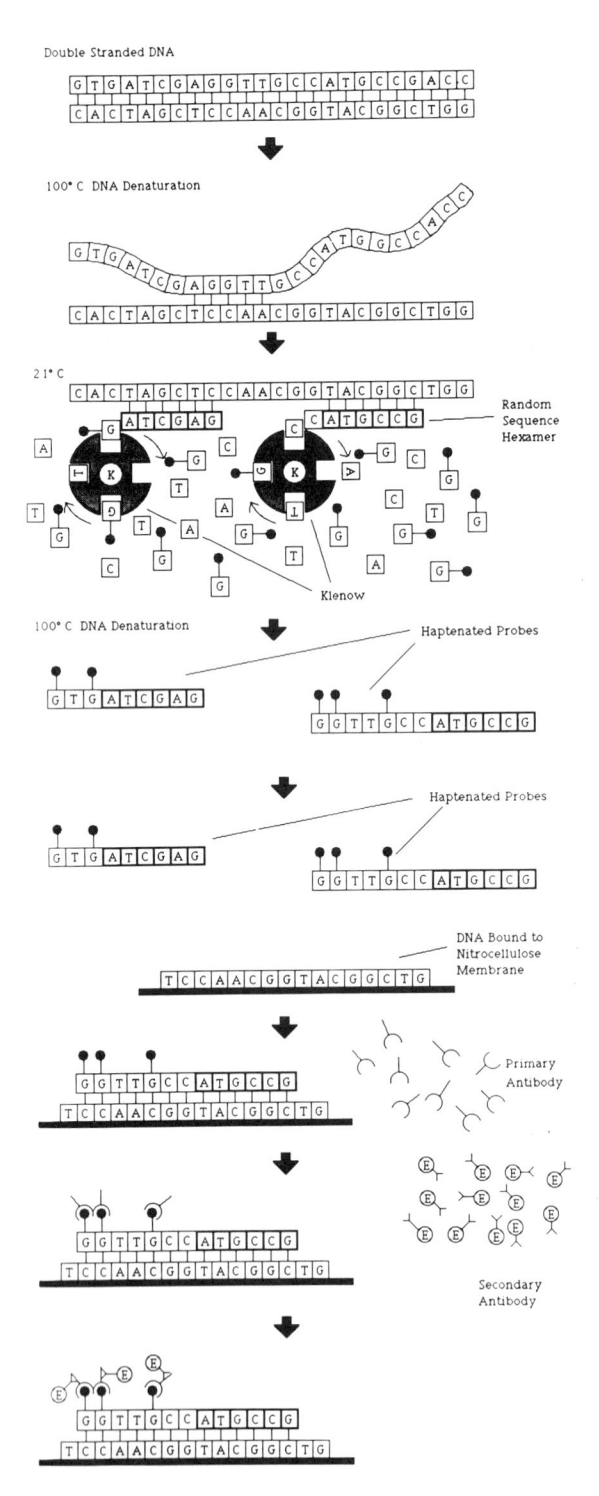

Figure 10 Multiprime assay incorporating into DNA probes a haptenated base: guanine (G). The revealing system proposed here consists of a primary antibody that is antihapten specific (e.g., rabbit anti-GUO-AAF) plus a second (revealing) antibody (e.g., enzymatically [alkaline phosphatase] labeled [E] goat antirabbit IgG).

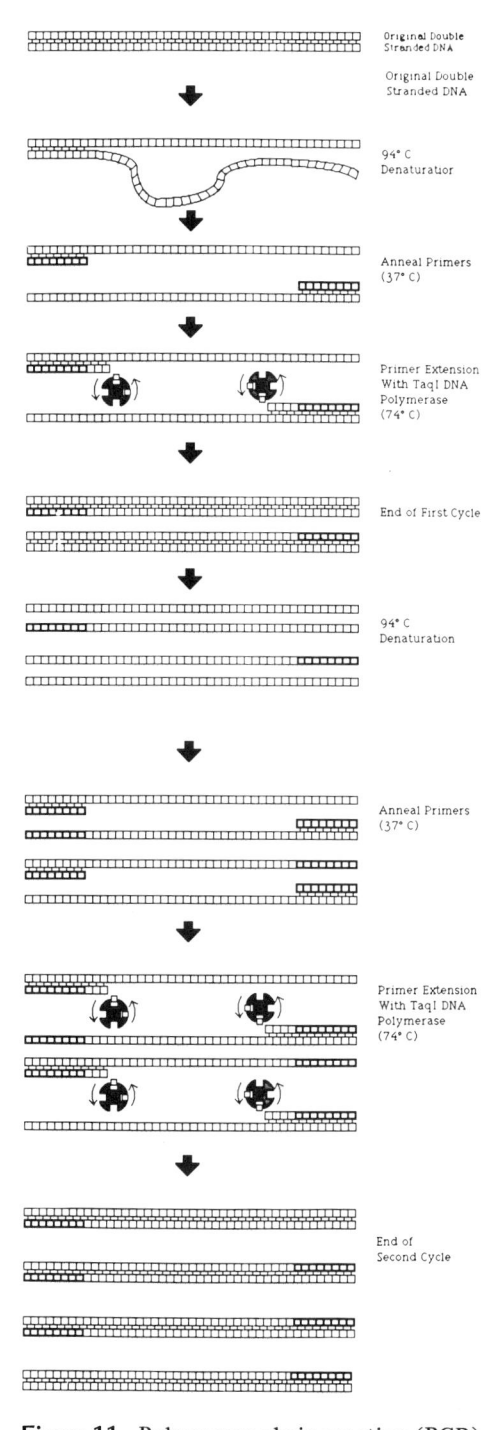

Figure 11 Polymerase chain reaction (PCR). A pair of oligonucleotide primers, complementary to sequences flanking a particular region of interest, are used to guide DNA synthesis in opposite and overlapping directions. Repeated cycles of DNA denaturation, primer renaturation, and DNA synthesis are possible when a thermostable polymerase (Taq polymerase) is used. In 30 cycles, the DNA segment can be accurately amplified 100,000 times [20].

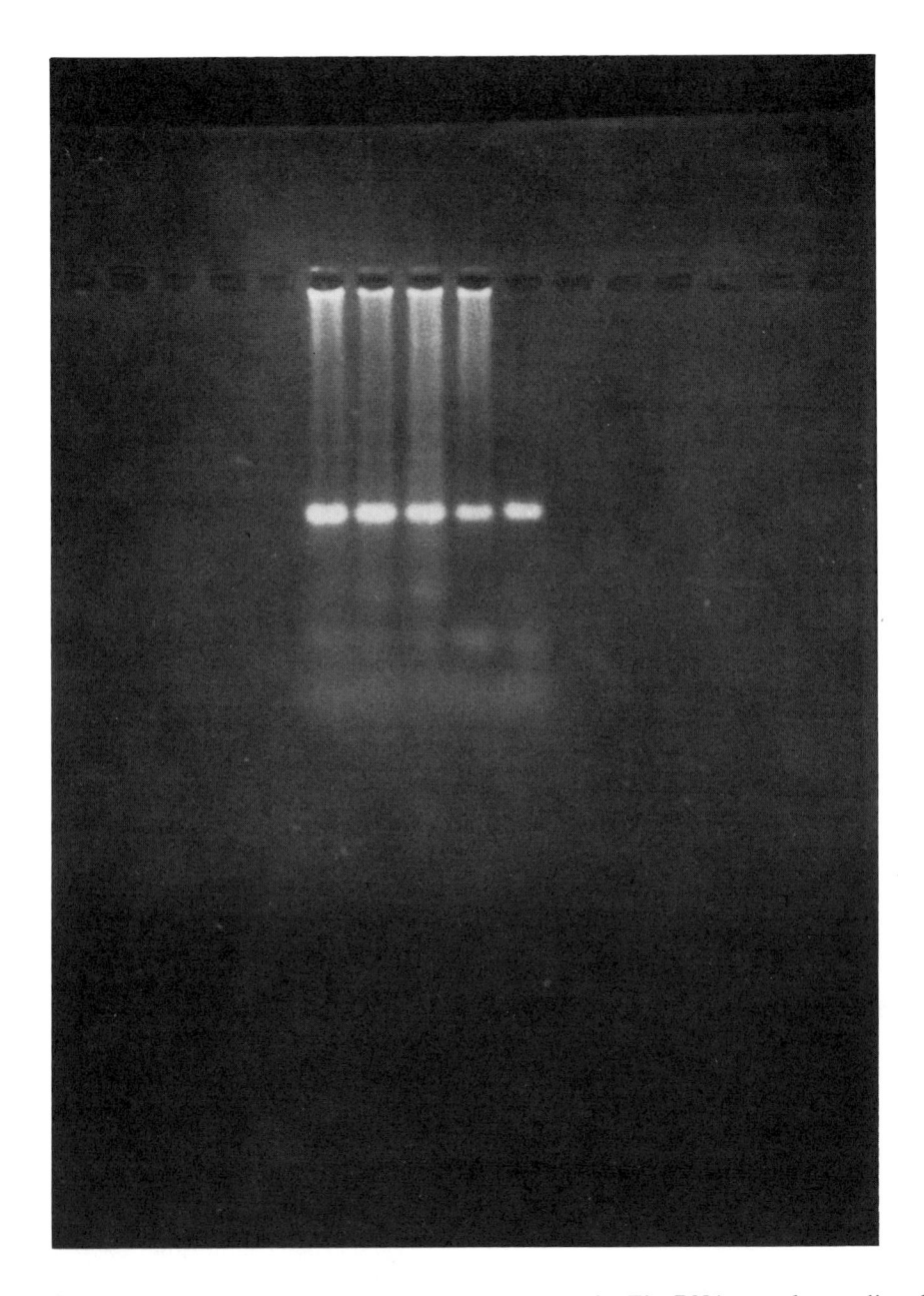

Figure 12 Agarose gel with an amplified DNA stretch. The DNA stretch encoding the first domain of the HLA DQ-beta chain has been amplified from the genomic DNA from the members of a family [22]. In slot number 1, as a control, the same amplified DNA stretch from a cDNA clone containing an HLA-DQ-beta chain gene was added [23].

instead on the specific activity of the various restriction enzymes. Although its application is for the moment quite limited to some well-known genetic disorders, we mention it here as an example of the "state of the art" of molecular techniques.

This procedure is based on the enzymatic amplification of short segments of human genomic DNA. The amplification can be obtained by polymerase chain reaction (PCR) [20] (Fig. 11) that offers at least two advantages. The first is that a high number of copies of a preselected stretch of DNA can be obtained in a limited amount of time. The number of correct copies of the specific sequence, in fact, is multiplied 100,000 times after 30 cycles, so that only 1 μg of genomic DNA as a starting amount is sufficient to permit the detection of different restriction-site polymorphisms. The second benefit of this technique takes advantage of the thermostable DNA polymerase *Taq* to allow repeated rounds of DNA synthesis without the addition of fresh enzyme at the beginning of each cycle. *Taq* polymerase is not inactivated by the high temperature used to denature the newly synthesized double-stranded DNA at each cycle.

The amplified stretch of DNA (Fig. 12) can easily be analyzed, after endonuclease digestion, on a gel stained with ethidium bromide under UV light; the presence of a single amplified band will be interpreted as the absence of a particular restriction site, while the presence of two bands indicates the presence of the site.

This very powerful technique already has been used successfully in the prenatal diagnosis of hemophilia. In this case the amplified sequence contained factor VIII intragenic polymorphisms identified by the restriction enzymes Bc/I and XbaI [21]. In addition to being extremely quick and completely specific, the technique has numerous other advantages. It uses only small amounts of genomic DNA, and requires neither the transfer of DNA to a matrix nor the preparation or use of any radioactive or nonradioactive probes. In the near future this technique should also be applicable to cancer diagnosis.

ACKNOWLEDGMENTS

Sharon Rosenshine kindly helped us by editing the manuscript. This work was supported in part by grant CA-44977, AI-23963, and DK-24021 from the National Institutes of Health, Bethesda, Maryland.

REFERENCES

1. Knudson AG. Hereditary cancer, oncogenes, and antioncogenes. Cancer Res 1985; 45:1437-1443.
2. Bishop JM. The molecular genetics of cancer. Science 1987;235:305-310.
3. Land H. Chen AC, Morgenstern JP, Parada LF, Weinberg RA. Behavior of myc and ras oncogenes in transformation of rat embryo fibroblasts. Mol Cell Biol 1986;6: 1917-1925.
4. Cleary ML, Warnke R, Sklar J. Monoclonality of lymphoproliferative lesions in cardiac transplant recipients: clonal analysis based on immunoglobulin-gene rearrangemens. N Engl J Med 1984;310:477-481.
5. Kimura N, Du RP, Mac TW. Rearrangement and organization of T cell receptor gamma chain genes in human leukemic T cell lines. Eur J Immunol 1987;17:1653-2656.
6. Watson J, Tooze J, Kunz D. Recombinant DNA. A short Course. New York: Scientific American Books, 1983:13.

7. Taub R, Kirsch I, Morton C, Lenoir G, Swan D, et al. Translocation of the c-myc gene into immunoglobulin heavy chain locus in human Burkitt lymphoma and murine plasmacytoma cells. Proc Natl Acad Sci USA 1982;79:7837-7841.
8. Southern EM. Detection of specific sequences among DNA fragments separated by gel electrophoresis. J Mol Biol 1975;98:503-508.
9. Little PRF. DNA analysis and antenatal diagnosis of hemoglobinopathies. In: Williamson R, ed. Genetic engineering 1. New York: Academic Press, 1981:62-100.
10. Santos E, Martin-Zonca D, Reddy EP, et al. Malignant activation of a K-ras oncogene in lung carcinoma but not in normal tissue of the same patient. Science 1984;223:661-665.
11. Sklar J. DNA hybridization in diagnostic pathology. Hum Pathol 1985;16:654-658.
12. Brodeur GM, Sieger RC, Schwab M, et al. Amplification of N-myc in untreated human neuroblastoma correlates with advanced disease stage. Science 1984;224:1121-1125.
13. Grady-Leopardi EF, Schwab M, Ablin AR, Rosenau W. Detection of N-myc oncogene expression in human neuroblastoma by in situ hybridization and blot analysis: relationship to clinical outcome. Cancer Res 1986;46:3196-3199.
14. Kelly RB, Cozzarelli NR, Deutscher MP, Lehman IR, Kornberg A. Enzymatic synthesis of deoxyribonucleic acid by polymerase at a single strand break. J Biol Chem 1970;245:39-45.
15. Feinberg AP, Vogelstein B. A technique for radio labeling DNA restriction endonuclease fragments to high specific activity. Anal Biochem 1983;132:6-13.
16. Langer P, Waldrop A, Ward D. Enzymatic synthesis of biotin-labeled polynucleotides: novel nucleic acid affinity probes. Proc Natl Acad Sci USA 1981;78:6633-6637.
17. Green NM. Avidin. Adv Protein Chem 1975;29:85-133.
18. Tchen P, Fuchs R, Sage E, Leng M. Chemically modified nucleic acids as immunodetectable probes in hybridization experiments. Proc Natl Acad Sci USA 1984;84:3466-3470.
19. Spodheim-Maurizot M, Saint-RUF G, Leng M. Conformational changes induced in DNA by in vitro reaction with N-hydroxy-N-2-aminofluorene. Nucleic Acids Res 1979;6:1683-1694.
20. Saiki RK, Scharf S, Faloona F, et al. Enzymatic amplification of beta-globin genomic sequences and restriction site analysis for diagnosis of sickle cell anemia. Science 1985;230:1350-1354.
21. Kogan S, Doherty M, Gitschier J. An improved method for prenatal diagnosis of genetic diseases by analysis of amplified DNA sequences. Application to hemophilia. A. N Engl J Med 1987;317:985-990.
22. Morel P, Dorman J, Todd J, McDevitt H, Trucco M. Aspartic acid at position 57 of HLA-DQ beta chain protects against type I diabetes: a family study. Proc Natl Acad Sci USA 1988;85:8111-8116.
23. Turco E, Care A, Compagnone-Post P, Robinson C, Cascino I, Trucco M. Allelic forms of the alpha- and beta-chain genes encoding DQw1-positive heterodimers. Immunogenetics 1987;16:282-290.

III
Organ Site-Associated Tumor Markers

15
Breast Epithelial Antigens in the Circulation of Breast Cancer Patients

ROBERTO L. CERIANI

John Muir Cancer and Aging Research Institute, Walnut Creek, California

ERNEST H. ROSENBAUM

Mt. Zion Hospital and University of California, San Francisco School of Medicine, San Francisco, California

I. INTRODUCTION

This review describes the use of newer immunoassays for circulating antigens (Ags) that have a direct relationship to breast cancer disease. With the increased sophistication now available to create and use reagents for assays (hybridoma production, genetic engineering of Ags, conjugation techniques, etc.), and our knowledge of the special characteristics of the breast (antigenicity corresponding to a skin appendage, milk production capability, etc.), the most successful and insightful approaches in immunoassays will be those taking advantage of these considerations to their full extent. Therefore, this succinct overview will discuss the different laboratory techniques and their participating components applicable to the detection of circulating breast epithelial markers as well as new knowledge that supports their use and optimal performance.

II. BREAST ANTIGENS

Breast cancer markers for clinical purposes can be classified into those used for the the follow-up of alternatives of the disease (breast cancer relapse or remission) and those used for screening it (detecting breast cancer among the general population). The latter will require high sensitivity and specificity, since the incidence per year for first diagnosis of breast cancer in the adult female population of the United States (25 years old and above) is approximately 0.5:1,000. Therefore, both the sensitivity and specificity will have to be above 99%. None of the markers proposed to date meets these expectations; however, panels of these markers could eventually approximate the desired characteristics. In contrast, several immunoassays are proposed today for the follow-up of the disease.

Among the Ags used for the follow-up of breast cancer with serum assays, two types have been focused on: those already characterized and to which antibodies (Abs) were created against their purified preparations, and those to which Abs were created after immunizations with human milk fat globule (HMFG), whole cells, cell homogenates, and/ or cell components. These latter Ags were then identified as new, previously unknown

Ags. Among those of the first type, carcinoembryonic antigen (CEA) has a very important place. In the second type are found those Ags discovered in the process of randomly creating Abs to different components of the breast cell. This is a large group of usually membrane-associated Ags that will generically be called breast epithelial Ags (BrE-Ags).

With regard to specificity, BrE-Ags can be organized into five distinct classes [1]: class I, normal Ags shared with cells of most other tissues, class II, Ags shared with cells of a few other tissues but with a preponderance in the breast, which thus could be referred to as characteristic of the breast; class III, Ags present only in the gland itself, being truly breast-tissue specific such as casein and alpha-lactalbumin; class IV, breast tumor Ags shared with other tumors arising in the organism; and, finally, class V, specific breast epithelial tumor Ags. Which Abs will be created and with what specificities depends on the intentions and rigorousness of the screening or absorption steps [2,3].

An alternative antigenic source is immunoaffinity-purified Ags used to obtain what could be called second-generation monoclonal Abs (MAbs) [4], which can be used to map the structure of the antigen. With the introduction of the newer techniques of genetic engineering, the synthesis of the polypeptide of the different glycoproteins, or fragments of it, has created a newer source of what could be called synthetic Ags for the preparation of anti-breast Abs [5]. They could have more specificity than those created against intact glycoproteins, which run the risk of generating Abs either totally or partially against the sugar moieties. When appropriate screening is performed for MAbs specific to breast epithelium, most if not all MAbs obtained are against the polypeptide region and not to the sugar moieties of the Ag [6,7]. MAbs such as Mc1 and Mc5 obtained after fastidious screening against different nonbreast epithelial cell lines [7] fulfill these characteristics.

III. ANTIBREAST ANTIBODIES

Different attributes are required for Abs to be useful in immunoassays of BrE-Ags in circulation. Among these attributes are lack of reactivity with serum proteins or with cells circulating in blood; specificity either for breast tumors and/or breast epithelial tissue; being able to withstand immunochemical manipulations (conjugations, solid phase binding, etc.); and the more attractive MAbs are those corresponding to epitopes found in multiple copies on the Ag (polyepitopic Ags). Polyepitopic Ags of the breast epithelial cells have been identified, and sandwich immunoassays have been successfully prepared using the same MAb for both catcher and tracer antibody [8].

Less than total specificity of BrE-Ags has proven not to be a complete drawback in serum assays, since panepithelial types of MAbs (as most if not all of those used to date are) have been useful in the follow-up of breast cancer. Although panepithelial MAbs cannot identify the tissue of origin of the carcinoma of the patient, when the patient is known to have breast cancer an increase in the circulating panepithelial Ag adequately reflects breast tumor presence and evolution.

Most MAbs used to date in breast cancer assay have been IgGs, but some assays have used IgM with success [8]. The immunoglobulin class could then obviously modify the configuration of the assay. Although MAbs have binding constants of 1×10^{-7} to 5×10^{-10} [9], it has been possible to create appropriate assays with them, although information in this regard is usually lacking in most published studies.

IV. TECHNIQUES TO BE USED

Three approaches have emerged as the most popular for present day immunoassays: the sandwich [10], the double-determinant [11], and the Ag displacement assays [12]. All these approaches are possible when using polyclonal Abs. When using MAbs in the assay, sandwich assays require Ags to be polyepitopic; in double determinant assays at least one of the epitopes detected on the Ag could be unique.

A sandwich immunoassay was used for detection of BrE-Ags in the original report [10]. In this report, the Ab was bound to a solid phase, the Ag in question was harvested from the serum by the solid-phase-bound Ab, and then the immobilized Ag was recognized by a tracer layer of the Ab (in the case of MAbs, both the tracer and catcher MAbs bound to two identical epitopes of the immobilized polyepitopic Ag [8]). To obtain a measurable signal, the tracer Ab is linked to a radioactive isotope or to an enzymatic system such as horseradish peroxidase.

When the epitope on the Ag for at least one of the MAbs is unique (or, if not, for assay configuration requirements), two different MAbs (double determinant assays) have been used: one MAb on the solid phase as a catcher of one epitope of the Ag and another MAb as a tracer binding a different epitope of the immobilized Ag [11]. In other cases, striving for higher specificity and sensitivity, immunoassays called displacement or otherwise Ab consumption techniques have been used [12]. In these displacement assays for BrE-Ags, stoichiometric concentrations of Ag and Ab are mixed with the serum and the competition for binding of the Ab by circulating Ag is then measured. However, the double determinant or sandwich assays are usually preferred for their ease of performance and short running time, while displacement assays confer more specificity but require longer incubation times and are more difficult to perform.

V. BREAST EPITHELIAL Ags FROM HUMAN MILK FAT GLOBULE AND BREAST EPITHELIAL CELLS

A. Background

Different systems of breast Ags have been known for quite some time. The breast epithelial cells synthesize specific glycoproteins (caseins) and a protein (alpha-lactalbumin), and a specific disaccharide, lactose. Casein was considered nonimmunogenic until polyclonal Abs were produced against it [13] after the hyperimmunization of rabbits. This approach opened the door to the creation of different anticasein Abs [14,15] that were used in the clinical setting with moderate success [15]. It was soon recognized that hormonal stimulation was required for casein synthesis by the mammary epithelium [16,17] and that breast tumors fail to synthesize it [18]. Also, Abs against alpha-lactalbumin were created [19]; however; for reasons similar to those for casein, attempts to use alpha-lactalbumin levels for breast cancer diagnosis proved fruitless. As for lactose, work with it proved elusive due to the difficult analytical procedures required for the isolation of lactose from body fluids.

A breast epithelial cell membrane marker studied later was a glycosyltransferase of breast epithelial cells [20]. Enzymatic values of sialyltransferase were found to be elevated in patients with large tumor burden [20,21], and large increases were reported for patients with terminal disease, but serum levels in patients with primary breast tumors were usually low or absent. A further problem was that glycosyltransferases, being nonspecific enzymes present in most cell surfaces, were not specific enough. So investigators continued their search for more specific components of the breast epithelial cell.

Starting in the earlier portion of the last decade, the literature dealt with so-called neoantigens such as CEA, originally described in colon carcinomas [22] but with a low expression in breast cancer [23]. In patients with breast tumors expressing CEA, it was useful in their follow-up [24] and in certain cases to indicate prognosis [25]. However, CEA is not ideal since responses to changes in breast tumor load are usually sluggish [24]. Despite these drawbacks, it is currently being used in breast cancer by many oncologists and clinical investigators, and a solid understanding of its uses and deficiencies has been established at this date (see [26]). In addition, CEA has found a valuable application as a benchmark against which to compare other breast-cancer-directed assays [27].

B. The Milk Fat Globule

The qualified ability of CEA to monitor breast cancer follow-up fostered the search for newer markers. Thus, in recent years the field of circulating breast cancer markers has been radically altered by the introduction of different antigenic markers [28] for which, unlike many of the previous ones, there is hardly any evidence for their metabolic and/or physiological function in the breast epithelial cell. These BrE-Ags have been found as a result of production of Abs directed against epithelial cells or their components, either normal or tumoral. BrE-Ags are easily detectable in both normal and neoplastic breast tumor cells and are characteristic, if not specific, for them [28].

At first, technical problems also encountered with other epithelial tissues had precluded the isolation of purified fractions of breast epithelial cell membrane [29], especially the abundant fat in the breast cells that interferes with the preparation of purified fractions of membrane. As an alternative, the milk fat globule of the cream fraction of human milk can be used as a source of membranous material [28]. The milk fat globule is the membrane of the apical region of the breast epithelial cell enveloping droplets of milk fat that is released during lactation to render milk fat in its emulsified state. The milk fat globule membrane can easily be separated from the fat in the globule itself by churning, which is the process used in the dairy industry to produce butter. The disrupted membrane sac, thus emptied of its fat load, is left void and can be easily centrifuged to a pellet [30], while butter floats above the whey. The milk fat globule can also be rid of its lipids by organic solvent extraction that leaves proteins, glycoproteins, and saccharides in the insoluble pellet [28]. Therefore, a successful approach was to use breast epithelial cell membrane obtained from the fat globule of milk for the immunization of hosts [28].

Since every component of the cell surface sought in the sera of breast cancer patients has been found in it, it is now expected that BrE-Ags with a high expression and prevalence in breast tumors will be found in high titers in the sera of most breast cancer patients. Thus, intense attention has been focused on the principal antigens on the milk fat globule membrane of different mammals. The milk fat globule is composed of a lipid bilayer plus an array of protiens and glycoproteins [31], which, upon radioiodination, can produce at least 25 reproducible bands in 7% polyacrylamide gel electrophoresis [32]. Five components are the most prevalent and have apparent molecular weights of 400,000, 150,000, 70, and 46-48,000 daltons [28]. The human milk fat globule (HMFG) membrane contains these molecular markers and also many enzymatic activities. It contains virtually all glycosyltransferases needed for glycoprotein synthesis [30].

When immunizing different species (R. Ceriani, 1986, unpublished results) with a delipidated HMFG preparation, the most immunogenic of these components is the one with the highest molecular weight. This is a mucin-like glycoprotein(s) first identified with the use of hybridoma techniques and called nonpenetrating glycoprotein(s) (NPGP) [7,33], and also identified by biochemical means [34,35]. High immunological recognition of this Ag complex by mice used as donors of stimulated spleen cells for fusions has resulted in the fact that most of the Abs obtained to HMFG had been directed against the NPGP complex. Even in cases where other preparations had been used to immunize, such as whole cells, antibodies have again been created almost exclusively to this molecule [36]. In contrast, the rabbit and rat react equally to other different sets of breast epithelial antigens ([7,28]; R. Ceriani, unpublished results); however, the full exploration of other species than mouse for immunizations as a means to identify useful antigens of HMFG is only beginning.

C. Assays with Polyclonal Antibodies Against HMFG

Originally, by injection of HMFG membranes, polyclonal Abs called antihuman mammary epithelial (anti-HME) were obtained and after repeated absorptions they bound breast epithelial cells selectively [28]. The discovery of this breast epithelial system of Ags opened many new immunological opportunities in immunohistopathology [37], serum assays [38], radioimaging [39], and eventually immunotherapy [40,41].

Anti-HME bound to breast epithelial cell lines, as well as normal breast cells, but did not bind nonbreast cell lines, fibrocytes, vascular cells, or blood cells [28,32]. The Ags of HMFG, a special group among BrE-Ags originally called human mammary epithelial antigens (HME-Ags), bound by this absorbed anti-serum (anti-HME) were created in the rabbit and are 150, 70, and 45-48Kd as established by affinity chromatography experiments [28] and double Ab immunoprecipitation [32].

A similar system exists in the mouse, where mouse mammary epithelial Ags are also detected by absorbed rabbit polyclonal antisera [42,43]. These antisera also identify in the mouse mammary cell membrane, whether normal or neoplastic, components of 150, 70, and 45-48 Kd, and their expression is not modified by mammary tumor virus infection [42]. These Ags are not detected in other normal tissue cells of mice [43].

Other authors later published reports of other polyclonal antisera against a step purification of HMFG [44]. These antisera, panepithelial in nature [45], were reactive only against the NPGP complex [46,47], in contrast to the original anti-HME [28]. Although anti-HME binds before absorptions to the NPGP complex, anti-HME final preparations did not recognize it as a result of absorptions with nonbreast epithelial cells to render anti-HME specific [28].

HME-Ags could be quantitated by an immunoassay in different human breast and nonbreast cell lines and in normal breast epithelial cells [32]. High concentrations were found in the normal breast epithelial cells, while the neoplastic cells had lower levels [32]. An important point established in this investigation was that protease treatment of the live breast epithelial cell surface releases most of these Ags from it [32]. Similar results in terms of cell surface digestion have been demonstrated previously [48], showing a 48-72 hr lapse for full reconstitution of the normal breast epithelial cell membrane after digestion [48].

With these polyclonal antibodies against breast epithelial cells or anti-HME, the search went on for the presence of these Ags in circulation. In nude mice carrying transplantable human breast tumors, high levels of HME-Ags were found that could be abolished by surgical removal of the breast tumor [49]. These anti-HME antisera had certain specificity since other transplantable human tumors (colon, lung, melanoma) did not increase HME-Ags values of such mice in serum [49].

The specificity of serum determinations of HME-Ags for breast tumors was tested and compared to sialyltransferase levels, which are also considered a marker for breast cancer and are present on the breast epithelial cell membrane together with HME-Ags [50]. They were measured simultaneously in the sera of nude mice grafted with human breast and nonbreast tumors [50]. Breast-tumor-bearing mice had elevated levels of both serum markers; however, sialyltransferase levels were also elevated in nonbreast tumors while levels of HME-Ags were not [50]. Upon surgical removal of all tumors, HME-Ags levels declined precipitously in the breast-tumor-bearing nude mice, while sialyltransferase levels remained elevated in both breast and nonbreast tumors [50], possibly due to surgical trauma and wound healing. At least in regards to sialyltransferase, which is a nonspecific cohabitant of the cell membrane together with HME-Ags, the specificity of the latter was proved. This indicates again that most, if not all, components of the breast epithelial cell are released into circulation, and that assay specificity (such as obtained with HME-Ags) is required to keep concurrent ailments or reactions in the tumor host from interfering with the diagnostic values obtained in the sera.

Determinations of HME-Ags levels in the sera of breast cancer patients were obtained in 1982 using a slightly different radioimmunoassay [10]. Beads coated with the polyclonal Ab were incubated with the patients' serum, then the immobilized antigen was detected by the polyclonal Ab labeled with biotin, and the latter detected by radiolabeled avidin. The assay was specific in that the sera of normal subjects, both male and female, with benign diseases of the breast, carcinomas of lung and colon, and neuroblastomas and melanomas were negative. In contrast, 25% of primary breast carcinomas stage I and more than 75% of disseminated breast cancer were above the cutoff line [10]. These results, in terms of their sensitivity for the detection of early and late breast cancer, are not very different from those obtained with the current assays based on MAb reagents. However, it should be noted that the specificity of the polyclonal is higher [10].

The only complete proof existing to date of existence of HME-Ags (or any other BrE-Ag) in all the human sera determined to have elevated values of these breast tumor markers is that provided by a very sensitive technique using radioiodination in situ of the HME-Ags bound to an immobilized antibody [10]. In contrast, other authors reported that only a small fraction of their breast cancer patients (most of whom were shown to have elevated values of BrE-Ags by serum immunoassays) gave positive results when stringent criteria to detect BrE-Ags in those sera were used, such as Western blotting [51,52]. When this in situ radioiodination approach [10] was used, the three Ags already detected in HMFG [10] by anti-HME were found in circulation (molecular weight 150, 70, and 45-48 Kd) in all breast cancer patients with elevated values. Control sera from normal subjects and patients with colon and lung carcinoma were negative [10]. In this same study the Ag corresponding to a MAb (Mc3) [7] was also found in the sera of these patients by the in situ labeling technique [10]. In later work the Mc3 antigen was found associated to immune complexes in the sera of breast cancer patients [53].

In brief, this polyclonal Ab sandwich serum test established for the first time the presence of circulating of BrE-Ags in breast cancer and led to the development of a new era in breast cancer serum diagnosis. Improved methodology has since increased our understanding of this system.

D. Assays with MAbs Against HMFG and Breast Cells

Once MAbs could be prepared, they were used for serum assays of breast cancer, especially due to the inexhaustible amounts of MAb that can be obtained after a stable hybridoma is produced. They have, as a drawback, their low binding constant and their unique binding, which introduces restricted specificity (polyclonal antibodies combine specificity for several epitopes of the same antigen). Thus, MAbs were prepared originally against HMFG [6,7] and also breast tumor cells [36,54]. As mentioned above, the most immunogenic Ag of both HMFG and breast cells is the NPGP complex, described for its binding to MAbs, Mc1, (also called HMFG-2 [6]), and Mc5 [7,33]. It is the only Ag thus far extensively quantitated in serum assays, with few exceptions [53]. Thus, despite the theoretical considerations mentioned above regarding the choice of a preferred BrE-Ag marker in circulation for immunoassays, the MAbs used to date for breast cancer serological diagnosis are those haphazardly obtained after immunization with injections of cell or membrane mixtures of breast epithelial cells and as the result of unplanned screenings. The hope is that the next generation of MAbs will take into consideration their ultimate use (serum assays, radioimaging, therapy, etc.) and specific characteristics of antibreast MAbs needed for those purposes.

After the original MAbs were created against HMFG [6,7], newer sets were created in different laboratories. They were used in immunoassays in which serum values were obtained for patients with primary breast tumor and disseminated disease. Overall, results gave partially positive values in patients with primary breast tumors and small tumor loads; as the tumor load increased, more sera became positive.

The original anti-HMFG MAbs created, HMFG-1 and HMFG-2 (the latter also called Mc1 [7]) that bind to the NPGP complex of the HMFG, were used in an assay that detected their corresponding antigen in 30% and 53% of cases of advanced cancer, respectively [51]. These percentages are low, possibly as a result of the configuration of the assay [55]. In addition, the authors were able to detect by means of Western blotting, in a few of the sera of all the positive sera detected by the assay, antigenic components with varying molecular weights (280-320 Kd) [51]. This finding could mean that either fragments of the native Ag were found or that these represent different polymorphic molecules for this Ag [52]. No positives were detected with the immunoblotting in any of the normal sera, although threshold values were detected by immunoassay [51].

Using MAb DF3 [56], which also binds the NPGP complex of HMFG, a comparison was made between a radioimmunoassay (RIA) and an enzyme-linked immunosorbent assay (ELISA) procedure. The results tended to agree: increased levels over the cutoff lines were obtained in disseminated breast cancer in more than 70% of the patients. In contrast, slightly more than 5% of the control women had values above the cutoff level. Patients with ovarian and pancreatic carcinoma and with melanoma had elevated values above the cutoff line in 47, 40, and 27% of cases [56], and 10 of 66 patients with benign liver disease also had elevated values [56]. Patients with visceral breast cancer involvement had a higher frequency of elevated values than those with local or skin recurrences. In this same study [56], Western blotting gave results similar to those outlined above for HMFG-1 and HMFG-2; few but not all patients with high immunoassay serum values of the Ag showed positive immunoblots.

Another MAb, also against the NPGP complex, called 115D8, was utilized in a sandwich serum assay [57] in which the same MAb was used for both layers. Similar to the results above, levels in about 5% of normal and benign diseases of the breast were above the cutoff line. Breast cancer patients were positive in 24% of stage I, and 21, 43, and 79% of stages II, III, and IV, respectively. There were also positive levels in patients with benign liver and kidney disease and in pregnant women (78%). Ovarian, colorectal, prostate, and lung carcinomas resulted in elevated values in a high percentage of cases, as did melanoma and lymphoma [57]. Correlation of the marker with progression or regression of the disease was reported in 95% of cases [57]. Another early attempt to measure the NPGP complex in circulation with MAb F36/22 [54] also obtained a similar percentage of positives that increase with the severity of disease, as reported in other assays [58].

A series of MAbs were created by another laboratory [59]. Two were selected, W1 and W9, that apparently also bind the NPGP complex. Elevated levels were found in 47% of breast cancer patients with visceral metastases (these were favored over localized metastases). This assay was also positive in 4% of normal subjects. Other carcinomas (colorectal, lung, ovarian, and prostate) showed elevated values in 12-60% of cases [59]. Another assay [60] using MAb AB13 recently detected the NPGP complex in approximately half of the patients with advanced breast cancer.

MAbs DF3 [52] and 115D8 [57], which were discussed above, have been used in a double determinant assay called CA 15-3 that is available commercially. The originators of the assay reported [61] levels above the cutoff line in appproximately 80% of the advanced cancer patients and in only approximately 30% of patients with primary breast cancers. These results match those of a similar study [62]. In contrast, other authors [63] report that only 13% of the primary breast tumors and 72% of the disseminated ones were positive. In a more detailed study [64], other authors find that only 24% of breast cancer patients have elevated CA 15-3, while in patients with disseminated disease almost 70% showed positive results. Comparisons were made between CEA, in both primary and disseminated breast cancer, and CA 15-3 proved the latter to be more sensitive [65]. In all, this commercial assay as well as all the others discussed above against the high-molecular-weight component of HMFG or NPGP attain at best, percentages of positives (sensitivity) similar to those reported in the original report that established the presence of BrE-Ags in circulation in breast cancer using polyclonal antibodies [10]. However, their specificity for tissue of origin of the tumor is very low (the antigen[s] is almost panepithelial) and their specificity for disease condition is hampered by high values in hepatic and kidney disease, pregnancy, polymorphic expression of the antigen(s), and other conditions.

A common feature of all these immunoassays (using either enzyme or radioassays to measure results) is that they indicate a positive correlation between increasing tumor load and higher serum levels. However, it would be desirable to increase the sensitivity to improve the detection in the early stages of the disease. In this regard, an assay using the NPGP complex as a marker that has shown particular sensitivity in the authors' hands [8,66] uses MAb 3E1.2. Up to 68% of early stage breast cancer patients are claimed to be positive by this assay in a limited number of patients, while CA15-3 only detected 3% of them [66]. However, the 3E1.2 assay results in 18% positives among patients with benign breast disease.

MAbs to be used in immunoassays should be created against BrE-Ag epitopes expressed in breast neoplasia. It has been shown that alterations in glycoprotein antigens on breast tumor cell membranes involve changes in the glycosylation pattern [67]; monosaccharide substitutions or elongation of oliposaccharide chains without modification of the core monosaccharide sequences [67]. A step in the right direction was the use of MAb B72-3 [36], which was claimed to be carcinoma-specific, although further investigations showed binding to NPGP in benign breast disease tissues [68] and some normal breast tissue [68]. The qualified carcinoma selectivity of B72.3 could have been conferred on an oligosaccharide alteration related to changed blood group expression in neoplasia [69,70]. Immunoassays have been created [71] with B72-3 and tested in breast cancer patients' sera; unfortunately they had low sensitivity perhaps due to the low prevalence of B72-3 in breast cancer tissue [72].

All immunoassays using MAbs described so far for breast cancer follow-up rely on serum levels of the NPGP complex. Two other antigens have been explored: GP-15, Mc3/Mc8 antigen. The former is a low-molecular-weight BrE-Ag (15 Kd) present in the cell membrane [73]. It detects mainly, if not exclusively, breast epithelium that has undergone apocrine metaplasia. It has been found, possibly as a result of its selectivity, in approximately 40% of the sera of breast cancer patients. In other work [53], a low-molecular-weight antigen (46 Kd) of the HMFG [7], already detected by radioiodination in situ in breast cancer patients' sera [10], was measured in sera by an immunoassay in a detailed study [53]. Using the configuration of sandwich assay previously used [10], levels of this Ag, identified by MAbs Mc3/Mc8, were detected in breast cancer patients and not in normal subjects, or in those with ovarian and colon carcinomas or osteosarcomas [53]. Levels of the Mc3/Mc8 Ag, however, were inversely related to the tumor load: small tumor load showed 95% positive, and high tumor load 65% positive. This fact was explained by the presence of immune complexes against this Ag that increased in titer in the group with high tumor load. The presence of higher-level immune complexes could have accelerated the clearance of Ag from blood.

In summary, as seen above, almost all assays using MAb have detected the NPGP complex of the HMFG [8,12,27,51,52,54-56,58-61,71], however, it is possible that most of them bind different epitopes. Heterogeneity of epitopic expression may have created the relatively small differences seen among the different assays. From all these studies, the diffusely panepithelial nature of NPGP complex as a marker is established, as shown by high circulating levels found in other carcinomas, and even its detection in melanomas and leukemias [57]. The levels obtained vary depending on the units used in the different immunoassays, however, the percentages positive at different stages of breast cancer are similar to those originally reported with the polyclonal assay to other components of the HMFG [10]. One drawback is that the present assays detecting NPGP complex lack the specificity of the original polyclonal assay [10].

To test the specificity issue, a comparison [12] was made among CEA; an assay detecting the NPGP complex using MAb Mc5 [7] in an Ag displacement assay configuration; and the original polyclonal assay against HME-Ags [10]. The polyclonal assay showed a very high specificity showing negative values for colon, ovarian, pancreatic, laryngeal, and endometrial carcinoma as well as for lymphomas, myelomas, melanomas, and leukemias. Only one case of lung carcinoma had an elevated value [12]. All normal serum controls were negative, thus giving this assay high specificity [12]. Positive serum values for both the NPGP complex and CEA assays were not restricted to breast tumors [12]. When a comparison was made with the other assays, a sensitivity and specificity above that of the MAb assay for the NPGP complex and CEA was found for the polyclonal assay for HME-Ags [12].

These three assays were also compared for their follow-up ability. An example of such studies in the breast cancer case is shown in Figure 1. There the response of the polyclonal antibody assay for HME-Ags (cutoff 100 μg/ml) to breast cancer relapse and tumor mass change shows, as demonstrated before [12], a very sensitive response far above the

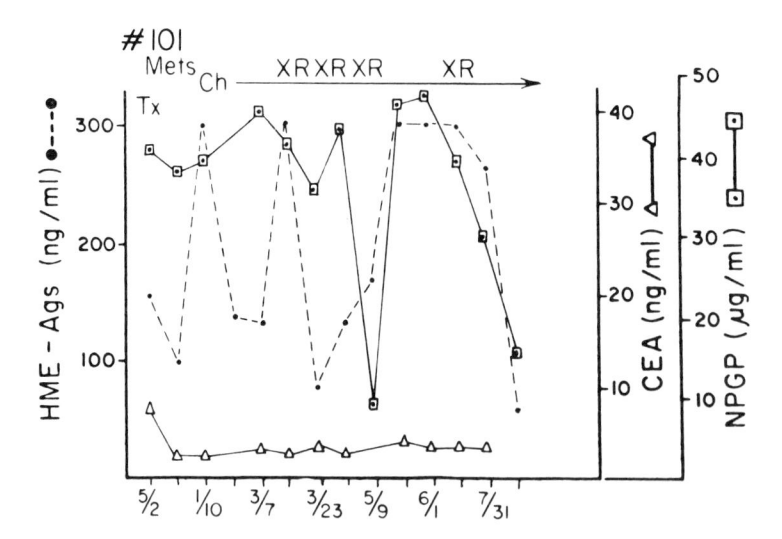

Figure 1 Simultaneous serum determinations of HME-Ags (cutoff value 100 μg HMFG protein equivalent/ml); NPGP complex (cutoff value 10 μg HMFG protein equivalent/ml); and CEA (Abbott CEA assay). Mets, metastases; ch, chemotherapy; XR, radiotherapy. Assay values shown in ordinate, time lapeses in abscissa.

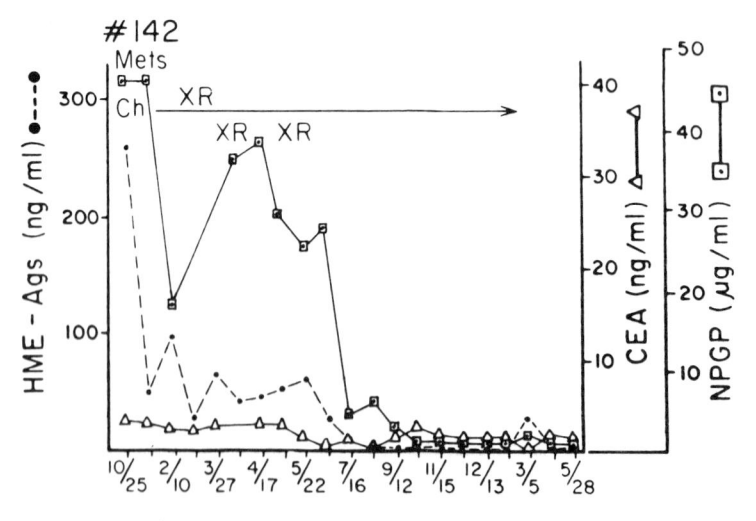

Figure 2 Simultaneous serum determinations of HME-Ags (cutoff value 100 μg HMFG progein equivalent/ml); NPGP complex (cutoff value 10 μg HMFG equivalent/ml); and CEA (Abbott CEA assay). Mets, metastases; ch, chemotherapy; XR, radiotherapy. Assay values shown in ordinate, time lapses in abscissa (from ref. [10]).

Table 1 Comparison of the Ability to Predict Relapse of HME-Ags (Polyclonal Assay), NPGP (MoAb NPGP-Complex-Directed Assay Using MoAb Mc5), CEA (Abbott Assay)

Ag	n[a]	Increased[b]	Predictive Ability (%)
HME-Ags	15	11/15	73.3
NPGP	15	7/15	47.2
CEA	15	7/15	47.2

[a]Breast cancer patients in disease-free period with no evidence of disease were followed for up to 1 year after the increase in the NPGP complex serum level was detected.
[b]An elevation of marker level of 50% or more in the NED breast cancer patient over the baseline level before any overt clinical finding of disease was considered an increase.

one demonstrated for the Mc5 assay (cutoff 10 $\mu g/ml$) for the NPGP complex. In contrast, CEA either responds slowly or not at all. In another case shown in Figure 2, where measurable breast tumor mass shrinkage was obtained, there was a fast decrease of HME-Ags corresponding to a decreased tumor mass brought about by irradiation, while levels of the NPGP complex remain high and CEA levels are unresponsive. In summary, the polyclonal assay for HME-Ags is much quicker to respond to changes in tumor mass and more accurate in predicting objective changes in tumor mass than are the other two assays (CEA and NPGP complex).

Previous studies showed a prognostic power for BrE-Ags [57] up to 90%. In another study, a comparison of the ability of these three assays (CEA, NPGP complex, and HME-Ags) to predict relapse was performed [74] and later extended. Table 1 depicts part of these studies and shows a comparison of the ability to detect relapse within the year in breast cancer patients with no evidence of disease (NED) after an increase of 30% in the serum marker baseline by any of the three assays. From these results, the HME-Ags technique could be the method of choice to establish prognosis due to its higher predictive ability and ability to detect a change in tumor mass earlier due to its sensitiveness, as shown in Figures 1 and 2.

VI. RELEASE, METABOLISM, AND CATABOLISM OF BrE-Ags

So far, with a few exceptions, the BrE-Ags identified in circulation in breast cancer patients have been Ags present on the apical surface of normal breast epithelial cells away from the vascular system. However, in tumors this apical polarization is lost and BrE-Ags are distributed throughout the cell membrane and, in most instances, are expressed in the cytoplasm. Little is known, however, about the mechanism for the release of BrE-Ags into the circulation by tumors. The loss of polarity of the antigens on the cell surface of the breast epithelial cell in breast cancer and alterations in its contact with the basement membrane could be responsible for their accessibility and the easy release in the stroma, and later for their availability to the capillary circulation. Tumor cell death, which is very conspicuous in breast cancer, could be another factor. Antigenic release in breast tumors is clearly demonstrated in histopathological sections: large clouds of BrE-Ags can be seen in immunoperoxidase staining surrounding the breast tumor islands dispersed in the thick stroma (R. Ceriani, 1988, unpublished observation) and not around normal lobules. This accumulation in the stromal tissue, which is notoriously poor in capillary circulation, could create deposits that could justify the slow release of this antigen into the circulation after tumor removal and the long turnover times reported by some authors [75]. In addition, availability of serum levels of these markers could be manipulated as shown by an increase in levels of certain BrE-Ags stimulated by hormones [76], which suggests interesting potential diagnostic approaches.

Our knowledge about the metabolism and catabolism of BrE-Ags is also scant. Some information exists about the rate at which BrE-Ags are regenerated after digestion on breast mammary cells and on the magnitude of this effect [48]. This synthesis could be a rate-limiting factor in the availability of the BrE-Ags in circulation. This production and release of BrE-Ags by tumors and posterior catabolic mechanisms must govern BrE-Ag levels in circulation. Using a very sensitive in situ radioiodination technique, HME-Ags (molecular weight 150, 70, 45 Kd) were never detected in normal sera and always in breast cancer patients with elevated values documented by immunoassay [10]. Other authors claim to find in normal subjects (even males) NPGP complex in circulation using Western blotting techniques [52]; however, others [51] report that only a small fraction of the positive sera were detected. Thus, still there is no complete certainty as to whether some values obtained by immunoassays in the sera of normal subjects could represent only vestigial noise in the immunoassay employed and also why the NPGP complex cannot be detected in all positive sera.

So far, the NPGP complex (approximately 400,000 dalton molecular weight) [51] and a 46-48 Kd HME-Ag [10] have been clearly identified in the sera of breast cancer patients. The latter can also be measured by assays using MAbs Mc3 and Mc8 [7], and has been found to integrate immune complexes in sera of breast cancer patients [53]. Although this Mc3-Mc8 antigen is present in the milk fat globule and the normal breast epithelial cell, breast cancer patients seem to respond immunologically to it. Thus, this Ag is found to integrate immune complexes in the sera of these patients [53] which could be dissociated by varying concentrations of MAbs Mc3 and Mc8 [53]. Also, when this BrE-Ag was added to the sera of breast cancer patients, it incorporated into immune complexes depending on the level of the free antigen already present in the sera [53]. As expected, results of immunoassays for the Mc3-Mc8 were shown to be influenced by circulating immune complex levels (which are in turn related to tumor burden) [53]. The presence of this HMFG Ag associated to circulating immune complexes identifies a mechanism that should accelerate its clearance. This is demonstrated by lower serum values of this antigen in patients with terminal stage breast cancer when circulating immune complexes were at their highest [53]. Thus, this Ag serum concentration is postulated to be balanced by the amount of circulating immune complexes present, which in turn influenced their catabolism [53].

No information is available on the mechanisms or site of catabolism of BrE-Ags. In recent studies (Fig. 3) in which affinity-purified NPGP complex from HMFG was produced by passage and ulterior release from MAb Mc5 columns, tagged with radioiodine and injected intravenously (i.v.) into mice, an extremely fast clearance of NPGP complex from circulation was found, if the injected i.v. dose was below 1 μg. Almost immediately after this injection, and while the tagged NPGP complex was cleared, a fragment with molecular weight of approximately 45 Kd started to accumulate in the serum (Fig. 3). This fragment had a much slower clearance from serum than NPGP complex. In addition, a larger fragment of apparent molecular weight 70K appears in a lesser amount as a faint band above.

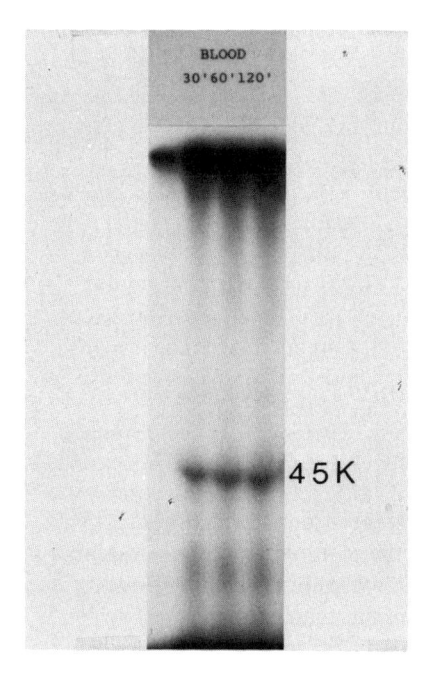

Figure 3 Polyacrylamide gel electrophoresis and autoradiograph of ^{125}I-labeled NPGP complex (first lane starting from left) before injection, and of sera of BALB/c mice at 30, 60, and 120 min after the i.v. injection of 12 × 10^6 cpm of the ^{125}I-NPGP complex. Direction of the migration: downwards; NPGP complex in first lane to the left upper band >400 Kd m.w.; and in 30, 60, and 120 min lanes upper bands (s) > 4000-280 Kd m.w. The lower intermediary band observed in lanes corresponding to 30, 60, and 120 min approximately 45 Kd m.w. (molecular weight standards not shown).

In the experiments above, a fast turnover time (T 1/2 = 250 min) was measured after the injection into BALB/c mice of radioiodinated NPGP complex. Most of the label accumulated soon in the liver (R. Ceriani, 1988, unpublished results), while, as shown above, in the serum an approximately 45 Kd molecular weight fragment appeared that had a much longer T 1/2. If the values obtained in the nude mouse (body weight 20 g, tumor weight 100 mg, NPGP 1 μg/ml [HMFG protein equivalent]) are transposed to the human situation, parameters (Table 2) can be calculated for the metabolic clearance of the NPGP complex from circulation in breast cancer patients in steady state of NPGP complex entrance into circulation (the following assumptions for the human subject are made: body weight, 60 kg; NPGP in sera, 30 μg [HMFG protein equivalent] ml; serum compartment, 4% of body weight).

Table 2 Parameters for Metabolic Clearance of NPGP Complex

NPGP level	30 μg (HMFG protein equivalent)/ml
Tumor load	300 g
T 1/2$_{biol}$	250 min
Cl$_{NPGP[equil]}$	0.06 μg (HMFG protein equivalent)/ml/min

Values such as that given for the clearance of the NPGP complex are not in agreement with those reported by other authors [75], who report 45-60 days. These latter results could be due to possible bodily storage of the Ag.

Thus, in view of the demonstrated rapid experimental clearance of the NPGP complex and the appearance of a smaller fragment in blood, the question to be asked regards the contribution of this fragment to the immunoassays. With an appropriate search for metabolic and catabolic phenomena connected to BrE-Ags, it may be possible to clarify this, and thus qualify judgment on the meaning of a level of circulating antigens in a given breast cancer patient, which will possibly depend on the clinical stage, tumor burden, and other unknown conditions of this patient.

VII. CONCLUSIONS AND NEWER AREAS OF STUDY

Enough information exists on antibody-detected BrE-Ags serum markers for us to highlight the following observations:

1. Among existing assays for breast cancer markers, those identifying HMFG and BrE-Ags seems to be the most attractive due to the prevalence of these antigens in the breast tumors themselves; their higher percentage of positives when compared to other assays; the ease of their assay; and the abundance of MAbs against them.
2. The profusion of assays aginst the NPGP complex (very immunogenic in the mouse) show, however, the limitations of this marker for breast cancer follow-up. Only some advanced breast cancer patients (50-70%) have detectable levels in circulation according to different assays. These percentages of positives in advanced (as well as low percentages for early stage) breast cancer (approximately 25%) are still at the same level as those in the original report of BrE-Ags with a polyclonal antibody [10]. The search for better assays using any other of the constitutively synthesized BrE-Ags in breast tumors is long overdue. The use of the NPGP complex (in fact, the largest component of HMFG) as a breast cancer marker was the fortuitous result of its high immunogenicity that translated into easy-to-obtain MAbs. However, due to its large size this molecule could have a reduced permeation into the circulation and a very fast clearance from serum and would not be the BrE-Ag of choice for serum immunoassay determinations in breast cancer.
3. The absence of NPGP in circulation in most patients with smaller breast tumors and in many patients with advanced disease, and its appearance in patients with hepatic disease, could indicate that levels in circulation are the result of a not yet understood release mechanism of the antigen into the circulation and a very active clearance mechanism, possibly of hepatic nature ([57] and above), which could be further interfered with by the presence of circulating immune complexes as is the case for other BrE-Ags [53]. Future research in the metabolism of the NPGP complex could bring answers to this question, but it could also indicate that other breast epithelial antigens with a more stable release mechanism, slower and better regulated serum clearance, imperviousness to other changes brought about by concomitant disease of other organs, and other characteristics, could be the marker(s) of choice. Other important issues are that the presence of genetic polymorphism of the components of the NPGP complex, and also heterogeneity of their antigenic expression, limits their usefulness in immunoassays. These two issues point to the

need to select for these tests, if possible, an invariable epitope expressed in all breast tumors or, in its absence, a panel of MAbs.
4. It will be most desirable that immunoassays express BrE-Ags values in concentration of weight units per ml (i.e., μ/ml) instead of arbitrary units to allow for comparisons among different assays.
5. None of the markers presently available are valuable breast cancer screening tools. Higher specificity and increased sensitivity will be needed for such assays.
6. In several studies the prognostic value of these assays approaches 75% levels. Although significant, this percentage does not yet warrant strong consideration. Improvements as mentioned above for BrE-Ag selection should also apply in this case.

Although present studies point to the limitations of the available serum markers for breast cancer in the last decade, they also show that a new breakthrough in the search for circulating breast cancer markers has occurred. Exploration of the many yet undiscovered antigenic possibilities of breast epithelial cells, rather than repetitious reevaluation of an existing NPGP complex, is where a wealth of new information and possibly better methods for breast cancer detection must be sought.

ACKNOWLEDGMENTS

I wish to acknowledge the helpful criticism of this overview by Dr. J. A. Peterson and Mr. E. W. Blank. This work was funded by NIH-NCI grants CA39932, 39933, and 1PO CA42767, and BRSG Grant S07-RR05929.

REFERENCES

1. Ceriani RL, Blank E. Experimental therapy of human breast tumors with ^{131}I-labeled monoclonal antibodies prepared against the human milk fat globule. Cancer Res 1988;48:4664-4672.
2. Fisher Ch, Lampson LA. Immunoblots and autoradiography used to analyze the proteins recognized by monoclonal antibodies. In: Kennet RH, Bechtol KB, McKearn TJ, eds. Monoclonal antibodies and functional cell lines. New York: Plenum Press, 1984:383-387.
3. Weir DM. Handbook of experimental immunology, 3rd ed. Oxford: Blackwell Scientific Publishers, 1978.
4. Griffiths AB, Burchell J, Gendler S, Lewis A, Blight K, Tilly R, Taylor-Papadimitriou J. Immunological analysis of mucin molecules expressed by normal and malignant mammary epithelial cells. Int J Cancer 1987;40:319-327.
5. Gendler SJ, Burchell JM, Duhig T, Lamport D, White R, Parker M, Taylor-Papadimitriou J. Cloning of partial cDNA encoding differentiation and tumour-associated mucin glycoproteins expressed by human mammary epithelium. Proc Natl Acad Sci USA 1987;84:6060-6064.
6. Taylor-Papadimitriou JT, Peterson JA, Arklie J, Burchell J, Ceriani RL, Bodmer WF. Monoclonal antibodies to epithelium-specific components of the human milk fat globule membrane: production and reaction with cells in culture. Int J Cancer 1981;28:17-21.
7. Ceriani RL, Peterson JA, Lee JV, Moncada R, Blank EW. Characterization of cell surface antigens of human mammary epithelial cells with monoclonal antibodies prepared against human milk fat globule. Somatic cell Genet 1983;9:415-427.

8. Stacker SA, Sacks NPM, Thompson CH, Smart C, Burton R, Bishop J, Golder J, Xing P-X, McKenzie IFC. A serum test for the diagnosis and monitoring of the progress of breast cancer. In: Ceriani RL, ed. Immunological approaches to the diagnosis and therapy of breast cancer. New York: Plenum Press, 1987:217-228.
9. Steward MW, Steensgaard J. Antibody affinity. Boca Raton, FL: CRC Press, 1983.
10. Ceriani RL, Sasaki M, Sussman H, Wara WM, Blank EW. Circulating human mammary epithelial antigens in breast cancer. Proc Natl Acad Sci USA 1982;79:5420-5424.
11. Matsuoka Y, Nakashima T, Endo K, Yoshida T, Kunimatsu M, Sakahara H, Koizumi M, Nakagawa T, Yamaguchi N, Torizuka K. Recognition of ovarian cancer antigen CA125 by murine monoclonal antibody produced by immunization of lung cancer cells. Cancer Res 1987;47:6335-6340.
12. Ceriani RL, Rosenbaum EH, Chandler M, Trujillo TT, Myers B, Sakada M. Role of circulating human mammary epithelial antigens (HME-Ags) as serum markers for breast cancer. In Ip C, ed. Tumor markers and their significance in the management of breast cancer. New York: Alan R. Liss, 1986:3-19.
13. Feldman MK, Ceriani RL. A comparative immunologic and electrophoretic analysis of rat and mouse caseins. Comp Biochem Physiol 1970;37:421-423.
14. Hencrick JC, Franchimont P. Radioimmunoassay of casein in the serum of normal subjects and of patients with various malignancies. Eur J Cancer 1974;10:725-730.
15. Franchimont P, Kendrick JC, Thirion A, Zangerle PF. Kappa casein: an index of normal mammary function and a tumor-associated antigen. In: Herberman RB, McIntire KR, eds. Immunodiagnosis of cancer, part 1. New York: Marcel Dekker, 1979:499-513.
16. Ceriani RL. Fetal mammary gland differentiation in vitro in response to hormones. II. Biochemical findings. Dev Biol 1970;21:506-529.
17. Ceriani RL. Hormone induction of specific protein synthesis in mid-pregnant mouse mammary cell cultures. J Exp Zool 1977;196:1-12.
18. Monaco MF, Bronzert DA, Tormey DC, Waalkes P, Lippman ME. Casein production by human breast cancer. Cancer Res 1977;37:749-751.
19. Kleinberg DL. Human alpha-lactalbumin: measurement in serum and in breast cancer organ cultures by radioimmunoassay. Science 1975;190:276-278.
20. Ip C, Dao TL. Alterations in serum glycosyltransferases and 5'-nucleotidase in breast cancer patients. Cancer Res 1978;38:723-728.
21. Dao TL, Ip C, Patel J. Serum sialyltransferase and 5'-nucleotidase as reliable biomarkers in women with breast cancer. J Natl Cancer Inst 1980;65:529-534.
22. von Kleist S, Burtin P. Cellular localization of an embryonic antigen in human colonic tumors. Int J Cancer 1969;4:874-879.
23. Bocker W, Schweikhart G, Pollow K, Kreienberg R, Klaubert A, Schroder S, Mitze M, Bahnsen J, Stegner HE. Immunohistochemical demonstration of carcinoembryonic antigen (CEA) in 120 mammary carcinomas and its correlation with tumor type, grading, staging, plasma-CEA, and biochemical receptor status. Pathol Res Pract 1985;180:490-497.
24. Schatz C, von Lieven H, Mulders M, Rowold J, Stracke H, Muller H, Grebe SF, Schatz H. Die tumormarker CEA, TPA und CA 19-9 sowie ferritin und osteocalcin in der verlaufskontrolle bei mammakarzinom. Wien Klin Wochenschr 1985;97:873-879.
25. Silva JS, Leight GS, Haagensen DE, Tallos PB, Cox EB, Dilley WG, Wells SA. Quantitation of response to therapy in patients with metastatic breast cancer by serial analysis of plasma gross cystic disease fluid protein and CEA. Cancer 1982;49:1236-1247.

26. Sugarbaker PH. Role of carcinoembryonic antigen assay in the management of cancer. Ad Immun Cancer Ther 1985;1:167-193.
27. Hilkens J, Bonfrer JM, Kroezen V, van Eykeren M, Nooyen W, de Jong-Bakker M, Bruning PF. Comparison of circulating MAM-6 and CEA levels and correlation with the estrogen receptor in patients with breast cancer. Int J Cancer 1987;39:431-435.
28. Ceriani RL, Thompson K, Peterson JA, Abraham S. Surface differentiation antigens of human mammary epithelial cells carried on the human milk fat globule. Proc Natl Acad Sci USA 1977;74:582-586.
29. Keenan TW, Morre DJ, Olson DE, Yunghans WN, Patton S. Biochemical and morphological comparisons of plasma membrane and milk fat globule membrane from bovine mammary gland. J Cell Biol 1970;44:80-93.
30. Parodi AJ, Blank EW, Peterson JA, Ceriani RL. Glycosyl transferases in mouse and human milk fat globule membranes. Mol Cell Biochem 1984;58:157-163.
31. Kobilka D, Carraway KL. Proteins and glycoproteins of the milk fat globule. Biochem Biophys Acta 1972;288:282-295.
32. Sasaki M, Peterson JA, Ceriani RL. Quantitation of human mammary epithelial antigens in cells cultured from normal and cancerous breast tissues. In Vitro 1981; 17:150-158.
33. Peterson JA, Selection of monoclonal antibodies for characterizing normal breast epithelium. Workshop on immunological markers of human mammary epithelial cells. Imperial Cancer Research Foundation, London, March, 1981.
34. Shimizu M, Yamauchi K. Isolation and characterization of mucin-like glycoprotein in human milk fat globule membrane. J Biochem 1982;91:515-524.
35. Shimizu M, Yamauchi K, Miyauchi Y, Sakurai T, Tokugawa K, McIlhinney RAJ. High molecular weight glycoprotein profiles in human milk serum and fat globule membrane. Biochem J 1986;233:725-730.
36. Schlom J, Wunderlich D, Teramoto YA. Generation of human monoclonal antibodies reactive with human mammary carcinoma cells. Proc Natl Acad Sci USA 1980;77:6841-6845.
37. Ceriani RL, Hill DL, Osvaldo L, Kandell C, Blank EW. Immunohistochemical studies in breast cancer using monoclonal antibodies against breast epithelial cell components and with lectins. In: Russo J, ed. Immunohistochemistry in tumor diagnosis. Boston: Martinus Nijhoff, 1985:233-261.
38. Ceriani RL, Orthendahl D, Sasaki M, Kaufman L, Miller S, Wara W, Peterson JA. Use of mammary epithelial antigens (MHE-Ags) in breast cancer diagnosis. In: Nieburgs HE, ed. Cancer detection and prevention 1981. New York: Marcel Dekker, 1981:603-609.
39. Wilbanks T, Peterson JA, Miller S, Kaufman L. Ortendahl D, Ceriani RL. Localization of mammary tumors in vivo with ^{131}I-labeled Fab fragments of antibodies against mouse mammary epithelial (MME) antigens. Cancer 1981;48:1768-1775.
40. Ceriani RL, Blank EW, Peterson JA. Experimental immunotherapy of human breast carcinomas implanted in nude mice with a mixture of monoclonal antibodies against human milk fat globule components. Cancer Res 1987;47:532-540.
41. Ceriani RL, Blank EW. Experimental therapy of breast cancer with anti-breast epithelial radioimmunoconjugates. In: Ceriani RL, ed. Immunological approaches to the diagnosis and therapy of breast cancer. New York: Plenum Press, 1987:229-243.
42. Ceriani RL, Peterson JA, Abraham S. Immunologic methods for the identification of cell types. II. Expression of normal mouse mammary epithelial (MME) cell antigens in mammary neoplasia. J Natl Cancer Inst 1978;61:747-752.
43. Ceriani RL, Peterson JA. Characterization of differentiation antigens of the mouse mammary epithelial cell (MME antigens) carried on the mouse milk fat globule. Cell Differ 1978;7:355-366.

44. Heyderman E, Steele K, Ormerod MG. A new antigen on the epithelial membrane: its immunoperoxidase localization in normal and neoplastic tissue. J Clin Pathol 1979;32:35-39.
45. Sloane JP, Ormerod MG. Distribution of epithelial membrane antigen in normal and neoplastic tissues and its value in diagnostic tumor pathology. Cancer 1981;47:1786-1795.
46. Ormerod MG, McIlhinney J, Steele K, Shimizu M. Glycoprotein PAS-O from the milkfat globule membrane carries antigenic determinants for epithelial membrane antigen. Mol Immunol 1985;22:265-269.
47. Ormerod MG, Bussolati G, Sloane JP, Steele K, Gugliotta P. Similarities of antisera to casein and epithelial membrane antigen. Virchows Arch [Pathol Anat] 1982;397:327-333.
48. Ceriani RL, Peterson JA, Abraham S. The removal of cell surface material by enzymes used to dissociate mammary gland cells. In Vitro 1978;14:887-894.
49. Sasaki M, Peterson JA, Wara W, Ceriani RL. Human mammary epithelial antigens (HME-Ags) in the circulation of nude mice implanted with breast and non-breast tumors. Cancer 1981;48:2204-2210.
50. Sasaki M, Barber S, Ceriani RL. Breast cancer markers: comparison between sialyltransferase and human mammary epithelial antigens (HME-Ags) for the detection of human breast tumors grafted in nude mice. Breast Cancer Res Treat 1985;5:51-56.
51. Burchell J, Wang D, Taylor-Papadimitriou J. Detection of the tumour-associated antigens recognised by the monoclonal antibodies HMFG-1 and 2 in serum from patients with breast cancer. Int J Cancer 1984;34:763-768.
52. Hayes DF, Sekine H, Marcus D, Alper C, Kufe D. Genetically determined polymorphism of the circulating human breast cancer-associated DF3 antigen. Blood 1988;71:436-440.
53. Salinas FA, Wee KH, Ceriani RL. Significance of breast carcinoma-associated antigens as a monitor of tumor burden: characterization by monoclonal antibodies. Cancer Res 1987;47:907-913.
54. Papsidero LD, Croghan GA, Capone PM, Johnson EA. Ductal carcinoma antigen: characteristics, tissue distribution and capacity to represent a target for monoclonal antibody therapy. In: Ceriani RL, ed. Monoclonal antibodies and breast cancer. Boston: Martinus Nijhoff, 1985.
55. Dhokia B, Pectasides D, Self C, Habib NA, Hershman M, Wood CB, Munro AJ, Epenetos AA. A low pH enzyme linked immunoassay using two monoclonal antibodies for the serological detection and monitoring of breast cancer. Br J Cancer 1986;54:885-889.
56. Hayes DF, Sekine H, Ohno T, Abe M, Keefe K, Kufe DW. Use of a murine monoclonal antibody for detection of circulating plasma DF3 antigen levels in breast cancer patients. J Clin Invest 1985;75:1671-1678.
57. Hilkens J, Kroezen v, Buijs F, Hilgers J, Van Vliet M, De Voogd W, Bonfrer J, Bruning PJ. MAM-6, a carcinoma associated marker: preliminary characterization and detection in sera of breast cancer patients. In: Ceriani RL, ed. Monoclonal antibodies and breast cancer. The Hague: Martinus Nijhoff, 1985:28-42.
58. Iacobelli S, Arno E, D.Orazio A, Coletti G. Detection of antigens recognized by a novel monoclonal antibody in tissue and serum from patients with breast cancer. Cancer Res 1986;46:3005-3010.
59. Linsley PS, Ochs V, Laska S, Horn D, Ring DB, Frankel AE, Brown JP. Elevated levels of a high molecular weight antigen detected by antibody W1 in sera from breast cancer patients. Cancer Res 1986;46:5444-5450.
60. Bray KR, Koda JE, Gaur PK. Serum levels and biochemical characteristics of cancer-associated antigen CA-548, a circulating breast cancer marker. Cancer Res 1987;47:5853-5860.

61. Hayes DF, Zurawski VR Jr, Kufe DW. Comparison of circulating CA15-3 and carcinoembryonic antigen levels in patients with breast cancer. J Clin Oncol 1986;4: 1542-1550.
62. Pons Anicet DMF, Krebs BP, Mira R, Namer M. Value of CA15-3 in the follow-up of breast cancer patients. Br J Cancer 1987;55:567-569.
63. Komuro K, Watanabe K, Ishida T, Sakurai M, Amano R, Hagiwara H, Hara Y, Hanzawa T, Itsubo K. Clinical evaluation of a new tumor marker, CA 15-3, in breast cancer: a comparative study with CEA. Gen To Kagaku Ryoho 1986;13:3145-3149.
64. Fujino N, Haga Y, Sakamoto K, Egami H, Kimura M, Nishimura R, Akagi M. Clinical evaluation of an immunoradiometric assay for CA15-3 antigen associated with human mammary carcinomas: comparison with carcinoembryonic antigen. Jpn J Clin Oncol 1986;16:335-346.
65. Hayes DF, Zurawski VR Jr, Kufe DW. Comparison of circulating CA15-3 and carcinoembryonic antigen levels in patients with breast cancer. J Clin Oncol 1986;4: 1542-1550.
66. Sacks NP, Stacker SA, Thompson CH, Collins JP, Russell IS, Sullivan JA, McKenzie IF. Comparison of mammary serum antigen (MSA) and CA15-3 levels in the serum of patients with breast cancer. Br J Cancer 1987;56:820-824.
67. Parodi AJ, Blank EW, Peterson JA, Ceriani RL. Dolichol-bound oligosaccharides and the transfer of distal monosaccharides in the synthesis of glycoproteins by normal and tumor mammary epithelial cells. Breast Cancer Res Treat 1982;2:227-237.
68. Castagna M, Nuti M, Squartini F. Mammary cancer antigen recognized by monoclonal antibody B672 in apocrine metaplasia of the human breast. Cancer Res 1987; 47:902-906.
69. Springer GF. Triple role of T and Tn antigens in cancer: as universal clonal carcinoma markers, in cancer cell adhesion, and as autoimmunogens. Surv Synth Pathol Res 1983;2:141-164.
70. Kjeldsen TK, Clausen H, Hirohashi S, Ogawa T. Iijima H, Hakomori S. Preparation and characterization of monoclonal antibodies directed to the tumor-associated O-linked sialosyl-2 → 6 α-N-acetylgalactosaminyl (Sialosyl-Tn) epitope. Cancer Res 1988;48:2214-2220.
71. Thor A, Ohuchi N, Szpak CA, Johnston WW, Schlom J. Distribution of oncofetal antigen tumor-associated glycoprotein-72 defined by monoclonal antibody B72.3. Cancer Res 1986;46:3118-3124.
72. Horan-Hand P, Nuti M, Colcher D, Schlom J. Definition of antigenic heterogeneity among human mammary carcinoma cell population using monoclonal antibodies to tumor associated antigens. Cancer Res 1983;43:728-735.
73. Haagensen DE, Wells SA, Haider M. Analysis of breast gross cystic disease fluid protein (GCDFP) by isotope dilution RIA with a solid phase second antibody. Clin Chem 1980;26:980.
74. Ceriani RL, Mammary epithelial antigens as a breakthrough in breast cancer research: Now where to? In: Rich MA, Hager JC, Lopez DM, eds. Breast cancer: scientific and clinical progress. Boston: Kluwer Academic Pub., 1988:223-261.
75. Sekine H, Hayes DF, Ohno T, Keefe KA, Schaetzl E, Bast RC, Knapp R, Kufe DW. Circulating DF3 and CA125 antigen levels in serum from patients with epithelial ovarian carcinoma. J Clin Oncol 1985;3:1355-1363.
76. Dilley WG, Haagensen DE Jr, Leight GS Jr, Ammirata S, Davis SR, Silva JS, Zamcheck N, Lokich JJ, Wells SA Jr. Fluoxymesterone stimulation of tumor marker secretion in patients with breast carcinoma. Breast Cancer Res Treat 1986;8:205-215.

16
Lymphocytic Leukemia and Lymphomas

HANS G. DREXLER*

Royal Free Hospital School of Medicine, University of London, London, United Kingdom

JUN MINOWADA

Fujisaki Cell Center, Fujisaki, Okayama, Japan

I. IMMUNOPHENOTYPING AS AN ALTERNATIVE APPROACH TO HISTOMORPHOLOGICAL CLASSIFICATIONS

The differentiation of acute lymphoblastic leukemia (ALL) from acute nonlymphocytic (ANLL)/acute myeloid (AML) leukemia is of great importance because these diseases are treated with different therapeutic regimens and have quite different prognoses. The differentiation of non-Hodgkin's lymphomas (NHL) from Hodgkin's disease (HD) is likewise important. The classic distinction of ALL from AML and of NHL from HD and the further subclassification of these disease entities is based on their cytomorphological, cytochemical, and histopathological appearance. The subclassifications of ALL and NHL are of further interest not only to hematopathologists but also to clinicians who must determine prognosis, treatment, and appropriate stratification for clinical trials.

Before the introduction of the French-American-British (FAB) classification of ALL and AML [1], no uniform morphological system was in use and concordance rates of diagnostic types and subtypes between independent hematopathologists were low [2]. By defining morphological parameters and introducing relatively stringent criteria, the FAB classification contributed significantly to an improvement in reproducibility of acute leukemia classification, with the distinction of ALL from AML in the range of 95% concordance rate [2]. The FAB system differentiates three morphologically defined ALL subtypes [1]. However, interobserver predictability of the FAB L1-L3 subtypes remained only in the range of 60-70% [3].

While FAB L3 morphology conveys a very poor prognosis, L2 morphology appears to predict in children a worse prognosis than L1 morphology [4]. Other studies failed to identify FAB subtypes as long-term significant risk factors [5]. No commonly accepted classification systems of the chronic lymphoid malignancies based on morphocytology have been proposed.

Present affiliation: DSM (German Collection of Microorganisms and Cell Cultures), Braunschweig, Federal Republic of Germany

Table 1 Markers Useful for Immunophenotyping of Lymphocytic Leukemias and Lymphomas

Antigen (Cluster)[a]	Molecular Weight	Reactivity	Comments	Selection of MAbs[b]
CD1a-c	43-49	Lymphoid (T)	T6	OKT-6 (O), NU-T2 (N), NA1/34 (D)
CD2	50	Lymphoid (T)	T11 (E-receptor)	OKT-11 (O), Leu-5 (BD), NU-T1 (N)
CD3	20,26	Lymphoid (T)	T3 (T-cell receptor-associated)	OKT-3 (O), Leu-4 (BD), UCHT1 (S)
CD4	60	Lymphoid (T)	T4 (T-helper/inducer)	OKT-4 (O), Leu-3a (BD)
CD5	67	Lymphoid (T)	T1	OKT-1 (O), Leu-1 (BD)
CD7	40	Lymphoid (T)		Leu-9 (BD), 3A1, WT-1
CD8	32	Lymphoid (T)	T8 (T-suppressor/cytotoxic)	OKT-8 (O), Leu-2a (BD)
CD10	100	Lymphoid (pre-B)	cALLA	BA-3 (H), J5 (C), NU-N1 (N)
CD11c	150/95	Myeloid (Mo)	HCL	Leu-M5 (BD)
CD13	150	Myeloid (G, Mo)		MY7 (C), MCS-2 (N)
CD14	55	Myeloid (Mo)		Mo2 (C), MY4 (C), UCHM1 (S), Leu-M3 (BD)
CD16	50-60	Granulocytes, NK cells	Fc-receptor	Leu-11 (BD)
CD19	95	Lymphoid (B)		B4 (C), Leu-12 (BD), HD37 (D)

CD20	32, 37	Lymphoid (B)	B1 (C), Leu-16 (BD)
CD22	130, 140	Lymphoid (B)	To15 (D), HD6, RFB-4, Leu-14 (BD)
CD23	45	Lymphoid (B)	MHM6, Blast-2
		IgE-Fc receptor	
CD33	67	Myeloid (progenitor)	MY9 (C)
CD34	115	Myeloid + lymphoid (progenitor)	MY10, 3C5, HPCA-1 (BD)
CD38	45	Myeloid/lymphoid	OKT-10 (D)
		Plasma cells	
SmIg		Lymphoid (B)	
		Ig (M/D/G/A/E; kappa/lambda)	
TdT		Lymphoid (progenitor)	TdT (BRL)
		Nuclear enzyme	
HLA-DR	28-33	Myeloid/lymphoid	OKIa (D), HLA-DR (BD)
		MHC class II	
FMC7	105	Lymphoid (B)	FMC-7 (S)
		Mature/activated B-cells	
PCA1	24	Lymphoid (B)	PCA-1 (C)
		Plasma cells	

[a] CD nomenclature as defined at the First, Second, and Third International Workshops on Human Leukocyte Differentiation Antigens (Paris, 1982; Boston, 1984; Oxford, 1986) [14-16]

[b] Commercial sources, where available: BD, Becton-Dickinson Monoclonal Center (Mountain View, CA); BRL, Bethesda Research Lab (Gaithersburg, MD); C, Coulter Immunology (Hialeah, FL); D, Dakopatts (Glostrup, Denmark); H, Hybritech (San Diego, CA); O, Ortho Diagnostic Systems (Raritan, NJ); N, Nichirei (Tokyo, Japan); S, Sera-Lab (Sussex, UK).

The multiplicity of histopathological classification systems involved in the categorization of NHL has resulted in both controversy and confusion [6,7]. Over the last 20 years several major classifications have been proposed, including these major ones: working classification, British National Lymphoma, Kiel, WHO, Lukes-Collins, Rappaport, NCI Working Formulation. Because the terminology differs strikingly between the different classification systems, it has been difficult to compare studies based on different classification schemes. No major prognostic differences could be demonstrated among the different systems. While distinct histopathological features could be detected with a 89-95% probability of concurrence among observers, the reproducibility for individual pathologists varied from 0.53 to 0.93. Comparisons of intersystem predictability in terms of diagnosis demonstrated that no one system predicted for any of the other systems within any major degree of reliability [7]. This inability to "translate" from one system to another is thought to be a reflection of the problems of both interpathologist agreement and intrapathologist reproducibility. These significant differences of interpretation were also encountered in subsequent studies on comparisons between and reproducibility of histopathological classification systems [8-10].

An alternative approach to histomorphological assessments of lymphocytic leukemias and lymphomas has evolved with the tremendous progress seen in the area of immunophenotyping of normal and neoplastic cells in suspension, on slides, or in tissue preparations. By exploiting recent advances in immunology it is now possible to define precisely stages of human lymphocyte differentiation using specific monoclonal antibodies (MAbs) that identify characteristic cellular antigens. These can be combined with more traditional cell markers such as surface membrane immunoglobulin (SmIg) on B lymphocytes and sheep erythrocyte receptors on T lymphocytes, and with cytochemical stains, morphological assessment, and genotyping. We summarize here the progress in the classification of the lymphoid leukemias and of lymphomas in the context of normal lymphoid differentiation and its implication for therapy and prognosis.

II. IMMUNOMARKER ANALYSIS

The detection of SmIg on B cells and the discovery that sheep red blood cells (SRBC) form rosettes with T cells led in the early 1970s to the description of a dichotomy in lymphoid differentiation. Hence, T- or B-cell malignancies could be discerned. However, a large percentage of cases (particularly ALL samples) could not be positively identified by these markers and therefore, by negative definition, they could only be termed non-T/non-B ALL [11]. Further advancement in the positive recognition of lymphoid neoplastic and normal cells was achieved with the identification of the cALL antigen [12]. The most crucial step towards modern immunophenotyping was the introduction of hybridoma technology, which provided for the unlimited and reproducible production of "clean" and specific reagents [13]. Before MAbs became available, heterologous, polyclonal antisera were used for the analysis of lymphocyte subsets and also for the differential diagnosis of leukemia and lymphoma. Careful, painstaking adsorption steps were required to eliminate unwanted reactivities and render these reagents diagnostically useful. Therefore, the most powerful impact on research and diagnosis has been made by the simplification of the production of specific reagents.

The subsequent abundance of these new reagents necessitated a detailed categorization and a nomenclature of the MAbs. In three past International Workshops these MAbs were assigned to clusters of differentiation (CD) on the basis of the biochemical nature of the antigen and/or similarities in their reactivity patterns on normal and malignant hematopoietic cell populations [14-16].

A panel of MAbs that are relevant for the immunophenotyping of lymphoid malignancies is listed in Table 1. The two basic techniques for the demonstration of surface and intracellular antigens are immunofluorescence staining of cells in suspension or of fixed cells or tissue preparations using fluorescence microscopy or flow cytometry for analysis and immunocytological labeling (peroxidase or alkaline phosphatase-antialkaline phosphatase) of fixed single cells or tissue using light microscopy. Immunofluorescence and immunocytology (also termed immunohistochemistry) can be used as complementary approaches; each method has its own unique advantages and disadvantages. The advantages of immunofluorescence are speed, double/triple labeling of the same cell, quantitation of positive cells, and staining intensity; those of immunocytology are simultaneous staining for morphology and immunomarkers, long-term storage, small sample quantity, and enhanced sensitivity [17].

III. HUMAN LEUKEMIA-LYMPHOMA CELL LINES

A significant step in the study of human leukemia-lymphoma cells was the identification of an in vitro milieu where malignant hematopoietic cells could be (initially only) maintained and then (via immortalization) grown indefinitely. In the 1960s and particularly in the 1970s tissue culture conditions were created that allowed for the establishment of human leukemia-lymphoma cell lines [18]. A large and diverse panel of different leukemia-lymphoma cell lines is now available (Fig. 1). These cell lines became invaluable resources in the analysis of leukemia-lymphoma pathological specimens and for the comparison of normal and malignant cells.

A number of findings are of particular significance: individual cell lines are monoclonal and their marker profiles remain stable in long-term culture for years; there is no leukemia-lymphoma-specific antigen and the majority of surface and intracellular antigens are not absolutely lineage specific [12,18]; the same marker profile found in each cell line can also be detected on the fresh original leukemia cells from which the cell lines were established (i.e., in vitro leukemia-lymphoma cell lines are approximate and not absolute replicas of in vivo leukemia-lymphoma cells); individual markers and marker profiles detected on malignant in vitro as well as on in vivo cells can equally be recognized on normal hematopoietic cells (in bone marrow, peripheral blood, lymphoid or fetal tissue), indicating that to each leukemia-lymphoma cell there is generally a normal counterpart (for discussions of abnormal gene expression and lineage infidelity, see below); and the heterogeneity of the immunoprofiles present both in leukemia-lymphoma cell lines and in large series of fresh leukemia-lymphoma cells suggest that this variety might reflect different but defined stages of normal human hematopoietic cell differentiation [18].

Due to their unlimited cell growth and monoclonality, such leukemia-lymphoma cell lines have been the preferential targets for any kind of leukemia-lymphoma-related research. Among other applications, cell lines have been successfully used as sources of immunogens for the production of MAbs. Innumerable contributions to progress in the understanding of the pathophysiology of leukemia-lymphoma, to knowledge of normal hematopoietic cell differentiation and to advances in leukemia-lymphoma immunophenotyping have resulted from these studies.

No.	Cell Line	Origin	No.	Cell Line	Origin	No.	Cell Line	Origin	No.	Cell Line	Origin
	T-cell		31	CTCL-2	SS	60	DAUDI	BL	89	BV-173	CML
			32	KH-2	ATL	61	NAMALVA	BL	90	Tahr 87	AUL
1	P30/OHKUBO	ALL	33	H9	ALL	62	RAMOS	BL	91	MR-87	AUL
2	MOLT-10	ALL	34	IZ-86	ATL	63	BJAB	BL			
3	RPMI 8402	ALL	35	ED-S⁻	ATL	64	DG-75	BL		Myelomonocyte-cell	
4	CCRF-CEM	ALL				65	CHEVALLIER	BL			
5	DND-41	ALL		B-cell		66	DND-39	BL	92	KG-1	AML
6	HPB-ALL	ALL				67	NK-9	BL	93	ML 1-3	AML
7	HPB-MLT	LY	36	NALM-1	CML	68	B46M	BL	94	HL-60	APL
8	HD-Mar-2	HD?	37	NALM 6-15	ALL	69	OGUN	BL	95	PL-21	APL
9	MOLT 13,14	ALL	38	NALM 17,18	ALL	70	AL-1	BL	96	JOSK-I	AMML
10	JM	ALL	39	LAZ-221	ALL	71	SL-1	BL	97	CTV 1,2	AMOL
11	TALL-1	ALL	40	KOPN 1-8	ALL	72	RIVA	LB	98	THP-1	AMOL
12	MOLT-3	ALL	41	HPB-Null	ALL	73	BALL-1	ALL	99	JOSK-K	AMOL
13	MOLT-4	ALL	42	KLM-2	AMOL	74	RPMI 8226	MM	100	JOSK-S	AMOL
14	P12/Ichi.	ALL	43	BALM 1,2	ALL	75	U-266	MM	101	JOSK-M	CML
15	KOPT-K1	LS	44	U-698-M	LS	76	ARH-77	MM	102	GDM-1	CML
16	MOLT-11	ALL	45	TANAKA	LB	77	Hs	MM	103	KCL-22	CML
17	MOLT 16,17	ALL	46	MANACA	LB	78	Nak-Pc2	MM	104	U-937	LY
18	MKB-1	AML	47	JOK-1	HCL	79	KMM-1	MM	105	EoL-1,3	EoL
19	Jurkat	ALL	48	SKW-4	LB	80	Hair-M	HCL			
20	CCRF-HSB-2	ALL	49	BALM 3-5	LB	81	JC-1	HCL		Non-L, Non-M-cell	
21	PEER	ALL	50	SU-DHL-4	LY	82	Hair-MT1	HCL			
22	KE-37	ALL	51	SCOTT	LB				106	K-562	CML
23	SKW-3	CLL	52	MLB-1084	LB		Non-T, Non-B cell		107	HEL	EL
24	A3/Kawa.	LY	53	BAL-KH	ALL				108	SPI 801,802	ALL
25	MOLT-15	AMOL	54	ABL-2	BL?	83	REH	ALL	109	SU-DHL-1	LY
26	MAT	LY	55	EB-3	BL	84	KM-3	ALL	110	HDLM 1-3	HD
27	ALL-Sil	ALL	56	RAJI	BL	85	NALL-1	ALL	111	L-428	HD
28	MT-1	ATL	57	HRIK	BL	86	NALM-16	ALL	112	KS-1	ALL
29	HUT-78	SS	58	B35M	BL	87	KOPN-K	ALL			
30	HUT-102	MF	59	KOBK-101	BL	88	NALM-19	ALL			

Figure 1 Panel of human leukemia-lymphoma cell lines (n = 112; July, 1988). All cell lines are original cell lines that represent the respective parental in vivo leukemia-lymphoma cells. The monoclonal cell lines are permanently established in culture with stable surface marker profiles.

IV. IDENTIFICATION OF T AND B LYMPHOCYTES

Membrane surface markers were first organized into the categories of receptors and antigens [18,19]. Receptors included the sheep erythrocyte receptor (on T cells), the Fc and complement receptor (on B cells, monocytes, granulocytes), the mouse erythrocyte receptor (on B cells), and the surface membrane immunoglobulin (on B cells). Surface antigens were identified with the use of first polyclonal and then MAbs (see above).

As already pointed out, the classic identification of T lymphocytes is the spontaneous rosette formation of human T cells with SRBC. This widely used marker of T cells has permitted decisive progress in the knowledge of T-cell physiology and in the understanding of T-lineage leukemia. The nature of this seemingly fortuitous interaction between SRBC and T cells has been clarified by the characterization of a T-cell surface molecule defined as CD2. This molecule appears to be involved in both T-cell activation (an alternative pathway to the T-cell receptor) and T-cell adhesion processes (via the leukocyte function antigen-3 ligand).

The use of T-cell lineage-associated MAbs has rendered the identification of T cells more sensitive and discriminatory. Besides the CD2 antigen, further T-cell lineage-specific or associated markers are CD1 (thymocytes), CD3 (pan-T; in functional linkage with the T-cell receptor), CD4/CD8 (T helper or suppressor), CD5 (pan-T), and CD7 (pan-T)

Lymphocytic Leukemia and Lymphomas

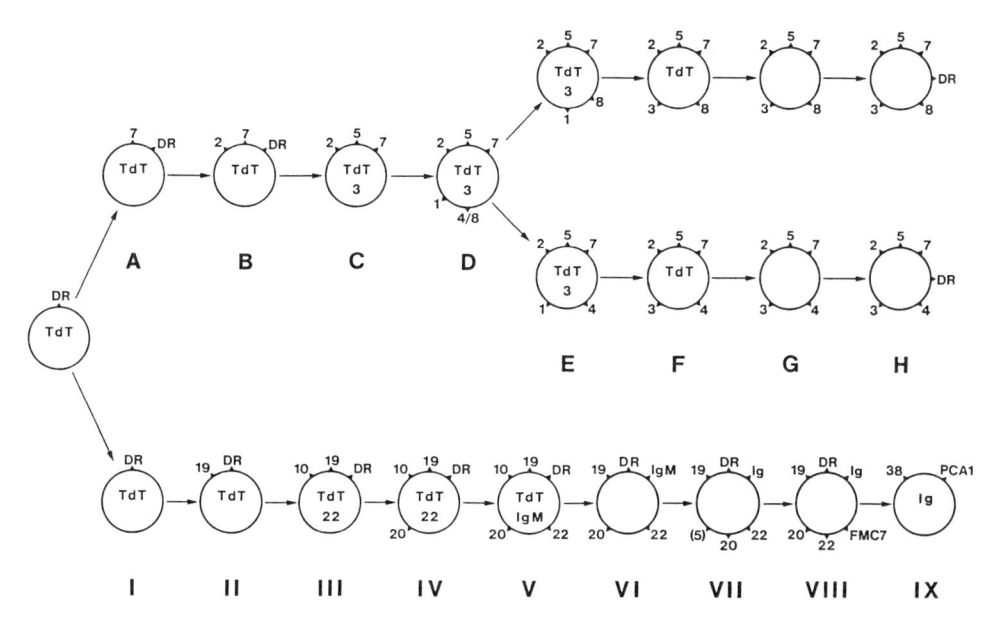

Figure 2 Schematic representation of human lymphoid differentiation. The T-lineage pathway includes the types A-H: type A; physiological stage committed progenitor; malignancy "pre T-ALL" or "early precursor T-ALL"; type B; prothymocyte; T-ALL; type C: immature thymocyte; T-ALL; types D and E: common thymocyte; T-ALL; type F: Mature thymocyte; T-ALL, T-LL; types G and H: peripheral blood T-helper/suppressor cell; T-ALL, T-CLL, T-PLL, ATL, CTCL, PTCL, T-LGL. Characteristic T-cell surface markers are CD7 (pan–T), CD2, and CD3 (the latter preceded by cytoplasmic CD3 expession).

The B-lineage pathway describes the types I-IX: type I: committed progenitor; morphologically and immunologically AUL, but genotypically belonging to the B cell due to Ig heavy chain rearrangement status; type II: early B-precursor; AUL or "early B-precursor ALL"; types III and IV: prepre B-cell; cALL; type V: pre B-cell; pre B-ALL; type VI: immature B-cell; B-ALL, Burkitt Ly; type VII: immature "resting" B-cell; B-CLL, B-NHL; type VIII: mature "activated" B-cell; B-PLL, HCL, B-NHL; type IX: plasma cell; multiple myeloma. Characteristic B-cell surface markers are CD19 (pan-B except for plasma cells), CD20, CD22, and SmIg (the latter two are first detectable in the cytoplasm). FMC7 (not clustered) is a marker of "activated," mature B cells; PCA1 (not clustered) is a marker of plasma cells.

The putative lymphoid stem cell or lymphoid progenitor cell is devoid of any lineage-specific surface or intracytoplasmic immunomarkers and has germline status for TCR and Ig gene rearrangements.

Figures relate to CD nomenclature; DR, HLA-DR (modifed after Refs. [20-27]).

(Table 1). Figure 2 is a schematic representation of the human lymphoid differentiation pathway. It is apparent that at distinct stages characteristic surface antigens are expressed or repressed (or even lost).

While T-cell precursors and prothymocytes express the human leukocyte antigen-(HLA) DR major histocompatibility complex class II antigen, this molecule is no longer detectable on more mature T cells. It should, however, be noted that activated mature T cells reexpress the class II antigens. The earliest characteristic T-cell marker is CD7 which subsequently can be found at all stages. However, this component is not entirely T-cell-specific: about 10% of AML cases have been described as expressing the CD7 antigen [28]. Other molecules acquired at the different thymocyte stages are CD2, CD5, and the simultaneous expression of CD4 and CD8. Mature thymocytes and peripheral blood T cells belong either to the T-helper/inducer (CD4) or T-suppressor/cytotoxic (CD8) category. The nuclear enzyme terminal deoxynucleotidyl transferase (TdT) (not specific for T cells but an indicator of maturity) is not displayed by peripheral blood T cells. Of interest is that CD3 can first be detected cytoplasmically, before its expression becomes apparent on the cell surface. This feature appears to be a truly specific T-cell parameter [23].

The presence of surface immunoglobulin (SmIg) is theoretically the most definitive indicator of B cells. However, SmIg is expressed at somewhat late stages of the B-cell differentiation pathway, and labeling with anti-Ig reagents is prone to nonspecific binding by other cell types (unless care is taken to avoid possible binding via the Fc-receptor). The surface component defined as CD19 appears to be the "best" pan-B cell marker. However, neither the latest stage of B-cell differentiation, plasma cells (which can be identified as such by other markers and morphological appearance), nor the seemingly earliest stage (characterized by genotyping as B-cell progenitor having rearranged Ig heavy chain genes) show the CD19 molecule. The HLA-DR antigen is found along the whole B-cell axis with the exception of plasma cells. Other markers are CD10 (which is expressed intermittently at the pre-pre-B and pre-B-cell stage), CD20, and CD22 (which are acquired sequentially). Mature granulocytes are also, positive for CD10, albeit weakly. The intracytoplasmic demonstration of IgM or CD22 is specific for B cells [21]. As indicated earlier, TdT is a differentiation rather than a lineage marker. The presumed counterpart to B CLL cells, which expresses uniquely the otherwise T-cell-associated antigen CD5, might be an immature, but resting B cell. Mature/activated B cells strongly display SmIg and the particular, but not entirely defined, FMC7 molecule. Plasma cells that are negative for SmIg (but cytoplasmic Ig-positive) and for most other antigens previously expressed in the B-cell pathway (such as CD19, CD20, CD22) can be labeled by CD38 and PCA1 MAbs. The latter MAbs are, however, not specific for plasma cells. Other surface structures which, although they were also found on other types of cells, have been regarded as B-cell markers are CD9, CD21, CD23, CD24, complement- and mouse erythrocyte recep-

Table 2 Basic Immunomarker Panel for the Differentiation of Monomyeloid from T and B Cells

Cell Type	CD13/33	POX	CD7	cCD3	CD19	cCD22
Monomyeloid	+	+	−	−	−	−
T	−	−	+	+	−	−
B	−	−	−	−	+	+

The combinations of CD13 or CD33 surface staining + cytoplasmic POX staining, surface CD7 + cytoplasmic cCD3, and surface CD19 + cytoplasmic cCD22 are definitive markers of monomyeloid, T and B cells, respectively [29].
cCD3 and cCD22, cytoplasmic expression of these antigens; POX, cytochemically detectable myeloperoxidase.

tor [22,25]. The putative lymphoid progenitor (or stem) cell appears to be immunophenotypically "naked" probably expressing intracellular TdT and surface HLA-DR.

A basic panel of immunomarkers for the differentiation of monomyeloid from lymphoid and of B-cells from T-cells is shown in Table 2. The CD13 and CD33 antigens are panmonomyeloid markers, which are, however, not absolutely specific since they can be found on some cases of ALL (see below); a specific cytoplasmic feature of monomyeloid cells is the cytochemically detectable enzyme myeloperoxidase. The T- and B-cell markers CD7 and CD19 are expressed at all or most stages of T- and B-cell differentiation, respectively; specific cytoplasmic markers of T and B cells are cCD3 and cCD22, respectively.

V. MAJOR IMMUNOLOGICALLY DEFINED LYMPHOID LEUKEMIA AND NHL SUBTYPES

A. Acute Lymphoblastic Leukemias

ALL is a heterogeneous disease that includes phenotypically and prognostically different subtypes. SRBC rosetting and SmIg staining allowed for the distinction of T-ALL, B-ALL, and non-T/non-B ALL. Using the common ALL (cALL) antiserum, the non-T/non-B ALL cases could be further subclassified into cALL and "null-ALL" (this group did not have distinctive immunological features). This immunologically defined ALL classification system discerning the four major subtypes has been widely accepted and used in a number of large clinical studies over the last 10 years. With the use of testing for cytoplasmic IgM expression, a subset termed pre B-ALL has been identified. The incidence of these ALL subtypes in a population of 3832 children and adults from 10 large series was as follows: cALL, 71%; T-ALL, 16%; Null-ALL, 12%; B-ALL, 1% [27].

With the development of stage- and lineage-specific MAbs and the complementary application of molecular genetic analysis, it became evident that further T-cell and B-cell subsets of ALL could be identified. Tables 3 and 4 list the distinct immunoprofiles of seven T-lineage ALL and six B-lineage ALL subsets (also compare with Fig. 2). The most common subtypes appear to be type C among the T-lineage ALL cases and type IV among the B-lineage ALL cases [20,24,30]. While detailed subclassifications are certainly of great importance for basic leukemia research, such compartmentalizations into smaller entity groups appear to be impractical for clinical application. Therefore, immunological classification systems described above that identify four or five subtypes (T-, B-, cALL, pre B-, and null-ALL) seem to represent, at least for the moment, the most relevant and practical scheme.

B. Chronic Lymphoid Leukemias and NHL

The main diagnostic instruments for the classification of the chronic lymphoid leukemias and related disorders are certain morphological analysis and clinical data. Nevertheless, different subtypes display characteristic marker profiles. Differentiation of T- or B-type leukemia can be achieved using either SRBC rosetting and SmIg labeling or staining with T- (CD2, CD3, CD7) or B- (CD19, CD20, CD22) associated markers. CD5 cannot be used as a discriminatory marker because B-CLL and other chronic B-cell leukemias give equally positive results. B-CLL, B-PLL, HCL, and multiple myeloma are monoclonal proliferations of Ig-positive B lymphocytes. Clonality is here demonstrated by the expression of a single Ig light chain per clone on the cell surface or in the cytoplasm (multiple myeloma).

Table 3 Immunoprofiles of T-Lineage ALL

Group	Leukemia Category	HLA-DR	TdT	CD7	CD2	CD5	CD1	cCD3	CD3	CD4/8	Physiological Stage
A	Early T-precursor ALL[a]	+	+	+	−	−	−	−	−	−	Committed T progenitor
B	T-ALL	+	+	+	+	−	−	−	−	−	Prothymocyte
C	T-ALL	−	+	+	+	+	−	+	−	−	Immature thymocyte
D	T-ALL	−	+	+	+	+	+	+	−	+	Common thymocyte
E	T-ALL	−	+	+	+	+	+	+	−	+[b]	Common thymocyte
F	T-ALL	−	+	+	+	+	−	+	+	+[b]	Mature thymocyte
G	T-ALL	−	−	+	+	+	−	+	+	+[b]	Peripheral blood T cell

[a] Also termed pre-T ALL.
[b] No longer simultaneous expression as in group D, but either CD4 or CD8.
To qualify as T-lineage ALL, group A type leukemic blasts have to display rearranged TCR genes (about 10% of AML cases are HLA-DR+, CD7+, and/or TdT+).
Coexpression of surface CD7 and cytoplasmic cCD3 is a definitive indication of T-cell lineage.
Source: Modified after Refs. [23,24,26]; also see Figure 2.

Table 4 Immunoprofiles of B-Lineage ALL

Group	Leukemia Category	HLA-DR	TdT	CD19	CD10	CD20	cCD22	CIgμ	SmIg	Physiological stage
I	Early B-precursor ALL[a]	+	+	−	−	−	−	−	−	Committed B-progenitor
II	Early B-precursor ALL[a]	+	+	+	−	−	−	−	−	Early B precursor
III	cALL	+	+	+	+	−	−	−	−	Pre-pre B cell
IV	cALL	+	+	+	+	+	+	−	−	Pre-pre B cell
V	Pre-B ALL	+	+	+	+	+	+	+	−	Pre B cell
VI	B-ALL	+	−	+	−	+	+	+	+	Immature B cell

[a]Lacking distinguishing morphological features, these groups have also been called AUL by morphologists. Group I ALL blasts cells can only be assigned to the B-cell lineage on the basis of a Ig heavy chain gene rearrangement status. Coexpression of surface CD19 and cytoplasmic cCD22 is a specific B-cell criterion.

Source: Modified after Refs. [20,21,25,30]; see also Figure 2.

Table 5 Distinctive Immunological Features of T-Cell Malignancies

Malignancy	HLA-DR	TdT	CD10	CD1	CD7	CD2	CD3	CD5	CD4	CD8	CD16
T-ALL	+/−	+	+/−	+/−	+	+	+	+	+/− and/or	+/−	−
T-LL	+/−	+	+/−	+/−	+	+	+	+	+/− and/or	+/−	−
T-CLL	−	−	−	−	+	+	+	+	+ or	+	−
T-PLL	−	−	−	−	+	+	+	+	+ or	+	−
ATL	+/−	−	−	−	+/−	+	+	+	+	−	−
CTCL	+/−	−	−	−	+/−	+	+	+	+	−	−
PTCL	+/−	−	−	−	+	+	+	+	+	−	−
T-LGL	+/−	−	−	−	+	+	+	+	−	+	+

LL, lymphoblastic (malignant) lymphoma; CLL, chronic lymphocytic leukemia; PLL, prolymphocytic leukemia; ATL, adult T-cell leukemia; CTCL, cutaneous T-cell lymphoma (mycosis fungoides, Sézary's syndrome); PTCL, peripheral T-cell lymphoma; LGL, large granular lymphocytosis or Tγ-lymphoproliferative disease.

Source: Data compiled from Refs. [22,31-33]. See also Table 3.

Table 6 Distinctive Immunological Features of B-Cell Malignancies

Malignancy	SmIg	k/λ[a]	HLA-DR	CD19	CD20	CD22	TdT	CD10	CD5	FMC7	CD11c	PCA1	CD38
cALL/pre-B ALL	–	–	+	+	+/–	+/–	+	+	–	–	–	–	–
B-ALL	+	+	+	+	+	+	–	+/–	–	–	–	–	–
Burkitt	+	+	+	+	+	+	–	+/–	–	–	–	–	–
B-CLL	+	+	+	+	+	+/–	–	–	+	+/–	–	–	–
B-NHL[b]	+	+	+	+	+	+	–	+/–	+/–	+/–	–	–	–
B-PLL	+	+	+	+	+	+	–	–	+/–	+	–	–	–
HCL	+	+	+	+	+	+	–	–	–	+	+	+/–	–
Myeloma[c]	–	+[d]	–	–	–	–	–	–	–	–	–	+	+

[a]Clonal expression of either kappa or lambda Ig light chain.
[b]This category includes a variety of histopathologically defined entities (see [31,34]).
[c]Multiple myeloma; other plasma cell dyscrasias: Waldenström's macroglobulinemia, plasma cell leukemia, plasmacytoma.
[d]Cytoplasmic expression.
CLL, chronic lymphocytic leukemia; PLL, prolymphocytic leukemia; HCL, hairy cell leukemia.
Source: Data compiled from Refs. [22,25,31,35,36]. See also Table 4.

The distinctive immunological features of the T- and B-cell malignancies in relation to their acute and chronic forms and in comparison with the different lymphoma variants are presented in Tables 5 and 6, respectively. Individual markers that are characteristic for a specific subset are the CD11c, CD16, CD38, FMC7, and PCA1 antigens (Tables 1, 5, 6). In some leukemias, quantitative differences (such as intensity of SmIg expression; e.g., HCL/PLL > CLL) can be exploited for the differential diagnosis. The majority of chronic lymphoid leukemias (>95%) belong to the B-cell lineage.

As indicated previously, the NHL are a diverse group of neoplasms whose histopathological classifications are controversial. In some of these schemes (e.g., the Kiel classification), CLL, HCL, mycosis fungoides, and Sézary's syndrome are listed in the low-grade lymphomas category. It is even more difficult to correlate histopathological descriptions with immunological marker patterns. The main contribution immunophenotyping can make is to the differentiation of T- and B-cell-type malignancy. Within cell lineages, however, the differentiation between leukemia and lymphoma is, for the most part, immunologically not apparent. A T-LL (lymphoblastic malignant lymphoma) is immunologically distinguishable barely or not at all from a T-ALL, particularly in its leukemia phase [33,34] (Table 5). The same applies for B-ALL vs. Burkitt's lymphoma (in children most B-ALL cases are probably leukemic phases of Burkitt's lymphomas) and B-CLL vs. B-NHL (in its different forms and leukemic phases). Some B-NHL express CD10, but are negative for CD5, in contrast to the reverse profile on B-CLL cells [36].

Peripheral and cutaneous T-cell lymphomas are identified by their morphology, clinical history, and their immunoprofile, which typically is that of mature T-helper cells (CD4, CD2, CD3, CD5 positive) [32].

VI. PROGNOSTIC SIGNIFICANCE OF ALL IMMUNOPHENOTYPING

The prognostic significance of immunophenotypes has been examined in a number of clinical studies. The majority of these series analyzed the prognostic impact of immunomarkers for patients with ALL. Comparatively little, if any, information is available for other types and subtypes of the lymphoid leukemias and lymphomas with regard to their prognostic importance. In contrast to the lack of prognostic significance of the diagnostic categories of the FAB system, it very soon became apparent that surface markers define both biological and prognostic characteristics in ALL [37,38]. T-ALL has been recognized since the mid-1970s as a poor-prognosis group of ALL. Even after other risk factors such as age or white blood cell (WBC) were adapted for, differences in clinical outcome between the various immunologically defined ALL subtypes remained significant [5]. In two large independent series on a total of more than 750 cases of ALL, it became apparent that a rank order (in terms of prognosis) exists between the identifiable ALL clones in children: namely cALL > null-ALL > T-ALL > B-ALL [5,39]. Prognosis is defined here as response to treatment, referring to achievement of complete remission, duration of remission, and overall survival.

Data on the prognostic significance of immunophenotypes in adult ALL indicate also a ranking order that is, however, different from that reported for children: T-ALL > cALL > null-ALL > B-ALL. In the largest series on 250 adult patients with ALL the immunoprofile clearly had a significant predictive impact on prognosis and was described to be independent by multivariate analysis [40].

VII. ACUTE MIXED LINEAGE AND UNDIFFERENTIATED LEUKEMIA

Current models of leukemic cell differentiation postulate the principle of "lineage fidelity." This concept proposes that leukemic cells arise from the clonal expansion of a transformed hematopoietic progenitor cell with preservation of either lymphoid or myeloid features characteristic for its normal counterpart at the stage at which maturation arrest had occurred [41]. A growing body of findings on leukemic blasts (mainly acute leukemias) that express simultaneously both lymphoid and myeloid features has recently suggested that lineage fidelity is violated in a significant number of cases [42]. These phenomena have been interpreted as *lineage infidelity* and *abnormal gene expression*; descriptive terms used are *acute hybrid, mixed lineage* and *biphenotypic* leukemias [43,44].

Most cases are of anecdotal nature and many reports are suspect on technical grounds [45]. Even though individual examples of truly aberrant gene expression may occur in leukemia, since they are found with insufficient regularity and at a low incidence they do not seem to be of general significance [45].

Other examples of apparent lineage infidelity might not reflect genetic misprogramming, but rather represent instead leukemic blasts arising from a progenitor cell that expresses markers of two or more cell lineages in a transient phase of limited promiscuity. The malignant counterpart to this normal bipotential or multipotential progenitor might preserve the unusual biphenotypic profile as a relic (*lineage promiscuity*) [45].

In the two largest series, 13% of childhood and adult ALL or AML (345 cases) showed evidence of "aberrant" marker expression [42,46]. Other studies claim an incidence of 30% or higher in adult ALL [47,48]. However, it should be stressed that the initial classification of these cases was based solely on morphological assessment [49]. In the absence of a gold standard, the application of absolutely stringent and objective criteria for the demonstration of lymphoid and myeloid features is mandatory. The actual incidence of hybrid acute leukemia might very well be lower than previously reported (possibly in the range of 10-20%) [50,51].

Lineage switches (conversion from a lymphoid to a myeloid phenotype or the reverse) and phenotype shifts (significant changes of immunophenotypes) have been reported for patients with acute leukemias after relapses. The incidences appear to be about 10% and 15%, respectively [42,52].

A certain percentage (2-10%) of acute leukemias are not classifiable according to the FAB system [27]. Morphology alone cannot provide conclusive evidence on the nature of the undifferentiated blast cells termed *acute undifferentiated/unclassified leukemia* (AUL). With the use of immunocytological methods in some of these cases, cytoplasmic staining with anti-myeloid MAbs (CD13) was detected [53]. Other specimens could be assigned to the B lineage on the basis of immunophenotyping and genotyping [54].

VIII. PRACTICAL PANELS FOR ROUTINE IMMUNOPHENOTYPING

The diagnostic precision of leukemia phenotyping has now been refined by the introduction of standard procedures and highly purified, "clean" reagents [17,55]. Flow cytometric evaluation has become a widely used, important diagnostic tool [56]. The gradual simplification and standardization of diagnostic panels for leukemia diagnosis allows for the processing of large numbers of samples and for the comparison of different studies.

Table 7 Practical Panels for Diagnostic Immunophenotyping

Primary panel for chronic lymphoid leukemias ("lymphocyte panel")

- T-lineage associated
  - CD2 — T cells
  - CD3 — T cells
  - CD5 (+ IgM)[a] — T cells (+ B-CLL)
  - CD7 — Pan-T
  - CD4 — T-helper cells
  - CD8 — T-suppressor cells
- B-lineage associated
  - HLA-DR — B cells, activated T cells, monocytes
  - CD19 — Pan-B
  - CD20 — B cells
  - CD22 — B cells
  - k + λ[b] — B cells
- Control
  - Control MAb

Primary panel for acute myeloid and acute lymphoid leukemias ("acute leukemia panel")

- Stem cell-associated
  - HLA-DR — Lymphoid/myeloid precursors, B cells, monocytes
  - CD34 — Lymphoid/myeloid precursors
- Myeloid-associated
  - CD13 — Panmonomyeloid
  - CD14 — Monocytes
- T-lineage associated
  - CD2 — T cells
  - CD3 — T cells
  - CD7 — Pan-T
- B-lineage associated
  - CD10 — Pre B cells
  - CD19 — Pan-B
  - CD20 — B cells
  - IgM — B cells
- Control
  - Control MAb

[a] Double labeling with anti-IgM for the detection of B-CLL cells.
[b] Double labeling with directly conjugated anti-k and anti-λ reagents to demonstrate monoclonality in B-cell disorders.
Source: Adopted and modifed from Refs. [17,55,57]. See also Table 1 for frequently used reagents.

Routine leukemia phenotyping now concentrates on the diagnostic enquiry using only a few crucial reagents with clearly established reactivity patterns [17]. The respective antibodies are available from commercial sources. An example of a standard primary panel of MAbs used in the microtiter plate assay [57] for chronic lymphoid disorders is shown in Table 7. The majority of chronic T- or B-cell malignancies can be recognized and classified with these markers (Lymphocyte Panel). Other reagents (e.g., CD38 for

Table 8 MIC Working Classification of ALL

Category	Morphology: FAB	CD7	CD2	TdT	CD19	DR	CD10	cIg	SmIg	Karyotype
Early T-precursor ALL	L1, L2	+	−	+						t or del(9p)
T-cell ALL	L1, L2	+	+	+						t(11;14), 6q−
Early B-precursor ALL	L1, L2			+	+	+	−	−	−	t(4;11), t(9;22)
Common ALL	L1, L2			+	+	+	+	−	−	6q−, near haploid, t or del(12p), t(9;22)
Pre B-ALL	L1			+	+	+	+	+	−	t(1;19), t(9;22)
B-cell ALL	L3			−	+	+	+/−	−/+	+	t(8;14), t(2;8), t(8;22), 6q−

DR, HLA-DR; cIg, intracytoplasmic immunoglobulin; SmIg, surface membrane immunoglobulin.
Source: Modified from Ref. [61].

multiple myeloma, CD11c for HCL) might be substituted or used in a secondary panel for further subtyping. Two-color staining with anti-kappa and anti-lambda light chain reagents permits documentation of the monoclonality of B-cell disorders. From the standpoint of differential diagnosis, the negative markers on a given specimen are as important as the positive criteria and are able to confirm or refute the diagnosis by exclusion in the context of a positive marker.

The primary panel for the acute leukemias (Acute Leukemia Panel) contains stem cell-, myeloid-, T-lineage, and B-lineage-associated markers (Table 7). Again, further tests might be required and other reagents might be included immediately. One of the most important additional markers is TdT, which might be employed in a double-staining assay. A "second round" panel can include additional myeloid (e.g., CD33 or CD15), erythroid (anti-glycophorin A), megakaryocytic (anti-gp IIb/IIIa and anti-gp Ib), antikappa + antilambda, or anti-IgM/IgD reagents (Table 1).

The microtiter plate method appears to be particularly practical and economical (in terms of time, sample material, and reagents) for routine leukemia-lymphoma phenotyping [17,57]. Another scheme for immunophenotyping of acute leukemias has been developed by Neame et al. [46], who use successive panels to screen and subclassify common and rare leukemias.

Similar panels of reagents have been described for the immunological staining of leukemias-lymphomas on cytospins, slides, or tissue preparations using labeling with enzymes, fluorochromes, or visible particles [17,31,58,59]. The most commonly applied techniques are immunoperoxidase and immunoalkaline phosphatase-antialkaline phosphatase (APAAP) stainings. Although immunofluorescence techniques are more frequently used by immunologists, and immunocytological methods appear to be preferred by hematopathologists, in many centers these staining procedures are used in parallel [17].

IX. COMBINED APPLICATION OF MORPHOLOGY, IMMUNOLOGY, AND CYTOGENETICS

In an effort to find an optimal method for typing acute leukemia, a combined classification of the immunophenotypes and FAB morphology-cytochemistry was devised [46]. It was reported that this combination of morphology, cytochemistry, and immunophenotyping improved considerably the diagnostic concordance of leukemic subtypes, that is, the reliability and validity of the methods used [60].

This approach to a multiparameter analysis was extended by including karyotyping in the assessment of the detailed cellular phenotype of acute leukemias. A morphological, immunological, and cytogenetic (MIC) Working Classification of ALL was proposed with the intention of correlating the findings of these three disciplines [61] (Table 8). This classification, which combines the above three major diagnostic approaches into a single and flexible classification, is proposed to be open-ended so that new entities can be entered as they become defined [61].

X. MULTIPLE MARKER ANALYSIS

The above mentioned MIC Workshops grew out of the realization that a large amount of data on different aspects of the biology of leukemias-lymphomas had accumulated during the last 10 years. This information, which was either acquired along parallel avenues or in the combination of different disciplines, is definitely of diagnostic and prognostic significance. Advances in immunology, cytogenetics, and molecular biology in particular have contributed to the exact definition of subtypes within previous entities and to the general expansion of knowledge of the different clinically relevant disease categories. This "multiple marker analysis" [18], which ultimately means the combination of a whole battery of disciplines will undoubtedly provide new insights into the pathobiology of leukemias and lymphomas, which in turn will be of great value for the study of events occurring during normal hematopoietic cell differentiation.

From a practical point of view, a combination of major diagnostic approaches can provide clinicians with a series of guidelines for the identification of leukemia-lymphoma subtypes; these have become of great importance for the management of patients with these neoplasms. Because genotyping (using molecular genetic techniques), and karyotyping (using traditional methods in cytogenetics) are more time-consuming, morphological-histopathological analysis and immunophenotyping will be the first-line and major diagnostic tools. Therefore, immunological phenotyping will definitely play an important part in the multiple marker analysis of leukemia and lymphoma and will clearly continue to make important contributions to basic and applied research.

ACKNOWLEDGMENTS

The authors thank Dr. Dario Campana for stimulating advice and Ms. Suzanne Gignac for help in the preparation of the manuscript.

REFERENCES

1. Bennett JM, Catovsky D, Daniel MT, et al. Proposals for the classification of the acute leukaemias. Br J Haematol 1976;33:451-458.
2. Head DR, Savage RA, Cerezo L, et al. Reproducibility of the French-American-British classification of acute leukemia: the Southwest Oncology Group experience. Am J Hematol 1985;18:47-57.
3. Bennett JM, Catovsky D, Daniel MT, et al. The morphological classification of acute lymphoblastic leukaemia: concordance among observers and clinical correlations. Br J Haematol 1981;47:553-561.
4. Miller DR, Leikin S, Albo V, Sather H, Hammond D. Prognostic importance of morphology (FAB classification) in childhood acute lymphoblastic leukaemia (ALL). Br J Haematol 1981;48:199-206.
5. Pullen DJ, Boyett JM, Crist WM, et al. Pediatric Oncology Group utilization of immunologic markers in the designation of acute lymphocytic leukemia (ALL) subgroups: influence on treatment response. Ann NY Acad Sci 1983;428:26-48.
6. The Non-Hodgkin's Lymphoma Pathologic Classification Project. National Cancer Institute sponsored study of classifications of non-Hodgkin's lymphomas: summary and description of a working formulation for clinical usage. Cancer 1982;40:2112-2135.
7. NCI Non-Hodgkin's Classification Project Writing Committee. Classification of Non-Hodgkin's lymphomas. Reproducibility of major classification systems. Cancer 1985;55:91-95.
8. Whitcomb CC, Crissman JD, Flint A, Cousar JB, Collins RD, Byrne GEJ. Reproducibility in morphologic classification of Non-Hodgkin's lymphomas using the Lukes-Collins system. Am J Clin Pathol 1984;82:383-388.

9. Ersboll J, Schultz HB, Hougaard P, Nissen NI, Hou-Jensen K. Comparison of the Working Formulation of Non-Hodgkin's lymphoma with the Rappaport, Kiel, and Lukes & Collins Classifications. Cancer 1985;55:2442-2458.
10. Lieberman PH, Filippa DA, Straus DJ, Thaler HT, Cirrincione C, Clarkson BD. Evaluation of malignant lymphomas using three classifications and the Working Formulation. Am J Med 1986;81:365-380.
11. Minowada J, Tsubota T, Nakazawa S, et al. Cell surface markers of human leukemia-lymphoma cells lines: Analysis of the marker expression on fresh leukemia cells from 527 patients. In: Bentvelzen P, Hilgers P, Yohn D, eds. Advances in comparative leukemia research. Amsterdam: Elsevier/North Holland Biomedical Press, 1978: 363-367.
12. Greaves MF, Janossy G. Patterns of gene expression and the cellular origin of human leukaemais. Biochim Biophys Acta 1978;516:193-230.
13. Köhler G, Milstein C. Continuous cultures of fused cells secreting antibodies of predefined specificity. Nature (London) 1975;256:495-497.
14. Bernard A, Boumsell L, Dausset J, Milstein C, Schlossman SF, eds. Leucocyte typing. Berlin: Springer, 1984.
15. Reinherz EL, Haynes BF, Nadler LM, Bernstein ID, eds. Leukocyte typing II. Berlin: Springer, 1986.
16. McMichael AJ, ed. Leucocyte typing III. White cell differentiation antigens. Oxford: Oxford University Press, 1987.
17. Janossy G, Amlot P. Immunofluorescence and immunohistochemistry. In: Klaus GGB, ed. Lymphocytes—a practical approach. Oxford: IRL Press, 1987: 67-108.
18. Minowada J. Immunology of leukemic cells. In: Gunz FW, Henderson ES, eds. Leukemia, 4th ed. New York: Grune & Straton, 1982: 119-139.
19. Bloomfield CD, Gajl-Peczalska KJ. The clinical relevance of lymphocyte surface markers in leukemia and lymphomas. Curr Top Hematol 1980;3:175-240.
20. Nadler LM, Korsmeyer SJ, Anderson KC, et al. B cell origin of non-T cell acute lymphoblastic leukemia. J Clin Invest 1984;74:332-340.
21. Campana D, Janossy G, Bofill M, et al. Human B cell development. I. Phenotypic differences of B lymphocytes in the bone marrow and peripheral lymphoid tissue. J Immunol 1985;134:1524-1530.
22. Foon KA, Todd RF III. Immunologic classification of leukemia and lymphoma. Blood 1986;68:1-31.
23. Campana D, Thompson JS, Amlot P, Brown S, Janossy G. The cytoplasmic expression of CD3 antigens in normal and malignant cells of the T lymphoid lineage. J Immunol 1987;138:648-655.
24. Dongen JJM van, Quertermous T, Bartram CR, et al. T cell receptor-CD3 complex during early T cell differentiation. Analysis of immature T cell acute lymphoblastic leukemias (T-ALL) at DNA, RNA, and cell membrane level. J Immunol 1987;138: 1260-1269.
25. Zola H. The surface antigens of human B lymphocytes. Immunol Today 1987;8: 308-315.
26. Dongen JJM van, Krissansen GW, Wolvers-Tettero ILM, et al. Cytoplasmic expression of the CD3 antigen as a diagnostic marker for immature T-cell malignancies. Blood 1988;71:603-612.
27. Drexler HG, Scott CS. Morphological and immunological aspects of leukaemia diagnosis. In: Scott CS, ed. Leukaemia cytochemistry. Chichester: Ellis Horwood, 1989: 13-67.
28. Vodinelich L, Tax W, Bai Y, Pegram S, Capel P, Greaves MF. A monoclonal antibody (WT-1) for detecting leukemias of T-cell precursors (T-ALL). Blood 1983;62: 1108-1113.

29. Campana D, Coustan-Smith E, Janossy G. The reliability of membrane and cytoplasmic markers of immature B and T lymphoid cells in acute leukemia diagnosis. Leukemia 1989;3:170-181.
30. Chen Z, Sigaux F, Miglierina R, et al. Immunological typing of acute lymphoblastic leukemia: concurrent analysis by flow cytofluorometry and immunocytology. Leuk Res 1986;10:1411-1417.
31. Stein H, Mason DY. Immunological analysis of tissue antigens in the diagnosis of lymphoma. In: Hoffbrand AV, ed. Recent advances in haematology, vol. 4. Edinburgh: Churchill Livingstone, 1985:127-169.
32. Knowles DM II. The human T-cell leukemias: clinical, cytomorphologic, immunophenotypic and genotypic characteristics. Hum Pathol 1986;17:14-33.
33. Weiss LM, Bindl JM, Picozzi VJ, Link MB, Warnke RA. Lymphoblastic lymphoma: an immunophenotype study of 26 cases with comparison to T cell acute lymphoblastic leukemia. Blood 1986;67:474-478.
34. Sloane JP. Histopathology and immunobiology of lymphomas. Clin Haematol 1987;1-44.
35. Jansen J, Ottolander GJ den, Schuit HRE, Waayer JLM, Hijmans W. Hairy cell leukemia: its place among the chronic B cell leukemias. Semin Oncol 1984;11:386-393.
36. Anderson KC, Bates MP, Slaughenhoupt BL, Pinkus GS, Schlossman SF, Nadler LM. Expression of human B cell-associated antigens on leukemias and lymphomas: a model of human B cell differentiation. Blood 1984;63:1424-1433.
37. Sallan SE, Ritz J, Pesando J, et al. Cell surface antigens: prognostic implications in childhood acute lymphoblastic leukemia. Blood 1980;55:395-402.
38. Sen L, Borella L, Clinical importance of lymphoblasts with T markers in childhood acute leukemia. N Engl J Med 1975;292:828-832.
39. Greaves MF, Janossy G, Peto J, Kay H. Immunologically defined subclasses of acute lymphoblastic leukaemia in children: their relationship to presentation features and prognosis. Br J Haematol 1981;48:179-197.
40. Hoelzer D, Thiel E, Löffler H, et al. Prognostic factors in a multicenter study for treatment of acute lymphoblastic leukaemia in adults. Blood 1988;71:123-131.
41. Greaves MF. "Target" cells, cellular phenotypes, and lineage fidelity in human leukemia. J Cell Physiol 1982;1:113-125 (suppl).
42. Stass SA, Mirro J Jr. Lineage heterogeneity in acute leukaemia: acute mixed-lineage leukaemia and lineage switch. Clin Haematol 1986;15:811-827.
43. Gale RP, Ben Bassat I. Hybrid acute leukaemia. Br J Haematol 1987;65:261-264.
44. Hoffbrand AV, Leber BF, Browett PJ, Norton JD. Mixed acute laukaemias. Blood Reviews 1988;2:9.
45. Greaves MF, Chan LC, Furley AJW, Watt SM, Molgaard HV. Lineage promiscuity in hemopoietic differentiation and leukemia. Blood 1986;67:1-11.
46. Neame PB, Soamboonsrup P, Browman GP, et al. Classifying acute leukemia by immunophenotyping: a combined FAB-immunologic classification of AML. Blood 1986;68:1355-1362.
47. Sobol RE, Mick R, Royston I, et al. Clinical importance of myeloid antigen expression in adult acute lymphoblastic leukemia. N Engl J Med 1987;316:1111-1117.
48. Hirsch-Ginsberg C, Childs C, Chang KS, et al. Phenotypic and molecular heterogeneity in Philadelphia chromosome-positive acute leukemia. Blood 1988;71:186-195.
49. Drexler HG. Classification of acute myeloid leukemias—a comparison of FAB and immunophenotyping. Leukemia 1987;1:697-705.
50. Ludwig WD, Bartram CR, Ritter J, Raghavachar A, Hiddemann W, Heil G, Harbott J, Seibt-Jung H, Teichmann JV, Riehm HJ. Ambiguous phenotypes and genotypes in 16 children with acute leukemia as characterized by multiparameter analysis. Blood 1988;71:1518-1528.

51. Chen SJ, Flandrin G, Daniel MT, Valensi F, Baranger L, Grausz D, Bernheim A, Chen Z, Sigaux F, Berger R. Philadelphia-positive acute leukemia: lineage promiscuity and inconsistently rearranged breakpoint cluster region. Leukemia 1988;2:261-273.
52. Greaves MF, Paxton A, Janossy G, Pain C, Johnson S, Lister TA. Acute lymphoblastic leukemia associated antigen. III. Alterations in expression during treatment and in relapse. Leuk Res 1980;4:1-14.
53. Matutes E, Oliveira MSP de, Foroni L, Morilla R, Catovsky D. The role of ultrastructural cytochemistry and monocloanl antibodies in clarifying the nature of undifferentiated cells in acute leukaemia. Br J Haematol 1988;69:205-211.
54. Raghavachar A, Bartram CR, Ganser A, Heil G, Kleihauer E, Kubanek B. Acute undifferentiated leukemia: implications for cellular origin and clonality suggested by analysis of surface markers and immunoglobulin gene rearrangement. Blood 1986;68:658-662.
55. Janossy G, Campana D, Bollum FJ. Immunofluorescence studies in leukaemia diagnosis. In: Beverley PCL, ed. Methods in haematology—monoclonal antibodies. Edinburgh: Churchill Livingstone, 1986:97-131.
56. Loken MR. Cell surface antigen and morphological characterization of leucocyte populations by flow cytometry. In: Beverley PCL, ed. Methods in haematology—monoclonal antibodies. Edinburgh: Churchill Livingstone, 1986:132-144.
57. Campana D, Janossy G. Leukemia diagnosis and testing of complement fixing antibodies for bone marrow purging in ALL. Blood 1986;68:1264-1271.
58. Erber WN, Mynheer LC, Mason DY. APAAP Labeling of blood and bone-marrow samples for phenotyping leukaemias. Lancet 1986;1:761-765.
59. Mason DY, Erber WN, Falini B, Stein H, Gatter K. Immunoenzymatic labeling of haematological samples with monoclonal antibodies. In: Beverley PCL, ed. Methods in haematology—monoclonal antibodies. Edinburgh: Churchill Livingstone, 1986:145-181.
60. Browman GP, Neame PB, Soamboonsrup P. The contributions of cytochemistry and immunophenotyping to the reproducibility of the FAB classification in acute leukemia. Blood 1986;68:900-905.
61. First MIC Cooperative Study Group. Morphologic, Immunologic, and Cytogenetic (MIC) Working Classification of acute lymphoblastic leukemias. Cancer Genet Cytogenet 1986;23:189-197.

17
Immunodiagnosis of Acute Nonlymphocytic Leukemia

CRAIG A. HURWITZ*, LEWIS C. STRAUSS, and CURT I. CIVIN
The Johns Hopkins Oncology Center, Baltimore, Maryland

I. INTRODUCTION

The French-American-British (FAB) classification of acute leukemia, designed in 1976, helped to standardize the morphological and cytochemical classification of acute leukemia [1-3]. Difficulties with subjective interpretation of leukemia subtypes within the FAB system [4,5] and identification of leukemia-associated antigens with monoclonal antibodies [6] stimulated interest in leukemia subclassification by immunophenotype. Although an historical objective was to identify cell surface antigens uniquely expressed by leukemia cells, no such "leukemia-specific" surface markers have yet been found. Significant advances, however, have been made in the development and characterization of normal differentiation antigens that are highly restricted in their expression to normal lymphoid and monomyeloid cell subsets. Careful use of monoclonal antibodies directed toward such normal myeloid (and lymphoid) -associated differentiation markers significantly helps in the diagnosis and classification of acute nonlymphocytic leukemia (ANLL). The flood of monoclonal antibodies prompted in the International Workshops on Leukocyte Differentiation Antigens [7-9], from which a cluster designation (CD) nomenclature for antigens recognized by specific antibodies was developed.

This chapter will review monoclonal antibodies that are particularly useful in the immunodiagnosis of ANLL. The application of these reagents for ANLL diagnosis will then be discussed, including a description of leukemic subtypes that could not have been identified without monoclonal antibodies.

II. USEFUL ANTIGENS IN THE IMMUNODIAGNOSIS OF ANLL

Brief synopses of the cell surface markers that aid in the diagnosis of ANLL are presented below. These markers extend beyond monomyeloid-associated antigens. Particular B- and T-lymphoid-associated antigens will also be reviewed, since they are also important for defining ANLL. The normal hematopoietic and leukemic cell-binding specificities of these antigens are summarized in Tables 1 and 2.

Present affiliation: St. Jude Children's Research Hospital, Memphis, Tennessee

Table 1 Frequently Used Monoclonal Antibodies in Leukemia Immunodiagnosis

Cluster Design	Common Hybridoma	Molecular Weight (kD)	CFU-Blast/ pre-CFU	CFU-Mix	Day 14 CFU-GM	Day 7 CFU-GM	Myelo-blast	Pro	Meta	Gran	Mono	BFUe	CFUe
			Hematopoietic Specificity										
CD33	My9,L4F3,UB2	67	−	+	+	+	+	+	−	−	+	+	+/−
CD13	My7,DuHL60-4	150	−	−	−	+	+	+	+	+	+	−	−
−	HLA-DR	28,34	+	+	+	+	+	−	−	−	+	+	−
CD15	LeuM1,My1	180	−	−	−	−	−	+	+	+	+/−	−	−
CD34	My10,HPCA1,12.8	115	+	+	+	+	+	−	−	−	−	+	−
CD11b	Leu15,Mol	170	−	−	−	−	−	+	+	+	+	−	−
CD14	LeuM3,Mo2	55	−	−	−	−	−	−	−	+	+	−	−
CD19	Leu12,B4	95	−	−	−	−	−	−	−	−	−	−	−
CD20	Leu16,B1	35	−	−	−	−	−	−	−	−	−	−	−
CD22	Leu14,SHC11	135	−	−	−	−	−	−	−	−	−	−	−
CD5	Leu1,T1	67	−	−	−	−	−	−	−	−	−	−	−
CD3	Leu4,T3	25	−	−	−	−	−	−	−	−	−	−	−
CD8	Leu2,T8	30	−	−	−	−	−	−	−	−	−	−	−
CD4	Leu3,T4	62	−	−	−	−	−	−	−	−	+	−	−
CD2	Leu5	50	−	−	−	−	−	−	−	−	−	−	−
CD7	Leu9,3A1,WT1	40	−	−	−	−	−	−	−	−	−	−	−
CD10	CALLA,J5,BA3	120	−	−	−	−	−	−	−	+	−	−	−
CD56	Leu19,NKH1	220	−	−	−	−	−	−	−	−	−	−	−
CD41	−	−	−	−	−	−	−	−	−	−	−	−	−
CD42	−	150	−	−	−	−	−	−	−	−	−	−	−
−	10F7MN	−	−	−	−	−	−	−	−	−	−	−	−

CFU-Meg	Platelet	Mature B Lymph	Mature T Lymph	NK Cell	Mature Erythro	Comments	Reference
−	−	−	−	−	−		10-13
−	−	−	−	+	−		10-12
−	−	+	−	−	−		21-23
−	−	−	−	+	−	X hapten	24-31
+	−	−	−	−	−		32-40
−	−	−	−	+	−	CBbiR (CR3)	41-43
−	−	−	−	−	−		44-46
−	−	+	−	−	−		40,50;79,81
−	−	+	−	−	−		6,40,51,79
−	−	+	−	−	−		6,40,52,53
−	−	−	+	−	−		55-58
−	−	−	+	+	−	Linked to T cell receptor	55,59-61
−	−	−	+	+	−		63
−	−	−	+	−	−	AIDS virus receptor	64-67
−	−	−	+	+	−	E rosette receptor	55,68
−	−	−	+	+	−		55,82
−	−	−	−	−	−		39-40.103
−	−	−	−	+	−	NCAM-1	61
+	+/−	−	−	−	−	gpIIb/IIIa	73-76
−	+	−	−	−	−	gpIb	73-77
−	−	−	−	−	+	Glycophorin	69-72

Table 2 Leukemic Cell-Binding Specificities of Antigens

Morphology (FAB classification)	ALL (L$_1$ - L$_3$) B Lineage	ALL (L$_1$ - L$_3$) T Lineage	ANLL (M$_1$ - M$_7$)
Cytochemical Analysis			
PAS[b]	+/− (Punctate)	+/− (Punctate)	+ (M$_6$)
Peroxidase	−	−	+ (M$_1$,M$_2$,M$_3$,M$_4$,M$_5$)
Sudan Black	−	−	+ (M$_1$,M$_2$,M$_3$,M$_4$,M$_5$)
TdT	+	+	− (∼5% may be +)
Specific esterase	−	−	+ (M$_2$,M$_3$,M$_4$)
NSE (ANA)[a]	−	−	+ (M$_4$,M$_5$,M$_7$)
Acid phosphatase	−	+ (Focal)	− (M$_7$ may be diffusely +)
Immunological Methods			
Primary Antigens			
HLA-DR	∼100	<1	∼80
CD5	<1	>95	<1
CD19	>95	<1	<10
CD20	>40	<1	<1
CD22	>65	<1	<1
CD33	<5	<5	∼80
CD13	<5	<5	∼65
Secondary antigens			
CD15	<5	<5	∼40
CD14	<1	<1	∼15
CD34	∼65	<1	∼50
CD11b	<5	<5	∼25
Leu19	<1	<1	∼15
CD41	−	−	<5
CD42	−	−	<5
Glycophorin	−	−	<5
CD3	<1	∼25	<1
CD8	<1	∼60	<1
CD4	<1	∼60	∼25
CD7	<1	>95	∼25
CD10	∼70	∼20	∼5
CD2	<1	∼90	∼10
Genetic probes			
Ig H[c]	+	Rare	Rare
Ig L[d]	+ (Usually)	−	−
TCR[e] gamma	Rare+	+	Rare
TCR beta	Rare+	+	Rare

[a] Nonspecific esterase (alpha naphthal acetate)
[b] Periodic acid-Schiff.
[c] Immunoglobulin heavy chain.
[d] Immunoglobulin light chain.
[e] T-cell receptor.

A. CD33

CD33 monoclonal antibodies detect a 67 kilodalton (kD) glycoprotein [10-12]. The CD33 gene was recently localized to the q13 region of chromosome 19 [13]. There is evidence that the earliest assayable colony-forming units (CFU) are CD33- [14,15]. In normal myelopoiesis, CD33 is first expressed on early multipotent progenitors (CFU-Mix). The CD33 antigen continues to be expressed on all committed myeloid progenitors (CFU-GM), and on a fraction of erythroid burst forming units (BFU-E). CD33 expression decreases during later granulopoiesis, and is only weakly expressed beyond the morphological promyelocytic stage of development. Monocytes, however, express the CD33 antigen more intensely as they mature. Platelets, B and T lymphocytes, and their committed precursors do not express the CD33 antigen, nor is the CD33 antigen detected on natural killer (NK) cells. The CD33 antigen is detected on fewer than 5% of B- and T-lineage ALL cases, but is present on approximately 80% of ANLL cases. Therefore, CD33 is most useful in the identification of ANLL.

B. CD13

Monoclonal antibodies within the CD13 cluster recognize a 150 kD glycoprotein [10-12]. This antigen is first expressed during normal hematopoiesis on late CFU-GM. Expression of the CD13 antigen continues throughout the remainder of myelomonocytic differentiation, increasing in concentration on the surface membrane of maturing monocytes, and gradually decreasing on maturing granulocytes [12,16-17]. It is not detected on early multipotent progenitors (e.g., CFU-Mix), nor on erythroid, platelet, or lymphoid precursor cells. Normal B and T cells do not express CD13; however, the antigen is present on some NK cells. Since CD13 is expressed on less than ~5% of B- and T-lineage ALL cases, but is present on approximately 65% of ANLL cases, it is important in the immunological differentiation of ANLL from ALL.

C. HLA-DR

The class II major histocompatibility antigen, HLA-DR (Ia), is expressed on stromal dependent pre-CFU [18], which are very early multipotent hematopoietic progenitors. It is assumed that the CFU-blast [19,20] also expresses HLA-DR. HLA-DR continues to be expressed on maturing granulocytic cells to about the promyelocytic stage of development, but is not expressed during the later stages of granulocyte development. HLA-DR is present on BFU-E, but is not expressed on CFU-E or later maturing erythrocytes. Neither platelets, NK cells, nor resting T cells express HLA-DR. HLA-DR is, however, expressed throughout monocytic and B lymphoid maturation, and activated T and NK cells also express HLA-DR [21-23]. Essentially all B-lineage ALL cases, and approximately 80% of ANLL cases, express HLA-DR. T-cell ALL, however, is HLA-DR- [21]. Thus, detection of HLA-DR expression on acute leukemia cells eliminates the diagnostic possibility of T-cell ALL (except for "activated, mature" Sézary cell leukemia, or mycosis fungoides, or HTLV-1+ T-cell leukemia, all of which are usually more chronic or subacute leukemias of adults).

D. CD15

The antigen recognized by CD15 monoclonal antibodies is an "X-hapten" with a carbohydrate sequence of lacto-N-fucopentaose-III (3-alpha-fucosyl-N-acetyllactosamine) [24].

The CD15 antigen is present on several proteins [25]. Antibodies directed against this X-hapten do not detect myeloid colony-forming cells, and label mature monocytes very weakly. CD15 antibodies do, however, bind at high levels to maturing cells of the granulocyte series. CD15 is first detected at the promyelocytic stage of granulocyte development, and it increases in cell surface expression as the granulocyte matures [14,26-28]. Some NK cells express CD15 weakly, but CD15 is not expressed on mature erythrocytes, platelets, B or T lymphocytes, or their precursors. CD15 is detected on approximately 40% of ANLL cases, but is expressed on <5% of B and T lineage ALL cases. However, the X-hapten is apparently present on many lymphoid cells in a cryptic, heavily sialylated form [29]. Treatment of normal lymphoid or ALL cells with neuraminidase exposes this cryptic form of the antigen, allowing its detection by CD15 monoclonal antibodies [29-31].

E. CD34

The CD34 antigen is a 115 kD glycoprotein that is selectively expressed on morphological blast cells in marrow, including multipotent progenitors (CFU-blast, pre-CFU, CFU-mix) and lineage committed progentors (BFU-E, CFU-meg, CFU-GM) [18-20,32-38]. Although mature B and T lymphoid cells are CD34-, approximately half of the CD34+ bone marrow cells coexpress CD19 (a B-lymphoid-restricted antigen). These CD34+, CD19+ cells are also positive for terminal deoxynucleotidyl transferase (TdT) and CD10 (CALLA), which further suggests that they are committed B-lymphoid progenitors [14, 39,40]. The CD34 antigen, however, is not found on further differentiated cells of any lineage; CD34+ cells make up only 1.5% of normal marrow cells [32-38]. Thus, CD34 is not specific for myeloid vs. lymphoid cells, but identifies very immature normal hematopoietic cells. CD34 antigen is found on less than 1% of T-lineage ALL cases, but is detected on approximately 70% of B-lineage ALL cases and 50% of ANLL cases.

F. CD11b

CD11b antibodies recognize CR3, the receptor for the C3bi component of complement [41,42]. The CR3 molecule is a heterodimer with a common 95 kD beta subunit (CD18), and one of three differing alpha subunits, recognized by antibodies designated CD11a, CD11b, and CD11c. The CD11b alpha subunit is a 170 kD glycoprotein [43]. CD11b monoclonal antibodies do not recognize multipotent or lineage-committed progenitors. CD11b is, however, detected on maturing granulocytic (from the myelocyte stage) and monocytic cells, as well as on a subset of NK cells. The CD11b antigen is found on about 25% of ANLL cases, but <5% of B- or T-lineage ALL cases, making it useful in diagnosing ANLL.

G. CD14

The CD14 differentiation antigen is a 55 kD glycoprotein selectively expressed on mature monocytes (although granulocytes are very weakly positive compared to lymphocytes, which are not detectably labeled) [44,45]. The gene encoding CD14 was recently localized to the q22-q32 region of chromosome 5 [46]. This region also encodes the monocyte colony stimulating factor (M-CSF) receptor (c-fms) and several hematopoietic growth factors, including interleukin 3 and granulocyte-macrophage colony stimulation factor (GM-CSF) [47-49]. The CD14 antigen is detected on fewer than 1-2% of B- or T-lineage ALL cases, but is expressed on approximately 15% of ANLL cases, making it relatively useful in ANLL immunophenotyping.

H. CD19

CD19 antibodies detect a 95 kD glycoprotein restricted to the B-lymphoid lineage. The antigen is first expressed on the earliest identifiable committed B-cell precursor, and continues to be present on the B-cell surface throughout B-lymphoid maturation [40, 50]. CD19 is not expressed on any of the myeloid lineages, nor is it detected at any stage of NK cell or T-lymphocyte development. Over 95% of B-lineage ALL cases express the CD19 antigen, but it is detected on <10% of T-lineage or ANLL cases. Because of this high specificity, CD19 is useful in acute leukemia immunophenotyping.

I. CD20

The CD20 antigen is a B-cell-restricted 35 kD phosphoprotein. CD20 is expressed only on relatively mature B lymphocytes, and is not expressed in hematopoietic cells of any other lineage. The CD20 antigen may be involved in receiving an early signal in B-cell activation, making a resting B cell "competent" and responsive to later signals [51]. CD20 is expressed on approximately 40% of B-lineage ALL cases, but is present on fewer than 1% of T-lineage ALL or ANLL cases, making it useful in differentiating B-lineage ALL from ANLL.

J. CD22

CD22 antibodies bind two highly glycosylated proteins of 135 and 140 kD. Cytoplasmic CD22 expression can be detected in very early (TdT+) B-cell precursors [52,53], but normal maturing B lymphocytes express cell surface CD22 only very late in development (i.e., after CD20 expression is evident) [14,40]. Expression of the CD22 antigen is detected on approximately 65% of B-lineage ALL cases, but CD22 is expressed on <1% of T-lineage ALL and ANLL cases [54]. Thus, this antigen, in conjunction with CD19 and CD20, is important in differentiating B-lineage ALL from T-lineage ALL or ANLL.

K. CD5

The CD5 antigen is a 67 kD glycoprotein expressed on T lymphocytes. CD5 is present on all but the earliest thymic T cells, and is also present on a rare subpopulation of B lymphocytes [55,56]. CD5 is not expressed on myelomonocytic cells, erythroid cells, platelets, or NK cells. The antigen is present on >90% of T-lineage ALL cases, but is expressed on fewer than 1% of B-lineage ALL and ANLL cases. Although CD5 is extremely useful in identifying T-lineage ALL, the antigen is also expressed on B-lineage chronic lymphocytic leukemia (CLL) [57,58].

L. CD3

The CD3 antigen, a 20-28 kD glycoprotein complex, is closely associated in the cell membrane with the T-cell receptor [59,60]. The CD3 antigen consists of three polypeptides which, together with the T-cell receptor, make up the T-cell receptor complex. CD3 expression on the cell surface is detected only on mature immunocompetent T cells, and on a subset of NK cells [55,61]. However, even very early T cells express cytoplasmic CD3 [62]. The CD3 antigen is present on the cell surface in about 25% of T-lineage ALL cases, but is detected on <1% of B-lineage ALL and ANLL cases.

M. CD8

Antibodies of the CD8 cluster detect a 32 kD glycoprotein on cytotoxic and suppressor T cells [63]. The antigen is also expressed on a subset of NK cells. CD8 is not present on cells of other lymphohematopoietic lineages. CD8 is detected on approximately 60% of T-cell ALL but on <1% of B-cell ALL and ANLL cases.

N. CD4

The CD4 antigen is a 62 kD glycoprotein shown to be the receptor for the acquired immunodeficiency syndrome (AIDS) virus [64-66]. It is a marker for helper T cells and is also present (at lower density) on monocytes [67]. Antibodies of the CD4 cluster recognize approximately 60% of T-cell ALL and 25% of ANLL cases. The CD4 antigen is not, however, present on B-lineage ALL (<1% of cases).

O. CD7

The CD7 antigen is another T-lymphocyte cell surface marker. It is a "pan T-cell" antigen, and is probably the earliest marker (along with cytoplasmic CD3) of T-cell lineage commitment [55]. CD7 is also present on a subset of NK cells, but is not expressed on hematopoietic cells of other lineages. CD7 is detected on >95% of T-lineage ALL, but is also found on a significant number (~25%) of ANLL cases. B-lineage ALL cases, however, rarely express CD7 (<1% of cases).

P. CD2

Monoclonal antibodies of the CD2 cluster detect a 50 kD glycoprotein that is the sheep erythrocyte receptor [68]. CD2 antigen is expressed on most, but not all, early thymocytes, and continues to be expressed throughout the remainder of T-cell differentiation [55]. CD2 is also expressed on a subset of NK cells, but is not expressed on normal myeloid cells or B lymphocytes. CD2 antigen is present on 90% of T-lineage ALL, and is also detected on ~10% of ANLL cases. B-lineage ALL cases rarely express the CD2 antigen (<1% cases).

Q. CD10

CD10 (CALLA) is a 120 kD glycoprotein found on the cell surface of B-lymphocyte precursors. The antigen is intensely expressed early in B-cell maturation, and decreases in expression at the next stage of B-lymphocyte development [39,40]. It is not detectable on mature B cells. CD10 is not expressed on normal T cells, but is found on some granulocytes. CD10 is present on ~70% of B-lineage ALL, about 20% of T-cell ALL, and ~5% of ANLL cases.

R. Leu19

The Leu19 (NKH-1) antigen (CD56 in the Fourth International Leukocyte Differentiation Antigen Workshop) is a specific marker of NK cells [61]. The 220 kD glycoprotein (NCAM-1) detected by Leu19 is not expressed on other hematopoietic cells, except perhaps a small T-cell subset. Leu19 antigen is rarely, if ever, detected on B-lineage ALL cases, and is found on <5% of T-lineage ALL cases. However Leu19 has been detected on a significant percentage of ANLL cases (~15%).

S. Glycophorin

Glycophorin is a major transmembrane sialoglycoprotein of the human erythrocyte [69, 70]. Glycophorin is restricted to the erythroid lineage, and is first detected at the morphologically identifiable erythroblast stage of development [71]. It is not present on erythroid colony-forming cells [70,72]. Antibodies to glycophorin recognize <5% of ANLL cases (these glycophorin+ ANLL cases are FAB type M6, erythroleukemias), but do not recognize either B- or T-lineage ALL cases. Thus, the glycophorin molecule is highly specific for ANLL, but is expressed only on a rare subtype.

T. CD41, CD42

Monoclonal antibodies within the CD41 and CD42 clusters recognize distinct differentiation antigens found only on megakaryocytes and platelets [73,74]. The CD41 antigen consists of platelet glycoproteins IIb and IIIa (gpIIb and gpIIIa) that form a Ca^{+2}-dependent complex in the platelet membrane [73]. The GpIIb/IIIa antigen complex is first detected on promegakaryoblasts, and expression of this CD41 antigen decreases with megakaryocytic maturation [75,76]. Antibodies within the CD42 cluster label glycoprotein Ib (gpIb), a 150 kD platelet membrane antigen [77]. CD42 is a late marker of megakaryopoiesis, first detected on megakaryocytes, and expressed at increasingly higher levels during megakaryocyte maturation. Either or both CD41 and CD42 antigens are specifically detected on FAB type M7 (megakaryocytic) ANLL cases, and may therefore be useful in identifying rare ANLL cases that are otherwise difficult to diagnose.

III. TECHNICAL PROBLEMS

Monoclonal antibodies that recognize the antigens described above are used to phenotype leukemias. Attention to characteristics of many of the antibodies and the subjectivity in interpretation of immunofluorescence assays help the investigator avoid misdiagnosis. For instance, the percentage of antigen+ cells in the sample required for a leukemic population to be considered "positive" for a particular antigen is somewhat arbitrary. Most studies have traditionally considered expression of an antigen by ≥20% of the cells within a leukemic population as a "positive" case [78]. Bone marrow samples with fewer than 20-30% leukemic blasts are especially difficult to analyze, because there are no known leukemia-specific markers, that is, all antigens detected on leukemic blasts are also expressed on normal hematopoietic cells. "False-positive" results, therefore, may occur from antibody recognition of normal contaminating hematopoietic cells.

HLA-DR antigen can sometimes be useful in locating blasts on a flow cytometer [78]. This is because HLA-DR is positive on <10-20% of normal marrow cells (monocytes [<5-10%], B cells [<5-10%], and progenitor cells [<1%] are HLA-DR+), but is detected on most B-lineage ALL cases and ~80% of ANLL cases. A marrow sample with >10-20% HLA-DR+ cells is abnormal, and the HLA-DR+ cell population is likely to contain the leukemic blasts. Double-label immunofluorescence assays using HLA-DR as a second marker to identify the leukemia cells can thus be performed to phenotype a case with a low percentage of blasts. Similar multiparameter analyses using other antibodies can often help in such cases.

Problems may also arise in specimens with large number of necrotic cells. Dying or dead cells tend to "autofluoresce" and to bind antibodies nonspecifically, resulting in "false-positive" observations. Density gradient (e.g., Ficoll-Hypaque, or Percoll centrifugation) should be used prior to antibody labeling of cells to remove dead cells.

False-positive results may also be caused by binding of monoclonal antibodies nonspecifically to Fc receptors on monocytic and granulocytic cells. In addition, IgM monoclonal antibodies bind nonspecifically to myeloid cells, for unknown reasons. Although this binding may be impossible to eliminate completely, certain steps during an immunofluorescence assay can decrease nonspecific antibody binding. First, proper negative control monoclonal antibodies (isotype matched) should be tested in every assay to determine the amount of nonspecific cell binding. Second, preincubation of cells with (1%) normal human AB+ serum results in coating of receptors (particularly Fc receptors) with human immunoglobulin in the serum, and thus decreases later nonspecific binding of mouse monoclonal antibody. Third, preincubation of cells with (1%) normal serum from the species of the second antibody (for indirect immunofluorescence, usually goat or rabbit serum) saturates cells with species-specific immunoglobulin, thus reducing nonspecific binding of the second antibody. It is beneficial to include these "blocking" sera throughout the procedure.

IV. IMMUNOLOGICAL DEFINITION OF ANLL

Table 1 lists the monoclonal antibodies that are useful in the immunodiagnosis of both lymphoid and myeloid leukemias. Unfortunately, no cell surface marker has been found to be absolutely myeloid-specific. All myeloid-associated antigens have been detected on at least a small percentage of ALL cases. The immunological definition of ANLL is improved by simultaneous use of B- and T-lymphoid-associated markers. Antibodies to certain of these lymphoid antigens recognize <5% of ANLL cases. The CD19, CD20, and CD22 antigens are highly specific for B-lineage ALL [6,50,79-81], and CD5 is present on >90% of cases of T-cell ALL but on <1% of B-cell ALL and ANLL cases [82]. An acute leukemia, therefore, that displays one or more of these lymphoid-restricted antigens is *not* likely to be ANLL. In fact, one could use these antigens to construct an immunological definition of ANLL (i.e., ANLL cases could be defined as those cases expressing *neither* CD19, CD20, CD22, *nor* CD5).

In conjunction with the *lack* of membrane expression of specific lymphoid surface antigens, the presence of myeloid cell surface markers should be assessed for each case of leukemia. As described earlier, CD33 is expressed in approximately 80% of ANLL cases. However, ~20% of CD5-, CD19-, CD20-, and CD22- cases still lack expression of CD33 (Fig. 1).

CD13 antibodies recognize approximately 65% of ANLL cases [83]. Almost 90% of CD5-, CD19-, CD20-, CD22- (ANLL) cases are positive for *either* CD33 *or* CD13 (Fig. 1). The ability to diagnose ANLL cases immunophenotypically is improved about 10% by using CD13, compared to the use of CD33 alone. However, one must be careful in interpretation, since CD33 and CD13 are each detected on some (<5%) ALL cases.

CD11b is a third myeloid-associated antigen that (slightly) increases the sensitivity of positive identification of ANLL cases. The CD11b antigen improves the ability to detect CD5-, CD19-, CD20-, CD22- (ANLL) cases (over the use of CD33 and CD13) by

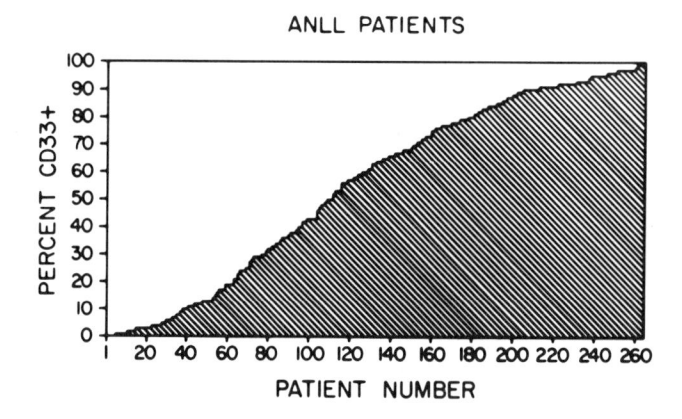

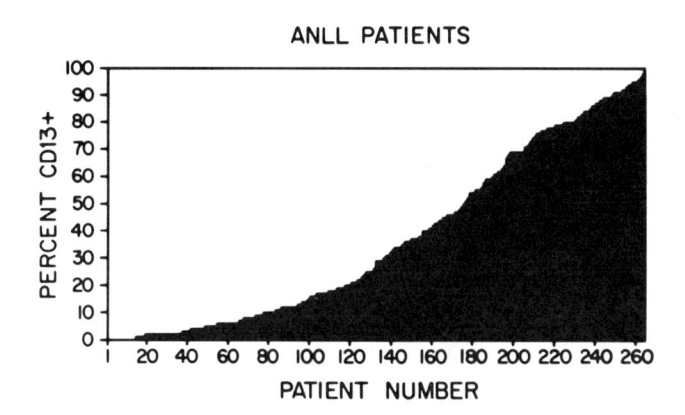

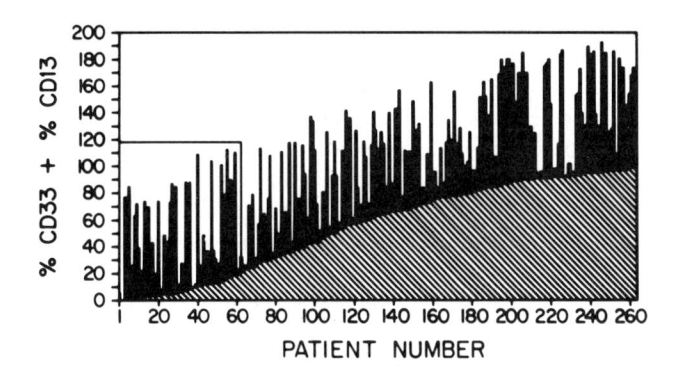

Figure 1 Frequency of CD33 and CD13 expression compared to the frequency of CD33 plus CD13 expression in ANLL cases. Hatched bars represent percentage CD33 expression; solid bars represent percentage CD13 expression of patient samples as assayed by one-color immunofluorescence. Samples with ≥20% antigen expression are considered positive. Cases in each graph are ordered by increasing CD33 (hatched bars) or increasing CD13 (solid bars) expression. The graph depicting %CD33 + %CD13 has the %CD13 expression superimposed on the %CD33 expression, for each particular case. Thus, %CD33 + %CD13 could theoretically equal 200%. CD33- cases are enclosed within the solid lines. About half of the CD33- cases are CD13+, thus improving the overall immunophenotypic diagnosis of ANLL (cases expressing either CD33 or CD13) by approximately 10%.

only about 2%, since CD11b is rarely expressed in CD33-, CD13- ANLL cases. In similar fashion, the mature myeloid markers, CD14 and CD15, are each positive on some of ANLL cases, but improve the sensitivity for immunophenotypic diagnosis of ANLL by only 1-2% each.

Antiglycophorin and the antiplatelet antibodies (CD41 and CD42) are found on fewer than 5% of ANLL cases, and are almost never expressed in the absence of other myeloid markers. Their major advantage in the immunodiagnosis of ANLL is their high specificity for FAB M6 and M7 subtypes.

Thus, using a small panel of well-characterized monoclonal antibodies, about 95% of ANLL cases can be effectively differentiated from ALL cases. A lineage diagnosis of the remaining questionable cases may tentatively be resolved by cytochemical stains, or by the finding of immunoglobulin or T-cell receptor gene rearrangements in the leukemic cells (Table 2), although these molecular markers are not absolutely specific either (see Section IX). Other troublesome cases will express "inappropriate" antigens associated with, but not necessarily restricted to, B- or T-lineage ALL (e.g., HLA-DR or CD10 in B-lineage ALL, and CD3, CD4, CD2, or CD7 in T-cell ALL). The remainder of "difficult to diagnose" specimens (~1%) may include cases of "mixed-lineage" or "undifferentiated" leukemia (discussed in Sections VII and VIII).

V. CORRELATION BETWEEN IMMUNOPHENOTYPE AND FAB CLASSIFICATION OF ANLL

Cell surface antigen expression differentiates cases of ANLL from ALL, but, in general, does not recapitulate the FAB morphological-cytochemical classification system. One study demonstrated a relationship between FAB subtype and surface immunophenotype in approximately 80% of 138 cases studied (84). In this study, ANLL cases with an M1 morphology were often positive for CD33 and CD13, but did not express either CD14 or CD15. The more mature granulocytic M2 cases tended to express CD33, CD13, and CD15, but were negative for the CD14 antigen. The monocytic leukemias (M4, M5) tended to express CD14 and HLA-DR. However, most M4 and M5 cases also expressed the primarily granulocyte-associated antigen CD15. In contrast, several other studies have not shown any correlation between FAB subtype and cell surface immunophenotype (83,85-88). Specifically, most investigations demonstrated a significant percentage of CD14+ M1 and M2 myeloid leukemias. A recent prospective analysis of 196 patients demonstrated CD14 antigen on 20% of M1 cases, 25% of M2 cases, and 50-75% of M4 and M5 cases (89). Since most available monoclonal antibodies bind to both granulocytes and monocytes (or subsets of precursors of both lineages), it may be unreasonable to expect sharp lineage (monocyte vs. granulocyte) distinctions to be demonstrated by these antibodies.

Difficulty in correlating FAB morphology with immunophenotype is compounded by the subjective nature of morphological classification. Several investigators have shown inter- or intraobserver agreement on FAB subtype in only 60-70% of cases studied (90). Although revisions have been made (3) to improve reproducibility within the FAB classification system, inherent problems of subjective interpretation still remain (91). Thus, studies attempting to correlate leukemic immunophenotypes with assigned FAB classifications are somewhat difficult to interpret. Nevertheless, monoclonal antibodies that recognize the erythroid antigen glycophorin and the platelet glycoproteins CD41 and CD42 are highly specific for M6 (erythroleukemia) and M7 (megakaryocytic) morphologies, respectively (92,93).

VI. ANTIGENIC HETEROGENEITY IN ANLL

The majority of studies that have looked at cell surface immunophenotypes in ANLL have shown substantial degrees of heterogeneity, not only among different cases but also among cells in a single ANLL case (12,86,94-96). Several investigators compared cell surface immunophenotypes of CFU-L, the ANLL clonogenic (i.e., self-renewing) cell population, to the total leukemic population (96,97-100). In general, antigen expression within the CFU-L population in many patients was heterogeneous, which suggested that ANLL cases may have more than one CFU-L subpopulation. CFU-L in many cases expressed antigens that were not found on the predominant leukemic cell population. The antigens were often those found on very early normal progenitor cells. These data have been interpreted to mean that some form of abnormal differentiation is proceeding in vivo in many patients with ANLL.

Furthermore, preliminary results from our laboratory indicate not only that individual ANLL cases are composed of antigenically heterogeneous populations, but also that sequential expression of antigens on individual cells does not always correspond to normal

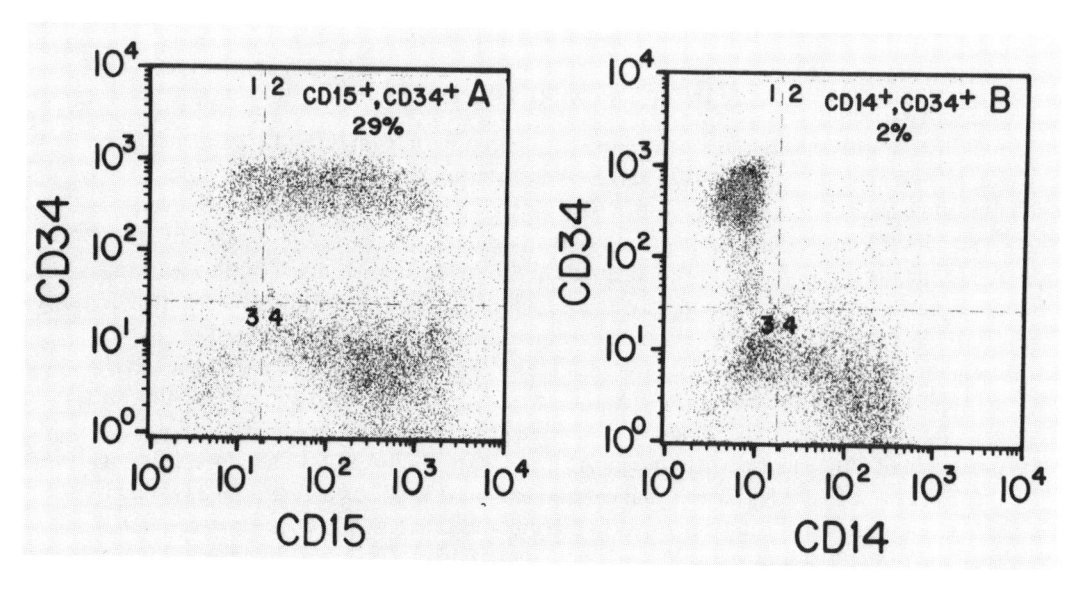

Figure 2 Expression of myeloid-associated antigens in a patient with ANLL. Cytograms show labeling of single cells with antigen combinations not found on normal marrow myeloid cells. CD34 is shown on the ordinate; CD15 (Left) and CD14 (Right) are shown on the abscissas. Region 1 contains single-labeled cells (red fluorescence); region 2 contains double-labeled cells (red plus green fluorescence); region 3 contains unstained cells (based on staining with labeled irrelevant control antibodies); region 4 contains only single-labeled cells (green fluorescence). Percentages represent the percentage of cells within a particular region.

myeloid differentiation pathways (Hurwitz et al., 1988, unpublished results). "Asynchronous" expression of differentiation antigens has been described and characterized in B-lineage ALL (31,101-103); B-lineage ALL cells were shown to coexpress early and mature differentiation antigens in combinations not found in normal lymphohematopoietic cells.

Two-color immunofluorescence assays indicate that distinct subsets of ANLL cells in many cases express both early and late antigens in combinations not found on normal myeloid cells. It is interesting that "antigenic asynchrony" seems to be more common with particular antigen pairs. For example, most CD34+, CD14+ ANLL cases contain distinct subpopulations of cells expressing CD14 or CD34, but not both, thus remaining faithful to normal monocytic differentiation. In contrast, the mature myeloid CD11b and CD15 antigens appear to be expressed on a significant percentage of CD34+ ANLL blasts. In normal myeloid differentiation, CD34 is not known to be simultaneously expressed with either CD11b or CD15 antigens. Double-label cytograms of an ANLL sample with a CD34+, CD14+, CD15+ immunophenotype are shown in Figure 2. The blast cell count on this ANLL sample was 88%. Twenty-nine percent of the leukemic cells were found to express simultaneously the CD15 and CD34 antigens. Only 2% of the leukemia cells (over background control), however, were CD34+, CD14+. Thus, on at least some ANLL subpopulations, CD15 (and CD11b) are expressed "asynchronously" simultaneously with CD34. The heterogeneity within ANLL cell populations and the presence of these "asynchronous" phenotypes may further reflect the abnormal differentiation patterns within acute leukemia. The relevance and prognostic implications of such heterogeneity and asynchrony within ANLL populations have yet to be determined.

VII. T- AND B-LINEAGE ANTIGEN EXPRESSION IN ANLL

Cell surface antigens clasically associated with either B- or T-lineage ALL cases are occasionally detected in patients with ANLL. For instance, the "helper T-cell antigen," CD4, is positive on approximately 15% of ANLL cases. This might be expected, however, since CD4 is known to be expressed on normal monocytes. Therefore, a case of CD4+ ANLL should not be taken to reflect aberrant differentiation or mixed lineage leukemia (31). Surface markers commonly associated with B-lymphocyte differentiation include HLA-DR, CD34, and CD10. As previously mentioned, HLA-DR is expressed on all B lymphocytes and B-lineage ALL cases, but is also expressed on normal monocytes, early granulocyte precursors, and 80% of ANLL cases. The earliest known differentiation antigen of progenitor cells, CD34, labels CFC of all cell lineages (18-20,32-38). It is expressed in 65-70% of cases of B-lineage ALL and 45-50% of ANLL. Therefore, it is not inconsistent to find ANLL cases that express either HLA-DR or CD34. CD10 is known best as a differentiation antigen of early B lymphoid cells (39,40,104). However, CD10 is also expressed on some normal granulocytes (105,106). Thus the observation of CD10+ ANLL in ~5% of cases is not surprising.

Other lymphoid differentiation markers have not (yet) been identified on normal monocytic or granulocytic cells. The T-lymphoid-associated antigen, CD7, is a sensitive indicator of T-cell ALL, detecting >95% of T-lineage ALL cases (82). Though CD7 is essentially never expressed in B-lineage ALL, CD7 labels 5-50% of ANLL cases (107-110). CD2 is likewise expressed on most normal T lymphocytes but not on normal myeloid cells. It is surprising that approximately 10% of ANLL cases express CD2.

This has been taken as evidence for "mixed (T and myeloid) lineage leukemia." An intriguing alternative possibility is that at least some CD2+, and/or CD7+ ANLL cases represent a previously undescribed leukemia subtype. Both of these surface markers have been detected on normal NK cell subpopulations (111-114). Leu19 (CD56), a surface marker restricted in its expression to NK cells and a minor T cell subset (61), is also positive on about 20% of ANLL cases. Our experience has been that CD2 and/or CD7 is expressed in 40% of Leu19+ ANLL cases (Civin, et al., unpublished observations) and 25% of those cases express all three antigens. Perhaps, then, CD56+, CD7+, CD2+ ANLL is a leukemia of NK cells.

VIII. "MIXED LINEAGE" LEUKEMIAS AND ANLL

Other rare cases (<10% of ANLL) coexpress myeloid antigens and lymphoid-restricted antigens (e.g., CD3, CD5, CD8, CD19, CD20, or CD22). Certain cases may be explained by contaminating normal cells, but other cases have been shown by multicolor analyses to be composed of cells with "mixed lineage" phenotypes. Thus, such cases may reflect true mixed lineage leukemias, that is, leukemias whose cells simultaneously express antigens normally restricted to different hematopoietic lineages (109,115). Other studies have demonstrated cases of beta and gamma chain T-cell receptor gene rearrangements in B-lineage ALL, and heavy-chain immunoglobulin gene rearrangements in T-cell ALL as well as ANLL (116-118).

The significance of these findings is unclear. Two theories have been developed that offer plausible explanations for these "mixed lineage" leukemias. The difference in these two views centers around the relationship of leukemic clones to normal hematopoietic differentiation. The first theory, referred to as "lineage infidelity," is based on the study of leukemic blasts in ANLL (119,120). According to this theory, the abnormal "mixed lineage" leukemia cell phenotypes result from expression of aberrant genetic programs within the self-renewing leukemic blast cell population. Thus, genetic misprogramming in the leukemic cell population causes cellular expression of antigens normally restricted to cells of another lineage. The alternative explanation, "lineage promiscuity," argues that *normal* multipotential stem cells may "transiently" coexpress antigens characteristic of later cells in more than one hematopoietic lineage (31). According to this view, "mixed lineage" leukemia cells reflect normally developing cells "frozen" in differentiation at a stage that is too rare to be detected in samples of normal cells.

The resolution of this controversy awaits further study of stem cell commitment and differentiation. Understanding of the relationship of proto-oncogene expression to normal stages of differentiation, the roles of colony-stimulating factors in leukemogensis and outgrowth of the leukemic clone, and the isolation of hematopoietic progenitors (14) with subsequent immunophenotyping of rare progenitor cell populations should aid in our understanding of the origin of "mixed lineage" leukemias.

IX. USE OF GENETIC PROBES IN THE DIAGNOSIS OF ANLL

Demonstration of immunoglobulin gene rearrangements and T-cell receptor gene rearrangements have been useful in the identification of B-lineage and T-lineage ALL, respectively (116-118,121-123). Since B lymphocytes are the only cells that produce immunoglobulin, and T lymphocytes (and some NK cells) are the only cells known to express the T-cell receptor, it was initially expected that rearrangements of these genes would be found solely in B or T lymphoid cases, respectively. However, heavy-chain immunoglobulin rearrangement has been observed in some ANLL and T-cell ALL cases (116,117). T-cell receptor gamma and beta chain gene rearrangements have been found in a low percentage of cases of B-lineage ALL (118) and in ANLL (123). However, T-cell receptor rearrangements have been reported only rarely in ANLL, and light-chain immunoglobulin gene rearrangements and expression of functional heavy chains have been documented *only* in B-lineage leukemias (124,125).

Unfortunately, then, genetic probes for immunoglobulin and T-cell receptor genes cannot guarantee lineage identification of leukemic cases in which immunophenotyping data are unclear or morphological classification is inconsistent with the leukemic cell phenotype. Used with caution, however, genetic probes for gene rearrangements provide presumptive additional evidence pointing to a lineage diagnosis.

X. FUTURE APPLICATIONS IN LEUKEMIA IMMUNOPHENOTYPING

As discussed, cell surface immunophenotyping enhances the morphologic and cytochemical classification of leukemia in approximately 95% of cases. The lineage diagnosis of about 5% of acute leukemia cases, however, remain questionable. These cases can be classified as "undifferentiated" leukemias (leukemias that fail to express any B-restricted, T-restricted, or myeloid-associated antigens [e.g., rare [<1%] leukemia cases that are HLA-DR+ only]), and "mixed lineage leukemias." Laboratories are now capable, through detailed assays, of probing for the genes of various markers. Genetic probes have been recently described for CD33, CD14, and for the myeloperoxidase gene (13,46,126). In the future, messenger RNA (mRNA) may be detectable by biotinylated cDNA probes and two- and three-color immunofluorescence assays may be designed that can look at surface and cytoplasmic or nuclear markers simultaneously. For example, high levels of myeloperoxidase mRNA have been observed in the promyelocytic cell line HL60, and decrease by >95% when differentiation of the cell line is induced. Assays designed to probe for mRNA levels may therefore provide a means by which lineage commitment of a cell can be ascertained without the actual cellular expression of the given protein marker.

In summary, our ability to diagnose ANLL is >95% accurate, if knowledge of cell surface markers is carefully applied. Still, a small number of cases (1-5%) remain problematic; our current understanding of such "mixed lineage" and "undifferentiated" leukemias is limited. The availability of new techniques should help to resolve these issues,, to improve clinical immunodiagnosis of leukemia, and to lead to a new understanding of leukemogenesis and normal hematopoiesis.

ACKNOWLEDGMENTS

This work was supported in part by grants from the National Cancer Institute, National Institutes of Health: CA-32318, CA-28476, CA-09071, CA-06973, and grant CH390 from the American Cancer Society; and grants from the Becton Dickinson Monoclonal Center and the Leukemia Research Foundation. Dr. Civin is a Scholar of the Leukemia Society of America, Inc.

REFERENCES

1. Bennett JM, Catovsky D, Daniel M-T, Flandrin G, Galton DAG, Gralnick HR, Sultan C. Proposals for the classification of acute leukaemias French-American-British (FAB) Co-operative group. Br J Haematol 1976;33:451-458.
2. Bennett JM, Catovsky D, Daniel MT, Flandrin G, Galton DAG, Gralnick HR, Sultan C. The morphological classification of acute lymphoblastic leukaemia: concordance among observers and clinical correlations. Br J Haematol 1981;47:553-561.
3. Bennett JM, Catovsky D, Daniel MT, Flandrin G, Galton DAG, Gralnick HR, Sultan C. Proposed revised criteria for the classification of acute myeloid leukemia. A report for the classification of acute myeloid leukemia. A report of the French-American-British co-operative group. Ann Intern Med 1985;103:620-625.
4. Head DR, Cerezo L, Savage RA, Craven CM, Bickers JN, Hartsock R, Hosty TA, Saiki JH, Wilson HE, Morrisong FS, Cottman CA, Hutton JJ. Institutional performance in application of the FAB classification of acute leukemia. The Southwest Oncology Group experience. Cancer 1985;55:1979-1986.
5. Head DR, Savage RA, Cerezo L, Craven CM, Bickers NJ, Hartsock R, Hosty TA, Saiki JH, Wilson HE, Morrison FS, Coltman CA, Hutton JJ. Reproducibility of the French-American-British classification of acute leukemia. The Southwest Oncology Group experience. Am J Hematol 1985;18:47-57.
6. Foon KA, Todd RF. Immunologic classification of leukemia and lymphoma. Blood 1986;68:1-31.
7. Bernard A, Boumsell L, Dausset J, Milstein C, Schlossman SF, eds. Leukocyte typing. New York: Springer-Verlag, 1984.
8. Reinherz E, Haynes BF, Nadler LM, Bernstein ID, eds. Leukocyte typing II. New York: Springer-Verlag, 1986.
9. McMichael AJ, et al., eds. Leukocyte typing III: White cell differentiation antigens. Oxford: Oxford University Press, 1987.
10. Griffin JD, Sullivan R, Beveridge RP, Larcom P, Schlossman SF. Induction of proliferation of purified human myeloid progenitor cells: a rapid assay for granulocyte colony-stimulating factors. Blood 1984;63:904-911.
11. Griffin JD, Beveridge RP, Schlossman SF. Isolation of myeloid progenitor cells from peripheral blood of chronic myelogenous leukemia patients. Blood 1982;60:30-37.
12. Griffin JD, Ritz J, Nadler LM, Schlossman SF. Expression of myeloid differentiation antigens on normal and malignant myeloid cells. J Clin Invest 1981;68:932-941.
13. Peiper SC, Ashmun RA, Kaufman S, Look TA. Molecular cloning, expression and chromosomal localization of a human gene encoding a 67 kD myeloid differentiation antigen (CD33). Blood 1987;70:195 (abst).
14. Civin CI, Loken MR. Cell surface antigens on human marrow cells: dissection of hematopoietic development using monoclonal antibodies and multiparameter flow cytometry. Int J Cell Clon 1987;5:267-288.
15. Andrews RG, Torok-Storb B, Bernstein ID. Myeloid-associated differentiation antigens on stem cells and their progeny identified by monoclonal antibodies. Blood 1983;62:124-132.
16. Griffin JD, Ritz J, Beveridge RP, Lipton JM, Daley JF, Schlossman SF. Expression of My7 antigen on myeloid precursor cells. Int J Cell Cloning 1983;1:33-48.
17. Griffin JD, Mayer RJ, Weinstein HJ, Rosenthal DS, Coral FS, Beveridge RP, Schlossman SF. Surface marker analysis of acute myeloblastic leukemia: Identification of differentiation-associated phenotypes. Blood 1983;62:557-563.
18. Watt SM, Katz FE, Davis L, Capellaro D, Gorden MY, Tindle RW, Greaves MF. Expression of HPCA-1 and HLA-DR on growth factor- and stroma-dependent colony forming cells. Br J Haematol 1987;66:153-159.

19. Rowley SD, Sharkis SJ, Hattenburg C, Sensenbrenner LL. Culture from human bone marrow of blast progenitor cells with an extensive proliferative capacity. Blood 1987;69:804-808.
20. Leary AG, Ogawa M. Blast cell colony assay for umbilical cord blood and adult bone marrow progenitors. Blood 1987;69:953-956.
21. Schlossman SF, Chess L, Humphreys RE, Strominger JL. Distribution of Ia-like molecules on the surface of normal and leukemic human cells. Proc Natl Acad Sci USA 1976;73:1288-1292.
22. Beckman IGR, Bradley J, Brooks DA, Kupa A, McNamara PJ, Thomas ME, Zola H. Human lymphocyte markers defined by antibodies derived from somatic cell hybrids. II. A hybridoma secreting antibody against an antigen expressed by human B and null lymphocytes. Clin Exp Immunol 1980;40:593-601.
23. Sieff C, Caine BG, Robinson J, Lam G, Greaves MF. Changes in cell surface antigen expression during hemopoietic differentiation. Blood 1980;60:703-713.
24. Huang LC, Civin CI, Magnani JL, Shaper JL, Ginsburg V. My-1, the myeloid-specific antigen detected by mouse monoclonal antibodies: is a sugar sequence found in lacto-N-fucopentaose III. Blood 1983;61:1020-1023.
25. Van Der Schoot CE, Visser FJ, Bos MJE, Tetteroo PAT, von dem Borne AEGKr. Serological and biochemical characterization of myeloid-specific antibodies of the myeloid panel. In: McMichael AJ, et al., eds. Leukocyte typing III: White cell differentiation antigens. Oxford: Oxford University Press, 1987:603-611.
26. Civin CI, Mirro J, Banquerigo ML. My-1, a new myeloid specific antigen identified by a mouse monoclonal antibody. Blood 1981;58:670-674.
27. Strauss LC, Stuart RK, Civin CI. Antigenic analysis of hematopoiesis. I. Expression of the MY-1 granulocyte surface antigen on human marrow cells and leukemic cell lines. Blood 1983;61:1222-1231.
28. Strauss LC, Skubitz KM, August JT, Civin CI. Antigenic analysis of hematopoiesis. II. Expression of human neutrophil antigens on normal and leukemic marrow cells. Blood 1984;63:574-578.
29. Tetteroo PAT, van't Veer MB, Tromp JF, von dem Borne AE. Detection of the granulocyte-specific antigen 3-fucosyl-N-acetyl-lactosamine on leukemic cells after neuraminidase treatment. Int J Cancer 1984;33:355-358.
30. Stockinger H, Majdic O, Liszka K, Aberer W, Bettelheim P, Lutz D, Knapp W. Exposure by desialylation of myeloid antigens on acute lymphoblastic leukemia cells. J Natl Cancer Inst 1984;73:7-11.
31. Greaves MF, Chan LC, Furley AJW, Watt SM, Molgaard HV. Lineage promiscuity in hemopoietic differentiation and leukemia. Blood 1986;67:1-11.
32. Leary AG, Ogawa M, Strauss LC, Civin CI. Single cell origin of multilineage colonies in culture: evidence that differentiation of multipotent progenitors are stochastic processes. J Clin Invest 1984;74:2193-2197.
33. Civin CI, Strauss LC, Brovall C, Fackler MJ, Schwartz JF, Shaper JH. Antigenic analysis of hematopoiesis. III. A hematopoietic progenitor cell surface antigen defined by a monoclonal antibody raised against KG-1a cells. J Immunol 1984;133:157-165.
34. Strauss LC, Rowley SD, LaRuss VF, Sharkis SJ, Stuart RK, Civin CI. Antigenic analysis of hematopoiesis. V. Characterization of My-10 antigen expression by normal lymphohematopoietic progenitor cells. Exp Hematol 1986;14:878-886.
35. Civin CI, Banquerigo ML, Strauss LC, Loken MR. Antigenic analysis of hematopoiesis. VI. Flow cytometric characterization of My-10-positive cells in normal human bone marrow. Exp Hematol 1987;15:10-17.
36. Leary AG, Strauss LC, Civin CI, Ogawa M. Disparate differentiation in hemopoietic colonies derived from human paired progenitors. Blood 1985;66:327-332.

37. Katz FE, Tindle R, Sutherland DR, Greaves MF. Identification of a membrane glycoprotein associated with haemopoietic progenitor cells. Leuk Res 1985;9: 191-198.
38. Andrews RG, Singer JW, Bernstein ID. Monoclonal antibody 12-8 recognizes a 115-kD molecule present on both unipotent and multipotent hematopoietic colony-forming cells and their precursors. Blood 1986;67:842-845.
39. Ryan D, Kossover S, Mitchell S, Frantz C, Hennessy L, Cohen H. Subpopulations of common acute lymphoblastic leukemia antigen-positive lymphoid cells in normal bone marrow identified by hematopoietic differentiation antigens. Blood 1986; 68:417-425.
40. Loken MR, Shah VO, Dattilio KL, Civin CI. Flow cytometric analysis of human bone marrow. II. Normal B lymphocyte development. Blood 1987;70:1316-1324.
41. Arnaout MA, Todd RF, Dana N, Melamed J, Schlossman SF, Colten HR. Inhibition of phagocytosis of complement C3- or immunoglobulin G-coated particles and of C3bi binding by monoclonal antibodies to a monocyte-granulocyte membrane glycoprotein (Mo-1). J Clin Invest 1983;72:171-179.
42. Wright SD, Rao PE, van Voorhis WC, et al. Identification of the C3bi receptor of human monocytes and macrophages by using monoclonal antibodies. Proc Natl Acad Sci USA 1983;80:5699-5703.
43. Hogg N, Horton MA. Myeloid antigens: new and previously defined clusters. In: McMichael AJ, et al., eds. Leukocyte typing III: White cell differentiation antigens. Oxford: Oxford University Press, 1987:576-602.
44. Loken MR, Shah VO, Civin CI. Characterization of myeloid antigens on human bone marrow using multicolor immunofluorescence. In: McMichael AJ, et al., eds. Leukocyte typing III: White cell differentiation antigens. Oxford: Oxford University Press, 1987:630-636.
45. Dimitriu-Bona A, Burmester GR, Waters SJ, Winchester RJ. Human mononuclear phagocyte differentiation antigens. I. Patterns of antigenic expression on the surface of human monocytes and macrophages defined by monoclonal antibodies. J Immunol 1983;130:145-152.
46. Goyert SM, Ferrero E, Rettig WJ, Yenamandra AK, Obata F, Le Beau MM. The CD14 monocyte differentiation antigen maps to a region encoding growth factors and receptors. Science 1988;239:497-499.
47. Le Beau MM, Epstein ND, O'Brien SJ, Nienhuis AW, Yang YC, Clark SC, Rowley JD. The interleukin 3 gene is located on human chromosome 5 and is deleted in myeloid leukemias with a deletion of 5q. Proc Natl Acad Sci USA 1987;84:5913-5917.
48. Le Beau MM, Westbrook CA, Diaz MO, et al. Evidence for the involvement of GM-CSF and FMS in the deletion (5q) of myeloid disorders. Science 1986;231:984-987.
49. Pettenati MJ, Le Beau MM, Lemons RS, Shima EA, Kawasaki ES, Larson RA, Sherr CJ, Diaz MO, Rowley JD. Assignment of CSF-1 to 5q33.1: evidence for clustering of genes regulating hematopoiesis and for their involvement in the deletion of the long arm of chromosome 5 in myeloid disorders. Proc Natl Acad Sci USA 1987;84: 2970-2974.
50. Nadler LM, Anderson KC, Marti G, Batyes M, Park M, Daley JF, Schlossman SF. B4, a human B lymphocyte-associated antigen expressed on normal, mitogen-activated, and malignant B lymphocytes. J Immunol 1983;131:244-250.
51. Clark EA, Ledbetter JA. Activation of human B cells mediated through two distinct cell surface differentiation antigens, Bp35 and Bp50. Proc Natl Acad Sci USA 1986;83:4494-4498.
52. Dorken B, Moldenhauer G, Pezzuto A, Schwartz R, Feller A, Kiesel S, Nadler LM.

CD39 (B3), a B lineage-restricted antigen whose cell surface expression is limited to resting and activated human B lymphocytes. J Immunol 1986;136:4470-4479.
53. Dorken B, Pezzutto A, Kohler M, Thiel E, Hunstein W. Expression of cytoplasmic CD22 in B-cell ontogeny. In: McMichael AJ, et al., eds. Leukocyte typing III: White cell differentiation antigens. Oxford: Oxford University Press, 1987:474-476.
54. Boue DR, LeBien TW. Expression and structure of CD22 in acute leukemia. Blood 1988;72:1480-1487.
55. Reinherz EL, Schlossman SF. The differentiation and function of human T lymphocytes. Cell 1980;19:821-827.
56. Engleman EG, Warnke R, Fox RI, Levy R. Studies of a human T lymphocyte antigen recognized by monoclonal antibody. Proc Natl Acad Sci USA 1981;58:1791-1795.
57. Martin PJ, Hansen JA, Siadak AW, Nowinski RC. Monoclonal antibodies recognizing normal human T lymphocytes and malignant human B lymphocytes: a comparative study. J Immunol 1981;127:1920-1923.
58. Boumsell L, Coppen H, Pham P, Raynal B, Lemerle J, Dausset J, Bernard A. An antigen shared by a human T cell subset and B-CLL cells: distribution on normal and malignant lymphoid cells. J Exp Med 1980;152:299-234.
59. Reinherz EL, Meuer S, Fitzgerald KA, Hussey RE, Levine H, Schlossman SF. Antigen recognition by human T lymphocytes is linked to surface expression of the T3 molecular complex. Cell 1982;30:735-743.
60. Meuer SC, Acuto O, Hussey RE, Hodgdon JC, Fitzgerald KA, Schlossman SF, Reinherz EL. Evidence for the T3-associated 90 kD heterodimer as the T-cell antigen receptor. Nature 1983;303:808-810.
61. Lanier LL, Le AM, Civin CI, Loken MR, Phillips JH. The relationship of CD16 (Leu-11) and Leu19 (NKH-1) antigen expression on human peripheral blood NK cells and cytotoxic T lymphocytes. J Immunol 1986;136:4480-4486.
62. Furley AJ, Mizutani S, Weilbaecher K, et al. Developmentally regulated rearrangement and expression of genes encoding the T cell receptor-T3 complex. Cell 1986;46:75-87.
63. Evans RL, Wall DW, Platsoucas CD, et al. Thymus-dependent membrane antigens in man: inhibition of cell-mediated lympholysis by monoclonal antibodies to the TH^2 antigen. Proc Natl Acad Scu USA 1981;78:544-548.
64. Maddon PJ, Dalgleish AG, McDougal JS, Clapham PR, Weiss RA, Axell R. The T4 gene encodes the AIDS virus receptor and is expressed in the immune system and the brain. Cell 1986;47:333-348.
65. Sattentau QJ, Weiss RA. The CD4 antigen: physiological ligand and HIV receptor. Cell 1988;52:631-633.
66. Fields AP, Bednarik DP, Hess A, May WS. Human immunodeficiency virus induces phosphorylation of its cell surface receptor. Nature 1988;333:278-280.
67. Ledbetter JA, Frankel AE, Herzenberg LE, Herzenberg LA. Human Leu T cell differentiation antigens: quantitative expression on normal lymphoid cells and cell lines. In: Hammerling G, Kearney J, eds. Monoclonal antibodies and T cell hybridomas: Perspectives and technical notes. New York: Elsevier/North Holland, 1981:16-22.
68. Howard FD, Ledbetter JA, Wong J, Bieber CP, Stinson EB, Herzenberg LA. A human T lymphocyte differentiation marker defined by monoclonal antibodies that block E-rosette formation. J Immunol 1981;126:2117-2122.
69. Marchesi VT, Tillack TW, Jackson RL, Segrest JP, Scott RE. Chemical characterization and surface orientation of the major glycoprotein of the human erythrocyte membrane. Proc Natl Acad Sci USA 1972;69:1445-1449.

70. Gahmberg CG, Jokinen M, Andersson LC. Expression of the major sialoglycoprotein (glycophorin) on erythroid cells in human bone marrow. Blood 1978;52:379-387.
71. Loken MR, Shah VO, Dattilio KL, Civin CI. Flow cytometric analysis of human bone marrow: I. normal erythroid development. Blood 1987;69:255-263.
72. Robinson J, Sieff C, Delia D, Edwards PAW, Greaves M. Expression of cell surface HLA-DR, HLA-ABC, and glycophorin during erythroid differentiation. Nature 1981;289:68-71.
73. Kieffer N, Lewis J, Henri A, et al. Reactivity of monoclonal anti-platelet workshop antibodies in immunofluorescence and immunochemical techniques. In: McMichael AJ, et al., eds. Leukocyte typing III: White cell differentiation antigens. Oxford: Oxford University Press, 1987:769-772.
74. Pilkington GR, Thorne HJ, Wilde TE, Jose DG. Analysis of expression, modulation, and synthesis of platelet-associated antigens by human blood monocytes. In: McMichael AJ, et al., eds. Leukocyte typing III: White cell differentiation antigens. Oxford: Oxford University Press, 1987:772-775.
75. Vainchenker W, Deschamps JF, Bastin JM, et al. Two monoclonal antiplatelet antibodies as markers of human megakaryocyte maturation: immunofluorescent staining and platelet peroxidase detection in megakaryocyte colonies and in vivo cells from normal and leukemic patients. Blood 1982;59:514-521.
76. Tomer A, Harker LA, Burstein SA. Flow cytometric analysis of normal human megakaryocytes. Blood 1988;72:1244-1252.
77. McMichael AJ, Ruse NA, Pilch JR, et al. Monoclonal antibody to human platelet glycoprotein I. I. Immunological studies. Br J Haematol 1981;43:501-509.
78. Griffin JD. The use of monoclonal antibodies in the characterization of myeloid leukemias. Hematol Pathol 1987;1:81-91.
79. Nadler LM. B cell/leukemia panel workshop. In: Reinherz EL, Haynes BF, Nadler LM, Bernstein ID, eds. Leukocyte typing II, vols. 2. New York: Springer-Verlag, 1986:3-42.
80. Stashenko P, Nadler LM, Hardy R, Schlossman SF. Characterization of a human B lymphocyte-specific antigen. J Immunol 1980;125:1678-1685.
81. Meeker TC, Miller R, Link M, Bindl J, Warnke R, Levy R. A unique human B lymphocyte antigen defined by a monoclonal antibody. Hybridoma 1984;3:305-320.
82. Borowitz MJ, Dowell BL, Boyett JM, et al. Monoclonal antibody definition of T cell acute leukemia: a Pediatric Oncology Group Study. Blood 1985;65:785-788.
83. Griffin JD, Mayer RJ, Weinstein HJ, et al. Surface marker analysis of acute myeloblastic leukemia: identification of differentiation-associated phenotypes. Blood 1983;62:557-563.
84. Neame PB, Soamboonsrup P, Browman GP, et al. Classifying acute leukemia by immunophenotyping: a combined FAB-immunologic classification of AML. Blood 1986;68:1355-1362.
85. Majdic O, Bettelheim P, Stockinger H, et al. M2, a novel myelomonocytic cell surface antigen and its distribution on leukemic cells. Int J Cancer 1984;33:617-623.
86. Ball ED, Fanger MW. The expression of myeloid-specific antigens on myeloid leukemia cells: correlations with leukemia subclasses and implications for normal myeloid differentiation. Blood 1983;61:456-463.
87. Linch DC, Allen C, Beverley PCL, Bynoe AG, Scott CS, Hogg N. Monoclonal antibodies differentiating between monocytic and nonmonocytic variants of AML. Blood 1984;63:566-573.
88. van der Reijden HJ, van Rhenen D, Lansdorp PM, et al. A comparison of surface marker analysis and FAB classification in acute myeloid leukemia. Blood 1983;61:443-448.

89. Griffin JD, Davis R, Nelson DA, et al. Use of surface marker analysis to predict outcome of adult acute myeloblastic leukemia. Blood 1986;68:1232-1241.
90. Browman GP, Neame PB, Soamboonsrup P. The contribution of cytochemistry and immunophenotyping to the reproducibility of the FAB classification in acute leukemia. Blood 1986;68:900-905.
91. Bloomfield CD, Brunning RD. The revised French-American-British classification of acute myeloid leukemia: is new better? Ann Intern Med 1985;103:614-616.
92. Greaves MF, Sieff C, Edwards PAW. Monoclonal antiglycophorin as a probe for erythroleukemias. Blood 1983;51:628-639.
93. Vainchenker W, Deschamps JF, Bastin JM, et al. Two monoclonal antiplatelet antibodies as markers of human megakaryocyte maturation: immunofluorescent staining and platelet peroxidase detection in megakaryocyte colonies and in vivo cells from normal and leukemic patients. Blood 1982;59:514-521.
94. Todd RF, Nadler LM, Schlossman SF. Antigens of human monocytes identified by monoclonal antibodies. J Immunol 1981;126:1435-1442.
95. Pessano S, Palumbo A, Ferrero D, et al. Subpopulation heterogeneity in human acute myeloid leukemia determined by monoclonal antibodies. Blood 1984;64:275-281.
96. Sabbath KD, Ball ED, Larcom P, Davis RB, Griffin JD. Heterogeneity of clonogenic cells in acute myeloblastic leukemia. J Clin Invest 1985;75:746-753.
97. Griffin JD, Linch D, Sabbath K, Larcom P, Schlossman SF. A monoclonal antibody reactive with normal and leukemic human myeloid progenitor cells. Leuk Res 1984;8:521-534.
98. Lange B, Ferrero D, Pessano S, et al. Surface phenotype of clonogenic cells in acute myeloid leukemia defined by monoclonal antibodies. Blood 1984;64:693-700.
99. Griffin JD, Larcom P, Schlossman SF. Use of surface markers to identify a subset of acute myelomonocytic leukemia cells with progenitor cell properties. Blood 1983;62:1300-1303.
100. Lowenberg B, Bauman JGJ. Further results in understanding the subpopulation structure in AML: clonogenic cells and their progeny identified by differentiation markers. Blood 1985;66:1225-1232.
101. Hurwitz CA, Loken MR, Graham ML, et al. Asynchronous antigen expression in B lineage acute lymphoblastic leukemia. Blood 1988;72:286-300.
102. Greaves MF, Delia D, Katz F, et al. Biological diversity of acute lymphoblastic leukemia. In: Murphy SB, Gilbert JR, eds. Leukemia research: Advances in cell biology and treatment. New York: Elsevier Science, 1983:97-112.
103. Greaves MF. Immunobiology of lymphoid malignancy. In: Neth R, Gallo RC, Greaves MF, Moore MAS, Winkler K, eds. Hematology and blood transfusion, vol. 28. Modern trends in human leukemia B. Berlin: Springer-Verlag, 1983:3-10.
104. Greaves MF, Hariri G, Newman RA, Southerland DR, Ritter MA, Ritz J. Selective expression of the common acute lymphoblastic leukemia (gp 100) antigen on immature lymphoid cells and their malignant counterparts. Blood 1983;61:628-639.
105. Cossman J, Neckers L, Leonard W, Greene W. Polymorphonuclear neutrophils express the common acute lymphoblastic leukemia antigen. J Exp Med 1983;157:1064-1069.
106. Braun MP, Martin P, Ledbetter J, Hansen J. Granulocytes and cultured human fibroblasts express common acute lymphoblastic leukemia-associated antigens. Blood 1983;61:718-25.

107. Ben-Ezra J, Winberg CD, Wu A, Rappaport H. Leu9 (CD7) positivity in acute leukemias: a marker for T-cell lineage? Hematol Pathol 1987;1:147-156.
108. Sutherland DR, Rudd CD, Greaves MF. Isolation and characterization of a human T lymphocyte-associated glycoprotein (gp40). J Immunol 1984;133:327-333.
109. Chan LC, Pegram SM, Greaves MF. Contribution of immunophenotype to the classification and differential diagnosis of acute leukaemia. Lancet 1985;1:475-479.
110. Sobel RE, Royston I, LeBien TW, et al. Adult acute lymphoblastic leukemia phenotypes defined by monoclonal antibodies. Blood 1985;64:730-735.
111. Lanier LL, Loken MR. Human lymphocyte subpopulations identified by using three color immunofluorescence and flow cytometry analysis: correlation of Leu2, Leu3, Leu7, Leu8, and Leu11 cell surface antigen expression. J Immunol 1984; 132:151-156.
112. Vodinelich L, Tax W, Bai Y, Pegram S, Capel P, Greaves MF. A monoclonal antibody (WT1) for detecting leukemias of T-cell precursors. Blood 1983;62:1108-1113.
113. Haynes BF, Eisenbarth GS, Fauci AS. Human lymphocyte antigens: production of a monoclonal antibody that defines functional thymus-derived lymphocyte subsets. Proc. Natl Acad Sci USA 1979;76:5829-5833.
114. Sobol RE, Rosemarie M, Royston I, et al. Clinical importance of myeloid antigen expression in adult acute lymphoblastic leukemia. N Engl J Med 1987;316:1111-1117.
115. Mirro J, Zipf TF, Pui C-H, et al. Acute mixed lineage leukemia: clinicopathologic correlations and prognostic significance. Blood 1985;66:1115-1123.
116. Rovigatti U, Mirro J, Kitchingman GR, et al. Heavy chain immunoglobulin gene rearrangement in acute nonlymphocytic leukemia. Blood 1984;63:1023-1027.
117. Kitchingman GR, Rovigatti U, Mauer AM, Melvin S, Murphy SB, Stass S. Rearrangement of immunoglobulin heavy chain genes in T cell acute lymphoblastic leukemia. Blood 1985;65:725-729.
118. Goorha R, Bunin N, Mirro J, et al. Provocative pattern of rearrangements of the genes for the gamma and beta chains of the T-cell receptor in human leukemias. Proc Natl Acad Sci USA 1987;84:4547-4551.
119. McCulloch EA. Stem cells in normal and leukemic hemopoiesis (Henry Stratton Lecture, 1982). Blood 1983;62:1-13.
120. Smith LJ, Curtis JE, Messner HA, Senn JS, Furthmayr H, McCulloch EA. Lineage infidelity in acute leukemia. Blood 1983;61:1138-1145.
121. Quertermous T, Murre C, Dialynas D, et al. Human T-cell gamma chain genes: organization, diversity, and rearrangement. Science 1986;231:252-255.
122. Sangster RN, Minowada J, Suciu-Foca N, Minden MD, Tak TW. Rearrangement and expression of the alpha, beta, and gamma chain T cell receptor genes in human thymic leukemia cells and functional T cells. J Exp Med 1986;163:1491-1508.
123. Cheng GY, Minden MD, Toyonaga B, Mak TW, McCulloch EA. T cell receptor and immunoglobulin gene rearrangements in acute myeloblastic leukemia. J Exp Med 1986;163:414-424.
124. Kemp DJ, Harris AW, Cory S, Adams JM. Expression of the immunoglobulin C_{mu} gene in mouse T and B lymphoid and myeloid cells lines. Proc Natl Acad Sci USA 1980;77:2876-2880.
125. Walker ID, Harris AW. Immunoglobulin C_{mu} RNA in T lymphoma cells is not translated. Nature 1980;288:290-293.
126. Tobler A, Miller C, Johnson K, Selsted M, Rovera G, Koeffler HP. Regulation of gene expression of myeloperoxidase (MPO) during myeloid differentiation. Blood 1987;70:196 (abst.)

18
Serum Tumor Markers for Pancreatic Carcinoma

HERBERT A. FRITSCHE, Jr.

University of Texas M. D. Anderson Cancer Center, Houston, Texas

FRANK B. GELDER

Louisiana State University Medical Center, Shreveport, Louisiana

I. INTRODUCTION

Considerable research effort continues to be directed to the identification of tumor markers for adenocarcinoma of the pancreas that might be helpful in the detection of this tumor when it is resectable. For symptomatic patients, tumor markers are needed that can aid in the differential diagnosis of pancreatic carcinoma and nonmalignant pancreatic disease. For those patients who are successfully treated, serum tumor markers can be used to monitor for recurrence of the disease. However, the monitoring of pancreatic cancer patients is only of limited prognostic value, since these patients cannot be helped with additional therapy. In this report, we review the progress of tumor marker research and the application of serum tumor markers to the diagnosis and clinical management of pancreatic cancer.

II. CARCINOEMBRYONIC ANTIGEN

Although carcinoembryonic antigen (CEA) continues to be the most widely used serum tumor marker test, it has not been demonstrated to be useful as a diagnostic aid in pancreatic cancer. While serum CEA abnormalities occur in a high percentage of patients, these abnormalities generally occur when metastatic disease is present. The high false-positive rate of CEA further limits its usefulness in establishing a diagnosis of pancreatic cancer [1].

III. PANCREATIC ONCOFETAL ANTIGEN

Pancreatic oncofetal antigen (POA), also termed pancreatic cancer-associated antigen (PCAA), was first reported in 1974 [2] and subsequently purified and characterized [3-7]. It is a high-molecular-weight glycoprotein that is produced by fetal pancreatic ductal epithelium and most pancreatic cancers, with little or no production by the normal adult pancreas. Immunohistochemical studies [7-10] have demonstrated the pres-

Table 1 Serum POA Concentrations in Individuals with Various Conditions

Diagnosis	N	Range (μg/ml)	% Elevated (>18 μg/ml)
Normal	114	1.5-15	0
Pancreatic cancer	49	3.5-206	63.0
Breast cancer	1087	1.5-48	16.5
Prostate cancer	323	2.6-29	16.5
Colon cancer	500	1.9-37	8.0
Lung cancer	48	3.4-31	6.3
Gastric cancer	36	5.2-27	13.9
Pancreatitis	116	2.1-22	15.5
Benign breast cancer	88	2.3-19	1.1
Biliary stone, cirrhosis, cholecystitis	16	2.9-28	13.5

ence of POA in ductal cells of the fetal pancreas, pancreatic cancer, and some other cancers, including those of the breast, stomach, lung, and prostate. In normal tissue, POA is located in those cells morphologically classified as mucin-positive goblet cells. Despite this morphological association, POA is biochemically and immunochemically distinct from mucin [11]. Unlike colon mucin, which is rich in threonine, serine, and proline and contains 60-75% carbohydrate by weight, POA isolated from pancreatic cancer, fetal pancreas, and normal colon mucosa is comparatively low in these amino acids and contains 19-30% carbohydrate. In addition, polyclonal antiserum POA does not react with purified colon mucin.

Studies have demonstrated both size and charge heterogeneity for POA isolated from different sources [10,11]. In particular, POA from pancreatic cancer, fetal pancreas, or colon mucosa exhibits large differences in molecular weight. Two-dimensional cross-immunoelectrophoretic analysis of POA recently demonstrated antigenic differences between pancreatic POA and POA from gastric cancer, prostate cancer, breast cancer, and normal colon mucosa [10]. In contrast, antigenic heterogeneity was not observed among POA preparations from eight different pancreatic cancer sources. It may be possible to exploit these differences and develop specific immunoassays for organ-associated POA epitopes.

With conventional polyclonal immunoassays, elevations of serum POA levels are observed in 56-98% of patients with pancreatic carcinoma (3-8). However, POA is not specific for pancreatic cancer. The incidence of POA abnormality ranges from 17 to 67% for other cancers, and a false-positive rate of 14-16% has been observed in nonmalignant diseases of the gastrointestinal tract (Table 1).

IV. CA 19-9

CA19-9 is a high-molecular-weight mucin defined by its reactivity to an IgG murine monoclonal antibody elicited against the SW 1116 colorectal carcinoma cell line [12]. The monoclonal antibody detects a carbohydrate epitope similar to that of the Lewis-a

Table 2 Incidence of Serum CA 19-9 Elevation In Individuals with Various Conditions[a]: Percentage of Patients with Serum CA 19-9 >37.0 U/ml

Author and Reference	Malignancy							Nonmalignant Disease		
	Pancreas	Gastric	Biliary	Colorectal	Hepatoma	Other	Liver	Pancreas	Other
Ritts (16)	72	62	67	19	—	6	3	0	—
Andruilli (17)	79	42	36	45	71	34	7	10	10
Satabe (18)	71	24	83	18	25	20	—	3	—
Safi (19)	92	38	—	32	—	17	—	28	14
Encabo (20)	—	—	—	—	—	21	21	14	6
Yoskikawa (21)	93	31	38	40	33	21	0	5	0

[a]Other malignant conditions include cancer of the breast, lung, ovary, and prostate. Other malignant conditions include various gastrointestinal, respiratory, metabolic, and infectious diseases.

antigen [13]. Cancer patients who are Lewis-a⁻, b⁻ do not synthesize the CA 19-9 antigen [14,15]. Various clinical studies (Table 2) have confirmed the high incidence of serum CA 19-9 abnormality in pancreatic carcinoma patients and defined the incidence of CA 19-9 elevation in patients with other cancers and nonmalignant disease [16-21].

Several groups have addressed the role of CA 19-9 in the differential diagnosis of pancreatic disease [22-25]. Since elevated values of serum CA-19-9 are usually associated with carcinoma and borderline elevations are generally observed in nonmalignant pancreatic disease, the diagnostic accuracy of CA 19-9 will depend on the degree of elevation observed in a particular patient. The diagnostic accuracy of CA 19-9 might be improved with the measurement of pancreatic fluids [26-28]. Also, the sensitivity of diagnostic ultrasound for pancreatic carcinoma might be enhanced when used with the CA 19-9 test [29,30].

In pancreatic cancer patients, the degree of CA 19-9 elevation may be predictive of tumor resectability. In one report, only 2 of 14 patients who had resectable disease demonstrated CA 19-9 values greater than 500 U/ml [31]. Approximately 50% of the patients with nonresectable or metastatic disease had CA 19-9 values greater than 500 U/ml. In another study, approximately 10% of patients with tumors smaller than 3 cm diameter had elevated CA 19-9 values, while 91% of the patients with larger tumors had CA 19-9 values greater than 37 U/ml [32]. When CA 19-9 values are elevated, a postsurgical decrease to normal values might have prognostic significance [33]. Glenn et al. reported a minimum 18 month survival time in 7 of 8 patients in whom the postsurgical CA 19-9 value normalized, whereas none of the 6 patients whose values did not normalize survived longer than 12 months [34]. Postsurgical increases in CA 19-9 are reported to reflect disease recurrence or progression [33,34].

V. OTHER BLOOD-GROUP-RELATED ANTIGENS

The cancer-associated expression of altered blood group antigens is well established and has been observed in pancreatic tumor tissue [35,36]. CA 19-9 is the first blood-group-related antigen to have application as a serum tumor marker. Recently, Itai and co-workers have reported that pancreatic tumor tissue can produce a sialylated Lewis-a substance that differs from the substance produced by nonmalignant pancreatic tissue [37]. They postulate that the ratio of these two substances might permit a more specific assessment of the presence of cancer. Other blood-group-related antigens that have recently been identified and investigated for use as serum tumor markers are CA 195, CA-50, and the fetoacinar protein.

A. CA 195

CA 195 is a tumor-related epitope defined by its reactivity with a murine monoclonal antibody elicited against purified cell membranes of a colonic tumor [38]. The monoclonal antibody is reactive to the sialylated and nonsialylated Lewis-a substance. Elevated concentrations of CA 195 are observed in the sera of patients with cancers of the colon, ovary, lung, breast, and pancreas [39,40]. One report suggests that the incidence of elevated CA 195 levels in patients with pancreatic cancer, as well as the false-positive rate in those with nonmalignant disease, are similar to those of CA 19-9, which detects only the sialylated Lewis-a epitope [40]. If this finding is confirmed, it seems unlikely that the combined use of CA 195 and CA 19-9 will result in improved sensitivity and specificity for pancreatic cancer.

B. CA-50

The CA-50 glycoprotein antigen is defined by its reactivity to an IgM murine monoclonal antibody elicited against a colorectal carcinoma cell line [41]. The antibody detects the epitope defined by the CA 19-9 antibody and a related epitope that lacks the fucose residue [42]. Preliminary studies have demonstrated elevated serum CA-50 concentrations in the majority of patients with pancreatic cancers, but the false-positive rate is also high in those with nonmalignant disease of the pancreas, liver, and biliary tract [43].

C. Fetoacinar Protein

Fetoacinar protein (FAP) is a fetal glycoprotein with a molecular weight of 110 KD expressed by pancreatic acinar cells. In a recent report, FAP was elevated in a significant proportion of cancer patients and showed a high positive rate in pancreatic cancer patients who did not express CA 19-9 [44]. The false-positive rate of FAP in patients with nonpancreatic benign disease of the gastrointestinal tract was equivalent to that for CA 19-9, but it was higher than for CA 19-9 in patients with nonmalignant pancreatic diseases.

D. Du-Pan-2

Du-Pan-2 is an IgM class monoclonal antibody elicited against the HPAF human pancreatic tumor cell line [45]. The antibody recognizes a tissue mucin(s) expressed by fetal cells and pancreatic ductal cells [46]. The Du-Pan-2 antigen is present in the circulation of normal individuals at concentrations ranging up to 300 units/ml [47]. In the initial study, the Du-Pan-2 antigen concentration exceeded the cutoff value of 400 units/ml in 68% of pancreatic cancer patients, 40% of gastric cancer patients, and 10% of colorectal cancer patients. Only 3 of 58 patients who had nonmalignant diseases of the gastrointestinal tract had false-positive values [47]. Confirmatory data have been reported by others who have also shown that 40-50% of patients with biliary tract cancer, hepatocellular carcinoma, hepatitis, and cirrhosis will also have significantly elevated concentrations of serum Du-Pan-2 [48-51].

The combined use of Du-Pan-2 with CA 19-9 appears to improve the true-positive rate for pancreatic cancer [52]. There is also some evidence to suggest that, in some cases, the CA 19-9 and Du-Pan-2 epitopes may be expressed by the same mucin molecule [53]. It is not yet clear whether the Du-Pan-2 epitope is a blood-group-related species, but it does not appear to be related to Lewis-a, since it can be detected in the sera of patients who are Lewis-a^-,b^-.

VI. SUMMARY

Various biochemical substances are being evaluated for use as serum tumor markers of adenocarcinoma of the pancreas. The currently established markers, CA 19-9 and POA, have an important but limited role in the diagnosis of pancreatic carcinoma. The role could be expanded if the specificity of these tests for pancreatic cancer could be increased, and this may be possible for each of these tests if several recent findings can be confirmed [10,37].

Most of the new tumor markers are blood-group-related substances, in that the cancer-associated substances share epitopes that are similar to those of the Lewis blood group system. It seems likely that a "panel" of these markers could improve the specificity of these tests for pancreatic carcinoma. However, improvement of the clinical specificity appears unlikely since each of these markers has a high false-positive rate in patients with other cancers, liver diseases, and nonmalignant diseases of the pancreas. Additional study will be required to determine the optimal group of these tests to be used as a pancreatic cancer test panel. Also, more emphasis should be directed to the identification of tumor markers that can be used to detect pancreatic cancer at a stage when it can be treated effectively. The use of tumor markers for monitoring patients does not result in longer patient survival times or a higher survival rate because salvage therapies for this disease are ineffective. If effective salvage therapies can be developed, monitoring with serum tumor markers will become more significant. Thus, continued emphasis should be given to the development of serum tumor markers that have diagnostic utility.

REFERENCES

1. Kam V, Tsao D, Itzkowitz S, Kim Y. Production, characterization and clinical utility of carcinoembryonic antigen. Adv Biotech Proc 1985;4:151-181.
2. Banwo O, Versey J, Hobbs J. New oncofetal antigen for human pancreas. Lancet 1974;1:643-645.
3. Gelder FB, Moossa AR, Hall T, Hunter R. Purification, partial characterization and clinical evaluation of a pancreatic oncofetal antigen. Cancer Res 1978;38;312-324.
4. Gelder FB, Reese C, Moossa AR, Hunter. Pancreatic oncofetal antigen. In: Herberman R, McIntyre R, eds. Immunodiagnosis of cancer. New York: Marcel Dekker, 1979:357-368.
5. Shimano T, Loor RM, Chu TM, et al. Isolation, characterization and clinical evaluation of a pancreas cancer associated antigen. Cancer 1981;47:1602-1613.
6. Schmiegel WH, Arndt R, Thiele HG, et al. Pancreatic oncofetal antigen in pancreatic juices: partial chemical characterization and diagnostic application of a pancreatic cancer-associated antigen. Scand J Gastroenterol 1981;16:1033-1040.
7. Hobbs JR, Knapp ML, Branfoot AC. Pancreatic oncofetal antigen (POA): its frequency and localisation in humans. Oncodev Biol Med 1980;1:37-48.
8. Kawa S, Homma T, Oguchi H, et al. Clinical application of the enzyme immunoassay for pancreatic oncofetal antigen. Ann NY Acad Sci 1983;417:400-409.
9. Muruyama H, Mori T, Shimano T, et al. Differential distribution of the pancreatic cancer associated antigen (PCAA) and the pancreatic tissue antigen (PaA) in pancreatic and gastrointestinal cancer tissue. Ann NY Acad Sci 1983;417:240-250.
10. Gelder FB, Morris DM, Goodwill WJ, Fowler MR. Evaluation of the pancreatic oncofetal antigen (POA) in patients with carcinoma and benign disease. Tumor Biol 1984;5 (abstracts issue), 98.
11. Shimano T, Mori T, Kitada M, et al. Purification and characterization of a pancreatic cancer associated antigen (PCAA) from normal colonic mucosa. Ann NY Acad Sci 1983;417:97-104.
12. Del Villano B, Brennan S, Brock P, et al. Radioimmunometric assay for a monoclonal antibody defined tumor marker, CA 19-9. Clin Chem 1983;29:549-552.
13. Magnani J, Steplewski Z, Koprowski H, Ginsberg V. Identification of the gastrointestinal and pancreatic cancer associated antigen detected by monoclonal antibody CA 19-9 in the sera of patients as a mucin. Cancer Res 1983;43:5489-5492.
14. Tempero M, Ucida E, Takasaki H, et al. Relationship of carbohydrate antigen CA 19-9 and Lewis antigens in pancreatic cancer. Cancer Res 1987;47:5501-5503.

15. Hirano K, Kawa S, Oguchi H, et al. Loss of Lewis antigen expression on erythrocytes in some cancer patients with high serum CA 19-9 levels. J Natl Cancer Inst 1987;79:1261-1268.
16. Ritts R, Del Villano B, Go V, et al. Initial clinical evaluation of an immunoradiometric assay for CA 19-9 using the NCI serum bank. Int J Cancer 1984;33:339-345.
17. Andriulli A, Gindro T, Piantino P, et al. Prospective evaluation of the diagnostic efficacy of CA 19-9 assay as marker for gastorintestinal cancers. Digestion 1986;33:26-33.
18. Satake K, Kanazawa G, Kho I, et al. A clinical evaluation of carbohydrate antigen 19-9 and carcinoembryonic antigen in patients with pancreatic carcinoma. J Surg Oncol 1985;29:15-21.
19. Safi F, Roscher R, Bittner R, Beger H. CA 19-9 as a marker for pancreatic cancer. J Tumor Marker Oncol 1987;2:187-194.
20. Encabo G, Ruibal A. Seric CA 19-9 levels in patients with nontumoral pathologies. Our experience in 892 cases. Bull Ca (Paris) 1986;3:256-259.
21. Yoshikawa T, Nishida K, Tanigawa N, et al. Carbohydrate antigenic determinant CA 19-9 and other tumor markers in gastrointestinal malignancies. Digestion 1985;31:67-76.
22. Farini R, Fabris C, Bonvicini P, et al. CA 19-9 in the differential diagnosis between pancreatic cancer and chronic pancreatitis. Eur J Cancer Clin Oncol 1985;21:429-432.
23. Piantino P, Andriulli A, Gindro T, et al. CA 19-9 assay in differential diagnosis of pancreatic carcinoma from inflammatory pancreatic diseases. Am J Gastroenterol 1986;81:436-439.
24. Venot J, Vincent D, Gainant A, et al. Carbohydrate antigen CA-19-9: value in pathology. Gastroenterol Clin Biol 1986;10:208-210.
25. Safi F, Roscher R, Bittner R, et al. High sensitivity and specificity of CA 19-9 for pancreatic carcinoma in comparison to chronic pancreatitis. Serological and immunohistochemical findings. Pancreas 1987;2:398-403.
26. Tappero G, Piantino P. Value of CA 19-9 assay in serum and duodenal juice in the differential diagnosis of pancreatic disease. Int J Biol Markers 1987;2:191-196.
27. Malesci A, Tommasini M, Bonato C, et al. Determination of CA 19-9 antigen in serum and pancreatic juice for differential diagnosis of pancreatic adenocarcinoma for chronic pancreatitis. Gastroenterology 1987;92:60-67.
28. Nishida K, Tasaki N, Miyagawa H, et al. Estimation of carbohydrate antigen CA 19-9 levels in pure pancreatic juice of patients with pancreatic cancer. Am J Gastroenterol 1988;83:126-129.
29. Iishi H, Yamamura H, Tatsuta M, et al. Value of ultrasonographic examination combined with measurement of serum tumor markers in the diagnosis of pancreatic cancer of less than 3 cm in diameter. Cancer 1986;57:1947-1951.
30. Wang T, Lin J, Chen D, et al. Noninvasive diagnosis of advanced pancreatic cancer by real-time ultrasonography, carcinoembryonic antigen and CA 19-9. Pancreas 1986;1:219-232.
31. Steinberg W, Gelfand R, Anderson K, et al. Comparison of the sensitivity and specificity of the CA 19-9 and CEA assays in detecting cancer of the pancreas. Gastroenterology 1986;90:343-349.
32. Sakahara H, Endo K, Nakajima K, et al. Serum CA 19-9 concentrations and computed tomography findings in patients with pancreatic carcinoma. Cancer 1986;57:1324-1326.
33. Beretta E, Malesci A, Zerbi A, et al. Serum CA 19-9 in the post surgical follow-up of patients with pancreatic cancer. Cancer 1987;60:2428-2431.

34. Glenn J, Steinberg W, Kurtzman S, et al. Evaluation of the utility of a radioimmunoassay for serum CA 19-9 in patients before and after treatment of carcinoma of the pancreas. J Clin Oncol 1988;6:462-468.
35. Kuhn W, Primus F. Alteration of blood groups and blood group precursors in cancer. Prog Clin Biochem 1985;2:49-95.
36. Itzkowitz S, Yuan M, Ferrell L, et al. Cancer associated alterations of blood group antigen expression in the human pancreas. J Natl Cancer Inst 1987;79:425-434.
37. Itai S, Tobe R, Kitahara A, et al. Significance of 2-3 and 2-6 sialylation of Lewis-a antigen in pancreas cancer. Cancer 1988;61:775-787.
38. Fukuta S, Magnani J, Gaur P, Ginsburg V. Monoclonal antibody CC3C 195, which detects cancer associated antigens in serum, binds to the human Le^a blood group antigen and to its sialylated derivative. Arch Biochem Biophys 1987;255:214-216.
39. Bhargava AK, Petrelli NJ, Karna A, et al. Circulating CA-195 in colorectal cancer. J Tumor Marker Oncol 1987;2:319-327.
40. Gupta M, Arciago R, Bukowski R, Gaur P. CA 195: a new sensitive monoclonal antibody defined tumor marker for pancreatic cancer. J Tumor Marker Oncol 1987; 2:201-206.
41. Lindholm L, Holmgren J, Svennerholm L, et al. Monclonal antibodies against gastrointestinal tumor associated antigens isolated as monosialogangliosides. Int Arch Allergy Appl Immunol 1983;71:178-181.
42. Mansson J, Fredman P, Nilsson O, et al. Chemical structure of carcinoma ganglioside antigens defined by monoclonal CA-50 and some allied gangliosides of human pancreatic adenocarcinoma. Biochem Biophys Acta 1985;834:110-117.
43. Harmenberg U, Wahren B, Wiechel K. Tumor markers carbohydrate antigens CA 19-9 and CA-50 and carcinoembryonic antigen in pancreatic cancer and benign diseases of the pancreatobiliary tract. Ca Res 1988;48:1985-1988.
44. Fujii Y, Albers G, Carre-Llopis A, Escribano M. The diagnostic value of the fetoacinar pancreatic (FAP) protein in cancer of the pancreas; a comparative study with CA 19-9. Br J Cancer 1987;56:495-500.
45. Metzgar R, Gaillard M, Levine S, et al. Antigens of human pancreatic adenocarcinoma cells defined by murine monoclonal antibodies. Ca Res 1982;42:601-608.
46. Lan M, Finn O, Fernstein P, Metzgar R. Isolation and properties of a human pancreatic adenocarcinoma associated antigen, DU-PAN-2. Ca Res 1985;45:305-310.
47. Metzgar R, Rodriquez N, Finn O, et al. Detection of a pancreatic cancer associated antigen (DU-PAN-2) in serum and ascites of patients with adenocarcinoma. Proc Natl Acad Sci 1984;81:5242-5246.
48. Sawabu N, Toya D, Takemori Y, et al. Measurement of a pancreatic cancer associated antigen (DU-PAN-2) detected by a monoclonal antibody in sera of patients with digestive cancers. Int J Ca 1986;37:693-696.
49. Ishii M, Urabe T, Seino Y. Clinical significance of serum DU-PAN-2 surveyed by solid phase sandwich EIA. Gan to Kaagaku Ryoho 1987;14:1860-1865.
50. Takemori Y, Sawabu N, Satomura Y, et al. Determination of serum DU-PAN-2 by enzyme immunoassay in patients with various digestive cancers. Gan to Kagaku Ryoho 1987;14:119-126.
51. Haviland AE, Borowitz MJ, Killenberg PG, et al. Detection of an oncofetal antigen (DU-PAN-2) in sera of patients with non-malignant hepatobiliary diseases and hepatomas. Int J Ca 1988;41:789-793.
52. Takasaki H, Uchida E, Tempero M, et al. Correlative study on expression of CA 19-9 and DU-PAN-2 in tumor tissue and in serum of pancreatic cancer patients. Ca Res 1988;48:1435-1438.
53. Lan M, Bast R, Colnaghi M, et al. Co-expression of human cancer associated epitopes on mucin molecules. Int J Ca 1987;39:68-72.

19

Melanoma: The Development of Immunoconjugates from a Preclinical Viewpoint

PAUL L. BEAUMIER and ALTON C. MORGAN, Jr.

NeoRx Corporation and University of Washington Schools of Medicine and Pharmacy, Seattle, Washington

CLIVE S. WOODHOUSE and KEVIN S. MARCHITTO

NeoRx Corporation, Seattle, Washington

I. INTRODUCTION

A. Melanoma Incidence, Mortality, and Characteristics

Statistics from the National Cancer Institute show that the incidence of malignant melanoma has increased from 2.9:100,000 people in 1950 to 9.8:100,000 people in 1985 [1]. Over this same period, the survival rate rose from 40% to 80%. In the 1 year period 1984-1985, the incidence of melanoma increased 10% in white males and females. Among all cancers in all races and both sexes, melanoma is fifteenth in terms of years of life lost [1]. In 1987, new cases of melanoma reached 25,800, with 7,800 deaths [1]. In spite of its relatively low incidence, melanoma remains a high-mortality cancer and continues to have a poor prognosis.

In some ways melanoma is a prototypic cancer; it is highly aggressive, refractory to radiation and drug treatments, and strikes any age group. It generally arises from pigmented cells on the body surface, is markedly heterogeneous, highly invasive, and spreads via the blood and lymphatics. Melanoma has a poor prognosis once it has penetrated the dermis. Rapid progression leads to metastases in organs and tissues throughout the body. Once spread beyond the dermis (stage II disease), advanced melanoma is treated by wide local excision; aggressive chemotherapy regimens provide marginal additional benefit [2].

B. Melanoma: A Model Disease

Since current treatment for disseminated disease is ineffective, research in the treatment of human melanoma has been vigorously pursued not only because of the impact on patients with the disease but also because in many respects, melanoma represents a model disease [3]. One treatment approach that has been pursued is immunological. Despite early anecdotal evidence of occasional spontaneous regressions, the disease has only recently been shown to exhibit special sensitivity to immunological therapy, as evidenced by a higher rate of response to interleukin-2 (IL-2) than other solid tumors [4]. Other immunological approaches have also been used to treat the disease successfully, including passive therapy with IL-2 and lymphokine-activated killer (LAK) cells [5].

With this background of success using immunotherapeutic strategies, immunoconjugates may represent another biological approach for advancing the treatment of the disease. As targets for antibodies, melanoma-associated antigens are among the best characterized in the field of tumor immunology. An extensive panel of monoclonal antibodies has been generated to multiple epitopes of these antigens. The potential of antibodies as carriers in clinical applications is finally being realized as the first antibody-based imaging products become commercially available.

It might appear inconsistent that melanoma, which has a relatively low incidence compared to lung, breast, or colorectal cancer, should be the first addressed by the research community using monoclonal antibody technology. Several aspects of the disease, however, have fostered a strong interest in melanoma as an initial target for immunoconjugates. Tissue samples can be readily obtained by biopsy and the availability of large numbers of cultured cell lines has facilitated the biochemical identification and characterization of melanoma antigens. The same cell lines have also provided an important and reliable source of immunogens for the generation of an extensive panel of monoclonal antibodies as well as in vitro test systems. Furthermore, melanoma grows well as a xenograft in nude mice and rats and offers a valuable and relevant preclinical in vivo model for testing immunoconjugates. In addition, the disease can be evaluated by histological examination of subcutaneous nodules to assess the localization of immunoconjugates and monitor antitumor effects clinically.

C. Melanoma: Antigen Target Selection

Passive immunological treatment strategies rely on selective delivery to tumor. Although a great number of antigens have been detected and identified as melanoma-associated, only a few have been considered appropriate as clinical targets. In an informative and concise review, Hersey catalogued several melanoma antigens and the monoclonal antibodies raised to them [6]. Of the more than 20 melanoma-associated antigens and epitopes listed, a few have been selected as targets in clinical trials: p97, the chondroitin sulfate proteoglycan antigen, and the GD2 and GD3 ganglioside antigens.

D. Antimelanoma Immunoconjugates: Preclinical Evaluation Process

It is important to recognize that monoclonal antibody immunoconjugates require extensive preclinical evaluation before being used in clinical trials. The generation and screening of antibody hybrids begin a laborious process focused on identifying a reagent that is highly reactive with tumor targets and yet shows minimal cross-reactivity with normal tissues. Biochemical identification and characterization contribute much to the understanding of the nature of the antigen and how to optimize targeted delivery. Once selected, the monoclonal antibody must be defined by a series of in vitro tests including immunological isotyping, Scatchard analysis, and isoelectric focusing to determine affinity, estimated sites/cell, and protein characteristics. In vitro immunoreactivity assays must be established to assess damage to the antibody resulting from purification, digestion to fragments, or chemical modification. Typical immunoreactivity assays include direct cell binding and competitive enzyme-linked immunosorbent assay (ELISA).

A series of in vitro studies in animal models are required to establish the capacity of the antibody to localize to tumor, the linker stability, and the kinetic impact on biodistribution produced by conjugation. For a number of immunoconjugate applications, antibody fragments are the delivery system of choice. Fragments of murine monoclonal antibodies may be prepared by enzymatic digestion with papain to generate Fab, or pepsin to generate F(ab')2 fragments. Fragmentation to Fab and F(ab')2 is subclass dependent since the IgG1 or IgG2a subclass can be cleaved enzymatically to produce both types of fragments whereas F(ab')2' is difficult to prepare from IgG2b. Immunoreactivity may be compromised after enzymatic cleavage. Recombinant fragments produced in bacteria have recently been reported [7]. Genetically engineered antibodies and fragments may provide a method for overcoming individual murine antibody distinctions, leading eventually to more homogeneous and less immunogenic antibody carriers.

After labeling, conjugation, or chemical modification, antibody conjugates must be tested for pyrogenicity, purity, and stability in serum at 37°C as well as for retention of the potency of the conjugated agent. Based on numerous published reports, minimal acceptable immunoreactive properties for the antibody include an affinity of 1×10^8 L/mole and 50,000 target sites/cell.

In this chapter we will review the process of immunoconjugate development exemplified by agents directed to the p97, proteoglycan, and GD3 ganglioside melanoma-associated antigens, with special emphasis on preclinical in vitro and in vivo evaluation. Selected examples of specific preclinical assays will be discussed for each antigen to illustrate the process. Clinical studies and therapeutic applications evolving from these preclinical evaluations will be detailed. When these three human melanoma-associated antigens were first identified, their clinical relevance could only be conjectured. Since that time, the evolution of clinically useful antibodies to these markers and their development as pharmaceuticals have followed distinctly separate paths in diagnostic as well as therapeutic applications.

II. p97 HUMAN MELANOMA-ASSOCIATED ANTIGEN: CHARACTERIZATION, EVALUATION, AND APPLICATION

A. p97 Antigen: Description and Characterization

The p97 antigen was first described by Woodbury as a 97 KD cell surface sialoglycoprotein [8]. An IgG1 monoclonal antibody 4.1 was produced using SK-MEL-28, a cultured human melanoma cell line, as the immunizing cell line. With this antibody used as a probe, p97 was detected in 90% of the melanomas and about 55% of the other tumors tested but not in fibroblasts or lymphoblasts. To assay the very low levels of p97 present in normal and nonmelanoma tumors, Brown et al. developed two sensitive in vitro assays to quantitate p97 on cultured cells and in biopsy materials: a live cell binding assay and a double determinant immunoassay [9]. When an improved monoclonal antibody, 96.5, was used, p97 was indeed detected (but at 100-fold lower levels) in normal adult and fetal tissues and in normal serum. Using a panel of monoclonal antibodies, Brown identified five epitopes of p97: p97a, p97b, and p97c present on a 40,000 molecular weight fragment, and p97d, p97e located on the residual portion of the molecule [10]. The antigen was shown to be a nonmodulating, integral sialoglycoprotein with intrachain disulfide bonds.

In 1983, p97 was purified and found to share N-terminal sequence homology with transferrin and lactoferrin [11]. It was also found that antisera to transferrin and lactoferrin cross-reacted with denatured p97, and that p97 exhibited iron-binding properties. The gene for p97 was assigned to chromosome 3, which also carries the gene for transferrin, its role in iron transport and metabolism may be different. With considerable metabolism. Since p97 is membrane bound, unlike the serum iron-binding protein transferrin, their role in iron transport and metabolism may be different. With considerable effort, the entire sequence and genomic organization of p97 was determined by Rose et al. [13]. The molecule consisted of two roughly homologous extracellular domains of 342 and 352 residues with a 25-residue hydrophobic C-terminus thought to serve as a membrane anchor. Of all the human tumor-associated antigens, p97 is the most thoroughly characterized molecularly.

B. p97 as a Target

1. Preclinical in Vitro Evaluation

The clinical application of anti-p97 antibodies was well grounded in basic research. Two monoclonal antibodies developed for clinical use were 96.5, an IgG2a specific for p97a, and 8.2, an IgG1 specific for p97c. Brown's original live cell-binding assay for p97 reactivity was modified to use a fixed cell line to simultaneously evaluate immunoreactivity of labeled antibodies directed to the p97, proteoglycan and GD3 ganglioside antigens [14]. For this purpose, several cloned melanoma cell lines were screened using immunofluorescence microscopy to select a cell line with high-density expression of all three antigens. After recloning, the M 2669 CL 13 cell line was selected by staining with combinations of fluorescein-labeled antimelanoma antibodies. It was important to detach cultured cells with buffered ethylenediamino tetra acetic acid (EDTA) rather than trypsin and to use paraformaldehyde rather than other fixatives to preserve antigen expression. Two control experiments established the degree of immunospecific binding to target cells: the use of an isotype-matched irrelevant antibody or fragment and the use of an antigen-negative cell line. An additional control was intrinsically or metabolically labeled antibody. Antibody labeled in this way can serve as a "gold standard" and can be used to define the limits of the assay. It was found that intrinsically labeled MAb and Fab showed the same percentage bound as the corresponding chloramine T trace-labeled material. Ferens et al. demonstrated that there was a differential sensitivity of the Fab 8.2 and Fab 96.5 to chloramine T iodination in this assay [15]. As the I/Fab molar ratio increased to 1.5:1, the immunoreactivity of Fab 96.5 was dropped by half but the immunoreactivity of Fab 8.2 was almost entirely lost.

To perform reproducibly, the cell-binding assay must be conducted under conditions of antigen excess. This condition was determined by increasing the cell number and measuring the percentage of counts bound. It was found that about 5 million cells were required to achieve maximal binding (antigen excess) consistently, which is typical for many other antibody/antigen systems. An alternative approach was developed by Lindmo [16]. This technique relies on a series of increasing cell (antigen) concentrations and a double inverse plot extrapolated to infinite antigen excess. Lindmo noted that the determination of immunoreactivity is really a twofold problem: to determine the fraction or percentage of immunoreactive radiolabeled antibody and to determine if the avidity of the radiolabeled antibody is equivalent to "native" antibody. Each preparation

Table 1 In Vitro Immunoreactivity Assessment for Anti-97 Antibodies 96.5 and 8.2.

MAb	Form	Cell Binding Percentage	Association Constant (L/mole)	Estimated Sites/Cell
96.5	MAb	79.5[a]	1.0×10^{10} [b]	360,000[c]
96.5	Fab	74.4[a]	2.4×10^{9} [a]	460,000[a]
8.2	MAb			300,000[c]
8.2	Fab	68[d]		

[a]Ref. [14].
[b]Ref. [10].
[c]Ref. [9].
[d]Ref. [18].

of radiolabeled antibody in reality consists of multiple components ranging from unlabeled antibody to a series of species labeled at increasing ratios of label to antibody. Thus the percentage bound reflects the average properties of multiple antibody populations.

A micro cell-binding assay using an oil separation technique has also been used as a rapid method for estimating the immunoreactive fraction of a labeled antibody preparation [14]. This technique is convenient for performing the series of binding assays required for Scatchard analysis of radiolabeled antibody.

Some in vitro characteristics of the anti-p97 antibodies 8.2 and 96.5 are listed in Table 1. Scatchard analysis provides a quantitative estimation of immunoreactivity: affinity, (avidity) and sites/cell. The equilibrium association constants and epitope density for 8.2 and 96.5 are also listed in Table 1. Software packages are available for the convenient analysis of the data using a microcomputer [17]. From Table 1, it can be seen that for 96.5 MAb and Fab results from the cell-binding assay, a relative and qualitative test, correlate reasonably well with the association constant from the more precise Scatchard analysis.

2. Preclinical In Vivo Evaluation

The preclinical in vivo evaluation of Fab fragments 8.2 and 96.5 was undertaken in nude mice bearing the subcutaneous human melanoma xenograft #2169 [18]. These anti-p97 Fabs were labeled with ^{125}I by the chloramine T method and each was coinjected with an irrelevant control Fab, 1.4, labeled with I-131. Blood clearance rates were similar for all three Fabs (T½ = 3 hr). This is consistent with predominant renal clearance of the low-molecular-weight Fab. Fab 8.2 achieved a peak tumor uptake of 3.5 ± 0.8% injected dose/g (ID/g) at 4 hr and Fab 96.5, 2.0 ± 0.3% ID/g. The irrelevant Fab, 1.4, distributed with similar kinetics but achieved lower levels and was completely cleared from tumor by 16 hr postinjection while the p97-specific 8.2 and 96.5 fragments were retained. Deiodination probably increased the apparent rate of clearance of Fab from the tumor. Immunospecific localization was critically evaluated using a localization index (LI) first mentioned by Moshakis [19], and calculated as the tumor/blood ratio for specific antibody divided by the ratio for irrelevant antibody. This parameter is useful since it controls for label instability and nonspecific localization into tumor extravascular space by normalizing to blood background. At 24 hr, the mean of 8.2/1.4 LI was 25 ± 5 (n = 5) and 19 ± 3 for the 96.5/1.4 combination (Table 2). The LIs for a panel of normal tissues ranged from 0.5 to 2.5.

Table 2 In Vivo Assessment of Anti-p97 Fab 8.2 and Fab 96.5 Labeled with ^{125}I by the Chloramine t Method

Time Post-injection (hr)	Mean % Injected D/G ± SD in Tumor	Mean Tumor Blood Ratio ± SD	Localization Index (L.I.)
Fab 8.2			
1	2.3 ± 0.6	0.3 ± 0.1	1.3
4	3.5 ± 0.8	1.1 ± 0.6	2.2
16	1.4 ± 0.7	4.5 ± 2.1	14.8
24	1.5 ± 0.5	7.5 ± 1.9	24.6
Fab 96.5			
1	2.1 ± 0.6	0.4 ± 0.0	2.1
4	1.8 ± 0.3	0.8 ± 0.1	2.6
15	1.0 ± 0.3	2.9 ± 0.1	12.9
24	0.8 ± 0.3	4.9 ± 2.2	18.9

Localization to human melanoma xenograft #2169 was determined in groups of five nude mice, at 1, 4, 16, and 24 hr postinjection at a dosage of 80 μg/mouse for Fab 8.2 and 10 μg/mouse for Fab 96.5. Localization index is the % D/G in tumor/blood for specific antibody divided by the same ratio for nonspecific antibody.
Source: Ref. (18).

The in vitro cell-binding assay contributed to predicting the quality of in vivo tumor localization, since a correlation could be made between in vitro immunoreactivity (IR) and in vivo targeting. Low IR preparations (<20%) produced lower LI (<5) whereas high IR preparations (>60%) produced higher LI (>12). The variability of the LI for different tumors may be related to tumor variables such as the degree of vascularization, necrosis, and antigen expression or differences between the antibodies. Sakahara described a similar correlation among three antiosteosarcoma antibodies; in this system affinity was shown to vary directly with tumor localization [20].

3. Clinical Evaluation

Larson was among the first to recognize the clinical utility of the Fab fragment for imaging [21]. The first clinical studies were performed at a low dose (30-50 μg) of chloramine T iodinated Fab 8.2 or 96.5, administered as an intravenous (i.v.) bolus. A large fraction of the dose accumulated in the liver and kidneys. It was subsequently found that a higher protein dose enhanced tumor localization, apparently saturating nonspecific binding sites. Clinical studies were pursued using this higher dose of anti-p97 Fabs and an irrelevant control, Fab 1.4. The latter antibody was included to assess antigen-specific localization and pharmacokinetics. Thirty-three patients with advanced malignant melanoma were studied [22]. Blood clearance for the specific Fab was much more rapid than for the nonspecific Fab and the specific/nonspecific ratios measured from biopsies were about 3.5:1 at 2-3 days [22]. The uptake in tumor could be correlated with the level of p97 in the biopsy sample. Results from this trial indicated that 88% of the known metastatic sites greater than 1.5 cm in diameter could be successfully imaged [22]. Whole body images after i.v. administration showed that the initial distribution was to the liver and the kidney, with the liver removing 40% of the injected dose at the earliest imaging time. Liver uptake was thought to result from a low-level p97 expression in the tissue.

Tumor uptake was estaimted in a 450 g portion of a tumor to be 17.5% of the injected dose or ~0.04% injected dose/g. However, typically, maximal values for Fab tumor uptake have been found to be typically 10 times lower. Dosimetry calculated from imaging data estimated that for 100 mCi of ^{131}I Fab administered, the tumor would receive 1,040 rad, the liver 325 rad, and marrow, the dose-limiting organ, 30 rads. A tumor to marrow ratio of 30:1 was estimated [22]. Studies since with the same antibody fragment have not produced such favorable dosimetry.

Murray et al. performed clinical imaging studies of melanoma patients using [^{111}In]-DTPA chelate conjugated to 96.5 MAb [23]. Thirty-one patients were imaged with doses ranging from 0.5 to 20 mg labeled with up to 5 mCi of ^{111}In. Tumors were successfully imaged in half the patients and better images were seen with higher antibody doses. A major limitation in the application of ^{111}In96.5 MAb was the high background observed in the liver, spleen, and marrow compared to iodinated antibody. A combination of factors contributed to this, including the slow clearance of whole antibody from circulation and the instability of the [^{111}In] in the DTPA complex compared to the highly avid metal-binding capacity of circulating transferrin. ^{111}In lost from the chelate would be quickly taken up by transferrin, which would be removed from circulation by transferrin receptors present in the reticuloendothelial system. The cross-reactivity of 96.5 with diffuse p97 antigen in the liver also further degraded tumor targeting and tumor/liver ratios.

Using subcutaneously administered ^{131}I-labeled Fab 96.5, Nelp et al. [24] evaluated the potential of lymphoscintigraphy for identifying regional metastatic deposits. An irrelevant nonspecific control, Fab 1.4 labeled with ^{125}I, was coadministered to assess immunospecific localization. Antigen expression in nodes was confirmed by immunohistological examination and autoradiography was performed to characterize the microdistribution of labeled Fab. It was found that specific and control Fab distributed within diseased nodes in essentially the same way and clearance from the lymphatic chain was slow and apparently nonspecific. Nelp noted that penetration of antibody from lymphatic vessels into tumorous nodes was largely incomplete and suggested that a higher dose might improve localization. Strategies producing better penetration into involved nodes and accelerated clearance from lymphatic circulation would contribute to improved targeting.

4. Future Prospects

The most widely used anti-p97 antibody, 96.5, has contributed much to the development of radiolabeled immunoconjugates. Since then, however, second generation anti-p97 antibodies superior to 96.5 have been reported. MAb 96.5 has been extensively studied, as mentioned above, preclinically and in several clinical trials both with iodine and indium labels. However, several difficulties frustrate the continued clinical application of anti-p97 immunoconjugates. Perhaps most significant is the inconsistent and heterogeneous expression of p97 in lesions in an individual patient and among patients. Also, extraction of a major part of the administered dose by normal liver appreciably reduces the fraction bioavailable to the tumor. This difficulty, as noted above, may be partially overcome by increasing the protein dose. Efforts are now being directed to the development of a melanoma vaccine [25] using a cloned sequence of p97 for the manipulation of the host immune response.

III. CHONDROITIN SULFATE PROTEOGLYCAN HUMAN MELANOMA-ASSOCIATED ANTIGEN: TARGET CHARACTERIZATION, EVALUATION, AND APPLICATION

A. Proteoglycan Antigen: Description and Characterization

The melanoma-associated proteoglycan antigen consists of a high-molecular-weight complex: a core glycoprotein of 240-260 kD with N- and O-linked carbohydrate giving a complete proteoglycan of about 450 Kd [26]. The complete proteoglycan consists of the core protein to which chondroitin sulfate chains are attached. The biological function of the proteoglycan antigen has not been completely elucidated; however, proteoglycans in general are components of the extracellular matrix and have been implicated in the loss of cell-cell contact inhibition, substrate adhesion, and metastatic potential of tumor cells [27].

The proteoglycan antigen is expressed on nearly all (>90%) melanoma tissue specimens and cultured cell lines [28]. It is shed into culture media but is detected in patients' serum in only low levels. Normal tissue distribution is limited to smooth muscle associated with blood vessels, basal cells in the skin, and kidney collecting tubules [29]. The proteoglycan antigen shows no evidence of modulation [30].

B. Proteoglycan as a Target

1. Antiproteoglycan Antibodies

The properties of high-level expression, homogeneous distribution within lesions, and prevalence on most melanomas have made the proteoglycan the most frequently targeted melanoma-associated antigen. More than 20 different monoclonal antibodies directed to the proteoglycan antigen have been detailed in published reports from various laboratories [31]. Among the first were 9.2.27, an IgG2a, developed by Morgan et al. in 1981 [32] and 225.8S developed in Ferrone's laboratory [33]. The antibodies recognize epitopes on the core glycoprotein and do not react with most carcinomas, lymphoid, or fibroblast cells. Another anti-proteoglycan MAb, 48.7, an IgG1, was developed in the Hellstrom laboratory [34].

An antibody with good characteristics is also extremely helpful in the production of second generation antibodies to the same target with improved properties. As an example, Woodhouse et al. developed second generation MAbs to the proteoglycan. One, NR-ML-05, an IgG2b, was derived from an immunogen of an immune complex of antigen with 9.2.27 [31]. The protocol consisted of immunization with intact cells followed by boosting with the immune complex: 65 clones were generated from 35 independent hybrids [31]. Four clones were selected for detailed analysis. One in particular, NR-ML-05, had properties that made it an attractive alternative to 9.2.27. The two antibodies differed in these respects: 9.2.27 showed fourfold greater Fc binding to cultured human monocytes than NR-ML-05, which is a subclass-dependent property; the two antibodies recognize different determinants; 9.2.27 recognizes a trypsin-sensitive epitope and the epitope recognized by NR-ML-05 is trypsin insensitive; and NR-ML-05 demonstrated increased tumor uptake in nude mice and reduced nonspecific retention in kidney compared to 9.2.27. These factors are important since lower NR-ML-05 binding to Fc receptor would be advantageous in reducing nonspecific uptake in normal tissues. NR-ML-05 recognition of a trypsin-insensitive epitope may be important since it would target higher-density and more homogeneously distributed epitopes throughout a tumor, including the necrotic regions where proteolytic activity might be expected to degrade 9.2.27 epitopes.

2. Preclinical In Vitro Evaluation

As described above for the anti-p97 antibodies, the immunoreactivity assessment of selected antiproteoglycan antibodies included the direct cell-binding assay and Scatchard analysis. A comparison of cell-binding assay results for 48.7, 9.2.27, and NR-ML-05 is presented in Table 3. Fixed cells were utilized in this comparison with antibody bound under conditions of antigen excess. The assay method for 48.7 used a single cell concentration, 5×10^6, which assumed that this cell number provided antigen excess, whereas the method for 9.2.27 and NR-ML-05 used multiple cell concentrations and extrapolated to antigen excess, using the double reciprocal analysis method described by Lindmo [16]. Affinity determinations conducted with radiolabeled antibodies and fragments were performed in essentially the same way for all three antibodies. Table 3 shows that the cell bound percentage for the 48.7 Fab (~20%) is much lower than values obtained for 9.2.27 and NR-ML-05 (~80%). This reflects the difference in affinity for 48.7, which is 2×10^7, compared to about 5×10^8 for the 9.2.27 and NR-ML-05 Fabs. It is also notable that Fab and MAb binding are essentially equivalent for 9.2.27 and NR-ML-05 (~80-90%) despite the lower affinity of monovalent Fab than divalent intact MAb. Above a certain threshold of cell number, a maximum appears to be reached and the percentage bound no longer correlates with affinity.

Table 4 presents an example of how immunoreactivity changes as a result of increasing levels of ligand substitution (35). NR-ML-05 MAb was derivatized with a diamide dimercaptide ^{99m}Tc chelate via lysine acylation. As the substituent to antibody ratio was increased, immunoreactivity deteriorated as shown by both a declining cell-binding percentage and a reduced affinity. In fact, percentage cell bound and affinity show a linear, parallel decline.

Table 3 In Vitro Immunoreactivity Assessment for Antichondroitin Sulfate Proteoglycan Antibodies

MAb	Form	Cell binding Percentage	Association Const (L/mol)	Estimated Sites/cell
48.7	MAb	52.6[a]		
48.7	Fab	20.4[a]	1.7×10^{7}[a]	420,000[a]
9.2.27	MAb	83.9[b]	2.0×10^{9}[b]	220,000[b]
9.2.27	Fab'2	88.9[c]		
9.2.27	Fab	85.2[b]	5.6×10^{8}[b]	430,000[b]
NR-ML-05	MAb	72.0[c]	7.6×10^{9}[b]	170,000[b]
NR-ML-05	Fab	87.5[c]	3.5×10^{8}[b]	310,000[b]

[a]Ref. [14].
[b]Ref. [31].
[c]Ref. [35] glutaraldehyde-fixed human melanoma A375 mixed met; average of 10 determinations.

Table 4 In Vitro Assessment of Antiproteoglycan MAb NR-ML-05 as a Function of the Degree of Substitution with Diamide Dimercaptide Technetium-99m Complex

Tc-Ligand: MAb Actual Substitution Ratio	Immunoreactivity % Cell Bound	Association Constant (L/mole)
0.01	83	4.4×10^9
1.25	76	1.1×10^9
4.32	59	5.3×10^8
9.60	17	9.2×10^7

Scatchard analysis and direct cell bound percentages were measured at four escalating antibody:ligand ratios.
Source: Ref. [35].

3. Preclinical In Vivo Evaluation

A xenograft model for in vivo evaluation of antiproteoglycan antibody was initially explored by Hwang et al. [36] using Bolton-Hunter iodinated 9.2.27 MAb and an isotype-matched irrelevant control antibody, RPC-5. Immunospecific localization was tested in melanoma xenografts in nude mice bearing SESX (high antigen expression), FMX-MetII (medium antigen expression), and LOX-L (little or no antigen expression) [35]. In vitro cell-binding assay results using 2×10^6 FMX-MetII cells indicated 48% cell bound with 9.2.27, with background binding of 3.8% with RPC-5. The percentage of the retained dose localizing to tumors (about 1.5 cm in diameter) correlated with antigen expression: 12% in SESX, 7.4% in FMX-MetII, and 0.2% in LOX-L.

Comparative biodistribution data for the 9.2.27 and NR-ML-05 intact antibody and Fab fragments was compared in subcutaneous human melanoma xenografts. Antibody, radiolabeled with diamide dimercaptide (N_2S_2) ^{99m}Tc or paraiodophenyl (PIP)-^{125}I, at a dose of 10-50 µg, was administered intravenously and %ID/g and tumor/blood (T/B) ratios determined at 20-24 hr. The data suggest that tumor localization with the NR-ML-05 Fab (2.8 %ID/g) is superior to the uptake achieved with 9.2.27 (2.1 %ID/g). The corresponding T/B ratio for NR-ML-05 also appears to be superior: 24 vs. 11. Compared to the Fab fragment, the %ID/g achieved with whole NR-ML-05 antibody is 10 times higher. This appreciably greater uptake results from the longer circulation half-life of intact MAb (more cycles through the tumor) as well as its higher avidity for antigen. Whole

Table 5 In Vivo Melanoma Xenograft Localization for Antichondroitin Sulfate Proteoglycan Antibodies 9.2.27 and NR-ML-05.

Antibody	Form	Mean % Injected D/G ± SD in Tumor	Mean Tumor/ Blood ± SD	Number of Determinations
9.2.27	Fab	2.08 ± 0.55[a]	10.83 ± 2.87[a]	16[a]
NR-ML-05	MAb	19.53 ± 3.62[b]	1.06 ± 0.20[b]	8[b]
NR-ML-05	Fab	2.75 ± 0.74[a]	23.96 ± 16.83[a]	113[a]

[a]Beaumier, unpublished data at 20 hr postinjection
[b]Beaumier, unpublished data at 24 hr postinjection

Table 6 Correlation of In Vitro with In Vivo Assessment of Antiproteoglycan Fab NR-ML-05 as a Function of the Degree of Substitution with Diamide Dimercaptide Technetium-99m Complex.

Tc-Ligand:MAb Substitution Ratio		Immunoreactivity % Cell Bound	% Injected Dose/g ± SD in Tumor (n =4)	Tumor/Blood Ratio
Offered	Actual			
0.01	0.01	83	3.32 ± 0.59	16.8 ± 7.0
3	1.4	78	2.85 ± 0.51	20.7 ± 5.5
16	4.8	59	2.45 ± 0.72	18.0 ± 3.6
70	7.0	42	1.30 ± 1.19	16.1 ± 7.5

Cell bound percentages were determined by the method of Lindmo [19]. Localization to 375-mm human melanoma xenografts was determined in groups of four nude mice, at 20 hr postinjection at a dosage of 10 µg/mouse.
Source: Ref. [37].

antibody's long half-life in blood reduces the T/B ratio to 1 at early time points, as detailed in Table 5. Although this ratio improves with time, it is typically low for MAb throughout the entire distribution time course.

One of the unstated goals in the preclinical evaluation process of monoclonal antibodies is to develop a correlation between in vitro and in vivo results. Table 6 presents such a correlation [37]. Following the experimental design of the study presented in Table 5, the in vitro immunoreactivity, in vivo tumor uptake, and T/B ratio are correlated with the ligand to antibody ratio. Increased substitution resulted in reduced immunoreactivity and decreased %ID/g localizing to tumor. These data suggest that even substitution of a single lysine residue reduces immunoreactivity and localization in an antibody fragment. This is supported by the tabulated differences between Fab derivatized at a 0.01:1 ratio compared to Fab derivatized at a 1.4:1 ratio. Since these ratios represent an average degree of substitution and since only radioactively labeled fragments are being studied, the difference between the two preparations is due to the proportion of antibody with more than one chelate/antibody. For example, from Poisson distribution calculations, at a ratio of 1.4:1 ligand/antibody, 16% of the antibody molecules are predicted to be derivatized with at least three chelates. In contrast, at tracer level derivation (0.01:1), only 1% of the antibody molecules are predicted even to be monosubstituted.

To validate this experimental prediction statistically, a biodistribution was repeated with NR-ML-05 Fab derivatized at 0.01:1 and at 1.5:1 ratios in larger groups of mice. Tumor uptake was 3.3 ± 0.8 %ID/g (n = 15) at the minimum substitution ratio, compared to 2.4 ± 0.8 %ID/g (n = 16) at the 1.5:1 ratio, which is a statistically significant difference at p = 0.004 [37]. To realize the maximal potential of the antibody carrier, care must be paid to minimizing the degree of substitution.

4. Clinical Evaluation

Antibodies directed to the proteoglycan antigen have been extensively evaluated in clinical trials. Since the preclinical evaluation outlined in the previous section was restricted mostly to technetium-labeled 9.2.27 and NR-Ml-05, the discussion of the clinical findings will focus on these antibodies labeled with technetium. It should be mentioned that antiproteoglycan monoclonal antibody 225.28S (ZME-018) labeled with chelated ^{111}In also has been studied in numerous clinical trials [38-40].

In an early study evaluating the clinical utility of the antiproteoglycan antibody, 9.2.27, Schroff et al. conducted clinical trials using escalating doses of antibody [30]. They found that the intratumor localization of the 9.2.27 antibody was dose dependent. Distribution was followed by immunoperoxidase staining of subcutaneous lesions, and immunofluorescent flow cytometry. They were able to show that low doses (1-10 mg) resulted in little or no localization, whereas higher doses (200-500 mg) produced saturation of tumor sites in most patients. It is likely that an antibody dose sufficient to overcome nonspecific binding in liver and spleen was needed to permit the remaining freely circulating antibody to localize to tumor. As evidence of this, higher serum levels of 9.2.27 were seen with the higher doses. In comparison to the p97 melanoma marker, flow cytometry demonstrated that 9.2.27 consistently stained more melanoma cells than did the anti-p97 antibody, 96.5. In addition, most melanoma cells in specimens were stained with 9.2.27, in contrast to the more heterogeneous expression noted with p97 [30].

Eary et al. detailed the application of the F(ab')2 and the Fab fragments of 9.2.27 labeled with ^{99m}Tc for the diagnostic imaging of melanoma [41]. Thirty-five patients received 10 mg of the radiolabeled specific 9.2.27 Fab or F(ab')2 and 50 mg of whole irrelevant antibody intravenously 1 hr before as a nonspecific blocker. This strategy, along with the administration of a small amount of unlabeled specific antibody, enhanced imaging. When the Fab fragment with a average serum half-life of 2 hr was used, optimal imaging was achieved at 6-9 hr postinjection. More than 80% (43/53) of the known metastases were visualized. Using the Fab or Fab' fragments, tumor uptake averaged 0.004% of the injected dose/g in image-positive tumors. The antiglobulin response in 4 of 25 (15%) of the patients was directed to 9.2.27 fragments; in contrast 54% of the patients developed an antiglobulin response to the irrelevant blocking antibody.

A melanoma imaging kit has been developed using the diamide dimercaptide ^{99m}Tc chelate conjugated to NR-ML-05 Fab and used for radioimmunodetection in a multicenter phase III clinical trial [42]. Data collected from a group of 99 patients demonstrated that 80% of known noncutaneous lesions were imaged, with an additional 108 previously undetected sites identified.

5. Future Prospects

The demonstrated clinical success with the melanoma imaging agent has provided a good foundation for the development of therapeutic isotope, drug, and toxin conjugates. Conjugation with multiple drug and toxin molecules is more of a challenge to the antibody delivery system than the tracer substitution levels required in imaging applications. The process of preclinical evaluation of different combinations of agents, linkers, and antibodies, the development of novel delivery strategies, and the demonstration of efficacy in animal models is underway. Given the resistance of melanoma to radiation and chemotherapy, successful treatment of the disease would be a major advance.

IV. GD3 GANGLIOSIDE HUMAN MELANOMA-ASSOCIATED ANTIGEN: TARGET CHARACTERIZATION, EVALUATION, AND APPLICATION

A. GD3 Antigen: Description and Characterization

Gangliosides are glycosphingolipids containing sialic acid residues. Four of the most commonly expressed gangliosides of human melanoma are GM3, GD3, GM2, and GD2 [43]. Among these glycolipids, the diasialoganglioside GD3 is probably the best expressed of the glycolipid antigens in melanoma and is only weakly expressed in normal melanocytes [43]. Gangliosides are characteristically found in neural tissues. These melanoma-associated antigens have received much attention as targets for antibody-based agents because of their restricted expression, immunogenicity, and potential susceptibility to antibody-directed cell-mediated cytotoxicity. The GD3 ganglioside has been investigated by several groups and its structure is known: NeuAcα2 → 8Neuαc2 → 3Galβ1 → 4Glcβ1 → (1Cer) [44].

It is surprising that although the GD3 carbohydrate structure is identical to that of the normal brain ganglioside, only the melanoma ganglioside is recognized by antibody. Nudelman hypothesized that cell surface GD3 is identified because it may be present in unusually high concentrations and/or its lipid ceramide chains are somewhat longer than the naturally occurring C18:0 fatty acid, and this may in some way make it more exposed or antigenic [44]. Tsuchida et al. surveyed ganglioside expression in a panel of largely metastatic biopsied melanomas using tissue extraction and thin layer chromatography methods [45]. They found GD3 present in 95% of the specimens tested, although the quantity and the distribution were markedly heterogeneous in both biopsied tissue specimens and cultured cells [45]. Using R24 antibody as a probe for GD3, others confirmed the restricted expression of GD3 in melanomas and nevi [46,47]. Wide variability in GD3 expression was noted, especially in specimens of metastatic disease and even among lesions biopsied from different sites in the same patient. GD3 is also weakly expressed in normal cells of neuroectodermal origin: melanocytes, adrenal medulla, glia, neurons, and pancreatic islet cells [48]. It has been reported that GD3 is shed into culture media and has been detected in the serum of melanoma patients [49]. Hersey noted unexpectedly high concentrations of GD3 in normal skin bordering nevi and primary melanoma, which he postulates may be shed and then diffuse into adjacent tissues [47]. The heterogeneous expression of GD3 prompted both Hoon and Tsuchida to stress the importance of characterizing a patient's ganglioside profile as a critical step in the design of a therapy protocol [43,45]. Many biological roles have been suggested for gangliosides: receptor for hormones, toxins, viruses, growth factors, and possible involvement in cell substrate attachment and infiltration [50]. It has been suggested that gangliosides may also operate in cell signaling, recognition, and the growth and differentiation of neuroectodermally derived cells [51]. Gangliosides and antibodies directed to them may interact with cells of the immune system [51]. More specifically, MAbs to GD3 may act as immunopotentiators by stimulating T-lymphocyte proliferation [51,52].

B. GD3 as a Target

1. Anti-GD3 Antibodies

R24, a murine IgG3, was first identified by Dippold et al. as recognizing a heat-stable glycolipid melanoma cell-surface antigen [53]. It was subsequently shown to react specifically with the disialoganglioside, GD3 [54]. R24 was also shown to strongly mediate multiple biological effector functions including antibody-dependent cell cytotoxicity and human complement-mediated cytotoxicity. Two other IgG3 MAbs, MG-21 and MG-22 (2B2), also specific for GD3, both mediate antibody-dependent cell cytcotoxicity with human lymphocyte effector cells as well [55]. It is not coincidental that most of the antibodies mediating effector functions are IgG3, since Ortaldo et al. reported that this murine isotype more frequently showed stronger cytotoxic activity than IgG1 or IgG2 isotypes [56]. They further suggested that the biochemical properties of the GD3 antigen or its location at the cell surface may be especially suited to coordinated interaction between antibody and effector cell.

2. Preclinical In Vitro Evaluation

In one example of in vitro assessment of radiolabeled anti-GD3 antibody, chloramine-T-labeled MG-22, showed a fixed cell-binding percentage of 69% for whole MAb (Table 7) [14]. It is interesting to note that the binding percentage using live M 2669 cells was considerably lower: only 28%. This difference may be related to modification or shrinkage of the cell membrane, leading to the exposure of additional epitopes of membrane-bound GD3.

Using metabolically labeled R24, Welt assayed the number of GD3-binding sites for each of 22 melanoma cell lines [57]. The average number of antibodies bound/cell ranged from 1×10^7 to 5×10^4. Tritiated R24 was more than 90% immunoreactive with appropriate target cells and its binding could be specifically blocked by addition of unlabeled R24 competitor. R24 was shown to induce the aggregation of cultured melanoma cells and also block the attachment of cells if added before plating. This antibody very effectively mediated complement cytolysis of melanoma target cells in the presence of rabbit, guinea pig, rat, or human sera but not mouse serum. In addition, R24 was very effective in inducing cell-mediated cytotoxicity with several sources of effector cells. Welt convincingly demonstrated that all of these processes—aggregation, complement-mediated lysis, or cell-mediated lysis—could be linearly correlated to the average number of MAb bound/cell [57].

3. Preclinical In Vivo Evaluation

Welt also reported that I-125 labeled R24 selectively localized to established tumors in vivo [57]. Inhibition studies were conducted to determine the therapeutic effect of

Table 7 In Vitro Immunoreactivity Assessment for Anti-GD3 Antibody MG-22 (2B2)

MAb	Form	Cell Binding Percentage	Association Constant (L/mol)	Estimated Sites/Cell
MG-22	MAb	68.5	3.9×10^8	7.4×10^4

Source: Ref. [4].

intraperitoneally administered R24 on xenograft progression in nude mice. Nude mice were implanted subcutaneously with 5×10^6 cells and treated with 150 μg intraperitoneally (i.p.) three times/week for a total of eight injections. It was shown that growth could be controlled if treatment was initiated within 3 days after implantation. However, if treatment was begun on day 7 postimplantation, no effect on tumor growth could be demonstrated. Welt et al. concluded that antibodies demonstrating strong in vitro effector function may be potent clinically despite variable or poor results in preclinical in vivo studies in [57]. Complement may play an important role since it is known that murine IgG3s effectively destroy target cells in conjunction with human serum and other heterologous sources of complement. However, mouse complement functions poorly if at all in this role. Reisfeld observed that successful immunotherapy of xenografts in an animal model system can be achieved using lymphokine-activated human effector cells prearmed and targeted by IgG3 antiganglioside monoclonal antibodies such as MAb 3.6 and R24 [58].

Using a similar antibody/animal model system, Hellstrom et al. were able to demonstrate the suppression of tumor growth in nude mice implanted subcutaneously with 1 mm cubes of the melanoma xenograft, M-2586 [55]. Treatment consisted of intravenous injections of MG-22 beginning on the day after implantation and every 3 days for a total of six injections. Tumor outgrowth in the treatment group occurred in only 1 of 10 instances, whereas in the control group, 18 of 20 tumors progressed. It was also reported that control MAb IF4, which did not mediate ADCC with mouse effector cells, did not inhibit tumor growth as progression was noted in 8 of 10 mice. Antibodies that can fix complement and direct cell-mediated cytotoxicity may use these immune mechanism synergistically to kill target cells [58].

4. Clinical Evaluation

Houghton et al. conducted a phase I clinical trial using R24, an anti-GD3 ganglioside monoclonal antibody demonstrated to have potent effector functions in vitro [59]. Escalating dosages of 8, 80, and 240 mg/m^2 were infused over a 2 week period. Patients receiving the two higher dosages experienced pruritus and urticaria, which typically developed 2-4 hr after treatment was begun, and inflammatory response around cutaneous tumor sites. No hepatic, renal, hematopoietic system, or neurological toxicities were noted despite the extensive expression of GD3 in these normal tissues. Three of the 12 patients treated showed significant tumor regression. Mixed responses were observed in two other patients. A human antimouse antiglobulin response was noted in all patients 2-5 weeks after the start of treatment. The mechanism for the therapeutic activity of R24 was not precisely known but may include both complement and cell-mediated cytotoxicity, which initiate inflammatory response in tumor sites.

5. Future Prospects

Antiganglioside antibodies offer a unique approach to melanoma therapy through the antibody-host cell and antibody-complement interaction. The demonstration of objective clinical responses by Houghton is a remarkable and highly encouraging finding [59]. Strategies need to be developed to make IgG3s easier to work with to avoid their tendency to aggregate at higher concentrations [57]. Perhaps the application of genetic engineering and the generation of chimeric antibodies with good specificity, optimized Fc effector function, and improved solubility could contribute to the design of ideal effector function antibodies. A better understanding of the multiple immune mechanisms involved in cytotoxicity must be developed. Welt noted that failure of R24 to con-

V. MELANOMA: FUTURE PROSPECTS FOR ANTIMELANOMA IMMUNOCONJUGATES

Several challenges remain to be resolved before antibody conjugates can achieve their full clinical potential in the diagnosis and treatment of melanoma as well as other cancers and medical conditions. Specific problems include the antiglobulin response to repeated administrations of murine protein, the low %ID/g delivered to tumor vs. nontarget tissues, the limited number of highly tumor-specific antibodies with minimal cross-reactivity to nontarget tissues, and chemically optimized linkages, chelates, and conjugates. Progress is being made regarding the antiglobulin response and some new strategies are being evaluated. These include the preadministration of nonspecific irrelevant antibody to divert immune response away from the specific antibody, the concept of selectively eliminating B cells that generate antiglobulin to the monoclonal antibody, and the development of human/murine chimeric antibodies [60,61]. Several possibilities need to be evaluated relating to achieving better uptake in tumor including modulation of tumor vasculature. Rosenblum et al. reported that preadministration of human alpha interferon resulted in a threefold enhancement in MAb 96.5-^{111}In uptake in tumor compared with untreated matched controls [62]. Conjugates produced from highly potent agents such as alpha emitters or toxins may take best advantage of the antibody-targeted delivery. Combinations of agents, which are a traditional approach in cancer chemotherapy, will also be important in newly developed regimens. Progress is clearly being made in the development of better monoclonal antibodies, as evidenced by the development of antibodies such as NR-ML-05 which is mentioned above [31]. Steady progress is being made in the chemistry of linkages, chelates, and conjugates as demonstrated by the development of the diamide dimercaptide chelated ^{99m}Tc radiopharmaceutical kit [42].

Success in the application of antibody-based agents to cancer requires a multidisciplinary approach integrating immunology, chemistry, and oncology and a systematic long-term commitment. Creative and innovative ways must be explored to apply rapidly developing technologies such as genetic engineering and lymphokine research to enhance the efficacy of established antibody carriers.

ACKNOWLEDGMENTS

We thank Drs. Robert Schroff and Everett Nichols for helpful discussions and members of the immunological assessment and immunopharmacology sections of NeoRx for providing some of the data presented.

REFERENCES

1. National Cancer Institute. 1987 Annual cancer statistics review. Drug Res Rep 1988; February 3:2-3.
2. Mastrangelo MJ, Baker AR, Katz HR. Cutaneous melanoma. In: DeVita VT Jr, Hellman S, Rosenberg SA, eds. Cancer. Principles and practice of oncology, 2nd ed. Philadelphia: JB Lippincott, 1985:1371-1422.

3. Macher E. Malignant melanoma as cancer model Hautarzt 1987;38:489-494.
4. Mitchell MS, Kempf RA, Harel W, et al. Effectiveness and tolerability of low-dose cyclophasphamide and low-dose intravenous interleukin-2 in disseminated melanoma. J Clin Oncol 1988;6:409-424.
5. Topalian SL, Solomon D, Avid FP, et al. Immunotherapy of patients with advanced cancer using tumor-infiltrating lymphocytes and recombinant interleukin-2: a pilot study. J Clin Oncol 1988;6:839-853.
6. Hersey P. Review of melanoma antigens recognized by monoclonal antibodies (MAbs). Their functional significance and applications in diagnosis and treatment of melanoma. Pathology 1985;17:346-354.
7. Skerra A, Pluckthum A. Assembly of a functional immunoglobulin Fv fragment in *Escherichia coli*. Science 1988;240:1038-1041.
8. Woodbury RG, Brown JP, Yeh M-Y, Hellstrom I, Hellstrom KE. Identification of a cell surface protein, p97, in human melanomas and certain other neoplasms. Proc Natl Acad Sci USA 1980;77:2183-2187.
9. Brown JP, Nishiyama K, Hellstrom I, Hellstrom KE. Structural characterization of human melanoma-associated antigen p97 with monoclonal antibodies. J Immunol 1981;127:539-546.
10. Brown JP, Woodbury RG, Hart CE, Hellstrom I, Hellstrom KE. Quantitative analysis of melanoma-associated antigen p97 in normal and neoplastic tissues. Proc Natl Acad Sci USA 1981;78:539-543.
11. Brown JP, Hewick RM, Hellstrom I, et al. Human melanoma-associated antigen p97 is structurally and functionally related to transferrin. Nature 1982;296:171-173.
12. Plowman GD, Brown JP, Enns CA, et al. Assignment of the gene for human melanoma-associated antigen p97 to chromosome 3. Nature 1983;303:70-72.
13. Rose TM, Plowman GD, Teplow DB, et al. Primary structure of the human melanoma-associated antigen p97 (melanotransferrin) deduced from the mRNA sequence. Proc Natl Acad Sci USA 1986;83;1261-1265.
14. Beaumier PL, Neuzil D, Yang HM, et al. Immunoreactivity assay for labeled antimelanoma monoclonal antibodies. J Nucl Med 1986;27:824-828.
15. Ferens JM, Krohn KA, Beaumier PL, et al. High-level iodination of monoclonal antibody fragments for radiotherapy. J Nucl Med 1984;25:367-370.
16. Lindmo T, Boven E, Cuttitta, Fedorko J, Bunn PA Jr. Determination of the immunoreactive fraction of radiolabeled monoclonal antibodies by linear extrapolation to binding at infinite antigen excess. J Immunol Methods 1984;72:77-89.
17. McPherson GA. Analysis of radioligand binding of experiments: a collection of computer programs for the IBM PC. J Pharmacol Methods 1985;14:213-228.
18. Beaumier PL, Krohn KA, Carrasquillo JA, et al. Melanoma localization in nude mice with monoclonal Fab against p97. J Nucl Med 1985;26:1172-2279.
19. Moshakis V, McIlhinney RAJ, Raghavan D, et al. Localization of human tumor xenografts after iv administration of radiolabeled monoclonal antibodies. Br J Cancer 1981;44:91-99.
20. Sakahara H, Endo K, Koizumi M, et al. Relationship between in vitro binding activity and in vivo tumor accumulation of radiolabeled monoclonal antibodies. J Nucl Med 1988;29:235-240.
21. Larson SM, Brown JP, Wright PW, et al. Imaging of melanoma with I-131-labeled monoclonal antibodies. J Nucl Med 1983;24:123-129.
22. Larson SM, Carrasquillo JA, Krohn KA, et al. Localization of 131-I-labeled p97-specific Fab fragments in human melanoma as a basis for radiotherapy. J Clin Invest 1983;72:2101-2114.
23. Murray JL, Rosenblum MG, Sobel RE, et al. Radioimmunoimaging in malignant

melanoma with In-111 labeled monoclonal antibody 96.5. Cancer Res 1985;45: 2376-2381.
24. Nelp WB, Eary JF, Jones RF, et al. Preliminary studies of monoclonal antibody lymphoscintigraphy in malignant melanoma. J Nucle Med 1987;28:34-41.
25. Estin CD, Stevenson US, Plowman GD, et al. Recombinant vaccinia virus vaccine against the human melanoma antigen p97 for use in immunotherapy. Proc Natl Acad Sci USA 1988;85:1052-1056.
26. Bumol TF, Reisfeld RA. Unique glycoprotein-proteoglycan complex defined by monoclonal antibody on human melanoma cells. Proc Natl Acad Sci USA 1982;79: 1245-1249.
27. Garriques JH, Lark MW, Lara S, et al. The melanoma proteoglycan: restricted expression on microspikes, a specific microdomain of the cell surface. J Cell Biol 1986;103:1699-1710.
28. Morgan AC, Woodhouse C, Bartholemew R, Schroff. Human melanoma-associated antigens: analysis of antigenic heterogeneity by molecular, serologic and flow-cytometric approaches. Mol Immunol 1986;23:193-200.
29. Woodhouse CS, Schroff RW, Beaumier PL, Morgan AC Jr. Monoclonal antibodies to a human melanoma-associated proteoglycan: efficient generation of antibodies with novel properties (submitted).
30. Schroff RW, Woodhouse CS, Foon KA, et al. Intratumor localization of monoclonal antibody in patients with melanoma treated with antibody to a 250,000 dalton melanoma-associated antigen. J Natl Cancer Inst 1985;74:299-306.
31. Woodhouse CS, Bordonaro JP, Beaumier PL, Morgan AC Jr. Second generation monoclonal antibodies to a human melanoma-associated proteoglycan. In: Ferrone S, ed. Human melanoma. From basic research to clinical application. New York: Springer-Verlag, 1990: 151-163.
32. Morgan AC Jr, Galloway DR, Reisfeld RA. Production and characterization of monoclonal antibody to a melanoma specific glycoprotein. Hybridoma 1981;1:27-36.
33. Buraggi GL, Callegaro L, Mariani G, et al. Imaging with 131-I-labeled monoclonal antibodies to a high-molecular-weight melanoma-associated antigen in patients with melanoma: efficacy of whole immunoglobulin and its F(ab')2 fragments. Cancer Res 1985;45:3378-3385.
34. Hellstrom I, Garriques J, Cabasco L, et al. Studies of a high molecular weight human melanoma-associated antigen. J Immunol 1983;130:1467-1472.
35. Marchitto KS, Kindsvogel WR, Beaumier PL, Finesk, Gilbert T, Levin SD, Woodhouse CS, Morgan AC Jr. In: Immunity to cancer II. New York: Alan R Liss, 1989:101-105.
36. Hwang KM, Fodstad O, Oldham RK, Morgan AC Jr. Radiolocalization of xenografted human malignant melanoma by a monclonal antibody (9.2.27) to a melaoma-associated antigen in nude mice. Cancer Res 1985;45;4150-4155.
37. Rao TN, Vanderheyden J-L, Kasina S, Beaumier P, Berninger R, Fritzberg AR. Dependence of immunoreactivity and tumor uptake on the ratio of Tc and Re N2S2 complexes per antibody Fab fragment. J Nucl Med 1988; abstract (in press).
38. Fawwaz RA, Wang TS, Estabrook A, et al. Immunoreactivity and biodistribution of indium-111-labeled monoclonal antibody to a human high molecular weight-melanoma associated antigen. J Nucl Med 1985;26:488-492.
39. Murray JL, Rosenblum MG, Lamki L, et al. Clinical parameters related to optimal tumor localization of indium-111-labeled mouse antimelanoma monoclonal antibody ZME-018. J Nucl Med 1987;28:25-33.
40. Taylor A, Milton W, Eyre H, et al. Radioimmunodetection of human melanoma with indium-111-labeled monoclonal antibody. J Nucl Med 1988;29:329-337.

41. Eary JF, Schroff RW, Abrams PG, et al. Successful imaging of malignant melanoma with Tc-99m labeled monoclonal antibodies. J Nucl Med 1989;30:25-32.
42. Schroff RW, Johnson SL, Eary JF, et al. Radioimaging of melanoma with Tc-99m labeled monoclonal antibodies. In: Ferrone S, ed. Human melanoma. From basic research to clinical application. New York: Springer-Verlag, 1990:344-354.
43. Hoon DSB, Irie RF, Cochran AJ. Gangliosides from human melanoma immunomodulate response of T cells to interleukin-2. Cell Immunol 1988;111:410-419.
44. Nudelman E, Hakomori S, Kannagi R, et al. Antigen defined by a monoclonal antibody, 4.2. J Biol Chem 1982;237:12752-12756.
45. Tsuchida T, Saxon RE, Morton DL, Irie RF. Gangliosides of human melanoma. J Natl Cancer Inst 1987;78:45-54.
46. Dippold WG, Dienes HP, Knuth A, zum Buschenfelde K-HM. Immunohistochemical localization of ganglioside GD3 in human malignant melanoma, epithelial tumors, and normal tissues. Cancer Res 1985;45:3699-3705.
47. Hersey P, Jamal O, Henderson C, Zardawi I, D'Alessandro G. Expression of the gangliosides GM3, GD3, and GD2 in tissue sections of normal skin nevi, primary and metastatic melanoma. Int J Cancer 1988;41:336-343.
48. Real FX, Houghton AN, Albino AP, Cordon-Cardo C, et al. Surface antigens of melanomas and melanocytes defined by mouse monoclonal antibodies: Specificity analysis and comparison of antigen expression in cultured cells and tissues. Cancer Res 1985;45:4401-4411.
49. Portoukalian J, Zwingelstein G, Abdul-Malak N, Dore JF. Alteration of gangliosides in plasma and red cells of humans bearing melanoma tumors. Biochem Biophys Res Commun 1978;85:916-920.
50. Cheresh DA, Pierschbacher MD, Herzig MA, Mujoo K. Disialogangliosides GD2 and GD3 are involved in the attachment of human melanoma and neuroblastoma cells to extracellular matrix proteins. J Cell Biol 1986;102:688-696.
51. Welte K, Miller G, Chapman PB, et al. Stimulation of T lymphocyte proliferation by monoclonal antibodies against GD3 ganglioside. J Immunol 1987;139:1763-1771.
52. Hersey P, MacDonald M, Burns C, Cheresh DA. Enhancement of cytotoxic and proliferative responses of lymphocytes from melanoma patients by incubation with monoclonal antibodies against ganglioside GD3. Cancer Immunol Immunother 1987;24:144-150.
53. Dippold WG, Lloyd KO, Li LT, Ikeda H, Oettgen HF, Old LJ. Cell surface antigens of human malignant melanoma: definition of six antigenic systems with mouse monoclonal antibodies. Proc Natl Acad Sci USA 1980;77:6114-6118.
54. Pukel CS, Lloyd KO, Travassos LR, et al. GD3, a prominent ganglioside of human melanoma. Detection and characterization by mouse monoclonal antibody. J Exp Med 1982;155:1133-1147.
55. Hellstrom I, Brankovan V, Hellstrom KE. Strong antitumor activities of IgG3 antibodies to human melanoma-associated ganglioside. Proc Natl Acad Sci USA 1985;82:1499-1502.
56. Ortaldo JR, Woodhouse C, Morgan AC, Herberman RB, Cheresh DA, Reisfeld R. Analysis of effector cells in human antibody-dependent cellular cytotoxicity with murine monoclonal antibodies. J Immunol 1987;138:3566-3572.
57. Welt S, Carswell EA, Vogel C-W, Oettgen HF, Old LJ. Immune and nonimmune effector functions of IgG3 mouse monoclonal antibody R24 detecting the disialoganglioside GD3 on the surface of melanoma cells. Clin Immunol Immunopathol 1987;45:214-229.
58. Reisfeld RA. Immunotherapy of cancer: a perspective view. J Clin Lab Anal 1988;2:124-130.

59. Houghton AN, Mintzer D, Cordon-Cardo C, et al. Mouse monoclonal IgG3 antibody detecting GD3 ganglioside: A phase I trial in patients with malignant melanoma. Proc Natl Acad Sci USA 1985;82:1242-1246.
60. Beaumier PL, Vanderheyden J-L, Woodhouse CS, Morgan AC Jr, Fritzberg AR. A comparison of radiation dosimetry estimates for MAb, F(ab')2 and Fab carriers of Rhenium-188 in xenografted nude mice. Third International Conference on Monoclonal Antibody Immunoconjugates for Cancer. San Diego, February 4-6, 1988.
61. Braslawsky GR, Ligato P, Kaneko T, Greenfield R. Specific immunosuppression of mice by conjugating adriamycin (ADM) to a foreign serum protein. Third International Conference on Monoclonal Antibody Immunoconjugates for Cancer. San Diego, February 4-6, 1988.
62. Rosenblum MG, Lamki LM, Murray JL, Carlo DJ, Gutterman JU. Interferon-induced changes in pharmacokinetics and tumor uptake of 111-In-labeled antimelanoma antibody 96.5 in melanoma patients. J Natl Cancer Inst 1988;80:160-165.

20
Markers of Central Nervous System Tumors

GEORG D. BIRKMAYER

Laboratory for Bio-Analytics and Med-Info, Vienna, Austria

A marker is any measurable parameter by which a transformed cell can be differentiated from its correspondent progenitor cell. This could be a nucleic acid, a protein, enzyme, metabolic product, or any altered cellular component. There is ample evidence that cancer cells differ from their normal counterparts in their biochemical and biological behavior [1]. It has therefore been postulated that the development and the maintenance of the transformed state of a neoplastic cell is based on the production of certain macromolecules [2]. Studies with oncogene probes have revealed that this is not so much a qualitative as a quantitative change [3]. This means that cancer cells produce certain components and macromolecules to a greater extent than do normal cells. Due to these predominantly quantitative changes, it seems very difficult, if not impossible, to find markers that are absolutely tumor specific. This is also true for markers of central nervous system tumors.

The most frequent tumors of the central nervous system are gliomas, followed by neuroblastomas, medulloblastomas, ependymomas, and menigiomas. This short review will focus mainly on gliomas and neuroblastomas.

I. CIRCULATING TUMOR-ASSOCIATED ANTIGENS

Gliomas are among the human tumors that express tumor-associated antigens [4] that differentiate them from their tissue of origin [5]. Monoclonal antibodies raised against human glioma cells were able to define at least three classes of glioma-associated membrane antigens [6] : individual or autologous tumor-associated antigens, which showed absolute restrictions to autologous tumor cells; glioma-associated antigens detectable on all glioma but also on neuroblastoma and melanoma cell lines; and antigens widely distributed in malignant as well as normal cells from human origins. In a recent report a monoclonal antibody produced against human glioma cell lines designated J-22 reacted with 18 of 20 glioma cell lines, but not with 39 cell lines derived from sarcoma, carcinoma, or hematopoietic tumors. The antigen was expressed in the brain of human embryos in early gestation (9 weeks) but not in late gestation (8 months) or in normal adult brain, which suggested that the antibody recognizes a neural differentiation antigen expressed by a neuroectodermal origin. The antigen defined by the J-22 monoclonal antibody could be regarded as a glioma-associated antigen because it has not been observed in other neural tumors such as ependymomas, meningiomas and neuroblastomas. This antigen could be detected in the cerebrospinal fluid (CSF) of patients with glioma [7]. It is a cell membrane component with an estimated molecular weight of 67 Kd. The authors claim that the monoclonal antibody J-22 is useful for the diagnosis of gliomas.

A number of other antigens have been detected in human gliomas and neuroblastomas that were claimed to be tumor associated. However, it turned out that they represent markers for cells of the central nervous system. This holds for the S-100 protein, glial fibrillary acidic protein (GFAP), neuron-specific enolase (NSE), myelin basic protein (MBP), neurofilaments, and vimentin. These antigens are detectable in almost all neuroectodermal tumors, especially those derived from the neuroepithelium, which include the various types of gliomas [8,9]. All these proteins are useful markers for neurogenic tumors in general, but also in the differentiation of real CNS tumors from brain metastasis of carcinomas. Another marker is D2 glycoprotein, which is found in gliomas and medulloblastomas but is absent in other brain tumors [10]. A further marker could be sialyllactotetraosylceramid, a recently described ganglioside marker for human malignant gliomas [11].

Certain tumor-associated antigens have been detected in the sera of patients with neuroblastoma. One of these is the ganglioside GD2. Sera of neuroblastoma patients showed significantly higher GD2 levels than did normal children ($p < 0.001$) and children with other tumors ($p < 0.001$), as determined by a quantitative competitive enzyme-linked immunosorbent assay. Furthermore, the GD2 serum level of one neuroblastoma patient, when evaluated serially, was found to correlate with progression of disease, thus suggesting the potential usefulness of this assay for the diagnosis and monitoring of neuroblastoma [15,16]. Highly selective recognition of human neuroblastoma cells by mouse monoclonal anitbody to a cytoplasmic antigen has been reported. Immunization of a BALB/c mouse with cells from the human neuroblastoma line LAN-1 and fusion of the spleen cells with mouse myeloma cells, NS-1, led to the production of a monoclonal antibody (Mab) with a unique reactivity for neuroblastoma. This Mab, named 5 A7, detects an antigen of an apparent molecular weight 65,000-67,000, localized mainly in the cytoplasm and released into the culture medium [12].

II. SINGLE-MARKER CLINICAL STUDY

The tissue polypeptide antigen (TPA) detected by Björklund in 1957 [13] has also been used as a marker for brain tumors [14]. According to these authors, TPA is a valid serological adjunct for the diagnosis of brain tumors.

III. MULTIPLE-MARKER CLINICAL STUDIES

In addition to TPA, a number of other established tumor markers such as carcinoembryonic antigen (CEA), ferritin, and NSE have been used for evaluating patients with brain tumors [17]. A definitive conclusion about the usefulness of these markers has not been reached. Long-term studies have not yet been evaluated.

IV. CIRCULATING TUMOR-ASSOCIATED ANTIBODIES AND IMMUNE COMPLEXES

To find tumor-associated antibodies in the serum of cancer patients, two criteria must be met: the particular neoplasm must express the tumor-associated antigen and the patient must produce antibodies against this marker to allow its detection.

Gliomas have been shown to elicit an immune response [18]. By a variety of methods, antigens directed partly against tumor-associated antigens and partly against glial-cell could be demonstrated in patients with glioma. An antigen was solubilized from the cell membrane of human gliomas. Its molecular weight was estimated at 55 kD. Cross-reacting antibodies could be detected in 79% of the glioma patients' whose sera have been tested so far [19]. The presence of two additional membrane-bound antigens in human glioma cells was substantiated by Solheid et al. [20] using autologous and heterologous patient sera. These findings were confirmed by Hitchcock et al. [21]. Using sera of patients with brain tumors, they were able to identify two tumor-associated antigens, one of which predominated in many glioma cells, and two tissue-associated antigens specific for normal white brain matter. Sheikh and co-workers [22] tried to differentiate between the various glioma-associated antigens. Sera from 17 patients with histologically proven malignant brain neoplasms were tested. Two were found to be positive for autologous antigen and five were positive for the cytoplasmic antigen component of tumor cells from different patients. No reaction was found with normal brain tissue. This study implies that certain patients produce circulating antibodies against a common cytoplasmic tumor component.

V. IN VIVO LOCALIZATION STUDIES

The development of monoclonal antibodies, with their high degree of specificity, should allow tumor localization. Therefore, a number of studies have been performed to try to localize specifically tumor tissue in vitro. The localization for imaging properties of two monoclonal antibodies specific for human glioma-associated extracellular matrix glycoprotein MA, 81C6 have been reported. This was compared with the nonspecific control monoclonal antibody MA, 45.6, raised against a normal human cell. It could be shown that the antiglioma antibody was able to localize specifically an intracranial xenograft of a human glioma cell line inoculated into athymic rats [23,24].

In another study, human monoclonal antibody prepared by fusing lymphocytes from glioma patients with a human lymphoma cell line was used to localize human tumor xenografts. The results clearly demonstrate the potential application of these antibodies for in vivo localization of these tumors [25].

The radioimmunodetection of human glioma xenografts by monoclonal antibodies to an epidermal growth factor represents a different approach. The epidermal growth factor receptor is expressed by human gliomas and tumors of other tissue origin but rarely by normal brain tissue. Monoclonal antibody did detect glioblastoma xenografts but not a xenograft of colorectal cancer tumor, which does not express the epidermal growth factor receptors. These data suggest that monoclonal antibodies against growth factor receptors are likely candidates for the clinical diagnosis of human gliomas [26]. The immunolocalization of human brain tumors has been reviewed recently by Bleehan [27]. The role of certain monoclonal antibodies in the diagnosis and therapy of brain tumors has been elucidated in a preliminary study [28]. In twelve patients with malignant gliomas the distribution of the antibody UJ-13A labeled with ^{131}J has been examined. The ratio of radiolabel in normal to malignant tissue ranged from 3.1 to 12.8; however, the absolute amount of labeled antibody reaching the tumor was approximately 0.001% of the injected dose. A phase I trial is presently being conducted with intrathecal radio-antibody therapy in cases of leptomeningeal malignancies resulting from various types of neoplasia. A clinical improvement was seen in four of five patients, without serious side effects. Such antibody-guided irradiation would be most suitable for the treatment of radiosensitive targets such as medulloblastomas, neuroblastomas, and some tumors of the pineal region. The rapid advances currently being made in our understanding of immunoregulation and T-cell function, as well as new technologies to gain human monoclonal antibodies by B-cell immortalization and recombinant DNA technology, and our preliminary results, justify the great hope we place in immunotherapy.

This chapter has provided a few important glimpses of the present state of the art. For more detailed coverage, the reader is referred to the reviews of monoclonal antibodies and tumor-associated antigens as reagents for brain tumor diagnosis (29-32). Enormous progress has been made in the last few years in the identification and application of tumor-associated antigens of CNS neoplasms, particularly through the use of monoclonal antibodies. To a certain extent, they contribute to the diagnosis and localization of brain tumors. However, they cannot now compete with the radiological techniques such as computed tomography (CT) or magnetic resonance imaging (MRI). Markers for CNS tumors may also provide a tool for therapeutic approaches, such as drug targeting. Phase I studies have already been performed. However, there is still a long road ahead of us before these tools will yield beneficial effect for patients.

REFERENCES

1. Birkmayer GD. Tumorbiologie. München/New York: Springer Verlag, 1984.
2. Todaro GJ, Huebner RJ. The viral oncogene hypothesis: new evidence. Proc Natl Acad Sci USA 1972;69:1009-1015.
3. Brodeur GM, Seeger RC, Schwab M, Varmus VE, Bishop IM. Amplification of N-mye in untreated human neuroblastoma with advanced disease stage. Science 1984; 224:1121-1124.
4. Cravioto H. Immunology of human brain tumors: a review. In Waters H, ed. The handbook of cancer immunology. New York: Garland STPM, 1978;3:365-397.
5. Neuwelt EA, Clark NK. Clinical aspects of neuroimmunology. Baltimore: Williams & Wilkins, 1978:73-119.
6. Pfreundschuh M, Shiku H, Takahashi T, et al. Serological analysis of cell surface antigens of human malignant brain tumors. Proc Natl Acad Sci USA 1978;75: 5122-5126.
7. Wakabayashi T, Yoshida J, Seo H, et al. Characterization of neuroectodermal antigen by a monoclonal antibody and its application in CSF diagnosis of human glioma. Neurosurgery 1988;68:449-455.
8. Reifenberger G, Szymas J, Wechsler W. Differential expression of glial- and neuronal-associated antigens in human tumors of the central and peripheral nervous system. Acta Neuropathol (Berl) 1987;74:105-123.
9. Weller RO. The immunopathology of brain tumours. In: Bleehan NM, ed. Tumours of the brain. New York: Springer Verlag, 1986:19-33.
10. Zhmareva EN, Brodskaia IA, Berezin, Schevchenko GM. Neurospecific proteins: potentials and prospects for their use in morphological research on tumors. Arkh Patol 1987;49:87-91.

11. Fredman P, von Holst H, Collins VP, Granholm L, Svennerholm L. Sialyllactotetraosylceramide, a ganglioside marker for human malignant gliomas. J Neurochem 1988;50:912-919.
12. Schulz G, Cheresh DA, Varki NM, Yu A, Staffileno LK, Reisfeld RA. Detection of ganglioside GD2 in tumor tissues and sera of neuroblastoma patients. Cancer Res 1984;44:5914-5920.
13. Kenshead JT, Lashford LS, Garson, Clayton J. Neuroblastoma; new approaches to diagnosis. Cancer Surv 1987;6:227-246.
14. Gross N, Beck D, Carrel S, Munoz M. Highly selective recognition of human neuroblastoma cells by mouse monoclonal antibody to a cytoplasmic antigen. Cancer Res 1986;46:2988-2994.
15. Björklund B, Björklund V. Antigenicity of pooled human malignant and normal tissues by cyto-immunological technique: presence of an insoluble, heat-labile tumor antigen. Int Arch Allergy 1957;10:153-184.
16. Sonoda H, Matsukado Y, Uemura S, Kaku M. Tissue polypeptide antigen as a tumor marker of brain tumors. Neurol Med Chir (Tokyo) 1984;24:655-662.
17. Klavins JV. Tumor markers. Clinical and laboratory studies. New York: Alan R. Liss, 1985.
18. Wood WC, Kornblith PL, Quindlen EA, Pollack EA. Detection of humoral immune response to human brain tumors. Specificity and reliability of microcytotoxicity assay. Cancer 1979;43:86-90.
19. Birkmayer GD. Tumor marker of the central nervous system. Cancer Detect Prev 1981;4:231-238.
20. Solheid C, Lauro G, Paladini G. Two separate membrane bound antigens on human glioma cells in tissue culture detected with sera from glioma patients by immunofluorescence. J Neurol Sci 1976;30:55-64.
21. Hitchcock MH, Hollinshead AC, Chretien P, Rizzoli HV. Soluble membrane antigens of brain tumors. I. Controlled testing for cell-mediated immune responses in a long surviving glioblastoma multiforme patient. Cancer 1977;40:660-666.
22. Sheikh KM, Apuzzo ML, Kochsiek KR, Weiss MH. Malignant glial neoplasmas: definition of a humoral host response to tumor-associated antigen(s). Yale J Biol Med 1977;50:397-403.
23. Bullard DE, Adams CJ, Coleman RE, Bigner DD. In vivo imaging of intracranial human glioma xenografts comparing specific with nonspecific radiolabeled monoclonal antibodies. J Neurosurg 1986;64:257-262.
24. Blasberg RG, Nakagawa H, Bourdon MA, Groothuis DR, Patlak CS. Regional localization of a glioma-associated antigen defined by monoclonal antibody 81C6 in vivo: kinetics and implications for diagnosis and therapy. Cancer Res 1987;47:4432-4443.
25. Sikoar K, Alderson T, Nethersell A, Smedley H. Tumour localization by human monoclonal antibodies. Med Oncol Tumor Pharmacother 1985;2:77-86.
26. Takahashi H, Herlyn D, Atkinson B, Powe J, Rodeck U, Alavi A. Radioimmunodetection of human glioma xenografts by monoclonal antibody to epidermal growth factor receptor. Cancer Res 1987;47:3847-3850.
27. Davies AG, Richardson RB, Bourne SP, Kemshead JT, Coakham HB. Immunolocalisation of human brain tumours. In: Bleehan NM, ed. Tumours of the brain. New York: Springer-Verlag, 1986:83-99.
28. Coakham HB, Kemshead JT. The present role of monoclonal antibodies in the diagnosis and therapy of brain tumours. Wien Klin Wochenschr 1987;99:372-378.
29. Coakham HB, Garson JA, Brownell B, Kemshead JT. Monoclonal antibodies as reagents for brain tumour diagnosis: a review. J R Soc Med 1984;77:780-778.
30. Schneider SL, Sasaki F, Zeltzer PM. Normal and malignant neural cells: a comprehensive survey of human and murine nervous system markers. CRC Crit Rev Oncol Hematol 1986;5:199-234.

31. Lee Y, Bigner DD. Aspects of immunobiology and immunotherapy and uses of monoclonal antibodies and biologic immune modifiers in human gliomas. Neurol Clin 1985;3:901-917.
32. Fischer DK, Chen TL, Narayan RK. Immunological and biochemical strategies for the identification of brain tumor associated antigens. J Neurosurg 1988;68:165-180.

21
Immunodiagnosis of Ovarian Tumors

IAN JACOBS

The London Hospital, Whitechapel, London, United Kingdom

ROBERT C. BAST, Jr.

Duke University Medical Center, Durham, North Carolina

During the last two decades intensive efforts have been made to improve the surgical and chemotherapeutic management of ovarian cancer and to determine the causes of this disease. As a result, some significant advances have been made. The importance of thorough surgical staging has been established, new chemotherapeutic regimens introduced, and the risk factors for ovarian cancer defined. Although these developments have been encouraging, they have not produced a dramatic improvement in ovarian cancer statistics. The incidence of ovarian cancer has steadily increased during the last 20 years and the bleak outlook for the disease has remained largely unchanged. Overall 5 year survival is still only 30% and cancer of the ovary is now responsible for more than 50% of deaths due to malignancies of the female reproductive tract.

Against this background, recent advances in the immunological diagnosis of malignancy appear to be of great potential in the search for methods to improve the management of ovarian cancer. Investigation of the value of immunological techniques in the diagnosis of ovarian cancer has concentrated largely on the measurement of serum levels of tumor-associated antigens and, more recently, their use in imaging by radioimmunoscintigraphy. The role of immunological techniques in histopathological and cytological diagnosis has been less extensively investigated.

I. CIRCULATING TUMOR MARKERS

Advances in the technology of antigen purification and radioimmunoassay development during the 1960s and 1970s led to the production of polyclonal antisera to oncofetal proteins, which aided in monitoring several malignancies but were of limited value in the management of epithelial ovarian cancer. During the last decade the introduction of technology for the production of monoclonal antibodies has led to a rapid increase in knowledge about antigens expressed by ovarian malignancy and to the production of monoclonal antibodies for their evaluation (Table 1). At present the most widely used immunological marker for epithelial ovarian cancer is CA 125. CA 125 is an antigenic determinant on a high-molecular-weight glycoprotein recognized by a monoclonal anti-

Table 1 Antigens Associated with Epithelial Ovarian Cancer

	Author and Reference
Oncofetal antigens	
Carcinoembryonic antigen	Stall and Martin, 1981 [1]
Human chorionic gonadotrophin	Donaldson, 1980 [2]
Alphafetoprotein	Donaldson, 1980 [2]
Tissue polypeptide antigen	Nakajima et al., 1984 [3]
Antigens defined by polyclonal antisera	
OCAA	Bhattacharya and Barlow, 1978 [4]
OCA	Knauf and Urbach, 1978 [5]
NB/70K	Knauf and Urbach, 1981 [80]
OVC-1	Inamura et al., 1978 [6]
OVC-2	Inamura et al., 1978 [6]
Antigens defined by monoclonal antibodies	
OC 125 (O)	Bast et al., 1981 [8]
ID 3 (O)	Bhattacharya et al., 1982 [9]
MOV-2 (O)	Tagliabue et al., 1985 [10]
OC 133 (O)	Berkowitz et al., 1983 [11]
MD 144, MF 61, MF 116 (O)	Mattes et al., 1984 [12]
632 (O)	Fleuren et al., 1984 [13]
OVTL-3 (O)	Poels et al., 1986 [104]
4F$_4$, 7A$_{10}$ (O)	Bhattacharya et al., 1984 [15]
WB12123 (O)	Knauf et al., 1986 [16]
MH55, MH94 (E)	Mattes et al., 1984 [12]
CA 19-9 (C)	Kaprowski et al., 1979 [7]; Charpin et al., 1982 [14]
DUPAN-2 (P)	Metzgar et al., 1982 [107]
F36/22 (B)	Papsidero et al., 1983 [111]; Croghan et al., 1984 [17]
DF3 (B)	Kufe et al., 1984 [18]; Sekine et al., 1985 [19]
B72.3 (B)	Colcher et al., 1983 [20]; Johnston et al., 1985 [21]
260FS, 280D11, 245E7 (B)	Frankel et al., 1985 [22]; Pirker et al., 1985 [23]
HMFG1, HMFG2, AUA1 (M)	Epenetos et al., 1982 [24]
CA-1 (L)	Ashall et al., 1982 [25]; Woods et al., 1982 [26]
NDOG2 (Pl)	Sunderland et al., 1984 [27]

Source of immunogen in parentheses: O, ovarian cancer; E, endometrial cancer; C, colorectal cancer; P, pancreatic cancer; B, breast cancer; M, milk fat globulin; L, laryngeal cancer; Pl, placental alkaline phosphatase.

Table 2 Preoperative Serum CA 125 Level in Ovarian Cancer Patients by FIGO Stage

Author and Reference	FIGO Stage I	II	III	IV	Total
Bast et al., 1983 [30]	1/1	2/2	15/16	3/3	21/22
Brioschi et al., 1986 [31]	4/13	3/3	29/30	22/23	58/69
Canney et al., 1984 [32]	–	–	–	–	48/58
Crombach et al., 1985 [33]	4/7	6/8	–	–	32/41
Cruickshank et al., 1987 [34]	3/12	2/3	15/16	10/10	31/42
Fuith et al., 1987 [35]	4/6	4/5	16/18	8/10	32/39
Heinonen et al., 1985 [36]	0/3	–	–	–	9/12
Kaeseman et al., 1986 [37]	–	31/46 –	–	–	166/199
Kivinen et al., 1986 [38]	3/3	10/10	15/15	1/1	29/29
Krebs et al.,[a] 1986 [39]	3/4	8/8	25/25	7/8	43/45
Li-Juan et al., 1986 [40]	1/2	3/3	22/23	–	26/28
Scilthius et al., 1987 [41]	6/8	5/5	20/20	13/13	44/46
Ricolleau et al., 1985 [42]	–	–	–	–	35/38
Zanaboni et al., 1987 [43]	8/15	3/4	29/34	3/4	43/57
Zurawski et al., 1988 [44]	12/24	10/12	–	–	–[b]
Total	49/98 (50.0%)	56/63 (90.0%)	186/197 (94.4%)	67/72 (93.1%)	617/725 (85.1%)

[a]Upper limit 25 U/ml, remaining studies 35 U/ml.
[b]Not included in total since this was a study of stage I and II disease.

body (OC 125) that was raised using an ovarian cancer cell line as an immunogen [8]. The CA 125 determinant is expressed by epithelial ovarian tumors as well as by other normal and abnormal tissues derived from the coelomic epithelium [28,29]. Serum levels of CA 125 are elevated preoperatively in more than 80% of patients with ovarian cancer (Table 2) but are also elevated in patients with a number of other pathological and physiological states (Table 3).

All of the antigenic markers described to date are expressed by normal or inflamed tissues and, consequently, are tumor-associated rather than tumor-specific antigens. Despite this limitation, several antigens have a potential role in screening for early-stage ovarian cancer, in the preoperative evaluation of pelvic masses, and in the detection of residual or recurrent disease during treatment. Each of these aspects of immunodiagnosis will be discussed below.

II. SCREENING

The rationale of screening for early-stage ovarian cancer is the well-documented observation that 5 year survival for this disease is closely correlated with stage at presentation. Ovarian cancer is characterized by its late and nonspecific symptoms. This feature, combined with the inaccessibility of the ovaries to inspection or thorough examination, ac-

Table 3 Benign and Malignant Conditions Associated with Elevation of Serum CA 125 Levels (>35 U/ml)

Malignancy	n	%	Reference
Endometrial	85/267	31.8	35,45-48
Breast	33/187	17.6	30,33,42,46,49,50
Colorectal	36/239	15.1	30,32,49,50,51
Pancreas	113/215	52.6	30,50-53
Lung	38/129	29.5	30,49,51
Gastric	25/81	30.9	50,51
Biliary tract	11/24	45.8	50
Liver	25/51	49.0	50
Esophageal	2/19	10.5	50
Benign ovarian tumors			
Serous	7/57	12.3	51,54-57
Mucinous	3/35	8.6	51,54-57
Benign cystic teratoma	5/74	6.8	51,54-57
Fibroma	1/7	14.3	40,55,57
All benign tumors	31/299	10.4	40,42,46,51,54-60
Other benign disorders of the reproductive tract			
Endometriosis			
Stage I + II	47/277	17.0	61-65
Stage III + IV	58/115	50.4	61-65
Acute salpingitis	19/47	40.4	48,57,66
Chronic salpingitis	2/24	8.3	61,66
Uterine myoma	12/122	9.8	40,51,57,58,61,67
Disorders of the digestive tract			
All cirrhosis	98/146	67.1	50,68-70
Cirrhosis + ascites	24/24	100.0	68
Chronic active hepatitis	3/33	9.1	50,69,70
Acute pancreatitis	20/62	32.2	50,53,69
Chronic pancreatitis	1/54	1.9	50,53
Physiological conditions			
1st trimester pregnancy			71-75
Menstruation			76-77

counts for the fact that 75% of patients have stage II, III, or IV disease at presentation [78]. Although 5 year survival for stage I and II disease is relatively good, that of women with disease that has spread beyond the pelvis at presentation is extremely poor. On the basis of such data it is reasonable to hypothesize that diagnosis of a greater proportion of cases at stage I would dramatically improve the overall prognosis for ovarian cancer.

Identification of the CA 125 determinant and the availability of an assay for its measurement have raised the possibility that it might provide an effective method for identification of early-stage ovarian cancer. This possibility has been dismissed by a number of authors in view of the elevation of serum CA 125 levels (>35 U/ml) in a number of benign conditions [35,41,46,79]. Specificity of the assay may, however, still be sufficient for screening since the majority of nonmalignant conditions associated with raised serum CA 125 levels occur in premenopausal women, whereas the logical target group for a screening program for ovarian cancer is the population of women who have passed menopause. As the threshold value of CA 125 considered "normal" is raised, specificity is improved with a more modest decrease in sensitivity. The incidence of ovarian cancer in women over 45 years of age in the United States and United Kingdom is approximately 40/100,000/year. Since positive results of a screening test for ovarian cancer is likely to prompt diagnostic surgery, it is unlikely that a test with a positive predictive value less than 10% would be acceptable in clinical practice. To achieve a positive predictive value of 10%, a screening test for ovarian cancer (even with 100% sensitivity) would require a minimum specificity of 99.6%. This level of specificity has been attained in a study of nuns over 40 years of age using 65 U/ml as the level for normal values. Whether this will permit sufficient sensitivity remains to be determined.

Jacobs et al. [81] have reported the results of the first phase of a study of screening for ovarian cancer in postmenopausal women using a cutoff level of 30 U/ml. The specificity of a single serum CA 125 measurement among the first 1010 women screened was high (97%) but would result in approximately 75 false-positive results for each case of ovarian cancer identified. Zurawski et al. (Gynecol Oncol, in press) have described similar levels of specificity in a 2 year study of 1082 women in Stockholm. In this trial, specificity was improved significantly by the sequential monitoring of women whose initial CA 125 values exceeded 35 U/ml. The only patient whose levels rose progressively was found to have ovarian cancer.

An alterantive approach to improving the specificity of CA 125 as a screening test has utilized secondary confirmatory tests. The role of two other antigens has been investigated in this context. TAG 72 [20,21,82] is expressed on a high-molecular-weight glycoprotein associated with breast, lung, colon, and ovarian cancer. Elevated serum levels are found in patients with colorectal, lung, and ovarian cancer. CA 15-3 is an antigen detected by an assay system using two monoclonal antibodies (115D8 [83] and DF3 [18]). Elevated serum levels of CA 15-3 are associated with breast, lung, prostate, and ovarian cancer. CA 15-3 and TAG 72 levels were elevated in 60% and 47%, respectively, of serum samples from ovarian cancer patients. All patients with serum elevation of either of these markers also had an elevated serum CA 125 level. In 50 serum samples from patients with benign disease who also had an elevated serum CA 125 level, TAG 72 and CA 15-3 levels were only elevated in 6% and 2%, respectively (Bast et al., in preparation). To investigate the use of these assays as confirmatory tests Jacobs et al. (in preparation) have measured CA 15-3 and TAG 72 levels in serum samples from patients recruited to the London Hospital screening project [81] who had a CA 125 level >30 U/ml. Definition of a positive test as a serum CA 125 level >30 U/ml in combination with either a serum CA 15.3 level >30 U/ml or a serum TAG 72 level >10 U/ml increased specificity to 98.9%. Specificity could also be improved by measurement of serial serum CA 125 levels. Definition of a positive test as a serum CA 125 level >30 U/ml at initial test and after a 3 month interval increased specificity to 98.4%. If this was combined with a requirement for coordinate elevation of either CA 15-3 or TAG 72, a specificity of 99.6% was achieved. The role of pelvic examination and ultrasound scanning as confirmatory tests for serum CA 125 measurement has also been investigated [81]. The combination of CA 125 measurement with ultrasound scanning achieved a specificity of 99.6% and the combination of CA 125 with vaginal examination a specificity of 100%.

The sensitivity of serum CA 125 measurement alone or in combination with other tests for preclinical early stage ovarian cancer is unknown. However, it is unlikely to be higher than the figure of 53% established for clinically diagnosed stage I disease (Table 1).

Retrospective analysis of serum samples stored in the JANUS serum bank has recently provided encouraging evidence that serum CA 125 measurement may be able to detect ovarian malignancy prior to clinical diagnosis [84]. Serum samples from 105 women who subsequently developed ovarian malignancy (interval, 1-143 months) were significantly greater than those of 323 matched controls. Six of 12 samples collected within 18 months of diagnosis had a serum CA 125 level >30 U/ml and 4 of 12 levels were >65 U/ml. Fourteen of 59 samples collected more than 60 months before diagnosis had a serum CA 125 level >30 U/ml. In another study, a single patient was monitored, in retrospect, for 3 years prior to the diagnosis of a stage III epithelial ovarian cancer [85]. CA 125 level was >35 U/ml 12 months prior to diagnosis and >65 U/ml 10 months prior to diagnosis.

Combined measurement of tumor markers may prove to be an effective method for improving sensitivity for early-stage disease if serum expression of the different antigens is independent. HMFG1 and HMFG2 are antigens recognized by antibodies raised against a preparation of human milk fat globule [86]. In the series of ovarian cancer patients reported by Dhokia et al. [87], serum levels of HMFG1 and HMFG2 were elevated in 56% and 65%, respectively, and CA 125 was elevated in 85%. Combined use of the three markers increased the sensitivity to 95% compared to 7% of controls. CA 125 and NB 70/K levels [80] are both elevated in the serum of approximately 80% of ovarian cancer patients but when used in combination up to 98% of patients have an elevated level of at least one marker [80,88,89]. There is evidence from a small series of patients of complementarity between CA 125 and DF-3 [19], placental alkaline phosphatase [70,90], lipid-associated sialic acid [91], galactosyl transferase [92], and MOV-2 [93].

Any increase in sensitivity achieved by combining tumor markers will be at the expense of some sacrifice in specificity. It may, however, be possible to achieve a high level of sensitivity for early stage disease using a panel of tumor-associated antigens and to increase specificity using ultrasound scanning as a secondary investigation. To define an appropriate panel of tumor markers, a large series of preoperative serum samples from patients with accurately staged early ovarian cancer is required. Such samples are difficult to obtain. A sufficiently large series can only be generated by a multicenter study.

III. DIFFERENTIAL DIAGNOSIS

Although screening for early-stage ovarian cancer remains the subject of research protocols, initial laparotomy will continue to provide the best opportunity to influence the natural history of this disease. The principles of surgical treatment of ovarian cancer have changed dramatically in recent years, but it remains the cornerstone of management for established disease. The objectives of the primary surgical procedure are to determine the extent of disease and reduce the tumor burden. Despite knowledge of the benefits of accurate surgical staging and cytoreductive surgery, many patients do not receive appropriate primary operations. A thorough surgical procedure for ovarian cancer often requires persistent, time-consuming, and aggressive surgery that requires the expertise of a gynecological oncologist. Patients with ovarian malignancy should be referred for primary laparotomy by a surgeon with appropriate operative experience. In practice, the diagnosis of ovarian cancer is not usually made preoperatively and consequently patients must often be referred to another institution for a second laparotomy if cytoreductive surgery is to be performed. An accurate method for preoperative diagnosis of ovarian cancer would provide a basis for appropriate referral before initial laparotomy.

Several studies suggest that preoperative serum CA 125 measurement is of value in the differential diagnosis of benign and malignant pelvic masses. Einhorn et al. [58] measured serum CA 125 levels preoperatively in 100 patients undergoing diagnostic laparotomy for palpable adnexal masses, 23 of whom were subsequently found to have a malignancy. Using an upper limit of 35 U/ml, serum CA 125 measurement had a sensitivity for malignant disease of 78% (18/22), a specificity of 95% (73/77), and a positive predictive value of 82% (18/22), and a predictive accuracy of 91% (91/100). Serum CA 125 levels >65 U/ml were detected in 13 of 23 patients with malignancy and 1 patient with benign disease (positive predictive value 93%), The results of this study suggested that preoperative CA 125 measurement may be of value in the differential diagnosis of benign and malignant pelvic masses.

Malkasian et al. [94] have obtained similar results. The test was of greater value for differentiating benign from malignant disease in postmenopausal patients with pelvic masses; CA 125 levels >65 U/ml identified malignant disease with a specificity of >90%. Vasilev et al. [57] conducted a similar study involving 182 gynecological patients with pelvic masses, but interpretation of their results is difficult because the study population were not consecutive admissions and included a large proportion of women without an adnexal mass. Seventy-one patients in this study had uterine fibroids alone and it is likely that the possibility of ovarian malignancy in most of these cases would be ruled out in routine clinical practice by ultrasound scanning. This study reported a sensitivity for malignant disease of 78% (14/18), a specificity of 78% (128/164), a positive predictive value of 28% (14/50), and a predictive accuracy of 78% (142/182). In two separate studies, a combination of CA 125, CA 15-3, and TAG 72 appeared to increase specificity over that observed with CA 125 alone (Soper et al., Obstet Gynecol, in press; Einhorn et al., Acta Oncologia, in press).

Clinical criteria and ultrasound may improve the utility of serum markers for differentiating benign from malignant disease. In a study of 142 patients admitted consecutively for diagnostic laparotomy with a suspected adnexal mass, Jacobs et al. (in preparation) compared clinical criteria, ultrasound scanning, and serum measurement of CA 125, CA 15-3, and TAG 72. The accuracy of a serum CA 125 measurement >30 U/ml and 70 U/ml in differentiation the 42 patients with malignant disease from the 101 patients with benign disease was 77% and 84%, respectively. This was similar to ultrasound scanning (79%) and greater than the criteria of age over 60 years (69%), postmenopausal status (61%), and clinical impression (69%). The predictive accuracy of serum CA 15-3 and TAG 72 measurement was 78% and 72%, respectively. Classification of patients as positive if their serum level of *any two* of CA 125, CA 15-3, or TAG 72.3 was elevated did not significantly improve upon the accuracy of CA 125 alone (predictive accuracy 85.3%). The combination of CA 125 with ultrasound also achieved a predictive accuracy (85%). The highest predictive accuracy (91.4%) was achieved by stepwise discriminant analysis involving CA 125, ultrasound, and clinical impression.

IV. DIAGNOSIS OF RESIDUAL AND RECURRENT DISEASE

A considerable number of studies have now correlated the course of ovarian cancer with serum CA 125 level. In cases where CA 125 levels are elevated in preoperative serum samples, serum CA 125 levels have correlated with clinical disease status in over 90% of cases. A persistently rising serum CA 125 level is consistently associated with progression of disease and is frequently evident several months prior to clinical evidence of

Table 4 Sensitivity, Specificity, and Overall Accuracy of Preoperative Serum CA 125 Measurement in Prediction of Findings at Second-Look Laparotomy for Ovarian Cancer

Author and Reference	CA 125 Raised/ Disease Present n	%	Ca 125 Normal/ Disease Absent n	%	Overall Accuracy n	%
Atack et al. [95]	3/9	33	8/8	100	11/17	65
Bast et al. [30]	17/28	61	4/4	100	21/32	66
Berek et al. [96]	12/31	39	24/24	100	36/55	65
Brioschi et al. [31]	17/27	63	13/13	100	30/40	75
Fish et al. [97]	8/12	67	2/2	100	10/14	71
Fuith et al. [35]	7/10	70	14/14	100	21/24	88
Khoo et al. [98]	11/22	50	19/20	95	30/42	71
Krebs et al. [39][a]	7/13	54	18/20	90	25/33	76
Lavin et al. [99]	12/19	63	9/10	90	21/29	72
Meier et al. [46]	12/16	75	6/6	100	18/22	82
Kaeseman et al. [37][b]	18/60	30	51/51	100	69/111	62
Niloff et al. [100]	7/25	28	9/10	90	16/35	46
Rome et al. [101]	13/31	42	18/18	100	34/49	69
Schwartz et al. [102]	3/9	33	9/9	100	12/18	67
Scilthius et al. [41]	12/39	31	4/4	100	12/18	67
Zanaboni et al. [43]	11/13	85	15/17	88	26/30	87
Total	170/364	47	223/230	97	393/594	66

Upper limit 35 U/ml
[a]Upper limit 25 U/ml.
[b]Upper limit 65 U/ml.

progression. A doubling of CA 125 levels outside the normal range of 35 U/ml is a reliable indicator of recurrent disease. The most accurate method of assessing disease status available is second-look laparotomy. Several studies have investigated the accuracy of CA 125 measurement as an indicator of disease status, using second-look findings as a "gold standard" (Table 4). An elevated serum CA 125 level prior to second-look was in all of these studies an extremely good indicator of the presence of disease. Residual disease was detected in 170 of 177 patients with an elevated serum CA 125 level (positive predictive value 96.1%). However, a serum CA 125 level within the normal range did not exclude the presence of disease (disease was present in 194/417 women with a normal serum CA 125 level). The overall accuracy of serum CA 125 measurement was 65%. It is worth noting that most of the patients undergoing second-look laparotomy in these studies were clinically free of disease. Serum CA 125 level was therefore considerably more sensitive as a detector of disease than was clinical assessment. False-negative CA 125 results were generally associated with small-volume disease, reflecting the fact that a minimum tumor volume is necessary to cause elevation of serum CA 125 levels. Serum CA 125 levels are elevated in only a small proportion of patients with microscopic tumor but in 70% or more with tumor >1-2 cm in diameter (Table 5).

Table 5 Proportion of Patients with Elevated Serum CA 125 Levels (> 35 U/ml) Related to Tumor Diameter

Author and Reference	Tumor Size at Laparoscopy/Laparotomy (cm)					
	Microscopic	<1	<2	>1	>2	>10
Canney et al. [32]				10/16 ─ ─ ─ ─ ─ ─ ─ 13/17[a] ─ ─		24/24
Brioschi et al.]31]			3/8 ─ ─ ─ ─ ─ ─ ─ ─ 14/19			
Berek et al. [96]	2/7 ─ ─ ─ ─ ─ ─ ─ ─ 3/13[b] ─ ─ ─ ─ ─ ─ 6/11[b]					
Atack et al. [95]	1/4 ─ ─ ─ ─ ─ ─ ─ ─ 1/3 ─ ─ ─ ─ ─ ─ ─ ─ ─ 1/2					
Fish et al. [97]				6/7 ─ ─ ─ ─ ─ ─ ─ ─ 5/6 ─ ─ ─ 15/15		
Khoo et al. [98]	2/6 ─ ─ ─ 3/7 ─ ─ ─ ─ ─ ─ ─ ─ 6/9					
Niloff et al. [100]			8/30 ─ ─ ─ ─ ─ ─ ─ 10/10			
Rome et al. [101]	1/4 ─ ─ ─ ─ ─ ─ ─ ─ 1/7[b] ─ ─ ─ ─ ─ ─ ─ 14/20[b]					
Schilthius et al. [41]	2/17					
Total	8/38 (21%)	14/45 (31%)	21/46 (46%)	30/38 (79%)	39/56 (70%)	39/39 (100%)

[a]>2 cm and < 10cm.
[b]>1.5 cm or < 1.5 cm.

A rising serum CA 125 level is extremely reliable evidence of recurrent or progressive disease. On the other hand, a CA 125 level falling to or staying within the normal range may be associated with small-volume residual disease (in up to 40% of cases) or a tumor not expressing the CA 125 determinant (up to 20% of cases). No study has yet addressed the question of whether the earlier detection of recurrent disease permitted by CA 125 results in an improved outlook for the patient. As methods for treating residual or recurrent disease are improved, the clinical utility of the marker should increase.

V. RADIOIMMUNOSCINTIGRAPHY

Radionuclide imaging of ovarian cancer using labeled polyclonal and monoclonal antibodies may have a role in the preoperative differential diagnosis of ovarian cancer and assessment of the extent and site of residual disease during follow-up. It is also possible that this approach will be of value as a further investigation in screening programs using serum tumor markers or ultrasound scanning.

Successful imaging of ovarian tumors has been achieved with antibodies to carcinoembryonic antigen (CEA) [103], HMFG2 [24,104,105], placental alkaline phosphatase [106], and CA 125 [107,108]. The anatomical location of the ovaries away from the central blood pool makes interpretation of images easier than in other sites. Labeled HMFG2 has been used to identify metastases as small as 0.8 cm in diameter. However there are limitations of specificity inherent in the use of tumor-associated rather than tumor-specific antigens as targets for radioimmunoscintigraphy. Although primary ovarian cancer could be localized in 19 of 20 patients imaged with HMFG2, scans were also positive in 5 of 10 patients who did not have ovarian cancer [105]. Granowska et al. [105] concluded that although radioimmunoscintigraphy with ^{123}I-labeled HMFG2 may be valuable in monitoring tumor response to treatment, it was not sufficiently specific to be useful in the differential diagnosis of ovarian malignancy. The use of radioimmunoscintigraphy for screening or differential diagnosis will require antibodies of greater specificity. Recent evidence suggests that the OVTL3 monoclonal antibody may have a high degree of specificity for ovarian carcinoma tissue. This antibody was produced using cells from an endometrioid ovarian carcinoma as immunogen. OVTL3 has been reported to bind strongly to ovarian carcinoma cells and weakly to a small proportion of endometrial carcinomas but not to other nonovarian tumors or normal tissues [106]. Epenetos et al. [109] have reported encouraging preliminary results using indium-^{111}I-labeled OVTL3 in three patients with ovarian cancer.

VI. IMMUNOCYTOCHEMISTRY AND IMMUNOHISTOCHEMISTRY

Ovarian cancer patients often present with ascites and are extensively investigated by physicians or general surgeons before the gynecological nature of their disease is established. Antibodies able to differentiate benign from malignant ascites and indicate the site of origin of a malignancy would therefore be of clinical value. Cross-reactivity of many antibodies with normal mesothelial cells limits their usefulness [110]. However, Epenetos et al. [111] identified malignant cells in 28 of 34 effusions that were histologically positive using the HMFG2 and AUA1 antibodies, while the absence of malignant cells was confirmed in 27 of 28 specimens.

The main potential role of immunohistochemical diagnosis is in the differential diagnosis of unknown primaries. Individual antibodies are of limited value since they generally interact with tumors from a number of tissues. Nevertheless, it may be possible to obtain useful results by using a panel of antibodies with different affinities to a number of primary sites.

REFERENCES

1. Stall KE, Martin EW Jr. Plasma carcinoembryonic antigen levels in ovarian cancer patients: A chart review and survey of published data. J Reprod Med 1981;26:73-79.
2. Donaldson ES, van Nagell JR Jr, Pursell S, Gay EC, Meeker WR, Kashmiri R, Van de Voorde J. Multiple biochemical markers in patients with gynecologic malignancies. Cancer 1980;45:948-953.
3. Nakajima H, Hirata T, Norisugi T, Furuno K, Ikehata N, Utsunomiya A, Kobayashi K, Negishi Y, Akiya K. A study on tumor marker of tissue-polypeptide-antigen (TPA) in gynecologic malignancies. Nipon Sanka Fujinka Gakkai Zasshi 1984;36:1877-1883.
4. Bhattacharya M, Barlow JJ. Ovarian tumor antigens. Cancer 1978;42:1616-1620.
5. Knauf S, Urbach GI. The development of a double-antibody radioimmunoassay for detecting ovarian tumor-associated antigen fraction OCA in plasma. Am J Obstet Gynecol 1978;131:780-787.
6. Imamura N, Takahashi T, Lloyd KO, Lewis JL Jr, Old LJ. Analysis of human ovarian tumor antigens using heterologous antisera: detection of new antigenic systems. Int J Cancer 1978;21:570-577.
7. Koprowski H, Steplewski Z, Mitchell K, Herlyn M, Herlyn D, Fuhrer P. Colorectal carcinoma antigens detected by hybridoma antibodies. Somatic Cell Genet 1979;5:957-971.

8. Bast RC Jr, Feeney M, Lazarus H, Nadler LM, Colvin RB, Knapp RC. Reactivity of a monoclonal antibody with human ovarian carcinoma. J Clin Invest 1981;68:1331-1337.
9. Bhattacharya M, Chatterjee SK, Barlow JJ, Fuji H. Monoclonal antibodies recognizing tumor-associated antigen of human ovarian mucinous cystadenocarcinomas. Cancer Res 1982;42:1650-1654.
10. Tagliabue E, Menard S, Della Torre G, Barbanti P, Mariani-Costantini R, Porro G, Colnaghi MI. Generation of monoclonal antibodies reacting with human epithelial ovarian cancer. Cancer Res 1985;45:379-385.
11. Berkowitz RS, Kabawat S, Lazarus H, Colvin RC, Knapp RC, Bast RC Jr. Comparison of a rabbit heteroantiserum and a murine monoclonal antibody raised against a human epithelial ovarian carcinoma cell line. Am J Obstet Gynecol 1983;146:607-612.
12. Mattes MJ, Cordon-Cardo C, Lewis JL Jr, Old LJ, Lloyd KO. Cell surface antigens of human ovarian and endometrial carcinoma defined by mouse monoclonal antibodies. Proc Natl Acad Sci USA 1984;81:568-572.
13. Fleuren GJ, Coerkamp E, Nap M, Warnar SO. Immunohistological characterization of monoclonal antibody directed against nonmucinous ovarian carcinomas. International meeting on monoclonal antibodies in oncology: Clinical applications, Nantes 1984.
14. Charpin C, Bhan AK, Zurawski VR Jr, Scully RE. Carcinoembryonic antigen (CEA) and carbohydrate determinant 19-9 (CA 19-9) localization in 121 primary and metastatic ovarian tumors: an immuno-histochemical study with the use of monoclonal antibodies. Int J Gynecol Pathol 1982;1:231-245.
15. Bhattacharya M, Chatterjee SK, Barlow JJ. Identification of a human cancer-associated antigen defined with monoclonal antibody. Cancer Res 1984;44:4528-4534.
16. Knauf S, Kalwas J, Helmkamp BF, Harwell LW, Beecham J, Lord EM. Monoclonal antibodies against human ovarian tumor associated antigen NB/70K: preparation and use in a radioimmunoassay for measuring NB/70K in serum. Cancer Immunol Immunother 1986;21:217-225.
17. Croghan GA, Papsidero LD, Valenzuela LA, Nemoto T, Penetrante R, Chu TM. Tissue distribution of an epithelial and tumor associated antigen recognized by monoclonal antibody F36/22. Cancer Res 1983;43:4980-4988.
18. Kufe D, Inghirami G, Abe M, Hayes D, Justi-Wheeler H, Schlom J. Differential reactivity of a novel monoclonal antibody (DF3) with human malignant versus benign breast tumors. Hybridoma 1984;3:223-232.
19. Sekine H, Hayes DF, Ohno T, Keefe KA, Schaetzl E, Bast RC Jr, Knapp RC, Kufe DW. Circulating DF3 and CA125 antigen levels in serum from patients with epithelial ovarian carcinoma. J Clin Oncol 1985;3:1355-1363.
20. Colcher D, Hand PH, Nuti M, Schlom J. Differential binding to human mammary and nonmammary tumors of monoclonal antibodies reactive with carcinoembryonic antigens. Cancer Invest 1983;1:127-138.
21. Johnston WW, Szpak CA, Lottich SC, Thor A, Schlom J. Use of a monoclonal antibody (B72.3) as an immunocytochemical adjunct to diagnosis of adenocarcinoma in human effusions. Cancer Res 1985;45:1894-1900.
22. Frankel AE, Ring DB, Tringale F, Hsieh-Ma ST. Tissue distribution of breast cancer-associated antigens defined by monoclonal antibodies. J Biol Response Mod 1985;4:273-286.
23. Pirker R, FitzGerald DJ, Hamilton TC, Ozols RF, Laird W, Frankel AE, Willingham MC, Pastan I. Characterization of immunotoxins active against ovarian cancer cell lines. J Clin Invest 1985;76:1261-1267.

24. Epenetos AA, Britton KE, Mather S, Shepherd J, Granowska M, Taylor-Papadimitriou J, Nimmon CC, Durbin H, Hawkins LR, Malpas JS, Bodmer WF. Targeting of iodine-123-labeled tumour-associated monoclonal antibodies to ovarian, breast and gastrointestinal tumours. Lancet 1982;2:999-1005.
25. Ashall F, Bramwell ME, Harris H. A new marker for human cancer cells. I. The CA antigen and the Ca1 antibody. Lancet 1982;2:1-6.
26. Woods JC, Spriggs AI, Harris H, McGee J O'D. A new marker for human cancer cells. 3. Immunocytochemical detection of malignant cells in serous fluids with the Ca-1 antibody. Lancet 1982;2:512-514.
27. Sunderland CA, Davies JO, Stirrat GM. Immunohistology of normal and ovarian cancer tissue with a monoclonal antibody to placental alkaline phosphatase. Cancer Res 1984;44:4496-4502.
28. Kabawat SE, Bast RC Jr, Bhan AK, Welch WR, Knapp RC, Colvin RB. Tissue distribution of a coelomic-epithelium-related antigen recognized by the monoclonal antibody OC125. Int J Gynecol Pathol 1983;2:275-285.
29. Kabawat SE, Bast RC Jr, Welch WR, Knapp RC, Colvin RB. Immunopathologic characterization of a monoclonal antibody that recognizes common surface antigens of human ovarian tumors of serous, endometrioid and clear cell types. Am J Clin Pathol 1983;79:98-104.
30. Bast RC Jr, Klug TL, St John E, Jenison E, Niloff J, Lazarus H, Berkowitz RS, Leavitt T, Griffiths CT, Parker L, Zurawski VR, Knapp RC. A radioimmunoassay using a monoclonal antibody to monitor the course of epithelial ovarian cancer. N Engl J Med 1983;309:883-887.
31. Brioschi PA, Irion O, Bischof P, Bader M, Forni M, Krauer F. Serum CA 125 in epithelial ovarian cancer. A longitudinal study. Br J Obstet Gynaecol 1987;94:196-201.
32. Canney PA, Moore M, Wilkinson PM, James RD. Ovarian cancer antigen CA 125: a prospective clinical assessment of its role as a tumour marker. Br J Cancer 1984;50:765-769.
33. Crombach G, Wurz H. CA 125, TPA and CEA in ovarian cancer—a critical evaluation of single and combined determinations of serum markers. In: Geten H, Klapdor RG, eds. New tumour associated antigens. Stuttgart-New York: Georg Thieme Verlag 1984:134-144.
34. Cruickshank DJ, Fullerton WT, Klopper A. The clinical significance of pre-operative serum CA 125 in ovarian cancer. Br J Obstet Gynaecol 1987;94:692-695.
35. Fuith LC, Daxenbichler G, Dapunt O. CA 125 in the serum and tissue of patients with gynecological disease. Arch Gynecol Obstet 1987;241:157-164.
36. Heinonen PK, Tontti K, Koivula T, Pystynen P. Tumour-associated antigen CA 125 in patients with ovarian cancer. Br J Obstet Gynaecol 1985;92:528-531.
37. Kaesemann H, Caffier H, Hoffmann FJ, Crombach G, Wurz H, Kreienberg R, Mobus V, Schmidt-Rhode P, Sturm G. Monoclonal antibodies in the diagnosis and follow-up of ovarian cancer. CA 125 as tumor marker. Klin Wochenschr 1986;64:781-785. 64:781-785.
38. Kivinen S, Kuoppala T, Leppilampi M, Vuori J, Kauppila A. Tumor-associated antigen CA 125 before and during the treatment of ovarian carcinoma. Obstet Gynecol 1986;67:468-472.
39. Krebs HB, Goplerud DR, Kilpatrick SJ, Myers MB, Hunt A. Role of CA 125 as tumor marker in ovarian carcinoma. Obstet Gynecol 1986;67:473-477.
40. Li-Juan L, Xiu-Feng H, Wen-Shu L, Ai-Ju W. A monoclonal antibody radioimmunoassay for an antigenic determinant CA 125 in ovarian cancer patients. Chin Med J 1986;99:721-726.

41. Schilthuis MS, Aalders JG, Bouma J, Kooi H, Fleuren GJ, Willemse PH, De Bruijn HW. Serum CA 125 levels in epithelial ovarian cancer: relation with findings at second-look operations and their role in the detection of tumor recurrence. Br J Obstet Gynaecol 1987;94:202-207.
42. Ricolleau G, Chatal JF, Fumoleau P, Kremer M, Douillard JY, Curtet C. Radioimmunoassay of the CA 125 antigen in ovarian carcinomas: advantages compared with CA 19-9 and CEA. Tumour Biol 1984;5:151-159.
43. Zanaboni F, Vergadoro F, Presti M, Gallotti P, Lombardi F, Bolis G. Tumor antigen CA 125 as a marker of ovarian epithelial carcinoma. Gynecol Oncol 1987;28:61-67.
44. Zurawski VR Jr, Knapp RC, Einhorn N, Kenemans P, Mortel R, Ohmi K, Bast RC, Ritts RE Jr, Malkasian G. An initial analysis of preoperative serum CA 125 levels in patients with early stage ovarian carcinoma. Gynecol Oncol 1988;30:7-14.
45. Niloff JM, Klug TL, Schaetzl E, Zurawski VR Jr, Knapp RC, Bast RC Jr. Elevation of serum CA 125 in carcinomas of the fallopian tube, endometrium and endocervix. Am J Obstet Gynecol 1984;148:1057-1058.
46. Meier W, Stieber P, Fateh-Moghadam A, Eiermann W, Hepp H. CA 125 in gynecological malignancies. Eur J Cancer Clin Oncol 1987;23:713-717.
47. Duk JM, Aalders JG, Fleuren GJ, de Bruijn HW. CA 125: a useful marker in endometrial carcinoma. Am J Obstet Gynecol 1986;155:1097-1102.
48. Patsner B, Mann WJ, Cohen H, Loesch M. Predictive value of preoperative serum CA 125 levels in clinically localized and advanced endometrial carcinoma. Am J Obstet Gynecol 1988;158:399-402.
49. Kawahara M, Terasaki PI, Chia D, Johnson C, Hermes M, Tokita K. Use of four monoclonal antibodies to detect tumor markers. Cancer 1986;58:2008-2012.
50. Haga Y, Sakamoto K, Egami H, Yoshimura R, Mori K, Akagi M. Clinical significance of serum CA 125 values in patients with cancers of the digestive system. Am J Med Sci 1986;292:30-34.
51. Kuzuya K, Nozaki M, Chihara T. Evaluation of CA 125 as a circulating tumor marker for ovarian cancer. Nippon Sanka Fujinka Gakkai Zasshi 1986;38:949-957.
52. Dietel M, Arps H, Klapdor R, Muller-Hagen S, Sieck M, Hoffmann L. Antigen detection by the monoclonal antibodies CA 19-9 and CA 125 in normal and tumor tissue and patients' sera. J Cancer Res Clin Oncol 1986;111:257-265.
53. Haglund C. Tumour marker antigen CA 125 in pancreatic cancer: a comparison with CA 19-9 and CEA. Br J Cancer 1986;54:897-901.
54. Pansini F, Bellinazzi A, Rainaldi V, Santoiemma M, Lenzi B, Jacobs M, Mollica G, Bagni B. Serum CA 125 in ovarian pathology and its variation in ovarian carcinoma after integrated therapy. Gynecol Obstet Invest 1986;21:47-51.
55. Shi QY, Sakawaki T, Kasamatsu T, Ohkura H. Clinical significance of CA 125 and its assessment with an enzyme immunoassay. Asia Oceania J Obstet Gynaecol 1987; 13:325-333.
56. Takahashi K, Shibukawa T, Moriyama M, Shirai T, Kijima S, Iwanari O, Matsunaga I, Kitao M. Clinical usefulness and false-positive results of CA 125 as a tumor marker of ovarian cancer—a study on 674 patients. Jpn J Surg 1986;16:305-310.
57. Vasilev SA, Schlaerth JB, Campeau J, Morrow CP. Serum CA 125 levels in preoperative evaluation of pelvic masses. Obstet Gynecol 1988;71:751-756.
58. Einhorn N, Bast RC Jr, Knapp RC, Tjernberg B, Zurawski VR Jr. Preoperative evaluation of serum CA 125 levels in patients with primary epithelial ovarian cancer. Obstet Gynecol 1986;67:414-416.
59. Fleuren GJ, Nap M, Aalders JG, Trimbos JB, de Bruijn HW. Explanation of the limited correlation between tumor CA 125 content and serum CA 125 antigen levels in patients with ovarian tumours. Cancer 1987;60:2437-2442.

60. Takahashi K, Yamane Y, Kijima S, Yoshino K, Shibukawa T, Kitao M. CA 125 antigen is an effective diagnostic for external endometriosis. Gynecol Obstet Invest 1987;23:257-260.
61. Barbieri RL, Niloff JM, Bast RC Jr, Schaetzl E, Kistner RW, Knapp RC. Elevated serum concentrations of CA 125 in patients with advanced endometriosis. Fertil Steril 1986;45:630-634.
62. Kauppila A, Telimaa S, Ronnberg L, Vuori J. Placebo-controlled study on serum concentrations of CA 125 before and after treatment of endometriosis with danazol or high-dose medroxyprogesterone acetate alone or after surgery. Fertil Steril 1988; 49:37-41.
63. Patton PE, Field CS, Harms RW, Coulam CB. CA 125 levels in endometriosis. Fertil Steril 1986;45:770-773.
64. Fedele L, Vercellini P, Arcaini L, Grazia da Dalt M, Candiani GB. CA 125 in serum, peritoneal fluid, active lesions, and endometrium of patients with endometriosis. Am J Obstet Gynecol 1988;158:166-170.
65. Pittaway DE, Fayez JA. The use of CA 125 in the diagnosis and management of endometriosis. Fertil Steril 1986;46:790-795.
66. Halila H, Stenman UH, Seppala M. Ovarian cancer antigen CA 125 levels in pelvic inflammatory disease and pregnancy. Cancer 1986;57:1327-1329.
67. Yabushita H, Masuda T, Ogawa A, Noguchi M, Ishihara M. Combination assay of CA 125, TPA, IAP, CEA and ferritin in serum for ovarian cancer. Gynecol Oncol 1988;29:66-75.
68. Bergmann JF, Bidart JM, George M, Beaugrand M, Levy VG, Bohuon C. Elevation of CA 125 in patients with benign and malignant ascites. Cancer 1987;59:213-217.
69. Ruibal A, Encabo G, Martinez-Miralles E, Murcia C, Capdevial JA, Salgado A, Martinez-Vasquez JM. CA 125 seric levels in nonmalignant pathologies. Bull Cancer 1984;71:145-146.
70. Eerdekens MW, Nouwen EJ, Pollet DE, Briers TW, De Broe ME. Placental alkaline phosphatase and cancer antigen 125 in sera of patients with benign and malignant diseases. Clin Chem 1985;31:687-690.
71. Seki K, Kikuchi Y, Uesato T, Kato K. Increased serum CA 125 levels during the first trimester of pregnancy. Acta Obstet Gynecol Scand 1986;65:583-585.
72. Niloff JM, Knapp RC, Schaetzl E, Reynolds C, Bast RC Jr. CA 125 antigen levels in obstetric and gynecologic patients. Obstet Gynecol 1984;64:703-707.
73. Haga Y, Sakamoto K, Egami H, Yoshimura R, Akagi M. Evaluation of serum CA 125 values in healthy individuals and pregnant women. Am J Med Sci 1986;292: 25-29.
74. O'Brien TJ, Hardin JW, Bannon GA, Norris JS, Quirk JG Jr. CA 125 antigen in human amniotic fluid and fetal membranes. Am J Obstet Gynecol 1986;155:50-55.
75. Jacobs IJ, Fay TN, Stabile I, Bridges JE, Oram DH, Grudzinskas JG. The distribution of CA 125 in the reproductive tract of pregnant and non pregnant women. Br J Obstet Gynaecol, 1988;95:1190-1194.
76. Mastropaolo W, Fernandez Z, Miller EL. Pronounced increases in the concentration of an ovarian tumor marker, CA 125, in serum of a healthy subject during menstruation. Clin Chem 1986;32:2110-2111.
77. Pittaway DE, Fayez JA. Serum CA 125 antigen levels increase during menses. Am J Obstet Gynecol 1987;156:75-76.
78. Kottmeier H (Ed). Annual report on the results of treatment in gynaecological cancer, vol 18. FIGO, Stockholm, 1982.
79. Lambert J. The value of CA 125 serum assay in the management of ovarian cancer. Br J Obstet Gynaecol 1987;94;193-195.

80. Knauf S, Urbach GI. Identification, purification and radioimmunoassay of NB/70K, a human ovarian tumor-associated antigen. Cancer Res 1981;41:1351-1357.
81. Jacobs I, Stabile I, Bridges J, Kemsley P, Reynolds C, Grudzinskas J, Oram D. Multimodal approach to screening for ovarian cancer. Lancet 1988;1:268-271.
82. Thor A, Gorstein F, Ohuchi N, Szpak CA, Johnston WW, Schlom J. Tumor-associated glycoprotein (TAG-72) in ovarian carcinomas defined by monoclonal antibody B72.3. J Natl Cancer Inst 1986;76:995-1006.
83. Hilkens J, Buijs F, Hilgers J, Hageman P, Calafat J, Sonnenberg A, Van Der Valk M. Monoclonal antibodies against human milk-fat globule membranes detecting differentiation antigens of the mammary gland and its tumors. Int J Cancer 1984;34:197-206.
84. Zurawski VR Jr, Orjaseter H, Andersen A, Jellum E. Elevated serum CA 125 levels prior to diagnosis of ovarian neoplasia: relevance for early detection of ovarian cancer. Int J Cancer, 1988;42:677-680.
85. Bast RC Jr, Siegel FP, Ronowicz C, Klug TL, Zurawski VR Jr, Schonholz D, Cohen CJ, Knapp RC. Elevation of serum CA125 prior to diagnosis of epithelial ovarian carcinoma. Gynecol Oncol 1985;22:115-120.
86. Taylor-Papadimitriou J, Peterson JA, Arklie J, Burchell J, Ceriani RL, Bodmer WF. Monoclonal antibodies to epithelium-specific components of the human milk fat globule membrane: Production and reaction with cells in culture. Int J Cancer 1981;28:17-21.
87. Dhokia B, Canney PA, Pectasides D, Munro AJ, Moore M, Wilkinson PM, Self C, Epenetos AA. A new immunoassay using monoclonal antibodies HMFG1 and HMFG2 together with an existing marker CA 125 for the serological detection and management of epithelial ovarian cancer. Br J Cancer 1986;54:891-895.
88. Knauf S, Anderson DJ, Knapp RC, Bast RC Jr. A study of the NB/70K and CA 125 monoclonal antibody radioimmunoassays for measuring serum antigen levels on ovarian cancer patients. Am J Obstet Gynecol 1985;152:911-913.
89. Dembo AJ, Chang D-L, Malkin A, Urbach GI. Plasma ovarian cancer antigen NB/70K. Clinical correlations. Proc Am Soc Clin Oncol 1985;6:30.
90. van den Burg MEL, van Putten WL, Cox PH, Haije WG. A panel of tumor markers for monitoring the epithelial ovarian carcinoma. ECCO 3, Stockholm 1985:135.
91. Schwartz PE, Chambers SK, Chambers JT, Gutmann J, Foemmel RS, Behrman HR. Circulating tumor markers in gynecologic malignancies: A preliminary experience using lipid associated sialic acid (LSA), CA 125 and carcinoembryonic antigen (CEA). Proc Am Soc Clin Oncol 1985;4:118.
92. Gauduchon P, Goussard J, Heron JF, Bouvet C, Tillier C, Bar E, Le Talger JY. Galactosyltransferase activity (GT) and CA 125 antigen in ovarian carcinoma. Ovarian Cancer Sumposium, Glasgow, February 24, 1985.
93. Colnaghi MI, Canevari S, Menard S, Miotti S, Rilke F. Tumor detection by immunocytochemical and radioimmunologic methods. Monoclonal Antibodies, 1984, Florence L24.
94. Malkasian GD Jr, Knapp RC, Lavin PT, Zurawski VR Jr, Podratz KC, Stanhope CR, Mortel R, Berek JS, Bast RC Jr, Ritts RE. Preoperative evaluation of CA 125 in premenopausal and postmenopausal patients with pelvic masses: Discrimination of benign from malignant disease. Obstet Gynecol 1988;159:341-346.
95. Atack DB, Nisker JA, Allen HH, Tustanoff ER, Levin L. CA 125 surveillance and second-look laparotomy in ovarian carcinoma. Am J Obstet Gynecol 1986;154:287-289.
96. Berek JS, Knapp RC, Malkasian GD, Lavin PT, Hacker NF, Whitney C, Niloff JM, Bast RC Jr. CA 125 serum levels correlated with second-look operations among ovarian cancer patients. Obstet Gynecol 1986;67:685-689.

97. Fish RG, Shelley MD, Maughan T, Rocker I, Adams M. The clinical value of serum CA 125 levels in ovarian cancer patients receiving platinum therapy. Eur J Cancer Clin Oncol 1987;23:831-835.
98. Khoo SK, Hurst T, Webb MJ, Dickie GJ, Kearsley JH, Mackay EV. Predictive value of serial CA 125 antigen levels in ovarian cancer evaluated by second-look laparotomy. Eur J Cancer Clin Oncol 1987;23:765-771.
99. Lavin PT, Knapp RC, Malkasian G, Whitney CW, Berek JC, Bast RC Jr. CA 125 for monitoring of ovarian carcinoma during primary therapy. Obstet Gynecol 1987;69:223-227.
100. Niloff JM, Bast RC Jr, Schaetzl EM, Knapp RC. Predictive value of CA 125 antigen levels in second-look procedures for ovarian cancer. Am J Obstet Gynecol 1985; 151:981-986.
101. Rome RM, Koh H, Fortune D, Cauchi M. CA 125 serum levels and secondary laparotomy in epithelial ovarian tumours. Aust NZ J Obstet Gynaecol 1987;27: 142-145.
102. Schwartz PE, Chambers SK, Chambers JT, Gutmann J, Katopodis N, Foemmel R. Circulating tumor markers in the monitoring of gynaecologic malignancies. Cancer 1987;60:353-361.
103. van Nagell JR Jr, Kim E, Casper S, Primus FJ, Bennett S, DeLand FH, Goldenberg DM. Radioimmunodetection of primary and metastatic ovarian cancer using radiolabeled antibodies to carcinoembryonic antigen. Cancer Res 1980;40:502-506.
104. Epenetos AA, Shepherd J, Britton KE, Mather S, Taylor-Papadimitriou J, Granowska M, Durbin H, Nimmon CC, Hawkins LR, Malpas JS, Bodmer WF. ^{123}I radioiodinated antibody imaging of occult ovarian cancer. Cancer 1985;55:984-987.
105. Granowska M, Britton KE, Shepherd JH, Nimmon CC, Mather S, Ward B, Osborne RJ, Slevin ML. A prospective study of 123
RJ, Slevin ML. A prospective study of ^{123}I-labeled monoclonal antibody imaging in ovarian cancer. J Clin Oncol 1986;4:730-736.
106. Poels LG, Peters D, van Megen Y, Vooijs GP, Verheyen RN, Willemen A, van Niekerk CC, Jap PHK, Mungyer G, Kenemans P. Monoclonal antibody against human ovarian tumor-associated antigens. J Natl Cancer Inst 1986;76:781-791.
107. Hnatowitch DJ, Griffin TW, McGann J, Rusckowski M, Gionet M, Hunter R, Doherty PW. Radioimmunodetection of ovarian and other tumours in patients using indium labeled OC 125 antibody; A comparison with the 19-9 antibody. International conference on Monoclonal Antibody Immunoconjugates for Cancer, San Diego, March 6-8, 1986, p. 40.
108. Haisma HJ, Battaile A, Knapp RC, Zurawski VR Jr. Immune complexes of OC 125-CA 125 following intravenous injection of radiolabeled monoclonal antibody into ovarian cancer patients. International Conference on Monoclonal Antibody Immunoconjugates for Cancer, San Diego, March 6-8, 1986, p. 44.
109. Epenetos AA, Lavender JP, Kenemans P, Poels L. Early results of the monoclonal antibody OVTL3 in specific detection of ovarian cancer (letter). J Clin Oncol 1987;5:160.
110. Ghosh AK, Mason DY, Springgs AI. Immunocytochemical staining with monoclonal antibodies in cytologically "negative" serous effusions from patients with malignant disease. J Clin Pathol 1983;36:1150-1153.
111. Epenetos AA, Canti G, Taylor-Papadimitriou J, Curling M, Bodmer WF, Use of two epithelium-specific monoclonal antibodies for diagnosis of malignancy in serous effusions. Lancet 1982;2:1004-1006.

22
Prostate Cancer–Associated Markers

T. MING CHU
Roswell Park Memorial Institute, New York State Department of Health, Buffalo, New York

When compared with most other tumor systems, immunodiagnosis of prostate cancer is better defined, in a more developed stage, and has a more significant role in the management of patients. This is primarily due to the fact that there are two immunochemically well-defined and clinically fully evaluated tumor markers: prostatic acid phosphatase (PAP) and prostate-specific antigen (PA or PSA). Acid phosphatase activity has been used for over five decades as an aid in serodiagnosis of advanced prostatic cancer [1]. Clinical utilization of this enzyme marker, including its limitations, is well known by urologists and oncologists. Realization of the difficulties in searching for a prostate tumor-specific marker and the focused investigation on a prostate-specific marker, an organ-site and cell-type specific antigen, have paved the way for dramatic advances in immunodiagnosis of prostatic cancer in recent years [2].

The established clinical application of prostate specific antigen and prostatic acid phosphatase as well as the potential of recently reported prostate tumor markers will be the subject of this chapter. Basic chemical and biological characteristics of these clinically important markers are also reviewed.

I. PROSTATE-SPECIFIC ANTIGEN

The existence of prostate-specific antigens has been reported for some time [3,4]. Although the identification of these prostatic-specific components had been indicated in the literature, its clinical potential was not appreciated until 1979 when a prostate-specific antigen, initially abbreviated as PA and now commonly known as PSA (the use of PA or PSA in this chapter follows its citation in the original references), was purified and characterized [5].

A. Molecular Characteristics

Purified PA is a glycoprotein (7% carbohydrate, 93% protein) of molecular weight 33 Kd and isoelectric point 6.9 [5-7]. The carbohydrate content of PA includes common hexoses, hexosamines, and 1% of sialic acid. The complete sequence of 240 amino acid residues in purified PA is also known, with isoleucine and proline as the N- and C-terminal amino acid, respectively [8]. The possible carbohydrate side chains have been postulated to be on asparagine-45 for N-linkage and on serine-69, threonine-70, and serine-71 for O-linkage [8]. The sequence of oligosaccharides and exact location of monosaccharide-amino acid linkages remain to be determined.

Other characteristics of PA include a Stokes' radius of 24 Å, partial specific volume of 0.73 ml/g, and a sedimentation coefficient of 3.1 S [7]. Immunochemically identical isomeric forms of PA have been reported, probably due to a microheterogeneity of the carbohydrate moiety [9].

PA is a highly immunogenic protein, although it is a comparatively small molecule. Immunization of animals, such as rabbit, goat, mouse, and baboon, with purified PA easily generates polyclonal antisera [5,9]. Upon absorption with femal normal serum and other tissue extracts, these polyclonal antisera are fairly specific for human prostate. Only normal, benign, and malignant prostate glands, and seminal plasma, have been found to contain PA reactive with the antisera. Monoclonal antibodies also can be easily generated. It should be emphasized that PA is not a prostate-tumor-specific antigen (prostate neoantigen), since a similar concentration (per unit of protein) is present in normal, benign hypertrophic and cancerous prostate [10]. Also, biochemical and immunochemical data available have clearly indicated that it is different from and not related to PAP [9, 11].

B. Initial Immunoassay System and Clinical Evaluation

With the use of a rabbit polyclonal anti-PA antiserum reagent, a sensitive and reproducible sandwich-type enzyme immunoassay was developed in 1980 shortly after the report on PA [12]. This assay system, with a sensitivity of 0.10 ng/ml, used CNBr-activated Sepharose 4B conjugated anti-PA IgG as a solid-phase and anti-PA IgG conjugated horseradish peroxidase as the quantitative marker enzyme-linked immunosorbent assay (ELISA). In this initial report, serum PA levels from normal male controls were found to range from <0.10 to 2.6 ng/ml, with a mean of 0.47. No serum PA was detectable (<0.10) from normal females or female patients with cancer. Male patients with cancer of nonprostatic origin were found to show serum PA levels similar to those of normal male controls.

Significant findings from this initial evaluation of serum PA included the detection of elevated PA levels in patients with prostate adenocarcinomas. The mean values for various clinical stages were: 4.8 ng/ml for stage A, 5.0 for stage B, 10.1 for stage C, and 24.2 for stage D. Highly elevated serum PA levels (>50 times the normal range) were found in many patients with stages C and D prostate cancer. In addition, patients with benign prostatic hyperplasia (BPH) exhibited a mean of 3.4 ng/ml, which is statistically higher than that of normal male controls and patients with nonprostatic cancer. No statistically significant difference was found between stage A and BPH, a suggestive difference was revealed between stage B and BPH, and a highly significant difference was shown between stage C or D and BPH. In comparison with normal controls, stages B, C, and D showed a highly significant difference: a quantitatively different serum PA level was present in patients with prostate cancer from normal men. These results also suggested that PA assay by this ELISA format alone was of limited use as a screening test or for early detection of prostatic cancer, since some degrees of overlap were observed between early prostate carcinoma and BPH.

Shortly after this initial ELISA for serum PA was established, its potential value in monitoring the disease status and treatment response became apparent [13]. In a pilot and double-blind study, a small number of carefully documented and evaluated patients with stage D_2 prostate cancer in whom previous hormonal therapies had failed were randomized to receive adjuvant chemotherapies (National Prostatic Cancer Project Treatment Group). The data revealed an astonishing prognostic value for patients' survival. Regardless of treatment regimen, the survival was found to be inversely and proportionally related to the pretreatment PA levels. An excellent correlation between PA level and clinical status was also noted: PA levels increased as disease progressed, decreased as disease regressed, and remained fluctuating in patients with stable disease.

These two initial and classic clinically significant reports, along with the original report on its restricted prostate tissue specificity, provided the basis for eventual use of PA as the most promising prostate tumor marker available to date [5,12,13]. Since 1980, improved immunoassays, especially with the use of monoclonal antibodies, have become commercially available. The assay kit manufactured by Hybritech Corp. has been approved by the Food an Drug Administration. Many others are awaiting approval.

C. Immunoassay Systems

The assay system developed by Hybritech Corp. is basically a two-site immunoradiometric procedure [14]. Two monoclonal antibodies, each directed against a distinct epitope on PSA molecule, were used in the kit. The first antibody immobilized on plastic beads serves as a "captor" to bind circulating PSA and separate it from other serum proteins and components; the second antibody, labeled with ^{125}I, serves as the "probe" to quantify the amount of the antigen.

The methodology of this two-site immunoradiometric assay has been most extensively evaluated (for review, see 15). In a recent report the detection sensitivity was shown to be 0.1 ng/ml, with assay coefficient variations ranging from 1.3 to 4.9%. A level of 0-2.8 ng/ml was established as a "reference interval" for healthy men [16]. Serum PSA levels greater than the reference interval were found in 50% of patients with stage A prostate cancer, 80% of those with stage B, and 100% for patients with both stages C and D. Follow-up PSA levels for most of the patients after prostatectomy were within 0-0.2 ng/ml. Most patients exhibiting abnormal PSA values were shown to have clinically detectable and residual disease.

In addition to Hybritech's assay, there are at least two other systems commonly available. One is the conventional competitive-inhibition-type polyclonal double-antibody radioimmunoassay, which is also commercially available (e.g., Diagnostic Products Corp., Yang Laboratories) [17,18]. Another is a monoclonal-antibody-based, two-site immunoenzymometric assay, such as that developed by Cetus, but this is not yet available [19]. This assay basically is similar to Hybritech's immunoradiometric assay, however, instead of labeling ^{125}I to the second monoclonal antibody, horseradish peroxidase is labeled and used as the quantifying marker. Evaluation of monoclonal antibody- and polyclonal antibody-based procedures is available, especially Hybritech's monoclonal immunoradiometric assay and Yang's polyclonal based radioimmunoassay. Comparative data for these two assays are available [16]. In a most recent report comparing PSA-Hybritech and PSA-Yang assays, both were reported to detect a similar percentage of abnormal values in patients with cancer of the prostate. However, controls showed a higher percentage abnormal (primarily BPH patients) by PSA-Yang assay (3% Hybritech vs. 17% Yang). Interassay and intra-assay variabilities were also higher with PSA-Yang, as reported previously by others, as was cost per patient. Based upon ease of assay performance, reliability, level of false-positive results, and cost, PSA-Hybritech was reported to be the preferred method [20].

D. Stability of Circulating Prostate-Specific Antigen

Stability of serum PSA as measured by the monoclonal antibody-based immunoradiometric assay of Hybritech has been investigated [21]. As expected, serum PSA is more stable at −20°C or −80°C than at 4°C. Serum specimens containing various levels of PSA (3.7-321.9 ng/ml) were stored at 4°C, −20°C, and −80°C for different periods of time. Aliquots were measured for PSA levels. Results indicated that no significant difference in PSA level was found between initial specimens and last aliquots stored for 7 days at 4°C, or between those stored at −20°C and at −80°C for up to 9 months. Thawing/freezing (at −20°C and −80°C) for eight times over a 6 month period did not affect the PSA level. Acidification of specimens also has no effect on the serum PSA levels. Therefore, with or without acidification, serum PSA is relatively stable and can be reliably quantified within 7 days if stored at 4°C or after long-term storage at −80°C [21]. Serum PSA also is stable for 48 hr at room temperature [22].

E. Serodiagnosis and Staging of Prostate Cancer

Since PSA levels are elevated in some patients with BPH, to eliminate this diagnostic difficulty some investigators have used a higher cutoff point. When an operational limit of upper normal, set at 10 ng/ml, was used, serum PSA levels were found to be elevated in 43% of 91 untreated patients with localized prostate cancer, 92% of 60 untreated patients with advanced disease, and none of 10 patients with BPH. Thus, the use of such an "operational" limit may be an approach to resolving the "false" PSA positive in BPH patients. Of course, ultimate diagnosis should be confirmed with biopsy. Elevation of PSA levels in BPH patients usually was slight and generally would not interfere with clinical interpretation [23]. From a large number of reports available, it is evident that an elevated serum PSA level is found in a great number of patients with an early stage of prostate cancer (10-50%) and in a significant majority of patients with metastatic disease (>90%).

Serum PSA levels appear to be associated with stage of prostate cancer. In a recent study involving 60 patients with prostatic carcinoma, 24 had localized disease (M0) and 36 had metastatic spread (M1) as judged by bone scan. An elevated PSA level (>10 ng/ml) was found in 16 (67%) of those with M0 disease and 34 (95%) of those with M1 disease [24]. However, data from this study revealed that PSA levels reflected neither the histological grade nor the local stage of the tumor, and were of little value in estimating tumor burden. In the study mentioned above involving 74 patients undergoing radical prostatectomy, PSA levels were elevated (>10 ng/ml) in 59% (26/44) with estracapsular disease and in only 7% (2/30) without extracapsular disease, that is, in the 26 of 28 patients (93%) with extracapsular disease [23]. Overall, an elevated PSA level was found in stage B_2 (2/6), C_1 (2/13), C_2 (4/6), C_3 (8/8), and D_2 (12/17). None of the 10 patients with A_1, 4 with A_2, or 10 with B_1 level tumors was found to have a PSA level >10 ng/ml.

F. Monitoring Prostate Cancer

At the present stage of development, measurement of serum PSA is of most clinical value in monitoring treatment response and predicting disease recurrence of patients with cancer as initially reported [13]. In a study of 152 patients, serial serum measurements showed that PSA either reflected or predicted clinical status in more than 97% of the patients. PSA was shown to be of significant prognostic value before and during endocrine therapy of 49 patients with stage D_2 disease, and to be of predictive value in clinical disease in 68 patients with progression and 35 patients with no progression [23].

If one uses 4 ng/ml instead of 10 ng/ml, the significance of PSA levels following radical prostatectomy is also apparent [25]. Preoperative PSA values were correlated with clinical stage and tumor volume in clinical stage A or B patients along with 28 patients with inoperable clinical stage C or D_2, but it could not accurately predict capsular penetration. Postoperative PSA levels were correlated with pathological stage of disease, and frequently associated with tumor recurrence.

A "reference range" of serum PSA levels obtained from 473 normal men in a recent study was reported to be up to 3.9 ng/ml (97.5 percentile), with a mean of 1.02. Age appears to be a factor, since an upper limit of 1.8 was found for 26 younger men (<40 years old) and 5.7 for 208 older men (>40 years old). In this study, a total of 174 serum samples from 80 patients with stage D metastatic prostate carcinoma over a 9 month period were analyzed for the association between disease activity and frequently of PSA elevation. PSA levels consistently correlated with disease status or predicted recurrence before clinical disease was evident [16].

The initial report that serum PSA level can be used to detect residual disease also has been confirmed by many investigators [26,27]. In a recent report PSA was used in the preoperative and postoperative evaluation of localized prostatic cancer treated with radical prostatectomy. Preoperatively, PSA levels were found to correlate directly with capsular penetration, lymph node involvement, and seminal vesicle involvement, although the diagnostic accuracy of an elevated PSA level on an individual basis was only 55% for capsular penetration and 50% for seminal vesicle involvement and lymph node involvement. Postoperatively, 127 patients were followed up for 2 months to 8.6 years, with a mean follow-up of 2 years. Of the 101 paitients who had favorable pathological findings at operation, 91% (15/20 patients with capsular penetration and 77/81 organ-confined cancer) had a follow-up PSA level in the range of 0.0-0.2, whereas only 19% (5/26) with either seminal vesicle involvement or lymph node involvement had a PSA level <0.2 ng/ml. All patients with documented clinical recurrence had an elevated follow-up serum PSA level. Also, the half-life of circulating PSA was calculated to be 3.15 days. PSA can therefore be used to detect residual cancer during the postoperative period.

In another recent study that also involved a large group of controls and patients, the normal range (mean ±2 SD) was found to be 0-2.5 ng/ml from 157 subjects [28]. Age variation in PSA level was not detected in this report using polyclonal radioimmunoassay as that in previous reports using monoclonal radioimmunometric assay [16,24]. For 127 patients with untreated prostate cancer, an elevated PSA level was found in 122 patients: 7/12 of clinical stage A (after transurethral resection), and 115/115 in stages B, C, and D. An unusual number of patients with BPH (63/73) also exhibited elevated levels. PSA level increased with advancing clinical stage. Through a series of multiple regression analysis, PSA was shown to be associated with estimated tumor volume in 45 patients undergoing radical prostatectomy. After the surgery, PSA routinely fell to undetectable levels, with a half-life of 2.2 days. In 6 patients evaluated postoperatively with serial PSA assays, PSA appeared to be useful in detecting residual and early recurrence of tumor and in monitoring response to radiation therapy [28].

G. Circulating Prostate-Antigen-Binding Globulin

Since PA is a prostate-organ- and epithelial cell-type-specific product, the formation of autoantibodies by the host would be a natural phenomenon according to classic immunology. Using a reverse immunoassay modified from the ELISA for serum PA initially described [12], circulating PA-binding globulin levels were measured in 167 serum specimens, including those of 18 normal men, 14 normal women, 10 male patients each with cancer of the lung, pancreas, and colon-rectum, 25 BPH patients, and 80 patients with various clinical stages of prostate cancer [29]. More than 10 μg/ml of PA-binding globulin was detected in at least 5-10% of the first four groups, even in the sera of normal women, and most profoundly in patients with stage D metastatic prostate carcinoma, in whom markedly elevated PA-binding globulin levels (>50 μg/ml) were detected in more than 50% of patients.

The data from the two initial clinical reports suggest that the presence of this circulating PA-binding globulin does not affect an accurate measurement of serum PA level [12,13]. However, one recent letter from Yang Laboratories has reported, based on very crude data from one single experiment, that the zero diluent in two PSA-Hybritech kits possibly containing a trace of prostate-antigen-binding globulin would interfere with Yang-polyclonal radioimmunoassay kit for PSA [30]. In response, Hybritech has presented data in detail showing otherwise, and indicating additionally that in one PSA-Yang kit tested, two extra ^{125}I-labeled tracers (a major one at 25 KDa and a minor one at 32 KDa), in addition to PSA were detected, and this may be the cause of nonspecific binding observed in the PSA-Yang kit [31].

Despite the possible analytical interference of circulating PA-binding globulin in one radioimmunoassay (RIA) PSA system [32], the potential biological significance of PA-binding globulin should be emphasized. PA-binding globulin isolated from sera of stage D patients was shown to be IgG, in both free form and PA-complexed form [29]. One working hypothesis derived from this finding is that cancer of the prostate is an autoimmune disease (as far as PA/PA binding globulin is concerned), and it raised the possibility of in vitro generation of human monoclonal antibodies directed against PA. Indeed, by fusing a lymphocyte preparation from lymph nodes taken from a patient with early stage of prostate cancer with human lymphoblastoid cells, monoclonal antibodies of human origin reactive with PA have been successfully prepared [33]. This result confirms the hypothesis that PA is an autoantigen of human prostate and that the lymphocytes in patients with prostate cancer are sensitized in vivo.

H. Biological Function

Little information on the biological function of PA is available and its important is yet to be fully appreciated. Purified PA has been shown to exhibit a mild protease activity [34]. This mild enzyme activity of PA examined with various natural and synthetic protease substrates and inhibitors has been characterized as distinct from that of any other known protease.

When the complete amino acid sequence of PA was used and compared structurally with those of the known enzymes, a high degree of sequence homology was observed with serine proteases of the kallikrein family [8]. "Chymotrypsin-like" and "trypsin-like" activity of PA have been detected. The finding that protease activity of PA could be inhibited by Zn^{++} and spermidine, among others, is most intriguing [8,34]. It is well known that normal prostate contains a relatively high level of Zn, and a reduced Zn level has been reported in malignant prostate. The reduced Zn level may lead to an increased proteolytic activity of PA in prostate tumor. Proteases are also associated with invasive behavior of tumor cells. The exact biological rule, if any, of PA in prostate adenocarcinoma will be of fundamental and clinical interest. Also, spermidine is found in abundant amounts in prostatic fluid, and plays a central role in the growth regulation of cells. A recent report has associated the proteolytic activity of PA with liquefaction of seminal clots and with the structural protein of human seminal coagulum (the primary secretory protein from seminal vesicle) as a possible in vivo physiological substrate [35]. These observations all suggest a potential important role that would have to be played by PA in biology and/or function of the prostate.

One aspect of PA's protease activity may be of interest in a PA assay system, that is, the autohydrolysis of PA, as in autolysis in chymotrypsin. Analysis of the amino-terminal sequence of purified PA has defined the possible localization of three endoproteolytic cleavages: lysine-148, lysine-185, and arginine-85 [8]. Similarly to those of other serine proteases, these cleavages may be autocatalytic. To avoid or minimize this potential autohydrolysis, the addition of protease inhibitors to PA assay reagents or PA specimens should be considered whenever necessary.

I. Immunohistopathology

In addition to being used as an effective serum marker for prostate cancer, PA is a most reliable marker for the immunohistopathological examination of tumors involving the prostate, and for identification of the prostate origin of metastases in various organs and sites. Due to its unique and stable antigenic determinants on the molecule, the expression of PA was preserved in conventionally prepared formalin-fixed, paraffin-embedded surgical and autopsy tissue specimens. The characteristic of PA expression specific for the prostate epithelial cells and in related components makes immunohistochemical staining a simple and effective tool for identifying the distribution of PA in various cells and tissues commonly examined in pathology laboratories.

The technique of immunohistochemistry have been described in detail in other chapters, and will not be repeated here. It should be emphasized once more that the specificity of the primary antibody reagent is the most critical factor in immunohistopathological evaluation. With its defined and restricted expression on prostatic epithelial cells, PA has been found to be the most useful histological marker in the recent history of the immunohistopathology of prostate neoplasms.

The first of such reports was published in 1981, and used a selected rabbit polyclonal antiserum reagent and a large panel of human tissue specimens [36]. In formalin-fixed paraffin-embedded tissue sections, 19 primary prostate carcinomas and 49 metastatic prostate carcinomas all showed positive staining for PA. The intensity of staining reaction was found to vary from cell to cell, from area to area, and from primary to metastatic tumor in the same patient. The positive immunoenzyme stain was confined to cytoplasm of epithelial, with greater concentration of brown granules in the paranuclear area of the luminal aspects of the cells. A most important finding was that none of the 78 nonprostate neoplasms was stained positively for PA. Of particular clinical interest was that 17 tumors of the urinary bladder with extensions to the prostate gland all stained negatively.

It should be noted that normal and hyperplastic glands of the prostate do express PA. Positive PA stains were found in the epithelial lining of acini and ducts as well as in secretion and concretions. However, the epithelia of periurethral glands, seminal vesicles, vasa deferentia, transitional epithelia of the urinary bladder and prostatic urethra, the glandular lining of von Brunn's nest, and areas of cystitis glandularis were negatively stained for PA. It was also interesting to observe a transition between the positive staining of prostatid ducts and the negative staining of the metaplastic squamous epithelium in prostatic infarcts. All other normal tissues, such as kidney, testis, stomach, liver, colon, rectum, pancreas, lung, breast, and salivary gland were all negative [36].

This report, with an extensive series of normal and pathological tissue specimens, revealed the universal expression of PA in primary and metastatic prostate carcinomas and the consistently negative findings in nonprostatic tumors or other normal tissues. This report also established the initial clinical application of PA in immunohistopathological examinations, providing a simple tool for the differential diagnosis of metastic adenocarcinoma in male subjects (e.g., tumor cells in bone marrow, lymph nodes, and distant organs). It also serves as a tissue marker for the histological classification of tumor involving prostate gland and adjacent tissues (e.g., bladder and rectum) and resolves one of the most common diagnostic problems in pathology.

J. Differential Diagnosis of Bladder from Prostate Carcinomas

This clinical application was best exemplified in a study in which 15 urinary bladder adenocarcinomas (11 in men, 4 in women) and 9 bladder tumors (5 in men, 4 in women) with mixed glandular and transitional features were examined with commercially available polyclonal reagents [37]. None of the tumors seen in either the male or female groups was positively stained for PA, although 3 of the 11 adenocarcinomas and 1 of the 5 tumors with mixed glandular and transitional features in men were positive for PAP, and 2 from each group in women also were positive for PAP. Rare isolated cells in two adenocarcinomas were reported to be slightly stained for PA.

A prospective clinical study involved tissue diagnosis in 21 patients in whom the tumors of the head and neck of the bladder could not be categorized as prostate or urothelial in origin by clinical and endoscopic assessment or by conventional histopathological examination [38]. In 8 of the 21 patients with lesions of the bladder neck, 4 had a past history of prostate tumor, 2 with urothelial tumors, and the other 2 with both. PSA was positive in three of the four patients with a previous prostate neoplasm and in none of the two with a previous history of transitional cell carcinomas. In the two patients who had had both prostatic and transitional cell carcinomas in the past, the bladder neck lesion was PSA-positive in one and negative in the other. All patients with PSA-positive lesions were treated as having prostate carcinomas and all negative ones as transitional cell carcinomas of the bladder. The tumors of the bladder neck occurred de novo in the remaining 13 patients, of which 8 were PSA-positive, indicating a prostate origin, and 7 were treated for prostate carcinoma. In the remaining patient, PSA stain was focal and only weakly positive. He underwent radiotherapy and radical cystoprostatectomy, since the endoscopic appearances and histological results, showing an anaplastic carcinoma, were considered more indicative of urethelial than a prostate tumor. Subsequent examination of the surgical specimens confirmed this to be a prostate tumor. In five patients the bladder neck tumor was PSA-negative. Four were treated appropriately for uroethlial tumor and one, with history of rectal cancer, was treated for recurrence of rectal carcinoma. In two additional patients, the prostate origin of lymph node metastases with atypical tumor morphology (cervical nodes in one and retroperitoneal nodes in the other) was detected by PSA-positive stain in the absence of a clinically apparent primary. The primary carcinomas were subsequently confirmed on prostate biopsies [38].

K. Identification of Metastatic Prostate Carcinoma

The application of PA to identification of metastatic origin of the prostate was further established with the use of a murine anti-PA monoclonal antibody F5 [39]. The expression of PA was examined in a panel of metastatic tumors, including 25 of prostatic origin and 73 of nonprostate carcinoma origin. Regardless of the site of dissemination or the malignancy grade, all secondary metastases originating from the prostate were positively stained. In contrast, nonprostatic metastases were all negative, including those primary tumors of other genitourinary tracts. Again, such promising clinical results were due primarily to the fine specificity of the F5 monoclonal antibody to the prostate, which had been characterized previously [40].

As commercial reagents become readily available, immunohistochemical staining of PSA has been more commonly practiced in pathology laboratories than in the past. In general, most of the recently published results have confirmed the original findings reported in 1981 [36]. However, several reports on rare and difficult types of prostate neoplasms are of interest. Positive PSA staining has been observed in endometrioid prostate adenocarcinoma and mucinous prostate adenocarcinoma; adenoid cystic carcinoma of the prostate showed negative findings [41-43]. Heterogeneity of small-cell carcinoma of the prostate also can be better appreciated than from morphological examination alone, since adenocarcinoma stained positively for PSA while the small-cell component stained negatively [44].

II. PROSTATIC ACID PHOSPHATASE

The measurement of acid phosphatase has been used for over five decades as an aid in the diagnosis of metastatic prostatic cancer [1]. The uses and limitations of this prostate tumor marker are well known. Several reviews are available for reference [45-48]. Only recent developments in its use will be discussed here. These include the comparison of PAP with PA in serodiagnosis of prostate cancer, in vivo radioimmunodetection of prostate cancer, and biological function of PAP.

A. Prostatic Acid Phosphatase Assay vs. Prostate-Specific Antigen Assay

At the present time, with data available from the extensive clinical evaluation of serum PA assay, it is generally concluded that PA is more sensitive and reliable than PAP as a prostate carcinoma marker. Abundant reports are available documenting the comparative values of PAP vs. PA in prostate cancer. Only a few most recent references are discussed. In a report dealing with comparison of monitoring and staging values between the two markers in patients with regionally confined prostate cancer with extracapsular disease, 93% were shown to have an elevated PSA level (>10 ng/ml), whereas elevated PAP values (>4 ng/ml) were detected in only 59% (using 97th percentile among BPH patients for both assays). In 86 patients with disseminated disease, 98% were documented with an elevated PSA level while 78% showed elevated PAP levels. Moreover, in 76% of patients who had elevated PSA and PAP levels, the PSA level was more elevated in 94%. Thus, PSA levels were almost always elevated in clinically active metastatic carcinomas of the prostate and were more clearly elevated than PAP in those patients who also had an elevated PAP level [23].

In another report of 127 patients with newly diagnosed prostate cancer, serum PSA level was elevated (>2.5 ng/ml) in 96% including 7 of 12 with stage A and all 115 with stage B-D [28]. In contrast, serum PAP levels were elevated (>2.1 ng/ml) only in 45% of the patients: none in 12 with stage A, 9% in stage B_1, 39% in stage B_2, 40% in stage B_3, 64% in stage C, and 96% in stage D_2. In a prospective comparative study of 80 patients with metastatic stage D cancer, while the overall frequency of abnormal PSA levels was only 76%, the PAP elevation occurred in even fewer patients (60%) [26].

Serial measurements for both markers in 69 patients receiving hormonal therapy were also available. Levels of PSA and PAP reflected the clinical progress in 70% (48/68), PSA alone in 27%, and PAP alone in 1.5%. In 33 patients for whom serial sera were available at least 6 months prior to clinical progression, elevated PSA and PAP were found in 30%, PA alone in 64%, and PAP alone 0% [23].

B. Radioimmunodetection of Prostate Cancer

No report is yet available for PA in in vivo radioimmunodetection of prostate cancer. However, initial evaluation of the use of anti-PAP antibodies has been reported both in animal and humans. The feasibility of using anti-PAP monoclonal antibody in the radioimmunodetection of human prostate carcinoma xenograft has been reported [49]. In a human prostate tumor (LNCaP)-nude mouse model and using ^{125}I-labeled anti-PAP monoclonal antibodies, the tumor/blood localization ration, 7 days after administration of antibodies, was six times higher in the anti-PAP antibody group than in the control groups.

Two previous reports involving a limited number of patients have shown that radioiodine-labeled polyclonal anti-PAP antiserum can be applied to the radioimmunodetection of metastatic prostate cancer [50,51]. A more systematic study has become available most recently, which evaluated 25 patients with stage D_2 disease using ^{111}In-labeled monoclonal antibody (PAY 276) [52]. The patients were evenly classified into five groups. Patients received an infusion of 5, 10, 20, 40, and 80 mg antibody, respectively, in each group, 1 mg of which was labeled with 5 mCi ^{111}In. Imaging was performed at 24, 72, and occasionally 120 hr after infusion. No adverse reactions (skin rash, bronchospasm, tachycardia, febrile reaction, or hypotension) were noted in the patients. No significant changes in serial hemoglobin levels, platelet count, blood chemistry profile, or urinalysis were noted. As with other clinical trials with murine monoclonal antibody, some patients did develop human antimouse antibodies (HAMA; 8 of 16 tested, 6 of whom received 40 mg or more PAY 276).

When compared with conventional radiographs and bone scintigraphy, which were considered in the study to be the definitive indicators of metastatic disease, the monoclonal antibody scan visualized at least 1 lesion in 19 of 25 patients: 4 in groups 1 and 2, and all 15 in groups 3, 4, and 5. Antibody scans detected 7 of 32 metastases in group 3, 31 of 58 in group 4, and 101 of 135 metastases in group 5. An increase in the detection of metastatic lesions was apparent as a greater concentration of unlabeled monoclonal antibody was administered. This increase in detection rate can be attributed to the decrease in the liver-to-heart distribution ratio of ^{111}In-labeled antibody. This approach to the increase in radioimmunodetection of tumor with the use of "cold" unlabeled monoclonal antibody has been reported with other monoclonal antibodies and tumor systems [53]. Nevertheless, this initial trial with 25 patients with stage D_2 disease does provide encouraging results for future investigations.

C. Biological Function

Although PAP has been known for a long time, its basic characteristics are still relatively unknown. Data available indicate that PAP is a glycoprotein of 100 kD, perhaps in a dimeric form, consisting of 87% peptide and 13% carbohydrate in various isomers (pI 4.8-53.) [54]. Only a partial amino acid sequence with lysine as the N-terminal (arginine in one report) has been determined [55,56]. The PAP molecule contains several antigenic epitopes, all of which are shared in various degrees (from 5% to more than 30%) with other organs [57]. Therefore, it is not as specific for prostate as PA.

Extensive investigations have recently been conducted to explore the biological function of PAP. Preliminary results, as yet few, are of potential interest. Available evidence suggests that, as for the growth rate of the prostate itself, PAP is regulated by androgens. Its physiological function has been suggested recently to be that of phosphotyrosyl-protein phosphatase, since PAP has been shown to exhibit the activity of this enzyme as well, that is, it hydrolyzes or removes the phosphate group linked to tyrosine residue from phosphotyrosine-containing proteins [58,59]. Since tyrosine phosphorylation of proteins is generally believed to play a role in the control of cell proliferation, PAP is thus speculated to act in a reverse role of phosphotyrosine kinase. In addition, epidermal growth factor receptor has been mentioned as the major substrate for tyrosine kinase activity in prostate carcinoma cells [60]. These initial observations and speculations have generated renewed interest in the biological investigation of PAP. We hope that gene cloning and other molecular biological tools may provide a clearer insight into the role of PAP in the physiology of the prostate.

III. MULTIPLE MARKERS IN IMMUNODIAGNOSIS OF PROSTATE CANCER

A. Prostate-Specific Antigen and Prostatic Acid Phosphatase

Based on the working hypothesis that PA and PAP are two distinct biochemical and immunological secretory products, a combination of these two markers may provide increased specificity and/or sensitivity in the serodiagnosis of prostate carcinoma. One approach would be to adjust the upper normal range in both assays to eliminate the BPH specimens as much as possible. An initial evaluation reported a few years ago did produce such an additive clinical value when both markers were assayed simultaneously [61]. In this study, serum specimens from 22 normal healthy men, 30 patients with advanced nonprostate adenocarcinomas, 29 patients with BPH, and 192 patients with prostate adenocarcinomas (including 12 with clinical stage A, 36 B, 28 C, and 116 stage D), were assayed by polyclonal antiserum-based ELISA for serum PA and PAP levels.

When upper cutoff values of 7.5 ng/ml for PA and 15.5 ng/ml for PAP (both were 95th percentile of BPH values) were used, the combination test resulted in a positive detection rate of 58% for prostate cancer stages A and B, 68% for stage C, and 92% for stage D, and only 10% for BPH. None of 52 healthy controls and other cancers registered a positive response. In light of the lack of prostate-tumor-specific seromarkers, the use of these two common markers in a simultaneous test may be an effective approach to the immunodiagnosis of prostate cancer. However, recent evidence discussed previously, as obtained from monoclonal-antibody-based reagents, would suggest that PSA assay alone may be as good as PSA and PAP combined.

B. PA, PAP, and Phosphatase Activities

Comparative data on the prognostic value of PA, PAP, and other commonly used enzyme markers associated with the prostate carcinoma have also been reported. Instead of analyzing a limited number of patients and comparing on an individual basis, a double-blind study was performed systematically on a large number of patients and specimens with well-documented clinical information (National Prostatic Cancer Project) [62]. A total of 1,065 serial serum specimens from 79 patients with prostate cancer stages B_2-D_1 (regionally confined disease) and 51 patients with newly diagnosed stage D_2 collected during an 8-year period were analyzed statistically. Forty of the 79 patients with B_2-D_1 disease and 21 of the 51 with D_2 disease presented a clinical progression of disease during follow-up period. In addition to PA and PAP, total acid phosphatase activity, total alkaline phosphatase, and bone isoenzyme of alkaline phosphatase were evaluated for their relative reliability in predicting clinical progression of disease by a series of statistical analyses. For patients with regionally confined disease, only PA and PAP were found to be prognostically important markers, with PA being more significantly reliable than PAP (p values 0.0052 vs. 0.0359). For metastatic prostate cancer, PA ($p < 0.0001$), bone alkaline phosphatase isoenzyme ($p = 0.0007$), and PAP ($p = 0.0206$) were the most reliable markers. Furthermore, after adjustment for the effect of PA, multivariate analyses revealed that no other marker was significantly related to the risk of disease progression. Elevated PA levels were predictive of increased risk even 6 months before disease progression in patients with B_2-D_1 prostate cancer. For all patients, the apparent overall order of prognostic reliability for disease progression was found to be PA > PAP > bone alkaline phosphatase isoenzyme > acid phosphatase > total alkaline phosphatase [62]. This information should be of value as a guide in selecting assays for monitoring disease progression in patients with prostate cancer.

IV. POTENTIAL IMMUNODIAGNOSTIC MARKERS

Several antigenic markers mainly recognized by monoclonal antibodies related to prostate cancer have been reported in recent years. None has reached the stage at which they can be considered for clinical use. Some of these potential markers are briefly described here.

A. KR-P8: Prostate-Specific Marker

KR-P8 is a monoclonal antibody originally prepared against the human prostate cell line PC-3 [63]. By immunostaining, KR-P8 reacted with the glandular epithelium of all specimens of normal, benign hypertrophic, and malignant prostates tested. No reactivity was detected with numerous other human tissues, normal and malignant, including bladder, kidney, and testis, which indicated that KR-P8 recognizes an antigen apparently specific for the prostate. The antigen recognized by the antibody and present in PC-3 cells and in seminal plasma is a glycoprotein with the molecular weight range of 48-75

Kd [64]. The epitope appears to be associated with carbohydrate units, since periodate oxidation resulted in the loss of antigenicity. The potential for using KR-P8 as a marker lies in its secretory characteristics, apparently from the prostate to seminal plasma, and from PC-3 cells to its culture medium. Neither the assay procedure nor its possible presence in patients sera has yet been evaluated. KR-P8 antigen is different from PA biochemically and immunologically; but like PA, it is a secretory product of the prostate, localized in the prostatic epithelia, and apparently prostate specific.

B. 7E11-C5: Prostate-Specfic Marker

A recently reported monoclonal antibody 7E11-C5, generated against the human prostate cell line LNCaP, also exhibited a restricted specificity only to the prostate epithelia [65]. In a panel of 175 human specimens, 7E11-C5 stained only with all 11 specimens of prostate carcinoma, 7 normal, and 7 benign hypertrophic prostate glands. None of the 26 nonprostate tumors nor 120 of the 122 other normal tissue reacted with 7E11-C5. Two of the 14 normal kidney specimens examined were shown to be stained positively. It was interesting that none of 32 cell lines of human normal or neoplastic cells tested was reactive with the antibody. In fact, LNCaP is the only cell line examined to date to react with 7E11-C5. The nature of the antigen remains unknown. Competitive binding inhibition assay revealed the presence of serum antigen in 20 of 43 patients, mostly with advanced prostatic cancer; and in 3 of 66 sera from patients with nonprostatic malignancies. None of the sera from 30 normal blood donors nor 7 with BPH sera was positive. These serum assay results remain to be confirmed by additional investigation with a more sensitive and quantitative assay. It should be noted that PA and KR-P8 are cytoplasmic proteins, while 7E11-C5 antigen is a plasma membrane protein.

C. Turp-27 and PD-41: Prostate-Specific and Prostate Tumor-Specific Markers

Like PA, KR-P8, and 7E11-C5, Turp-27 is a prostate-specific antigen, present in BPH and adenocarcinoma of the prostate and, in reduced concentration, in normal prostate. The monoclonal antibody Turp-27 was generated against a membrane preparation derived from a pool of transurethral resection specimens, which included three BPH and one adenocarcinoma [66]. Therefore, Turp-27 also cannot differentiate BPH from carcinoma. Using a membrane extract prepared identically from a poorly differentiated carcinoma of the prostate, a monoclonal antibody PD-41 was reported most recently [67]. Reactivity of PD-41 appeared to be restricted to prostate carcinoma. PD-41 was not reactive with PA, PAP, normal prostate, BPH, or a variety of normal and nonprostate tumor tissue. Positive immunostain of PD-41 was detected in 83% of poorly differentiated tumor and in only 33% of well-differentiated tumors; thus, PD-41 may be a grade-related and prostate-tumor-specific marker. If the nature of tumor specificity can be confirmed, PD-41 indeed will be one of the most significant prostate tumor markers to become available in recent years, and its clinical potential could be substantial.

D. 83.21 and Other Prostate Tumor-Associated Markers

One of the most extensively characterized monoclonal antibodies associated with prostate cancer is 83.21. This antibody was generated against the human prostate tumor line DU-145 [68]. Initially 83.21 was shown to react with human prostate and bladder tumor cell lines, and not with 38 other normal or malignant human cell lines. On human prostate tissue sections, reactivity was found in 7 of 11 undifferentiated prostate carcinoma, 4 of 5 poorly differentiated, 1 of 3 moderately differentiated, and none of 6 well-differentiated prostate adenocarcinomas. No reactivity was found with normal prostate or BPH specimens. Reactivity was also shown with one of four bladder tumor specimens, and no stain with two benign bladder tumors or three normal bladder specimens. Of 15 other adenocarcinomas and 30 normal tissues examined, positive stain was found in only 2 normal kidneys at the area of proximal convoluted tubules. A particular feature of this antibody is its reactivity with cytomegalovirus-transformed human embryonic lung cell line. The 83.21 antibody initially was reported to direct against a surface glycoprotein of 180 kD, but recent data revealed that two glycopeptides of 60 kD and 28 kD were isolated from both human prostate tumor and cytomegalovirus-in transformed cells by 83.21 antibody affinity chromatography [69]. 83.21 could be of use in the immunohistopathological characterization of urogenital tumors and in the study of a possible link between cytomegalovirus and transformation of prostate malignancy.

Recent reports have indicated the potential application of anti-Leu-7 (natural killer cells) and anti-Leu 4 (pan T cell) in the immunohistochemical examination of prostate tumor [70,71]. Other potential monoclonal antibodies related to prostate neoplasm include P6.2, αPro-3, αPro-13, αPro-15, Pr-1, and Pr-2 (see Ref. 2 for review). Clinical evaluation of these markers in immunodiagnosis is not yet available.

V. SUMMARY

Immunodiagnosis of prostate cancer is at a more advanced stage than that of most other tumors. Two well-known markers, prostatic acid phosphatase and prostate-specific antigen, have been used in the clinical management of patients. Prostate-specific antigen is a more sensitive and reliable marker than prostatic acid phosphatase. Serum prostate-specific antigen is effective in monitoring disease status, predicting recurrence, and detecting residual disease. Prostate-specific antigen is a tool for the histological differential diagnosis of metastatic carcinomas, especially in the identification of metastatic prostate tumor cells in distant organs and in the differentiation of primary prostate carcinoma from poorly differentiated transitional cell carcinoma of the bladder. Few data on biological function are available. Prostatic acid phosphatase functions as a phosphotyrosyl-protein phosphatase and prostate-specific antigen as a protease. Physiological function in the prostate remains to be elucidated. Several of the prostate-specific and prostate-tumor-associated antigens, as well as a putative prostate tumor-specific antigen, as recognized by monoclonal antibodies are available. Clinical evaluation of these potential markers is not yet available.

ACKNOWLEDGMENTS

This chapter is dedicated to my present and former colleagues and students at Roswell Park Memorial Institute and elsewhere who contributed most diligently to our research projects on prostatic acid phosphatase and prostate-specific antigen. Grant and fellowship support from the National Cancer Institute, the American Cancer Society, and international agencies is also acknowledged.

REFERENCES

1. Gutman EB, Sproul EE, Gutman AB. Increased phosphatase activity of bone at site of osteoplastic metastases secondary to carcinoma of prostate gland. Am J Cancer 1936;28:485-488.
2. Chu TM. Monoclonal antibodies to human prostate cancer-related antigens. In: Sell R, Reisfeld R, eds. Monoclonal antibodies in cancer. Clifton NJ: Humana Press, 1985:309-324.
3. Flocks RH, Urich VC, Patel CA, Opitz JM. Studies on the antigenic properties of prostate tissue. J Uron 1960;84:134-143.
4. Ablin RJ, Bronson P, Soanes WA, Witebsky E. Tissue- and species-specific antigens of normal human prostatic tissue. J Immunol 1970;104:1329-1339.
5. Wang MC, Valenzuela LA, Murphy GP, Chu TM. Purification of a human prostate specific antigen. Invest Urol 1979;17:159-163.
6. Wang MC, Valenzuela LA, Murphy GP, Chu TM. A simplified purification procedure for human prostate antigen. Oncology 1982;39:1-5.
7. Wang MC, Loor RM, Li SL, Chu TM. Physico-chemical characterization of prostate antigen purified from human prostate gland and seminal plasma. IRCS Med Sci 1983;11:327-328.
8. Watt KWK, Lee PJ, M'Timkulu T, Chan WP, Loor R. Human prostate-specific antigen: structural and functional similarity with serine proteases. Proc Natl Acad Sci USA 1986;83:3166-3170.
9. Wang MC, Kuriyama M, Papsidero LD, Loor RM, Valenzuela LA, Murphy GP, Chu TM. Prostate antigen of human cancer patients. Methods Cancer Res 1982;19:179-197.
10. Papsidero LD, Kuriyama M, Wang MC, Horoszewicz JS, Leong SS, Valenzuel KA, Murphy GP, Chu TM. Prostate antigen: a marker for human prostate epithelial cells. N Natl Cancer Int 1981;66:37-42.
11. Papsidero LD, Wang MC, Valenzuela LA, Murphy GP, Chu TM. A prostate antigen in sera of prostatic cancer patients. Cancer Res 1980;40:2428-2432.
12. Kuriyama M, Wang MC, Papsidero LD, Killian CS, Shimano T, Valenzuela LA, Nishiura T, Murphy GP, Chu TM. Quantitation of prostate-specific antigen in serum by a sensitive enzyme immunoassay. Cancer Res 1980;40:4658-4662.
13. Kuriyama M, Wang MC, Lee CL, Papsidero LD, Killian CS, Inaji H, Slack NH, Nishiura T, Murphy GP, Chu TM. Use of human prostate-specific antigen in monitoring prostate cancer. Cancer Res 1981;41:3874-3876.
14. Myrtle JF, Schackelford W, Bartholmew RM, Wampler J. Prostate-specific antigen: quantitation in serum by immunoradiometric assay. Clin Chem 1983;29:1216.
15. Wang MC, Killian CS, Lee CL, Chu TM. Monoclonal antibody for prostate cancer. In: Kupchik H, ed. Cancer diagnosis in vitro using monoclonal antibodies. New York: Marcel Dekker, 1987;195-219.
16. Chan DW, Bruzek DJ, Osterling JE, Walsh PC. Prostate-specific antigen as a marker for prostatic cancer: a monoclonal and a polyclonal immunoassay compared. Clin Chem 1987;33:1916-1920.
17. Killian CS, Emrich LJ, Constantine RI, Kubasik N, Chu TM. Evaluation of prostate-specific antigen (PSA) for follow-up of prostate cancer (PCa) using two commercially available kits. Clin Chem 1987;33:928.
18. Liedtke RJ, Batjer JD. Measurement of prostate-specific antigen by radioimmunoassay. Clin Chem 1984;30:649-652.
19. Loor R, M'Timkulu T, Jung N, DeWitt SD, Leffland R, Soloway MS, Guinan P. Measurement of prostate tissue-specific markers: prostate-specific antigen and prostatic acid phosphatase. J Clin Immunoassay 1986;9:153-158.

20. Williams RD, Mardan A. Comparison of two commercially available immunoassays for prostate-specific antigen. J Urol 1988;139:192A.
21. Killian CS, Vargas FP, Chu TM. Long-term stability of prostate-specific antigen in native and acidified serum. Clin Chem 1988;34:1296.
22. Huben RP, Schnell Y, Hering F, Rutishauser G. Prostate specific antigen: experimental and clinical observations. Scand J Urol Nephrol 1987;104:33-39.
23. Ercole CJ, Lange PH, Mathiesen M, Chiou RK, Reddy PK, Vessella RL. Prostate specific antigen and prostatic acid phosphatase in the monitoring and staging of patients with prostatic cancer. J Urol 1987;138:1181-1184.
24. Ferro MA, Barnes I, Roberts JB, Smith PJ. Tumor markers in prostatic carcinomas. A comparison of prostate-specific antigen with acid phosphatase. Br J Urol 1987;60:69-73.
25. Hudson MA, Catalona WJ, Bahnson RR. Prostate specific antigen levels following radical prostatectomy. J Urol 1988;139:314A.
26. Ahmann FR, Schifman RB. Prospective comparison between serum monoclonal prostate specific antigen and acid phosphatase measurement in metastatic prostatic cancer. J Urol 1987;137:431-434.
27. Oesterling JR, Chan DW, Epstein JI, Limball AW, Bruzek DJ, Rock RC, Brendler CB, Walsh PC. Prostate specific antigen in the preoperative and postoperative evaluation of localized prostate cancer treated with radical prostatectomy. J Urol 1988;139:766-772.
28. Stamey TA, Yang N, Hay AR, McNeal JE, Freiha FS, Redwine E. Prostate-specific antigen as a serum marker for adenocarcinoma of the prostate. N Engl J Med 1987;317:909-916.
29. Chu TM, Kuriyama M, Johnson E, Papsidero LD, Killian CS, Murphy GP, Wang MC. Circulating antibodies of prostate antigen in patients with prostate cancer. Transplan Proc 1984;16:481-485.
30. Yang N. Endogenous antibody to prostate-specific antigen in women. Clin Chem 1988;34:647-648.
31. Wolfert RL, Peterson C, Hussa RO, Kabakoff DS. Endogenous antibody to prostate-specific antigen in women: in reply. Clin Chem 1988;34:648-649.
32. Chan DW. Endogenous antibody to prostate-specific antigen in women: in reply. Clin Chem 1988;34:648.
33. Janvier M, Diakun K, Chu TM. Production of human monoclonal antibodies reactive with prostate specific antigen. J Tumor Marker Oncol 1987;2:217-224.
34. Ban Y, Wang MC, Watt KW, Loor RM, Chu TM. The proteolytic activity of human prostate-specific antigen. Biochem Biophys Res Commun 1984;123:482-488.
35. Lilja H. A kallikrein-like serine protease in prostatic fluid cleaves the predominant seminal vesicle protein. J Clin Invest 1985;76:1899-1903.
36. Nadji M, Tabei SZ, Castro A, Chu TM, Wang MC, Murphy GP, Morales AR. Prostatic specific antigen: an immunohistologic marker for prostatic noeplasms. Cancer 1981;48:1229-1232.
37. Epstein JI, Kuhajda FP, Lieberman PH. Prostate specific acid phosphatase immunoreactivity in adenocarcinoma of the urinary bladder. Hum Pathol 1986;17:939-942.
38. Ford TF, Butcher D, Masters JRW, Parkinson MC. Immunocytochemical localization of prostate-specific antigen: specificity and application of clinical practice. Br J Urol 1985;57:50-55.
39. Papsidero LD, Croghan GA, Asirawatham J, Gaeta J, Abenoza P, Englander L, Valenzuela L. Immunohistochemical demonstration of prostate-specific antigen in metastases with the use of monoclonal antibody F5. Am J Pathol 1985;121:451-454.
40. Papsidero LD, Croghan GA, Wang MC, Kuriyama M, Johnson EA, Valenzuela LA, Chu TM. Monoclonal antibody (F5) to human prostate antigen. Hybridoma 1983;2:139-147.

41. Epstein JI, Woodruff JM. Adenocarcinoma of the prostate with endometroid features. Cancer 1986;57:111-119.
42. Epstein JI, Lieberman PH. Mucinous adenocarcinomas of the prostate gland. Am J Surg Pathol 1985;9:299-308.
43. Kuhajda FP. Adenoid cystic carcinoma of the prostate. Am J Clin Pathol 1984;81: 257-260.
44. Ghandur-Mnaymneh L, Satterfield S, Block NL. Small cell carcinoma of the prostate gland with inappropriate antidiuretic hormone secretion: morphological, immunohistochemical and clinical expressions. J Urol 1986;135:1263-1266.
45. Chu TM, Wang MC, Lee CL, Killian CS, Murphy GP. Prostatic acid phosphatase in human prostate cancer. In: Chu TM, ed. Biochemical markers for cancer. New York: Marcel Dekker, 1987:17-136.
46. Yam LT. Clinical significance of the human acid phosphatase. Am J Med 1974;56: 604-616.
47. Kuriyama M, Loor RM, Wang MC, Lee CL, Killian CS, Papsidero LD, Inaji H, Nishiura T, Slack NH, Murphy GP, Chu TM. Prostatic acid phosphatase and prostate-specific antigen in prostate cancer. Int Adv Surg Oncol 1982;5:29-49.
48. Ban Y, Wang MC, Chu TM. Immunologic markers and the diagnosis of prostate cancer. Urol Clin North Am 1984;11:269-276.
49. Lee CL, Kawinski E, Horoszewicz JS, Chu TM. Radioimmunodetection of human prostate adenocarcinoma xenografts using monoclonal antibodies to prostatic acid phosphatase. J Tumor Marker Oncol 1986;1:47-59.
50. Golbenberg DM, DeLand FH, Bennet SJ, Primus FJ, Nelson MD, Hanigan RC, McRoberts JW, Bruces AW, Mahan DE. Radioimmunodetection of prostatic cancer. JAMA 1983;250:630-631.
51. Vihko P, Heikkila J, Knotturi M, Walhberg L, Vihko R. Radioimaging of prostate and metastases of prostatic carcinoma with ^{99}Tc-labeled prostatic acid phosphatase specific antibodies and their Fab fragment. Ann Clin Res 1984;16:5152-5154.
52. Babain RJ, Murray JL, Lamki LM, Haynie TP, Hersh EM, Rosenblum MG, Glenn HJ, Unger MW, Carlo DJ, von Eschenbach AC. Radioimmunological imaging of metastatic prostatic cancer with ^{111}Indium-labeled monoclonal antibody PAY 276. J Urol 1987;137:439-443.
53. Eger RR, Covel DG, Carrasquillo JA, Abrams PG, Food KA, Reynolds JC, Schroff RW, Morgan AC, Larson SM, Weinstein JN. Kinetic model for the biodistribution of an In-111 labeled monoclonal antibody in humans. Cancer Res 1987;47:3328-3336.
54. Lin MF, Lee CL, Li SL, Chu TM. Purification and characterization of a new human prostatic acid phosphatase isoenzyme. Biochemistry 1983;22:1055-1062.
55. Lin MF, Lee CL, Sahrief FS, Li SL, Chu TM. Glycoprotein exhibiting immunological and enzymatic activities of human prostatic acid phosphatase. Cancer Res 1983;43: 3841-3846.
56. Choe BK, Rose NR. Prostatic acid phosphatase. Methods Cancer Res 1982;19:199-232.
57. Lee CL, Li CY, Jou YH, Murphy GP, Chu TM. Immunochemical characterization of prostatic acid phosphatase with monoclonal antibodies. Ann NY Acad Sci 1982; 390:52-61.
58. Lin MF, Clinton G. Human prostatic acid phosphatase has phosphotyrosyl protein phosphatase activity. Biochem J 1986;235:351-357.
59. Li HC, Chernoff J, Chen LB, Kirschonbaum A. A phosphotyrosyl protein phosphatase activity associated with acid phosphatase from human prostate gland. Eur J Biochem 1984;138:45-51.

60. Lin MF, Clinton GM. Human prostatic acid phosphatase and its phosphotyrosyl-protein phosphatase activity. Adv Protein Phosphatase 1987;4:199-228.
61. Kuriyama M, Wang MC, Lee CL, Killian CS, Papsidero LD, Inaji H, Loor RM, Lin MF, Nishiura T, Slack NH, Murphy GP, Chu TM. Multiple marker evaluation in prostate cancer using tissue specific antigens. J Natl Cancer Inst 1982;68:99-105.
62. Killian CS, Emrich LJ, Vargas FP, Yang N, Wang MC, Priore RL, Murphy GP, Chu TM. Relative reliability of five serially measured markers for prognosis of progression in prostate cancer. J Natl Cancer Inst 1986;76:179-185.
63. Raynor RH, Hazra TA, Moncure CW, Mohanakumar T. Characterization of a monoclonal antibody KR-P8 that detects a new prostate-specific marker. J Natl Cancer Inst 1984;73:617-625.
64. Raynor RH, Hazra TA, Moncure CW, Mohanakumar T. Biochemical nature of the prostate associated antigen identified by the monoclonal antibody KR-P8. Prostate 1986;9:21-31.
65. Horoszewicz JS, Kawinski E, Murphy GP. Monoclonal antibodies to a new antigen marker in epithelial prostatic cells and serum of prostatic cancer patients. Anticancer Res 1987;7:927-935.
66. Starling JJ, Sieg SM, Beckett ML, Wirth PR, Wahab L, Schellhammer PF, Ladaga LE, Polesksic S, Wright GL. Human prostate tissue antigens defined by murine monoclonal antibodies. Cancer Res 1986;46:367-374.
67. Wright GL, Beckett ML, Schellhammer PF, Roland W, Wirth PR, Lipford G, Polesksic L, Ladaga LE. Monoclonal antibody PD-41 identifies a marker specifically expressed on prostatic carcinoma. J Urol 1988;139:174A.
68. Starling JJ, Sieg SM, Beckett ML, Schellhammer PF, Ladaga LE, Wright Gl. Monoclonal antibody to human prostatic and bladder tumor-associated antigen. Cancer Res 1982;42:3084-3089.
69. Campbell AE, Beckett ML, Starling JJ, Sieg SM, Wright GL. Antiprostatic carcinoma monoclonal antibody (D83.21) cross-reacts with a membrane antigen expressed on cytomegalovirus-transformed human fibroblasts. Prostate 1985;6:205-215.
70. Wahab ZA, Wright GL. Monoclonal antibody (anti-Leu 7) directed against natural killer cells reacts with normal, benign and malignant prostatic tissues. Int J Cancer 1985;36:677-683.
71. Grignon DJ, Banerjee D. Cross reactivity of a monoclonal pan T-cell antibody (anti-leu 4) with prostate epithelium. J Urol 1987;137:330-332.

23
Tumor Markers of Head and Neck Carcinoma

DAVID E. EIBLING

U.S. Air Force Medical Center, Lackland Air Force Base, Texas

ROBIN L. WAGNER and JONAS T. JOHNSON

Eye and Ear Institute, Pittsburgh, Pennsylvania

The development of a reliable circulating tumor marker for the presence of carcinoma of the head and neck would represent a major advance in the evaluation of patients with these malignancies. Such a marker could enhance the diagnosis of occult malignancy, be used in follow-up after therapy, be adapted for therapeutic interventions, or ultimately be used for the screening of susceptible populations.

An ideal tumor marker (Fig. 1) would be found within tumor tissue but not normal tissues and released into the circulating blood, where it would be present in concentrations proportional to the amount of tumor present. Serum concentrations of the marker would correlate with tumor burden as well as clinical control or progression of disease. Tumor marker levels would return to normal with successful eradication of disease and reappear prior to the clinical manifestations of recurrence. This ideal marker would remain stable both in vivo and in vitro for a sufficient period of time to allow distribution and sampling. Finally, an ideal marker would be detectable in low concentrations by a reliable and sensitive assay of reasonable cost. Useful clinical applications of such a hypothetical tumor marker are readily apparent in patients with carcinoma of the head and neck.

I. DIAGNOSTIC AID

The vast majority of head and neck malignancies develop within the mucosa of the aerodigestive tract and are visible by direct examination. However, due to the "hidden" nature of much of this mucosa, early detection is contingent upon the early development of symptoms for the fortuitous identification of a mucosal abnormality during a routine examination. Early mucosal lesions tend to be nearly symptom free with only a mild sore throat, otalgia, or a sensation of a foreign body or "mucus." Detection of malignancy is frequently delayed until advanced-stage disease is heralded by pain, deformity, weight loss, or regional metastatic disease. Among the exceptions are glottic lesions, which may present with hoarseness when the tumor mass is quite small. As a consequence, 5 year disease-free survival rates for head and neck carcinoma (with the excep-

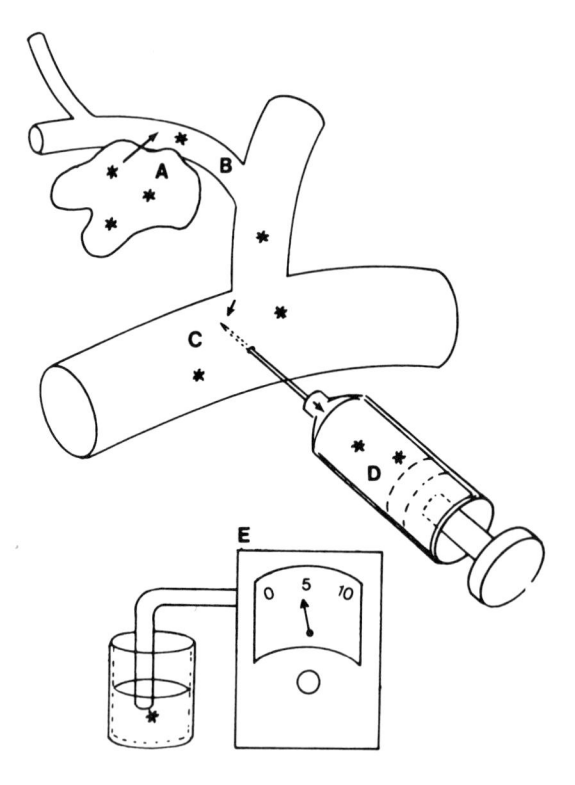

Figure 1 Diagrammatic representation of an ideal tumor markers: A. Found within tumor tissue but not normal tissue. B. Released into the circulating blood. C. Circulates and remains stable in vivo. D. Remains stable in vitro for a sufficient amount of time to allow for sampling. E. Detectable in low concentrations to allow for measurement by a sensitive and reliable assay at reasonable cost.

tion of glottic carcinoma) average less than 55% [1]. The development of a reliable and sensitive tumor marker may increase the number of patients in whom early stage disease is detected at initial presentation and, thereby, potentially improve survival rates.

II. EVALUATION OF THE ASYMPTOMATIC NECK MASS

Many patients with malignancies of the upper aerodigestive tract present with an asymptomatic neck mass. Proper evaluation and subsequent therapy require an extensive search for the primary mucosal lesion by a combination of physical examination, flexible endoscopy, and direct examination under general anesthesia. Such an exhaustive search will reveal a primary mucosal lesion in all but a small percentage of patients. Clinically useful tumor markers may be developed that could be used in the evaluation of these patients.

III. SELECTION OF PATIENTS FOR ADJUVANT THERAPY

Helmuth Geopfert [2], in an editorial for the March/April 1988 edition of *Head and Neck Surgery* entitled "Subclinical Cancer," eloquently presents another impetus in the search for clinically reliable markers for the detection of subclinical malignancy. He suggests that the ability to answer accurately the question "Doctor, did you get it all?" will enable the clinician not only to predict prognosis but also to use objective evidence of persistent subclinical disease when selecting patients for further adjuvant therapy. This objective selection process will eliminate unnecessary adjuvant radiotherapy or chemotherapy and will not only benefit the patient but also reduce the cost of medical care.

IV. FOLLOW-UP

Advanced-stage head and neck malignancies are characteristically treated by multimodality therapy, including surgery, radiotherapy, and chemotherapy. Nevertheless, 50-75% of these patients can be expected to develop recurrence within 2 years of therapy [3]. Although the salvage of these patients with recurrence is difficult and usually unsuccessful, it is nevertheless clear that the earlier the recurrence is detected, the greater the likelihood of successful salvage. Unfortunately, the therapy used at the time of primary treatment frequently produces such scarring and deformity that early detection of recurrent disease is difficult or impossible. A reliable tumor marker could be used in the follow-up of these patients to increase the likelihood of earlier identification of recurrence.

V. MONITORING THERAPY

A reliable tumor marker could monitor patients receiving adjuvant therapy by periodic sampling. Evaluation of serial samples obtained from patients undergoing adjuvant treatment may provide insight into the efficacy of the treatment. For example, an ineffective treatment could be stopped and the treatment plan reassessed if levels of a reliable marker continued to rise during therapy. That would avoid the time wasted on ineffective treatment regimens and allow earlier institution of alternative modalities.

VI. TUMOR CELLULAR PHYSIOLOGY

Investigation into the pathophysiology of marker production and release may provide insight into the peculiar biochemical properties of tumor tissue. Further understanding of the biochemical precursors of the marker may permit the development of improved therapeutic modalities. Markers may represent a portion of the host response to tumors and improved understanding of host responses may suggest additional avenues of investigation.

A variety of tumor markers have been investigated for a possible role in head and neck malignancies. To date, none of these has been ideal and none has been universally acclaimed or utilized in the clinical diagnosis or treatment of head and neck carcinoma. Markers that have been investigated in head and neck carcinoma are listed in Table 1.

A. Oncofetal Antigens

1. Carcinoembryonic Antigen

Carcinoembryonic antigen (CEA) is an oncofetal antigen first described by Gold and Freeman [4] in 1965 as a tumor-specific antigen in colorectal cancer. Later, this protein complex was also demonstrable in patients with inflammatory bowel disease [5],

Table 1 Tumor Markers Studied for Head and Neck Carcinoma

Oncofetal antigens
  CEA
  Alpha fetoprotein

Proteins
  Ferritin
  Glycoproteins
    Alpha-1-antitrypsin
    Alpha-1-acid glycoprotein
    Alpha-2 HS glycoproteins
    Prealbumin
    Albumin
  Beta protein
  Tissue polypeptide antigen
  Beta-2- microglobulin

Enzymes
  Alkaline phosphatase
  Placental alkaline phosphatase
  Lactate dehydrogenase
  Allesterase
  Phosphohexose isomerase
  Adenosine deaminase
  Neuron-specific enolase

Hormones: Calcitonin

Metabolic byproducts: Polyamines

Immune parameters
  Immunoglobulins
  Immune complexes

Lipids
  Lipid sialic acid
  "Oncolipids"

Viral Markers
  Epstein-Barr virus
  Herpes simplex virus
  Papilloma virus

Tumor-associated antigens
  SCC-associated antigen
  SCC antigen

Multiple markers
  SCC and LSA
  CEA and Alpha-1-acid glycoprotein
  B2M, IgE, Ferritin, LSA, and PHI

chronic smokers [6,7], and normal tissues [8] as well as in patients with other types of cancer [9-17]. Silverman et al. [18] studied CEA levels in 439 patients with head and neck squamous cell carcinoma and 276 non-tumor-bearing controls, of whom 154 were smokers and 122 were nonsmokers. While CEA levels in only 5% of the nonsmoker control patient population exceed 5 ng/ml, 36% of tumor-bearing patients had levels greater than 5 ng/ml. Five percent of control patients who were smokers had levels greater than 7 ng/ml, while 17% of tumor-bearing patients had levels greater than 7 ng/ml. They demonstrated that although CEA levels correlated with stage of disease, if patients with advanced disease were excluded, tumor-bearing patients, tumor-free previously treated patients, and healthy controls who smoked all had similar levels.

Schneider et al. [19] reported that 47% of 85 patients with head and neck carcinoma had CEA levels greater than 5 ng/ml, but they were unable to correlate elevated levels with site or stage of disease. Al-Sarraf [20] demonstrated CEA levels greater than 2.5 ng/ml in 80% of patients with untreated carcinoma of the head and neck. The CEA levels greater than 5 ng/ml were present in 32% of that population. In a group of patients with recurrent disease, 53% had levels greater than 5 ng/ml. He was unable to correlate CEA levels with site or stage of disease, but did demonstrate correlation between posttherapy levels and clinical response.

Alexander et al. [21] revealed similar values in patients with carcinoma of the esophagus, but noted a higher percentage of positive levels in patients with adenocarcinoma of the esophagus. These studies seem to indicate that CEA is not a particularly useful diagnostic marker for carcinoma of mucosal origin.

2. Alpha Fetoprotein

Alpha fetoprotein was first associated with malignancy in 1963 by Abelov and co-workers [22]. It is the major protein found in the fetus and is normally found in low concentrations in the serum of adult patients [23,24]. Alpha fetoprotein has been demonstrated to show a high correlation with germ-cell tumors [25,26] and liver cancer [27,28] and is not particularly useful as a tumor marker for carcinoma of the head and neck.

B. Proteins

1. Ferritin

Ferritin is a storage and major iron-binding protein that is normally found in small quantities in human serum. High serum ferritin levels were originally noted in patients with Hodgkin's disease and acute leukemia [29], but have been demonstrated in those with other neoplastic processes as well [30-34].

Maxim and Veltri [35] investigated serum ferritin as a potential marker in carcinoma of the head and neck. Patients and controls were classified into several groups: cigarette smokers, healthy controls, a group of Latter Day Saints, a group of treated cancer patients who showed no evidence of disease (NED) for ≥5 years, and patients with newly diagnosed head and neck carcinoma.

One hundred and eighteen normal volunteers had a mean ferritin level of 32.2 ng/ml. High levels were seen in controls ≥32 years old compared to either younger normals (<32 years) or the whole normal population control group.

No significant differences were found between groups of smokers and the group of Latter Day Saints. When normals were compared to untreated head and neck cancer patients, there was a significant difference at the level of $p < .001$. A comparison of men and women showed no difference in incidence between the groups.

Attempts to correlate ferritin levels with stage of head and neck cancer patients did not show any trends. The group of previously treated cancer patients who were NED ≥5 years showed statistically increased levels of ferritin compared to normal controls, but the groups mean levels were also significantly lower than the levels in untreated head and neck carcinoma patients.

Serial ferritin levels were obtained in a group of head and neck cancer patients. Samples were obtained monthly from patients following surgical extirpation or combination therapy with surgery and radiotherapy. Levels did not decrease until 5-6 months after initial treatment. The investigators were able to classify head and neck carcinoma patients into two groups based on serum ferritin levels. Patients with poor outcome and survival of <12 months had serum ferritin levels that remained consistently elevated during therapy. However, patients with a favorable outcome and survival of >12 months had ferritin levels that decreased during therapy. They were unable to prove statistical significance because of the small sizes studied in the two groups.

Although some ferritin levels seem to correlate with prognosis, further investigation is needed to corroborate these findings.

2. Glycoproteins

Glycoproteins are normally synthesized in the liver and are thought to increase in response to inflammatory stimuli [36]. Glycoproteins have been studied in various types of malignancies [37-41], including those of the liver [42]. Changes in glycoprotein levels in carcinogenesis have been linked to altered liver synthesis or lymphoid tissue of normal circulating proteins.

Wolfe et al. [43] studied six circulating glycoproteins: haptoglobin, alpha-1-antitrypsin, alpha-1-acid glycoprotein, alpha-2-HS glycoprotein, prealbumin, and albumin. They demonstrated correlation between levels of alpha-1-antitrypsin and alpha-2-acid glycoprotein and tumor burden. Haptoglobin levels were elevated significantly in a group of tumor-bearing patients, but no correlation between levels and tumor burden could be demonstrated. Tumor-bearing patients had levels of alpha-2-HS glycoprotein, prealbumin, and albumin that were significant lower than those in normal subjects. Non-tumor-bearing smokers had elevated alpha-1-trypsin levels and lower albumin and prealbumin levels compared to nonsmokers. The authors were unable to demonstrate any correlation between tumor site and any of the glycoproteins studied. Treated patients had levels of haptoglobin and alpha-1-acid glycoprotein that were lower than in tumor-bearing patients, but still higher than non-tumor-bearing controls. Alpha-2-HS glycoprotein, albumin, and prealbumin levels were significantly higher in disease-free previously treated patients than in tumor-bearing patients. The usefulness of glycoproteins as tumor markers for squamous cell carcinoma of the head and neck needs further investigation before firm recommendations regarding their clinical utility can be formulated.

3. Beta Protein

A chemical assay for the detection of cancer was reported by Bucovaz et al. in 1976 [44]. Beta protein is a nonspecific biological marker found in the serum of cancer patients. The assay is based on a reaction between a radioactive labeled protein fractioned from bovine, liver, or baker's yeast and a protein present in the serum of cancer patients [45]. The specific protein is thought to be a substructure of coenzyme A-synthesizing protein complex of baker's yeast.

Two types of assays have been developed: a quantitative measure to monitor patients before and after therapy and a screening test developed for use in the clinician's office that detects the presence of beta protein very rapidly.

The investigators have studied more than 2,657 patients diagnosed with cancer and approximately 87% have shown elevated beta protein levels [47]. Beta protein levels have also been determined in 3,879 controls, both healthy normals and those with benign medical conditions. Approximately 9% of the control group have tested positive. Benign medical conditions shown to elevate beta protein levels include severe burns, Crohn's disease, and the third trimester of pregnancy. Study into the possible correlations between the above conditions, cancer, and beta protein assay is underway. Long-term follow-up will be necessary to determine if some of the false-positive results are indicative of occult neoplastic processes.

Eighty-eight percent (61/69) of patients with head and neck carcinoma studied by Bucovaz demonstrated seropositivity. Continued investigation will possibly answer the questions regarding beta protein, but its lack of sensitivity hinders its utility as a marker for patients with head and neck carcinoma.

4. Tissue Polypeptide Antigen

Over 30 years ago, Bjorklund and co-workers [47] isolated an antigenic polypeptide from anaplastic tumor tissue. Early studies demonstrated this substance termed tissue polypeptide antigen (TPA) could be identified in the serum of patients with carcinoma [48-50]. With use of a hemagglutin assay, elevated TPA levels were identified in four of seven patients with head and neck carcinoma [51]. More recent studies utilizing a radioimmunoassay have demonstrated elevated levels in 50-80% of patients with malignancies [52]. It is also identified in 69% of patients with benign inflammatory processes. Levels of TPA correlate closely with CEA levels in patients with breast carcinoma and correlate with recurrent disease in patients with a variety of tumors.

There are no recent studies of TPA levels in head and neck carcinoma patients utilizing this radioimmunoassay. The high false-positive rate would suggest that this marker may not be useful in studies of carcinoma of the head and neck.

5. B_2 Microglobulin

B_2 microglobulin is a light chain of the human leukocyte antigen and is of low molecular weight. It is filtered and metabolized within the kidney and its serum concentrations are dependent on renal function [53]. B_2 microglobulin levels have been measured in various malignancies [55-57] as well as in patients with various benign medical disorders [58,59]. Vinzenz et al. [54] studied a group of 206 patients with known head and neck carcinoma and 50 age-matched healthy controls. One hundred and one patients had had no prior therapy and 105 patients had recently been treated for head and neck carcinoma. Elevated levels of B_2 microglobulin were found in approximately 50% of the patients with head and neck carcinoma.

C. Enzymes

1. Alkaline Phosphatase

Alkaline phosphatase is an enzyme required for bone mineralization. Although the majority of this enzyme is produced in bone, it is also produced in liver, intestinal mucosa, placenta, and breast [60].

Katz et al. [61] correlated alkaline phosphatase levels with response rates in 55 patients undergoing induction chemotherapy with *cis*platin and bleomycin for carcinoma of the head and neck. Of the 40 patients who responded to chemotherapy, pretreatment alkaline phosphatase levels showed a mean of 82 µg/ml, while alkaline phosphatase in the nonresponder group showed a mean of 66 µg/ml. Ninety percent of patients with alkaline phosphatase levels >100 µg/ml responded to chemotherapy, while only 50% with alkaline phosphatase levels <60 µg/ml were responsive to therapy ($p < 0.05$). They concluded that alkaline phosphatase may be a potentially useful marker in determining chemotherapy response prior to therapy.

In a subsequent study, Coker et al. [62] retrospectively reviewed 51 patients with squamous cell carcinoma receiving chemotherapy containing *cis*platin and found no correlation between pretreatment alkaline phosphatase levels and response rates to chemotherapy treatment. In contrast to Katz et al.'s findings, their data demonstrated better responses in the group with lower alkaline phosphatase levels.

Burres et al. [63] were unable to demonstrate correlation between alkaline phosphatase levels in untreated patients and response rates in patients with squamous cell carcinoma of the head and neck undergoing induction chemotherapy.

Although early reports showed alkaline phosphatase to be a potential marker in the prediction of chemotherapeutic response in head and neck patients, follow-up studies failed to confirm this application.

2. Placental Alkaline Phosphatase

Placental alkaline phosphatase (PLAP) is an isoenzyme of alkaline phosphatase found in normal placental tissue. This isoenzyme usually makes up <1% of the total alkaline phosphatase level in the serum of normal adults. It was termed Regan isoenzyme in 1968 when detected by Fishman et al. [64] in tumor tissue and serum of a patient with lung carcinoma. Elevated levels of this isoenzyme were later demonstrated in various neoplastic states including head and neck carcinoma [65-67].

Muensch and colleagues [68] demonstrated a high correlation between smoking and PLAP levels. Levels in smokers were sometimes as high as 10-fold above the upper normal limit of nonsmokers. They also studied 286 patients with carcinoma of various sites and, although 23% of patients had an increased PLAP serum level, only 4.2% had levels that exceeded those found in apparently healthy smokers. Therefore, the Regan isoenzyme was deemed too nonspecific to be used as a tumor marker.

3. Lactate Dehydrogenase

Lactate dehydrogenase was reported a possible marker for malignant disease by Hill et al. [69] in 1954. It is an enzyme of the glycolytic pathway and is normally released after tissue damage [70]. Elevations in LDH levels have been demonstrated in all types of malignancies [71-76], but most studies have shown no significant differences from levels in patients with nonmalignant diseases [77-79]. LDH generally is neither specific nor sensitive as a tumor marker in malignancy.

4. Allesterase

Allesterase is an enzyme required for fat metabolism. In contrast to what is found in the enzymes required for glycolysis, allesterase levels have been shown to decrease in patients with malignancy. Lal et al. [80] studied allesterase levels in 38 patients with carcinoma of the head and neck and 25 control patients. Allesterase levels in controls averaged 2.38 ± 0.06 units/ml whereas the groups with disease had levels that averaged 1.11 ± 0.07 units/ml. Patients with ulcerative lesions had levels that were statistically lower than those with proliferative lesions. Serum levels varied inversely with tumor stage. Levels rose during radiotherapy, but did not reach control levels.

5. Phosphohexose Isomerase

Phosphohexose isomerase (PHI) is an enzyme of glycolysis that transforms glucose-6-phosphate. Aerobic glycolysis is increased in patients with malignant tumors and elevated levels of PHI have been noted in association with advanced tumors.

Goel et al. [80] studied PHI levels in 28 patients with head and neck malignancies and 25 healthy controls. PHI activity for the group with disease was 130 ± 5.67 mu/ml, whereas the serum levels of normal controls was 54 ± 2.52 mu/ml. Serum levels correlated with tumor bulk and decreased during radiotherapy in 9 of 10 (90%) patients.

Vinzenz et al. [54] noted elevated PHI levels in 62% of a group of 206 patients with carcinoma of the head and neck, with a strong correlation noted with stage of disease. However, only 43% of patients with recurrent disease had elevated levels. Those with stage I and II diseases demonstrated less than 50% sensitivity rate to PHI; thus PHI can be added to the list of markers that are not sensitive enough to be useful in carcinoma of the head and neck.

6. Adenosine Deaminase

Lal et al. [82] studied adenosine deaminase (ADA) levels in a group of 40 patients with carcinoma of the head and neck and 25 normal controls. The mean serum activity for the control group was 9.65 ± 0.64, whereas the mean activity for the group with disease was 25.52 ± 1.47, or a statistically significant increase. Mean levels for patients with the ulcerative lesions were greater than for the proliferative lesions in levels correlated with tumor stage. Levels fell during radiotherapy in 15 patients.

These enzymes all demonstrate a correlation with tumor burden and response to therapy. Unfortunately, these changes are not sufficiently specific to warrant use as a diagnostic tumor marker. Measurement of serum enzyme levels may have a place in evaluating the effects of therapy. Evaluation of the mechanism of increased or decreased serum levels may provide insight into tumor cell biochemistry and suggest possible additional therapeutic modalities.

7. Neuron-Specific Enolase

Neuron-specific enolase (NSE) is an isoenzyme of the glycolytic enzyme enolase. This protein was first described by Moore and McGregor in 1965 [83], but its application as a tumor marker is more recent [84]. It was first isolated from bovine brain tissue and has been demonstrated to be found in neuronal and neuroendocrine cells of the central and peripheral nervous system [85,86]. Neuron-specific enolase levels between 4.3 and 9.6 ng/ml have been measured in normal adults and in infants. Elevated levels have been described in patients with tumors of neural crest origin such as neuroblastoma [87-92], small-cell carcinoma of the lung [93-97], and melanoma [98].

Although NSE is not useful as a marker for mucosal malignancies, it may nevertheless represent a valuable marker for the patient with a head and neck tumor of neural crest origin.

D. Hormones: Calcitonin

Calcitonin is a polypeptide hormone produced by the C cells of the thyroid. To date, calcitonin appears to be specific for medullary carcinoma of the thyroid [99-105], but it has been investigated in various other carcinomas as well [106-113].

In 1971, Melvin et al. [114] first described the clinical usefulness of calcitonin in medullary thryoid carcinoma. Twelve patients with a positive family history of medullary thyroid carcinoma had abnormal concentrations and were found to have medullary thyroid carcinoma demonstrable microscopically [115]. Eleven of 12 patients had no evidence of disease clinically, however.

Wolfe et al. [116] subsequently studied family members of patients with medullary thyroid carcinoma using yearly and/or biyearly serum calcitonin levels. Three patients with initially normal levels demonstrated subsequent elevations in calcitonin levels. After thyroidectomy was performed, histological examination demonstrated foci of hyperplastic C cells. The pathological findings were postulated to represent a premalignant lesion. Calcitonin levels normalized postoperatively and remained low in these individuals.

Other studies have confirmed these findings and, therefore, calcitonin is thought to be useful in screening for medullary carcinoma of the thyroid.

E. Metabolic Byproducts: Polyamines

Polyamines such as purtescine, spermidine, and spermine have been found in all types of cells and are organic cations involved in tissue growth [117]. Previous studies have indicated that polyamines may be linked to the control of RNA metabolism. In 1971, Russell et al. [118] studied urinary polyamine levels in patients with malignancy and demonstrated increased levels in patients with malignancy compared to normal controls.

Polyamines have been measured in urine [118-122], cerebral spinal fluid [123], and blood products [124-126]. Shideler et al. [127] studied erythrocyte polyamines in patients with head and neck carcinoma. Twenty-nine previously untreated patients with head and neck carcinoma were studied, of whom nine (31%) had increased levels of spermidine and/or spermine concentrations that were considered above normal. Mean spermidine levels correlated with tumor burden in these patients. In a smaller subgroup of 12 patients, pretreatment and posttreatment samples were obtained and levels decreased after surgery or irradiation in all but one case.

This study of head and neck carcinomas demonstrated that polyamines were not sensitive enough to function as tumor markers and confirm other studies suggesting that polyamine excretion may not be specific to the neoplastic process.

F. Immune Parameters

Early investigations into the systemic indications of head and neck carcinoma focused on studies of the immune system and its function [128]. Investigation revealed decreased levels of T-cell activity as well as circulating immunoglobulins. Abnormalities were noted in measurements of both broad forms of immune response [129-137], although early investigation failed to indicate whether the source of the abnormality lay within the direct effect of the tumor or in secondary effects on the patient such as malnutrition and weight loss. Further research into the cellular response has, however, suggested a close correlation between patient host response and cancer progression [138-

141]. The current status of research into the cellular immune response is beyond the scope or intent of this chapter.

G. Lipids

1. Lipid Sialic Acid

Sialic acid is a component of most glycoproteins. Winzler et al. [149] first recognized elevations in sialic acid levels in patients with carcinoma as well as patients with inflammatory disorders such as arthritis. Serum lipid sialic acid (LSA) levels have been shown to correlate with stage of disease, presence of recurrence, and tumor burden in patients with breast carcinoma [150] and malignant melanoma [151].

Katopodes et al. [152] evaluated LSA levels in 850 serum samples from patients with malignant disease, benign medical conditions, and normal controls. One hundred and eighty controls consisted of three different populations: normal persons, patients with systemic medical disorders, and surgical patients with benign medical disorders. The overall median LSA level for these thee subgroups was 17.8 mg/100 ml. No statistical difference was found between the three subgroups, but the patients with systemic medical disorders had the widest range of LSA levels (5-34.3 mg/100 ml).

LSA levels for the tumor patients had an overall median of 26.3 mg/100 ml. The difference from the normal group was statistically significant ($p < 0.001$). These authors' studies revealed the assay to be 93% specific in normal persons, 94% specific in benign surgical patients, and only 77% specific in patients with benign medical disorders. The test specificity of the cancer population ranged from 77 to 97%, with breast carcinomas being the least sensitive and sarcoma being the most specific.

Santamaria and colleagues [153] evaluated LSA in 35 patients with serial levels of laryngeal carcinoma, 27 with benign tumors of the head and neck, and 14 with lymph node metastases. Samples were obtained preoperatively, 2, 7, 21, 55, and 95 days postoperatively. Levels in patients with benign diseases remained within the normal range of 18-22 mg/100 ml, wherease the mean value in patients with laryngeal carcinoma was 27 mg/100 ml prior to surgery. Positive correlation between serum levels, tumor size, and extent of nodal disease was also evident.

LSA levels decreased rapidly after surgery in 29 of 35 patients. In six patients LSA levels increased dramatically immediately following surgery, before later decreasing. In all 35 patients with laryngeal carcinoma, LSA levels returned to normal by postoperative day 35.

In 14 cases of nodal recurrent disease that occurred after primary therapy, six patients demonstrated levels of LSA within a normal range even after histological confirmation of nodal disease. The investigators concluded that elevations in LSA levels were indicative of tumor, but also noted a high false-negative rate.

2. Oncolipids

In 1986, Fossel et al. [154] described a new nonspecific cancer marker involving the use of water-suppressed proton nuclear magnetic resonance (NMR) spectroscopy of plasma. Lipoprotein particles found in plasma of cancer patients have been termed "oncolipids" and preliminary tests have suggested that this assay may provide a universal cancer screening test [155,156].

The technique uses an NMR spectrometer to study the patient's plasma. By suppressing the bands of water molecules contained in the sample, the spectrometer reads the lipid portion of the lipoproteins.

NMR line widths of >33 Hz correlated with tumor-free status while values <33 Hz signified malignancy. Two different subgroups of patients had values that overlapped into the group of patients with malignancy: pregnant women and men with benign prostatic hypertrophy. The two subgroups demonstrated line widths similar to cancer patients, although the mechanisms for this are still unclear.

More studies are needed before this new modality can be accepted as a clinically useful marker. Clinical trials to examine more extensively the oncolipid marker are currently underway [157].

H. Viral Markers

1. Epstein-Barr Virus

Old et al. [158] first observed the relationship between nasopharyngeal carcinoma and the Epstein-Barr virus (EBV) antigen in 1966. Since that time, various investigations have studied EBV and its relationship to carcinomas of the head and neck [159-173].

Sako and colleagues [174] evaluated serum from 23 patients with nasopharyngeal carcinoma, 86 patients with head and neck carcinoma of other sites, and 222 age-matched controls. Anti-EBV capsid antibody titers were quantitatively determined. In the control group, the average anti-EBV titer was 1188, while the mean antititer in the nasopharyngeal carcinoma group was 2671. The mean anti-EBV titer in the group of patients with other head and neck carcinomas was 1400. The nasopharyngeal carcinoma group had statistically significant higher anti-EBV titers when compared to either of the other groups.

If an upper limit of normal of 1280 was used, 74% of the patients with nasopharyngeal carcinoma had titers greater than the upper normal limit, while only 23% of the patients with head and neck carcinoma had levels higher than the upper normal limit.

The elevated anti-EBV titer in nasopharyngeal carcinoma has been confirmed by reports. The association between the Epstein-Barr virus and nasopharyngeal carcinoma, WHO types II and III, has been well demonstrated.

EBV-associated antigens consist of the following: viral capsid antigen (VCA), early antigen (EA), and EBV antigens in the serum of IgA fraction. Coates et al. [175] studied EBV-associated antigens in patients with occult primary malignancies in the head and neck and in patients with treated and untreated nasopharyngeal carcinoma. Thirty-five patients had positive titers for viral capside antigen and 94% of the patients had positive EA titers.

Serum samples from 85 patients with other head and neck carcinomas, 80 patients with lymphoma, and 47 control patients were studied. Thirty-six of 47 (77%) control patients had elevated VCA titers, while 91% (77 of 85) of the patients with other head and neck carcinomas had VCA titers that were considered to be elevated. Lymphoma patients demonstrated a positive VCA rate of 88%.

The EA determinations demonstrated a statistically significant difference between the different study groups ($p < 0.001$). The rate of positive results was 94% in the group of patients with nasopharyngeal carcinoma. Those with head and neck carcinomas, lymphomas, and normal populations yielded positive EA titer percentages of 45%, 59%, and 36%, respectively.

The highest specificity for the nasopharyngeal carcinoma was demonstrated in the serum antibody to EBV-induced VCA in the IgA immunoglobulin fraction. Positive IgA titers were demonstrated in 17 of 19 (89%) of the patients with nasopharyngeal carcinoma. Six of 16 patients with nasopharyngeal carcinoma who had NED on clinical examination demonstrated negative titers. The remaining subgroup of 10 patients demonstrated low levels. The three control groups had positive titers in only 11% or fewer of patients tested and this was noted to be a statistically significant difference at $p < 0.001$.

The investigators also sought to determine whether occult nasopharyngeal carcinoma could be detected by examination of serum alone. Thirty-eight serum samples were examined blind and immunofluorescent testing was performed to determine EA titers and IgA titers to VCA. The EA titers were then plotted against IgA titers to VCA and serum was then categorized into two groups: that identifying patients likely to have nasopharyngeal carcinoma and that identifying patients unlikely to have nasopharyngeal carcinoma. The values used to determine patients with nasopharyngeal carcinoma were EA titers >40 and IgA titers to EBV >10. After patients were classified into two groups, the samples were identified. Thirty-seven of 38 samples, or 97%, correlated correctly.

The authors concluded that titers of EBV IgA antibody to VCA and EA are probably the most specific and reliable markers for nasopharyngeal carcinoma. EBV serological testing may be also helpful in the diagnosis of occult malignancy of the head and neck area, but further study is needed.

2. Herpes Simplex Virus

Hollinshead et al. [176] first described herpes simplex virus (HSV)-tumor-associated antigen (TAA) from squamous cell carcinomas in 1972. Since then, strong correlations have been seen in the antibodies to HSV in patients with squamous cell carcinoma and in patients who have been previously treated for squamous cell carcinoma of the head and neck [177-183].

Silverman et al. [184] studied 94 patients with squamous cell carcinoma of the head and neck, 46 patients who had no prior treatment, and 48 patients who had previously received treatment but had had NED for at least 4 years. Forty-one age- and sex-matched controls were also studied.

Antibodies of HSV-TAA were not demonstrable in control patients, but 89% of the tumor patients demonstrated HSV-TAA antibodies. The prior-treatment group and the no-prior-treatment group demonstrated similar levels: 90% and 88%, respectively. Statistically different percentages of HSV-TAA antibodies were noted in those with stage I carcinoma among the group with larynx cancer (63%) when compared to those with stages II-IV carcinoma of the larynx (94%) ($p < 0.05$) or patients with head and neck cancer of other sites (95%) ($p < 0.01$).

Smith et al. [185] studied HSV-induced antigens (HSV-IA) in normal populations of smokers and nonsmokers to investigate if a relationship between smoking, alcohol use, and HSVIA was evident. They concluded that IgG and IgM anti-HSVIA were found in both the control groups with equal frequency but that antibody titers were significantly higher in the smoking group compared to the nonsmoking group ($p < 0.05$). Cigarette smokers who drank alcohol also demonstrated higher IgA antibodies to HSVIA compared to smokers who used no alcohol ($p < 0.025$). They, therefore, concluded that smoking and alcohol use play a significant part in altering humoral immunity to HSV in populations at high risk for carcinoma of the head and neck.

Shillitoe et al. [186] studied neutralizing antibodies to HSV type I, type II, and measle virus in smokers, nonsmokers, patients with oral carcinoma, and patients with oral leukoplakia. Results demonstrated lower levels of neutralization titers to HSV-I in nonsmoking controls compared to smoking controls (p < 0.001). Significantly lower levels were found in nonsmokers compared to the group with oral leukoplakia (p < 0.02).

Patients with previously untreated oral carcinoma all had antibody titers to HSV-I within the range of the smoking control group. However, the highest levels of HSV-I were found in patients with larger tumor burden. After treatment, patients with a short-term disease-free interval had lower levels of HSV-I compared to patients with a long-term disease-free interval (p < 0.01).

HSV seems to have a role in the development of tumors of the head and neck, but the exact mechanisms are still unclear. The incidence of elevated antibody titers in normal controls would indicate insufficient specificity to serve as a clinically useful tumor marker.

3. Human Papilloma Virus

Much evidence has accumulated linking human papilloma virus (HPV) with squamous cell carcinomas of various sites including head and neck. To date, immunohistochemical, morphological, and DNA studies have suggested the role of HPV in squamous cell carcinomas of the head and neck [187-190].

Although the authors acknowledge that HPV may someday be investigated as a possible tumor marker for squamous cell carcinomas of the head and neck, no serological investigations have been performed to date, thus precluding its usefulness as a tumor marker.

I. Tumor-Associated Antigens

1. Squamous Cell Carcinoma-Associated Antigen

Saito et al. [191] have described a tumor antigen isolated from squamous cell carcinoma of the maxillary sinus. By using a combination of isoelectric focusing, electrophoresis, and chromatography techniques, they obtained a purified IgG component directed against this antigen. Twenty-nine tissue specimens were studied using indirect immunofluorescence and staining by the immunoperoxidase technique. The SCC-associated antigen was positive in 27 of 29 (93%) of the tissue samples tested. Serum studies are currently underway to investigate potential SCC-associated serum antigen levels in hopes of providing a diagnostic marker specific for squamous cell carcinoma of the head and neck.

2. SCC Antigen

Another squamous-cell-derived antigen that has recently been evaluated is tumor antigen TA-4, isolated and purified from squamous cell carcinoma of the uterine cervix by Kato and Torigoe [192] in 1977. They purified antigen obtained from fresh tumor tissue on Sephadex columns followed by electrophoresis and then prepared antibody to this purified antigen. Subfractions of this antigen were obtained by isolectric focusing and the most neutral subfraction was termed SCC antigen. This purified SCC antigen is utilized in a radioimmunoassay (SCC-RIA; Abbott Laboratories) as a research tool. The specificity of this assay has been quite high, with extremely low levels detectable in patients without carcinoma and with levels greater than 2.5 ng/ml detected in 54% of patients with known squamous cell carcinoma of the uterine cervix [193]. Kato and co-workers [194] demonstrated that serum levels of TA-4 correlated with extent of disease as well as successful therapy. Recurrent disease was associated with elevated levels in 15 of 17 patients and, in 11 of these, elevated serum levels were detected before recurrence was evident clinically. Patients with elevated levels (>15 ng/ml) had a significantly worse prognosis. Immunohistological studies have demonstrated the presence of this antigen in malignant epithelial cells from the uterine cervix. Direct correlation betwee tissue levels and serum levels of TA-4 could not be demonstrated.

Initial studies have demonstrated circulating SCC antigen in approximately one-half of patients with SCC of the uterine cervix [195]. Further investigation into the potential utility of this marker in squamous cell carcinoma of other sites has been undertaken. Studies of SCC antigen in patients with carcinoma of the esophagus [196], lung [197], and head and neck [198,199] have suggested a possible clinical role for this marker.

Eibling and colleagues [200] recently evaluated SCC antigen using retrospective sequential serum samples maintained in the department serum bank from patients with head and neck carcinoma. Three hundred and fifty-nine serum samples from 102 patients with malignancies of the head and neck were studied. Eighty-nine patients had a diagnosis of SCC from multiple head and neck sites. Two patients had verrucous carcinoma, three patients had anaplastic tumors, and eight patients had head and neck malignancies of non-squamous-cell origin. A minimum of 2 years of follow-up was available on all patients. Any SCC antigen levels above 2.0 ng/ml were considered abnormal.

The mean pretreatment SCC antigen level of all patients studied was 2.51 ng/ml. Among the 89 patients with head and neck SCC, the mean SCC antigen level was 2.57 ng/ml (range, 0.27-20.51 ng/ml). Thirty-nine of the 89 patients with head and neck SCC (44%) had elevated levels prior to treatment. In 50 (56%), pretreatment levels were less than 2.0 ng/ml.

Stage of disease did not correlate with SCC antigen levels. However, the mean SCC antigen level for each group correlated with increasing tumor stage. Mean SCC antigen levels were elevated in groups of patients with nodal disease without extracapsular spread (ECS) (2.60 ng/ml) and those with ECS (3.31 ng/ml) compared to levels in patients with negative nodes or in whom no neck dissection was performed (1.81 ng/ml).

Three hundred and fifty-nine serum determinations were organized into two groups depending on the patient's clinical status at the time the serum was drawn (Fig. 2). One

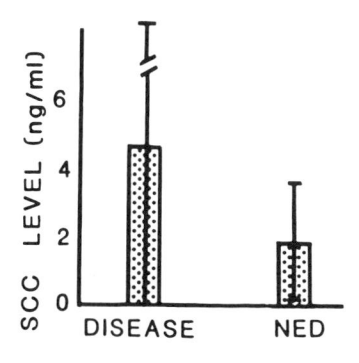

Figure 2 Mean SCC antigen levels grouped by presence or absence of clinical disease at the time of collection. (Reprinted with permission from Eibling DE, Johnson JT, Wagner RL, Su S. SCC-RIA in the diagnosis of squamous cell carcinoma of the head and neck. Laryngoscope 1988, accepted for publication.)

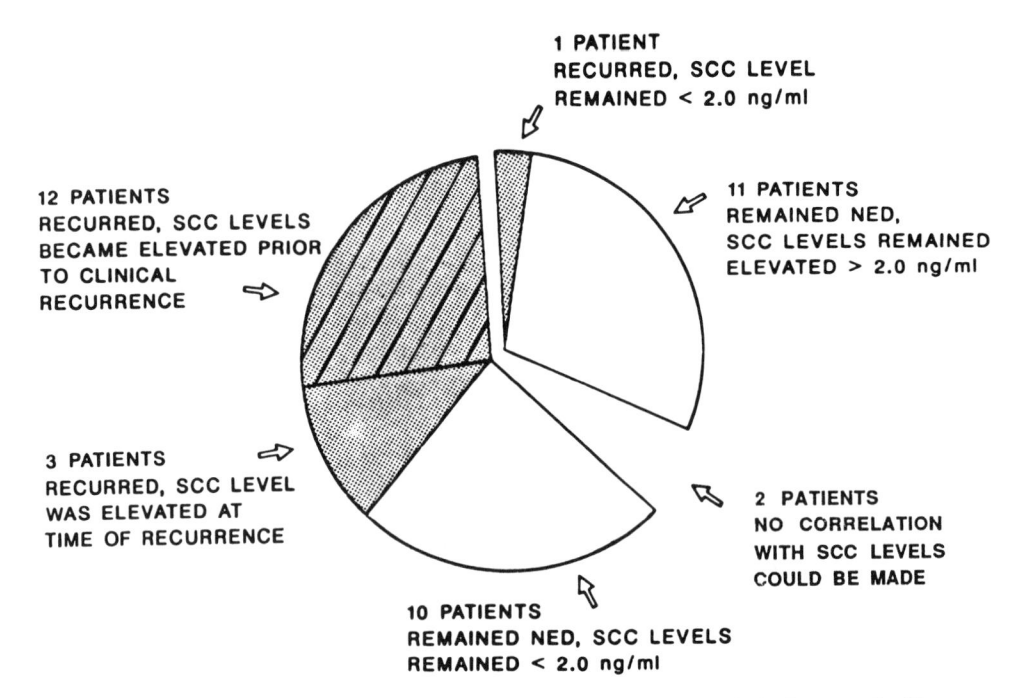

Figure 3 Pie chart shows correlation of SCC antigen levels with clinical course in 39 patients. Stippled area denotes patients in whom SCC recurred. Two patients were unable to be correlated with SCC levels. (Reprinted with permission from Eibling DE, Johnson JT, Wagner RL, Su S. SCC-RIA in the diagnosis of squamous cell carcinoma of the head and neck. Laryngoscope 1988, accepted for publication.)

hundred and forty-five (40%) of the total samples were obtained when the patients had clinical evidence of disease and 214 (60%) were obtained when the patients were considered to have NED. The mean value for the group with disease was 4.59 ng/ml (range, 0.27-150 ng/ml), whereas the mean SCC antigen level for the NED group was 1.90 (range, 0.20-17.67 ng/ml).

Thirty-nine patients with head and neck squamous cell carcinoma had elevated SCC antigen levels prior to therapy. In 25 (64%) of these patients, the serum SCC antigen levels reflected the clinical course (Fig. 3). The SCC antigen levels did not reflect the clinical course in the remaining 12 patients (31%). In 11 patients whose disease was controlled, SCC antigen levels remained elevated following therapy despite the lack of persistent or recurrent disease. Mean follow-up on these patients was 60 months, with a range of 33-76 months.

Of the 16 patients with elevated levels prior to therapy who developed recurrent cancer, serum SCC antigen levels greater than 2.0 ng/ml were noted at, or prior to, recurrence in 15 (93.5%). In 12 of these patients, the levels were elevated prior to clinical detection of recurrence by an average of 5.4 months (range, 3-11 months). The remaining three had simultaneous elevation of serum SCC antigen with detection of recurrent disease. One patients developed recurrent disease without demonstrating a concomitant rise in serum SCC antigen levels. Of the 21 patients with elevated SCC antigen levels prior to therapy who remained NED, 10 (48%) had a persistent drop of SCC antigen levels to normal values (<2 ng/ml) following surgical therapy.

Although SCC antigen levels increased with increasing tumor burden, only approximately 50% of each group of patients with squamous cell carcinoma had elevated pretreatment levels, regardless of their age, sex, tumor burden, or tumor differentiation.

Correlation with clinical course after definitive therapy was evident in the group of patients who demonstrated elevated serum SCC antigen levels prior to treatment. Only one patient from this group developed recurrent disease without a concomitant rise in serum SCC antigen levels. Moreover, elevated SCC antigen levels could be demonstrated prior to the clinical detection of recurrence in three-quarters of those patients whose cancer recurred.

The preliminary data suggest that SCC antigen deserves further investigation. Current information suggests that its greatest utility may be in the group of tumor-bearing patients with elevated levels before surgery.

J. Multiple Markers

1. SCC Antigen and LSA

Goodwin et al. [201] recently evaluated a combination of tumor markers in head and neck carcinoma using LSA and SCC antigen determinations. The investigators studied 144 patients with SCC of the head and neck, of whom 45 had SCC present at the time of sampling and 99 were undergoing routine follow-up after treatment for SCC. Seven age- and smoking-matched controls and 13 patients with malignancies of the head and neck other than SCC were also studied. They determined that the upper limit of normal for LSA was 20 mg/dl and 2.5 ng/ml for SCC antigen.

Specificity and sensitivity were calculated for four different groups: (1) if neither SCC antigen or LSA levels were elevated; (2) if only LSA levels were elevated; (3) if only SCC antigen levels were elevated; and, (4) if both SCC antigen and LSA levels were elevated. The calculated specificity was 71% when levels of either marker were elevated, whereas a specificity of 97% was obtained if levels of both markers were elevated. They concluded that the combined use of LSA and SCC antigen was more useful to the clinician than just a single marker and that serial determinations would likely prove to be more useful than single data points as studied above.

2. CEA and Alpha-1-Acid Glycoprotein

Al-Sarraf et al. [202] studed CEA levels and alpha-1-acid glycoprotein levels in patients with head and neck carcinoma. Levels of both markers were elevated in 44% of untreated patients and 59% of patients with recurrence. In this study, they concluded that CEA was sensitive for tumor progression while alpha-1-acid glycoprotein demonstrated sensitivity as an indicator for tumor remission.

3. B_2M, IgE, Ferritin, LSA, and PHI

Vinzenz et al. [54] studied 5 markers in 206 patients and 50 age-matched controls. Findings from each marker were reported individually and multivariant analysis combining the results of the battery has not yet been performed. Such statistical analysis of the group may provide a significant advantage over any of the markers when utilized individually.

VII. SUMMARY

Evaluation of the various tumor markers that have been studied in patients with SCC of the head and neck reveals no single marker or battery of markers that is sufficiently specific or sensitive to provide useful clinical data in the treatment of patients with these cancers. This finding reinforces previous investigative and clinical impressions that would suggest that carcinoma of the head and neck is a heterogeneous entity. These tumor marker studies provide additional support for the hypothesis that although various carcinomas of the head and neck may appear identical microscopically, their behavior both clinically and biochemically is nevertheless heterogeneous. Due to the wide variability of the behavior of these tumors, the search for the perfect tumor marker continues. The development of additional multiple-marker batteries may provide the closest approximation to a perfect marker that is feasible. Additional studies into the nature and function of oncogenes may provide additional clues to the variability of these tumors and the resultant heterogenicity of their marker expression.

REFERENCES

1. Cancer Statistics, 1988. CA 1988;22.
2. Geopfert H. Editorial; subclinical cancer. Head Neck Surg 1988;10:217-218.
3. Vikram B, Farr HW. Adjuvant radiation therapy in locally advanced head and jneck cancer. CA 1983;134-138.
4. Gold P, Freedman SO. Demonstration of tumor specific antigens in human colonic carcinomata by immunological tolerance and absorption techniques. J Exp Med 1965;121:439-459.
5. Lurie BB, Loewenstein MS, Zamcheck V. Elevated carcinoembryonic antigen levels and biliary tract obstruction. JAMA 1975;233:326-330.
6. Cullen KJ, Stevens DP, Frost MA, et al. Carcinoembryonic antigen (CEA), smoking and cancer in a longitudinal population study. Aust NZ J Med 1976;6:279-283.
7. Alexander JC Jr, Silverman NA, Chretien PB. Effect of age and cigarette smoking on carcinoembryonic antigen levels. JAMA 1976;235:1975-1979.
8. Costanza ME, Das S, Nathanson L, et al. Carcinoembryonic antigen. Cancer 1974; 33:583-590.
9. Zamcheck N. The present status of carcinoembryonic antigen (CEA) in diagnosis, detection of recurrence, prognosis and evaluation of therapy of colonic and pancreatic cancer. Clin Gastroenterol 1976;5:625-638.
10. Gold P, Shuster J, Freedman SO. Carcinoembryonic antigen (CEA) in clinical medicine: historical perspectives, pitfalls, and projections. Cancer 1978;42:1399-1405.
11. Wanebo HJ, Stearns M, Schwartz MK. Use of CEA as an indicator of early recurrence and as a guide to a selected second-look procedure in patients with colorectal cancer. Ann Surg 1978;188:481-493.
12. Wanebo HJ, Rao B, Pinsky CM, et al. Preoperative carcinoembryonic antigen level as a prognostic indicator in colorectal cancer. N Engl J Med 1978;299:448-451.
13. Shively JE, Spayth V, Chang F, et al. Serum levels of carcinoembryonic antigen and a tumor-extracted carcinoembryonic antigen-related antigen in cancer patients. Cancer Res 1982;42:2506-2513.
14. Concannon JP, Dalbow MH, Hodgson SE, et al. Prognostic value of preoperative carcinoembryonic antigen (CEA) plasma levels in patients with bronchogenic carcinoma. Cancer 1978;42:1477-1483.

15. Falkson HC, Falkson G, Portugal MA, et al. Carcinoembryonic antigen as a marker in patients with breast cancer receiving postsurgical adjuvant chemotherapy. Cancer 1982;49:1859-1865.
16. Goslin RH, Sharin AT, Zamcheck N. Carcinoembryonic antigen: a useful monitor of therapy of small cell lung cancer. JAMA 1981;246:2173-2176.
17. Vincent RG, Chu TM, Lane WW. The value of carcinoembryonic antigen in patients with carcinoma of the lung. Cancer 1979;44:685-691.
18. Silverman NA, Alexander JC, Chretien PB. CEA levels in head and neck cancer. Cancer 1976;37:2204-2210.
19. Schneider M, Demard F, Chauvel P, et al. Carcinoembryonic antigen determinations in head and neck cancer. In: Krebs BP, Lalanne CM, Schneider M, eds. Clinical applications of carcinoembryonic antigen. Amsterdam: Excerpta Medica, 1977: 384-387.
20. Al-Sarraf M, Chu C, Lai L, Carey MK. Tumor markers in patients with epidermoid cancer of the head and neck and the effect of therapy. Head Neck Surg 1981;3:346.
21. Alexander JC, Chretien PB, Deecon AL, Snyder J. CEA levels in patients with carcinoma of the esophagus. Cancer 1978;42:1492-1497.
22. Abelev GI, Perova SD, Khramkova NI, et al. Production of embryonal A-globulin by transplantable mouse hepatomas. Transplantation 1963;1:174-180.
23. Waldman TA, McIntyre KR. The use of a radioimmunoassay for alpha-fetoprotein in the diagnosis of malignancy. Cancer 1974;34:1510-1515.
24. Abelev GI. Alpha-protein in oncogenesis and its association with malignant tumors. Adv Cancer Res 1971;14:295-358.
25. Gold P, Labitan A, Wong HCG, et al. Physiochemical approach to the purification of human A-fetoprotein from the ascites fluid of a hepatoma-bearing patient. Cancer Res 1978;38:6-12.
26. Ishiguro J, Sugitoehi I, Sakaguchi H, et al. Serum alpha-fetoprotein subfractions in patients with primary hepatoma or hepatic metastasis of gastric cancer. Cancer 1985;55:156-159.
27. Purves LR, Bershon I, Path FC, et al. Serum alpha-fetoprotein and primary cancer of the liver in man. Cancer 1970;25:1261-1270.
28. Bloomer JR, Waldman TA, McIntire KR, et al. Alphafetoprotein in non-neoplastic hepatic disorders. JAMA 1975;233:38-41.
29. Jones PAE, Miller FM, Worwood M, et al. Ferritinaemia in leukemia and Hodgkin's disease. Br J Cancer 1973;27:212-217.
30. Lipschitz DA, Cook JD, Finch CA. A clinical evaluation of serum ferritin as an index of iron stores. N Engl J Med 1974;290:1213-1216.
31. Tanaka M, Kato K. The measurement of ferritin in the leukemic blasts with a "sandwich" type enzyme immunoassay method. Cancer 1983;51:61-64.
32. Tappin JA, George WD, Bellingham AJ. Effect of surgery on serum ferritin concentration in patients with breast cancer. Br J Cancer 1979;40:658-660.
33. Kew MC, Torrance JD, Derman D, et al. Serum and tumor ferritins in primary liver cancer. Gut 1978;19:294-299.
34. Hann HL, Levy HM, Evans AE. Serum ferritin as a guide to therapy in neuroblastoma. Cancer Res 1980;40:1411-1413.
35. Maxim PE, Veltri RW. Serum ferritin as a tumor marker in patients with squamous cell carcinoma of the head and neck. Cancer 1986;57(2):305-311.
36. Hammarstrom L, Fuchs T, Smith IE. The immunodepressive effect of human glucoproteins and their possible role in the nonrejection process during pregnancy. Acta Obstet Gynecol Scand 1979;58:417-422.
37. Harris CC, Primack A, Cohen MH. Elevated alpha-1-antitrypsin serum levels in lung cancer patients. Cancer 1974;34:280-281.

38. Baskies AM, Chretien PB, Weiss JF, et al. Serum levels of alpha 2 HS-glycoprotein and acute phase proteins correlate with cellular immunity in cancer patients. Proc Am Assoc Cancer Res 1978;19:221.
39. Baskies AM, Chretien PB, Yang C, et al. Serum glycoproteins and immunoglobulins in nasopharyngeal carcinoma. Am J Surg 1990:138:478-488.
40. Bradley WP, Blasco AP, Weiss JF, et al. Correlations among serum protein-bound carbohydrates, serum glycoproteins, lymphocyte reactivity, and tumor burden in cancer patients. Cancer 1977;40:2264-2272.
41. Baskies AM, Chretien PB, Weiss JF, et al. Serum glycoproteins in cancer patients: first report of correlations with in vitro and in vivo parameters of cellular immunity. Cancer 1980;45:3050-3060.
42. Chis LF, Oon CJ. Changes in serum α_1-antitrypsin, α_1-acid glycoprotein and β_2 glycoprotein in patients with malignant hepatocellular carcinoma. Cancer 1978;43: 596-604.
43. Wolf GT, Chretien PB, Elias EG, et al. Serum glycoproteins in head an neck squamous carcinoma: correlations with tumor extent, clinical tumor stage, and T-cell levels during chemotherapy. Am J Surg 1979;138:489-500.
44. Bucovaz ET, Schweikert AW. B-protein assay use in cancer detection. In: Nieburgs HE, ed. Third international symposium, detection and prevention of cancer, vol. 1. New York: Marcel Dekker, 1978:257-269.
45. Bucovaz ET, Schweikert AW. Cancer detection: assay procedures designed to detect B-protein. Am Clini Prod Rev 1986;11:10-16.
46. Bucovaz ET. An assessment of serological testing for cancer and the B-protein assay. IRCS Med Sci 1983;11:1-4.
47. Bjorklund B, Bjorklund V. Antibody of pooled human malignant and normal tissues by cyto-immunological techniques. Int Arch Allergy 1957;10:153.
48. Bjorklund B. Tissue polypeptide antigen (TPA): biology, biochemistry, improved assay methodology, clinical significance in cancer and other conditions, and future outlook. Antiboitics Chemother 1978;22;16-31.
49. Inoue M, Inoue Y, Hiramatsu K, et al. The clinical value of tissue polypeptide antigen (TPA) in patients with gynecological tumors. Cancer 1985;55:2618-2623.
50. Buccheri BF, Ferrigno D, Sartoris AM, et al. Tumor markers in bronchogenic carcinoma: superiority of tissue polypeptide antigen to carcinoembryonic antigen and carbohydrate antigenic determinant 19-9. Cancer 1987;60:42-50.
51. Menendez-Bolet CJ, Detegen HF, Pinsky CM, Schwartz MK. A preliminary evaluationof tissue polypeptide antigen in serum or urine (or both) of patients with cancer or benign neoplasms. Clin Chem 1978;24:868-872.
52. Mross KB, Wolfrom DI, Rauschecker H. Determination of tissue polypeptide antigen (TPA) levels in different cancer types an controls. Oncology 1985;42:288-295.
53. Cunningham BA, Berggard I. Structure, evolution, and significance of B_2 microglobulin. Transplant Rev 1974;21:3-14.
54. Vinzenz K, Schonthal E, Zekert F, Wunderer S. Diagnosis of head and neck carcinomas by means of immunological tumor markers (beta-2-microglobin, immunoglobulin E, ferritin, N-acetyl-neuraminic acid, phosphohexase-isomerase). J Craniofac Surg 1987;15:270-277.
55. Teasdal C, Mander AM, Fifield R, et al. Serum B_2 microglobulin in controls and cancer patients. Clin Chem 1973;78:135-142.
56. Hallgren R, Nou E, Lundquist G. Serum B_2 microglobulin in patients with bronchial carcinoma and controls. Cancer 1980;45:780-785.
57. Amlot PL, Adinolfi M. Serum β_2 microblogulin and its prognostic value in lymphomas. Eur J Cancer 1979;15:791-796.

58. Descos L, Andre C, Beorchia S, et al. Serum levels of β_2 microglobulin. A new marker of activity in Crohn's disease. N Engl J Med 1979;301:440-441.
59. Rashid SA, Cooper EH, Axon ATR, et al. Serum β_2 microglobulin in malignant and benign diseases of the stomach and pancreas. Biomedicine 1980;33:112-116.
60. Narayanan S. Alkaline phosphatase as a tumor marker. Ann Clin Lab Sci 1983;13:133-13.
61. Katz AE, Hong W, Bhutani R, et al. Prognostic indicators in chemotherapy for head and neck carcinoma: alkaline phosphatase levels. Laryngoscope 1980;90:924-929.
62. Coker DD, Morris D, Elias EG, et al. Head and neck cancer: relationship of the pre-chemotherapy serum alkaline phosphatase levels to response rate of induction chemotherapy. Arch Otolaryngol 1982;108:28-29.
63. Burres SA, Jacbos JR, Peppard SB, Al-Sarraf M. Significance of alkaline phosphate and chemotherapy for head and neck carcinoma. Otolaryngol Head Neck Surg 1982; 90:188-192.
64. Fishman WH, Inglis NI, Stolbach LL' et al. A serum alkaline phosphatase isoenzyme of human neoplastic cell origin. Cancer Res 1968;28:150-154.
65. Nathanson L, Fishman WH. New observations on the Regan isoenzyme of alkaline phosphatase in cancer patients. Cancer 1971;55:1388-1397.
66. Pollet DE, Nouwen E, Schilstraete FB, et al. Enzyme-antigen immunoassay for human placental alkaline phosphatase in serum and tissue extracts and its application as a tumor marker. Clin Chem 1985;31:41-45.
67. Eerdekens MW, Nouwen EJ, Pollet DE, et al. Placental alkaline phosphatase and cancer antigen 125 in sera of patients with benign and malignant disease. Clin Chem 1985;31:687-690.
68. Muensch HA, Maslow WC, Azama F, et al. Placental-like alkaline phosphatase: re-evaluation of the tumor marker with exclusion of smokers. Cancer 1986;58:1689-1694.
69. Hill BR, Levi C. Elevation of a serum component in neoplastic disease. Cancer Res 1954;14:513-515.
70. Vanderlinde RE. Measurement of total lactate dehydrogenase activity. Ann Clin Lab Sci 1985;15:13-31.
71. Schneider RJ, Seibert K, Passe S, et al. Prognostic significance of serum lactate dehydrogenase in malignant lymphoma. Cancer 1980;46:139-143.
72. Rotenberg Z, Weinberger I, Fuchs Y, et al. Elevation of serum lactic dehydrogenase levels as an early marker of occult malignant lymphoma. Cancer 1984;54:1379-1381.
73. Goldman R, Kaplan NO, Hall TC. Lactic dehydrogenase in human neoplastic tissues. Cancer Res 1964;24:389-399.
74. Kornberg A, Polliack A. Serum lactic dehydrogenase (LDH) levels in acute leukemia: marked elevations in lymphoblastic leukemia. Blood 1980;56:351-355.
75. Quinn JJ, Altman AJ, Frantz CN. Serum lactic dehydrogenase, an indicator of tumor activity in neuroblastoma. J Pediatr 1980;97:89-91.
76. Farley FA, Healey JH, Caparros, Sison B, et al. Lactase dehydrogenase as a tumor marker for recurrent disease in Ewing's sarcoma. Cancer 1987;59:1245-1248.
77. Ratcliff CR, Culp TW, Hall FF. Serum lactic dehydrogenase and other enzymes in malignant disease: comparison with an immunodiagnostic test for cancer. Am J Gastroenterol 1971;56:199-208.
78. Ridgway D, Smiley S, Neerhout RC. The prognostic value of presenting serum LDH in non-Hodgkin lymphoma. J Pediatr 1981;99:611-613.
79. Jacobs DS, Robinson RA, Clark GM, et al. Clinical significance of the isomorphic pattern of the isoenzymes of serum lactate dehydrogenase. Ann Clin Lab Sci 1977; 7:411-421.

80. Goel H, Kohli GS, Lal H. Serum phosphohexose isomerase levels in patients with head and neck cancer. J Laryngol Otol 1986;100:581-585.
81. Lal H, Madan HC, Kohli GS, Yadav SPS. Serum enzyme in head and neck cancer II. Allesterase. J Laryngol Otol 1987;101:819-822.
82. Lal H, Munjal SK, Wig U, Saini AS. Serum enzymes in head and neck cancer III. J Laryngol Otol 1987;101:1062-1065.
83. Moore BW, McGregor T. Chromatographic and electrophonetic fractionation of soluble proteins of brain and liver. J Biol Chem 1965;240:1647-1653.
84. Prinz RA, Marangos PJ. Use of neurospecific enolase as a serum marker for neuroendocrine neoplasms. Surgery 1982;97:887-889.
85. Schmechel D, Marangos PJ, Brightman M. Neuron specific enolase in a molecular marker for peripheral and central neuroendocrine cells. Nature 1978;273:834-836.
86. Marangos PJ, Schnechel DE, Parma M, et al. Measurement of neuron specific (NSE) and non-neuronal (NNE) isoenzymes of enolase in rat, monkey, and human nervous tissue. J Neurochem 1979;33:319-329.
87. Ishiguro Y, Kato K, Ito T, et al. Nervous system specific enolase in serum as a marker for neuroblastoma. Pediatrics 1983;72:696-700.
88. Beemer FA, Vlug A, van Vellen C, et al. Isoenzyme pattern of enolase of childhood tumors. Cancer 1984;54:293-296.
89. Zeltzer PM, Marangos PJ, Satter H, et al. Prognostic importance of serum neuron specific enolase in local and widespread neuroblastoma. Progr Clin Biol Res 1984; 17:319-329.
90. Notomi T, Morikawa J, Kata K, et al. Radioimmunoassay development for human neuron-specific enolase with some clinical results in lung cancers and neuroblastoma. Tumour Biol 1985;6:57-66.
91. Zeltzer PM, Marangos PJ, Evans AE, Schneider SL. Serum neuron-specific enolase in children with neuroblastoma. Cancer 1986;57:1230-1234.
92. Cooper EH, Pritchard J, Bailey CC, Ninane J. Serum neuron-specific enolase in children's cancer. Br J Cancer 1987;56:65-67.
93. Carney DN, Marangos PJ, Inde DC, et al. Serum neuron-specific enolase: a marker for disease extent and response to therapy of small cell lung cancer. Lancet 1982; 1:583-585.
94. Johnson DH, Marigos PJ, Forbes JT, et al. Potential utility of serum neuron-specific enolase levels in small cell carcinoma of the lung. Cancer Res 1984;44:5409-5414.
95. Akoun GM, Scarna HM, Milleron BJ, et al. Serum neuron-specific enolase: a marker for disease extent and response to therapy for small-cell lung cancer. Chest 1985; 87:39-43.
96. Natomi L, Morikawa J, Kato K, et al. Radioimmunoassay development for human neuron-specific enolasee with some clinical results in lung cancers and neuroblastoma. Tumour Biol 1985;6:57-66.
97. Fujita K, Haimoto H, Inaizumi M, et al. Evaluation of neuron-specific enolase as a tumor marker for lung cancer. Cancer 1987;60:362-369.
98. Kato K, Ishiguro Y, Suzuki F, et al. Distribution of nervous system-specific forms of enolase in peripheral tissues. Brain Res 1982;239:441-448.
99. Stepanas AV, Samoan NA, Hile CS, Hickey RC. Medullary thyroid carcinoma: importance of serial serum calcitonin measurement. Cancer 1979.43:825-837.
100. Sikri L, Varndell IM, Hamud JA, et al. Medullary carcinoma of the thyroid: an immunocytochemical and histochemical study of 25 cases using eight separate markers. Cancer 1985;56:2481-2491.
101. Palmer BV, Harmer CL, Shaw HJ. Calcitonin and carcinoembryonic antigen in the follow-up of patients with medullary carcinoma of the thyroid. Br J Surg 1984;71:101-104.

102. Graze K, Spiler IJ, Tashjian AH Jr, et al. Natural history of familial medullary thyroid carcinoma: effect of a program for early diagnosis. N Engl J Med 1978; 299:980-985.
103. Goltzman D, Potts JT, Ridgway EC, et al. Calcitonin as a tumor marker. Use of radioimmunoassay for calcitonin in the postoperative evaluation of patients with medullary thyroid carcinoma. N Engl J Med 1974;290:1035-1039.
104. DeLillis RA, Rule AH, Spiler I, et al. Calcitonin and carcinoembryonic antigen as tumor markers in medullary thyroid carcinoma. Am J Clin Pathol 1978;70:587-594.
105. Poston GJ, Seitz PK, Townsend CM, et al. Calcitonin gene-related peptide: possible tumor marker for medullary thyroid cancer. Surgery 1987;102:1049-1054.
106. Coombes RC, Hillyard C, Greenburg PB, et al. Plasma immunoreactive calcitonin in patients with non-thyroid tumors. Lancet 1974;1:1080-1083.
107. Silva OL, Becker KL, Primack A, et al. Ectopic secretion of calcitonin by oat-cell carcinoma. N Engl J Med 1974;290:1122-1124.
108. Silva OL, Broder LE, Doppmann JL, et al. Carcitonin as a marker for bronchogenic cancer: a prospective study. Cancer 1979;44:680-684.
109. Mueder H, Hacking WHL, Silberbusch J, et al. Value of serum calcitonin estimation in clinical oncology. Br J Cancer 1981;43:786-792.
110. Austin LA, Heath H. Calcitonin physiology and pathophysiology. N Engl J Med 1981;304:269-278.
111. Body JJ, Kellermann Gk Muquardt C, Brokowski G. Serum immunoreactive calcitonin: useful as a general tumor marker? Horm Metab Res 1983;15:624-625.
112. Cate CC, Douple EB, Andrews KM, et al. Calcitonin as an indicator of the response of human small cell carcinoma of the lung cells to drugs and radiation. Cancer Res 1984;44:949-954.
113. Hom D, Waxman K. Calcitonin as a tumor marker for nonthyroid neoplasia. J Surg Oncol 1985;29:59-60.
114. Melvin KEW, Miller HH, Tashjian AH. Early diagnosis of medullary carcinoma of the thyroid gland by means of calcitonin assay. N Engl J Med 1971;285:1115-1120.
115. Melvin KEW, Tashjian AH Jr, Miller HH. Studies in familial (medullary) thryoid carcinoma. Rec Prog Horm Res 1972;28:399-470.
116. Wolfe HF, Melvin KEW, Cervi-Skinner S, et al. C-cell hyperplasia preceding medullary thyroid carcinoma. N Engl J Med 1973;289:437-441.
117. Lipton A, Sheehan L, Kessler GAF. Urinary polyamine levels in human cancer. Cancer 1975;35:464-468.
118. Russell DH, Levy CC, Schimpff CC, Hawk IA. Urinary polyamines in cancer patients. Cancer Res 1971;31:1555-1558.
119. Sanford EJ, Drago JR, Rohner TJ, et al. Preliminary evaluation of urinary polyamines in the diagnosis of genitourinary tract malignancy. J Urol 1975;113:218-221.
120. Waalkes TP, Gehrke CW, Tormey DC, et al. Urinary excretion of polyamines by patients with advanced malignancy. Cancer Chemother Rep 1975;59:1103-1106.
121. Lipton A, Sheehan L, Mortel R, Harvey HA. Urinary polyamine levels in patients with localized malignancy. Cancer 1976;38:1344-1347.
122. Tormey DC, Walkes TP, Kuo KC, et al. Biologic markers in a breast carcinoma: clinical correlations with urinary polyamines. Cancer 1980;46:741-747.
123. Edwards MSB, Davis RL, Laurent JP. Tumor markers and cytologic features of cerebrospinal fluid. Cancer 1985;56:1773-1777.
124. Russell DH. Clinical relevance of polyamines as biochemical markers of tumor kinetics. Clin Chem 1977;23:22-27.

125. Cooper KE, Shukla JB, Rennert OM. Polyamine compartmentalization in various human disease states. Clin Chim Acta 1978;82:1-7.
126. Savory J, Shipe JR, Wills MR. Polyamines in blood cells as cancer markers. Lancet 1979;2:1136-1137.
127. Shideler CE, Johns ME, Cantrele RW, et al. Erythrocyte polyamine determination in patients with head and neck cancer. Arch Otolaryngol 1981;107:752-754.
128. Katz AE. Advances in the immunology of head and neck cancer. Otolaryngol Clin North Am 1980;13:431-436.
129. Catalona WJ, Sample WF, Chretien PB. Lymphocyte reactivity in cancer patients: Correlation with tumor histology and clinical stage. Cancer 1973;31:65-71.
130. Maisel RH, Ogura JH. Abnormal dinitrochlorobenzene skin sensation: a prognostic sign of survival in head and neck squamous cell carcinoma. Laryngoscope 1973; 83:2012-2019.
131. Twomey PL, Catalona WJ, Chretien PB. Cellular immunity in cured cancer patients. Cancer 1974;33:435-440.
132. Olivari A, Pradier R, Feierstein J, et al. Cell-mediated immune response in head and neck cancer patients. J Surg Oncol 1976;8:287-294.
133. Jenkins VK, Ray P, Ellis HN, et al. Lymphocyte response in patients with head and neck cancer. Arch Otolaryngol 1976;102:596-600.
134. Maisel RH, Ogura JH. Dinitrochlorobenzene skin sensitization and peripheral lymphocyte count: predictors of survival in head and neck cancer. Ann Otol 1976; 85:517-522.
135. Betto RM, Burger DR, Vandenbard AA, Finke PE. Changes in tumor immunity during therapy determined by leukocyte adherence inhibition and dermal testing. Cancer 1978;41:1034-1039.
136. Papenhausen PR, Kukwa A, Croft CB, et al. Cellular immunity in patient with epidermoid cancer of the head and neck. Laryngoscope 1979;89:538-549.
137. Menzio P, Cortesina G, Sartoris A, et al. Relationships between cervical node histological patterns and rosette test scores: possible prognostic value in laryngeal cancer. Laryngoscope 1980;90:1032-1038.
138. Eilber FR, Morton DL, Ketcham AS. Immunologic abnormalities in head and neck cancer. Am J Surg 1974;128:534-537.
139. Hilal EY, Wanebo HJ, Pinsky CM, et al. Immunologic evaluation and prognosis in patients with head and neck cancer. Am J Surg 1977;134:469-473.
140. Veltri RW, Sprinkle PM, Maxim PE, et al. Immune monitoring protocol for patients with carcinoma of the head and neck. Ann Otol 1978;87:692-700.
141. Zighelboim J, Dorey F, Parker NH, et al. Immunologic evaluation of patients with advanced head and neck cancer receiving weekly chemoimmunotherapy. Cancer 1970;44:117-123.
142. Mandel MA, Dvorak K, DeCosse JJ. Salivary immunoglobulins in patients with oropharyngeal and bronchopulmonary carcinoma. Cancer 1973;31:1408-1413.
143. Brown AM, Lally ET, Frankel A, et al. The association of the IgA levels of serum and whole saliva with the progression of oral cancer. Cancer 1975;35:1154-1162.
144. Vincenz K, Parelka R, Schonthal E, et al. Serum immunoglobulin levels in patients with head and neck cancer (IgE, IgA, IgM, IgG). Oncology 1986;43:316-322.
145. Katz AE, Yoo TJ, Harker LA. Serum immunoglobin A (IgA) levels in carcinoma of the head and neck. TAAOO 1976;82:131-137.
146. Katz AE, Nysather JO, Harker LA. Major immunoglobulin ratios in carcinoma of the head and neck. Ann Otol 1978;87:412-415.
147. Ockhuizen T, Pandey JP, Veltri RW, et al. Immunoglobulin allotypes in patients with squamous cell carcinoma of the head and neck. Cancer 1982;49:2021-2024.

148. Baseler NW, Maxim PE, Veltri RW. Circulating IgA immune complexes in head and neck cancer, nasopharyngeal carcinoma, lung cancer and colon cancer. Cancer 1987;59:1727-1731.
149. Winzler RJ. Determination of serum glycoprotein methods. Biochem Anal 1958; 2:279-311.
150. Dnistrian AM, Schwartz MK. Plasma lipid-bound sialic acid and carcinoembryonic antigen in cancer patients. Clin Chem 1981;27:1737-1739.
151. Silver HKB, Rangel DM, Morton DL. Serum sialic acid elevations in malignant melanoma patients. Cancer 1978;4:1497-1499.
152. Katopodis N, Hirshaut Y, Geller NL, Stock CC. Lipid associated sialic acid test for the detection of human cancer. Cancer Res 1982;42:5270-5275.
153. Santamaria L, Galioto GB, Benazzo M, et al. Serum lipid sialic acid: a valuable tumor marker for laryngeal carcinomas detection and clinical monitoring. In: Robietti G, Galioto G, Dall'Acqua F, et al. (eds). Medecine Biologie Environment, vol. 13(1), 1985; 287-291.
154. Fossel ET, Carr JM, McDonagh J. Detection of malignant tumors. Water-suppressed proton nuclear magnetic resonance spectroscopy of plasma. N Engl J Med 1986;315:1369-1376.
155. Schein PS. Oncolipids? A possible new approach to diagnosis of cancer. N Engl J Med 1986;315:1410-1411.
156. Zylke JW. Oncolipids offer promise as universal screening tool for cancer. JAMA 1987;257:1849.
157. VanDerWerf M. New diagnostic technique for cancer to enter clinical trials. Oncol Biotechnol News 1987;1(3):1, 18.
158. Old LJ, Boyse EA, Oehgen HF, et al. Precipitating antibody in human serum to an antigen present in cultured Burkitt's lymphoma cells. Proc Natl Acad Sci USA 1966;56:1699-1704.
159. Henle W, Henle G, Ho HC, et al. Antibodies to Epstein-Barr virus in nasopharyngeal carcinoma, other head and neck neoplasms and control groups. J Natl Cancer Inst 1979;44:225-237.
160. Miller D, Goldman JM, Goodman ML. Etiologic study of nasopharyngeal cancer. Arch Otolaryngol 1971;94:104-108.
161. DeSchryvev A, Klein G, Henle G, et al. EB-virus associated serology in malignant disease: antibody levels to viral-capsid antigens (VCA) membrane antigens (MA) and early antigens (EA) in patients with various neoplastic conditions. Int J Cancer 1972;9:353-364.
162. Henle W, Ho HC, Henle G, Kwan HC. Antibodies to Epstein-Barr virus related antigens in nasopharyngeal carcinoma. Comparison of active cases with long term survivors. J Natl Cancer Inst 1973;51:361-369.
163. Ho HC, Ng MH, Kwan HC, Chau JCW. Epstein-Barr virus specific IgA and IgG specific antibodies in nasopharyngeal carcinoma. Br J Cancer 1976;34:656-659.
164. Ho HC, Ng MH, Kwan HC. Immunohistochemistry of local immunoglobulin production in nasopharyngeal carcinoma. Br J Cancer 1987;37:514-519.
165. Ho HC, Ng MH, Kwan HC. Factors affecting serum IgA antibody to Epstein-Barr virus capsid antigens in nasopharyngeal carcinoma. Br J Cancer 1978;37:356-362.
166. Halili MR, Spigland I, Foster N, Ghossein NAA. Epstein-Barr virus (EBV) antibody in patients treated by radical radiotherapy for head and neck cancer. J Surg Oncol 1978;10:457-463.
167. Coates HL, Pearson GV, Bryan N, et al. An immunologic basis for the detection of occult primary malignancies of the head and neck. Cancer 1978;41:912-918.
168. Ho HC, Ng MH, Kwan HC. Factors affecting serum IgA antibody to Epstein-Barr viral capsid antigens in nasopharyngeal carcinoma. Br J Cancer 1978;37:356-362.

169. Neel HB, Pearson GR, Weiland LH, et al. Anti-EBV serologic tests for nasopharyngeal carcinoma. Laryngoscope 1980;90:1981-1989.
170. Neel HB, Pearson GR, Weiland LH, et al. Immunologic detection of occult primary cancer of the head and neck. Otolaryngol Head Neck Surg 1981;89:230-234.
171. Neel HB, Pearson GR, Weiland LH, et al. Immunologic detection of occult primary cancer of the head and neck. Otolaryngol Head Neck Surg 1981;89:230-234.
172. Ringhorgs, Henle W, Henle G, et al. Epstein Barr virus specific serodiagnostic tests in carcinomas of the head and neck. Cancer 1983;52:1237-1243.
173. Neel HB, Pearson GR, Weiland LH, et al. Application of Epstein-Barr virus serology to the diagnosis and staging of North American patients with nasopharyngeal carcinoma. Otolaryngol Head Neck Surg 1983;91:255-262.
174. Sako K, Minswada J, Marchetta F. Epstein Barr Virus antibodies in patients with carcinoma of the nasopharynx and carcinoma of other sites in the head and neck. Am J Surg 1975;130:437-439.
175. Coates HL, Pearson FR, Neel HB, et al. An immunologic basis for detection of occult primary malignancies of the head and neck. Cancer 1978;41:912-918.
176. Hollinshead A, Lee OB, McKelway W, et al. Reactivity between herpes virus type 2-related soluble cervical tumor cell membrane antigens and matched cancer control sera. Proc Soc Exp Biol Med 1972;141:688-693.
177. Hollinshead AC, Lee O, Chretien PB, et al. Antibodies to herpes virus nonvirion antigens in squamous carcinomas. Science 1972;182:713.
178. Hollinshead A, Tarro G, Foster WA, et al. Studies of tumor-specific and herpes non-virion antigens. Cancer Res 1974;34:1122-1125.
179. Balaram P, Pillai MR, Abraham T. Immunology of premalignant and malignant conditions of the oral cavity. II. Circulating immune complexes. J Oral Pathol 1987;16:389-391.
180. Shillitoe EJ, Greenspan D, Greenspan JS, Silverman S. Immunoglobulin class of antibody to HSV in patients with oral cancer. Cancer 1983;51:65-71.
181. Henderson BE, Louie E, Bogdanoff E, et al. Antibodies to herpes group viruses in patients with nasopharayngeal and other head and neck cancers. Cancer Res 1974; 34:1207-1210.
182. Smith HG, Horowitz N, Silverman NA, et al. Humoral immunity to herpes simplex viral-induced antigens in smokers. Cancer 1976;38:1155-1162.
183. Hollinshead A, Tarro G, Foster WA, et al. Studies of tumor-specific and herpes non-virion antigens. Cancer Res 1974;34:1122-1125.
184. Silverman NA, Alexander JC Jr, Hollinshead AC, Chretien PB. Correlation of tumor burden with in vitro lymphocyte reactivity and antibodies to herpes virus tumor-associated antigens in head and neck squamous carcinoma. Cancer 1976;37:135-140.
185. Smith HG, Chretien PB, Henson DE, et al. Viral specific humoral immunity to herpes induced antigens in patients with squamous carcinoma of the head and neck. Am J Surg 1976;132:541-548.
186. Shillitoe EJ, Greenspan D, Greenspan JS, Silverman S, Hansen LS. Neutralizing antibody to herpes simplex virus type I in patients with oral cancer. Cancer 1982; 49:2315-2320.
187. Syrjanen K, Syrjanen S, Pyrhonen S. Human papilloma virus (HPV) antigens in lesions of laryngeal squamous cell carcinomas. ORL 1982;44:323-334.
188. Syrjanen K, Syrjanen S, Lamberg M, Pyrhonen S, Nuutinen J. Morphological and immunohistochemical evidence suggesting human papilloma virus (HPV) involvement in oral squamous cell carcinogenesis. Int J Oral Surg 1983;12:418-424.

189. Loning T, Ikenberg H, Becker J, Gissmann L, Hoepfer, zur Hausen H. Analysis of oral papillomas, leukoplakias, and invasive carcinomas for human papilloma virus type related DNA. J Invest Dermatol 1985;84:417-420.
190. Schantz S, Dakmezian RH, Goepfert H, Batsakis JB, Katz RL. In situ hybridization for HPV subtypes in squamous carcinomas of the oral cavity (abstr). Arlington, VA: Second International Head and Neck Oncology Research Conference, September, 1987.
191. Sazito S, Komatsu N, Sakai M, et al. Specific antigens associated to squamous cell carcinoma isolated from maxillary cancer. In: Nieburgs HE, ed. Third international symposium. Detection and prevention of cancer, vol. 1. New York: Marcel Dekker, 1978:1467-1474.
192. Kato H, Torigoe T. Radioimmunoassay for tumor antigen of human cervical squamous cell carcinoma. Cancer 1977;40:1621-1628.
193. Kato H, Morioka H, Aramaki S, et al. Prognostic significance of the tumor antigen TA-4 in squamous cell carcinoma of the uterine cervix. Am J Obstet Gynecol 1983;145:350-354.
194. Kato H, Tamai K, Morioka H, et al. Tumor-antigen TA-4 in the detection of recurrence in cervical squamous cell carcinoma. Cancer 1984;54;1544-1546.
195. Maruo T, Shibata K, Kimura A, et al. Tumor-associated antigen, TA-4, in the monitoring of the effects of therapy for squamous cell carcinoma of the uterine cervix. Cancer 1985;56:302-308.
196. Gion M, Mione R, Dittadi R, Bruscagnin G. SCC antigen in patients with esophageal carcinoma. In: Kato H, de Bruijn HWA, Ebert W, Herberman RB, Johnson JT, eds. SCC antigen in the management of squamous cell carcinoma. New Jersey: Excerpta Medica, 1987:130-141.
197. Yasuda Y, Hanawa J, Miyamota Y, et al. SCC-antigen in patients with lung cancer. J Jpn Soc Cancer Ther 1986;21:1036-1048.
198. Johnson JT, Wagner RL, Eibling DE, Su S. Radioimmunoassay for SCC antigen in the diagnosis of squamous cell carcinoma of the head and neck: a preliminary report. In: Kato H, de Bruijn HWA, Ebert W, Herberman RB, Johnson JT, eds. SCC antigen in the management of squamous cell carcinomas. New Jersey: Excerpta Medica, 1987:112-123.
199. Fischbach W, Meyer T, Barthel K. Early results concerning the clinical usefulness of SCC antigen in oral and facial squamous cell carcinoma. In: Kato H, de Bruijn HWA, Ebert W, Herberman RB, Johnson JT, eds. SCC antigen in the management of squamous cell carcinoma. New Jersey: Excerpta Medica, 1987;124-129.
200. Eibling DE, Johnson JT, Wagner RL, Su S. SCC-RIA in the diagnosis of squamous cell carcinoma of the head and neck. Laryngoscope 1989;99(2):117-124.
201. Goodwin WJ, Kirchner JC, Sasaki CT. Plasma tumor markers associated with squamous carcinoma of the upper aerodigestive tract. In: Wolf GT and Carey TE (eds.) Head and neck oncology research. Kugler & Ghedini, California: 1988;69-73.
202. Al-Sarraf M, Chu CYT, Lai L, Carey MK, Drelichman A. Multiple tumor markers in monitoring patients with epidermoid cancer of the head and neck. Proc Am Assoc Cancer Res. 1981;22:285.

24
Uterine Tumors

HIROSHI KATO

Yamaguchi University School of Medicine, Ube, Japan

In gynecological practice, it is well documented that the sensitive radioimmunoassay using human chorionic gonadotrophin (HCG) has contributed greatly to the diagnosis and management of trophoblastic disease. Encouraged by the pioneering studies of Gold and Freedman who discovered carcinoembryonic antigen (CEA) [1], and more recently by the introduction of monoclonal antibody technology, a variety of tumor markers have been prepared for uterine cancer. These new tumor markers will be considered in two groups: tumor-associated antigens and viral antigens. Several tumor markers that have been used generally in other types of malignancies have also been applied to gynecological oncology: some of the oncofetal antigens (CEA, alpha-fetoprotein), placental antigens (HCG, Regan isoenzyme), tumor-associated antigens (CA 125), and other types of tumor markers (carbohydrate antigen). This chapter will review the recent experience/ studies of tumor markers in gynecological malignancies, focusing particularly on those prepared specifically for uterine cancer.

I. SPECIAL CONSIDERATIONS OF UTERINE TUMOR

The major malignancies of the uterus are cervical cancer and endometrial cancer. Approximately 90% of cervical cancer is squamous cell carcinoma, while more than 90% of endometrial cancer is adenocarcinoma (Table 1). Gestational trophoblastic disease, which includes hydatidiform mole, invasive mole, and choriocarcinoma, also appears in the uterus but the incidence rate of choriocarcinoma is less than 0.5 per 100,000 women.

The initial diagnosis of uterine cancer has been performed easily by use of cytological smear test, which is sensitive and specific enough to detect more than 90% of uterine cancers. In fact, the well-organized mass screening program using the Papanicolaou smear test has contributed to a reduction in the death rate from cervical and endometrial carcinoma. However, despite tremendous efforts to stress its effectiveness, the cytological test is used in less than 15% of the target population in Japan. This has resulted in high percentages of advanced stages of disease at initial diagnosis. Also, there are no reliable methods for monitoring disease progress or evaluating treatment efficacy.

Table 1 Histological Types of Uterine Tumor

Cervical tumor
    Squamous cell carcinoma
    Adenocarcinoma
    Undifferentiated carcinoma
Endometrial tumor
    Adenocarcinoma
        Adenocarcinoma, endocervical type
        Endometrioid adenocarcinoma
        Clear cell adenocarcinoma
        Adenoid cystic carcinoma
    Squamous cell carcinoma
    Adenosquamous carcinoma
    Undifferentiated carcinoma

The early diagnosis of recurrent tumor is still difficult. The major aims of the clinical application of tumor markers should be directed to improving these clinical problems.

II. TUMOR MARKERS SPECIFIC FOR UTERINE TUMORS

A. SCC

Squamous cell carcinoma antigen (SCC), formerly referred to as tumor antigen-4 (TA-4), is a protein originally purified squamous cell carcinoma of the uterine cervix [2]. Recent studies with isoelectric focusing showed that SCC was further divided into at least 14 subfractions. Each subfraction had almost the same molecular weight of 42,000-48,000 daltons, and has several antigenic determinants in common with the other subfractions. These subfractions were roughly classified into two subgroups: acidic SCC with pI of less than 6.25, and neutral SCC with pI of 6.25 or greater. Squamous cell carcinoma contains both acidic and neutral SCC. Acidic SCC is released into the surrounding tissue and appears in the blood circulation. SCC in the nonmalignant squamous epithelium is mainly neutral SCC [3]. SCC antigen is one of the TA-4 subfractions, which consists of 334 amino acids with pI of 6.5. The currently available radioimmunoassay kit for SCC antigen (SCC RIA Beads, DaiNabot Ltd Co, Tokyo, Japan) can detect all SCC subfractions with a minimum sensitivity of 0.2 ng/ml. The results of the assay are expressed in nanograms of SCC antign [4].

1. Tissue Localization of SCC

Immunohistochemical studies have shown the presence of SCC in the cytoplasm of squamous cell carcinoma, particularly in those of the keratinizing type and large-cell nonkeratinizing type [5-7]. Matsuta et al. [9] reported the intracellular localization of SCC by immunoelectron microscopic studies, in which SCC was detected in the cytosols, but not in tonofilaments, of the neoplastic cells [8]. SCC cells also appeared in the intermediate layer of the normal squamous epithelium [5,7,9]. Hoshina et al. therefore suggested that SCC might be associated with squamous maturation or metaplasia [7]. It was interesting that SCC levels were often increased in the nonmalignant cells when they were adjacent to the tumor nest [9]. It is not clear whether these cells with high SCC content will progress to malignant cells, but the expression of SCC appears to indicate some change in the nature of squamous cells. Also, the expression of acidic SCC was heterogeneous in the tumor cells, and increased particularly in those cells at the periphery of the tumor nest [10]. It remains to be verified whether the appearance of acidic SCC in these particular area of the tumor is due to the change in the cells or to some influences from the neighboring normal tissues. It is also unclear whether the appearance of acidic SCC is related to the malignant potential of tumor cells. However, it is clear that biochemical analysis of cells provides different information from morphological evaluation, and can be useful for studying the nature of malignant cells.

2. Clinical Values of SCC

With the recent development of a sensitive assay for SCC antigen, the initial cutoff value of circulating SCC antigen (2.0 ng/ml) has been reappraised. The new cutoff is 1.5 ng/ml. Approximately 51-67% of cervical squamous cell carcinomas showed elevated values of serum SCC antigen [11-14]. The positive rate was increased according to the clinical stages and high pretreatment values of circulating SCC antigen identified a "high risk group of patients" with poor prognosis [15].

Serum SCC levels were increased in 17-33% of patients with cervical adenocarcinoma [13,14] and 8-29% of those with endometrial carcinoma [12,13]. Kenemans et al. [16] reported that the combination assay using SCC antigen and CA 125 was more sensitive for detecting endometrial carcinoma than CA 125 alone. Other histological types of malignancy showed very low rates of circulating SCC. Some instances of false-positve results included patients with pulmonary diseases (13%), benign ovarian tumor (6.7%), myoma of the uterus (3-4%), and hepatitis (9.1%) [11].

Serum SCC values usually decrease to undetectable levels within at least 48 hr after surgery [6], and serial determinations of serum SCC levels are useful to evaluate rapidly the efficacy of therapy and disease progress.

In patients with recurrent cervical cancer, SCC antigen was found in 83-100% of cases, with a lead time of 3.6 months [12,17]. Serum SCC is particularly useful for

Table 2 Detection Rates of Circulating Tumor Markers in Uterine Tumor

Tumor Marker	Cervical Cancer	Endometrial Cancer	References
CEA	402/989 (40.6)	72/179 (40.2)	14,30,34-37,39-41
SCC	387/730 (53.0)	29/82 (35.4)	6,11-14
AFP	80/175 (45.5)	21/42 (50.0)	37
HCG	54/287 (18.8)	26/167 (15.6)	37,50
HSAP	46/233 (19.7)	4/62 (6.5)	34,52
CA 125	18/84 (21.4)	47/109 (43.1)	14,41,43,44

detecting distant recurrence, while an abnormal cytological appearance is often the first sign of vaginal recurrence. Combination testings with the vaginal cytological test and circulating marker values would be recommended for the early diagnosis of recurrent cervical carcinoma.

B. Viral Antigens

A number of reports have appeared regarding the association of herpes simplex virus (HSV) or human papilloma virus (HPV) infection with squamous cell carcinoma of the female genital tract.

The antibody against HSV type 2 is detected in 88-89% of invasive cervical carcinomas [18,19]. Tarro et al. [20] demonstrated with studies from a serum panel from NCI-Mayo Clinic Serum Bank, that seven of eight patients (87.5%) with cervical cancer showed elevated levels of antibody against HSV. These authors also noted that the changes in the antibody titers reflected the progress of the disease: they decreased after tumor resection and increased in patients with recurrent disease. A virus antigen was purified by Tarro et al., which was a glycoprotein with a molecular weight of approximately 70,000 daltons [20].

Among HPV, types 16 and 18 are reported to be high-risk types often associated with invasive cervical cancer [21]. Although the epidemiological data suggested that HSV and HPV would play important roles in the pathogenesis of cervical cancer, the clinical value of viral antigens as tumor markers for cervical cancer requires further extensive studies.

C. Other Tumor-Associated Antigens of Uterine Cancer

Several tumor-associated antigens have been demonstrated, mostly by immunohistochemistry, in cervical cancer [22-24]. Tsuji et al. also recently reported on a new tumor marker (C-12 antigen) for endometrial adenocarcinoma, which appeared in the circulation of patients with ovarian cancer, endometrial carcinoma, and cervical carcinoma [25]. Although very few markers have been used in clinical practice, their investigation should be encouraged, and will provide useful information that allows us to understand better the biology of tumors.

III. TUMOR MARKERS IN GENERAL USE

A. Carcinoembryonic Antigen

In immunohistochemical studies, carcinoembryonic antigen (CEA) is detected in 25% of patients with mild dysplasias [26], 60-83% of cervical squamous cell carcinoma [23,27,28], and in 77% of cases of cervical adenocarcinoma [29]. On the other hand, CEA is not common in the endometrial adenocarcinoma tissue [30,31]. CEA might, therefore, be useful for differentiating endocervical from endometrial adenocarcinoma [32].

There are a number of reports on the clinical values of circulating CEA in patients with uterine tumor. Van Nagell et al. have reported elevated plasma CEA values in 23% of patients with moderate dysplasia, 26% of severe dysplasia, and 33% of carcinoma in situ [33], while other investigators showed a low rate of appearance of circulating CEA in patients with carcinoma in situ of the uterine cervix [34-36]. In patients with invasive cervical cancer, 36-80% have shown elevated serum CEA values [34-40], and high pretreatment CEA values generally indicate a wide spread of tumor [35] and poor prognosis [38].

Senekjian et al. [14] recently demonstrated the pretreatment values of serum CEA and SCC antigen in 30 patients with cervical carcinoma, and found that 13% of cases with normal SCC antigen values showed abnormal CEA levels. CEA is also useful for monitoring disease progress and detecting recurrent disease [38] (Table 3). However, we must be aware of several noncancerous conditions (e.g., smoking or hepatic dysfunction) in which circulating CEA levels are often elevated. Although CEA activity is weak in endometrial carcinoma tissue, serum CEA values are elevated in 14-100% of cases of endometrial cancer [35,39-41].

B. CA-125

Since its first report in 1983, CA-125 has been used widely for monitoring epithelial ovarian cancer [42]. Several reports have recently indicated that CA-125 could also be useful for uterine cancer. Niloff et al. [43] reported that serum CA-125 levels were increased in 14 of 18 patients with stage IV or recurrent endometrial cancer, whereas none of the 11 patients with stage I or II disease showed positive values. Malkasian et al. [44] also showed an increase in serum CA-125 values in 6 of 34 patients with endometrial adenocarcinoma, 6 of 15 patients with recurrent endometrial carcinoma, and 2 of 13 patients with recurrent or active uterine sarcoma. Levels of CA-125 were elevated in 7-18.8% patients with cervical squamous cell carcinoma [41,43], 83% of those with cervical adenocarcinoma, and in 100% of cases of fallopian tube cancer [43]. Senekjian et al. reported a combination assay using measurement of serum CEA, SCC, and CA-125 in 30 patients with cervical carcinoma, which showed that all of three markers were elevated in 10% of patients, CEA alone in 10%, SCC alone in 36.7%, and CA-125 alone in none [14]. CA-125 may therefore be useful as a marker for endometrial carcinoma.

Immunohistochemical studies demonstrated the presence of CA-125 in a variety of noncancerous tissues, including endocervical and endometrial epithelium of the uterus, epithelium of the fallopian tube, and decidual tissue. The presence of CA-125 in these normal tissues may cause false-positive results under some circumstances [45-47]. Serum CA-125 levels are increased in early stages of pregnancy and also in patients with severe peritoneal or pleural lesions (e.g., peritonitis or pleuritis; serum CA-125 values may be several thousand U/ml). Malkasian [44] showed an increase in serum CA-125 levels in 13 of 31 patients with endometriosis and 5 of 24 patients with leiomyomas. Although the clinical value of CA-125 in ovarian cancer is already established, further investigations are necessary before this marker can be used for uterine cancer.

C. Human Chorionic Gondadotropin

Human chorionic gonadotropin (HCG) is undoubtedly the best tumor marker for trophoblastic disease. The sensitive radioimmunoassay of HCG, with the use of effective antitumor drugs such as methotrexate, has greatly reduced the death rate from choriocarcinoma. HCG is produced and released from the syncytial trophoblast. The production of HCG is so active that 10^4-10^5 tumor cells are enough to release detectable amounts of HCG into the blood circulation. Moreover, radioimmunoassay for the beta subunit of HCG makes it possible to differentiate HCG from other glycoprotein hormones [48].

Table 3 Circulting Tumor Markers Related to Clinical Stage in Cervical Carcinoma

Tumor Marker	Clinical State, No. Positive/No. Tested (%)					References
	0	I	II	III	IV	
SCC[a]	11/87 (12.6)	40/162 (24.7)	116/167 (69.5)	75/86 (87.2)	29/32 (90.6)	6,11-14
CEA	36/215 (6.7)	81/206 (39.3)	117/239 (49.0)	95/150 (63.3)	21/31 (67.7)	34-39
AFP	20/66 (30.3)	19/36 (52.8)	17/39 (43.6)	20/28 (71.4)	2/7 (28.6)	37
HCG	2/6 (33.3)	19/85 (22.4)	16/71 (22.5)	11/48 (22.9)	1/11 (9.1)	37,50

[a]Detection rates of SCC in cervical squamous cell carcinoma.

Since 50-60% of choriocarcinoma develops from hydatidiform mole, changes in serum or urinary HCG values after the evacuation of hydatidiform mole should be monitored carefully. Serum HCG levels usually decrease to less than 1,000 mIU/ml within 5 weeks and to normal levels within 12 weeks after the evacuation of hydatidiform mole. Any abnormal regression pattern strongly indicates reappearance of malignant cells. Molecular heterogeneity has recently been demonstrated in the carbohydrate structure of HCG of choriocarcinoma [49]. An assay system to identify this molecular heterogeneity might be useful to differentiate the malignant changes from those occurring in normal pregnancy.

HCG also appears in the presence of nontrophoblastic tumors, but the results are still inconsistent. Rutanen et al. reported that serum HCG levels were elevated in 21% of cases of cervical cancer, and in 14% of endometrial cancers [50]. Donaldson et al. [37] showed an elevation of serum beta HCG in 24% of cases of invasive cervical carcinoma and 21% of cases of endometrial carcinoma. However, Bychko et al. could not detect any HCG activity in any gynecological tumors [28].

D. Other Tumor Markers

Donaldson et al. [37] and Seppala et al. [40] have reported that serum alpha fetoprotein levels are elevated in 21% of patients with cervical carcinoma. Also, Regan isoenzyme, the heat-stable, placental alkaline phosphatase, appeared in 21-22% of cases of cervical carcinoma and in 25% of cases of endometrial carcinoma [51,52].

Pregnancy-specific beta 1 glycoprotein (SP1) has been found in the normal syncytiotrophoblast and in choriocarcinoma. An interesting finding was that SP1 levels were positive in 11% of patients with choriocarcinoma who showed negative serum HCG levels [53]. Although HCG would be the first choice as a marker for trophoblastic disease, SP1 might be an alternative for those patients who do not produce any HCG. Placental protein 5 (PP-5) is also a glycoprotein, produced by the syncytiotrophoblast. PP-5 appears in the circulation during normal pregnancy and in the presence of hydatidiform mole, but not in choriocarcinoma or invasive mole. PP-10, PP-11, and PP-12 are found in some endometrial adenocarcinoma tissues. However, the clinical significance of these proteins has not yet been established [54].

IV. OVERVIEW OF GYNECOLOGICAL TUMOR MARKERS

Since the determination of circulating tumor markers has not aided in the detection of early malignancies, attempts have been made to use cells or secretions of the local region for the diagnosis of cervical carcinoma. However, often the cells must be washed thoroughly to remove the mucus that contains high amounts of tumor markers and influences the results of immunostaining. Suehiro et al. used flow cytometry to analyze the SCC contents of cytological specimens from the uterine cervix, and found that 82% of cervical carcinoma showed abnormal histograms [9]. The results of the immunocytochemistry of cervical smears are so far encouraging, and this may be a useful approach for the development of an automated screening system.

McDicken et al. [55] reported that CEA concentrations in cervical mucus were elevated in 85% of patients with cervical cancer. SCC is also increased in the cervical mucus of patients with malignant disease. This may be applicable to the diagnosis of cervical carcinoma. However, the isoelectric focusing analysis of the SCC subfractions in the cervical mucus revealed that the patterns of SCC subfractions in the mucus were very similar to those found in the tissue. Therefore, the amounts of tumor marker in the mucus would not directly reflect the potential of the cells to produce the marker substances, but might be influenced by the number of cells, including normal cells, destroyed in the mucus.

Immunohistochemical techniques have generally been used to identify the origin of tumor markers. However, positive marker values in blood are not always associated with positive immunohistochemistry staining. For example, cellular SCC contents are usually high in large-cell nonkeratinizing or keratinizing types of squamous cell carcinoma but negligible in the small-cell type, whereas there is no difference in the appearance of SCC in circulating blood among these three histological types of squamous cell carcinoma. This is also the case in some other tumor markers. In our studies with six patients with endometrial carcinoma, sialyl Le^x-i [57] was detected in all tumor tissues but could not be detected in the circulation, even in patients with advanced disease (Kato, unpublished data). We have also recently found a cell line of squamous cell carcinoma that releases SCC into the incubation medium, but cannot be stained immunohistochemically by SCC antibodies. Thus, the intensity of immunostainings of tumor marker would not necessarily reflect the ability of the cells to produce it.

There has been a feeling of optimism in the past that there might be a tumor-specific substance that would be present in every malignant cell, but would not appear in normal cells. This has turned to be unlikely. Since all cells in the body originate from the single fertilized cell, every cell should possess the same information on its genes as well. The only difference among the cells would be in the control system to regulate gene expression, to avoid any confusion that might occur. Once the control is broken by malignant changes, for example, each cell would show random gene expression. It is unlikely, therefore, that any such substances appear in all tumor cells as well. Likewise, every normal cell would have the same gene information as the malignant cells, and once the normal cells received a pathological stimuli (e.g., inflammation), it is very possible that those normal cells could produce abnormal substances, that is, tumor markers. Thus, attention should be focused now on what the tumor marker means by its expression in the cell, rather than the traditional discussion of whether the tumor marker is strictly specific to the malignant changes. Tumor marker would then be promising aids for investigating cell biology.

REFERENCES

1. Gold P, Freedman SO. Demonstration of tumor-associated antigens in human colonic carcinomata by immunological tolerance and absorption techniques. J Exp Med 1965; 122:439-463.
2. Kato H, Torigoe T. Radioimmunassay for tumor antigen of human cervical squamous cell carcinoma. Cancer 1977;40:1621-1628.
3. Kato H, Nagaya T, Torigoe T. Heterogeneity of a tumor antigen TA-4 of squamous cell carcinoma in relation to its appearance in the circulation. Gann 1984;75:433-435.
4. Ikeda I. Two-site radioimmunometric (sandwich) assay of SCC antigen using monoclonal antibodies. In: Kato H, de Bruijn HWA, Ebert W, Herberman RB, Johnson JT, eds. SCC antigen in the management of squamous cell carcinoma. Princeton: Excerpta Medica, 1987:215-226.
5. Ueda G, Inoue Y, Yamasaki M, et al. Immunohistochemical demonstration of tumor antigen TA-4 in Gynecologic tumors. Int J Gynecol Pathol 1984;3:291-298.

6. Maruo T, Shibata K, Kimura A, Hoshina M, Mochizuki M. Tumor-associated antigen, TA-4 in the monitoring of the effects of therapy for squamous cell carcinoma of the uterine cervix. Serial determinations and tissue localization. Cancer 1985; 56:302-308.
7. Hoshina M, Kimura A, Shibata K, Maruo T, Mochizuki M. Immunocytological distribution of the tumor antigen TA-4: expression during carcinogenesis and maturation of squamous epithelium of human uterine cervix. Asia Oceania J Obstet Gynaecol 1986;12:119-126.
8. Matsuta M, Kagabu T, Nishiya I. Immunohistochemical studies on tumor antigen-4 in squamous cell carcinoma of uterine cervix. Acta Histochem Cytochem 1987;20: 75-85.
9. Suehiro Y, Kato H, Nagai M, Torigoe T. Flow cytometric analysis of tumor antigen TA-4 in cervical cytologic specimens. Cancer 1986;57:1380-1384.
10. Kato H, Suehiro Y, Morioka H, et al. Heterogeneous distribution of acidic TA-4 in cervical squamous cell carcinoma: immunohistochemical demonstration with monoclonal antibodies. Jpn J Cancer Res (Gann) 1987;78:1246-1250.
11. Kato H, Morioka H, Hashimoto K, et al. SCC antigen and its clinical applications. In: Kato H, de Bruijn HWA, Ebert W, Herberman RB, Johnson JT, eds. SCC antigen in the management of squamous cell carcinoma. Princeton: Excerpta Medica, 1987:1-15.
12. de Bruijn HWA, Bouma J, Krans M, et al. SCC antigen, a useful tumor marker in patients with squamous cell carcinoma of the cervix. In: Kato H, de Bruijn HWA, Ebert W, Herberman RB, Johnson JT, eds. SCC antigen in the management of squamous cell carcinoma. Princeton: Excerpta Medica, 1987:18-33.
13. Cromback G, Wuerz H. Serum concentrations of SCC antigen in patients with cervical cancer. In: Kato H, de Bruijn HWA, Ebert W, Herberman RB, Johnson JT, eds. SCC antigen in the management of squamous cell carcinoma. Princeton: Excerpta Medica, 1987:54-63.
14. Senekjian EK, Young JM, Weiser PA, et al. An evaluation of squamous cell carcinoma antigen in patients with cervical squamous cell carcinoma. Am J Obstet Gynecol 1987;157:433-439.
15. Kato H, Morioka H, Aramaki S, Torigoe T. Prognostic significance of the tumor antigen TA-4 in squamous cell carcinoma of the uterine cervix. Am J Obstet Gynecol 1983;145:350-354.
16. Kenemans P, Yedema KA, Wobbes T, et al. Assay of SCC antigen in patients with breast cancer and other adenocarcinoma. In: Kato H, de Bruijn HWA, Ebert W, Herberman RB, Johnson JT, eds. SCC antigen in the management of squamous cell carcinoma. Princeton: Excerpta Medica, 1987:86-95.
17. Kato H, Tamai K, Morioka H, et al. Tumor-antigen TA-4 in the detection of recurrence in cervical squamous cell carcinoma. Cancer 1984;54:1544-1546.
18. Aurelian L, Davis HJ, Julian CG. Herpesvirus type 2 induced, tumor-specific antigen in cervical carcinoma. Am J Epidemiol 1973;98:1-9.
19. Hollinshead AC, Lee O, Chretien PB, Tarpley JL, Rawls W. Antibodies to Herpesvirus nonvirion antigens in squamous carcinoma. Science 1973;182:713-715.
20. Tarro G, Flaminio G, Maturo S, Esposito C, Cocchiara R. Further studies on an immunoenzymatic assay for herpes simplex virus tumor-associated antigen. Clin Immunol Immunopathol 1982;25:126-132.
21. Koss LG. Cytologic and histologic manifestations of human papillomavirus infection of the female genital tract and their clinical significance. Cancer 1987;60:1942-1950.
22. Haines HG, McCoy JP, Hofheinz DE, et al. Cervical carcinoma antigens in the diagnosis of human squamous cell carcinoma of the cervix. J Natl Cancer Inst 1981; 66:465-474.

23. Harlozinska A, Kula J, Stepinska B, et al. Cervical carcinoma antigen, carcinoembryonic antigen (CEA), and nonspecific cross-reacting antigen (NCA) in appraisal of uterine cervix smears. Am J Clin Pathol 1985;83:301-307.
24. Koprowska I, Zipfel S, Ross AH, Herlyn M. Development of monoclonal antibodies that recognize antigens associated with human cervical carcinoma. Acta Cytol 1986;30:207-213.
25. Tsuji Y, Yoshioka M, Ogasawara T, Takemura T, Isojima S. Identification of an H antigen-like blood group antigen in sera of cancer patients using a novel monoclonal antibody raised against endometrial carcinoma. Cancer Res 1987;47:3543-3550.
26. Lindgren J, Wahlstrom T, Seppala M. Tissue CEA in premalignant epithelial lesions and epidermoid carcinoma of the uterine cervix: prognostic significance. Int J Cancer 1979;23:448-453.
27. van Nagell JR Jr, Donaldson ES, Gay EC, et al. Carcinoembryonic antigen in carcinoma of the uterine cervix. 2. Tissue localization and correlation with plasma antgen concentration. Cancer 1979;44:944-948.
28. Bychkov V, Rothman M, Bardawil WA. Immunocytochemical localization of carcinoembryonic antigen (CEA), alpha-fetoprotein (AFP), and human chorionic gonadotropin (HCG) in cervical neoplasia. Am J Clin Pathol 1983;79:414-420.
29. Hurlimann J, Gloor E. Adenocarcinoma in situ and invasive adenocarcinoma of the uterine cervix. An immunohistologic study with antibodies specific for several epithelial markers. Cancer 1984;54:103-109.
30. van Nagell JR Jr, Donaldson ES, Sharrey RM, Goldenberg DM. The prognostic significance of carcinoembryonic antigen in the plasma and tumors of patients with endometrial adenocarcinoma. Am J Obstet Gynecol 1977;128:308-313.
31. Cohen C, Shulman G, Budgeon LR. Endocervical and endometrial adenocarcinoma. An immunoperoxidase and histochemical study. Am J Surg Pathol 1982;6:151-157.
32. Warlstrom T, Lindgren J, Korhonen M, Seppala M. Distinction between endocervical and endometrial adenocarcinoma with immunoperoxidase staining of carcinoembryonic antigen in routine histological tissue specimens. Lancet 1979;2:1159-1160.
33. van Nagell JR Jr, Meeker WR, Parker JC Jr, Kashmiri R, McCollum V. Carcinoembryonic antigen in intraepithelial neoplasia of the uterine cervix. Am J Obstet Gynecol 1976;126:105-109.
34. Malkin A, Kellen JA, Lickrish GM, Bush RS. Carcinoembryonic antigen (CEA) and other tumor markers in ovarian and cervical cancer. Cancer 1978;42:1452-1456.
35. DiSaia PJ, Haverback BJ, Dyce BJ, Morrow CP. Carcinoembryonic antigen in patients with gynecologic malignancies. Am J Obstet Gynecol 1975;121:159-163.
36. Kjorstad KE, Orjaseter H. Carcinoembryonic antigen levels in patients with squamous cell carcinoma of the cervix. Obstet Gynecol 1978;51:536-540.
37. Donaldson ES, van Nagell JR Jr, Pusell S, et al. Multiple biochemical markers in patients with gynecologic malignancies. Cancer 1980;45:948-953.
38. van Nagell JR Jr, Donaldson ES, Gay EC, et al. Carcinoembryonic antigen in carcinoma of the uterine cervix. 1. The prognostic value of serial plasma determinations. Cancer 1978;42:2428-2434.
39. Barrelet V, Mack J-P. Variations of the carcinoembryonic antigen level in the plasma of patients with gynecologic cancers during therapy. Am J Obstet Gynecol 1975;121:164-168.
40. Seppala M, Pihko H, Ruoslahti E. Carcinoembryonic antigen and alpha fetoprotein in malignant tumors of the female genital tract. Cancer 1975;35:1377-1381.

41. Schwartz PE, Chambers SK, Chambers JT, et al. Circulating tumor markers in the monitoring of gynecologic malignancies. Cancer 1987;60:353-361.
42. Bast BC Jr, Klug TL, St John E, et al. A radioimmunoassay using a monoclonal antibody to monitor the course of epithelial ovarian cancer. N Engl J Med 1983; 309:883-887.
43. Niloff JM, Klug TL, Schaetzl E. Elevatin of serum CA125 in carcinomas of the fallopian tube, endometrium, and endocervix. Am J Obstet Gynecol 1984;148: 1057-1058.
44. Malkasian GD Jr, Podratz KC, Stanhope CR, Ritts RE Jr, Zurawski VR Jr. CA 125 in gynecologic practice. Am J Obstet Gynecol 1986;155:515-518.
45. Kabawat SE, Bast RC, Welch WR, Knapp RC, Colvin RB. Immunopathologic characterization of a monoclonal antibody that recognize common surface antigens of human ovarian tumors of serous, endometrioid, and clear cell types. Am J Clin Pathol 1983;79:98-104.
46. de Bruijn HWA, Calkoen-Carpay TVB, Jager Duk JM, Aalders JG, Fleuren GJ. The tumor marker of CA 125 is a common constituent of normal cervical mucus. Am J Obstet Gynecol 1986;154:1088-1091.
47. O'Brien TJ, Hardin JW, Bannon GA, Norris JS, Quirk JG Jr. CA 125 antigen in human amniotic fluid and fetal membranes. Am J Obstet Gynecol 1986;155:50-55.
48. Vaitukaitis JL, Braunstein GD, Ross GT. A radioimmunassay which specifically measures human chorionic gonadotropin in the presence of human luteinizing hormone. Am J Obstet Gynecol 1972;113:751-758.
49. Mizuochi T, Nishimura R, Derappe C, et al. Structures of the asparagine-linked sugar chains of human chorionic gonadotropin produced in choriocarcinoma: appearance of triantennary sugar chains and unique bianntenary sugar chains. J Biol Chem 1983;258:14126-14129.
50. Rutanen E-M, Seppala M. The HCG-beta subunit radioimmunoassay in nontrophoblastic gynecologic tumors. Cancer 1978;41:692-696.
51. Nathanson L, Fishman WH. New observations on the Regan isoenzyme of alkaline phosphatase in cancer patients. Cancer 1971;27:1388-1397.
52. Kellen JA, Bush RS, Malkin A. Placenta-like alkaline phosphatase in gynecological cancers. Cancer Res 1976;36:269-271.
53. Rutanen E-M, Seppala M. Pregnancy-specific β-1-glycoprotein in trophoblastic disease. J Clin Endocrinol Med 1980;50:57-61.
54. Bohn H. Systematic identification of specific oncoplacental proteins. In: Fishman WH, ed. Oncodevelopmental markers. New York: Academic Press, 1983:69-84.
55. McDicken IW, McMillan DL, Rainey M. Carcinoembryonic antigen levels in cervicovaginal fluid from patients with intraepithelial and invasive carcinoma of the cervix. Eur J Cancer Clin Oncol 1982;18:917-919.
56. Kannagi R, Fukushi Y, Tachikawa T, et al. Quantitative and qualitative characterization of human cancer-associated serum glycoprotein antigens expressing fucosyl or sialyl-fucosyl type 2 chain polylactosamine. Cancer Res 1986;46:2619-2626.

25
Testicular Tumors

DEBORAH J. LIGHTNER*

Veterans Administration Medical Center, Minneapolis, Minnesota

PAUL H. LANGE

University of Washington, Seattle, Washington

I. OVERVIEW

The management of patients with nonseminomatous germ cell tumors (NSGCT) has improved markedly since radioimmunoassays have allowed for the sensitive determination of serum levels of alpha fetoprotein (AFP) and beta human chorionic gonadotropin (B-HCG). These markers are exquisitely sensitive detectors of recurrent or residual NSGCT; they are sufficiently accurate that the decision to start or resume cytotoxic chemotherapy may be based solely on the finding of elevated serum levels.

Other tumor markers used with NSGCT are less specific, and therefore are less valuable in the treatment of patients than are serum determinations of AFP and B-HCG.

Unfortunately, there currently exists no marker that is useful in the routine management of patients with pure seminomatous disease; several serum markers have been evaluated but fail to achieve sufficient sensitivity to be clinically applicable.

Because an understanding of the biological nuances of each tumor marker is crucial to the proper interpretation of laboratory determinations, each of the tumor markers for testis tumor is reviewed here.

II. TESTICULAR TUMOR MARKERS

A. AFP

The production of AFP by the fetal yolk sac, liver, and gastrointestinal tract is a fact critical to the understanding of the biology of this marker. It will exist where these embryological elements are present and, therefore, is present only in trace amounts in the serum of adults. In the embryo, production of AFP by the yolk sac is gradually replaced by the liver, with serum levels peaking in the maternal circulation at approximately 12-14 weeks of gestation; thereafter, levels decline, and should become undetectable during the first year of life [1].

---

*Present affiliation: Abbott-Northwestern Hospital, Minneapolis, Minnesota

AFP is a glycoprotein with a molecular weight of 70 kD and exists in human serum with a half-life of 4.8 days. Slower than calculated falls in serum levels should alert the clinician to the possibility of residual disease. In addition, several heterogeneous forms of AFP exist, for example, AFP derived from yolk sac will bind to concanavalin A (ConA); liver-derived AFP will not bind to Con A.

AFP is synthesized by the endothelium lining of yolk sac elements of NSGCT, but this production is not limited to yolk sac tumors. While over 90% of children with yolk sac tumors will show serum AFP elevations [2], positive AFP immunohistochemical staining as well as serum AFP elevations can occur in tumors with yolk sac, embryonal or teratocarcinomatous elements [3]. Serum elevations occur in 53% of patients with NSGCT [4].

Neither positive histochemical staining nor serum elevations of AFP should occur with pure seminoma. This is of great importance in treatment planning; pure seminomatous tumor is exquisitely radiosensitive while nonseminomatous tumors require chemotherapy and/or surgical extirpation. Therefore, patients with elevated serum levels of AFP should be treated as having NSGCT.

Other clinical conditions, primarily of hepatic or gastrointestinal origin, can cause an elevation in AFP levels (see Table 1).

B. HCG

HCG is a glycoprotein of molecular weight 38 kD, which exists as two noncovalently linked polypeptide chains. The alpha chain is primarily a carrier, and is similar in sequence to the alpha subunits of both luteinizing hormone and thyroid-stimulating hormone. The beta chain defines the immunospecificity of HCG; 30 additional carboxy-terminal amino acids as well as minor variations in the remainder of the beta-chain sequence allow for the differentiation of HCG from the pituitary peptide hormones [6,7]; hence, the more sensitive assays of tumor expression of HCG measure the beta subunit. The half-life of HCG is 2.1 days.

HCG is normally produced by the syncytiotrophoblastic giant cell of the placenta; the same syncytiotrophoblast cell identified in NSGCT and in a small percentage of seminomas is responsible for elevated levels of HCG found in patients with these malig-

Table 1 Conditions Causing Elevations in AFP and HCG Levels

Serum Elevations of AFP	Serum Elevations of HCG
Liver regeneration, as seen in hepatitis, cirrhosis	Cross-reactivity to luteinizing hormone when LH is elevated (see text)
Breast cancer	
Gastric cancer	
Pancreatic cancer	Pancreatic cancer
Lung cancer	Lung cancer
Malignant hepatoma	Malignant hepatoma
Ataxia telangiectasia	
Hereditary tyrosinemia	

nancies. Elevated serum levels of HCG occur preoperatively in approximately 5% of pure seminomatous malignancies, whereas the incidence of NSGCT producing only HCG is 32% [4]. Since HCG elevation is five times more common with NSGCT. and significant treatment differences exist for these two malignancies, an elevation of HCG alone should always invite a search for NSGCT in what otherwise appears to be a case of pure seminoma. This recommendation is especially compelling when the HCG level is markedly elevated (greater than 1000 ng/ml).

III. VALUE OF MARKERS

NSGCT markers may be discordant: neither or only one marker may be elevated with a given tumor or with a recurrence. This occurs because the various cell lines making up a malignancy differ in their production of the tumor markers. The Danish Testicular Cancer study demonstrated that while 33% of nonseminomatous tumors will show a preoperative elevation of B-HCG, 53% will show an elevation in AFP levels [4]. Elevations of either AFP or B-HCG levels will occur in 86-94% of NSGCT, which underscores their high degree of combined sensitivity [8]. We assume that the varying propensity of each cell population to metastasize allows for the discordance seen with recurrences: levels of one or both tumor markers may be either elevated or normal. Twenty-eight percent of patients with recurrent NSGCT will have normal levels of serum tumor markers [9]. In simplistic terms, one can expect teratomatous elements to produce neither AFP nor B-HCG, while embryonal elements can produce both markers. Yolk sac malignancies, in general, produce AFP; as might be anticipated by its placental origin, choriocarcinoma produces HCG.

Other placental proteins can be demonstrated in tumors containing syncytiotrophoblasts or choriocarcinoma elements. For example, levels of seminoprotein-1 (SP-1) are elevated in some patients with choriocarcinoma; these elevations occur independently of B-HCG production [10,11]. Because of its low frequency of occurrence, this assay is not widely used in clinical practice.

Another placental-specific protein, PP-5, is of limited clinical value as a tumor marker, although it may prove useful occasionally in histological staining. Human placental lactogen (HPL) does occur in choriocarcimatous elements as well [12].

Although AFP and B-HCG are the most clinically useful tumor markers for testicular tumors, new tumor markers might complement or offset certain clinical deficiencies; for example, neither AFP nor B-HCG is useful in seminoma. The most promising of these markers for seminoma are human placental alkaline phosphatase (PLAP) and gamma-glutamyl transpeptidase (GGT).

PLAP is histologically expressed in all germ cell tumors [13]. However, current assays detect serum PLAP elevations in only 36% of seminomatous tumors, and in 15% of NSGCT [14].

According to one study, GGT levels were elevated in 10 of 33 patients with seminoma; when combined with PLAP and HCG determinations, marker elevations were found in 80% of patients [15]. Further confirmation of this work is needed to determine if multiple marker studies will prove useful in the diagnosis and prognosis of seminoma.

The production of most of these oncofetal hormones is not confined to the malignant testes; HCG is produced in small amounts by the normal tests [16]; HPL, SP-1, HCG, and PP-5 are all found in seminal plasma [17]. It is not yet known if these proteins are completely homologous with the normal placental products.

IV. NONSPECIFIC MARKERS

Nonspecific markers have also been investigated for their clinical utility. Lactate dehydrogenase (LDH) is not sufficiently specific for general application, but may prove useful in some patients with advanced disease and normal AFP and B-HCG levels, and during chemotherapy [18,19]. A recent report suggests that a new isoenzyme, LDHX, is not found in seminomatous tissue, and that LDH-1 and LDH-2 are found in higher concentrations in seminoma than in normal testicular tissue [20].

Carcinoembryonic antigen, alpha-1-antitrypsin, ferritin, and fibronectin have not proved to be clinically useful markers for testicular tumors [5].

In summary, the management of nonseminatous disease has been greatly simplified by the assiduous use of the tumor markers AFP and B-HCG, and occasionally LDH. The other tumor marker candidates are rarely if ever used, although PLAP, when it is more widely available, may prove valuable in some patients with seminomatous disease. Since patients with active germ cell tumors still present many clinical problems, the search for additional tumor markers should continue.

REFERENCES

1. Lange PH, Fraley EE. Serum alpha-fetoprotein and human chorionic gonadotropin in the treatment of patients with testicular tumors. Urol Clin North Am 1977;4: 393-406.
2. Abelev GI. Alpha-fetoprotein as a marker of embryo-specific differentiations in normal and tumor tissues. Transplant Rev 1974;20:3-37.
3. Kurman RJ, Scardino PT, McIntire KR, et al. Cellular localization of alpha-fetoprotein and human chorionic gonadotropin in germ cell tumors of the testis using an indirect immunoperoxidase technique: a new approach to classification utilizing tumor markers. Cancer 1977;40:2136-2151.
4. Norgaard-Pedersen B, Schultz H, Arends J, Brincker H, Krag Jacobsen G, Lindelov B, Rorth M, Svennekjaer IL, and the DATECA Study Group. Biochemical markers for testicular germ-cell tumors in relation to histology and stage: some experiences from the Danish Testicular Cancer (DATECA) Study from 1976 through 1981. Ann NY Acad Med 1983;417:390-399.
5. Sidi AA, Chiou R-K, Lange PH. Recent reflections on tumor markers. World J Urol 1984;2:18-25.
6. Birken S, Canfield RE. Isolation and amino acid sequence of COOH-terminal fragments from the B subunit of human choriogonadotropin. J Biol Chem 1977;252: 5386-5392.
7. Keutmann HT, Williams RM. Human chorionic gonadotropin; amino acid sequence of the hormone-specific COOH-terminal region. J Biol Chem 1977;252:5393-5397.
8. Bosl GJ, Lange PH, Nochomovitz LE, et al. Tumor markers in advanced nonseminomatous testicular cancer. Cancer 1981;47:572-576.
9. Bosl GJ, Geller NL, Cirrincione C, Misselbaum J, Vugrin D, Whitmore WF, Golbey RB. Serum markers in patients with metastatic germ cell tumors of the testis. Am J Med 1983;75:29-35.
10. Rutanen EM, Seppala M. Pregnancy-specific B-1-glycoprotein in trophoblastic disease. J Clin Endocrinol Metab 1980;50:57-61.
11. De Bruijn HWA, Suurmeijer AJH, Sleijfer TH, Schraffordt Koops H, Ockhuizen TH, Willemse PHB, Marrink J. Evaluation of pregnancy specific beta-1-glycoprotein in patients with nonseminomatous testicular germ cell tumours. Eur J Cancer Clin Oncol 1982;18:911-916.

12. Heyderman E. Multiple tissue markers in human malignant testicular tumours. Scand J Immunol 1978;8(Suppl 8):119-126.
13. Epenedos AA, Travers P, Gatter KC, Oliver RTD, Mason DY, Bodmer WF. An immunohistological study of testicular germ cell tumours using two different monoclonal antibodies against placental alkaline phosphatase. Br J cancer 1984;49:11-15.
14. Jeppsson A, Wahren B, Stigbrand T, Edsymr F, Andersson L. A clinical evaluation of serum placental alkaline phosphatase in seminoma patients. Br J Urol 1983;55: 73-78.
15. Javadpour N. Tumor markers in testicular cancer—an update. Prog Clin Biol Res 1985;203:141-154.
16. Braunstein GD, Rasor J, Wade ME. Presence in normal testes of a chorionic gonadotropin-like substance distinct from luteinizing hormone. N Engl J Med 1975;293: 1339-1343.
17. Salem HT, Menabawey M, Seppala M, Shaaban MM, Chard T. 1982.
18. Lippert MC, Javadpour N. Lactic dehydrogenase in the monitoring and prognosis of testicular cancer. Cancer 1981;48:2274-2278.
19. Von Eyben FE. Lactic dehydrogenase and its isoenzymes in testicular germ cell tumors: an overview. Oncodev Biol Med 1983;4:395-414.
20. Sakakibara N, Sakashita S, Koyanagi T. Total LDH activities and LDH isoenzyme patterns in the tissues of testicular tumors. Nippon Hinyokika Gakkai Zasshl 1987; 78:249-252.

26
Markers for Hepatocellular Carcinoma

ALAN H. B. WU and STEWART SELL
University of Texas Medical School, Houston, Texas

I. INCIDENCE AND CAUSES OF LIVER CANCER

As with all neoplastic processes, causative factors in human hepatocellular carcinoma are complex and involve a mix of genetic and environmental factors. Cancers that originate from parenchymal liver cells make up the majority of primary liver cancers and are termed hepatocellular carcinomas (HCC). Malignancy originating from the bile ducts is termed cholangiocellular carcinoma. In North America and western Europe, research in primary liver cancer received considerably less attention and support than that for malignancies of other organs because of the much lower incidence of this cancer compared with that of skin, lung, breast, and colon cancers. Figure 1 illustrates the differences in incidence of HCC from some geographic regions [1]. The range of values are 0.3 per 100,000 for women in São Paulo, Brazil, to 34.4 per 100,000 for men in Hong Kong. In general, individuals from Asia and Africa have a much higher incidence of liver cancers than those from western countries. The epidemiological data show that racial and probably genetic differences exist among populations within the same region. As shown in Figure 1, Asian populations have a higher incidence of HCC than blacks, who have a higher incidence than whites. It is also clear from these figures that male subjects have roughly a three to fivefold higher risk of liver cancer than female subjects.

Causative factors that must be considered include infection by the hepatitis B virus (HBV), the underlying presence of cirrhosis, exposure to dietary carcinogens, and the risks related to certain occupations and other environmental hazards.

A. Hepatitis B Virus

The evidence suggesting that chronic hepatitis B viral infection is one of the causative agents in primary liver cancer is impressive although not conclusive: (1) geographic areas with a high incidence of hepatitis B infection are coincident with areas with a high incidence of HCC; (2) a statistically higher percentage of HCC patients are infected with the B virus than controls matched for age and geographic region; and (3) the integration of hepatitis B viral DNA into the DNA of many but not all HCC tumors.

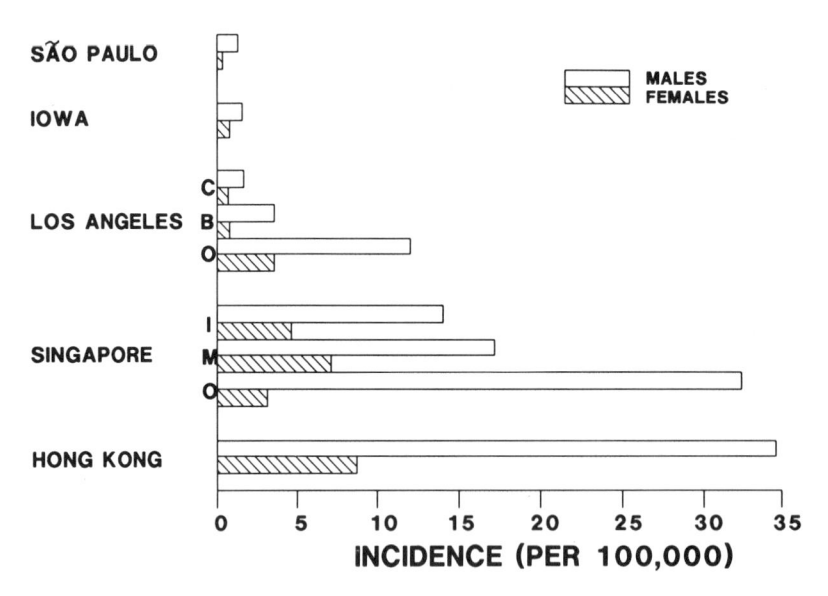

Figure 1 Incidence of primary liver cancer in selected regions of the world. Subpopulations: C, whites; B, blacks, O, Asians; I, Asian Indian; M, Malayan (from ref. [1]).

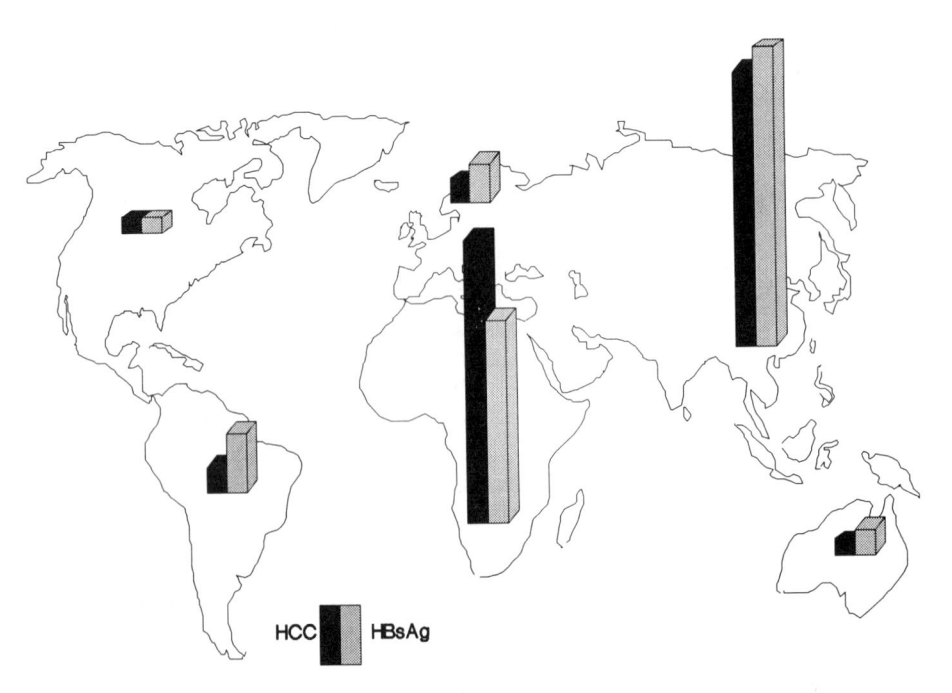

Figure 2 Relative world incidence of HCC for males and females (from ref. [1]) and HBsAg detection (from ref. [2]).

Figure 2 demonstrates that the geographic prevalence of hepatitis B infection matches with the incidence of HCC [2]. Such figures may be coincidental, however, since populations with high HCC incidence have lower standards of living and sanitation conditions, which may be more conducive for developing both diseases as compared to western societies. If these were truly unrelated statistical events, however, the chances of having *both* HCC and HBV infection should be only a small fraction of the chance of having either one. Table 1 shows, however, that the *majority* of HCC patients (as many as 80%) are chronically infected with hepatitis B surface antigen in most populations; that is, the chances of HCC occurring in patients who carry hepatitis B antigens (HBsAg) is 10- to 20-fold higher than in noninfected patients within the same general population. Furthermore, as shown in Table 1, HCC patients have a twofold lower incidence of HBV immunity, as evidence by positive antibodies to HBsAg, than the general population. These data suggest that patients who develop immunity to HBV are less likely to develop HCC than those who are not immune. Implementation of a world-wide hepatitis B vaccination program (currently underway as pilot projects) would be expected to reduce the incidence of liver cancer.

The incorporation of hepatitis B viral DNA into hepatoma DNA tissues and cell lines derived from HCC has been well documented using genetic probes and recombinant DNA techniques. Chen et al., for example, found integrated HBV DNA sequences in 80% of HCC tumors in Taiwan [3]. This is consistent with the fact that up to 80% of the general population in the Republic of China are carriers of HBsAg [4]. Data from Japan are similar: Hino found HBV DNA in eight of nine hepatomas [5]. The integration patterns by Southern blots are unique and often involve multiple sites. Such data suggest either that primary hepatomas are oncogenically transformed by HBV or that HBV is an associated "passenger virus" and not the causative agent. In either case, HBV qualifies as an "oncogenic virus."

B. Liver Cirrhosis

Several investigators have suggested that the successive sequelae of chronic hepatitis B virus infection begin with acute infection, followed by chronic active hepatitis, liver cirrhosis, and finally hepatocellular carcinoma [6]. Lin et al. reported that approximately 66% of HCC patients in Taiwan have cirrhosis [7]. Miyaji has found that as many as 40% of Japanese patients with liver cirrhosis will develop HCC [8]. In addition, cirrhosis from other causes such as alcohol ingestion, metabolic diseases such as inborn errors of metabolism, and biliary cirrhosis is also associated with an increased risk for HCC. American bartenders and wine makers, for example, are at increased risk for HCC, presumably

Table 1 Prevalence (%) of Primary Liver Cancer with HBsAg and Anti-HBs

	HBsAg		Anti-HBs	
Country	HCC	No HCC	HCC	No HCC
S. Africa (Bantu)	59.5	9.0	11.6	33.4
Japan	37.3	2.6	7.5	18.3
Uganda	78.9	7.0	16.0	45.4
Senegal	61.2	11.0	18.2	42.1

Source: Data modifed from ref. [2].

due to their increased alcohol consumption [9]. It may therefore be logical to assume that patients with HBV infections are at increased risk for HCC because they will eventually develop cirrhosis, and that cirrhosis itself is the actual risk factor. Liaw et al. however, showed that in patients with small HCC tumors, chronic hepatitis could progress directly to HCC before the development of cirrhosis [6]. This was concluded on the basis that patients with these small tumors have a lower incidence of cirrhosis and hepatitis Be antigenicity than patients with larger ones, and that cirrhosis and seroconversion of HBe antigen to antibody can occur after development of HCC. The development of cirrhosis and HCC from chronic hepatitis are therefore related but nevertheless independent events.

C. Dietary and Environmental Factors in Hepatic Carcinogenesis

Studies have clearly shown that certain substances, when given in high doses, can produce hepatocellular carcinoma in animals. Particularly potent carcinogens include N-acetyl-2-aminofluorene, diethylnitrosamine, and aflatoxin B_1. These studies are important for understanding the pathogenesis of primary liver cancer in humans. Aflatoxin, for example, has been implicated as a major cause of HCC in China [10] and in Africa [11] where food storage conditions may be conducive for growth of *Aspergillus flavus*, a natural source of aflatoxin. In the United States, a recent epidemiological study showed a slightly increased incidence of liver cancer among male subjects from the southeast (where the average aflatoxin B_1 intake was estimated to be 13-197 ng/kg body weight) compared to male subjects from the north and west (average intake 0.2-0.3 ng/kg bw) [12]. However the differences in mortality were not as high as what was expected based on results from experimental hepatocarcinogenesis and prior epidemiological studies done in Africa and Asia. Development of HCC in the United States is therefore likely linked to other factors.

Some of these other environmental influences have been studied on an epidemiological basis. Elevated risk for HCC has been shown for individuals working in laundries, dry cleaning services, and gasoline stations, presumably because of increased exposure to organic solvents [9]. Agricultural occupations are associated with about a twofold risk for HCC due to exposure to pesticides. Cigarette smoking in contrast is not associated with an increased risk, since no consistent dose-response relationship can be defined [13].

II. TUMOR MARKERS FOR HCC

A. Diagnosis and Screening

The development of HCC is associated with production of alpha fetoprotein, a glycosylated embryonic serum protein that migrates in the alpha-1 region by zone electrophoresis. AFP is synthesized by the fetal liver and by the yolk sac, is the major serum protein during the first 16 weeks of gestation, and is important for the maintenance of oncotic pressure since the concentration of albumin is low in early fetal circulation. In the second trimester, the fetal liver switches towards the production of albumin, and the concentration of AFP drops. The reference range of AFP in adults is dependent on the assay used, but usually ranges from 5 to 10 µg/L. Elevations in AFP secondary to HCC were first demonstrated in mice by Abelev et al. in 1963 [14] and in humans by Tatarinov in 1964 [15]. Although other approaches have been examined, AFP remains the single most useful marker for the diagnosis and management of HCC. The current analytical

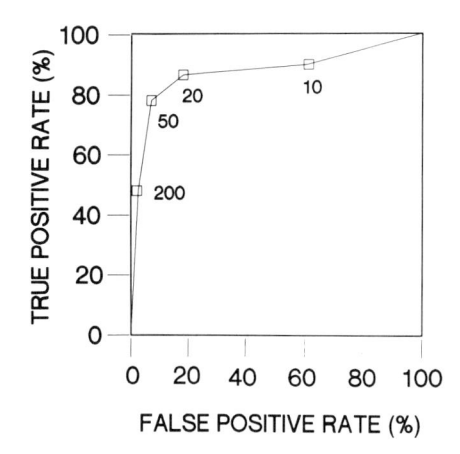

Figure 3 Receiver operating characteristic curves for the diagnosis of HCC using alpha fetoprotein from selected published reports [4,16,18,19]. Decision limits, (µg/L) are given with each point along with the reference.

methods include competitive radioimmunoassay and nonisotopic immunoenzymometric assays. The newest procedures make use of dual monoclonal antibodies, are highly automated, and are not interfered with by other proteins [16].

1. Diagnosis of Hepatocellular Carcinoma by AFP

Unlike most tumor markers for other cancers, AFP is highly specific for HCC when very high levels (e.g., >3000 µg/L) are present in the blood. Unfortunately, at these levels,

Table 2 Incidence of Alpha Fetoprotein Elevations in HCC, Liver Diseases, and Other Malignancies

Disease	% Incidence[a]
Hepatocellular carcinoma	90
Cirrhosis	34
Chronic active hepatitis	58
Chronic persistent hepatitis	42
Subacute hepatitis	69
Acute viral hepatitis	47
HBsAg carriers	23
Gastric cancer	18
Colorectal cancer	14
Pancreatic cancer	22
Miscellaneous other cancers	30

[a]Cutoff for AFP = 10 µg/L.
Source: Data modified from Ref. [4].

A. GALACTOSAMINE-INDUCED LIVER INJURY

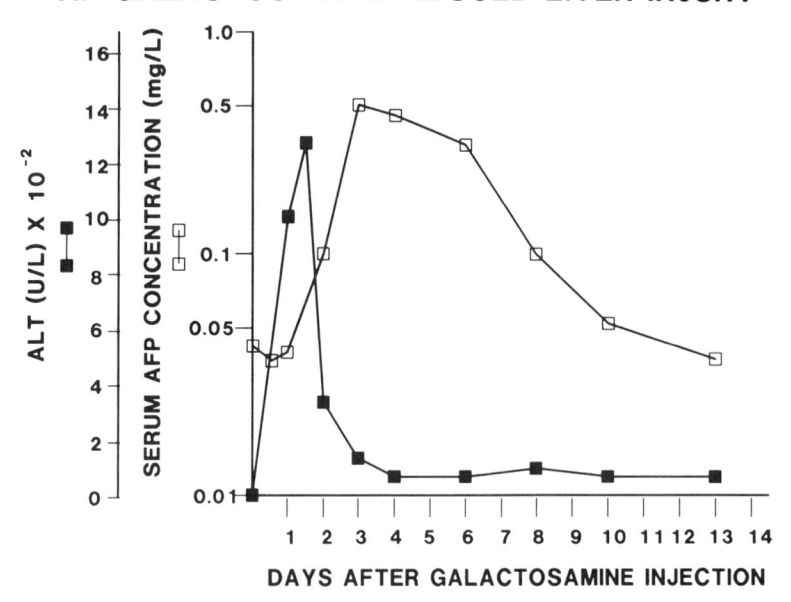

Figure 4 Serial determination of serum alpha fetoprotein (AFP) in rats following chemically induced liver injury (A) and growth of a transplantable hepatoma (B). A shows alanine aminotransferase (ALT) and AFP elevations occurring following a sublethal dose of galactosamine [21]. Within 12 hr ALT levels rise sharply, falling back to normal within 3-4 days. Elevation of serum AFP levels is first seen on day 2, reaches a maximum on day 3, and gradually falls to normal (<0.06 µg/L) during the next week. B shows AFP elevations following transplantation of Morris hepatoma 7777 [22]. The shaded area shows serum AFP concentration in eight rats inoculated intramuscularly with 5×10^6 hapatoma cells. The sigmoidal elevation reflects the growth kinetics of the tumor, with death (†) of all animals occurring by day 40. If the tumor is removed surgically, the serum AFP level falls with a half-life of approximately 1 day. Complete removal results in a return to normal levels, but recurrence or lung metastasis results in a re-elevation of AFP levels. The ordinate of panel A covers two logs for AFP; that of panel B covers six logs.

the clinical sensitivity of AFP for small tumors is poor. When the decision limits are lowered, the specificity decreases because cancers other than HCC and nonmalignant liver diseases can produce false-positive results. This relationship can be illustrated using standard receiver-operating characteristic curves (ROC), which plot clinical sensitivity against specificity at different decision limits [17]. Figure 3 shows the relationship of true with false-postive results for HCC using different cutoff values for AFP. ROC curves are not usually computed from different data sets and there is great variability in reported clinical sensitivity and specificity secondary to the analytical methods used, Figure 3 does illustrate that specificity rapidly decreases when detection limits are lowered. Many of these false-positive results are due to chronic liver disease. Table 2 lists benign conditions that have been reported to produce AFP [4].

B. GROWTH OF HEPATOMA 7777

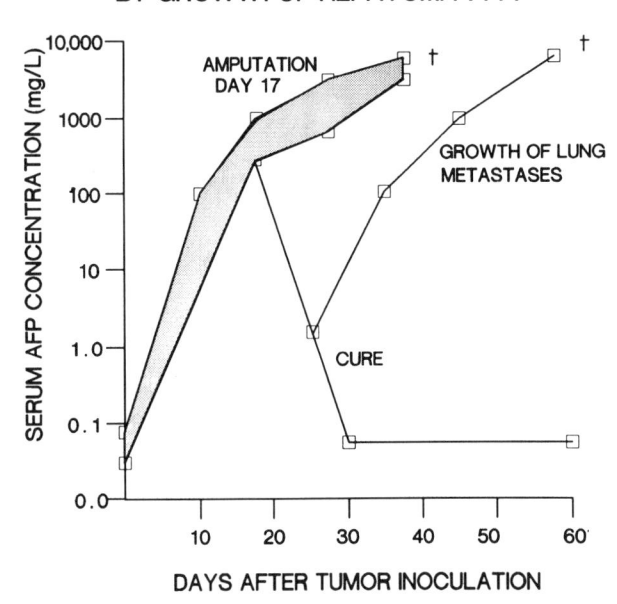

Figure 4 (continued)

Increased clinical specificity, however, can be achieved when serial determinations are made. Numerous investigators in both controlled animal studies and in humans have demonstrated that nonmalignant liver diseases such as hepatitis and cirrhosis generally produce transient levels of AFP [3]. Moreover, AFP levels are briefly elevated after acute liver injury, reflecting hepatic regeneration [20]. In contrast, patients with untreated HCC usually have AFP levels that are persistently high and may increase over time if the tumor volume increases. The relationship of serial AFP concentrations with experimental liver injury and transplantable carcinomas in rats is illustrated graphically in Figure 4 [22].

The major diagnostic limitation for AFP appears to be low clinical sensitivity for small tumors. Because the rate of AFP production can be variable, tumors that are <3 cm in diameter often do not produce detectable quantities of AFP. This limitation is critically important in these patients, since a good prognosis is dependent on early diagnosis and treatment. Occasionally even large and advanced tumors have been shown not to produce AFP. Most studies show that the clinical sensitivity of AFP for HCC is at best only 80-90% despite the use of decision limits that are set to a low value, for example, the upper limit of normal (10 μg/L, see Fig. 3).

2. Mass Screening for HCC by AFP

The low incidence of HCC in western societies coupled with the low clinical sensitivity of AFP for small tumors renders inefficient large-scale screening in these areas because of the high costs. However, since the prognosis associated with advanced HCC is poor, large-scale screening may be useful in areas where the incidence of HCC is high. The important criteria for screening cancer in the general population are (1) the prospects of detecting a high percentage of asymptomatic patients, (2) the availability of appropriate treatment regimens for asymptomatic patients to improve their prognosis and survival rates, (3) the availability of highly sensitive screening assays to minimize the number of false negatives and a highly specific confirmation assay to rule definitively out false positives, and (4) the availability of an assay that is inexpensive, standardized, and easy to use in the field [23]. AFP fulfills many of these criteria for screening HCC in that marginal elevations (50-300 µg/L) are observed in early hepatic tumors. In addition, reasonably sensitive and inexpensive screening assays based on reverse hemagglutination are available with detection limits of 50 µg/L. However, as discussed above, AFP alone is not sufficiently specific for early HCC detection at these decision limits. Moreover, use of more analytically sensitive procedures for AFP (e.g., radioimmunoassay) will not improve the clinical specificity. Other diagnostic methods are therefore required to determine the nature of moderate AFP elevations.

Nevertheless, screening programs have been instituted in the People's Republic of China (see the review by Xu [23]) as well as in other areas where the prevalence of HCC is much higher than in the United States. Tang found over 1000 cases of HCC when over 5 million Chinese residents were screened beginning in 1971 by using both agar gel diffusion (AGD, sensitivity 500 ng/L) and reverse hemagglutination [24]. When the latter, more sensitive, assay was used for screening, 61% of the positive cases were from patients with subclinical disease compared to 30% using AGD. The increase in sensitivity, however, was accompanied by an increase in the false-positive rate of 0.28-3.2%. Early detection of small hepatocellular tumors and subsequent treatment by surgical resection has improved the survival from 14.8 to 79.1% after 1 year, and from 5.5 to 61.6% after 3 years [24]. Such studies show that screening for HCC by AFP is important in reducing the mortality in areas that have a high incidence of liver cancer.

B. Methods for Confirmation of HCC

Because it is necessary to use a low decision limit to maximize clinical sensitivity, followup for patients who have elevated levels of AFP is necessary before the diagnosis of HCC can be made. This is particularly important in patients with cirrhosis since they have a high incidence of HCC and because low AFP elevations are not useful diagnostically. Among the several approaches that have been used to improve the diagnostic efficiency of AFP for the detection of HCC include (1) measurement of AFP subfractions, (2) detection of liver tumors using imaging techniques, and (3) use of AFP with other markers in multivariate analyses.

1. AFP Subfractions

There is considerable heterogeneity in human AFP based on differences in glycosylations resulting from specific enzymes present in the endoplasmic reticulum and Golgi apparatus that modify 4-6% of AFP [25]. One classification can be made on the basis of their affinities toward binding concanavalin A (ConA) [26-28]. As shown in Figure 5, hepatic tumors produce an AFP variant that binds to Con A whereas yolk-sac derived AFP has an additional N-acetylglucosamine that blocks binding to Con A. These subfractions can be analyzed using Con A affinity chromatography and electrophoresis. Current commercial immunoassays that make use of polyclonal antisera are not able to differentiate between these variants. Several investigators, however, have developed strategies to raise monoclonal antibodies specific to these subfractions [29,30].

Results from studies of AFP subfractions in HCC have shown that variant analysis can be used to differentiate patients with HCC from those with other AFP-producing

Markers for Hepatocellular Carcinoma

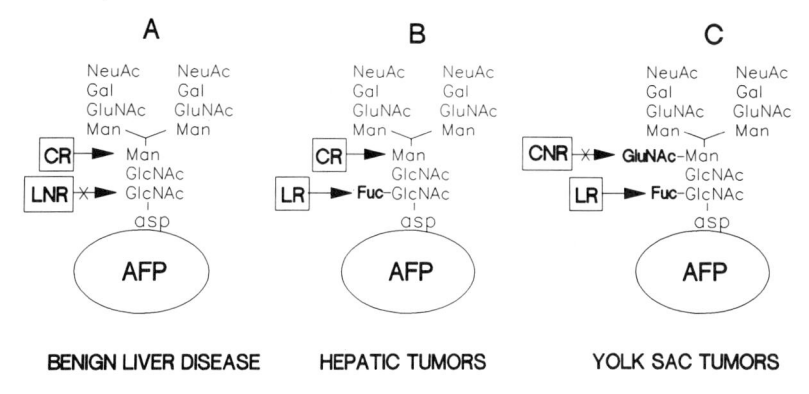

Figure 5 Possible structure of the predominant AFP subfraction from patients with benign liver disease (A), hepatic carcinoma (B), and yolk sac tumors (C), illustrating the sites of binding for con A and lentil lectins. CR, con A reactive; CNR, con A nonreactive; LR, lentil reactive; LNR, lentil nonreactive (from ref. [25]).

tumors [31]. Primary liver cancer patients typically have a 10% Con A nonreaction (CNR) component of total AFP, whereas patients with yolk sac tumors and secondary liver metastasis have AFP that is typically 50% CNR. Differentiation between primary HCC and benign liver diseases cannot be made on the basis of Con A nonreactivity, as both groups exhibit low Con A nonreactivity (Fig. 5). However, significant differences between these two groups may exist when affinity chromatography based on lentil lectin binding is used. Aoyagi et al. have shown that the lentil-lectin-reactive fraction contains carbohydrates of the fucosylated biantennary complex type [32]. Patients with HCC have a higher degree of fucosylated AFP (typically 45%) than patients with benign liver disease (typically only 5% fucosylated). Figure 5 shows the possible site of AFP fucosylation in patients with hepatomas and yolk sac tumors and its absence in AFP from patients with benign liver disease. Additional microheterogeneity may exist with other sites on the sugar moiety since AFP will react with other lectins such as *Ricinus communis, Phaseolus vulgaris*, and *Vicia faba* agglutinin [33]. A systematic nomenclature based on reactivity to specific lectins has been proposed to simplify the comparison of results [25]. Further study may lead to the development of appropriate highly specific immunoassays that will permit routine "fractionation" of AFP and will improve clinical diagnosis.

2. Special Diagnostic Procedures

The most sensitive method for noninvasive diagnosis of HCC involves imaging techniques such as radionuclide scintigraphy, computed tomography, angiography, and ultrasonography. Of these, real-time ultrasound appears the most promising, particularly for tumors that are <3 cm in diameter [34-36]. Reports show that the clinical sensitivity and specificity for HCC by ultrasound range from 86 to 99%. A comparable clinical sensitivity by AFP can only be achieved when a low decision limit is used, such as 20 µg/L. At this limit, however, the specificity decreases to 82%. Nevertheless, the higher costs and inconvenience of ultrasound limit its use to only high-risk patients, particularly patients with chronic active hepatitis and liver cirrhosis. In addition, because liver cancer has a high incidence in Asia and Africa, the availability of ultrasound will limit its use as a screening tool for HCC.

In cases where the suspicion of HCC is high and the evaluation of the patient's status is equivocal when both marker analysis and imaging techniques are used, invasive diagnostic techniques may be required. Percutaneous coarse-needle liver biopsy has a reported specificity of 100% and sensitivity of 85% [37]. The use of ultrasonically guided fine-needle biopsy can reduce the sampling error and increase the clinical sensitivity to 95% [38]. Use of the fine needle also reduces the incidence of bleeding and other complications. As diagnostic techniques continue to improve, situations requiring laparoscopic biopsy will become increasingly rare.

3. Other Markers and Approaches to Multiple Marker Studies

The limitation of sensitivity of AFP for early liver cancer have prompted many investigators to search for other markers to be used in conjunction with AFP. Considerable attention has been focused on the measurementof serum ferritin, an iron-storing protein found in high concentrations in the liver, spleen, and bone marrow. Low levels of ferritin generally indicate iron deficiency. Moderately high levels have been demonstrated in patients with chronic liver disease secondary to direct liver damage. Very high levels are associated with primary hepatomas, possibly due to its production by the tumor. Some investigators consider that ferritin measurements are superior to those of AFP for documenting HCC in terms of clinical sensitivity and specificity [39], whereas others have suggested that ferritin is not as useful as AFP [40]. Because ferritin levels are elevated in other malignant diseases (e.g., leukemia and breast cancer), as well as nonmalignant diseases, it is not likely that assays for ferritin will replace AFP for diagnosis. Some improvements in clinical specificity can be gained when ferritin levels are evaluated in conjunction with iron levels and a laboratory assessment of liver function (e.g., by aspartate and alanine aminotransferases) [41]. The greatest potential for ferritin may be when it is used in conjunction with AFP. The diagnostic efficiency is improved, especially for small tumors, over individual AFP or ferritin measurements [40]. It is unlikely, however, that the slight improvements seen with ferritin justify the extra costs for the analysis.

Numerous other biochemical markers have been examined for use in HCC with varying degrees of success. Tissue polypeptide antigen (TPA) is isolated from endoplasmic reticulum of malignant cells and has been found in 96% of HCC [42]. Unfortunately, TPA severely lacks specificity, since high blood concentrations are observed associated with a variety of malignant and nonmalignant conditions. Carcinoembryonic antigen (CEA), another oncofetal antigen, and CA-50 are other cancer markers found in high concentrations in patients with hepatocellular carcinoma. Both of these markers, however, are less specific than AFP, since elevations are observed in patients with other gastrointestinal tumors such as carcinoma of the colon (see Chapter 23) and pancreas (see Chapter 19). Other newly available markers include transforming growth factor [43], urinary pseudouridine [44], liver-specific F antigen [45], and thyroxine-binding globulin [46]. Whether any of these become clinically important for the detection of HCC will require further examination.

Many of the markers described above are not ideal complements to AFP, because they are elevated in patients with liver metastasis, other gastrointestinal tumors, and benign liver diseases. Measurement of des-γ-carboxy (abnormal) prothrombin (DCP), however, may provide the specificity that these other markers lack [47]. Prothrombin is normally carboxylated in the liver by a vitamin-K-dependent carboxylase. Normal individuals have undetectable levels of DCP. High concentrations are found in the blood of patients who have vitamin K deficiencies, those who are treated with anticoagulants (e.g., vitamin K inhibitors such as warfarin), and patients with HCC. In contrast to levels of alpha fetoprotein, patients with liver cancers other than HCC and other patients with benign liver disease have detectable but low levels of DCP. When the decision limit for DCP is set to exclude these groups, the resulting clinical sensitivity of DCP for HCC is 66%. When used in conjunction with AFP (using a conservative cutoff of 400 µg/L), the sensitivity increases to 84-87% with almost 100% specificity when patients with vitamin K deficiencies are excluded from the differential diagnosis. Current assay methods for DCP include radioimmunoassay and spectrophotometry with chromogenic substrates [48].

C. Monitoring HCC with AFP

Levels of AFP reasonably reflect tumor burden in both experimental animal models and humans. Kelsten et al. showed a correlation of 0.55 when the percentage tumor volume (as calculated from computerassisted tomography) is plotted against AFP concentrations [49]. Perhaps more significant is that the rate of serum AFP elevation is more closely related to the rate of tumor growth [50]. Successful surgical resection is characterized by declining AFP levels that eventually reach normal concentrations. If AFP level do not return to normal, incomplete tumor removal must be considered. If AFP levels return to a high concentration after a period of remission, metastasis and/or recurrence of tumors is likely. Figure 4B summarizes the changes in AFP in eight rats with transplanted hepatomas.

AFP monitoring is also useful for HCC patients treated with chemotherapy, such as Adriamycin, 5-fluorouracil, cyclophosphamide, and methotrexate. Although such therapy generally has limited success, a decrease in the serum AFP level is indicative of a positive response to treatment. Measurement of the kinetics of AFP provides a semiquantitative measure of the adequacy of therapy [51].

In contrast to levels of total AFP, measurement of AFP subfractions is not useful for monitoring treatment. Cells from primary hepatocellular carcinoma as well as from germ-cell tumors are "preprogrammed" to produce a reasonably consistent pattern of AFP glycosylations [52]. No correlation between minor fluctuations in percentage CNR-AFP and the clinical course of patients can be concluded [15]. Likewise, measurement of ferritin does not appear to be useful for the follow-up of treated HCC patents [53].

D. Markers for Routine Laboratory Tests of Liver Function

1. Total Enzyme Measurements

Measurement of enzyme activities routinely used in diagnosis of liver diseases include aspartate and alanine aminotransferase (AST and ALT), alkaline phosphatase (ALP), gamma-glutamyl transferase (GGT), and 5'nucleotidase (5NT). Presence of these enzymes in elevated concentrations in blood is indicative of nonspecific liver injury or obstruction and they are useful in hepatocellular carcinoma only to the extent that they reveal the presence of these processes secondary to the neoplastic process itself. Levels of AST and ALT, for example, are marginally elevated when hepatocytes are destroyed during the growth and invasion of the tumor. Levels of ALP and GGT are elevated secondary to obstruction by the "space-occypying" tumor. Measurement of these enzymes is not specific to primay cancer, since elevations are observed in secondary liver

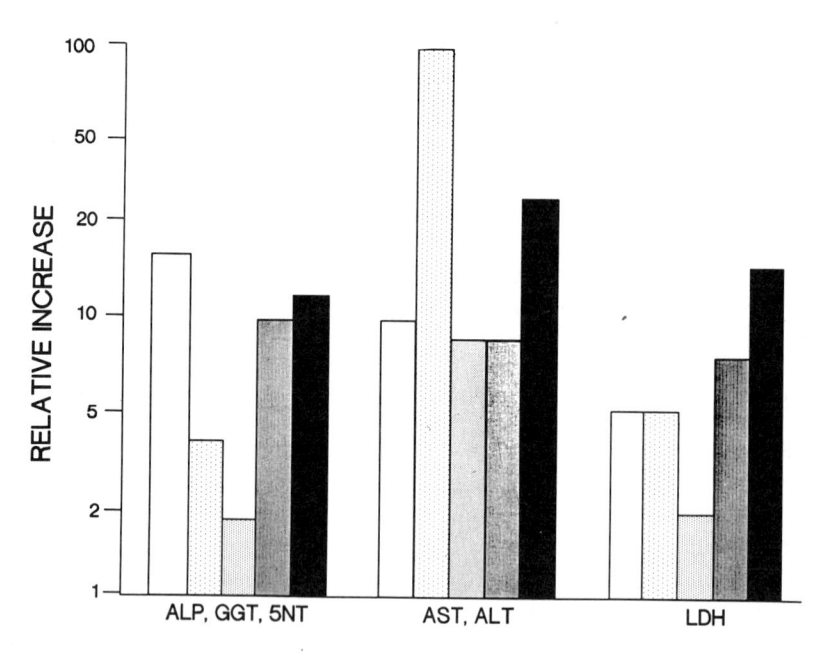

Figure 6 Typical results for enzymes in various types of liver disease. The ordinate represents the elevations observed for each class with the reference limit set to one. Clear bar, liver cancer; light shaded bar, obstruction; medium shaded bar, hepatocellular necrosis; dark shaded bar, alcoholic cirrhosis, black bar, other disease (for ALP, bone disease; for AST and ALT, muscle and myocardial disease; for LDH, muscle, myocardial diseases, hematological cancers, and hemolysis).

cancer as well as other diseases such as hepatitis, cirrhosis, and biliary obstruction (see Fig. 6). Use of these liver enzymes is somewhat helpful for the diagnosis of liver metastasis. For example, GGT has a clinical sensitivity of 97%. The specificity is low, however, since GGT levels are elevated in many other conditions, including chronic alcohol use. 5NT is more specific at 93% for secondary liver cancer, but has a low clinical sensitivity [54]. Although a combination of GGT and 5NT would appear to be the most efficient use of these markers, they are not currently used because they are not efficient in detecting early metastasis [55].

2. Isoenzyme Markers

Increased efficiency for secondary liver cancer has been observed when specific isoenzymes are measured. One of the first isoenzyme markers for cancer was an alkaline phosphatase variant identified by Stolbach et al. and subsequently termed the Regan isoenzyme [56]. This atypical isoenzyme is thought to be a placental antigen: it resembles placental ALP in terms of its resistance to denaturation at 56°C, electrophoretic migration, and immunological properties. It is not, however, an effective tumor marker because of its very low sensitivity (<25% for most cancers).

Other isoenzymes have demonstrated a higher degree of diagnostic efficiency, particularly for the detection of liver metastasis. The alpha 1-isoenzyme of alkaline phosphatase has been reported to have a clinical sensitivity 97% [57]. This isoenzyme is not normally present in healthy individuals and is termed the "fast liver" fraction because it migrates beyond the liver isoenzyme in the anodic direction on cellulose acetate electrophoresis. Levels of fast liver isoenzyme are, however, elevated in obstructive liver diseases, especially extrahepatic forms [58]. A fith enzyme of 5′-nucleotide phosphodiesterase (NPD) has been identified in serum and correlated with cancer [59]. As with other liver tumor markers, however, nonspecific liver disease produces false-positive results. Because of the limitations in sensitivity and specificity, particularly for early diagnosis, and the lack of convenient routine assays, isoenzymes of 5 NPD are not routinely used. Detection of an albumin-migrating GGT isoenzyme has been shown to have a clinical sensitivity of 75% and specificity of >90% when measured in patients with other nonhepatic malignancies, and other liver diseases [60]. Studies on larger numbers of patients are necessary before the clinical utility of these markers is known. To date, none of the isoenzymes of routine liver enzymes is being used for diagnosis of HCC.

3. Other Indicators of Liver Function

Nonspecificity is also a problem when other traditional nonenzyme markers of liver functin are examined for patients with HCC. Elevations in serum bilirubin and urinary urobilinogen levels reflect the extent of disruption in hepatic architecture secondary to liver cancer and range from normal to fivefold elevated. Total protein and albumin levels may be normal or decreased depending on the stage and aggressiveness of the tumor. Early tumor growth is likely to be associated with normal protein levels, whereas decreased levels are typically observed in the later stages due to the consumption of substrates by the tumor itself. Vitamin B_{12} levels are often elevated in patients with liver disease. Reports have shown that patients with HCC may also have high vitamin B_{12} levels secondary to the production of an abnormal serum-binding protein [61]. Although detection of this variant binding protein may have diagnostic implications, its production has not been demonstrated in certain populations such as South African blacks [62]. Levels of serum copper are nonspecifically elevated in patients with liver disease. Exceptionally high levels are observed in patients with Wilson's disease. Copper has also been described in patients with hepatomas as well as other carcinomas, but it is thought to reflect altered metabolism and is itself not carcinogenic [63].

Other liver function tests may be useful for predicting which of a population of diagnosed HCC patients are likely to survive long enough (e.g., 60 days) for the performance of valid chemotherapeutic trials. Of several laboratory factors examined, measurement of AST, blood urea nitrogn, and total bilirubin together with an assessment of alcohol consumption and the presence or absence of encephalopathy appear to be the most discrimminating factors for recruiting patients for drug studies in countries with a high incidence of HCC [64].

E. Immunodiagnosis and Immunotherapy

Immunodiagnosis and therapy require identification of specific antigens produced by the hepatoma. Because of their importance in serum, it is logical first to assess the serum-borne liver tumor markers that are found at high frequency in tumor tissues themselves. Such studies have shown that alpha-1-antitrypsin occurs in hepatoma tissue with a frequency of 83%, ferritin 70%, alpha fetoprotein 50%, and carcinoembryonic antigen 30% [65,66]. Differences in analytical sensitivity and study population may account for the differences in the reported incidence of these markers in tumor tissue compared to serum. When the same individual is studied, however, markers are invariably found in both [65].

Antibodies have been examined as potential markers for immunolocalization. One known as anti-PLC$_1$ appears to have a high specificity towards membranes of human hepatoma cells grown in culture [67].

Studies on the in vivo localization of HCC using monoclonal antibodies have been limited. In 1974, preliminary data showed that radiolabeled antimouse AFP preferentially localized in hepatocellular tumor tissue if the serum of tumor-bearing mice was cleared of circulating AFP [68]. Localization of transplanted AFP producing hepatocellular carcinoma was demonstrated in rats using ^{125}I horse F(ab′)2 to AFP [69]. When this technique was applied to human HCC, however, the identification rate was only 7 of 15 patients when using polyclonal radiolabeled anti-AFP and 2 of 9 using monoclonal labeled anti-AFP; all patients had elevated serum AFP concentrations and tumor masses that were detectable ty celiac angiography, abdominal computed tomographic (CT) scan, and ultrasonography [70]. In contrast, Kim et al. [71] examined 16 cases of malignant neoplasia with ^{131}I-labeled goat anti-alpha-fetoprotein antibody and demonstrated 100% sensitivity for detecting AFP-producing tumors and 80% specificity. Although sensitivity for HCC is inferior to localization by ultrasound.

Immunotherapy is based on the delivery of chemotherapeutic agents and/or radiation directly to neoplastic tissue through binding by specific antibodies to tumor-expressed proteins. Studies done on animal models in the 1970s suggested that anti-AFP administered to animals bearing AFP-producing transplatanble hepatomas might suppress tumor growth if given at the time of tumor transfer [20,72,73]. Hirai et al. have treated 23 patients with AFP-producing hepatomas [70,74]. Administratin of purified horse anti-AFP resulted in an immediate drop in the serum AFP concentration, but had little or no effect on liver enzyme markers [74]. Seven of the 23 patients had decreased serum AFP concentrations for long periods, but no definite conclusion could be made regarding possible therapeutic efficacy.

Tsukada et al. recently conjugated the cytotoxic drug daunomycin to anti-AFP in an attempt to direct its effect on AFP-producing tumors [75,76]. Administration of purified conjugated horse antirat AFP 3, 5, and 7 days after transplantation of hepatoma AH66 to Donryu rats produced a 50% cure rate (vs. 0% in controls). Daunomycin-conjugated monoclonal antibody to rat AFP injected 14-22 days after tumor transplantation resulted in prolongation of survival from 25 days in control animals to 60 days in the treated animals [73]. Removal of the tumor transplant surgically at 14 days followed by daunomycin-conjugated monoclonal antibody treatment produced a 60% cure rate compared to 0% with surgery alone. Daunomycin-conjugated anti-AFP also suppresses growth of human hepatomas transplanted into nude mice [77]. Clinical trials using daunomycin-conjugated anti-human AFP are now underway in Japan.

Order et al. has shown that antiferritin labeled with ^{131}I may be useful for patients with unresectable liver tumors [78]. Of 38 patients treated with conventional external radiation therapy, 23% went into partial remission. Of the 66 patients who received antiferritin treatment (with 25 of the 38 having also undergone irradiation treatment), 48% went into either partial or total remission, for a total remission rate of 50% for all 79 patients. It remains to be seen if this approach becomes a valuable member of the limited methods for treating liver cancer. Studies have shown that the use of interleukin-2 and lymphokine-activated killer cells for patients with disseminated cancer has produced little success [79,80].

III. FUTURE CONSIDERATIONS

Active investigation into new and more accurate markers for cancer include the use of cancer-related epitopes detected by monoclonal antibodies, chromosomal rearrangements, and oncogene expression [81]. These approaches have not as yet been developed to the point of being useful for the diagnosis or treatment of liver cancer. The study of early reversible changes that occur during the induction of hepatocellular carcinomas in animals by chemicals gives support for the possibility of detecting "precancerous" changes before overt malignancy develops [81-83]. If the causative agents are known, it is theoretically possible that the affected individual could be removed from the environment or his or her diet changed so that the ultimate development of cancer is prevented. Exposure of mice or rats to chemical hepatocarcinogens results in elevations of AFP that may occur well in advance of hepatocellular carcinoma [82,84]. Removal of the hepatocarcinogen from the diet of the animal is followed by reversal of "premalignant" changes in the liver and a decline in serum AFP levels to normal [85,86]. In humans, elevations of AFP associated with early carcinomas have been found in screening studies (see above and [87-89]), but convincing data indicating that serum AFP elevations may be used to detect hepatocarcinogen exposure prior to the appearance of cancer have not been reported. A possible exception is the report of Purves et al. [90], who examined Bantu men from Mozambique and Malaysia, who are known to have a higher incidence of HCC and higher baseline serum levels of AFP than their white counterparts. When these men went to work in mining camps and were given a diet of whites, their serum AFP dropped to control levels after 1 year. A possible explanation is the presence of an hepatocarcinogen in their native diet. Since it is known that the Bantu eat foods containing aflatoxin, it appears likely that their high serum AFP concentrations are caused by aflatoxin and that this effect may be reversed if aflatoxin is removed before HCC develops.

Monoclonal antibodies are now being used to identify early cellular changes during experimental hepatocarcinogenesis in rats. Much of this work is reviewed elsewhere [81]. Routine histopathological examination as well as enzyme histochemistry and tissue localization of AFP during experimental hepatocarcinogenesis have previously suggested that there are multiple possible celluar lineages to cancer [81-83]. Among the possibilities that must be considered are so-called neoplastic nodules, oval cells, zones of "atypical hyperplasia," as well as hepatocellular cancers arising directly from normal-appearing hepatocytes. For this chapter it is not appropriate to include a complete discussion of this system; the reader is referred to other works [81,82]. Monoclonal antibodies recognize epitopes on different cell types that appear during chemical hepatocarcinogenesis. One group of epitopes is shared by oval cells and tumors that arise from carcinogenic regimens that stimulate oval cell production. Another group of epitopes is found on microcarcinomas that appear to arise de novo using other regimens. Neither of these epitopes is seen on early nodular cells. Thus the phenotypes of putative premalignant cells found during experimental hepatocarcinogenesis are consistent with the multiple lineage hypothesis of liver cancer, most likely arising from a liver "stem cell."

IV. SUMMARY

Markers for hepatocellular cancer include the best and worst of cancer detection. Although hepatocellular cancer is relatively infrequent compared to other cancers in the western world, HCC has a very high incidence in parts of Asia and Africa. It is estimated to be one of the most common cancer worldwide. High risk factors for HCC include previous hepatitis B infection, heavy alcohol consumption, cirrhosis, and aflatoxin exposure. Alpha fetoprotein may be the best human cancer marker that appears in the serum, but levels of this marker are often not elevated until the tumor is beyond surgical treatment. No other serum or tissue marker is particularly useful. Screening of high-risk populations in China has detected previously undiagnosed HCC in 1,000 of 5 million individuals tested and has led to an increase in survival from 5.5 to 61.6% with surgical resection over those who are later diagnosed with HCC without screening. Elevations of AFP due to yolk sac tumors may be differentiated from those due to HCC on the basis of Concanavalin A reactivity. Immunodetection using radiolabeled anti-AFP and immunoscintigraphy have given inconsistent results that are not as sensitive as ultrasonography in detecting HCC in the liver. Various enzymes, isoenzymes, and other markers may be useful as adjuncts to diagnosis in selected cases, but are not generally as good as AFP alone. If a patient has an AFP-producing tumor, the serum levels of AFP provide an excellent means of monitoring its progression. If the serum AFP levels drop to normal and stay there, cure is almost certain. If, however, the serum AFP level does not fall at the normal catabolic rate after therapy, or subsequently rises, regrowth of metastases are indicated. Immunotherapy using anti-AFP has not been shown to induce remission, but experimental studies indicate that drug-conjugated anti-AFP is effective in inhibiting growth of AFP-producing tumors. Clinical trials using drug-conjugated anti-AFP are now underway. Monoclonal antibodies have not yet identified the "antigens" useful for the diagnosis or treatment of HCC, but epitopes identified by monoclonal antibodies have been studied experimentally in rats which indicate multiple cellular lineages to HCC in cases of experimental chemically induced hepatocarcinoma.

REFERENCES

1. Waterhouse J, Muir C, Shanmugaratnam K, Powell J, eds. Cancer incidence in five continents. V. Lyon: IV. World Health Organization, IARC Scientific Pub., 1982.
2. Szmuness W. Hepatocellular carcinoma and the hepatitis B virus: evidence for a causal association. Prog Med Virol 1978;24:40-69.
3. Chen JY, Harrison TJ, Lee CS, et al. Detection of hepatitis B virus DNA in hepatocellular carcinoma. Br J exp Pathol 1986;67:279-282.
4. Chen DS, Sung JL. Serum alpha-fetoprotein in hepatocellular carcinoma. Cancer 1977;40:779-783.
5. Hino O, Kitagawa T, Koike K, et al. Detection of hepatitis B virus DNA in hepatocellular carcinoma in Japan. Hepatology 1984;4:90-95.
6. Liaw YF, Tai DI, Chu CM, et al. Early detection of hepatocellular carcinoma in patients with chronic type B hepatitis. A prospective study. Gastroenterology 1986; 90:263-267.
7. Lin DY, Liaw YF, Chu CM, et al. Hepatocellular carcinoma in noncirrhotic patients; a laparoscopic study of 92 cases in Taiwan. Cancer 1984;54:1466-1468.
8. Mayaji T. Association of hepatocellular carcinoma with cirrhosis among autopsy cases in Japan (1858-1967). Cancer Res 1973;14:179-204.
9. Stemhagen A, Slade J, Altman R, Bill J. Occupational risk factors and liver cancer. A retrospective case-control study of primary liver cancer in New Jersey. Am J Epidemiol 1983;117:443-454.

10. Sun TT, Chu YY. Carcinogenesis and prevention strategy of liver cancer in areas of prevalence. J Cell Physiol 1984;3(Suppl):39-44.
11. Wogan GU. Dietary factors and special epidemiological situations of liver cancer in Thailand and Africa. Cancer Res 1975;35:3499-3502.
12. Stoloff L. Aflatoxin as a cause of primary liver cell cancer in the United States: a probability study. Nutr Cancer 1983;5:165-186.
13. Austin H, Delzell E, Grufferman S, et al. A case-control study of hepatocellular carcinoma and the hepatitis B virus, cigarette smoking, and alcohol consumption. Cancer Res 1986;46:962-966.
14. Abelev GI, Perova S, Khramkova N, et al. Production of embryonic α-globulin by transplantable mouse hepatoma. Transplantation (Baltimore) 1963;1174-1180.
15. Tatrinov Y. Detection of embryo specific α-globulin in the blood sera of patients with primary liver tumor. Vopr Med Khim 1964;10:90-94.
16. Chan DW, Kelsten M, Rock R, Bruzek D. Evaluation of a monoclonal immunoenzymometric assay for alpha-fetoprotein. Clin Chem 1986;32:1318-1322.
17. Winter J. Computer assessment of observer performance by receiver operating characteristic curve and information theory. Comput Biomed Res 1982;15:555-562.
18. Bell H. Alpha-fetoprotein and carcinoembryonic antigen in patients with primary liver carcinoma, metastatic liver disease, and alcoholic liver disease. Scand J Gastroenterol 1982;17:897-903.
19. Alpert E. Serum alpha-fetoprotein (AFP) in benign and malignant gastrointestinal diseases: Evaluation of an immunoenzymatic assay. Clin Chim Acta 1975;58:77-83.
20. Endo HY, Kanai K, Oda T, et al. Clinical significance of α-fetoprotein in hepatitis and liver cirrhosis. Ann NY Acad Sci 1975;259:234-238.
21. Sell S, Stillman D, Gochman N. Serum alpha-fetoprotein: a diagnostic and prognostic indictor of liver cell injury following experimental injury by galactosamine in rats. Am J Clin Pathol 1976;66:847-856.
22. Sell S, Wepsic HT, Nickel R, Nichols M. Rat alpha$_1$-fetoprotein IV. Effect of growth and surgical removal of Morris hepatoma 7777 on serum α-F concentration of buffalo rats. J Natl Cancer Inst 1974;52:133-137.
23. Xu KL. Large-scale AFP screening for hepatocellular carcinoma in China. In Fishman WH, ed. Oncodevelopmental markers, biologic, diagnostic, and monitoring aspects. Orlando, FL: Academic Press, 1983:395-408.
24. Tang Z. Screening and early treatment of primary liver cancer—with special reference to the east part of China. Ann Med Singapore 1980;9:203-205.
25. Breborowicz J. Microheterogeneity of human alphafetoprotein. Tumor Biol 1988; 9:3-14.
26. Chan DW, Miao YC. Affinity chromatographic separation of alpha-fetoprotein variants: development of a mini-column procedure, and application to cancer patients. Clin Chem 1986;32:2143-2146.
27. Vessella RL, Santrach MA, Bronson D, et al. Evaluation of AFP glycosylation heterogeneity in cancer patients with AFP-producing tumors. Int J Cancer 1984;34:309-314.
28. Xu KL, Yu EX, Lu LN, Zhou J. Alpha-fetoprotein variants and their clinical significance. Chin Med J 1984;97:538-542.
29. Brock DJH, Barron L, Van Heyningen V. Approaches to the production of monoclonal antibodies specific for concanavalin A binding and non-binding forms of alpha-fetoprotein. Tumour Biol 1984;5:171-178.
30. Kitagawa H, Oekouchi E, Hata JI, et al. Monoclonal antibodies with fine specificities distinguishing alpha-fetoproteins of hepatoma and yolk sac tumor origin. Jpn J Cancer Res (Gann) 1986;77:1012-1017.

31. Breborowicz J, Mackiewicz A, Breborawicz D. Microheterogeneity of alphafetoprotein in patient serum as demonstrated by lectin affino immunoelectrophoresis. Scand J Immunol 1981;14:15-20.
32. Aoyagi Y, Isemura M, Yosizawa Z, et al. Fucosylation of serum α-fetoprotein in patients with primary hepatocellular carcinoma. Biochim Biophys Acta 1985; 830:217-232.
33. Buamah PK, Cornell C, Cassells-Smith AJ, Harris AL. Fucosylation of α-fetoprotein in hepatocellular carcinoma. Lancet 1986;1:922-923.
34. Maringhini A, Cottone M, Sciarrino E, et al. Ultrasonographic and radionuclide detection of hepatocellular carcinoma in cirrhotics with low alpha-fetoprotein levels. Cancer 1984;54:2924-2926.
35. Zokazaki N, Yoshida T, Yoshino M, Matue H. Screening of patients with chronic liver disease for hepatocellular carcinoma by ultrasound. Clin Oncol 1984;10:241-246.
36. Sheu JC, Sung JL, Chen DS, et al. Early detection of hepatocellular carcinoma by real-time ultrasonography. Cancer 1985;56:660-666.
37. Chlebowski RT, Tong M, Weissman J, et al. Hepatocellular carcinoma. Diagnostic and prognostic features in North America patients. Cancer 1984;53:2701-2706.
38. Buscarini L, Sbolli G, Cavanna L, et al. Clinical and diagnostic features of 67 cases of hepatocellular carcinoma. Oncology 1987;44:93-97.
39. Giannoulis E, Arvanitakis C, Nikopoulos A, et al. Diagnostic value of serum ferritin in primary hepatocellular carcinoma. Digestion 1984;30:236-241.
40. Tatsuta M, Yamamura H, Iishi H, et al. Value of serum alpha-fetoprotein and ferritin in the diagnosis of hepatocellular carcinoma. Oncology 1986;43:306-310.
41. Nakano S, Kumada T, Sugiyama K, et al. Clinical significance of serum ferritin determination for hepatocellular carcinoma. Am J Gastroenterol 1984;79:623-627.
42. Kew MC, Berger EL. The value of serum concentrations of tissue polypeptide antigen in the diagnosis of hepatocellular carcinoma. Cancer 1986;58:127–301.
43. Yeh YC, Tsai JF, Chuang LY, et al. Elevation of transforming growth factor and its relationship to the epidermal growth factor and α-fetoprotein levels in patients with hepatocellular carcinoma. Cancer Res 1987;47:896-901.
44. Tamura S, Amuro Y, Nakano T, et al. Urinary excretion of pseudouridine in patients with hepatocellular carcinoma. Cancer 1986;57:1571-1575.
45. Mori Y, Iwama S, Mori T, et al. Liver-specific F antigen in the serum of patients with liver diseases and the detection of early stage hepatocellular carcinoma. Clin Chim Acta 1987;164:127-137.
46. Terui S. Clinical evaluation of thyroxine-binding globulin (TBG) as a marker of liver tumors. Eur J Nucl Med 1984;9:121-124.
47. Liebman HA, Furie BC, Tong MJ, et al. Des-γ-carboxy (abnormal) prothrombin as a serum marker of primary hepatocellular carcinoma. N Engl J Med 1984;310:1427-1431.
48. Soulier JP, Gozin D, Lefrere JJ. A new method to assay des-γ-carboxyprothrombin. Results obtained in 75 cases of hepatocellular carcinoma. Gastroenterology 1986; 91:1258-1262.
49. Kelstein ML, Chan DW, Bruzek DJ, Rock RC. Monitoring hepatocellular carcinoma by using a monoclonal immunoenzymometric assay for alpha-fetoprotein. Clin Chem 1988;34:76-81.
50. Sheu JC, Sung JL, Chen DS, et al. Growth rate of asymptomatic hepatocellular carcinoma and its clinical implications. Gastroenterology 1985;89:259-266.
51. Buamah PK, Skillen AW, Ward AW. Serum alpha fetoprotein kinetics in hepatocellular carcinoma. A case for cessation of therapy. Cancer Chemother Pharmacol 1986; 16:75-77.

52. Vessella RL, Santrach MA, Bronson D, et al. Evaluation of AFP glycosylation heterogeneity in cancer patients with AFP-producing tumors. Int J Cancer 1984; 34:309-314.
53. Nagasue N. Yukaya H, Chang YC, Ogawa Y. Serum ferritin level after resection of hepatocellular carcinoma. Correlation with alpha-fetoprotein level. Cancer 1986; 57:1820-1823.
54. Kim NK, Yasmineh WG, Freier EF, et al. Value of alkaline phosphatase, 5'-nucleotidase, γ-glutamyltransferase, and glutamate dehydrogenase activity measurements (single and combined) in serum in diagnosis of metastasis to the liver. Clin Chem 1977;23:2034-2038.
55. Beck PRl, Belfield A, Spooner RJ, et al. Serum enzymes in the diagnosis of hepatic metastatic carcinoma. Clin Chem 1978;24:839.
56. Stolbach LL, Krant MJ, Fishman WH. Ectopic production of an alkaline phosphatase isoenzyme in patients with cancer. N Engl J Med 1969;281:757-762.
57. Viot M, Thyss A, Viot G, et al. Comparative study of gamma glutamyl transferase, alkaline phosphatase, and its α1 isoenzyme as biological indicators of liver metastasis. Clin Chem Acta 1981;115:349-358.
58. Crofton PM, Elton RA, Smith AF. High molecular weight alkaline phosphatase: a clinical study. Clin Chim Acta 1979;98:263-275.
59. Tsou KC, Lo KW. Serum 5'-nucleotide phosphodiesterase isoenzyme-V test for human liver cancer. Cancer 1980;45:209-213.
60. Sacchetti L, Castaldo G Salvatore F. The gamma-glutamyltransferase isoenzyme pattern in serum as a signal discriminating between hepatobiliary diseases, including neoplasias. Clin Chem 1988;34:352-355.
61. Waxman S, Gilbert HS. A tumor-related vitamin B_{12} binding protein in adolescent hepatoma. N Engl J Med 1973;282:1053-1056.
62. Van Tonder S, Kew MC, Hodkinson J, et al. Serum vitamin B_{12} in South African Blacks with hepatocellular carcinoma. Cancer 1985;56:789-792.
63. Miatto O, Casaril M, Gabrielli GB, et al. Diagnosis and prognostic value of serum copper and plasma fibrinogen in hepatic carcinoma. Cancer 1985;55:774-778.
64. Attali P, Prod'Homme S, Pelletier G, et al. Prognostic factors in patients with hepatocellular carcinoma. Attempts for the selection of patients with prolonged survival. Cancer 1987;59:2108-2111.
65. Imoto M, Nishimura D, Fukuda Y, et al. Immunohistochemical detection of α-fetoprotein, carcinoembryonic antigen, and ferritin in formalin-paraffin sections from hepatocellular carcinoma. Am J Gastroenterol 1985;80:902-906.
66. Busachi CA, Ferrari S, Ballardini PL, et al. Tissue antigen distribution in hepatocellular carcinoma. Tumori 1986;72:1-5.
67. Shouval D, Eilat D, Carlson RT, et al. Human hepatom-associated cell surface antigen: identification and characterization by means of monoclonal antibodies. Hepatology 1985;5:347-356.
68. Parks LC, Baer A, Pollace M, Williams GM. Alpha-fetoprotein: an index of progression or regression of hepatoma, and a target for immunotherapy. Ann Surg 1974;180:599-605.
69. Koji T, Ishii N, Munehisa T, et al. Localization of radioiodinated antibody to alpha-fetoprotein in rats and a case report of alpha-fetoprotein antibody treatment of a hepatoma patient. Cancer Res 1980;40:3010-3015.
70. Ishii N, Nakata K, Muro T, et al. Radioimmunodetection of cancer using antibodies to alpha-fetoprotein and carcinoembryonic antigen. Ann NY Acad Sci 1983;417: 270-276.
71. Kim EE, DeLand FH, Nelson MO, et al. Radioimmunodetection of cancer with radiolabeled antibodies to α-fetoprotein. Cancer Res 1980;40:3008-3012.

72. Gousev A, Yazova A. Termination of natural tolerance to alpha-fetoprotein in rats: its effect on hepatoma growth and course of pregnancy. In: Masseyeff R, ed. Colloques L'Inserm l'alpha foetoproteine. Paris, France: Editions INSERM, 1974: 255-270.
73. Mizejewski GJ, Allen RP. Immunotherapeutic suppression in transplantable solid tumors. Nature 1974;250:50-52.
74. Hirai H, Taga T, Kaneda H, et al. Alpha-fetoprotein: its diagnostic specificity and therapeutic use of its antibody. In Cimino F, et al., eds. Human tumor marker. Berlin: de Gruyter, 1987:181-201.
75. Tsukada Y, Bischof WKD, Hibi N, et al. Effect of a conjugate of daunomycin and antibodies to rat alpha-fetoprotein on the growth of alpha-fetoprotein producing tumor cells. Proc Natl Acad Sci 1982;79:621-625.
76. Tsukada Y, Ohkawa K, Hibi N. Therapeutic effect of treatment with polyclonal or monoclonal antibodies to alpha-fetoprotein that have been conjugated to daunomycin via a dextran bridge. Studies with an alpha-fetoprotein producing rat hepatoma tumor model. Cancer Res 1985;47:4293-4295.
77. Tsukada Y, Ohakawa K, Hibi N. Suppression of human alpha-fetoprotein producing hepatocellular carcinoma growth in nude mice by an anti-alpha-fetoprotein antibody-daumycin conjugate with a poly L-glutamate acid derivative as intermediate drug carrier. Br J Cancer 1985;52:111-116.
78. Order SE, Stillwagen GB, Klein JL, et al. Iodine 131 antiferritin, a new treatment modality in hepatoma: a radiation therapy oncology group study. J Clin Oncol 1985;3:1573-1582.
79. Lotze MT, Chang A, Seipp CA, et al. High-dose recombinant interleukin 2 in the treatment of patients wtih disseminated cancer. Responses, treatment-related morbidity, and histologic findings. JAM 1986;256:3117-3124.
80. DiBisceglie AM, Rustgi VK, hoofnagle JH, et al. Hepatocellular carcinoma. Ann Intern Med 1988;108:390-401.
81. Sell S, Hunt JM, Knoll BJ, Dunsford HA. Cellular events during hepatocarcinogenesis in rats and the question of premalignancy. Adv Cancer Res 1987;48:37-111.
82. Sell S. Cancer markers: past, present and future. In Resifeld R, Sell S, eds. Monoclonal antibodies and cancer therapy. New York: Alan R. Liss, 1985:3-21.
83. Sell S. Leffert HL. An evaluation of cellular lineages in the pathogenesis of experimental hepatocellular carcinoma. Hepatology 1982;2:77-86.
84. Farber E, Cameron R. The sequential analysis of cancer development. Adv Cancer Res 1980;31:125-226.
85. Sell S, Becker FF. Alphafetoprotein. J Natl Cancer Inst 1978;60:19-26.
86. Becker FF, Sell S. Early elevation of alpha-fetoprotein in N-1-fluorenylacetamide hepatocarcinogenesis. Cancer Res 1974;34:2489-2494.
87. Teebor G. Sequential aspects of liver carcinogenesis. In Becker FF, ed. Cancer: a comprehensive treatise. New York: Plenum Press, 1975:345-352.
88. LeBlanc L, Tuyns AJ, Masseyeff R. Screening for primary liver cancer. The relationship between the time of appearance of α-fetoprotein in the serum and clinical symptoms in nine cases. Digestion 1973;8:8-14.
89. Lehmann FG, Wegener T. Alphafetoprotein in liver cirrhosis II. Early detection of hepatoma. In: Lehman FG, ed. Carcino-embryonic proteins. Vol 1. Amsterdam: Elseiver, 1979:233-245.
90. Purves LR, Branch WR, Geddes EW, et al. Serum alphafetoprotein VII. The range of apparent serum values in normal people, pregnant women, and primary liver cancer high risk populations. Cancer 1973;31:578-587.

27

Bone Tumors

MORTON K. SCHWARTZ

Memorial Sloan-Kettering Cancer Center, New York, New York

Bone cancers can be classified into two groups: primary bone cancer and cancer metastatic to bone. Primary bone cancers reflect the cellular and extracellular structure of the skeleton. They are derived histogenetically from the osteoclasts, whose origin is from hematopoietic cells and from osteoblasts that originate from stromal cells. Stromal cells may also differentiate embryologically into chondroblasts and fibroblasts. Bone cancers may also on occasion arise from other hematopoietic and neural cells.

I. BENIGN BONE TUMORS

There are a large variety of benign bone tumors, including osteochondromas, chondromas, benign giant cell tumors, unicameral bone cysts, osteoid osteomas, and fibromas. The diagnosis and monitoring during treatment of these tumors involves histopathological examination, bone scans, computed axial tomography, and conventional radiological studies [1]. Tumor markers are not useful in these diseases. Levels of the most of used bone tumor markers, alkaline phosphatase, are usually not elevated [2].

II. PRIMARY BONE TUMORS

Malignant bone tumors include tumors of bony origin, multiple myeloma and reticulum cell sarcoma, which are hematopoietic in origin and involve an influence on bone due to the pressure of the cancer within the bone marrow. The osteoid malignant bone tumors include chondrosarcomas, osteosarcomas, fibrosarcoma, Ewing's tumor, malignant giant cell tumors, and chondromas [1]. Because of their influence on calcium metabolism, parathyroid adenomas may be considered bone tumors.

A. Multiple Myeloma

Multiple myeloma is a malignancy of plasma cells, with an incidence in the United States of about 30 cases/10^6 population. The plasma cells within the bone marrow produce a unique and diagnostic amount of immunoglobulins and generate, for the most part, osteolytic bone changes. The bone change is probably due to production by the malignant plasma cells of a substance that stimulates osteoclast proliferation. Alkaline phosphatase levels are elevated in about 30% of the patients, but in most cases to levels less than four times the upper limit of normal [2]. Most patients with myeloma demonstrate an anemia but only 25% of patients with myeloma have hypercalcemia and 20% exhibit elevations of serum urea nitrogen and uric acid levels, presumably related to secondary renal damage. Ninety-eight percent of the patients exhibit a homogeneous (monoclonal) globulin peak on conventional electrophoresis of serum or urine. Immunoglobulin assays permit one to confirm the diagnosis and monitor treatment of the disease [1].

The single electrophoretic homogeneous peak on electrophoresis is usually an IgG or IgA heavy chain with either kappa or lambda light chains. IgD or IgE heavy chains are very rare. Bence Jones protein, which is a light chain component with a molecular weight of 20,000 Kd, is found in the urine of about 20% of patients with myeloma, primarily when renal failure inhibits normal catabolism. In about 20% of patients a monoclonal gammopathy is not observed in the serum or urine but in these patients immunofluorescent microscopy of bone marrow will identify malignant plasma cells. Once a diagnosis is made, serial monitoring of the myeloma proteins is essential in managing these patients [3,4].

After the immunoglobulin abnormality is identified by screening with serum protein electrophoresis, radial immunodiffusion or nephelometry is used to identify the class of immunoglobulins and to quantitate the amount. Immunoelectrophoresis and immunofixation are then used to type and classify the gammopathy. In nonsecretory multiple myeloma, severe anemia and multiple areas of bone destruction, are seen. However, unlike in typical myeloma, there is marked depression of normal immunoglobulins and no renal failure or hypercalcemia [5].

Flow cytometry is useful in predicting prognosis in myeloma. In one study, aneuploidy of DNA content was present in 80% of 115 patients with active myeloma, but was not found in any of 12 patients with disease in remission. A low thymidine labeling index ($<$1%) indicated longer survival [6].

B. Ewing's Sarcoma

Ewing's sarcoma is a rapidly growing, often lethal tumor in which routine follow-up with chest x-rays and bone scans is not sufficiently sensitive or specific to determine either tumor progression or cancer recurrence. Serum alkaline phosphatase levels are usually within normal limits or slightly elevated. In several early reports total serum lactate dehydrogenase (LDH) levels were reported to be of value in predicting survival in these patients [7,8]. In the study by Glabiger of 117 patients treated at the National Cancer Institute (NCI), survival was highly correlated with levels of serum LDH ($p < 0.0001$) [7]. When the results were adjusted for treatment protocol, the significance disappeared.

Two studies from Memorial Sloan-Kettering Cancer Center have not confirmed the prognostic role of LDH [9,10]. In 1981 Rosen and his associates reported that initial serum LDH values correlated with the bulkiness of the tumor but that enzyme activity had no prognostic value with post-1976 chemotherapy protocols [9]. In a more recent study of 133 patients studied between 1970 and 1985, 56 were evaluated for a minimum of 2 years (mean follow-up, 48 months) and were subject to retrospective analysis

[10]. Of 32 patients, 31% with initially normal LDH levels showed recurrence (follow-up 26 ± 19 months) whereas 5 of 15 (33%) patients with elevated LDH levels had a recurrence (follow-up, 29 ± 21 months). There was no statistical difference in the two groups. However, at recurrence the mean serum level in 20 patients was 330 ± 208 U/L compared to 192 ± 41 U/L in 31 patients who did not experience recurrence. These values were significantly different ($p < 0.001$); the sensitivity was 60% (12/20 patients) and the specificity 80% (25/31 patients). At a mean follow-up of 58 ± 24 months, the specificity was 94% (29/31 months). These studies suggest that total serum LDH levels are not a prognostic factor nor an appropriate monitor of therapy. However, a normal LDH level is consistent with the absence of recurrent disease. Measurement of serum LDH levels, in association with x-rays and bone scans, can be an important part of the long-range surveillance of patients with Ewing's sarcoma.

C. Osteosarcomas

Classic osteosarcomas are classified as true osteosarcomas, chondrosarcomas, and fibroblastic sarcomas. A number of markers have been described for use in management of these patients.

1. DNA Ploidy

All classic osteosarcomas appear to be hyperploid when evaluated by single-cell cytophotometry [11]. DNA content was measured by flow cytometry in 20 malignant osteosarcoma tissues after preoperative chemotherapy. Fourteen showed aneuploid DNA stem lines with a medium value of 1.67 and an S-phase proportion of 16.1%. These values were significantly higher than observations in untreated osteosarcomas. It appears that higher DNA indices as well as DNA aneuploidy and cells in S-phase indicate a poor regression of the tumor [12].

In a study of 158 tumors, all of 41 benign bone tumors were diploid whereas 92 of 96 high-grade osteosarcomas were hyperploid (high DNA content). In 17 cases in which the pathological diagnosis was indeterminate, those patients who experienced recurrence had hyperploid tumors and those who did not recur showed diploid tumors [13].

Hyperploid DNA distribution was observed in chondrosarcoma in 22 of 48 cases, whereas all 4 benign chondromas were diploid [14]. It was concluded that malignant disease could not be differentiated from benign tumors by DNA analysis nor could it establish tumor grade. Most grade III tumors were hyperploid and most grade I tumors were diploid, but there was sufficient overlap to prevent use of this indicator. However, whatever the grade diploid tumors exhibited significantly higher survival rates and lower metastatic rates in the hyperploid tumors. Ploidy assays may be very useful in predicting prognosis. In 10 year survival studies diploid patients had a better prognosis regardless of location, size, or grade of tumors. It was concluded that ploidy is probably the best predictor of the clinical course in patients with chondrosarcomas [15].

2. Carcinoembryonic Antigen

Plasma carcinoembryonic antigen (CEA) has been reported to be present in 17 of 21 patients with active osteosarcomas, but within normal limits in each of 9 patients in remission [16]. Longitudinal studies in nine patients indicated a decrease in levels in seven after successful surgery and in eight of the nine patients rises and falls in CEA levels followed the clinical course during chemotherapy. In two cases, clinical recurrence preceded or coincided with elevations in CEA levels.

3. Alkaline Phosphatase

Alkaline phosphatase is the traditional marker in classic osteosarcoma and levels are generally not elevated in patients with chondrosarcoma or fibrosarcomas. In tissues from twenty-six patients with osteosarcomas, elevations of alkaline phosphatase level above 0.6 U/min/mg tissue were associated with pulmonary metastases in 16 of 17 patients, whereas only 1 of 7 patients with levels less than 0.6 U/min/mg experienced metastases [17]. Preoperative serum alkaline phosphatase levels are also useful in predicting prognosis [18]. In 17 patients with elevated preoperative alkaline phosphatase levels, 12 experienced recurrence, compared to recurrence in 4 of 13 who had normal preoperative phosphatase activity ($p \ll 0.05$). Preoperative elevated alkaline phosphatase levels indicated poor prognosis [19]. In these tumors, serum alkaline phosphatase activity is elevated when the tumor is predominantly osteoblastic. In osteosarcomas of the osteoblastic type, the serum level appears to parallel the growth of the tumor. Levels as high as 40 times the upper limit of normal have been reported. Elevated preoperative levels are associated with a poor prognosis and elevated values that persist after amputation indicate an extremely poor prognosis. Enzyme activity is sensitive to therapeutic response, and rises and falls are indicative of the clinical status. The changes in the enzyme level occur prior to clinical evidence of response and permit the oncologist to institute changes in therapy earlier than would otherwise be possible [2,20].

The differentiation of bone from liver alkaline phosphatase has been difficult since both are probably products of the same structural gene and molecular differences are small, probably occurring only in carbohydrate side chains. Until recently immunochemical, chemical inhibition methods, and electrophoresis have not permitted a clear and distinct differentiation. Differentiation becomes an important diagnostic tool when there is an elevated serum alkaline phosphatase level and the question is whether the patient has beon metastases, liver metastases, or a combination. In our clinical laboratory we have combined the assay of 5'-nucleotidase with alkaline phosphatase to achieve this differentiation [2]. Since 5'-nucleotidase levels are elevated almost exclusively as a result of space-occupying lesions in the liver, such elevations indicate there is liver disease and the elevated alkaline phosphatase level must be further evaluated by bone scans and conventional skeletal surveys. An elevated alkaline phosphatase level in association with a normal 5'-nucleotidase level is clearly indicative of bone disease.

In 1984 Rosalki and Ying Foo found that wheat germ lectin reacted preferentially with bone alkaline phosphatase (80% precipitation and only minimal liver isoenzyme precipitation) and used this approach to achieve excellent differentiation and quantitation of bone from liver alkaline phosphatase levels [21]. In the cellulose actate membrane electrophoresis method the membrane is soaked in buffer containing the lectin before electrophoresis. It has been reported that with this method the bone isoenzyme made up about 70% of total serum alkaline phosphatase levels in normal young adult men and women (under 50 years of age). In older men the bone alkaline phosphatase level decreases to 60% of the total but the percentage does not decrease in women [22].

It is important to emphasize that in children elevated total serum alkaline phosphatase levels and the bone component must be evaluated with an understanding of the effects of bone growth on serum alkaline phosphatase. In children the reference level of total serum alkaline phosphatase activity is 400 U/L and the bone fraction is about 90%. The wheat germ lectin method must be thoroughly studied in patients with cancer.

Immunochemical and heat inactivation methods have presented diagnostic problems when there is only moderate involvement of either liver or bone. In a study by Moss, who used a heat inactivation technique, each of 20 patients with known bone malignancy had elevations of the bone isoenzyme, but 19 of the 20 demonstrated values only between one to two times the upper limit of normal [23]. This compared with 40 patients with liver involvement who all had elevations of the liver fractions; 37 of the 40 exhibited elevations greater than twice the upper limit of the liver isoenzyme.

4. Other Markers

We have evaluated a variety of tumor markers in a small number of patients with osteosarcomas [24]. The analytes included parathyroid hormone, osteocalcin, calcitonin, calcidiol, calcium, alkaline phosphatase, and the general tumor marker sialic acid (N-acetylneuraminic acid). Abnormal test values in the cancer patients were observed primarily for alkaline phosphatase and sialic acid. However, when compared to normal persons and individuals with benign bone disease, sialic acid exhibited the best combination of sensitivity (0.78) and specificity (0.70) for the detection of osteocarcinoma. Although the data suggest that sialic acid may be useful in evaluating bone cancers, it should be noted that the test is not cancer specific and elevated serum values may be seen in persons with infections or inflammatory disease, including osteoarthritis.

D. Parathyroid Adenomas

Alkaline phosphatase levels are markedly elevated in 95-100% patients with adenomas or carcinomas of the parathyroid. In almost two-thirds of these patients the levels are more than four times the upper limit of normal and the rises and falls in alkaline phosphatase levels reflect the course of the disease [2].

Serum osteocalcin and urinary hydroxyproline have been used as markers in parathyroid adenoma and carcinoma [25]. Osteocalcin is a noncollagen calcium-binding protein characterized by vitamin K dependently synthesized gamma carboxy glutamic acid. It is presumably a product of osteoblastic activity. Hydroxyproline is an amino acid constituent of collagen. Two-thirds of the body concentration of hydroxyproline is in bone and is measured in urine as an indicator of bone formation or resorption. Urinary measurements are normalized to creatinine concentrations. In cancer of the parathyroid, hypercalcemia is presumably due to increases in bone resorption and renal calcium resorption mediated by parathyroid hormone. It should therefore be anticipated that hydroxyproline turnover (a marker of resorption) should be increased and osteocalcin (a marker of osteoblastic activity) levels should remain unchanged in patients with parathyroid adenomas. In each of 21 patients the hydroxyproline excretion was significantly elevated (27.5 ± 6.2 µg/mg creatinine) in controls and (52.5 ± 27.6 µg/mg creatinine) in the patients ($p < 0.001$). In nine patients with hypercalcemia related to metastatic bone cancer the value is higher: 71.7 ± 58.4 µg/mg creatinine ($p < 0.005$) [25]. However, the osteocalcin concentration in the parathyroid patients was significantly reduced, whereas in the cancer patients it remained constant. It was suggested that in patients with hypercalcemia and elevated alkaline phosphatase levels, elevated hydroxyproline excretion and depressed osteocalcin levels are diagnostic of primary parathyroid adenomas whereas elevated hydroxyproline excretion and nondepressed osteocalcin levels are indicative of metastatic bone disease.

III. METASTATIC BONE TUMORS

Many cancers metastasize to bone and can create either osteoblastic or osteolytic lesions. Prostate cancer, Hodgkin's lymphoma, and carcinoid are primarily osteoblastic or osteolytic, or a combination. Thyroid, kidney, and colorectal cancer are primarily osteolytic when they metastasize to bone [2].

In patients with carcinoma metastatic to bone, elevations in serum alkaline phosphatase levels are observed when the metastatic lesion is osteoblastic. Since most cases of bone metastases in breast carcinoma are osteolytic, alkaline phosphatase levels are usually in the normal range or slightly elevated. In prostatic carcinoma, the opposite is true, and most bone metastatic lesions are osteoblastic. Hence, the serum alkaline phosphatase is usually elevated to values between 4 and 10 times the upper limit of activity.

In patients with carcinoma metastatic to bone and elevated alkaline phosphatase activity, there is occasionally a paradoxical increase in activity following favorable response to therapy. This is exemplified in sequential alkaline phosphatase determinations in a patient diagnosed as having carcinoma of the prostate 18 months prior to the study. In that period, he had undergone a transurethral resection and hormone ablation. On roentgenological examination, he showed generalized bone metastases to nearly all bones, primarily of an osteoblastic nature. He was admitted to the hospital because of weakness and increasing pain. In the next 2 months, the patient experienced clinical signs of remission, and, in general, the bone pain disappeared. During this period, the serum alkaline phosphatase level rose. After this the patient's condition worsened, he experienced severe pain, and the alkaline phosphatase level fell. The initial rise in alkaline phosphatase levels was interpreted as the initial phase of a reparative process of bone, and the eventual decrease was indicative of the latter stages of skeletal healing.

Attempts have been made to use the urinary hydroxyproline index (μg hydroxyproline/mg creatinine) in evaluating metastatic bone disease [26-28]. Many studies have reported that in patients with prostate cancer metastatic to bone and radiographic evidence of metastases, urinary hydroxyproline levels are elevated [28,29]. Men who did not exhibit radiographic evidence of bone metastases had values within the normal reference range. In a study of 35 untreated patients with positive bone scans, significant elevations of urinary hydroxyproline levels were noted. Patients who responded to hormonal therapy had a concomitant reduction in hydroxyproline excretion. One study of patients with metastatic prostate cancer reported that alkaline phosphatase was more sensitive to the presence of skeletal metastases than was hydroxyproline. However, other workers have reported that hydroxyproline is much more indicative of progressive or stable disease than is serum alkaline phosphatase. In 14 patients with progressive disease as evidenced by bone scans, hydroxyproline levels increased and in several the increases were observed 2-4 months before positive bone scan or clinical changes indicated recurrence. Clinical improvement was associated with decreases in hydroxyproline levels and in patients with stable disease there was no change.

In an outpatient population from whom urine was collected without dietary restrictions (a low collagen or a collagen-free diet and an overnight fast), the usefulness of urinary hydroxyproline assays was not confirmed [29]. Sixteen patients were evaluated for 3-18 months (55 assays). In seven patients with stable disease and nine with progressive disease the patterns of hydroxyproline excretion were variable. The conclusion was that urinary hydroxyproline levels did not establish a correlation with disease state and that the measurement of hydroxyproline is not useful in the management of patients with metastatic bone cancer.

IV. CONCLUSIONS

Except on a limited basis, tumor markers are not primary tools in the diagnosis and management of bone tumors. When markers such as alkaline phosphatase are used, immunochemical methods are not usually used. Several monoclonal antibodies to human osteosarcoma antigens have been described that react positively with osteosarcomas on immunostaining and have been proposed for possible use in imaging and therapy. They have not been used to evaluate circulating antigens. In one study the antibody reacted strongly with 15 of 17 fresh frozen samples of osteosarcoma as well as with neuroblastoma and rhabdomyosarcoma tissue. An undifferentiated sarcoma, fibrosarcoma, and Ewing's sarcoma tissue reacted weakly [30]. Bone scans and magnetic resonance imaging are at this time the techniques used for the initial diagnosis and for monitoring response to therapy.

REFERENCES

1. Isselbacher KJ, Adams RD, Braumwald E, Petersdorf RG, Walker JD. Harrison's Principles of Medicine. New York: McGraw-Hill, 1980.
2. Schwartz MK, Fleisher M, Bodansky O. Clinical application of phosphohydrolase measurements in cancer. Ann NY Acad Sci 1969;166:775-791.
3. Hobbs JR. Immunoglobulins in clinical chemistry. Adv Clin Chem 1971;14:291-317.
4. Seon BK, Pressman D. Immunoglobulins as cancer markers in humans. In: Chu TM, ed. Biochemical markers for cancer. New York: Marcel Dekker, 1982:301-320.
5. Carro M, Galieni P, Gobbi M, Baldrati L, Leardini L, Baccarani M, Tura S. Non secretory multiple myeloma. Presenting findings, clinical course and "prognosis." Acta Haematol 1985;74:27-30.
6. Latreille J, Barlogie B, Johnston D, Drewinko B, Alexanian R. Ploidy and proliferative characterization in monoclonal gammopathies. Blood 1982;59:43-51.
7. Glaubiger DL, Makuch R, Schwarz J, Levine AS, Johnson RE. Determination of diagnostic factors and their influence on therapeutic results in patients with Ewing's sarcoma. Cancer 1980;45:2213-2219.
8. Brereton HD, Simon R, Pomeroy TC. Pretreatment serum lactate dehydrogenase predicting metastatic spread in Ewings sarcoma. Ann Intern Med 1975;83:352-354.
9. Rosen G, Camparros B, Nirenberg A, Marcove RC, Huvos AG, Kosloff C, Lane J, Murphy ML. Ewings sarcoma: ten years experience with adjuvant chemotherapy. Cancer 1981;47:2204-2213.
10. Farley FA, Healey JH, Caparros-Sisson B, Godbold J, Lane JM, Glasser DB. Lactate dehydrogenase as a tumor marker for recurrent disease in Ewing's sarcoma. Cancer 1987;59:1245-1248.
11. Kreicbergs A, Bronstrom LA, Cewrien G, Einhorn S. Cellular DNA content vs. humas osteosarcoma. Aspects on diagnosis and prognosis. Cancer 1982;50:2476-2481.
12. Bosing T, Roessner A, Hiddemann W, Mellin W, Grundmann E. Cytostatic effects in osteosarcoma as detected by flow cytometric DNA analysis after preoperative chemotherapy according to the COSS 1980/1982 protocol. Cancer Res Clin Oncol 1987;113:369-375.

13. Baur HCF, Kreicbergs A, Silfversward C, Trubakait B. DNA analysis in differential diagnosis of osteocarcinoma. Cancer 1988;61:2532-2540.
14. Kreicbergs A, Zetterberg A, Soderberg G. The prognostic significance of nuclear DNA content in chondrosarcoma. Anal Quant Cytol 1980;2:272-279.
15. Kreicbergs A, Boquist L, Borssen B, Larsson SE. Prognostic factors in chrondrosarcoma. A comparative study of cellular DNA content and clinicopathologic features. Cancer 1983;50:577-583.
16. Contes EP, Chu TM, Wang JJ, Holyoke D, Wallace HJ, Murphy GP. Carcinoembryonic antigen in osteosarcoma. J Surg Oncol 1977;9:257-266.
17. Levine AM, Rosenberg SA. Alkaline phosphatase levels in osteosarcoma tissue are related to prognosis. Cancer 1979;44:2291-2293.
18. Thorpe WP, Reilly JJ, Rosenberg SA. Prognostic significance of alkaline phosphatase measurement in patients with osteogenic sarcoma receiving chemotherapy. Cancer 1971;43:2178-2181.
19. Posen S. Doherty E. The measurement of serum alkaline phosphatase in clinical medicine. Adv Clin Chem 1981;22:163-245.
20. Schwartz MK. Laboratory aid to diagnosis—enzymes. Cancer 1976;37:542-548.
21. Rosalki SB, Ying Foo A. Two new methods for separating and quantifying bone and liver alkaline phosphatase isoenzymes in plasma. Clin Chem 1984;30:1182-1186.
22. Kuwanta T, Sugita O, Yakata M. Reference limits of bone and liver alkaline phosphatase isoenzymes in the serum of healthy subjects according to age and sex as determined by wheat germ lectin affinity electrophoresis. Clin Chem Acta 1988;173:273-280.
23. Moss DW. Alkaline phosphatase isoenzymes. Clin Chem 1982;28:2007-2016.
24. Savoia M, Dnistrian AM, Piergentili C, Lane JM, Schwartz MK. Biochemical markers in bone cancer. Clin Chem 1987;33:929.
25. de la Piedra C, Tourul V, Rapado A. Osteocalcin and urinary hydroxyproline/creatinine ratio in the differential diagnosis of primary hyperparathyroidism and hypercalcemia of malignancy. Scand J Clin Lab Invest 1987;47:587-592.
26. Niell HB, Neely CL, Palmieri GM. The post absorptive urinary hydroxyproline (spot HYPRO) in patients with multiple myeloma. Cancer 1981;48:783-787.
27. Niell HB, Palmieri GM, Neely CL, McDonald MW. Postabsorptive urinary hydroxyproline test in patients with metastatic bone disease from breast cancer. Arch Intern Med 1981;141:1471.
28. Hopkins SC, Palmieri GM, Niell HB, Moinuddin M, Soloway MS. Total and nondialyzable hydroxyproline excretion in stage D2 prostate cancer. Cancer 1984;53:117-121.
29. O'Brien WM, Lynch JH. Hydroxyproline as a marker for following patients with metastatic prostate cancer. J Urol 1988;139:66-68.
30. Heiner JP, Miraldi F, Kallick S, Makley J, Neely J, Smith-Mensah WH, Cheung NV. Localization of G_{D2}-specific monoclonal antibody 3F8 in human osteosarcoma. Cancer Res 1987;47:5377-5381.

28
Thyroid Tumors

HINRICH G. SEESKO* and SAMUEL A. WELLS, Jr.
Washington University School of Medicine, St. Louis, Missouri

I. INTRODUCTION

Malignancies of the thyroid gland are relatively uncommon and each year fewer than 5,000 people in the United States of America die from this disorder. There is substantial controversy concerning the management of patients with the more common types of thyroid tumors (e.g., papillary and follicular carcinomas), since some clinicians believe that total thyroidectomy is the procedure of choice while others (probably the minority) believe that subtotal resection is satisfactory. The question of appropriate therapy will remain unsettled until clinical trials are conducted in which adequate numbers of patients are staged and prospectively randomized to various treatment arms and then followed postoperatively for a length of time sufficient to compare morbidity and mortality rates.

The diagnosis and operative therapy of patients with medullary thyroid carcinoma (MTC) must be considered separately primarily because of the availability of the plasma radioimmunoassay (RIA) for calcitonin (CT), the sensitive hormonal tumor marker for this disease. Furthermore, MTC is uniformly present in the multiple endocrine neoplasia type II (MEN-II) syndromes (MEN-IIa, MEN-IIb) and a great deal of effort is currently being directed toward identifying the genes responsible for these diseases.

The rare giant-cell and small-cell thyroid carcinomas represent the most virulent thyroid malignancies; no therapy has proved effective in the management of patients with these tumors. It is now recognized that some patients with apparent small-cell carcinomas (SCC) actually have thyroid lymphomas; the identification of these cases is important since these patients, if appropriately treated, have a much more favorable prognosis than patients with anaplastic SCC.

The various thyroid carcinomas differ in their frequency and prognosis. The international classification of thyroid tumors has provided a uniform nomenclature for epidemiological studies (Table 1) [1]. However, the development of new histological (especially immunohistochemical) techniques has served to modify the existing classification schema. For example, by means of immunohistochemistry, it has been demonstrated that SCCs of the thyroid exist as a specific entity and that at least some are

---

*Present affiliation: Phillips Universitaet, Marburg, Marburg, West Germany

Table 1 Partial Histological Classification of Thyroid Tumors: Malignant Epithelial Tumors

---
Follicular carcinoma
Papillary carcinoma
Squamous cell carcinoma
Undifferentiated (anaplastic) carcinoma
   Spindle cell type
   Giant-cell type
   Small-cell type
Medullary carcinoma

Source: Hedinger CH, Sobin LH (eds) Histological typing of thyroid tumours. Geneva: WHO, 1974.

variants of medullary thyroid carcinoma [2]. In addition, ultrastructural and immunohistochemical studies have provided evidence for the existence of true mixed follicular and medullary thyroid carcinomas in primary tumors as well as in metastases [3-6]. The discriminating power of immunohistochemistry was further shown in a study of 35 cases previously classified as undifferentiated thyroid carcinoma by routine histological examination. On further examination, one-third of the cases lacked the immunohistochemical findings typical of epithelial cells, thus raising the question of their being of nonepithelial origin [7].

In this chapter we will discuss the serum and cellular markers associated with thyroid malignancies and describe how they are useful clinically in the diagnosis and management of patients with papillary, follicular, or medullary tumors.

II. SERUM AND PLASMA MARKERS IN PATIENTS WITH THYROID MALIGNANCIES

A. Differentiated (Papillary and Follicular) Thyroid Malignancies

1. Thyroglobulin: Biochemistry and Physiology

Thyroglobulin (TG) is a glycoprotein molecule with a molecular weight of 660,000 daltons. It is synthesized in the rough endoplasmic reticulum of the thyroid follicular cells and secreted into the colloid, where it constitutes the stored form of thyroid hormone and the major component of the thyroid gland. Thyroglobulin reaches the circulation by the thyroid veins and extrathyroid lymphatics and is cleared from the circulation by the liver [8-10].

Under normal conditions thyroid-stimulating hormone (TSH) regulates serum TG (s-TG) levels and normal euthyroid persons have increased s-TG levels in response to the administration of bovine TSH [11,12]. Furthermore, it has been demonstrated that repeating the TSH injection after a 7 day administration of triiodothyronine (100 µg/day) results in clearly suppressed s-TG levels compared to basal conditions [11]. It has also been reported [13] that following the administration of thyrotropin-releasing hormone, there is an increase in serum TSH and subsequently an increase in s-TG levels. However, to establish reference values the need for an international TG reference standard has been noted [14].

2. Problems Associated with Assay of s-TG

Serum TG levels in euthyroid subjects range between 0 and 27 ng/ml [15-19], with mean values of 5.1-15 ng/ml. However, mean values of 32 ± 4 ng/ml ranging as high as 43 ng/ml [20,21] have been reported, and even higher values are found in women in the last trimester of pregnancy [19,22]. Serum TG levels are measured by RIA, and the sensitivity ranges between 0.8 and 2.5 ng/ml [16,19,22-24]. Undetectable s-TG levels are found in 15% of euthyroid persons [16,19,22]. This finding probably reflects assay insensitivity or day-to-day variation in s-TG levels [9].

Autoantibodies (AAB) directed against s-TG impair the usefulness of TG measurement since they may interfere with the RIA and confuse the interpretation of s-TG levels [25]. Autoantibodies are found in 4-15% of patients with differentiated thyroid cacinoma [14,18,20,26-28]. In patients with proven metastases and AAB against s-TG, low or undetectable s-TG levels have been reported [29]. Various factors such as AAB titer and specificity of the precipitating antiserum [25] influence the interference of AAB in the RIA for s-TG. Although some authors have de-emphasized the importance of the anti-TG-AAB titer [30], it is generally accepted that the presence and concentration of AAB should be evaluated for each patient [28,29,31,32].

The use of monoclonal or polyclonal antibodies has failed to enhance the specificity of the assay since the s-TG in patients with thyroid carcinoma may differ qualitatively from the TG in normal tissues: the TG in carcinoma tissue may variably express the epitopes present on TG purified from normal thyroid tissue. Conversely, the use of antibodies raised against normal thyroglobulin in the TG RIA could lead to false results in evaluating s-TG levels in patients with differentiated carcinoma.

The reported broad ranges in the levels of s-TG in normal subjects has led to an international cooperative study to evaluate s-TG standards [14]. Several participating laboratories evaluated standard lyophilized serum samples and reported a great variability in the s-TG levels. Although this could be related in part to the recovery of the lyophilized antigen, the wide interassay variation raised concern about the establishment of threshold s-TG levels that could serve as international standards in the postoperative evaluation of patients treated for differentiated thyroid carcinoma [14].

3. Serum Thyroglobulin Follow-up

Elevated s-TG levels have been reported in most patients with differentiated thyroid carcinoma or mixed papillary-follicular carcinoma [15,16,18,21,22,29,30,33,34]. However, patients with benign thyroid diseases have also been found to have s-TG levels that overlap with levels in patients with malignant disease (Figures 1,2) and thus, s-TG measurements cannot be used to differentiate benign from malignant disease [20-22,24,35]. The measurement of s-TG is important in the postoperative evaluation of patients undergoing total thyroidectomy for differentiated thyroid carcinoma, since an elevated level would indicate the presence of regional or distant tumor metastases. Patients with primary or metastatic anaplastic thyroid carcinoma or MTC reportedly have normal s-TG levels [15,16,18,21,22,29].

Usually, but not always, higher levels of s-TG are found in patients with follicular carcinoma than in patients with papillary carcinoma [16,17,22,27,33]. This variability appears to be attributed to the marked histological variation in the quantity of follicular areas within individual papillary carcinomas and within mixed carcinomas [33]. High levels of s-TG are frequently detected in patients with extensive metastases, indicating a

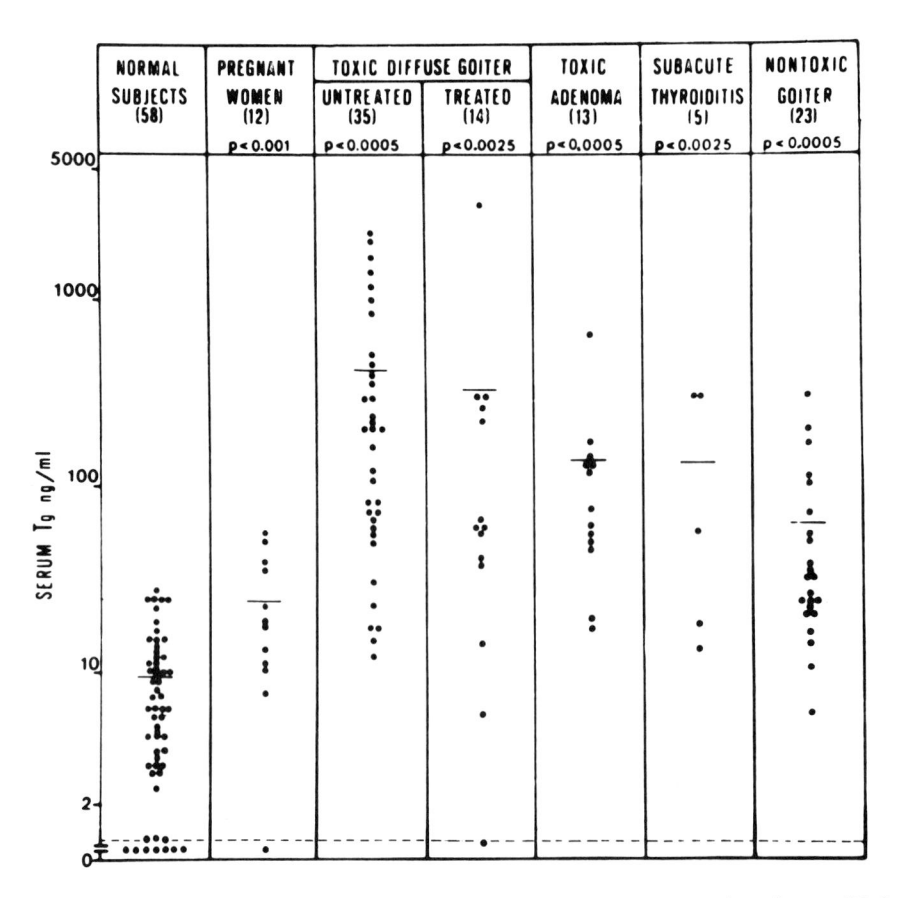

Figure 1 Serum TG levels in normal subjects, pregnant women and patients with benign thyroid disorders. The number of subjects is indicated in parentheses. The broken horizontal line represents the limit of sensitivity of the assay (1.25 ng/ml). The mean concentration of s-TG for each group is indicated by a solid horizontal line. The significance of differences with respect to normal subjects was assessed by Student's test (from Pacini F, Pinchera A, Giani C, et al. Serum thyroglobulin in thyroid carcinoma and other thyroid disorders. J Endocrinol Invest 1980; 3:283).

relationship between s-TG and total tumor mass [17,21,22,36,37]. The highest values of s-TG are seen in patients with osseous metastases [17,22,27,33] (Fig. 2).

The mechanisms of TG release from thyroid carcinoma cells are not completely understood. Release from the tumor tissue is seen in patients who have undergone total ablative therapy of the primary tumor but present with distant metastases and high s-TG levels [15,16,22]. In studying the iodine content of s-TG in normal individuals and in patients with benign and malignant thyroid tumors, Schneider and co-workers [38] concluded that the s-TG must come directly from the tumor tissue since the iodine levels were markedly decreased or absent in both the s-TG from tumor-bearing patients and the TG extracted from their tumor tissue. Even though the iodine content was low in s-TG from normal subjects, it was elevated in normal thyroid tissue. If the mechanism of elevated s-TG levels in patients with differentiated thyroid tumors was destruction

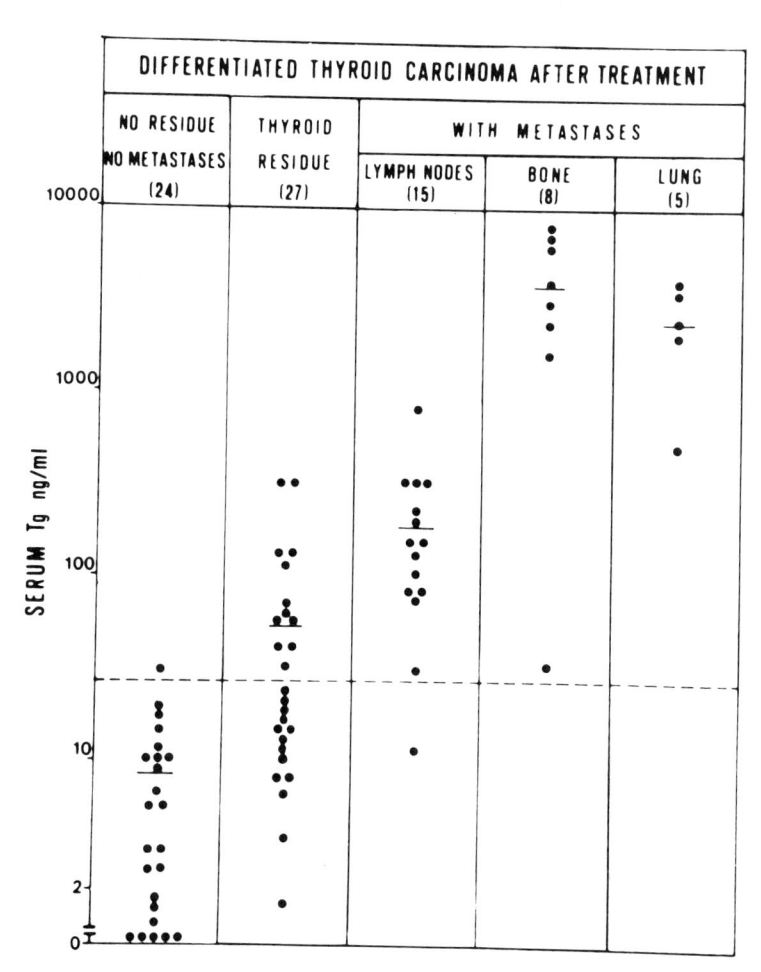

Figure 2 Serum TG levels in patients with differentiated thyroid carcinoma previously treated with total thyroidectomy, with or without subsequent radioiodine therapy. The number of subjects is indicated in parentheses. The solid horizontal lines indicate the mean concentration of s-TG for each group. The broken horizontal line represents the upper limit of the normal range (from Pacini F, Pinchera A, Giani C, et al. Serum thyroglobulin in thyroid carcinoma and other thyroid disorders. J Endocrinol Invest 1980; 3:283).

of adjacent normal thyroid tissue, one would have expected to see an increased iodine content in their s-TG, but this was not the case. The frequently normal levels of s-TG in patients with metastases of tumors of nonfollicular origin (i.e., medullary thyroid carcinoma or thyroidal metastases from lung carcinoma) add further evidence to this notion [15,16,18,20-22,30]. However, infiltration of normal thyroid tissue by nonthyroidal malignancies can also lead to increased s-TG levels, as was reported in patients with non-Hodgkin's lymphoma or epidermoid carcinoma [18,39].

The inability of administered thyroid hormone to suppress completely s-TG secretion in patients with thyroid carcinoma [27,30,35] suggests basal autonomous TG release from tumor tissue. However, in patients who have had ablative therapy for differentiated carcinoma, s-TG levels have been documented to rise after discontinuation of thyroid suppressive therapy [22] or prolonged TSH stimulation. Such findings are evidence of occult residual or metastatic disease even when there is no uptake of radioiodine to indicate metastases in these patients [35].

The direct correlation between s-TG levels and the quantity of differentiated thyroid carcinoma (i.e., a decrease in elevated s-TG levels after ablative therapy and a rise in the presence of relapse) [15,16,18,21-23,29,33,36,40,41], as well as the ability to indicate metastases not detectable by total body ^{131}I scans (TBS) [42] have established the clinical usefulness of s-TG in the follow-up of patients with this disease.

The rationale for posttherapeutic monitoring of patients with differentiated thyroid carcinoma using s-TG is the assumption that thyroid tissue (benign or malignant) is the only source of thyroglobulin. Therefore, after complete surgical and radioiodine ablative therapy s-TG levels should not be detectable unless thyroid tissue remains. However, up to 13% of patients who are clinically free of disease and with no signs of residual tissue on TBS have been shown to produce normal levels of s-TG [21,22,37,41-44]. Therefore, serial measurements are much more reliable than a single measurement since increasing levels of s-TG indicate relapse [9,36] and the observed trend is more meaningful than a single absolute value [45]. Patients who are clinically disease free for long periods following ablative therapy for thyroid carcinoma need to be monitored continuously since relapses may occur 30 or more years after the initial treatment [41].

Persistently elevated s-TG levels may herald the appearance of previously occult carcinoma or metastases not detectable by TBS [20,30,42]. In 13% of 284 patients with documented spread of well-differentiated thyroid carcinoma, metastases were not detected by TBS but were suggested by abnormal s-TG levels and/or clinical findings. Conversely, only 4.2% of patients with positive findings on TBS had undetectable s-TG levels [42]. That the s-TG levels may vary with time and assay sensitivity is of utmost importance as is the influence of TG AAB and the possibility of false-positive TBS. Clinically evident tumor growth associated with low or undetectable s-TG levels (with high assay sensitivity) has been described [17,37,46]. Very small tumors sometimes fail to produce detectable increases in s-TG in patients [21,24,43] and in rare instances undetectable s-TG levels might be related to tumor progression towards dedifferentiation.

In interpreting the results of the s-TG assay, it is important to know the degree of previous ablation. Thyroid remnants do alter the specificity of the test since the s-TG assay cannot differentiate between the presence of residual normal thyroid tissue and recurrent local carcinoma or metastatic disease [20,27]. Monitoring is obviously most useful in patients whose thyroid is totally ablated [27].

To refine and improve the monitoring of patients in apparent remission following ablative therapy for thyroid carcinoma, s-TG has been measured during thyroid hormone substitution and after withdrawal. In these studies TSH has been shown to result in an increase in s-TG levels [22,23,29,35,40]. It was demonstrated that s-TG levels increased following the administration of TSH even in patients whose metastases did not take up radioiodine. Thus, in the absence of normal thyroid tissue TSH increases the sensitivity of s-TG as a tumor marker in patients with negative TBS [35]. However, the potential risk of stimulating the growth of residual thyroid cancer cells with TSH must be considered a possibility and may obviate the use of this stimulation test [33].

Normal s-TG levels measured during thyroid hormone replacement therapy do not exclude the presence of metastases and therefore hormone administration should be stopped prior to s-TG measurement to minimize the number of false-negative results [22,23, 29,35,40,44,47,48].

Some studies have shown that the combined measurement of s-TG and TBS is superior to the measurement of either alone in detecting residual functioning thyroid tissue [35, 40]. The best time for s-TG measurement and TBS is 2 weeks after withdrawal of replacement therapy [20,35]. When s-TG remains undetectable following TSH stimulation it may be acceptable to omit or delay TBS [43]. Pacini and associates [49] studied 17 consecutive patients with s-TG levels above 15 ng/ml following thyroid ablative therapy. They were able to demonstrate metastases in 16 in whom conventional TBS (5 mCi tracer) was negative but became positive after a therapeutic dose of 131-iodine (75-140 mCi). In a study of 53 patients in whom s-TG levels were less than 10 ng/ml after withdrawal of T4 or were less than 1 ng/ml during T4 replacement therapy, total ablation of thyroid tissue was suggested and a TBS could be avoided [42].

In a most interesting prospective study reported by Thoresen and associates [50], serum samples stored in a biological serum bank over several years were evaluated for 43 patients who subsequently developed differentiated thyroid carcinoma, and for 128 case controls matched for age, sex, geographic region of Norway, and time of blood sampling. It was found that 9 (25%) of 36 patients with papillary carcinoma, 4 (100%) of 4 patients with follicular carcinoma, and 3 (100%) of 3 patients with anaplastic carcinoma had elevated s-TG levels (>90 μg/L) from 1 to 10 years before the cancer diagnosis. Similar levels of s-TG only occurred in 7 (5.4%) of the 128 control cases. Of three patients (all female, with s-TG levels above 150 μg/L) one was found to have undergone surgery for simple goiter years earlier and a second was recently operated on for this disease. Of the three individuals with anaplastic carcinoma whose elevated s-TG levels preceded cancer diagnosis by 4.5 and 7 years, presumably a differentiated thyroid carcinoma was present earlier that underwent dedifferentiation and rapid progression.

It can be concluded that the main indications for s-TG measurement clinically are:

1. Posttherapeutic follow-up in patients with confirmed differentiated thyroid carcinoma. (A significant increase in a previously low or undectable s-TG level can be considered highly suspicious for recurrence.)
2. Follow-up in patients with positive s-TG levels and negative TBS.
3. Differential diagnosis in patients with metastases of an unknown primary tumor, for example, the assay can be used to determine TG in body cavity effusions [9].

B. Medullary Thyroid Carcinoma

1. Plasma Calcitonin

Medullary thyroid carcinoma (MTC) accounts for 5-10% of all thyroid carcinomas and may present in either a familial or nonfamilial pattern. Familial MTC presents in an autosomal dominant pattern and includes the multiple endocrine neoplasia syndromes (type IIa or IIb) and non-MEN MTC. Medullary thyroid carcinoma is an endocrine tumor derived from the calcitonin (CT)-secreting thyroid C cells. These cells are of neuroectodermal origin and have been included in the amine precursor uptake and decarboxylation (APUD) system [51]. Tumors derived from C cells have great biosynthetic potential and may synthesize serotonin, calcitonin, neuron-specific enolase, histaminase, chromogranin, somatostatin, and dopa decarboxylase [52-62]. Calcitonin gene-related peptide (CGRP), which has been described recently [63,64] has been shown to be synthesized and released by MTC but, unlike calcitonin, appears to be unresponsive to calcium stimulation [65]. Out of this spectrum, diagnostic and prognostic relevance is mainly attributed to measurement of CT.

a. *Measurement of CT.* The main secretory product of the C cells is calcitonin, a 32-amino-acid peptide 3500 daltons in molecular weight. Plasma CT is measured by radioimmunoassay [66-69], and its quantitation (basally and after stimulation with CT secretagogues) has proved to be an excellent screening method both for the early diagnosis of MTC in MEN IIa or IIb kindred members at risk for MTC and for the detection of recurrence in patients with MTC following thyroidectomy.

In 1970, Tashjian and associates [70,71] demonstrated in patients with MTC that plasma CT levels increased above basal following a 4 hr calcium infusion (15 mg/kg). The synthetic pentapeptide pentagastrin was subsequently shown to be a potent CT secretagogue [68,72,73]. To determine the optimal test method for the diagnosis of MTC, Wells and associates [74] evaluated four different intravenous provocative regimens in 26 patients. They administered calcium gluconate (2 mg/kg/min), pentagastrin (0.5 µg/kg/5 sec), or calcium chloride (3 mg/kg/10 min), or the combination of calcium gluconate followed immediately by pentagastric (dosages as above). Blood samples were collected before and at 1, 2, 3, 5, and 10 min following the single or combined infusion. The combination of calcium gluconate and pentagastrin caused higher peak plasma CT levels than either one of the agents administered alone (Fig. 3). It was also found that the plasma CT response in some patients was substantially greater following the infusion of pentagastrin compared to calcium, while the converse was true in others. The combined infusion of calcium plus pentagastrin serves as the most reliable provocative test for the diagnosis of MTC. However, the identification of patients at risk depends on the establishment of a cut-off point between normal and abnormal plasma CT levels, both basally and after provocative testing.

b. *Problems Associated with Plasma CT Assay.* Various laboratories have reported basal plasma CT levels in normal subjects from 0 to 80 pg/ml and peak levels following provocative testing between 150 and 550 pg/ml [74-78]. Sex and age also affect the peak responses [32,50]. The differences between values from the various laboratories may be related to differences in the antisera used and variation in the testing methods. It is impossible to switch from one assay to another while monitoring a patient, unless both RIA methods are standardized on a set of common samples, because it is impossible to produce a conversion factor that is valid under all conditions from one assay to another [79].

A variety of nonthyroidal tumors (e.g., breast, pancreas, small-cell carcinoma of the lung, carcinoid, insulinoma, pheochromocytoma) [66,80-84] apparently secrete CT and are associated with elevated plasma CT levels; however, the CT levels observed in these patients are usually in the 500-1,000 pg/ml range and do not increase following provocation with calcium and/or pentagastrin.

c. *Management.* Stimulated peak plasma CT levels above 1000 pg/ml in a patient with normal basal levels who is at risk for MTC are virtually diagnostic and indicate the need for total thyroidectomy [85]. The diagnosis may not be certain in patients from MTC kindreds whose peripheral plasma levels are undetectable in the basal state but increase minimally (300-600 pg/ml) following provocative testing [86,87]. In a study of 25 subjects at direct risk for familial MTC with basal plasma CT levels below 240 pg/ml,

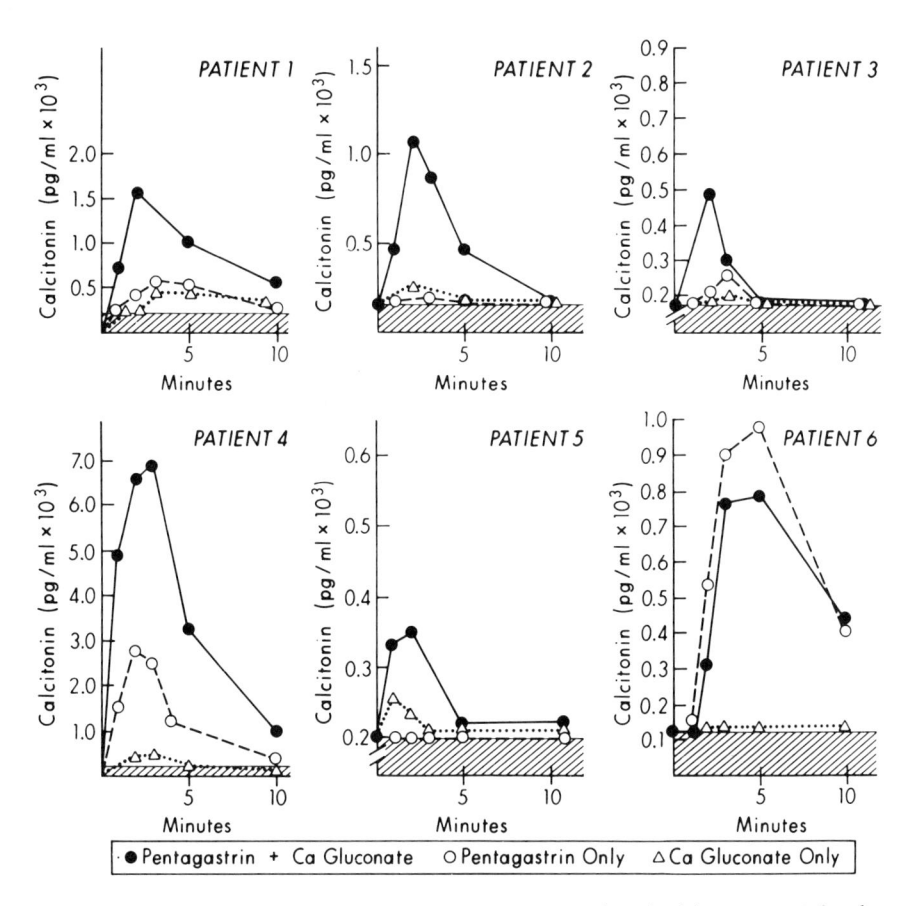

Figure 3 Plasma calcitonin levels in six patients stimulated with pentagastrin plus calcium gluconate, pentagastrin alone, or calcium gluconate alone. (*Source*: Wells SA, Baylin SB, Linehan WM, et al. Provocative agents and the diagnosis of medullary carcinoma of the thyroid gland. Ann Surg 1978;188:139-141.)

Table 2 Plasma Calcitonin and Prognosis

Preop CT (pg/ml)	RLNM[a] (%)	Postop CT[a] (%) (>300 pg/ml)	DM (%)	DTH (%)
1. 250-1,000 (n = 25)	1 (4)	1 (4)	0	0
2. 1,000-5,000 (n = 36)	3 (8.3)	6 (16.7)	0	0
3. 5,000-10,000 (n = 8)	2 (25)	1 (12.5)	0	0
4. >10,000 (n = 23)	13 (57)	14 (61)	4 (17)	2 (8.7)

Preop CT: preoperative stimulated plasma CT level; Postop CT: postoperative stimulated plasma CT level; RLNM: regional lymph node metastases; DM: distant metastases; DTH: death.
[a]Group 1 or Group 2 vs. Group 4, p < 0.001.

17 had peripheral plasma CT levels that peaked between 260 and 940 pg/ml after stimulation. All 17 patients underwent repeat provocative testing with simultaneous sampling from a peripheral vein and a selectively catheterized inferior thyroid vein. All patients demonstrated a marked increase in inferior thyroidal vein plasma CT levels, which were in almost all cases substantially higher than peripheral peak plasma levels. All 17 underwent thyroidectomy and were found to have either C-cell hyperplasia or MTC in the resected specimens [86].

The use of provocative testing in individuals at risk for MTC is well established and, furthermore, the stimulated preoperative plasma CT levels are of prognostic significance [87]. Ninety-two patients were assigned to four different groups according to their preoperative stimulated peak plasma CT level (Table 2). Of 25 patients with peak plasma CT levels between 250 and 1000 pg/ml, only 1 had metastases to regional lymph nodes and only 1 had elevated plasma CT levels following provocative testing in the postoperative period. By contrast, 60% of 23 patients with preoperative stimulated plasma CT levels greater than 10,000 pg/ml had regional lymph node metastases and elevated stimulated plasma CT levels in the postoperative period [87].

Kindred members at risk for familial MTC (MEN IIa, MEN IIb, or familial non-MEN MTC) should generally be screened beginning at 5 years of age and then subsequently on an annual or semiannual basis. Thyroidectomy should be advised if plasma CT levels exceed 1000 pg/ml (either basal or after provocative testing). If the stimulated plasma CT level is less than 300 pg/ml, the subject should be re-evaluated annually. In those individuals whose stimulated plasma CT levels are between 300 and 1000 pg/ml, provocative testing may be combined with selective inferior thyroid vein catheterization. Inferior thyroid vein plasma values exceeding 1000 pg/ml are virtually diagnostic of MTC or C-cell hyperplasia [73,74].

If patients have normal stimulated CT levels in the first month or two following surgery, they should be tested at yearly intervals. The management of patients who have an elevated plasma CT level, either in the immediate postoperative period or on subsequent testing, is unclear.

2. Carcinoembryonic Antigen

In 1976, Ishikawa and associates [88] reported elevated plasma carcinoembryonic antigen (CEA) levels in patients with MTC, whereas normal levels were seen in patients with other thyroid carcinomas except in one patient with papillary adenocarcinoma. Carcinoembryonic antigen levels in patients with MTC returned to normal after curative surgery. It was subsequently noted that plasma CEA levels were related to tumor burden [89-93].

Plasma CEA is less reliable than plasma CT as a tumor marker in patients with MTC since many patients with elevated plasma CT levels do not have elevated plasma CEA levels [93]. Also, plasma CEA levels do not increase above basal following a calcium or pentagastrin infusion. However, the measurement of plasma CEA may be useful prognostically to identify patients at risk for developing recurrence [89,91]. Saad and associates [92] reported a correlation between disease progression and rate of increase of plasma CEA levels as a function of time. The steepest slopes were found in patients with rapidly progressing disease, while flat slopes were seen in patients with nonprogressive metastases as well as in those apparently free of disease.

In a follow-up of 14 patients with MTC in apparent complete remission but with elevated plasma CT levels, Rougier and co-workers [91] reported 8 patients who also had

Thyroid Tumors

elevated plasma CEA levels. Six of these eight patients relapsed, whereas none of the patients with normal CEA levels relapsed. Furthermore, in 5 of the 6 patients with recurrent disease, the elevated plasma CEA levels heralded the clinical relapse from 6 to 36 months.

A nonparallel pattern of plasma CT and CEA levels (i.e., the rise or persistence of CEA in the presence of decreasing or normal CT levels) was reported by Busnardo and coworkers [94]. This correlated with immunohistochemical findings: in patients with aggressive MTC a diminishing or absent intracellular CT content is evident in the presence of positive homogeneous CEA staining [95]. Also, CEA has not been seen in normal or even hyperplastic C-cells [90].

III. CELL-BOUND MARKERS

A. Differentiated Thyroid Carcinoma

Most papillary and follicular thyroid carcinomas are capable of synthesizing TG. Biochemical studies have demonstrated that TG is present in tissue fragments of differentiated thyroid cancers in amounts one-half to one-third that of normal thyroid tissue [96]. The iodine content of TG from malignant tissue has been reported to be lower than that in normal thyroid tissue from the same patient [38]. Berge-Lefranc [97] demonstrated a direct correlation between the amount of TG-mRNA and the degree of cell differentiation in human differentiated thyroid carcinomas. In moderately differentiated follicular carcinomas, the TG-mRNA content was one-half to one-third as much as the amount in the well-differentiated follicular and papillary forms.

The use of monoclonal antibodies directed against TG may be helpful in allowing one to identify the thyroid origin of poorly differentiated primary tumors or their metastases [98]. However, there are about 40 antigenic sites on the thyroglobulin molecule and their expression may vary in different tumors. Thus, monoclonal antibodies generated against specific epitopes of the thyroglobulin molecule may not be useful in all patients [31,32]. Also, atypical iodoproteins may not give a positive reaction [99].

Non-thyroid-specific tissue antibodies may be helpful in the assessment of cell heterogeneity. It was recently demonstrated that the analysis of intermediate cytoskeleton filament proteins of the cytokeratin/prekeratin and vimentin type can be utilized to differentiate papillary from follicular thyroid carcinomas, benign from malignant papillary lesions, and anaplastic thyroid carcinoma from sarcomas or lymphomas [100]. Since undifferentiated thyroid carcinoma cells stain positively for keratin [11,101], anitkeratin antibodies can serve to detect small anaplastic foci in differentiated carcinoma. This finding would portend a worse prognosis [101].

Several authors have reported the existence of true mixed follicular and medullary carcinomas, the cells of which give positive immunohistochemical reactions to both thyroglobulin and calcitonin [3-6,101-103]. Therefore, the concept of different origin of C cells (neuroectoderm) and follicular cells (endoderm) has been questioned [102]. However, after neoplastic transformation, cells may exhibit morphological and functional features not observed in their nonneoplastic progenitors. Therefore, it is unreasonable to deduce embryonic derivations from characteristics expressed in neoplastic cells. In addition, it has not been determined so far whether the "thyroglobulin" seen in association with C cells is essentially identical to the 19s-TG, or whether it represents immuno-cross-reactive material that differs in structure and biological function [5]. The classification of these tumors as "differentiated thyroid carcinoma intermediate type" has been suggested [6]. The reported association of calcitonin and CEA in follicular thyroid carcinoma cells has raised the issue of screening tumors for these markers and eventually using them clinically for monitoring therapy [3].

B. Medullary Thyroid Carcinoma

Mendelsohn and co-workers [95] studied the tissue distribution of CEA and its relationship to CT in early, localized, or disseminated MTC. In early lesions, the distribution of CEA was similar to that of CT, being seen in virtually every cell. In virulent MTC tumors, however, the cellular distribution of CEA was abundant and inversely related to that of CT. Thus CEA is present in tumor cells at all developmental stages of MTC and retained even when CT is markedly diminished or absent as seen in virulent, widely disseminated disease. This finding may in some degree reflect a "maturation block" in aggressive MTC cells [95].

In an earlier study, a similar inverse relation was described for both L-dopa decarboxylase (DDC) and histaminase, or diamine oxidase (DA), and CT respectively. Staining of CT was markedly reduced in areas of MTC that simultaneously had increased contents of DDC and DA. This pattern was suggested to characterize a loss of or failure to attain the fully differentiated phenotype in tumor deposits in patients with virulent disease. It appears that CT immunostaining in primary MTC lesions is of prognostic importance. Patients harboring tumors with generally decreased CT staining or patchy CT cellular distribution appear to be at risk for widely metastatic MTC and a rapidly progressing downhill clinical course [57].

Neuron-specific enolase (NSE), the neuronal form of the glycolytic enzyme enolase, also can be found in APUD cells and neurons of the diffuse neuroendocrine system but not in other peripheral cells. Thus, its immunohistochemical presence in tumors derived from these cells can help to differentiate endocrine from nonendocrine malignancies [60,62]. It is a nonspecific marker for neuroendocrine tumors and its staining has been found to be less consistent than that of CEA or CT in MTC [58,103]. Therefore, it is only of limited value in the diagnosis or prognosis of patients with MTC.

Immunohistochemistry of sputum cytological specimens may be helpful in assessing the diagnosis in patients suspected of having pulmonary metastases of MTC [104].

IV. ONCOGENES IN THYROID CANCER

Activated cellular oncogenes are able to express gene products that may transform normal cells into neoplastic, malignant cells. Both the activated c-myc and ras oncogenes have been reported in thyroid malignancies [105,106,107]. The ras oncogene codes for a protein of 21,000 molecular weight (p-21), which has been found to be expressed weakly in normal thyroid tissue but more strongly in thyroid malignancies (MTC, anaplastic, follicular and papillary carcinomas) as well as inflammatory and nonneoplastic proliferative thyroid lesions. Thus, it is not helpful in differentiating benign from malignant lesions [105]. A recent study demonstrated that enhanced expression of ras p21 on the apical cell surface may be characteristic of papillary thyroid carcinoma [108].

The c-myc oncogene also is not specific for thyroid malignancy. It has been described in one case of anaplastic carcinoma and was overexpressed in the anaplastic component [107].

A new oncogene was recently reported in human papillary thyroid carcinoma. This transforming gene was described as different from any other known viral or human oncogene. It has been found activated in the primary tissue of five patients with papillary thyroid carcinoma and in metastases from two of the five patients. It was not observed in normal thyroid or lymphoid tissues obtained from the same patients [109,110].

V. OVERALL IMMUNE COMPETENCE

George and associates [111] demonstrated that patients with MTC had mononuclear cells whose migration was inhibited by MTC extracts but not by normal thyroid antigen.

Amino and co-workers [112] evaluated humoral and cellular-mediated immunity in 16 patients with thyroid carcinoma. Their studies indicated that spontaneous antitumor immunity is not common in thyroid carcinoma, but could be induced in one of the three patients studied by autologous, active immunization with tumor homogenate. In one patient, a transient clinical benefit was described. The existence of cell-mediated immunity to tumor-associated antigens (TAA) in patients with thyroid adenocarcinoma has been reported [113]. Fifty percent positive reactivity (leukocyte migration inhibition) was observed in patients with metastatic thyroid cancer when their leukocytes were exposed to TAA of a mixed papillary-follicular thyroid carcinoma. Inhibition was not seen in healthy controls or patients with Graves' disease. Furthermore, TAA was shown to be cancer tissue specific and distinct from thyroglobulin [113].

Balch and associates [114] reported depressed levels of granular lymphocytes with natural killer cell function in 247 patients with malignancy. This applied also to a group of 64 patients with "head and neck carcinoma," but thyroid carcinoma was not listed as a separate entity. Thyrocyte-specific killer cell activity was recently reported to be decreased in patients with thyroid carcinoma. When thyrocytes were used as target cells, consistently less lysis was observed with natural killer cells from patients with thyroid carcinoma than with those from normal subjects. This was not the case when a standard thyroid tumor cell line (K 562) was used as target. Thus, the authors postulate a defect in immune homeostasis against specific thyrocyte targets that may be important in mediating or aggravating proliferation of thyroid carcinoma [115].

A decrease in the number of killer cells in thyroid carcinoma tissue was observed when compared to tissue from functionally inactive thyroid adenomas. However, the overall number of immunocompetent cells in differentiated thyroid carcinomas was 2.5-8 times as much as in benign lesions. No major differences from normal ranges of killer or suppressor cells were seen in the peripheral blood [116]. Boros and co-workers [117] studied the postoperative activity of natural killer cytotoxicity of large granular lymphocytes from patients with thyroid carcinoma. The cytotoxic capacity was found to be elevated in tumor-free patients and in patients with highly differentiated tumor compared to patients with metastases or those with anaplastic tumors [117].

Hashimoto's (autoimmune) thyroiditis may be associated with various benign and malignant neoplasms of the thyroid. Also, lymphocytic infiltration of the thyroid without the classic features of Hashimoto's thyroiditis can be found in many types of thyroid disease [118]. Thus, it had been suggested that many reported cases of thyroid carcinoma associated with Hashimoto's thyroiditis, especially those that preceded modern diagnostic laboratory tests and expertise with fine-needle aspiration cytological studies, in fact represent a lymphoid reaction to tumor rather than a true autoimmune disorder. In a study of 112 cases of Hashimoto's thyroiditis treated over a 10-year period, the incidence of associated thyroid malignancy was 6.3%, which correlated well with the incidence of thyroid carcinoma seen at autopsy. Thus, it was concluded that patients with autoimmune thyroiditis are at no greater risk for thyroid carcinoma than the general population [119].

Protein-bound carbohydrate levels in the serum of patients with cancer have been shown to be related directly to tumor burden and primarily reflect changes in the acute-phase protein, found in the alpha 1 fraction of serum [120]. Tamura and associates [121] showed that the glycoprotein isolated from ascitic fluid of cancer patients was indistinguishable from normal alpha 1 acid glycoprotein by immunodiffusion and immunoelectrophoresis, but differed from the normal glycoprotein by gel-isoelectric focusing, molecular weight, and carbohydrate content.

"Immunosuppressive substance" (ISS) of similar molecular weight and isoelectric point to the glycoprotein isolated by Tamura and co-workers [121] has been purified from the serum of patients with thyroid cancer [122]. Significantly higher values of ISS were observed in patients whose thyroid carcinomas were invading surrounding organs than in those whose tumors were confined to the thyroid, and in patients lacking antithyroid autoantibodies compared to those with autoantibodies. In four patients with distant metastases and positive thyroid AAB, serum ISS levels were found to be low and the patients experienced a more favorable clinical course. It was therefore suggested that antitumor immunity existed in patients with concomitant positive antithyroid autoantibody [122].

Thyroid-stimulating antibodies (TSAB), which are known to be the cause of hyperthyroidism in Graves' disease, have also been described in patients with thyroid carcinoma [123,124]. A recent report of three patients with differentiated thyroid carcinoma and associated Graves' disease gave clinical and experimental evidence that TSAB may also stimulate the function and growth of thyroid carcinoma tissue. Thus, the authors conclude, in some patients TSAB may be responsible for tumor progression despite therapy directed to suppressing TSH secretion [123]. The occurrence of hyperthyroidism in patients with thyroid carcinoma is rare and it is not quite clear whether the observed hyperthyroidism in Valenta's series [124] could be related to the presence of long-acting thyroid stimulators (LATS) or whether it was due to the large tumor masses in the three patients. In seven patients with thyroid carcinoma but without any signs of hyperthyroidism, the presence of LATS was reported by the same authors [125]. In Hancock's series [126] of eight patients with thyroid carcinoma and hyperthyroidism, LATS was not found in any patient and LATS protector (LATS-P) activity was detectable in only one case. Edmonds and Tellez [127] reported a patient with high serum levels of thyrotropin-binding-inhibiting immunoglobulins (TBII). These were shown to stimulate thyroid cell function in vitro. Furthermore, radioiodine was clearly concentrated in pulmonary metastases despite undetectable TSH levels during thyroxin therapy.

In a series of 69 patients with thyroid carcinoma, a significant correlation has been found between the quantity of circulating immune complexes (CIC) and the histological type of carcinoma, with the highest levels being found in patients with papillary carcinoma. A positive correlation existed between the size of the papillary carcinoma and the preoperative levels of CIC. A significant decrease in CIC levels was seen following thyroidectomy [128].

Ruf and co-workers [32] used a monoclonal antibody (MAB)-binding inhibition assay to demonstrate TG autoantibodies in the serum of patients with Grave's disease, Hashimoto's thyroiditis, and thyroid carcinoma. Each serum was tested against 10 radiolabeled

MABs specific for different regions of the TG molecule. In the serum of cancer patients, they saw strong inhibition of all MABs tested, indicating a generalized and uniform increase in autoantibody, whereas anti TG autoantibody from patients with Graves' disease and, to a lesser degree, patients with Hashimoto's thyroiditis, exhibited specific and more restricted patterns of MAB binding inhibition. Therefore, the authors suggest that under pathological conditions the loss of tolerance and appearance of auto-AB may proceed differently in patients with thyroid carcinoma [86] than in those with Graves's disease and Hashimoto's thyroiditis.

VI. DNA ANALYSIS

In recent years, analysis of DNA content in thyroid malignancies has been investigated [129-132]. Backdahl [129] demonstrated in patients with follicular carcinoma a high correlation between the DNA profile of tumor cells and prognosis: 89% of patients who died of follicular thyroid carcinoma had a high proportion of aneuploid tumor cells, whereas 95% of patients with a good prognosis (10-21 year survival rate) had a polyploid DNA content or a low proportion of aneuploid cells. Similar data have been presented for patients with MTC and papillary carcinoma of the thyroid [130,131]. The measurement of DNA content apparently differentiates tumors with a low from those with a high degree of malignant behavior, regardless of their morphological classification. The prognostic power of the tumor cell DNA content equalled that of all other preoperative prognostic factors combined (age, sex, tumor size, preoperative tumor classification) for follicular and medullary carcinoma and exceeded it for papillary carcinoma [133].

REFERENCES

1. Hedinger CE, Sobin LH. Histological typing of thyroid tumors. In: International histological classification of tumors, No. 11. Geneva: World Health Organization, 1974.
2. Mendelsohn G, Bigner SH, Eggleston JC, Baylin SB, Wells SA Jr. Anaplastic variants of medullary thyroid carcinoma. Am J Surg Pathol 1980;4:333-341.
3. Calmettes C, Caillou B, Moukhtar MS, Milhaud G, Gerard-Marchant R. Calcitonin and carcinoembryonic antigen in poorly differentiated follicular carcinoma. Cancer 1982;49:2342-2348.
4. Fenoglio CM, Uribe M, Grimes M, Feind C. Medullary carcinoma of the thyroid gland: clinical, pathologic, and immunohistochemical features with review of the literature. Lab Invest 1985;52:21a-22a.
5. Hales M, Rosenau W, Okerlund MD, Galante M. Carcinoma of the thyroid with a mixed medullary and follicular pattern: morphologic, immunohistochemical and clinical laboratory studies. Cancer 1982;50:1352-1359.
6. Ljungberg O, Bondeson L, Bondeson A. Differentiated thyroid carcinoma, intermediate type: a new tumor entity with features of follicular and parafollicular cell carcinoma. Hum Pathol 1984;15:218-228.
7. Hedinger CE. Epidemiology of thyroid cancer. In: Jaffiol C, Milhaud G, eds. Thyroid cancer. Amsterdam: Excerpta Medica, 1985:3-9.
8. Gavin LA, Thyroid physiology and testing of thyroid function. In: Clark OH, ed. Endocrine surgery of the thyroid and parathyroid glands. St. Louis: C. V. Mosby, 1985:1-34.
9. Van Herle AJ. Measurement and clinical significance of thyroglobulin in serum and

body fluids. In: Ingbar SH, Braverman LE, eds. The thyroid. Philadelphia: JB Lippincott, 1986:534-545.
10. Van Herle AJ, Vassart G, Dumont JE. Control of thyroglobulin synthesis and secretion. N Engl J Med 1979;301:307-314.
11. Permanetter W, Nathrath WBJ, Loehrs U, Gaertner R. Immunohistochemical studies of keratin and thyroglobulin in thyroid tumors. Acta Endocrinol Suppl 1983;252: 40-41.
12. Uller RP, Van Herle AJ, Chopra IJ. Thyroidal response to graded doses of bovine thyrotropin. J Clin Endocrinol Metab 1977;45:312-318.
13. Uller RP, Van Herle AJ, Chopra IJ. Comparison of alterations in circulating thyroglobulin, triiodthyronine and thyroxine in response to exogenous (bovine) and endogenous (human) thyrotropin. J Clin Endocrinol Metab 1973;37:741-745.
14. Van Herle AJ, Van Herle IS, Greipel MA. An international cooperative study evaluating serum thyroglobulin standards. J Clin Endocrinol Metab 1985;60:338-343.
15. Pacini F, Pinchera A, Grasso L, Giani C, Doveri F, Baschieri L. Serum thyroglobulin in various thyroid disorders. Ann Endocrinol 1977;38:51A.
16. Pezzino V, Cozzani P, Filetti S, Galbiati A, Lisi E, Squatrito S, Vigneri R. A radioimmunoassay for human thyroglobulin: methodology and clinical applications. Eur J Clin Invest 1977;7:503-508.
17. Shah DH, Dandekar SR, Jeevanram RK, Kumar A, Sharma SM, Ganatra RD. Serum thyroglobulin in differentiated thyroid carcinoma: histological and metastatic classification. Acta Endocrinol 1981;98:222-226.
18. Van Herle AJ, Uller RP. Elevated serum thyroglobulin. A marker in differentiated thyroid carcinomas. J Clin Invest 1975;56:272-277.
19. Van Herle AJ, Uller RP, Matthews NL, Brown J. Radioimmunoassay for measurement of thyroglobulin in human serum. J Clin Invest 1973;52:1320-1327.
20. Lo Gerfo P, Colacchio T, Collacchio D, Feind C. Thyroglobulin in benign and malignant thyroid disease. JAMA 1979;241:923-925.
21. Schlossberg AH, Jacobson JC, Ibbertson HK. Serum thyroglobulin in the diagnosis and management of thyroid carcinoma. Clin Endocrinol 1979;10:17-27.
22. Pacini F, Pinchera A, Giani C, Grasso L, Doveri F, BAscieri L. Serum thyroglobulin in thyroid carcinoma and other thyroid disorders. J Endocrinol Invest 1980;3:283-292.
23. Barsano CP, Skosey C, DeGroot LJ, Refetoff S. Serum thyroglobulin in the management of patients with thyroid cancer. Arch Intern Med 1982;142:763-767.
24. Schneider AB, Murray JF, Stachura ME, Arnold JE, Yun Ryo U, Pinsky S, Colman M, Arnold MJ, Frohman LA. Plasma thyroglobulin in detecting thyroid carcinoma after childhood head and neck irradiation. Ann Intern Med 1977;86:29-34.
25. Schneider AB, Pervos R. Radioimmunoassay of human thyroglobulin: effect of antithyroglobulin autoantibodies. J Clin Endocrinol Metab 1978;47:126-137.
26. Rochman H, DeGroot LJ, Rieger CHL, Varnavides LA, Refetoff S, Joung JI, Hoye K. Carcinoembryonic antigen and humoral antibody response in patients with thyroid carcinoma. Cancer Res 1975;35:2689-2692.
27. Schlumberger M, De Vathaire F, Guermazi F, Fragu P, Parmentier C. Thyroglobulin measurement during the follow-up of patients with thyroid remnants. In: Jaffiol C, Milhaud G, eds. Thyroid cancer. Amsterdam: Excerpta Medica, 1985:273-278.
28. Van Herle AJ. Serum thyroglobulin: its diagnostic and prognostic value in patients with thyroid cancer. In: Jaffiol C, Milhaud G, eds. Thyroid cancer. Amsterdam: Excerpta Medica, 1985:237-244.
29. Feldt-Rasmussen U, Holten I, Sand Hanse H. Influence of thyroid substitution therapy and thyroid autoantibodies on the value of serum thyroglobulin in recurring thyroid cancer. Cancer 1983;51:2240-2244.

30. Black EG, Gimlette TMD, Maisey MN, Cassoni A, Harmer CL, Oates GD. Serum thyroglobulin in thyroid cancer. Lancet 1981;2:443-445.
31. Ruf J, Carayon P, Lissitzky S. Various expressions of a unique anti-human thyroglobulin antibody repertoire in normal state and autoimmune disease. Eur J Immunol 1985;15:268-272.
32. Ruf J, Henry M, De Micco C, Carayon P. The use of monoclonal antibodies to human thyroglobulin in the investigation of thyroid cancer. In: Jaffiol C, Milhaud G, eds. Thyroid cancer. Amsterdam: Excerpta Medica, 1985:253-260.
33. Ericsson UB, Tegler L, Lennquist S, Borup Christensen S, Stahl E, Thorell JI. Serum thyroglobulin in differentiated thyroid carcinoma. Acta Chir Scand 1984;150:367-375.
34. Lo Gerfo P, Li Volsi V, Colacchio D, Feind C. Thyroglobulin production by thyroid cancers. J Surg Res 1978;24:1-6.
35. Schlumberger M, Charbord P, Fragu P, Lumbroso J, Parmentier C, Tubiana M. Circulating thyroglobulin and thyroid hormones in patients with metastases of differentiated thyroid carcinoma: relationship to serum thyrotropin levels. J Clin Endocrinol Metab 1980;51:513-519.
36. Lo Gerfo P, Colacchio D, Stillman T, Feind C. Serum thyroglobulin and recurrent thyroid cancer. Lancet 1977;1:881-882.
37. Ng Tang Fui SC, Hoffenberg R, Maisey MN, Black EG. Serum thyroglobulin concentrations of whole-body radioiodine scan in follow-up of differentiated thyroid cancer after thyroid ablation. Br Med J 1979;2:298-300.
38. Schneider AB, Ikekubo K, Kuma K. Iodine content of serum thyroglobulin in normal individuals and patients with thyroid tumors. J Clin Endocrinol Metab 1983;57:1251-1256.
39. Shimaoka K, Van Herle AJ, Dindogru A. Thyrotoxicosis secondary to involvement of the thyroid with malignant lymphoma. J Clin Endocrinol Metab 1976;43:64-68.
40. Colacchio TA, Lo Gerfo P, Colacchio DA, Feind C. Radioiodine total body scan versus serum thyroglobulin levels in follow-up of patients with thyroid cancer. Surgery 1982;91:42-45.
41. Tubiana M, Schlumberger M, Rougier P, Laplanche A, Benhamou E, Gardet P, Caillou B, Travagli J, Parmentier C. Long-term results and prognostic factors in patients with differentiated thyroid carcinoma. Cancer 1985;55:794-803.
42. Ashcraft MW, Van Herle AJ. The comparative value of serum thyroglobulin measurements and iodine-131 total body scans in the follow-up studies of patients with treated differentiated thyroid cancer. Am J Med 1981;71:806-814.
43. Schlumberger M, Fragu P, Parmentier C, Tubiana M. Thyroglobulin assay in the follow-up of patients with differentiated thyroid carcinomas: comparison of its value in patients with or without normal residual tissue. Acta Endocrinol 1981;98:215-221.
44. Schneider AB, Line BR, Goldman JM, Robbins J. Sequential serum thyroglobulin determinations, 131-I scans, and 131-I uptakes after triiodothyronine withdrawal in patients with thyroid cancer. J Clin Endocrinol Metab 1981;53:1199-1206.
45. Botsch H, Glatz J, Schulz E, Wenzel KW. Long-term follow-up using serial serum thyroglobulin determinations in patients with differentiated thyroid carcinoma. Cancer 1983;52:1856-1859.
46. Delprat CC, Hoefnagel CA, Marcuse HR. Methodology of posttherapeutic control. In: Jaffiol C, Milhaud G, eds. Thyroid cancer. Amsterdam: Excerpta Medica, 1985:221-227.
47. Baschieri L, Giani C, Mariotti S, Pacini F, Busnardo B, Girelli ME, Pinchera A. Markers of thyroid tumours. In: Colnaghi MI, Buraggi GL, Ghione M, eds. Markers for diagnosis and monitoring of human cancer. London: Academic Press, 1982:259-270.

48. Mueller-Gaertner H, Schneider C. Clinical evaluation of tumor characteristics predisposing serum thyroglobulin to be undetectable in patients with differentiated thyroid cancer. Cancer 1988;61:976-981.
49. Pacini F, Lippi F, Formica N, Elisei R, Anelli S, Ceccarelli C, Pinchera A. Therapeutic doses of iodine-131 reveal undiagnosed metastases in thyroid cancer patients with detectable serum thyroglobulin levels. J Nucl Med 1987;28:1888-1891.
50. Thoresen SO, Myking O, Glattre E, Rootwelt K, Andersen A, Foss OP. Serum thyroglobulin as a preclinical tumor marker in subgroups of thyroid cancer. Br J Cancer 1988;57:105-108.
51. Pearse AGE. Common cytochemical and ultrastructural characteristics of cells producing polypeptide hormones (the APUD series) and their relevance to thyroid and ultimobranchial C-cells and calcitonin. Proc R Soc London [B] Biol Sci 1968;170:71-80.
52. Atkins FL, Beaven MA, Keiser HR. DOPA decarboxylase in medullary carcinoma of the thyroid. N Engl J Med 1973;289:545-548.
53. Baylin SB, Wells SA Jr. Management of hereditary-medullary thyroid carcinoma. J Clin Endocrinol Metab 1981;10:367-378.
54. Baylin SB, Beaven MA, Engelman K, Sjoerdsma A. Elevated histaminase activity in medullary carcinoma of the thyroid gland. N Engl J Med 1970;283:1239-1244.
55. Cox CE, Van Vickle J, Froome LC, Mendelsohn G, Baylin SB, Wells SA Jr. Carcinoembryonic antigen and calcitonin as markers of malignancy in medullary thyroid carcinoma. Surg Forum 1979;30:120-121.
56. Feldman JM, Wells SA Jr. Tissue levels of monoamines and monoamine metabolizing enzymes in medullary carcinoma and other thyroid diseases. Am J Med Sci 1981;282:34-40.
57. Lippmann SM, Mendelsohn G, Trump DL, Wells SA Jr, Baylin SB. The prognostic and biological significance of cellular heterogeneity in medullary thyroid carcinoma: A study of calcitonin, L-dopa decarboxylase, and histaminase. J Clin Endocrinol Metab 1982;54:233-240.
58. Lloyd RV, Sisson JC, Marangos PJ. Calcitonin, carcinoembryonic antigen and neuron-specific enolase in medullary thyroid carcinoma. Cancer 1983;51:2234-2239.
59. O'Connor DT, Burton D, Deftos LJ. Immunoreactive human chromogranin-A in diverse polypeptide hormone producing human tumors and normal endocrine tissues. J Clin Endocrinol Metab 1983;57:1084-1086.
60. Schmechel D, Marangos PJ, Brightman M. Neuron-specific enolase is a molecular marker for peripheral and central neuroendocrine cells. Nature 1978;276:834-836.
61. Sikri KL, Varndell IM, Hamid QA, Wilson BS, Kameya T, Ponder BAJ, Lloyd RV, Bloom SR, Polak JM. Medullary carcinoma of the thyroid: an immunocytochemical and histochemical study of 25 cases using eight separate markers. Cancer 1985;56:2481-2491.
62. Tapia FJ, Barbosa AJA, Marangos PJ, Polak JM, Bloom SR, Dermody C, Pearse AGE. Neuron-specific enolase is produced by neuroendocrine tumors. Lancet 1981;1:808-811.
63. Morris HR, Panico M, Etienne T, Tippins J, Girgis SI, MacIntyre I. Isolation and characterization of human calcitonin gene-related peptide. Nature 1984;308:746-748.
64. Steenbergh PH, Hoeppener JWM, Zandberg J, Lips CJM, Jansz HS. A second human calcitonin/CGRP gene. FEBS Lett 1985;183:403-407.
65. Poston GJ, Seitz PK, Townsend CM, Alexander RW, Rajaraman S, Cooper CW, Thompson JC. Calcitonin-gene related peptide: possible tumor marker for medullary thyroid cancer. Surgery 1987;102:1049-1054.
66. Coombes RC, Hillyard C, Greenberg PB, MacIntyre I. Plasma-immunoreactive-calcitonin in patients with non-thyroid tumours. Lancet 1974;1:1080-1083.

67. Dilley WG, Wells SA Jr, Cooper CW. Calcitonin radioimmunoassay. In: Rose NR, Friedman H, eds. Manual of clinical immunology. Washington, DC: Society for Microbiology, 1980:944-950.
68. Hennessy JF, Wells SA Jr, Ontjes DA, Cooper CW. A comparison of pentagastrin injection and calcium infusion as provocative agents for the detection of medullary thyroid carcinoma. J Clin Endocrinol Metab 1974;39;487-495.
69. Silva OL, Snider RH, Becker KL. Radioimmunoassay of calcitonin in human plasma. Clin Chem 1974;20:337-339.
70. Tashjian AH Jr, Melvin KEW. Medullary carcinoma of the thyroid gland: studies of thyrocalcitonin in plasma and tumor extracts. N Engl J Med 1968;279:279-283.
71. Tashjian AH Jr, Howland BG, Melvin KEW. Immunoassay of human calcitonin: clinical measurement, relation to serum calcium and studies in patients with medullary carcinoma. N Engl J Med 1970;283:890-895.
72. Cooper CW, Schwesinger WH, Mahgoub AM, Ontjes DA. Thyrocalcitonin: stimulation of secretion by pentagastrin. Science 1971;172:1238-1240.
73. Wells SA Jr, Ontjes DA, Cooper CW, Hennessy JF, Ellis GJ, McPherson HT, Sabiston DC Jr. The early diagnosis of medullary carcinoma of the thyroid gland in patients with multiple endocrine neoplasia type II. Ann Surg 1975;182:362-368.
74. Wells SA Jr, Baylin SB, Linehan WM, Farrell RE, Cox EB, Cooper CW. Provocative agents and the diagnosis of medullary carcinoma of the thyroid gland. Ann Surg 1978;188:139-141.
75. Deftos LJ, Weisman MH, Williams GE, Karpf DB, Frumar AM, Davidson BJ, Parthemore JG, Judd HL. Influence of age and sex on plasma calcitonin in human beings. N Engl J Med 1980;302:1351-1353.
76. Heath H III, Sizemore GW. Plasma calcitonin in normal man: differences between men and women. J Clin Invest 1977;60:1135-1140.
77. Wells DA Jr, Baylin SB, Gann DS, Farrell RE, Dilley WG, Preissig SH, Linehan WM, Cooper CW. Medullary thyroid carcinoma: relationship of method of diagnosis to pathologic staging. Ann Surg 1978;188:377-382.
78. Wolfe HJ, Melvin KEW, Cervi-Skinner SJ, Al-Saadi A, Juliar JF, Jackson CE, Tashjian AH Jr. C-cell hyperplasia preceding medullary thyroid carcinoma. N Engl J Med 1973;289:437-441.
79. Calmettes C. Problems associated with the assay of calcitonin in medullary cancer of the thyroid. In: Jaffiol C, Milhaud G, eds. Thyroid cancer. Amsterdam: Excerpta Medica, 1985:91-95.
80. Deftos LJ, McMillan PJ, Sartiano GP, Abuid J, Robinson AG. Simultaneous ectopic production of parathyroid hormone and calcitonin. Metabolism 1976;25:543-550.
81. Milhaud G, Calmette C, Taboulet J, Julienne A, Moukhtar MS. Hypersecretion of calcitonin in neoplastic conditions. Lancet 1974;1:462-463.
82. Schwartz KE, Wolfsen AR, Forster B, Odell WD. Plasma calcitonin, use as a tumor marker. Clin Res 1978;26:134A.
83. Silva OL, Becker KL. High plasma calcitonin levels in breast cancer (letter). Br Med J 1976;1:460.
84. Whitelaw AGL, Cohen SL. Ectopic production of calcitonin. Lancet 1973;2:443.
85. Wells SA Jr, Dilley WG, Farndon JA, Leight GS, Baylin SB. Early diagnosis and treatment of medullary thyroid carcinoma. Arch Intern Med 1985;145:1248-1252.
86. Wells SA Jr, Baylin SB, Johnsrude IS, Harrington DP, Mendelsohn G, Ontjes DJ, Cooper CW. Thyroid venous catheterization in the early diagnosis of familiar medullary thyroid carcinoma. Ann Surg 1982;196:505-511.
87. Wells SA Jr, Baylin SB, Leight GS, Dale JK, Dilley WG, Farndon JR. The importance of early diagnosis in patients with hereditary medullary thyroid carcinoma. Ann Surg 1982;195:595-599.

88. Ishikawa H, Hamada S. Association of medullary carcinoma of the thyroid with carcinoembryonic antigen. J Cancer Res Clin Oncol 1976;34:111-115.
89. Busnardo B, Girelli ME, Pelizzo MR, Zorat PL, De Besi P, Eccher C. Different diagnostic significance of carcinoembryonic antigen versus calcitonin as tumor marker of medullary thyroid carcinoma. Acta Endocrinol Suppl 1983;252:57-58.
90. DeLellies RA, Rule AH, Spiler I, Nathanson L, Tashijian AH Jr, Wolfe HJ. Calcitonin and carcinoembryonic antigen as tumor markers in medullary thyroid carcinoma. Am J Clin Pathol 1978;70:587-594.
91. Rougier P, Calmettes C, Laplanche A, Travagli JP, Lefevre M, Parmentier C, Milhaud G, Tubiana M. The values of calcitonin and carcinoembryonic antigen in the treatment and management of nonfamilial medullary thyroid carcinoma. Cancer 1983;51:855-862.
92. Saad MF, Fritsche HA J, Samaan NA. Diagnostic and prognostic values of carcinoembryonic antigen in medullary carcinoma of the thyroid. J Clin Endocrinol Metab 1984;58:889-894.
93. Wells SA Jr, Haagensen DE Jr, Linehan WM, Farrell RE, Dilley WG. The detection of elevated plasma levels of carcinoembryonic antigen in patients with suspected or established medullary thyroid carcinoma. Cancer 1978;42:1498-1503.
94. Busnardo B, Girelli ME, Simioni N, Nacamulli D, Busetto E. Nonparallel patterns of calcitonin and carcinoembryonic antigen levels in the follow-up of medullary thyroid carcinoma. Cancer 1984;53:278-285.
95. Mendelsohn G, Wells SA Jr, Baylin SB. Relationship of tissue carcinoembryonic antigen and calcitonin to tumor virulence in medullary thyroid carcinoma. Cancer 1984;54:657-662.
96. Thomas-Morvan C, Nataf B, Tubiana M. Thyroid proteins and hormone synthesis in human thyroid cancer. Acta Endocrinol 1974;76:651-669.
97. Berge-Lefranc J, Cartouzou G, De Micco C, Fragu P, Lissitzky S. Quantification of thyroglobulin messenger RNA by in-situ hybridization in differentiated thyroid cancers: difference between well-differentiated and moderately differentiated histologic types. Cancer 1985;56:345-350.
98. Bellet D, Schlumberger M, Bidard JM, Assicot M, Caillou B, Motte P, Vignal A, Bohoun C. Production and in vitro utilization of monoclonal antibodies to human thyroglobulin. J Clin Endocrinol Metab 1983;56:530-533.
99. Dralle H, Boecker W, Nielson G, Rehpenning W. Morphometric light microscopic and immunohistochemical analyses of differentiated thyroid carcinomas. Virchows Arch Pathol Anat Histopathol 1982;398:87-99.
100. Miettinen M, Franssila K, Lehto V, Paasivuo R, Virtanen I. Expression of intermediate filament proteins in thyroid gland and thyroid tumors. Lab Invest 1984;50:262-270.
101. Caillou B, Talbot M, Schlumberger M, Rougier P, Bellet D, Travagli JP. Usefulness of new methodological and conceptual approaches in thyroid carcinoma (especially in medullary carcinoma). Acta Endocrinol Suppl 1983;252:29-31.
102. Caillou B. Immunohistochemistry and electron microscopy in thyroid cancer. In: Jaffiol C, Milhaud G, eds. Thyroid cancer. Amsterdam: Excerpta Medica, 1985:11-14.
103. Houcke M, Delobelle A, Proye C, Mourot-Same C, Parent M, Bonniere X. Immunological study of thyroid medullary carcinomas. In: Jaffiol C, Milhaud G, eds. Thyroid cancer. Amsterdam: Excerpta Medica, 1985:299-300.
104. Hamilton CW, Bigner SH, Wells SA Jr, Johnston WW. Metastatic medullary thyroid carcinoma in sputum: a light and electron microscopic study. Acta Cytol 1983;27:49-53.
105. Johnson TL, Lloyd RV, Thor A. Expression of ras oncogene p21 antigen in normal and proliferative thyroid tissue. Am J Pathol 1987;127:60-65.

106. Lemoine NR. Activated oncogenes in human thyroid tumors. Ann Endocrinol 1987;48:81.
107. Terrier P, Douc-Rasy S, Schlumberger M, Fragu P, Tubiana M, Caillou B, Travagli JP, Riou G. Enhanced expression of the c-myc oncogene in a case of anaplastic carcinoma of the thyroid gland. In: Jaffoil C, Milhaud G, eds. Thyroid cancer. Amsterdam: Excerpta Medica, 1985:287-288.
108. Mizukami Y, Nonomura A, Hashimota T, Terahata S, Matsubara F, Michigishi T, Noguchi M. Immunohistochemical demonstration of ras p21 oncogene product in normal, benign, and malignant human thyroid tissues. Cancer 1988;61:873-880.
109. Fusco A, Grieco M, Santoro M, Berlingeri MT, Pilotti S, Pierotti MA, Della Porta G, Vecchio G. A new oncogene in human thyroid papillary carcinomas and their lymph-nodal metastases. Nature 1987;328:170-172.
110. Grieco M, Santoro M, Berlingieri MT, Donghi R, Pierotti MA, Della Porta G, Fusco A, Vecchio G. Identification of a new oncogene in human thyroid papillary carcinomas and their lymph-nodal metastases. Ann Endocrinol 1987;48:81.
111. George JM, Williams MA, Almoney R, Sizemore G. Medullary carcinoma of the thyroid: cellular immune response to tumor antigen in a heritable human cancer. Cancer 1975;36:1658-1661.
112. Amino N, Pysher T, Cohen EP, DeGroot LJ. Immunologic aspects of human thyroid cancer: humoral and cell-mediated immunity, and a trial of immunotherapy. Cancer 1975;36:963-973.
113. Aoki N, DeGroot LJ. Inhibition of leukocyte migration by human thyroid adenocarcinoma associated antigens. Acta Endocrinol 1982;99:56-63.
114. Balch CM, Tilden AB, Dougherty PA, Cloud GA, Abo T. Depressed levels of granular lymphocytes with natural killer (NK) cell function in 247 cancer patients. Ann Surg 1983;198:192-199.
115. Sack J, Baker JR Jr, Weetman AP, Wartofsky L, Burman KD. Thyrocyte specific killer cell activity is decreased in patients with thyroid carcinoma. Cancer 1987; 59:1914-1917.
116. Wenisch HJC, Schumm-Draeger PM, Beyer PA, Encke A. Immune system and thyroid carcinoma. Acta Endocrinol 1986;274:89.
117. Boros P, Balazs G, Szegedi G. Natural killer activity in thyroid cancer patients. Haematologic (Budapest) 1987;20:189-193.
118. Kini SR, Mueller JM, Hamburger JI. Problems in the cytologic diagnosis of the "cold" thyroid nodule in patients with lymphocytic thyroiditis. Acta Cytol 1981; 25:506-512.
119. Maceri DR, Sullivan MJ, McClatchney KD. Autoimmune thyroiditis: pathophysiology and relationship to thyroid cancer. Laryngoscope 1986;96:82-86.
120. Bradley WP, Blasco AP, Weiss JF, Alexander JC Jr, Silverman NA, Chretien PB. Correlations among serum protein-bound carbohydrates, serum-glycoproteins, lymphocyte reactivity, and tumor burden in cancer patients. Cancer 1977;40: 2264-2272.
121. Tamura K, Shibata Y, Matsuda Y, Ishida N. Isolation and characterization of an immunosuppressive acidic protein from ascitic fluids of cancer patients. Cancer Res 1981;41:3244-3252.
122. Kaneko G, Sugenoya A, Miyakawa M, Yoshitsugu Y, Fujii M, Makiuchi M, Iida F. Tumor immunity in thyroid carcinoma: correlative study between immunosuppressive substance and anti-thyroid antibodies. In: Ui N, Torizuka K, Nagataki S, Miyai K, eds. Current problems in thyroid research. Amsterdam: Excerpta Medica, 1982:537-540.
123. Belfiore A, Filetti S, Amir S, Daniels G, Ingbar SH, Vigneri R. The thyroid stimu-

lating antibodies (TSAbs) of Graves' disease may stimulate the function and growth of thyroid cancer. Ann Endocrinol 1987;48:83.
124. Valenta L, Lemarchand-Beraud T, Nemec J, Griessen M, Bednar J. Metastatic thyroid carcinoma provoking hyperthyroidism, with elevated circulating thyroid stimulators. Am J Med 1970;48:72-76.
125. Lemarchand-Beraud T, Valenta L, Vannotti A. Biochemical differences between normal and cancerous thyroid tissues. In: Hedinger C, ed. UICC monograph series, Vol. 12, thyroid cancer. Heidelberg/New York: Springer Verlag, 1969:205.
126. Hancock BW, Bing RF, Dirmikis SM, Munro S, Neal FE. Thyroid carcinoma and concurrent hyperthyroidism: a study of ten patients. Cancer 1977;39:298-302.
127. Edmonds CJ, Tellez M. Hyperthyroidism and thyroid cancer. Clin Endocrinol 1988;28:253-259.
128. Eto S, Shibata Y, Fujihara T, Suzuki H. Circulating immune complexes in autoimmune thyroid diseases and thyroid carcinoma. In: Ui N, Torizuka K, Nagataki S, Miyai K, eds. Current problems in thyroid research. Amsterdam: Excerpta Medica, 1982:553-556.
129. Backdahl M, Auer G, Forsslund G, Granberg P, Hamberger B, Lundell G, Loewhagen T, Zetterberg A. Prognostic value of nuclear DNA content in follicular thyroid tumours. Acta Chir Scand 1986;152:1-7.
130. Backdahl M, Tallroth E, Auer G, Forsslund G, Granberg P, Lundell G, Loewhagen T. Prognostic value of nuclear DNA content in medullary thyroid carcinoma. World J Surg 1985;9:980-987.
131. Cohn K, Backdahl M, Forsslund G, Auer G, Lundell G, Lowhagen T, Tallroth R, Willems JS, Zetterberg A, Granberg PO. Prognostic value of nuclear DNA content in papillary thyroid carcinoma. World J Surg 1984;8:470-474.
132. Ryan JJ, Hay ID, Goellner JR, Van Heerden JA, Grant CS. Prognostic value of flow cytometric DNA measurements in medullary thyroid carcinoma. Ann Endocrinol 1987;48:82.
133. Backdahl M, Wallin G, Loewhagen T, Auer G, Granberg P. Fine needle biopsy cytology and DNA analysis: their place in the evaluation and treatment of patients with thyroid neoplasms. Surg Clin North Am 1987;67:197-211.

29
Biological, Molecular, and Clinical Markers for the Diagnosis and Typing of Lung Cancer

ADI F. GAZDAR and JAMES L. MULSHINE

NCI-Navy Medical Oncology Branch, National Cancer Institute and Naval Hospital, Bethesda, Maryland

BARNETT S. KRAMER

NCI-Navy Medical Oncology Branch, National Cancer Institute and Uniformed Services University of the Health Sciences, Bethesda, Maryland

Lung cancer is the most common form of cancer as well as the leading cause of cancer deaths in the United States. Approximately 90% of patients eventually die from their disease [1]. Newer forms of therapy are therefore needed. In addition, improvements in our current diagnostic, staging, and typing methodologies would be of benefit. Lung cancer presents a pradoxical situation for the biologist and clinician. While numerous excellent biological and molecular markers are available for tumor tissue diagnosis, typing, and progression, few, if any, have successfully weathered the perilous journey from laboratory to bedside. In this chapter we will discuss the reasons for this paradox, and critically assess the more promising of the many markers clinicians have attempted to use in the management of this disease.

I. TYPES OF LUNG CANCER

A knowledge of the major forms of lung cancer and the principles of their therapeutic management are essential for understanding the role of markers. The World Health Organization [2] classifies lung cancers into four major types: squamous cell (or epidermoid), adenocarcinoma, large-cell, and small-cell carcinomas. In addition, mixed or combined forms may occur, as well as several rare types. However, for a number of clinical and biological reasons, we can classify lung cancers into two broad therapeutic categories: small-cell lung cancer (SCLC) and every other type into non-SCLC (NSCLC) cancer [1]. SCLC frequently metastasizes before it is diagnosed, and thus is seldom cured by surgery. However, it is initially responsive to chemotherapy and radiotherapy, and combination chemotherapy with or without radiotherapy is the treatment of choice at most institutions for patients with both extensive and limited disease. However, most SCLC tumors eventually recur, at which time they are resistant to further therapy. The long-term survival rate is between 5 and 10%, with patients

---

The opinions or assertions contained herein are the private views of the authors and are not to be construed as official or as reflecting the views of the Department of Navy or the Department of Defense.

with limited-stage disease having a better prognosis. By contrast, NSCLC usually is resistant to cytotoxic therapy, and curative resection, if possible, is the preferred treatment [1,3]. Thus, early diagnosis, accurate typing, and staging are important facets of the clinical management of lung cancer.

II. BIOLOGICAL AND MOLECULAR MARKERS FOR LUNG CANCER

A. SCLC and Other Neuroendocrine Tumors of the Lung

Because a large panel of well-characterized and studied cell lines exists, we know considerably more about the pathobiology of SCLC than other types of lung cancer [4,5]. The frequent association of SCLC with paraneoplastic syndromes, and the finding of dense core granules in the cytoplasm of tumor cells, gradually led to the concept that SCLC was not an "undifferentiated tumor," but belonged in a category of endocrine cells. These cells were originally characterized by their amine precursor uptake and decarboxylation (APUD) properties, but are now referred to as neuroendocrine (NE) cells because they share many properties with certain neural cells. The major function of NE cells is the synthesis, packaging, and secretion of specific peptide and amine products, such as calcitonin secretion by C cells of the thyroid, or catecholamines by adrenal medullary cells. These specific products identify the NE cell and tumor type. In addition to the specific properties, all NE cells express certain common properties [6]. These general NE markers categorize a cell or tumor as belonging in the NE cell group, but identification of the specific cellular products is necessary for further subtyping.

The general NE properties include the presence of dense core granules (the cytoplasmic storage site of the specific products), chromogranin A (CgA)(a matrix protein of the granules), L-dopa decarboxylase (an enzyme essential for biogenic amine production), and synaptophysin (a newly identified marker for neurons and NE cells). In addition, NE cells share certain other properties with neurons, including expression of an acidic form of enolase (neuron-specific enolase, NSE) and expression of certain surface markers initially identified on natural killer (NK) cells (Leu 7 and NKH1). Several other NE markers have been described, but because they have not yet been fully characterized or confirmed they will not be discussed in this chapter. Thus several excellent markers for NE cells exist, and the list may increase substantially in the not too distant future. The genes for CgA and synaptophysin have been cloned, and their role in diagnostic pathology is being evaluated [6]. In addition, excellent biochemical assays for L-dopa decarboxylase exist and NSE and synaptophysin can be identified or quantitated immunologically.

Identification of NE cell type by specific markers is more difficult to perform and interpret. More than 35 NE cell products have been identified, and the list grows longer every year. Without some clue as to which product to look for, one is reduced to looking for a needle in a haystack. In addition, many of these products are produced in relatively low concentrations (compared to expression of the general NE properties), thus requiring highly specific and sensitive assays for their detection. Also, tumors may secrete ectopic hormones (i.e., products that are not characteristic of the normal cell of origin).

A clue as to which specific NE products to look for in lung cancers is provided by studying the normal NE cells present in the lung. The lung is an endocrine organ, containing scattered NE cells in the bronchial mucosa. This is similar to the gastrointestinal (GI) tract, although the number and variety of NE cells are greater in most parts of the GI tract. Pulmonary endocrine cells are most numerous in late fetal and early neonatal life, although small numbers persist at all ages. Exposure to a variety of stimuli, including anoxia, chronic obstructive disease, and carcinogens, leads to a proliferation of pulmonary endocrine cells. While mature pulmonary endocrine cells have little ability to divide, it is presumed that NE lung cancers arise from mucosal cells committed to endocrine differentiation. Among the specific products of these cells are serotonin, calcitonin, and the mammalian homolog of amphibian bombesin, usually referred to as gastrin-releasing peptide [7]. Thus production of these substances by NE lung tumors should be considered as examples of eutopic secretion (i.e., secretion by a tumor of products produced by its normal precursor cells).

While eutopic secretion of GRP and calcitonin is frequently associated with SCLC [8,9], more than 15 NE cell products have been described in tumors, and individual cell lines may secrete 10 different peptides [10]. The ectopically secreted products include ACTH, arginine vasopressin, neurophysins, and neurotensin.

SCLC is a classic NE tumor, expressing all of the markers characteristic of this cell type [6] and secreting a variety of NE cell products. A closely related lung tumor is the bronchial carcinoid. Carcinoids are better differentiated than SCLC tumors, and express the general NE cell markers in higher concentrations. They are relatively slow growing, with low metastatic potential, and, in marked contrast to SCLC, are hightly chemoresistant [1]. Tumors intermediate in appearance and behavior between SCLC and the typical bronchial carcinoid occur and have been termed atypical carcinoids. While NSCLC characteristically lack NE markers, we and others have demonstrated expression of multiple NE markers in about 12% of otherwise typical appearing NSCLC tumors [5]. Of considerably interest is that cell lines established from such tumors are much more chemosensitive in vitro than other NSCLC, and of comparable sensitivity to SCLC cell lines established from untreated patients [11]. While this concept currently is being tested clinically, it is possible that expression of NE markers may identify a subset of NSCLC that are more responsive to cytotoxic therapy.

While it is not a NE cell marker, very high intracellular concentrations of the brain isoenzyme of creatine kinase (CK-BB) differentiate SCLC from NSCLC [12]. However, almost all cells contain some CK-BB, and thus it is a quantitative rather than a qualitative marker. Although CK-BB is not actively secreted, its use as a clinical marker has been explored [13].

B. Proto-oncogene Amplification and Overexpression in Lung Cancers

Proto-oncogene research represents one of the most exciting areas of tumor research. While comprehensive studies of tumor tissues are currently underway, some knowledge is currently available from both tumors and cell lines. As in other areas of lung cancer research, the lack of surgically resected SCLC specimens makes it mandatory to study cell lines. To date, most of the observations made on cell lines have been confirmed with tumor tissues, thus proving that cell lines established from lung cancer cell lines are representative of the tumors from which they were derived [14]. Cell lines offer another major advantage: they are free of contaminating stromal cells. The presence of stromal cells greatly complicates the interpretation of molecular genetic studies, especially those involving gene deletions (see below).

Both amplification and overexpression of several protooncogenes have been described in lung cancer [15]. These include the *myc* and *ras* families, c-*myb*, p53, c-*jun*, and c-*raf*-1. Frequent amplification of only one oncogene family has been associated with SCLC, namely the *myc* family. All of the three well-characterized members of this family, c-, N-, and L-, have been found to be amplified and overexpressed in tumors and corresponding cell lines, especially in samples taken after treatment [16-18]. It is not yet clear whether chemotherapy actually induces oncogene amplification or whether amplification, which is a tumor progression event, occurs at a certain frequency, and therapy permits the patient to live long enough for such an event to occur spontaneously. Of interest is that all of the *myc* genes are expressed in the normal fetal and adult lung. While the clinical significance and biological effects of N- and L-*myc* amplification, if any, are unknown, c-*myc* amplification results in profound biological changes. SCLC cell lines having c-*myc* amplification (usually about 25-100-fold) have marked phenotypic changes, and are referred to as variant cell lines [19,20].

Variant lines morphologically more closely resemble large cell undifferentiated carcinoma than SCLC, they have considerable growth advantages over typical or classic lines, and have partial loss of NE Markers. Of great interest is that variant lines exhibit considerable radioresistance, similarly to NSCLC lines. Patients with corresponding tumors (termed mixed small-cell-large-cell tumors by pathologists) have poorer responses to therapy and a shorter survival time.

We have recently found that c-*jun* is overexpressed in many SCLC and NSCLC lines (J. Schutte et al., unpublished observations). Also, this gene is expressed in relatively high amounts in normal lung tissue. It is a nuclear oncogene that apparently functions as a transcriptional activator. It is interesting to speculate that overexpression of c-*jun* may be associated with expression of the numerous endocrine and other markers characteristic of SCLC.

In contrast to SCLC, no frequent oncogene amplification pattern has been described in NSCLC. However, point mutations in the *ras* family oncogenes are relatively common in adenocarcinomas, and in some tumor systems they have been associated with metastatic behavior.

In summary, while none of the currently identified oncogenes has been implicated in the pathogenesis of lung cancer, they may influence its biology and clinical course.

C. Chromosome and Gene Deletions

Whang-Peng et al. described a specific chromosomal abnormality, deletion 3p, in SCLC [21]. While these findings initially were controversial, restriction fragment length polymorphism studies have confirmed that all or almost all SCLC tumors and cell lines have loss of genetic material on the short arm of chromosome 3 [22,23]. This is a finding of great interest and suggests that the deletion uncovers an otherwise recessive mutation on the cytogenetically normal chromosome 3. Similar mechanisms (i.e., homozygous loss of tumor suppressor genes) have been implicated in retinoblastoma, Wilms' tumor, and other cancers [24]. Another major recent finding is the demonstration that SCLC tumors and cell lines express abnormalities of the retinoblastoma *Rb* gene structure or expression at about the same frequency as in retinoblastoma [25]. Deletions may also be present on other chromosomes, such as 17. Thus the pathogenesis of SCLC may be complex, and involve homozygous loss of genetic material present on several chromosomes. In a limited study, similar abnormalities were noted in carcinoid tumors but not in NSCLC tumors with NE properties (author's unpublished data).

D. Growth Factors and Receptors

During the last few years it has gradually become apparent that tumor cells are capable of secreting many of the factors essential for their growth. One of the NE-related peptides secreted by SCLC, GRP, functions as an autocrine growth factor [26]. In addition, lung cancers of all types secrete insulin-like growth factor-1, and may also produce other essential factors such as transferrin [27,28].

A major difference between SCLC and NSCLC is the presence of receptors for epidermal growth factor (EGF) on the latter (and most epithelial cells) and absence of expression on SCLC [29]. In addition, squamous cell tumors are upregulated for EGF receptor expression [29].

These differences between SCLC and NSCLC may be of diagnostic significance and also offer potential new therapeutic approaches.

E. Differentiation Markers for NSCLC

Many of the biological studies in lung cancer have focused on the NE properties of SCLC. NSCLC remains relatively poorly characterized but forms a diverse group of tumors with multiple forms of differentiation.

While squamous differentiation is not present in the normal respiratory epithelium, squamous metaplasia follows a long list of acute and chronic lesions. Squamous tumors express all of the markers characteristic of squamous differentiation, including medium to high molecular weight keratins, involcurin, transglutaminase, and formation of cornified envelopes ([30]; M. Levitt et al., unpublished data). While these findings are of biological interest, they have not as yet had much impact on diagnostic pathological studies.

Adenocarcinomas are currently the most common form of lung cancer in the United States [5]. We have noted an apparent increase in bronchioloalveolar carcinomas, which are a subtype of adenocarcinomas arising from the peripheral airways. The origin of these tumors is diverse, but includes the progenitor cells of the peripheral airways, namely Clara cells and surfactant-producing type II pneumocytes. The major surfactant-associated protein, SAP-35, may be used as an immunohistochemical marker for many peripheral airway tumors [5]. We are currently exploring the use of other markers characteristic of type II cells or Clara cells.

III. APPLICATION OF MONOCLONAL ANTIBODIES TO LUNG CANCER

Hybridoma technology has permitted the generation of a vast number of new reagents that react with lung cancer cells (Table 1). The advantage of this approach over previous heteroantisera reagents is the potential for monospecificity and unlimited antibody production of consistently immunoreactive material. With the use of such reagents one can attempt to complement the routine histological classificatio of lung cancer or even reclassify lung cancer based on the expression of classes of antigens recognized by particular monoclonal antibodies. To date, the primary emphasis of monoclonal antibody generation has been on making reagents that improve the precision of light microscopic classification of lung cancer. Based on limited data, pathologists are beginning to look for expression of markers such as neuron-specific enolase, S-100, CgA, and synaptophysin as adjuncts in assisting diagnosis. We are currently evaluating the clinical significance

Table 1 Application of Monoclonal Antibodies in the Classification of Lung Cancer

Defined by Cluster Phenotype [33]	Defined by Differentiation
Cluster 1: Neural cross-reactivity Either a 124-165 Kd or a 25-40 Kd antigen which also clusters with expression. May be involved with neuropeptide synthesis or storage.	Neuroendocrine differentiation [6,19,26,31]: Leu 7, chromogranin A, NSE, GRP, other peptides. CK-BB, while not a NE marker, is a marker for SCLC [12,13].
Cluster 2: Epithelial antigen, 35-40 Kd	Squamous differentiation: medium-weight keratins (57-59 Kd), involucrin, cornified envelopes, transglutaminase [30,58]
Cluster 3: Cell-cycle-dependent antigen	Epidermal growth factor receptor: upregulated in squamous tumors; absent in SCLC [29].
Cluster 4: Less characterized antigen expressed on both neural and epithelial tissues	Papillary differentiation: surfactant-associated proteins [5]
Cluster 4a: Provisionally identified as a 100 Kd antigen with broad expression by both neural and endocrine tissues	Glandular differentiation: Tag 72 antigen, human milk fat globulin, CEA [44-53,58]

of the expression of neuroendocrine differentiation in NSCLC using three monoclonal antibodies. The monoclonal antibodies, Leu-7, antichromogranin A, and neuron-specific enolase were shown in a retrospective study using logistic regression analysis to be the most precise in the immunohistochemical identification of neuroendocrine differentiation as defined by a biochemical test for L-dopa decarboxylase activity [31]. Our hypothesis in evaluating NSCLC for this feature is that positive expression for neuroendocrine differentiation may correlate with a clinical course more similar to SCLC rather than NSCLC. We do not support the routine use of immunohistochemical analysis for neuroendocrine differentiation unless this identification provides clinically useful information.

In an attempt to organize research in this field, a cluster analysis of monoclonal antibody binding specifically was performed by the First International Monoclonal Antibody Workshop held in London in the spring of 1987 [32]. As with the previous leukocyte differentiation workshops, a group of lung cancer investigators exchanged reagents and tested a panel of antibodies on a variety of targets. These data were then subjected to a sophisticated dendrogram-cluster analysis, which suggested that 31 reagents defined 5 independent antigenic determinants [33]. Another 20 antibodies remained unclustered. This analysis provides a rational analysis for the systematic study of monoclonal-antibody-based lung cancer typing.

An unrealized potential provided by monoclonal antibodies is to reclassify lung cancer completely based on the pattern of reactivity with a panel of defined markers, independent of conventional morphologic criteria. In lung cancer one can argue that clinical course as well as appropriate therapeutic management are tied more closely to stage than to histological type. This is especially true if one considers NSCLC, in which histological type is rarely useful as a prognostic feature in large studies. It may soon be possible to determine if phenotyping will result in a classification of lung cancer that provides more useful clinical information that conventional light microscopic examination.

A preliminary stage of this approach would be to use monoclonal antibodies as adjuncts to conventional histological classifications. The bulk of the lung cancer monoclonal antibodies have been devoted to this pursuit and the results have been generally disappointing. One approach is to use cluster markers as defined by the previously mentioned Workshop, in much the same way as the leukocyte workshop classifications, which have greatly aided the diagnosis of leukemias and lymphomas.

One can use various antibody-defined biologically significant markers, such as oncogene expression, differentiation, or cell-cycle-dependent markers to classify lung cancer. A summary of the cluster- and differentiation-based markers currently available is shown in Table 1. The continuing work opened by the first Lung Cancer Workshop may result in more definitive answers following the second Lung Cancer Workshop, which convened in Cambridge, England, in the spring of 1990.

When investigators first generated monoclonal antibodies, it was widely believed that reagents with unique specificity for malignant antigens could be generated. Since then hundreds of reagents have been reported and, as documented by the International Lung Cancer Antibody Workshop, all lung cancer antibodies cross-react with normal tissues [33]. As suggested by Ginsberg, Magnani, and others [34,35], monoclonal antibodies to lung cancers, as with other solid tumors, are generally differentiation antigens that are overexpressed in the course of malignant progression. We have recently reported the results of a collaboration that attempts to exploit this for the early detection of lung cancer [36].

This analysis was possible since investigators at Johns Hopkins Hospital in a previous National Cancer Institute sponsored study of chest x-ray and sputum cytological examination as early detection tools [37] saved the sputum specimens from all the patients who had moderate to severe dysplasia. Clinical data on the long-term follow-up of all patients also were available. This archived material permitted the evaluation of two monoclonal antibodies, one to SCLC [38] and one to NSCLC [39], to determine if they could be used with an immunostaining technique [40] to identify patients at very high risk for lung cancer.

Representative samples of the available cytological material were immunostained with the two antibodies and evaluated in a blinded fashion. Immunostaining of the sputum specimens (available from approximately 5 years before the average patient developed lung cancer, as documented on chest x-ray or by sputum cytological studies) correlated with the eventual development of lung cancer with a 70% accuracy. A parallel analysis, performed on material obtained from the patients approximately 2 years before the diagnosis of lung cancer, was associated with a 91% prognostic accuracy. This correlation was highly significant. If this result is reproducible in another prospective study currently underway with the Lung Cancer Study Group, we may have a very useful method for the early detection of lung cancer based on surface changes associated with premalignant cells.

The two antibodies used for this application are 70304 [39] and 624H12 [38] (Table 2). The screening technique used to select 70304 involved screening specifically for antigens that were not expressed on SCLC. In contrast, 624H12 was selected for reactivity only with SCLC line and nonreactivity with an autologous B lymphoblastoid line. Both behave as typical tumor-associated antigens, so there is little to suggest that these

Table 2 Features of Monoclonal antibodies Used in Early Lung Cancer Detection Studies

Antibody	Class	Immunogen	Antigen	Reactivity	Reference
70304	IgG$_{2ak}$	NCI-H157	31 Kd protein	90% NSCLC 15-20% SCLC	39
624H2	IgM	NCI-H69	Unknown	75% SCLC 40% SCLC	38

antibodies have unique properties that will permit their use for early detection. We presume that other monoclonal antibodies may complement the original two markers in further improving the accuracy of this approach. Further information regarding the biology of lung cancer would greatly facilitate the appropriate selection of further targets for such a monoclonal-antibody-directed approach to early detection.

IV. MARKERS FOR LUNG CANCER IN SERUM AND BODY FLUIDS

A. Background

As mentioned earlier in this chapter, a number of histochemical, biochemical and molecular markers for bronchogenic carcinoma can play a role in the diganostic work-up of biopsy specimens from patients with bronchogenic carcinoma. Some of these markers, as well as others, may be found in soluble form in the blood or other body fluids of patients with lung cancer. A great body of literature exists on the use of these soluble markers in determining diagnosis, prognosis, and response to therapy. Over the years, there has been the recurring hope, as each new marker is discovered, that a single marker or panel of markers will find a major role in routine practice. In general, however, early hopes have not been fulfilled. This section will discuss some of the markers that have been studied and some of the reasons why early speculation about their utility has fallen short. A more detailed discussion of these concepts can be found elsehwere [41]. The ideal tumor marker would fulfill several criteria [42,43]. In addition to having high sensitivity and specificity, the level of the marker should closely follow the body burden of tumor, but its levels should be abnormally elevated even when there are very small amounts of malignancy present. To be truly useful in clinical practice, a tumor marker should provide information not obtainable through routine diagnostic, staging, and follow-up procedures; or it should be able to replace invasive or expensive procedures. To date, none of the soluble tumor markers clearly meet these criteria for bronchogenic carcinoma. This is not to say that the markers do not provide some interesting insights into the biology of lung tumors. The elaboration of substances from lung cancers that are not found in high concentrations in normal lung tissue raises interesting issues about tumor differentiation/dedifferentiaton pathways and about the ontogeny of lung cancers.

B. Serum Markers

Table 3 catalogs a variety of soluble serum markers that have been measured in patients with lung cancer. The list is partial, since a large number of other substances have been associated with lung cancer, including chorionic gonadotrophin, alpha fetoprotein, adrenocorticotropic hormone (ACTH), beta lipotropin, arginine vasopressin/vasotocin.

Table 3 Selected Tumor Markers for Lung Carcinoma in Serum

Marker	Comments	References
CEA	A family of cell surface glycoproteins; may be elevated in carcinomas of lung, GI tract, breast, as well as some benign conditions (inflammatory bowel disease, liver disease, renal disease, pulmonary inflammatory disease, connective tissue disease, smokers); level above 10 ng/ml usually indicate malignancy; elevated levels persisting more than 6 weeks beyond "curative" resection often indicate residual disease	44-53
NSE	An isoenzyme of the glycolytic enzyme enolase; found in neurons and APUD cells and tumors; reported in serum of about 70% of SCLC patients; levels roughly correlate with stage and response to treatment; may also be found in NSCLC	58-61
CK-BB	The isoenzyme of creatine kinase found in normal brain; elevated in a variety of tumors, including SCLC (about 25%), prostate cancer, breast cancer, lymphoma; levels in SCLC correlate with response to treatment and with survival	12,13,62-64
CgA	Protein found in catecholamine storage granules; elevated levels found in about 50-70% of SCLC patients (roughly correlating with disease extent); also found in NSCLC lung cancer, although generally in lower levels	66
GRP	A normal human peptide in the GI tract and central nervous system; has homology at its active carboxyl terminal with frog bombesin; an autocrine growth factor for SCLC cells in vitro; although found in many SCLC tumors, has little or no value as a serum marker because of its short half-life (less than 2 min)	67,68
Calcitonin	Increased levels originally described in patients with medullary thyroid carcinoma; subsequently found in patients with a variety of other malignancies (as well as in patients with benign pulmonary disease); elevated serum levels found in about 50% of patients with bronchgenic carcinoma (about 65% of patients with SCLC); lung cancer patients tend to have a higher-molecular-weight form of the circulating molecule than patients with medullary thyroid carcinoma	69-72

However, most of these occur with too low a frequency to play a significant role in the management of bronchogenic carcinoma.

A large volume of literature exists on the use of carcinoembryonic antigen (CEA) in the management of numerous epithelial tumors, especially of the colon, rectum, and breast. Since its first description by Gold and Freedman in 1965 [44], thousands of articles on its use in clinical practice have appeared [45-47]. A brief discussion of what can be concluded from the numerous studies can, perhaps, bring into focus the issues involved with all of the markers listed in Table 1. CEA is not actually a single compound but a family of antigenically related glycoproteins [46]. As such, the available commercial kits may not be measuring the identical molecule. Although there is generally good correlation between results obtained by different assays, there clinically important differences in results from the same sample have also been described [48]. Hence, in using serial CEA measurements in evaluating a patient, one must be sure that the same assay has been used. When initially reported, CEA was thought to be a sensitive as well as a specific marker of gastrointestinal malignancy. However, CEA has subsequently been found in other tumors such as breast and lung cancers. In fact, many normal people have detectable levels of circulating CEA, although nearly always less than 10 ng/ml [49,50]. It is clear from the published literature that CEA cannot be used as a screening test for bronchogenic carcinoma. As mentioned above, it is not specific for lung tumors, since it is found in other malignancies. Levels of CEA may also be elevated in benign conditions such as pulmonary inflammatory diseases, as well as in heavy smokers [47,51]. More important is that the specificity of a CEA assay is less than 90%. Given a prevalence of lung cancer in the population of less than 10:1000, there would be more than 30 false-positive tests for each patient who was correctly identified with lung cancer if measurement of CEA was used as a screening test [47]. Moreover, with a sensitivity of about 30% for detecting pulmonary cancer [52], about 70% of lung cancers would be missed.

It is also clear that the CEA assay cannot be used to stage a patient definitively once the diagnosis of lung cancer has been made. Although elevated CEA levels may be associated with a poor prognosis for bronchogenic carcinoma [53], the sensitivity and specificity of the test are not nearly good enough to preclude an attempt at curative surgery in the absence of definitive evidence of inoperability. Some have also proposed that serial CEA determinations be used to detect early recurrence of lung cancer after initial curative therapy [42]. It is certainly true that CEA levels roughly correlate with body burden of tumor and that rising levels sometimes occur before clinical recurrence is detected. However, it is less clear how frequently patients would benefit from the routine use of serial CEA determinations, especially since there is no evidence that detection of early recurrence can be put to advantage (that is, to prolong the patient's life). To address these issues in SCLC, a prospective study of serial CEA determinations has recently been performed [52]. The authors concluded that CEA determinations reliably indicate stage of disease and prognosis in less than 10% of patients with SCLC. Since it is still uncertain whether early detection of relapse in SCLC can be translated into better survival, the conclusion was made that the clinical relevance of serial CEA tests in patients with SCLC is quite low.

As noted in Table 1, a number of other serum markers have been studied in both SCLC and NSCLC. These markers include various peptides, CK-BB, and CgA. Compared to CEA, the data base for these markers is smaller. Nevertheless, many of the same concepts discussed above hold for these markers. Since experience with them is limited, their ultimate utility in the clinical arena is yet to be determined. Initial enthusiasm must be tempered until more information becomes available. In this regard, a careful reading of an important paper [54] describing the appropriate evaluation of diagnostic marker tests before admitting them into common usage is recommended. It points out that many marker tests have not been completely evaluated before they become "marketed" for routine clinical practice. Since it is certain that no single serum marker is ideal for the detection, staging, or follow-up of patients with lung cancer, some have tried to use a combination of markers [55]. A final note is in order about the cost of using serum marker studies in routine management. In most cases, cost-benefit analyses are not available. However, the issue of cost vs. benefit has been addressed for the use of serial

CEA tests to monitor for recurrence of colorectal carcinoma in patients who had undergone a primary resection. The hope was to detect local recurrences that could still be resected [55]. In that study, the cost per resectable recurrent tumor was nearly $25,000.

C. Marker Panels

Because no individual marker appears to be of major clinical importance, the use of panels of markers has been attempted by several groups. In perhaps the most comprehensive of these studies, Gail et al. [56] tested sera from patients with lung cancer, benign lung disease, and normal controls for the presence of 10 marker substances. Only two markers, CEA and ferritin, appeared to help differentiate lung cancer patients from normal controls. However, because of the high false-negative rate, a complete work-up would still be required of the marker-negative patients.

This study reflects on their earlier work, in which they used a logistic regression model to test the utility of a panel of 11 serum markers to differentiate advanced lung cancer from benign pulmonary disease. They found that a model incorporating the serum CEA and total sialic acid levels was most accurate, with 95% specificity and 54% sensitivity. The model gave better results than any of the other 11 markers used individually. As the authors of the article point out, however, it is not certain that the discrimination capabilities of a regression model add to conventional diagnostic tests. It is apparent that a combination of mediocre markers results in a mediocre panel, while greatly adding to the cost of screening.

D. Other Body Fluid Markers for Lung Carcinoma

Little information exists on the use of markers for lung carcinoma in body fluids other than blood. However, soluble markers for SCLC have been measured in the cerebrospinal fluid (CSF) to detect central nervous system metastases [57]. Patients with either parenchymal brain metastases or carcinomatous meningitis had elevations in CSF NSE. Likewise, most patients with meningeal disease had elevated CSF CK-BB or GRP (see Table 3 for references) levels. Again, the issues raised in the interpretation of serum tumor marker studies are important in the interpretation of these results. Specifically it is important to determine whether CSF markers add to the use of routine radiological and cytological detection studies, whether elevations in marker levels significantly precede the onset of symptoms, and, finally, whether their measurement can be used to improve survival in this subset of patients with a very poor prognosis.

As mentioned earlier, one of the characteristics of an ideal serum marker would be a near-perfect correlation between body tumor burden and level of circulating marker. In some cases, even if the rate of secretion of a soluble marker into the bloodstream correlated well with tumor burden, serum measurements could still correlate poorly if there were rapid metabolism or clearance of the marker. This may indeed be the case with serum GRP, since it is hydrolyzed in plasma. In such cases, measurements of urinary excretion may correlate better with tumor mass and clinical status. Indeed, there is preliminary evidence from studies in our branch that this may be true in the case of calcitonin and GRP (Johnson and Becker, unpublished data).

V. CONCLUSIONS AND SUMMARY

The multitude of NE cell markers expressed by SCLC and other NE lung cancers offers opportunities for their use in clinical medicine. While they may be used for tumor tissue typing, they have proved of limited clinical value. Why? The most useful clinical markers are actively secreted or shed by tumor cells. While the peptide products are actively secreted, their concentrations are relatively low, and some, such as GRP, are unstable in blood. Their detection in fluids with low protein levels, such as CSF, may be more sensitive and specific. Chromogranin is actively secreted along with the other contents of the dense core granules. It is relatively stable in blood, and may live up to its early promise as a useful marker. However, it cannot be used to differentiate between the various NE tumors. The other NE markers are not actively secreted, but presumably may be released by dead or damaged cells. Such markers may be used to detect tumor bulk, but are unlikely to detect early, small lesions.

Thus, despite the fact that numerous soluble markers have been identified in patients with lung cancer and that levels correlate to varying degrees with either histological subtype or clinical stage, it is by no means clear that routine measurement of any single marker or combination in body fluids can yet play a role in clinical decisions. It is not our current policy at the National Cancer Institute-Navy Medical Oncology Branch to use measurement of any of the body fluid tumor markers mentioned in this chapter in the management of patients with bronchogenic carcinoma.

Monoclonal antibodies have played a pivotal role in both clinical diagnosis and biological research. While their application to lung cancer has been disappointing, their original promise remains, and will proably be fulfilled in the future.

Molecular markers for the diagnosis and typing of lung cancers offer a promising new approach. They are currently useful for tissue diagnosis, but molecular techniques remain tools available predominantly at research institutions. In the not too distant future, they will become part of routine diagnostic procedures at most institutions. However, their application for early detection and staging procedures may be limited.

As is apparent, most clinicians still believe that no marker currently available offers much more information than a careful physical examination and routine staging procedures. However, we believe that the future for markers in lung cancer is promising. Major advances in our understanding of the biology of lung cancer have occurred during the last decade, and lung cancer continues to be intensively studied in many laboratories around the world. Large panels of well-characterized cell lines exist and remain the cornerstones of biological studies. Biological studies have yielded many new diagnostic and therapeutic avenues for us to explore.

REFERENCES

1. Minna JD, Higgins GA, Glatstein EJ. Cancer of the lung. In: DeVita VT, Hellman S, Rosenberg SA, eds. Cancer: principles and practice of oncology. Philadelphia: Lippincott, 1985:507-597.
2. The World Health Organization histological typing of lung tumours. Am J Clin Pathol 1982;77:123-136.
3. Mulshine JL, Glatstein E, Ruckdeschel JC. Treatment of non-small cell lung cancer. J Clin Oncol 1986;4:1704-1715.
4. Carney DN, De Leij L. Lung cancer biology. Semin Oncol 1988;15:199-214.

5. Gazdar AF, Linnoila RI. The pathology of lung cancer—changing concepts and newer diagnostic techniques. Semin Oncol 1988;15:215-225.
6. Gazdar AF, Helman L, Israel MA, Russell EK, Linnoila I, Mulshine J, Schuller H, Park JG. Expression of neuroendocrine cell markers L-dopa decarboxylase, chromogranin A, and dense core granules in human tumors of endocrine and non-endocrine origin. Cancer Res 1988;48:4078-4082.
7. Moody TW, Pert CB, Gazdar AF, Carney DN, Minna JD. High levels of intracellular bombesin characterize human small cell lung carcinoma. Science 1981;214:1246-1248.
8. Becker KL, Gazdar AF. The endocrine lung in health and disease. Philadelphia: WB Saunders, 1984.
9. Gazdar AF, Carney DN, Becker KL, Deftos LJ, Go LV, Marangos PJ, Moody TW, Wolfsen AR, Zweig MH. Expression of peptides and other markers in lung cancer cell lines. Adv Cancer Res 1985;99:168-174.
10. Sorenson GD, Pettengill OS, Brinck-Johnsen T, Cate CC, Mauer LH. Hormone production by cultures of small cell carcinoma of the lung. Cancer 1981;47:1289-1296.
11. Gazdar AF, Tsai CM, Park JG, Mulshine J, Linnoila RI, Yesner R, Helman L, Israel M, Whang-Peng J, Minna JD. Relative chemosensitivity of non-small cell lung cancers expressing neuroendocrine cell properties. Proc Am Soc Clin Oncol 1988;7:200 (abstr).
12. Gazdar AF, Zweig MH, Carney DN, Van Stierteghen AC, Baylin SB, Minna JD. Levels of creatine kinase and its BB isoenzyme in lung cancer specimens and cultures. Cancer Res 1981;41:2773-2777.
13. Carney DN, Zweig MH, Ihde DC, Cohen MH, Makuch RW, Gazdar AF. Elevated serum creatine kinase BB levels in patients with small cell lung cancer. Cancer Res 1984;44:5399-5403.
14. Gazdar AF, Minna JD. Cell lines as an investigational tool for study of the biology of small cell lung cancer. Eur J Clin Oncol 1986;22:909-911.
15. Birrer MJ, Minna JD. Molecular genetics of lung cancer. Semin Oncol 1988;15:226-235.
16. Johnson BE, Ihde DC, Makuch RW, Gazdar AF, Carney DN, Oie H, Russell E, Nau MM, Minna JD. myc Family oncogene amplification in tumor cell lines established from small cell lung cancer patients and its relationship to clinical status and course. J Clin Invest 1987;79:1629-1634.
17. Johnson BE, Makuch RW, Simmons AD, Gazdar AF, Burch D, Cashell AW. Myc family DNA amplification in small cell lung cancer patients' tumors and corresponding cell lines. Cancer Res 1988;48:5163-5166.
18. Nau MM, Brooks BJ, Carney DN, Gazdar AF, Battey JF, Sausville EA, Minna JD. Human small cell lung cancers show amplification and expression of the N-myc gene. Proc Natl Acad Sci USA 1986;93:1092-1096.
19. Carney DN, Gazdar AF, Bepler G, Guccion J, Marangos PJ, Moody TW, Zweig MH, Minna JD. Establishment and identification of small cell lung cancer cell lines having classic and variant features. Cancer Res 1985;45:2913-2923.
20. Gazdar AF, Carney DN, Nau MM, Minna JD. Characterization of variant subclasses of cell lines derived from small cell lung cancer having distinctive biochemical, morphological and growth properties. Cancer Res 1985;45:2924-2930.
21. Whang-Peng J, Kao-Shan CS, Lee EC, Bunn PA, Carney DN, Gazdar AF, Minna JD. A specific chromosome defect associated with human small cell lung cancer. Science 1982;215:181-185.
22. Brauch H, Johnson B, Hovis J, Yano T, Gazdar A, Pettengill OS, Graziano S, Sorenson GD, Poiesz BJ, Minna JD, Linehan M, Zbar B. Molecular analysis of the short

arm of chromosome 3 in small-cell and non-small cell carcinoma of the lung. N Engl J Med 1987;317:1109-1113.
23. Johnson BE, Sakaguchi AY, Gazdar AF, Minna JD, Burch D, Marshall A, Naylor SL. Restriction fragment length polymorphism studies show consistent loss of chromosome 3p alleles in small cell lung cancer patients' tumors. J Clin Invest 1988;82: 502-507.
24. Ponder B. Gene loss in human tumours. Nature 1988;401-403.
25. Harbour JW, Lai SL, Whang-Peng J, Gazdar AF, Minna JD, Kaye FJ. Abnormalities in structure and expression of the human retinoblastoma gene in small cell lung cancer. Science 1988;241:353-357.
26. Cuttitta F, Carney DN, Mulshine J, Moody TW, Fedorko J, Fischler A, Minna JD. Bombesin-like peptides can function as autocrine growth factors in human small cell lung cancer. Nature 1985;316:823-826.
27. Nakanishi Y, Milshine JL, Kasprzyk PG, Natale RB, Maneckjee R, Avis I, Treston AM, Gazdar AF, Minna JD, Cuttitta F. Insulin-like growth factor-1 can mediate autocrine proliferation of human small cell lung cancer cell lines. J Clin Invest 1988; 82:354-359.
28. Nakanishi Y, Cuttitta F, Kasprzyk PG, Avis I, Steinberg SM, Gazdar AF, Mulshine JL. Growth factor effects on small cell lung cancer cell lines using a colorimetric assay: can a transferrin like factor mediate autocrine growth? Exp Cell Biol 1988; 56:74-85.
29. Hendler FJ, Ozanne BW. Human squamous cell lung cancers express increased epidermal growth factor receptors. J Clin Invest 1984;74:647-651.
30. Banks-Schlegel SP, Gazdar AF, Harris CC. Intermediate filament and cross-linked envelope expression in human lung cancer cell lines. Cancer Res 1985;45:1187-1197.
31. Linnoila RI, Mulshine JL, Steinberg SM, Funa K, Matthews MJ, Cotelingam JD, Gazdar AF. Neuroendocrine differentiation in endocrine and non-endocrine lung carcinomas. 1988; in press.
32. Souhami RL, Beverley PC, Bobrow LG. Antigens of small-cell lung cancer: First International Workshop. Lancet 1987;2:325-326.
33. Souhami RL, Beverley PC, Bobrow LG. The First International Workshop on small cell lung cancer antigens. Lung Cancer 1988;4:1-4.
34. Hakomori S. Glycosphingolipids. Sci Am 1986;54:44-53.
35. Ginsburg V, Fredman P, Magnani JL. Cancer-associated carbohydrate antigens detected by monoclonal antibodies. In: Genes and antigens in cancer cells: the monoclonal antibody approach. Basel: Karger Verlag, 1988.
36. Tockman MS, Gupta PK, Myers JD, Frost JK, Baylin SB, Gold EB, Mulshine JL. Sensitive and specific monoclonal antibody recognition of human lung cancer antigen on preserved sputum cells. J Clin Oncol 1988; in press.
37. Frost JK, Fontana RS, Melamed MR, et al. Early lung cancer detection: Summary and conclusions. Am Rev Respir Dis 1984;130:565-570.
38. Rosen ST, Mulshine J, Cuttitta F, Fedorko J, Carney DN, Gazdar AF, Minna JD. Analysis of human small cell lung cancer differentiation antigens using a panel of rat monoclonal antibodies. Cancer Res 1984;44:2052-2061.
39. Mulshine JL, Cuttitta F, Bibro M, Fedorko J, Fargion S, Little C, Carney DN, Gazdar AF, Minna JD. Monoclonal antibodies that distinguish non-small cell from small cell lung cancer. J Immunol 1983;131:497-502.
40. Gupta PK, Myers JD, Baylin SB, Mulshine JL, Cuttitta F, Gazdar AF. Improved antigen detection in ethanol-fixed cytologic specimens: a modified avidin-biotin-peroxidase complex (ABC) method. Diagn Cytopathol 1985;1:133-136.

41. Soc HC. Probability theory in the use of diagnostic tests: an introduction to critical study of the literature. Ann Intern Med 1986;104:60-66.
42. Virgi MA, Mercer DW, Herberman RB. Tumor markers in cancer diagnosis and prognosis. CA 1988;38:104-126.
43. Coombs RC, Neville AM. Significance of tumor-index substances in management. In: Stoll BA, ed. Secondary spread in breast cancer. Chicago: William Heineman, 1977:113-138.
44. Gold P, Freedman SO. Demonstration of tumor-specific antigens in human colonic carcinomata by immunological tolerance and absorption techniques. J Exp Med 1965;121:439-462.
45. Ladenson JH, McDonald JM, Landt M, Schwartz MK. Colorectal carcinoma and carcinoembryonic antigen (CEA). Clin Chem 1980;26:1213-1220.
46. Beatty JD, Terz JJ. Value of carcinoembryonic antigen in clinical medicine. Prog Clin Cancer 1982;8:9-29.
47. Fletcher RH. Carcinoembryonic antigen. Ann Intern Med 1986;104:66-73.
48. Fleisher M, Nisselbaum JS, Loftin K, Smith C, Schwartz MK. Roche RIA and Abbott EIA carcinoembryonic antigen assays compare. Clin Chem 1984;30:200-205.
49. Herbeth B, Bagrel A. A study of factors influencing plasma CEA levels in an unselected population. Oncodev Biol Med 1980;1:191-198.
50. Tabor E, Gerety RJ, Needy CF, Elisberg BL, Colo AR, Jones R. Carcinoembryonic antigen levels in asymptomatic adolescents. Eur J Cancer Clin Oncol 1981;17:257-258.
51. Stevens DP, Mackay IR, Cullen KJ. Carcinoembryonic antigen in an unselected elderly population. Br J Cancer 1975;32:147-151.
52. Krischke W, Niederle N, Schutte J, Pfeiffer R, Hirche H. Is there any clinical relevance of serial determination of serum carcinoembryonic antigen in small cell lung cancer patients? Cancer 1988;62:1348-1354.
53. Dent PB, McCulloch PB, Wesley-James O, MacLaren R, Muirhead W, Dunnett CW. Measurement of carcinoembryonic antigen in patients with bronchogenic carcinoma. Cancer 1978;42:1484-1491.
54. Nierenberg AA, Feinstein AR. How to evaluate a diagnostic marker test. JAMA 1988;259:1699-1702.
55. Sanders RS, Freund DA, Herbst CA, Sandler DP. Cost effectiveness of postoperative carcinoembryonic antigen monitoring in colorectal cancer. Cancer 1984;53:193-198.
56. Gail MH, Muenz L, McIntire KR, Radovich B, Braunstein G, Brown PR, Deftos L, Dnistrian A, Dunsmore M, Elashoff R, Geller N, Go VL, Hirji K, Klauber MR, Pee D, Petroni G, Schwartz M, Wolfsen AR. Multiple markers for lung cancer diagnosis: Validation of models for localized lung cancer. J Natl Cancer Inst 1988;80:97-101.
57. Hansen M, Pedersen AG. Tumor markers in patients with lung cancer. Chest 1986;89:219S-224S.
58. Gazdar AF, McDowell EM. Pathobiology of lung cancers. In: Rosen ST, Mulshine JL, Cuttitta F, Abrams PG, eds. Biology of lung cancer: diagnosis and treatment. New York: Marcel Dekker, 1988:1-42.
59. Marangos PJ, Gazdar AF, Carney DN. Neuron-specific enolase in human small cell carcinoma cultures. Cancer Lett 1982;15:67-71.
60. Carney DN, Marangos PJ, Ihde DC, Bunn PA, Cohen MH, Minna JD, Gazdar AF. Serum neuron specific enolase: a marker for disease extent and response to therapy in patients with small cell lung cancer. Lancet 1982;1:583-585.
61. Johnson DH, Marangos PJ, Forbes JT, Hainsworth JD, Van Welsh R, Hande KR.

Potential utility of serum neuron-specific enolase levels in small cell carcinoma of the lung. Cancer Res 1984;44:5409-5414.
62. Zweig MH, van Steirteghem AC. Assessment of radioimmunoassay of serum creatine kinase BB (CK-BB) as a tumor marker: studies in patients with various cancers and a comparison of CK-BB concentrations to prostate acid phosphatase. J Natl Cancer Inst 1981;66:859-862.
63. Thompson RJ, Rubery ED, Jones HM. Radioimmunoassay of serum creatine kinase BB as a tumor marker in breast cancer. Lancet 1980;2:673-675.
64. Rubery ED, Doran JF, Thompson RJ. Brain-type creatine kinase BB as a potential tumor marker: serum levels measured by radioimmunoassay in 1015 patients with histologically confirmed malignancies. Eur J Cancer Clin Oncol 1979;44:1414-1418.
65. Carney DN, Zweig MH, Ihde DC, Cohen MH, Makuch RW, Gazdar AF. Elevated serum creatine kinase BB levels in patients with small cell lung cancer. Cancer Res 1984;44:5399-5403.
66. Sobol RE, O'Connor DT, Addison J, Suchocki K, Royston I, Deftos LI. Elevated serum chromogranin A concentrations in small cell lung carcinoma. Ann Intern Med 1986;105:698-700.
67. Pederson AG, Becker KL, Back F, Snider RH, Gazdar AF. Sorenson PS, Bunn PA, Hansen HH. Cerebrospinal fluid bombesin and calcitonin in patients with central system metastases from small cell lung cancer. J Clin Oncol 1989;4:1620-1627.
68. Wood S, Wood J, Ghatei M, Sorensen G, Bloom S. Is bombesin a tumor marker for small cell carcinoma? Lancet 1982;1:690-691.
69. Silva OL, Becker KL, Primack A. Ectopic secretion of calcitonin. Lancet 1973;2:317.
70. Silva OL, Becker KL, Primack A, et al. Ectopic secretion of calcitonin in oat cell carcinoma. N Engl J Med 1974;290:1122.
71. Becker KL, Snider RH, Silva OL. Calcitonin heterogeneity in lung cancer and medullary thyroid cancer. Acta Endorcinol 1978;89:89.
72. Coombes RC, Greenberg PB, Hillyard C. Plasma immunoreactive calcitonin in patients with non-thyroid tumours. Lancet 1974;1:1080.

30
Immunodiagnosis of Renal Cell Carcinoma

THOMAS EBERT and NEIL H. BANDER

The New York Hospital-Cornell University Medical Center and Memorial Sloan-Kettering Cancer Center, New York, New York

Renal cell carcinoma (RCC) is the most common malignant tumor of the kidney and the third most common urologic cancer, with approximately 18,000 cases occurring annually in the United States [1]. Since the onset of symptoms frequently occurs late and clinical signs may vary considerably, correct diagnosis is often established only when the disease is advanced. In fact, at initial presentation approximtely 50% of patients already have metastatic disease [2-4]. The median duration of survival in patients with overt distant disease at presentation is 10 months with a 10% 2 year survival. There is no effective systemic therapy, since renal cancer has proven refractory to chemotherapy. There are, however, recent preliminary reports of responses to immunotherapy [5] that need to be pursued.

Despite this rather grim background, there is reason for guarded optimism. Renal cancers are curable if diagnosed and excised prior to spread. Since they tend to be relatively large (>3 cm) prior to metastasizing [6] and since tumor growth rate is relatively slow [7], these lesions, in general, are present for years before spreading. This affords a broad "window" of time during which to diagnose and excise the lesion. These cancers are also likely to communicate with the urine, potentially allowing urinary assays as screening procedures for the disease. The multitude of active peptides inducing the paraneoplastic syndromes associated with this cancer provide numerous potential targets on which to focus recombinant DNA and monoclonal antibody technology for the development of assays designed to detect and monitor these cancers. While the immunodiagnosis of renal cancer is not yet a reality, the techniques necessary to reach this goal successfully are now at hand. This chapter outlines some of the potential targets for immunodiagnosis.

I. PARANEOPLASTIC PEPTIDES: POTENTIAL TUMOR MARKERS

About 30% of patients with RCC initially present without specific genitourinary symptoms [8]. In many cases, however, systemic manifestations mimicking other disorders may be observed. Indeed, because RCC often presents with a broad variety of nonuro-

Table 1 Paraneoplastic Syndromes in Renal Cell Carcinoma

Extrarenal Manifestation	Incidence (%)	Serological "Marker"	Reference
Hepatic dysfunction	10-40	Alakaline phosphatase	11
		Alpha-2-globulin	12
Anemia	25	Lactoferrin	13
		Iron	
Pyrexia	20	(endogenous pyrogens)	14
Hypercalcemia	15	PTH-like peptide	15
Erythrocytosis	2-10	Erythropoietin	16
Amyloidosis	3	?	17
Hypertension	?	Renin	18
Hypotension	1	Prostaglandin A	19
Cushing's syndrome	1	ACTH	20
Galactorrhea	1	Prolactin	21
?	1	Alpha fetoprotein	22
Enteropathy	1	Enteroglucagon	23
Hyperglycemia	1	?	24
Hypoglycemia	1	Insulin	25
Gynecomastia/hirsutism	1	Gonadotropin	26
Carcinoid-like syndrome	1	Prostaglandin E and F2a	27

logical symptoms (Table 1), it has been called "one of the great masqueraders" in medicine [9]. The most common paraneoplastic syndromes in RCC are hepatic dysfunction (Stauffer's syndrome), anemia, pyrexia, and hypercalcemia [10].

A. Pyrexia and Anemia

Fever and anemia occur in 20-25% of patients with renal cancer. In relatively few cases is the anemia secondary to hematuria [28]. Likewise, fever is rarely due to coexistent infections or other known disorders [29] and is believed secondary to release of endogenous pyrogen (interleukin-1).

Endogenous pyrogens (EP) have been found in renal cancer tissue extracts [14]. Furthermore, these pyrogens could be demonstrated only in tumor tissue of patients presenting with fever. Neither tumor tissue from patients without febrile episodes nor normal kidney tissue contained measurable amounts of these substances. Since leukocytes are a well-recognized source of EP, it is still unclear if EP is released by tumor-infiltrating leukocytes or produced and released by the tumor cells per se [30].

Endogenous pyrogen can, in turn, induce the release of lactoferrin [31], a glycoprotein present in most body fluids and in neutrophils [32]. Lactoferrin tightly binds two ferric ions and removes them from the circulation [33]. In RCC the release of lactoferrin mediated by endogenous pyrogen has been proposed as a possible explanation for the anemia that occurs in many patients [13].

B. Hepatic Dysfunction

Since Stauffer [11] published the first report on reversible hepatic dysfunction in patients with nonmetastatic RCC, several authors subsequently described prolonged sulfobromophthalein (BSP) retention, hypoproteinuria, elevated alpha-2-globulin levels, elevated indirect bilirubin levels, and increased alkaline phosphatase levels in patients with RCC. Although this syndrome is not specific for RCC [34,35], it can be found in up to 40% of patients [12,36].

Histological findings in liver biopsies include Kupfer's cell proliferation, hepatocellular degeneration, focal necrosis, and portal triaditis [12]. The pathogenesis of this syndrome is still unknown, however, the most likely cause is a hepatotoxic factor either released directly by the tumor or produced by the host in response to the presence of the tumor [37]. Although undetected hepatic metastasis or nonspecific reactive hepatitis may account for some of the impaired liver function, the theory of a circulating factor is strongly supported by the fact that the abnormalities disappear in the majority of patients following nephrectomy [38]. A human renal cancer cell line has recently been derived from a patient with Stauffer's syndrome. Conditioned medium from this cell line, when injected into mice, causes the animals to develop an identical clinical picture and pathological liver changes [39].

While the finding of hepatic dysfunction prior to therapy has no prognostic significance [12], its persistence after nephrectomy seems to indicate residual disease [40] and its reappearance might accompany recurrence [12]. Purification and sequencing of the responsible factor(s), cloning the gene(s), and development of a sensitive immunoassay may result in a greatly improved ability to diagnose and monitor renal cancer.

C. Hypercalcemia

Hypercalcemia in the absence of bone involvement (humoral hypercalcemia of malignancy [HHM]) has long been associated with a variety of malignant diseases. A number of factors, such as ectopic parathormone, vitamin D and its metabolites, osteoclast-activating factors, prostaglandins, transforming growth factors, and colony-stimulating factors have been cited as causative agents of elevated serum calcium levels [41,42]. Although hypercalcemia in patients with RCC occasionally might be induced by each of these substances, there is much recent evidence that parathyroid hormone (PTH)-like substances mediate this syndrome in many instances. Permanent tissue culture cell lines from RCC that induce hypercalcemia in nude mice have been established, and culture supernatant of these cells contains a substance(s) that binds to the human PTH receptor [43]. Several laboratories have recently reported the successful purification, cloning, and sequencing of a tumor-derived PTH-like peptide [15,44,45].

The sequenced peptide bears a 67% homology to human PTH within the first 13 N-terminal amino acids [15]. This particular region is believed to play an important role in the biological activity of the PTH molecule [46]. In vivo, PTH regulates serum calcium levels by enhanced reabsorption of calcium (distal tubule) and by diminished reabsorption of phosphate (proximal tubule). It also mediates the resorption of bone, which leads to the release of calcium and phosphate into the circulation [46]. These actions of PTH seem to be shared by the PTH-like peptide. However, distinctive biological actions in vitro have also been described [47,48]. It is believed that the PTH-like peptide may interact with PTH receptors for PTH-like actions and that it may react with its own class of receptors for distinctive actions [49].

Although serum calcium has been proposed [38] as a possible marker for monitoring patients after nephrectomy, no prospective studies reported have examined the value of this particular parameter. Since the regulation of serum calcium levels is controlled by various other factors, the secretion of a PTH-like protein by the tumor may not necessarily lead to hypercalcemia in all patients. Attempts, therefore, should be made to measure circulating PTH-like peptides directly rather than serum calcium levels. The development of monoclonal antibodies (MABs) against these peptides might result in a promising new assay for monitoring at least a subset of renal cancer patients.

Levels of other substances (Table 1) have been found to be elevated in some patients with RCC and have been probed as "tumor markers." Due to low specificity and sensitivity they are usually of no diagnostic value. In individual cases, however, persistence of or normalization of elevated levels may correlate with stage or clinical course of disease.

II. ANTIGENIC DETERMINANTS IN RCC

A. Steroid Receptors

Interest in steroid hormone receptors in renal cancer was prompted by the observation of Kirkman in 1959 [50] that kidney tumor development in a Syrian hamster model could be prevented by the administration of progesterone. Various studies have been performed to examine the presence of steroid receptors in renal carcinoma cells and its prognostic significance in this disease [51-55]. Although it has been proposed that the hormone receptor status of renal cancer would have a similar value to that in breast cancer, to date no correlation between survival, response to hormone therapy, or prognosis and expression of steroid receptors has been established.

B. Cell Surface Markers

In addition to secreted paraneoplastic factors, cell surface markers also represent a potential target for the immunodiagnosis of renal cancer. Mouse MABs have been generated and used in various laboratories to study the distribution of antigenic determinants in normal and malignant renal tissue (Table 2).

1. Immunopathological Findings

At Memorial Sloan-Kettering Cancer Center (MSKCC) a series of MABs have been developed that identify several surface antigens on human kidney cells [58,59,61]. Serological tests on cultured cells as well as immunohistochemical studies on frozen tissue sections were performed to characterize the antigenic phenotype of normal kidney and renal cell carcinoma [59,62]. The reactivity pattern of several MABs in normal kidney tissue is shown in Figure 1. Comparison of antigen expression of normal kidney and kidney tumor supports earlier electron microscopic findings that defined the proximal tubule as the site of origin of RCC. While none of the distal tubular markers could be detected in kidney tumor specimens, the vast majority of these tumors (93%) express one or more proximal tubular antigens [76].

Although a correlation between antigenic expression and variation in histological cell type has not yet been established, the presence or absence of some surface antigens seems to correlate with different clinical features. That is, S4$^+$/S23$^+$ tumors presented at a lower clinical stage and were associated with a significantly better survival (12 of 15; 80%) than S4$^-$/S23$^-$ tumors, which presented at a higher stage with poorer survival (1 of 11; 9%). Reciprocal expression (S4$^+$/S23$^-$ or S4$^-$/S23$^+$) was associated with intermediate results (Bander, unpublished observation, 1983).

MABs S6, S23, and S27 were found to react with different epitopes of the same antigen. The antigen (gp 120) is the adenosine-deaminase binding protein (ADAbp; [77]), which is present on a variety of normal and malignant human cells including normal kidney proximal tubular cells and renal cancer cells [61]. An immunoassay using these MABs has been formatted and is capable of detecting ADAbp in urine [63]. Although no ADAbp is present in normal urine, diseases in which there is proximal tubular injury such as acute tubular necrosis are associated with elevated urinary ADAbp levels. The assay is currently being evaluated in patients with renal cancer.

Oosterwijk et al. [75] described 4 MABs that were highly specific for kidney tissue. One of these antibodies, designated RC 38, was found in immunohistochemical studies to be positive on 95% of primary and 60% of metastatic RCC. It did not stain any of 179 tumors of various nonrenal origin. To assess the possible role of this MAB in the diagnosis of RCC, it was used to immunostain 18 malignant lesions for which RCC was considered only among other possible diagnoses. Six of 18 tissues finally were diagnosed as RCC (2 primary, 4 metastatic), while 12 specimens were considered to show malignancies distinct from RCC. The immunohistochemical procedure with MAB RC38 showed negative results in all 12 nonrenal tumors but was positive in 4 of 6 renal-cancer-derived lesions. Only two metastatic RCC specimens (femur, subcutis) were not detected by RC38. Based on the high specificity and fairly good sensitivity of this MAB, the authors concluded that RC38 might be of particular value when a diagnosis of metastatic lesions is difficult to establish by conventional histological techniques.

2. Immunoradioscinitigraphy

Attempts are being made with a wide variety of tumors to identify metastatic sites by radionuclide scans using systemically administered radiolabeled MAB to tumor-associated antigens. Human renal cancer xenografts transplanted into nude mice have been successfully imaged with labeled MAB [78,79].

A radioimmunolocalization study using ^{131}I-labeled MAB F23 has been performed in 22 patients with stage III/IV RCC at New York Hospital-Cornell Medical Center and MSKCC [80]. F23 (IgG2a) reacts with a cell surface glycoprotein of 140Kd expressed by the epithelium of the normal proximal tubule and by 90% of RCC. Only biliary tract epithelium, fibroblasts, and polymorphonuclear leukocytes have also been found to react with this antibody [59]. Two mCi of ^{131}I conjugated to MAB (1-10 mg) was given intravenously over 1 hr and scans were obtained daily for 5-7 days. In 5 of 19 evaluable patients, increased radioactivity at tumor sites was observed. Skeletal disease seemed to be imaged better than lung metastases. To determine the specificity of the increased radioactivity, three of those five patients received 2 mCi of ^{99}Tc-labeled human serum albumin (HSA). In two of three patients a similar increase of radioactivity at tumor sites was observed, which suggested that blood flow in the tumor might account for at least some of the observed increased activity. All patients tolerated MAB administration without side effects. Further studies are underway to address the specificity of radioactive uptake at tumor

Table 2 Monoclonal Antibodies to Kidney Antigens

Monoclonal antibodies (Ig Class)	Defined Antigen	Site of Expression	Laboratory	Reference
T138 (IgM)	gp25	Vascular endothelium	MSKCC	56
J143/URO-1 (IgG$_1$)	gp140,120,30	Glomerulus, ureothelium	MSKCC	56
AJ8 (IgG$_1$)	gp100 (CALLA)	Glomerulus, PT	MSKCC	57
S4/URO-2 (IgG$_{2a}$)	gp 160	Glomerulus, PT	MSKCC	58-62
T43/URO-10 (IgG$_1$)	gp85	PT	MSKCC	56
F23/URO-3 (IgG$_{2a}$)	gp140	PT	MSKCC	59-61
S23/URO-4 (IgG$_1$)	gp120 (ADAbp)	PT, LH	MSKCC	58-63
S27/URO-4 (IgG$_1$)	gp120 (ADAbp)	PT, LH	MSKCC	58-62
S6 (IgG$_1$)	gp120 (ADAbp)	PT, LH	MSKCC	58-62
F31/URO-8 (IgM)	glycolipid	PT	MSKCC	64
C26 (IgG$_{2a}$)	gp40	DT, CT, urothelium	MSKCC	61
M2 (IgM)	A blood group	CT, urothelium	MSKCC	58,61
S8 (IgM)	B blood group	CT, urothelium	MSKCC	58,61
T16 (IgG$_{2b}$)	gp48,42	DT, CT, urothelium	MSKCC	56,60,61
S22 (IgG$_1$)	gp115	Renal cancer only	MSKCC	58,59,61
10.32 (IgG$_1$)	gp90 (Tamm-Horsfall)	LH, DT	Cedarlane Labs	61
OKIa1 (IgG$_2$)	HLA-DR	Endothelium, mesangium	Ortho Pharm.	
Anti-SSEA-1 (IgM)	glycolipid	PT	Wistar Inst.	65
A6H (IgG$_1$)	N.D.	PT	U. Minn.	66,67
D5D (IgG$_1$)	N.D.	BC	U. Minn.	66,67

C5H (IgG$_1$)	p115	Glomerulus	U. Minn.	66,67
Anti-B1 (IgG$_{2a}$)	N.D.	Fetal ureteral bud	Dana Farber	68
BA-1 (IgM)	N.D.	BC,DT,CT	U. Minn.	68
BA-2 (IgG$_3$)	p24	BC,DT,CT	U. Minn.	68
E6 (N.D.)	N.D.	Renal cancer only	Mainz, W. Germany	69
B7 (N.D.)	N.D.	Renal cancer only	Mainz	69
C8 (N.D.)	N.D.	Renal cancer only	Mainz	69
D8 (N.D.)	N.D.	Renal cancer only	Mainz	69
DU-ALL-1	p24	Glomerulus, DT	Duke Univ.	70
DU-HL60-4	p160	PT,BC	Duke Univ.	70
DU-HL60-3 (IgM)	p15	PT	Duke Univ.	70
K2.7 (IgG$_3$)	gp43,60,63	PT,DT	Osaka, Japan	71
MKi-1 (IgG$_1$)	p140	PT	Milan, Italy	72
Da1-K20 (IgG$_1$)		PT,DT	Halifax, Canada	73
Da1-K29 (IgG$_1$)		Glomerular capillaries	Halifax	73
Da1-K45 (IgG$_1$)	p/gp150,177	Renal cancer only	Halifax	73
5F4 (IgG$_1$)	N.D.		Texas A & M	74
RC3 (IgG$_1$)	N.D.	PT	Leiden, The Netherlands	75
RC69 (IgG$_{2b}$)	N.D.	PT	Leiden	75
RC154 (IgG$_{2a}$)	N.D.	PT,DT,CT	Leiden	75
RC38 (IgG$_1$)	N.D.	Glomerulus, PT	Leiden	75

PT = proximal tubule; LH = loop of Henle; DT = distal tubule; CT = collecting duct; BC = Bowman's capsule.

	PT	LH	DT	CT
G

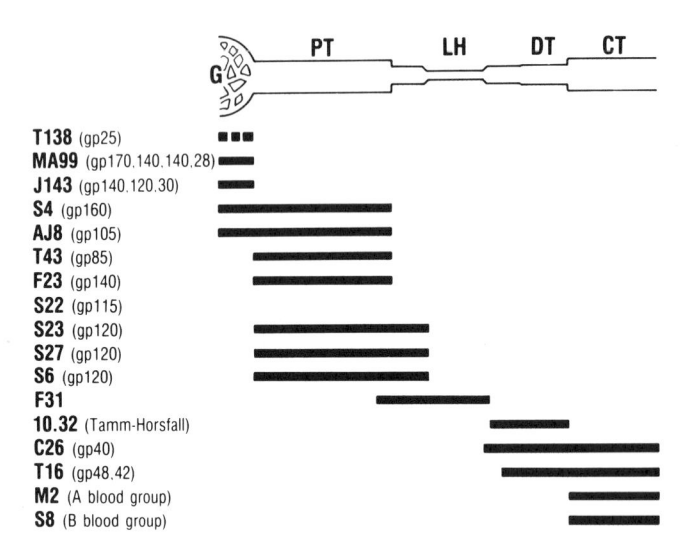

T138 (gp25)
MA99 (gp170,140,140,28)
J143 (gp140,120,30)
S4 (gp160)
AJ8 (gp105)
T43 (gp85)
F23 (gp140)
S22 (gp115)
S23 (gp120)
S27 (gp120)
S6 (gp120)
F31
10.32 (Tamm-Horsfall)
C26 (gp40)
T16 (gp48,42)
M2 (A blood group)
S8 (B blood group)

Figure 1 "Antigenic map" of the human nephron. The bar indicates the site of binding of each monoclonal antibody as determined by immunohistochemistry on frozen kidney sections. The immunochemical characterization of each defined antigen is indicated (e.g., gp25 is a glycoprotein with a molecular weight of 25,000). S22, for which no bar is shown, does not bind to normal kidney sections but does bind to some renal cancers.

sites by using a control isotype-matched unreactive antibody and by obtaining surgical biopsy specimens after MAB administration.

Another radioimmunolocalization study of RCC was presented by Vesella et al. [81]. With their MAB A6H, infrequent imaging was observed when 5 or 10 mg doses of ^{131}I-labeled antibody were given. Higher dosages caused fever and chills, which were found to be abolished by pretreatment with methylprednisolone and diphenhydramine hydrochloride (Benadryl). Administration of 20 or 40 mg of labeled MAB resulted in higher antibody serum levels and improved imaging of suspected tumor lesions.

Although these first attempts are somewhat encouraging, the results do not yet match the expected sensitivity of immunoimaging. Various factors might account for this: F23 and A6H bind to highly restricted antigens that are preferentially expressed on renal tissue. However, reactivity in the case of F23 was also observed with the above-mentioned cell types, which might explain the antibody uptake in liver and spleen. A6H seems to react with a circulating antigen that might be shed or secreted by tumor cells. Both events are presumed to compromise the quality and specificity of tumor imaging or might prevent imaging altogether. New antibodies without these problems need to be developed to improve radioimaging in RCC for diagnostic and therapeutic purposes.

III. CHROMOSOME CHANGES IN RCC

The first structural chromosomal abnormality reported in human malignant disease was the Philadelphia chromosome [82]. Since then, nonrandom chromosome changes have been observed in patients with many malignancies [83]. Genetic abnormalities that might be involved in RCC have been investigated in a few laboratories. In 1979 Cohen et al. [84] reported on a family with hereditary RCC in which a chromosome translocation between chromosome 3 and 8 was noted. This translocation was associated with a high probability of developing this disease by age 59 in those family members who inherited the translocation. Another translocation involving chromosome 3 and 11 was found in another family with RCC [85]. In both cases the breakpoint on chromosome 3 was located at the same region (3p 14-21). Abnormalities of this particular region have also been described in RCC from "sporadic," unrelated patients [86,87]. There appears to be little doubt that a chromosomal defect at this site plays a role in the induction and/or progression of RCC.

This has been supported by studies investigating the chromosomal location of proto-oncogenes. Little is known about the function of the protein products of many of these cellular oncogenes; however, there is evidence that several of these products mediate actions involved in cell differentiation and cell proliferation by controlling the expression of growth factors and/or their receptors [88]. Alteration of these genes might result in neoplastic growth [89]. Several of these proto-oncogenes have been found to be adjacent to the breakpoints in chromosomal changes of malignancies [90]. So far, in two human malignancies the link between chromsome change and the involvement of an oncogene has been established: chronic myeloblastic lymphoma (CML) for the C-erb and Burkitt's lymphoma for the C-myc oncogene [91,92]. In RCC a shift of the C-RAF1 locus on chromosome 3 [93] as well as a translocation of C-myc from chromosome 8 to chromosome 3 [94] have been described.

The C-ERB-B oncogene that has been linked to the expression of the EGF-receptor on human cells is located on chromosome 7. Although nonrandom changes on this chromosome have not been reported in patients with RCC, Real et al. [95] have examined the expression of this receptor on normal and malignant renal cell lines and tissue sections. Using MAB reacting with the EGF receptor, they found that this receptor in normal kidney tissue was located on the proximal, distal, and collecting tubules. It was also expressed in six of six renal cancer specimens. The authors also observed that the expression of the EGF receptor was clearly increased in RCC cell lines compared to short-term cultures of normal kidney epithelium.

In contrast to the oncogene activation hypothesis, more recent data suggest that RCC may arise by the deletion of genes [96,97]. In these studies restriction fragment-length polymorphisms (RFLPs) on the short arm of chromosome 3 were analyzed in renal cancer tissue and autologous normal kidney. Zbar et al. observed in all 11 evaluable patients a loss of alleles mapped to the 3p region of chromosome 3 in tumor specimens. Kovacs et al. used 4 different 3p-specific DNA probes to examine the specimens of 34 patients. The 3p-21-specific sequences were either completely lost or their signal was significantly decreased in 16 of 21 (76%) evaluable cancer tissues. These results support the previous findings underlining the importance of the 3p region in the origin or evolution of this tumor. Genes that enable the renal cell to respond appropriately to environmental stimuli might be located in this particular region and their loss might trigger intracellular events that result in malignant transformation.

To date, the final role of chromosome changes in diagnosis and prognosis cannot yet be defined. We hope that further studies will lead to a broader understanding of the molecular events occurring at the DNA level and their relationship to clinical features in patients with RCC.

IV. SUMMARY

Development of immunoassays for the diagnosis of RCC offers significant promise of improved clinical treatment of this disease. While this promise remains to be fulfilled, there is reason to believe that advances will soon be forthcoming. MABs have defined numerous kidney-associated antigens. In at least one instance (ADAbp), an immunoassay has already been established that aids in the diagnosis of benign renal diseases. A similar approach may also be useful in renal cancer. Likewise, in at least one instance a previously undefined peptide (the HHM factor) has been closed and sequenced. Synthesis of this peptide and production of antibodies to it cannot be far off. Measurement of this peptide and/or of the many other aberrantly produced peptides associated with this cancer may be anticipated to improve substantially our ability to diagnose early and, therefore, cure renal cancer.

REFERENCES

1. Paganini-Hill A, Ross RK, Henderson BE. Epidemiology of kidney cancer. In: Skinner DA, ed. Urological cancer. New York: Grune & Stratton, 1983:383-407.
2. Robson CJ, Churchill BM, Anderson W. The results of radical nephrectomy for renal cell carcinoma. J Urol 1969;101:297-301.
3. Giuliani L, Martorana G, Giberti C, Pescatore D, Magnani G. Results of radical nephrectomy with extensive lymphadenectomy for renal cell carcinoma. J Urol 1983;130:664-668.
4. Bassil B, Dosoretz DE, Prout GR Jr. Validation of the tumor, nodes and metastasis classification of renal cell carcinoma. J Urol 1985;134:450-454.
5. Rosenberg SA, Lotze MT, Muul LM, Chang AE, Avis FP, Leitman S, Linehan WM, Robertson CN, Lee RE, Rubin JT, Seipp CA, Simpson CG, White DE. A progress report on the treatment of 157 patients with advanced cancer using lymphokine-activated killer cells and interleukin-2 or high-dose interleukin-2 alone. N Engl J Med 1987;316:889-897.
6. Bell ET. Renal diseases. Philadelphia: Lea & Febiger, 1950.
7. Rabes HM, Carl P, Meister P, Rattenhuber U. Analysis of proliferative compartments in human tumors. I. Renal adenocarcinoma. Cancer 1979;44:799-813.
8. Holland JM. Cancer of the kidney—natural history and staging. Cancer 1973;32:1030-1042.
9. Marshall FF, Walsh PC. Extrarenal manifestations of renal cell carcinoma. J Urol 1977;117:439-440.
10. Chisholm GD, Roy RR. The systemic effects of malignant renal tumours. Br J Urol 1971;43:687-700.
11. Stauffer MH. Nephrogenic hepatosplenomegaly. Gastroenterology 1961;40:694.
12. Utz DC, Warren MM, Gregg JA, Ludwig J, Kelalis PP. Reversible hepatic dysfunction associated with hypernephroma. Mayo Clin Proc 1970;45:161-169.
13. Loughlin KR, Gittes RF, Partridge D, Stelos P. The relationship of lactoferrin to the anemia of renal cell carcinoma. Cancer 1987;59:566-571.
14. Rawlins MD, Luff RH, Cranston WI. Pyrexia in renal carcinoma. Lancet 1970;1:1371-1373.
15. Mangin M, Webb AC, Dreyer BE, Posillico JT, Ikeda K, Weir EC, Stewart AF, Bander NH, Milstone L, Barton DE, Francke U, Broadus AE. Identification of a cDNA encoding and parathyroid hormone-like peptide from a human tumor associated with humoral hypercalcemia of malignancy. Proc Natl Acad Sci USA 1988;85:597-601.

16. Murphy GP, Kenny GM, Mirand EA. Erythropoietin levels in patients with renal tumors or cysts. Cancer 1970;26:191-194.
17. Svane S. Hypernephroma and systemic amyloidosis. A report on 3 cases. Acta Chir Scand 1970;136:68-76.
18. Hollifield JW, Page DL, Smith C, Michelakis AM, Staab E, Rhamy R. Renin-secreting clear cell carcinoma of the kidney. Arch Intern Med 1975;135:859-864.
19. Zusman RM, Snider JJ, Cline A, Caldwell BV, Speroff L. Antihypertensive function of a renal-cell carcinoma. Evidence for a prostaglandin-A-secreting tumor. N Engl J Med 1974;290:843-844.
20. Riggs BL Jr, Sprague RG. Association of Cushing's syndrome and neoplastic disease. Arch Intern Med 1961;108:85-93.
21. Turkington RW, Ectopic production of prolactin. N Engl J Med 1971;285:1455-1458.
22. Morimoto H, Tanigawa N, Inoue H, Muraoka R, Hosokawa Y, Hattori T. Alpha-fetoprotein-producing renal cell carcinoma. Cancer 1988;61:84-88.
23. Gleeson NH, Bloom SR, Polak JM, Henry K, Dowling RH. Endocrine tumor in kidney affecting small bowel structure, motility, and absorptive function. Gut 1971;12:773-782.
24. Palgon N, Greenstein F, Novetsky AD, Lichter SM, Rosen Y. Hyperglycemia associated with renal cell carcinoma. Urology 1986;28:516-517.
25. Papaioannou AN. Tumors other than insulinomas associated with hypoglycemia. Surg Gynecol Obstet 1966;123:1093-1109.
26. Golde DW, Schambelan M, Weintrabu BD, Rosen SW. Gonadotropin-secreting renal carcinoma. Cancer 1974;33:1048-1053.
27. Plaksin J, Landau Z, Coslovsky R. A carcinoid-like syndrome caused by a prostaglandin-secreting renal cell carcinoma. Arch Intern Med 1980;140:1095-1096.
28. Loughlin KR, Gittes RF. Serum iron: a tumour marker in renal carcinoma. Br J Urol 1986;58:617-620.
29. Bodel P. Part I. Generalized perturbations in host physiology caused by localized tumors. Tumors and fever. Ann NY Acad Sci 1974;230:6-13.
30. Cranston WI, Luff RH, Owen D, Rawlins MD. Studies on the pathogenesis of fever in renal carcinoma. Clin Sci Mol Med 1973;45:459-467.
31. Klempner MS, Dinarello CA, Gallin JI. Human leukocytic pyrogen induces release of specific granule contents from human neutrophils. J Clin Invest 1978;61:1330-1336.
32. Masson PL, Heremans JF, Schonne E. Lactoferrin, an iron-binding protein in neutrophilic leukocytes. J Exp Med 1969;130:643-658.
33. Lukens JN. Iron deficiency and infection. Am J Dis Child 1975;129:160-162.
34. Abels JC, Rekers PE, Binkley GE, Pack GT, Rhoads CP. Metabolic studies in patients with cancer of the gastrointestinal tract. II. Hepatic dysfunction. Ann Intern Med 1942;16:221-240.
35. Nieburgs HE, Parets AD, Perez V, Boudreau C. Cellular changes in liver tissue adjacent and distant to malignant tumors. Arch Pathol 1965;80:262-272.
36. Boxer RJ, Waisman J, Lieber MM, Mampaso FM, Skinner DG. Non-metastatic hepatic dysfunction associated with renal carcinoma. J Urol 1978;119:468-471.
37. Walsh PN, Kissane JM. Nonmetastatic hypernephroma with reversible hepatic dysfunction. Arch Intern Med 1968;122:214-222.
38. Sufrin G, Golio A, Murphy GP. Serologic markers, paraneoplastic syndromes, and ectopic hormone production in renal adenocarcinoma. In: deKernion JB, Pavone-macaluso M, eds. Tumors of the kidney. Baltimore: Williams & Wilkins, 1986:51-71.
39. Sherwood ER, Keer HN, Fike W, Smith D, Kozlowski JM. The role of colony stim-

ulating factor(s) in the pathogenesis of nephrogenic hepatic dysfunction (Stauffer's) syndrome. J Urol 1988;139, 254A (abstract 368).
40. Lemmon WT Jr, Holland PV, Holland JM. The hepatopathy of hypernephroma. Am J Surg 1965;110:487-491.
41. Mundy GR, Ibbotson KJ, D'Souza SM, Simpson EL, Jacobs JW, Martin TJ. The hypercalcemia of cancer: clinical implications and pathogenic mechanisms. N Engl J Med 1984;310:1718-1727.
42. Tashjian AH. Tumor humors and the hypercalcemias of cancer. N Engl J Med 1974; 290:905-906.
43. Strewler GJ, Williams RD, Nissenson RA. Human renal carcinoma cells produce hypercalcemia in the nude mouse and a novel protein recognized by parathyroid hormone receptors. J Clin Invest 1983;71:769-774.
44. Burtis WJ, Wu T, Bunch C, Wysolmerski JJ, Insogna KL, Weir EC, Broadus AE, Stewart AF. Identification of a novel 17,000-dalton parathyroid hormone-like adenylate cyclase-stimulating protein from a tumor associated with humoral hypercalcemia of malignancy. J Biol Chem 1987;262:7151-7156.
45. Moseley JM, Kubota M, Diefenbach-Jagger H, Wettenhall REH, Kemp BE, Suva LJ, Rodda CP, Ebeling PR, Hudson PJ, Zajac JD, Martin TJ. Parathyroid hormone-related protein purified from a human lung cancer cell line. Proc Natl Acad Sci USA 1987;84:5048-5052.
46. Habener JF, Rosenblatt M, Potts JT Jr. Parathyroid hormone: biochemical aspects of biosynthesis, secretion, action, and metabolism. Physiol Rev 1984;64:985-1053.
47. Nissenson RA, Strewler GJ, Williams RD, Leung SC. Activation of the parathyroid hormone receptor-adenylate cyclase system in osteosarcoma cells by a human renal carcinoma factor. Cancer Res 1985;45:5358-5363.
48. Stewart AF, Vignery A, Silverglate A, Ravin ND, LiVolsi V, Broadus AE, Baron R. Quantitative bone histomorphometry in humoral hypercalcemia of malignancy: uncoupling of bone cell activity. J Clin Endocrinol Metab 1982;55:219-227.
49. Wu TL, Insogna KL, Hough LM, Milstone L, Stewart AF. Skin-derived fibroblasts respond to human parathyroid hormone-like adenylate cyclase-stimulating proteins. J Clin Endocrinol Metab 1987;65:105-109.
50. Kirkman H. Estrogen-induced tumors of the kidney in the Syrian hamster. Nat Cancer Inst Monogr 1959;1:59-91.
51. Concolino G, Marocchi A, Conti C, Tenaglia R, Di Silverio F, Bracci U. Human renal cell carcinoma as a hormone-dependent tumor. Cancer Res 1978;38:4340-4344.
52. Mukamel E, Bruhis S, Nissenkorn I, Servadio C. Steroid receptors in renal cell carcinoma: relevance to hormonal therapy. J Urol 1984;131:227-230.
53. Nakano E, Tada Y, Fujioka H, Matsuda M, Osafune M, Kotake T, Sato B, Takaha M, Sonoda T. Hormone receptor in renal cell carcinoma and correlation with clinical response to endocrine therapy. J Urol 1984;132:240-245.
54. Pizzocaro G, Piva L, Salvioni R, DiFronzo G, Ronchi E, Miodini P, The Lombardy Group. Adjuvant medroxyprogesterone acetate and steroid hormone receptors in category M0 renal cell carcinoma: an interim report of a prospective randomized study. J Urol 1986;135:18-21.
55. Pizzocaro G, Piva L, DiFronzo G, Giongo A, Cozzoli A, Dormia E, Minervini S, Zanollo A, Fontanella U, Longo G, Maggioni A. Adjuvant medroxyprogesterone acetate to radical nephrectomy in renal cancer: 5-year results of a prospective randomized study. J Urol 1987;138:1379-1381.
56. Fradet Y, Cordon-Cardo C, Thompson T, Daly ME, Whitmore WF Jr, Lloyd KO, Melamed MR, Old LJ. Cell surface antigens of human bladder cancer defined by mouse monoclonal antibodies. Proc Natl Acad Sci USA 1984;81:224-228.

57. Cairncross JG, Mattes MJ, Beresford HR, Albino AP, Houghton AN, Lloyd KO, Old LJ. Cell surface antigens of human astrocytoma defined by mouse monoclonal antibodies: identification of astrocytoma subsets. Proc Natl Acad Sci USA 1982; 79:5641-5645.
58. Ueda R, Ogata S-I, Morrissey DM, Finstad CL, Szkudlarek J, Whitmore WF, Oettgen HF, Lloyd KO, Old LJ. Cell surface antigens of human renal cancer defined by mouse monoclonal antibodies: identification of tissue-specific kidney glycoproteins. Proc Natl Acad Sci USA 1981;78:5122-5126.
59. Finstad CL, Cordon-Cardo C, Bander NH, Whitmore WF, Melamed MR, Old LJ. Specificity analysis of mouse monoclonal antibodies defining cell surface antigens of human renal cancer. Proc Natl Acad Sci USA 1985;82:2955-2959.
60. Cordon-Cardo C, Bander NH, Fradet Y, Finstad CL, Whitmore WF, Lloyd KO, Oettgen HF, Melamed MR, Old LJ. Immunoanatomic dissection of the human urinary tract by monoclonal antibodies. J Histochem Cytochem 1984;32:1035-1040.
61. Bander NH, Cordon-Cardo C, Finstad CL, Whitmore WF Jr, Vaughan ED Jr, Oettgen HF, Melamed M, Old LJ. Immunohistologic dissection of the human kidney using monoclonal antibodies. J Urol 1985;133:502-505.
62. Bander NH. Study of the normal human kidney and kidney cancer with monoclonal antibodies. Uremia Invest 1985;8:263-273.
63. Tolkoff-Rubin NE, Cosimi AB, Delmonico FL, Russell PS, Thompson RE, Piper DJ, Hansen WF, Bander NH, Finstad CL, Cordon-Cardo C, Klotz LH, Old LJ, Rubin RH. Diagnosis of tubular injury in renal transplant patients by a urinary assay for a proximal tubular antigen, the adenosine-deaminase-binding protein. Transplantation 1986;41:593-597.
64. Bander NH, Finstad C, Cordon-Cardo C, Ramsawak RD, Vaughan ED Jr, Whitmore WF Jr, Oettgen HF, Melamed MR, Old LJ. Mouse monoclonal antibody defines a specific region of the human proximal tubule and major subsets of renal cell carcinomas. Submitted.
65. Solter D, Knowles BB. Monoclonal antibody defining a stage-specific mouse embryonic antigen (SSEA-1). Proc Natl Acad Sci USA 1978;75:5565-5569.
66. Chiou R-K, Vessella RL, Elson MK, Clayman RV, Gonzalez-Campoy JM, Klicka MJ, Shafer RB, Lange PH. Localization of human renal cell carci-noma xenografts with a tumor-preferential monoclonal anti-body. Cancer Res 1985;45:6140-6146.
67. Lange PH, Vessella RL, Chiou R-K, Elson MK, Moon TD, Palme D, Shafer RK. Monoclonal antibodies in human renal cell carcinoma and their use in radioimmune localization and therapy of tumor xenograft. Surgery 1985;98:143-150.
68. Platt JL, LeBien TW, Michael AF. Stages of renal ontogenesis identified by monoclonal antibodies reactive with lymphohemopoietic differentiation antigens. J Exp Med 1983;157:155-172.
69. Schaerfe T, Becht E, Kaltwasser R, Thuroff J, Jacobi G, Hohenfellner R. Tumor-specific monoclonal antibodies for renal cell carcinoma. Eur Urol 1985;11:117-120.
70. Borowitz M, Weiss M, Bossen E, Metzgar R. Characterization of renal neoplasms with monoclonal antibodies to leukocyte differentiation antigens. Cancer 1986;57:251-256.
71. Kinouchi T, Nakayama E, Ueda R, Ishiguro S, Uenaka A, Oda H, Kotake T. Characterization of a kidney antigen defined by a mouse monoclonal antibody K2.7. J Urol 1987;137:151-154.
72. Tagliabue E, Canevari S, Menard S, Fossati G, Balsari A, Della Torre G, Colnaghi MI. Human renal antigen defined by a murine monoclonal antibody. J Natl Cancer Inst 1984;73:363-369.
73. Luner SJ, Ghose T, Chatterjee S, Cruz HN, Belitsky P. Monoclonal antibodies to kidney and tumor-associated surface antigens of human renal cell carcinoma. Cancer Res 1986;46:5816-5820.

74. Kochevar J. A renal cell carcinoma neoplastic antigen detectable by immunohistochemistry is defined by a murine monoclonal antibody. Cancer 1987;60:2031-2036.
75. Oosterwijk E, Ruiter DJ, Wakka JC, Huiskens VD, Meij JW, Jonas U, Fleuren G-J, Zwartendijk J, Hoedemaeker P, Warnaar SO. Immunohistochemical analysis of monoclonal antibodies to renal antigens. Am J Pathol 1986;123:301-309.
76. Bander NH. Monoclonal antibodies. A.U.A. Update Series. Vol 5, lesson 22.
77. Andy RJ, Finstad CL, Old LJ, Lloyd KO, Kornfeld R. The antigen identified by a mouse monoclonal antibody raised against human renal cancer cells is the adenosine deaminase binding protein. J Biol Chem 1984;259:12844-12849.
78. Bander NH, Welt S, Houghton AN, Lloyd KO, Grando R, Yeh S, Whitmore WF Jr, Vaughan ED Jr, Old LJ. Radionuclide imaging of human renal cancer with labeled monoclonal antibodies. Surg Forum 1984;35:652-655.
79. Chiou R-K, Vessella R, Moon T, Elson MK, Afrman EW, Gonzalez-Campoy JM, Lange PH. Study of monoclonal antibodies in diagnosis and therapy of renal cell carcinoma using the nude mouse model (abstr). J Urol 1985;133:156A.
80. Real FX, Bander NH, Yeh S, Cordon-Cardo C, Lloyd KO, Welt S, Wong G, Old LJ, Oettgen HF. Monoclonal antibody F23: radiolocalization and phase I study in patients with renal cell carcinoma. Proc Am Soc Clin Oncol 1987;6:240.
81. Vessella RL, Chiou R-K, Grund FM, Elson MK, Johnson TK, Nowak JA, Palme DF, Lightner DJ, Mathisen ML, Wessels BW, Olson RE, Rittenhouse HG, Stroup SD, Lange PH. Renal cell carcinoma (RCC) phase I-II trials with 131I-labeled monoclonal antibody A6H: imaging and pharmacokinetic studies. Proc Am Assoc Cancer Res 1987;28:385.
82. Nowell PC, Hungerford DA. Minute chromosome in human chronic granulocytic leukemia. Science 1960;132:1497.
83. Sandberg AA. A chromosomal hypothesis of oncogenesis. Cancer Genet Cytogenet 1983;8:277-285.
84. Cohen AJ, Li FP, Berg S, Marchetto DJ, Tsai S, Jacobs SC, Brown RS. Hereditary renal-cell carcinoma associated with a chromosomal translocation. N Engl J Med 1979;301:592-595.
85. Pathak S, Strong LC, Ferrell RE, Trindade A. Familial renal cell carcinoma with a 3:11 chromosome translocation limited to tumor cells. Science 1982;217:939-941.
86. Yoshida MA, Ohyashiki K, Ochi H, Gibas Z, Prout GR Jr, Pontes EJ, Huben R, Sandberg AA. Rearrangement of chromosome 3 in renal cell carcinoma. Cancer Genet Cytogenet 1986;19:351-354.
87. Carroll PR, Murty VVS, Reuter V, Jhanwar S, Fair WR, Whitmore WF, Chaganti RSK. Abnormalities at chromosome region 3p12-14 characterize clear cell renal carcinoma. Cancer Genet Cytogenet 1987;26:253-259.
88. Pimentel E. Oncogenes. Boca Raton, FL: CRC Press, 1986:121-140.
89. Bishop JM. The molecular genetics of cancer. Science 1987;235:305-311.
90. LeBeau MM, Rowley JD. Heritable fragile sites in cancer. Nature 1984;308:607-608.
91. Gale RP, Cannani E. The molecular biology of chronic myelogenous leukaemia. Br J Haematol 1985;60:395-408.
92. Croce CM, Nowell PC. Molecular basis of human B cell neoplasia. Blood 1985;65:1-7.
93. Teyssier JR, Henry I, Dozier C, Ferre D, Adnet JJ, Pluot M. Recurrent deletion of the short arm of chromosome 3 in human renal cell carcinoma: shift of the c-raf 1 locus. J Natl Cancer Inst 1986;77:1187-1195.
94. Drabkin HA, Bradley C, Hart I, Bleskan J, Li FP, Patterson D. Translocation of c-myc in the hereditary renal cell carcinoma associated with a t(3;8) (p14.2;q24.13) chromosomal translocation. Proc Natl Acad Sci USA 1985;82:6980-6984.

95. Real FX, Rettig WJ, Garin Chesa P, Melamed MR, Old LJ, Mendelsohn J. Expression of epidermal growth factor receptor in human cultured cells and tissues: relationship to cell lineage and stage of differentiation. Cancer Res 1986;46:4726-4731.
96. Zbar B, Brauch H, Talmadge C, Linehan M. Loss of alleles of loci on the short arm of chromosome 3 in renal cell carcinoma. Nature 1987;327:721-724.
97. Kovacs G, Erlandsson R, Boldog F, Ingvarsson S, Muller-Brechlin R, Klein G, Sumegi J. Consistent chromosome 3p deletion and loss of heterozygosity in renal cell carcinoma. Proc Natl Aca Sci USA 1988;85:1571-1575.

31
Recent Diagnostic Modalities for Bladder Cancer

JOEL SHEINFELD and CARLOS CORDON-CARDO

Memorial Sloan-Kettering Cancer Center, New York, New York

NEIL H. BANDER

The New York Hospital-Cornell University Medical Center and Memorial Sloan-Kettering Cancer Center, New York, New York

Approximately 45,000 new cases of bladder cancer will be diagnosed in the United States in 1988. Most (75%) of these will present as superficial disease confined to the mucosa or the submucosa (T1). Approximately 50-80% of these tumors will recur and 10-30% will show grade and stage progression [1]. Identifying those patients whose tumors will progress is critical since most of the 12,000 annual deaths resulting from bladder cancer are in patients with muscle wall invasion and metastatic disease [1].

The remaining 25% of bladder tumor patients present initially with muscle invasive disease. Although they form a high-risk group with aggressive disease, selecting appropriate therapy requires the ability to predict and detect micrometastatic disease and sensitivity to multidrug chemotherapy.

The optimal management of human bladder cancer requires early detection of the tumor and an accurate assessment of its biological potential. Although advances in radiology, ultrasonography, endoscopy, cytology, and cytometry have improved our ability to diagnose and stage bladder cancer accurately, many pitfalls still exist. The multicentricity and tumor heterogeneity of bladder cancer further complicate the issue.

Histopathological and radiographic staging and cytohistological grading have been the cornerstone of bladder cancer management to date. This chapter explores recent efforts to identify tumor markers that can potentially complement existing modalities and thus improve our understanding, diagnosis, and treatment of bladder cancer.

I. TUMOR-ASSOCIATED ANTIGENS

A number of specific monoclonal antibodies to bladder cancer related antigens have been produced [2-10]. In addition to providing fundamental insights into tumor biology, they have enormous clinical potential for diagnosis as well as treatment. The immunophenotyping of bladder cancer could potentially identify markers of tumor progression, invasion, and metastasis, thus allowing greater precision in predicting the course of individual patients and improved tailoring of therapy in each case. Detection of highly restricted antigens in exfoliated cells obtained from either voided or barbotage specimens could en-

Table 1 Monoclonal Antibodies to Bladder Cancer Related Antigens

| mAb (Ig Class) | Antigen | Urothelium |||| Outside Urothelium | Reference |
		Normal	Low-Grade Cancer	High-Grade Cancer	Cancer In Situ		
Om5 (γ_1)	N.D.[a]	47/81[b]	32/37	4/24	21/24		4
T43 (γ_1)	gp85	0/22	4/19	18/19	3/8	Proximal tubules, stratified epithelium (basal layer); lymphocytes	4
T138 (μ)	gp25	0/22	3/19	15/19	2/8	Vascular and lymphatic endothelium	4
T23 (μ)	N.D.	0/22	0/19	6/19	0/8	Astrocytes, melanocytes, fibroblasts, cartilage, smooth muscle, mature T cells	4
G4 (μ)	gp80	0/16	0/4	8/8	3/3	16 normal and 14 malignant tissues[c]	3
E7 (μ)	N.D.	0/16	0/4	8/8	3/3	16 normal and 14 malignant tissues[c]	3

Bladder Cancer

SK4H-12 (γ_{28})	p100	0/3		7/8	? (Limited panel)	Koho et al.
P7A5-4 (γ_1)	p92/23/17	0/3		7/8	Endothelium, BPH, prostate cancer	Koho et al.
3G2-C6 (γ_1)	p90	0/4	—	+	N.D.	10
HBA4 (μ)	gp130/78/40	0/3		3/5	N.D.	7
HBE10 (μ)	neutral glycolipid	0/3		3/5	N.D.	7
9A7 (?)	N.D.	0/4	4/4	3/3	N.D.	8
2E1 (?)	N.D.	0/3	2/2	2/2	N.D.	8
2A6 (?)	N.D.	0/3	2/2	1/1	N.D.	8
Mano 4/4 (γ_1)	p28	0/6	18/19	18/19	Epidermis, kidney, stomach, tonsil	11
486 P3/12 (μ)	p200	0/6[d]	17/19	17/19	Kidney, stomach, tonsil, lymph node, breast, endometrium, granulocytes	11

[a] N.D., not defined.
[b] Number positive/number tested.
[c] Tissues not specified.
[d] Occasional umbrella cells.

hance our ability to detect malignant and premalignant cells. Finally, these highly restricted antigens could be targeted for diagnostic imaging with radiolabeled antibodies or treatment using antibodies conjugated with chemotherapeutic drugs.

Table 1 lists a limited number of these antibodies. A number of these antigens are expressed by most transitional cell carcinomas but not normal urethelium. Fradet et al. [4] have identified 11 cell-surface antigens of human bladder cancer. There appears to be a dichotomy of antigen expression when one compares superifical and muscle-infiltrating bladder tumors. OM-5 is a highly restricted differentiation antigen detected in the normal urothelium of 50-60% of persons. OM-5 is also expressed in 88% of patients with superficial bladder tumors, but in only 20% of those with invasive or metastatic tumors [5]. Paired biopsies of normal urethelium and bladder tumors from the same individual have shown that OM-5 expression in superficial bladder tumors or lack of expression in invasive tumors is independent of OM-5 expression in the patient's normal urothelium [5]. Antigens M344 and 19A211 are also preferentially expressed in superficial bladder tumors rather than invasive tumors [5].

Conversely, antigens T43 and T138 are not expressed by normal urothelium or most superficial tumors, but are detected in a high proportion of patients with invasive and metastatic bladder tumors [5]. Sixteen of 19 invasive or metastatic lesions were Om-5-/T43+ while 20 of 27 superficial tumors were Om-5+/T43-. Do the seven superficial bladder tumors that expressed the more ominous T43 antigen represent the subgroup of superficial cancers with the biological potential to progress?

The gene locus encoding T43 has recently been assigned to chromosome 11. The expression of T138 in normal tissues is limited to vascular and lymphatic endothelium, thus raising the possibility that the T138 antigen gp25 may represent a molecule involved in a cancer cell's ability to invade blood or lymphatic vessels.

Arndt et al. produced two monoclonal antibodies, Mano 4/4 and 486 P 3/12, that identify bladder tumor-associated glycoprotein antigens of 28 kD and 200 kD, respectively [11]. Mano 4/4 reacted with 17 of 20 transitional cell carcinomas (TCC) while 486 P 3/12 reacted with 17 of 19 [11]. Huland et al. [12] performed immunocytological assays on exfoliated cells from bladder barbotage specimens using 486 P 3/12. By detecting 486-P-3/12-positive cells, they noted an improved sensitivity (90% vs. 43%) in bladder tumor detection compared to standard cytological studies.

Chopin et al. [3] isolated two IgM monoclonal antibodies, E7 and G4, which react preferentially with high-grade TCC. Using E7, Neuwirth et al. [13] developed a urine assay and found that urine from 24 control patients was nonreactive while 22 of 24 patients with TCC had detectable levels of the E7 antigen.

II. BLOOD-GROUP-RELATED ANTIGENS

Blood-group-related antigens are carbohydrate structures bound to membrane lipids or proteins of erythrocytes and certain epithelial tissues, including urothelium. The Lewis antigens are structures that are genetically and biochemically related to the ABO blood group antigens. The ABO and Lewis antigenic profile of red cells and urothelial cells is the result of complex interaction between three structural gene loci: ABO, Lele and Hh [14]. A fourth gene, the secretor gene (Sese), regulates the 2-fucosyltransferase encoded by the H gene [15]. The sequential addition of saccharide residues by gene-specific glycosyltransferases to precursor structures is shown in Figure 1. The 4-fucosyltransferase synthesized by the Le gene converts precursor type I chain to Le^a by addition of

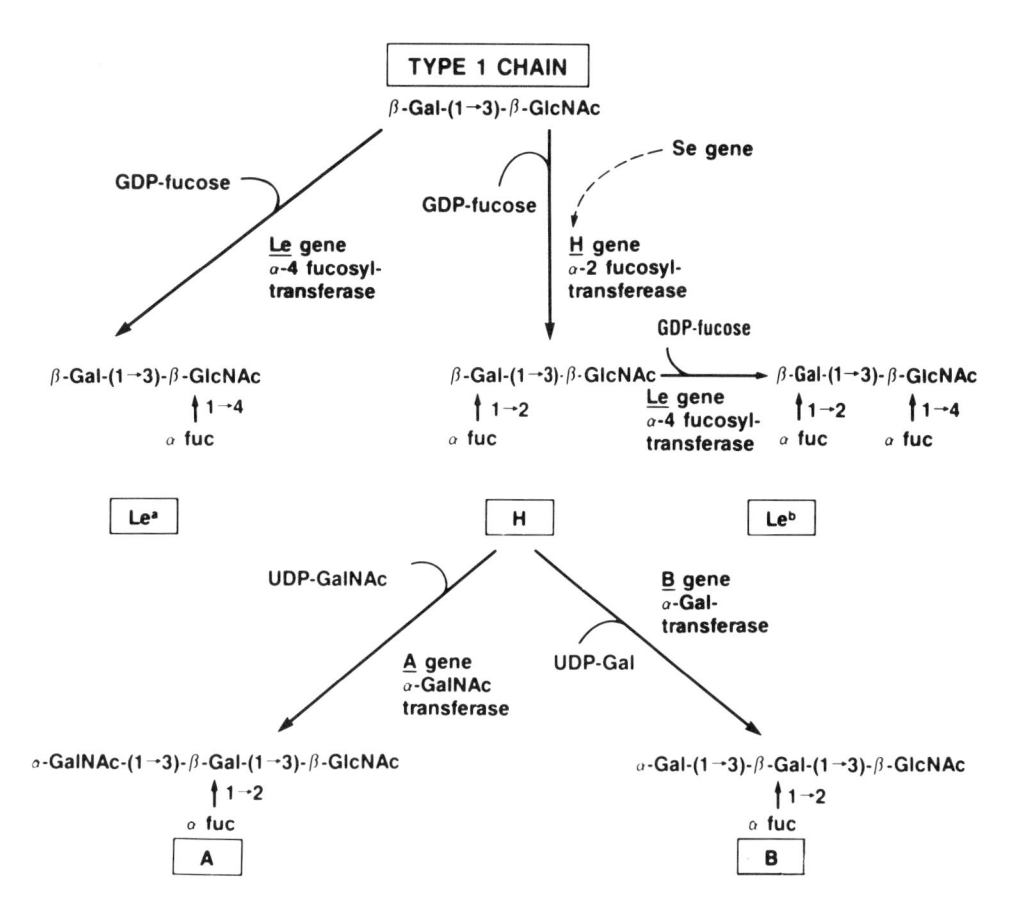

Figure 1 Genetic and biosynthetic pathways of type 1 and type 2 blood-group-related antigens.

a fucose residue. However, in the presence of the H gene and its regulatory Se gene, precursor type I chain is preferentially converted to H substance by the addition of a fucose via the H-controlled 2-fucosyltransferase. H substance can then be converted to Le^b by addition of a second fucose by the 4-fucosyltransferase controlled by the Le gene, or to A and B determinants in the presence of the appropriate genes and their specific glycosyltransferases (N-acetylgalactosamine transferase and D-galactose transferase, respectively).

The 3-fucosyltransferase synthesized by the X gene converts type 2 precursor chains to Le^x. However, in the presence of the H and Se genes, precursor type 2 chain is preferentially converted to type 2 H substance by the addition of fucose. Type 2 H substance can then be converted to A and B determinants by the addition of N-acetylgalactosamine and D-galactose in the presence of the A and B genes and their respective transferases. Type 2 H substance can also be converted to Le^y by the addition of a second fucose via the 3-fucosyltransferase controlled by the X gene [16].

The expression and modulation of blood group related antigens in normal, malignant, and premalignant urothelium have been the focus of considerable interest and have resulted in an extensive literature on the subject [17-23]. As early as 1961, Kay and

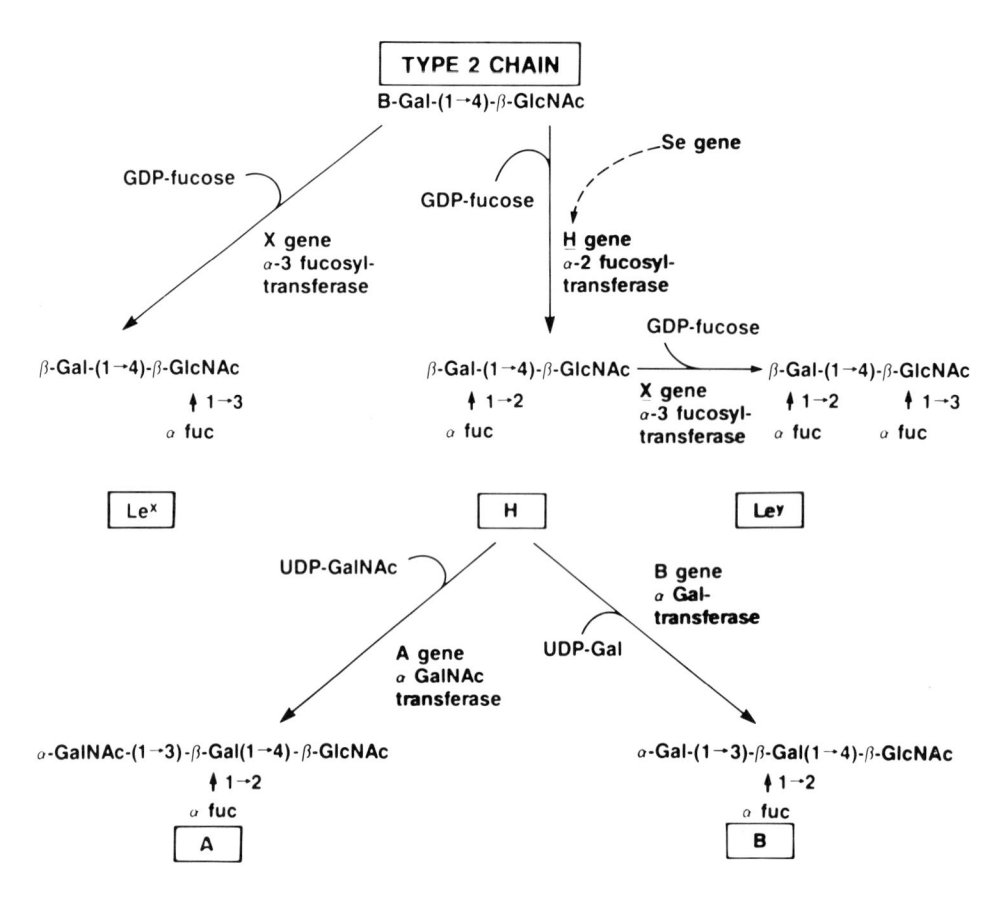

Figure 1 (continued)

Wallace [22] suggested that the absence of A and B blood group determinants correlated with an decreased degree of histological differentiation in TCC. However, the correlation was inconsistent since blood group antigens were present in 20 of 38 invasive high-grade tumors tested. This inconsistency has been attributed to the methodology used, since the study was done with agglutination of cell suspensions that may have contained patient erythrocytes or adjacent normal urothelium [22]. In 1975, DeCenzo et al. studied the expression of ABH antigens in 22 patients with stage A TCC of the bladder using a mixed cell agglutination test on paraffin-embedded formalin-fixed tissue sections. They noted that the absence of ABH antigen expression on the original tumor correlated with the development of invasive cancer. None of the 13 patients with ABH antigens on the original tumors had invasive tumors during a 5-14 year follow-up, while 8 of 9 patients without ABH antigens experienced an invasive recurrence [17]. This landmark observation was corroborated by other investigators (Table 2). Newman et al. [24] studied 322 patients with superficial bladder tumor for ABH deletion and found that 88% (71 of 80) of patients whose disease subsequently progressed to invasion showed ABH deletion, while 127 of 146 patients (87%) who had no recurrence of superficial tumor for at least 5 years retained ABH antigens. Of 96 patients with only superficial recurrences,

Table 2 Incidence of Invasive Recurrence Related to ABH(0) Status of Superficial Tumor

Reference	Number	ABH-Positive (%)	ABH-Negative (%)
17	22	0/13 (0)	8/9 (89)
24	322	9/223 (4)	71/99 (72)
18	40	4/25 (16)	11/15 (73)
20	37	2/16 (13)	16/21 (76)
21	60	5/26 (19)	21/34 (62)
25	23	1/9 (11)	13/14 (93)
23	34	0/22 (0)	14/26 (54)
19	30	0/15 (0)	9/15 (60)

90% retained their respective blood group antigens [24]. Using the red cell adherence (RCA) test on 37 patients with superficial bladder cancer, Lange et al. [20] found that 16 of 21 (76%) patients with RCA-negative tumors had a subsequent invasive tumor, while only 2 of 16 (13%) who were RCA positive showed disease progression [20]. Johnson and Lamm [19] studied 30 patients with superficial TCC of the bladder. With a minimum follow-up of 5 years, they found that 9 of 15 patients with a negative mixed cell agglutination test experienced invasive disease while none of 15 with a positive test did.

Das et al. [26] found that ABH antigen loss was the most significant predictor of tumor recurrence and that dysplastic urothelium with ABH deletion had a greater number of tumor recurrences than dysplastic urothelium with retained antigens. Weinstein and associates [27] found that in 50% of cystectomy specimens from patients with flat in situ carcinoma, histologically normal-appearing urothelium failed to have detectable levels of ABH antigens. This suggested that antigenic changes may precede overt histological changes in transformed urothelium. It is most likely that the absence of ABH antigens results from suppression or inactivation of normally active glycosyltransferases [28].

Recently, Cordon-Cardo et al. [15] have shown that an individual's genetically controlled secretor status has a profound impact on the antigenic profile of blood-group-related substances in normal urothelium. The normal urothelium of "nonsecretors" does not express A, B, or H determinants due to their inability to fucosylate type 1 and 2 precursor chains of type 1 and 2 H substances [15]. Therefore, deletion of ABH antigens can only be ascertained in "secretors." Although nonsecretors only make up approximately 20% of the population, this observation mandates a reexamination of earlier studies and conclusions and may explain some inconsistencies and false-negative results in previous reports.

Immunohistochemical analysis has demonstrated that Le^x is not detected in normal adult urothelium except for occasional umbrella cells [15]. However, papillomas, carcinoma in situ, and TCC expressed this blood-group-related antigen in 30 of 33 (91%) cases, regardless of grade, stage, blood type, or secretor status of the individual [15,28,29]. Sheinfeld et al. found that the presence of Le^x-positive cells (exclusive of umbrella cells) in bladder barbotage specimens identified bladder tumors in 76 of 89 (85%) cases compared to a 62% sensitivity for cytology alone [30]. The combination of either a positive Le^x level and/or a positive cytological examination yielded a sensitivity of 93% while the specificity in the study of 40 controls was 85% [30].

Although the H, Le^b, and Le^y antigens are present in the normal and malignant urothelium of secretor individuals, they are not detected in normal urothelium of nonsecretors [15]. Again, this is presumably due to the inability to fucosylate type 1 and 2 precursor chains to type 1 and 2 H substances, which are the necessary intermediates in the synthesis of Le^b and Le^y (Fig. 1). However, transformed urothelium in nonsecretors demonstrates neoexpression of H, Le^b, and Le^y [15].

Precursor type 1 chain has not been detected in normal urothelium or in low-grade or low-stage bladder tumors; however, it tends to accumulate in invasive TCC [29] as normally active glycosyltransferases are downregulated. Conversely, the neoexpression of Le^x in tumors of all individuals and H, Le^b, and Le^y in tumors of nonsecretors can be accounted for by the upregulation of normally inactive fucosyltransferases [15].

The T antigen is thought to represent a precursor of the human MN blood goup system [31,32]. Based on findings from lectin immunohistochemical studies, the T antigen is only detected in normal urothelium after neuraminidase digestion to cleave a terminal sialic acid residue, that is, cryptic T antigen-positive (C_rT+). TCC have usually been found to be T-antigen-positive (T+) without neuraminidase pretreatment or cryptic T-antigen-negative (C_rT-). Furthermore, patients with low-stage bladder tumors whose tumor cells were T+ or C_rT- had a higher rate of subsequent invasive recurrence [31,32].

III. CYTOGENETICS, ONCOGENES, AND GROWTH FACTORS

One of the earliest proponents of the genetic basis of neoplasia was Boveri [33], a zoologist, who in 1929 proposed the "chromosomal hypothesis of neoplasia." He suggested that chromosomal alterations in structure and number can result in neoplasia. While the development of karyotyping and improved banding techniques have identified chromosomal markers in many tumors, advances in molecular biology resulted in the identification, characterization, and mapping of oncogenes [34]. Recent evidence indicates that proto-oncogenes can be mutated or that specific chromosomal breakpoints and rearrangements can result in regulatory changes in oncogenes and thereby lead to altered cellular physiology, which in turn can result in neoplastic transformation [34].

TCC is a good example of the clinical implications of tumor karyotyping [34,35]. Several investigators have shown that the presence of marker chromosomes in superficial bladder tumors correlates with subsequent invasion and prognosis [35-37]. Summers and associates performed cytogenic analyses on 65 patients with superficial (Ta/T1) TCC and found that only 10% (2 of 20) patients without marker chromosomes experienced recurrences while 90% (18 of 20) with marker chromosomes did. Furthermore, of the 25 patients with stage T1 TCC, all had chromosome markers. Twenty (80%) experienced recurrences and 12 (50%) died from tumor progression [37].

In a separate study, Summers et al. analyzed the tumor karyotype of 39 patients with Ta/T1 superficial TCC who were evaluated for 3-11 years. Twenty-eight of 39 patients had chromosome markers. Twenty-four of 28 patients with chromosome markers had recurrences, as did all 13 patients who developed invasive tumors [36].

Several specific chromosomal abnormalities have been identified. Trisomy 7 has been observed in TCC of the bladder and ureter [38]. The implications of extra copies of chromosome 7 are interesting, since the c-erb-B oncogene that codes for the intracellular portion of the epidermal growth factor receptor (EGFr) is located at 7p12-13 [34,35, 39]. Neal and associates [40] were unable to detect EGFr in normal urothelium but found that poorly differentiated bladder cancers showed EGFr immunoreactivity more commonly than moderately differentiated tumors (85.7% vs. 30%). They also noted the presence of EGFr in a higher proportion of invasive tumors (87.5%) than noninvasive tumors (29.2%) [40]. Berger et al. [41] also showed a correlation between tumor grade and stage with EGFr expression but noted that in bladder tumors EGF receptors are not abnormal in size, autophosphorylation activity, or gene structure.

Messing et al. [42] noted minimal EGFr reactivity in normal superficial urothelial cells, but immunostaining of all 21 TCC studied. They observed that the staining intensity appeared to be inversely proportional to tumor differentiation. Messing et al. [42] recently identified EGF receptors in the basal layers of normal human urothelium and in superficial cells of premalignant and malignant urothelium. Furthermore, EGF, which is found in very high concentrations in urine, was noted in the rat model to induce in vivo activity of ornithine decarboxylase and DNA synthesis, as shown by thymidine incorporation in the basal epithelial cell layer [42].

Other chromosomal abnormalities noted in bladder cancer are monosomy 9, isochrome 5p (i5p), 3p duplication and rearrangement (3p14-3pter), and 11p deletion [35,38,39, 43]. One study reported tumor progression in all five patients with 11p deletion; three of these patients died of their disease. It appears that specific chromosome changes may herald a more ominous prognosis in bladder cancer, particularly 11p deletion, 3p duplication, and trisomy 7 [38].

Deletion of genetic information may result in oncogene activation in one of three ways. First, an inhibitory coding sequence upstream from an oncogene may be lost or, second, the deletion may place the oncogene next to a promoter sequence. Either case results in overproduction of the oncogene product. A third mechanism may be deletion of a tumor-suppressor gene (antioncogene) [34]. Fearson et al. found deletion of 11p in 5 of 12 patients with TCC and suggested that the 11p deletion appeared to confer a selective growth advantage on tumor cells [43]. Further experimental evidence suggesting the presence of antioncogenes on chromosome 11 comes from cytogenetic analysis of somatic cell hybrids. Loss of chromosome 11 is associated with neoplasia, which can then be reversed by reintroduction of a normal copy of chromosome 11 into the cell in a Wilms' tumor [44].

Also mapped to chromosome 11 is the transforming Harvey ras oncogene (c-Ha-ras), which has been associated with a variety of solid tumors [35]. It is present in the T24 bladder carcinoma cell line and was the first oncogene isolated by molecular cloning techniques [45]. Although the normal cellular gene does not have transforming activity, point mutations at codon 12 or 61 result in a single amino acid substitution (valine for glycine) in the protein product, p21. Monoclonal antibody assays have demonstrated increased p21 immunoreactivity in carcinoma in situ and high-grade tumors compared to normal urothelium and low-grade tumors [46]. Brosman and Liu [45] have recently identified a ras-related oncogene product (p55) in the urine of patients with TCC. The expression of p55 correlated with the stage and grade of the tumor [45] and also found within the cytoplasm of exfoliated cells obtained by bladder barbotage from patients with TCC.

The galactosyltransferases involved in the synthesis of ABH blood groups have been mapped to the 9q34 region of chromosome 9 and monosomy 9 may explain ABH antigen deletion in patients with TCC [35]. The oncogene c-fms, which codes for a macrophage-activating factor called CSF-1, is located in the 5q34 region of chromosome 5 [35] and could theoretically be activated by i5p changes; however, its relationship to transformed urothelium is unclear at this time.

IV. P-GLYCOPROTEIN

The P-glycoprotein is the 170,000 dalton protein product of the multidrug resistance (MDR 1) gene. Overproduction of this protein due to amplification and/or overexpression of the MDR 1 gene is thought to result in resistance to a number of drugs including vinblastine and Adriamycin [47], both of which are used in the treatment of invasive and/or metastatic bladder cancer [48]. Recent studies, using monoclonal antibodies to the P-glycoprotein with human cell lines [49] and renal cell carcinomas [50], indicate that the P-glycoprotein may play a critical role in multidrug resistance of these cancers. Current investigations in our laboratory indicate that normal urolthelium does not express the P-glycoprotein but that a number of TCC of the bladder do. Identification of patients whose tumors contain a population of tumor cells expressing the P-glycoprotein may become crucial in selecting a treatment modality since response to M-VAC chemotherapy may be suboptimal in these patients. Furthermore, the role of intravesical instillation of Adriamycin may need to be reassessed, since it may contribute to the development or selection of cells characterized by multidrug resistance.

V. INTERMEDIATE FILAMENTS

Intermediate filaments are a group of differentiation-related cytoskeletal proteins, so called because they are intermediate in diameter between microtubules and microfilaments. They are thought to be involved in cell motility, maintenance of cellular architecture, and regulation of cellular events [51].

Cytokeratins are the intermediate filaments of epithelial cells.. To date, 19 different cytokeratin structures have been identified. They can be classified as type 1 or acidic and type II or basic [51]. The keratin distribution in normal urothelium is highly specific and tumors tend to retain the keratin makeup of their cell of origin [52].

Investigators have shown the presence of cytokeratins 7, 8, 18, and 19 in normal urothelium; while others have noted cytokeratin 18 only in umbrella cells [52]. Furthermore, immunohistochemical studies on TCC have shown that low-grade (grades 0 and 1) tumors express cytokeratin 18 (RGE 53) only in superficial cells while higher-grade tumors show immunoreactivity throughout all layers [52].

Using AE-1, a high-molecular-weight cytokeratin, Huffman et al. [53] showed strong affinity for normal and malignant bladder epithelial cells by flow cytometry on both exfoliated cells and deparaffinized tissue section, and were thus able to exclude inflammatory cells from the DNA analysis.

VI. SUMMARY

Recent advances in hybridoma technology, immunohistochemistry, cytopathology, cytogenetics, and molecular biology have not only furthered our understanding of the biology of bladder cancer but are also beginning to find clinical applications in the early diagnosis of the disease. It is likely that further study and refinement of these tools could reliably identify patients with superficial bladder tumors who are at increased risk for subsequent muscle infiltrative disease, or patients whose tumors are not amenable to current frontline chemotherapy.

ACKNOWLEDGMENT

Dr. Sheinfeld is supported in part by a fellowship as an AUA/Bard Scholar and in part by NCI Grant CA47538 (CCC).

REFERENCES

1. Catalona WJ, Dresner SM, Haaff EO. Management and superficial bladder cancer. In: Skinner DG, Lieskovsky G, eds. Diagnosis and management of genitourinary cancer. Philadelphia: WB Saunders, 1988:281-284.
2. Baricordi OR, Sensi A, De Vinci C, Melchiorri L, Fabris G, Marchetti E, Corrado F, Mattiuz PL, Pizza G. A monoclonal antibody to human transition-cell carcinoma of the bladder cross-reacting with a differentiation antigen of neutrophilic lineage. Int J Cancer 1985;35:781.
3. Chopin DK, de Kernion JB, Rosenthal DL, Fahey JL. Monoclonal antibodies against transitional cell carcinoma for detection of malignant urothelial cells in bladder washing. J Urol 1985;134:260.
4. Fradet Y, Cordon-Cardo C, Thompson T, Daly M, Whitmore WF Jr, Lloyd KO, Melamde MR, Old LJ. Cell surface antigens of human bladder cancer defined by mouse monoclonal antibodies. Proc Natl Acad Sci 1984;81:224.
5. Fradet Y, Islam N, Boucher L, Parent-Vaugeois C, Tardiff M. Polymorphic expression of a human superficial bladder tumor antigen defined by mouse monoclonal antibodies. Proc Natl Acad Sci USA 1987;84:7227.
6. Grossman HB. Hybridoma antibodies reactive with human bladder carcinoma cell surface antigens. J Urol 1983;130:610.
7. Masuko T, Yagita H, Hashimoto Y. Monoclonal antibodies against cell surface antigens present on human urinary bladder cancer cells. J Natl Cancer Inst 1984; 72:523.
8. Messing EM, Bubbers JE, Whitmore KE, de Kernion JB, Nestor MS, Fahey JL. Murine hybridoma antibodies against human transitional carcinoma-associated antigens. J Urol 1984;132:167.
9. Trejdosiewicz LK, Southgate J, Donald JA, Masters JRW, Hepburn PJ, Hodges GM. Monoclonal antibodies to human urothelial cell lines and hybrids: production and characterization. J Urol 1985;133:533.
10. Young DA, Prout GR Jr, Lin CW. Production and characterization of mouse monoclonal antibodies to human bladder tumor-associated antigens. Cancer Res 1985;45: 4439.
11. Arndt R, Durkopf H, Huland H, Donn F, Loening T, Kalthoff H. Monoclonal antibodies for characterization of the heterogeneity of normal and malignant transitional cells. J Urol 1987;137:758.
12. Huland H, Arndt R, Loening T, Steffens M. Monoclonal antibody 486 P 3/12: a valuable bladder carcinoma marker for immunocytology. J Urol 1987;137:654.
13. Neuwirth H, Zhu L, Stock LM, Liu BCS, Fahey JL. Glycoprotein as a tumor marker in urine of patients with bladder cancer. Surg Forum 1986;37:657.
14. Juhl BR, Hartzen SH, Hainau B. Lewis antigen in transitional cell tumors of the urinary bladder. Cancer 1986;58:222.
15. Cordon-Cardo C, Lloyd KO, Finstad CL, McGroarty ME, Reuter VE, Bander NH, Old LJ, Melamed MR. Immunoanatomic distribution of blood group antigens in the human urinary tract: influence of secretor status. Lab Invest 1986;55:444.

16. Lloyd KO. Blood group antigens as markers for normal differentiation and malignant change in human tissues. Am J Clin Pathol 1987;87:129.
17. DeCenzo JM, Howard P, Irish CE. Antigenic deletion and prognosis of patients with stage A transitional cell bladder carcinoma. J Urol 1975;114:874.
18. D'Elia FL, Cooper HS, Mulholland G. ABH isoantigens in stage O papillary transitional cell carcinoma of the bladder: correlation with biological behavior. J Urol 1982;127:665.
19. Johnson JD, Lamm DL. Prediction of bladder tumor invasion with the mixed cell agglutination test. J Urol 1980;123:25.
20. Lange PH, Limas C, Fraley EE. Tissue blood-group antigens and low stage transitional cell carcinoma of the bladder. J Urol 1978;119:52.
21. Limas C, Lange P, Fraley EE, Vessella RL. A, B, H antigens in transitional cell tumors of the urinary bladder. Correlation with the clinical course. Cancer 1979; 44:2099.
22. Kay HF, Wallace DM. A and B antigens arising from urinary epithelium. J Natl Cancer Inst 1961;26:1349.
23. Wiley EL, Mendelsohn G, Droller MJ, Eggleston JC. Immunoperoxidase detection of carcinoembryonic antigen and blood group substances in papillary transitional cell carcinoma of the bladder. J Urol 1982;128:276.
24. Newman AJ, Carlton CE Jr, Johnson S. Cell surface A, B or O (H) blood group antigens as an indictor of malignant potential in stage A bladder carcinoma. J Urol 1980; 124:27.
25. Young AK, Hammond E, Middleton AW Jr. The prognostic value of cell surface antigens in low grade, non-invasive, transitional cell carcinoma of the bladder. J Urol 1979;122:462.
26. Das G, Buxton NJC, Stewart PAH, Glashan RW. Prognostic significance of ABH antigenicity of mucosal biopsies in superficial bladder cancer. J Urol 1986;136:1194.
27. Weinstein RS, Alroy J, Farrow GM, Miller AW III, Davidson I. Blood group isoantigen deletion in carcinoma in situ of the urinary bladder. Cancer 1979;43:661.
28. Cordon-Cardo C, Reuter VE, Lloyd KO, Sheinfeld J, Fair WR, Old LJ, Melamed MR. Blood group-related antigens in human urothelium: enhanced expression of precursor, Le-x and Le-y determinants in urothelial carcinoma. Cancer Res 1988;in press.
29. Cordon-Cardo C, Reuter VE, Hermanssen D, Fair WR, Melamed MR. Urinary bladder papilloma: heterogeneity by immunohistochemistry and flow cytometry analyses. Proc Am Assoc Cancer Res 1988;29:389 (abst 1550).
30. Sheinfeld J, Cordon-Cardo C, Reuter UE, Herr H, Whitmore WF Jr, Sogani PC, Morse M, Melamed MR, Fair WR. Enhanced bladder cancer detection utilizing the Le-x antigen as a marker of neoplastic transformation. 1988;submitted for publication.
31. Coon JS, Weinstein RS, Summers JL. Blood group precursor T-antigen expression in human urinary bladder carcinoma. Am J Clin Pathol 1982;77:692.
32. Limas C, Lange P. T-antigen in normal and neoplastic urothelium. Cancer 1986; 58:1236.
33. Boveri T. The origin of malignant tumors. London: Bailliere, 1929.
34. Klein EA, Fair WR, Chaganti RSK. Molecular and cytogenetic events in urologic tumors. Semin Urol 1988;6:2.
35. Sandberg A. Chromosome changes in bladder cancer: clinical and other correlations. Cancer Genet Cytogenet 1986;19:163.
36. Summers JL, Coon JS, Ward RM, Falor WH, Miller AW, Weinstein RS. Prognosis in carcinoma of the urinary bladder based upon tissue blood group ABH and Thomsen-Freidenreich antigen status and karyotype of the initial tumor. Cancer Res 1983; 43:934.

37. Summers JL, Falor WH, Ward R. A 10-year analysis of chromosomes in non-invasive papillary carcinoma of the bladder. J Urol 1981;125:177.
38. Babu VR, Lutz MD, Miles BJ, Farah RN, Weiss L, Van Dyke DL. Tumor behavior in transitional cell carcinoma of the bladder in relation to chromosomal markers and histopathology. Cancer Res 1987;47:6800.
39. Gibas Z, Prout GR Jr, Connolly JG, Pontes JE, Sandberg A. Nonrandom chromosomal changes in transitional cell carcinoma of the bladder. Cancer Res 1984;44:1257.
40. Neal DE, Bennett MK, Hall RR, Marsh C, Abel PD, Sainsbury JRC, Harris AL. Epidermal-growth-factor receptors in human bladder cancer: comparison of invasive and superficial tumors. Lancet 1985;1:366.
41. Berger MS, Greenfield C, Gullick WJ, Haley J, Downward J, Neal DE, Harris AL, Waterfield MD. Evaluation of epidermal growth factor receptors in bladder tumors. Br J Cancer 1987;56:533.
42. Messing EM, Hanson P, Ulrich P, Erturk E. Epidermal growth factor interactions with normal and malignant urothelium: in vivo and in situ studies. J Urol 1987;138:1329.
43. Fearson ER, Feinberg AP, Hamilton SH, Vogelstein B. Loss of genes on the short arm of chromosome 11 in bladder cancer. Nature 1985;318:377.
44. Weissman BE, Saxon PJ, Pasquale SR, Jones GR, Geiser AG, Stanbridge EJ. Introduction of a normal human chromosome 11 into a Wilms' tumor cell line controls its tumorigenic expression. Science 1987;236:175.
45. Brosman SA, Liu B. Oncogenes: their role in neoplasia. Urology 1987;30:1.
46. Viola MV, Froniowitz F, Oravez S. Ras oncogene P21 expression is increased in premalignant lesions and high grade bladder carcinoma. J Exp Med 1985;161:1213.
47. Kartner N, Riordan JR, Ling V. Cell surface P-glycoprotein associated with multidrug-resistance in mammalian cell lines. Science 1983;221:1285.
48. Scher HI, Sternberg CN. Chemotherapy of urologic malignancies. Semin Urol 1985;3:239.
49. Kartner N, Evernden-Porelle D, Bradley G, Ling V. Detection of P-glycoprotein in multidrug-resistant cell lines by monoclonal antibodies. Nature 1985;316:820.
50. Kakehi Y, Kanamaru H, Yoshida O, Ohkubo H, Nakanishi S, Gottesman MM, Pastan I. Measurement of multidrug-resistance messenger RNA in urogenital cancers; elevated expression in renal cell carcinoma is associated with intrinsic drug resistance. J Urol 1988;139:863.
51. Cordon-Cardo C, Finstad CL, Bander NH, Old LJ, Melamed MR. Immunoanatomic distribution of cytostructural and tissue-associated antigen in the human urinary tract. Am J Pathol 1987;126:269.
52. Feitz WFJ, Debruyne FMJ, Vooijs GP, Herman CJ, Ramaekers FCS. Intermediate filament proteins as tissue specific markers in normal and malignant urological tissues. J Urol 1986;136:922.
53. Huffman JL, Garin-Chesa P, Gay H, Whitmore WF Jr, Melamed MR. Flow cytometric identification of human bladder cells using a cytokeratin monoclonal antibody. Ann NY Acad Sci 1985;X:302.

IV
Broad or Non-Organ Site-Restricted Tumor Markers

32
Immunoglobulins in Diagnosis and Monitoring of Neoplasia

DANIEL E. BERGSAGEL

Ontario Cancer Institute and Princess Margaret Hospital, Toronto, Ontario, Canada

WALDEMAR PRUZANSKI

University of Toronto and Wellesley Hospital, Toronto, Ontario, Canada

I. INTRODUCTION

Homogeneous immunoglobulins (monoclonal [M] proteins) are produced by most plasma cell neoplasms, and by some of the other tumors arising from cells of the B-lymphocyte lineage, such as chronic lymphocytic leukemia and diffuse lymphomas. In addition, M-proteins have been detected in a variety of patients with other diseases, and even in 0.9% of asymptomatic adults over the age of 25 years in a screening study[1]. In this chapter we will review the diseases associated with M-proteins, the laboratory evaluation of patients with immunoglobulin-producing tumors, the use of M-proteins as an index of the response to therapy, and prognostic factors relating to M-proteins.

II. DISEASES ASSOCIATED WITH M-PROTEINS

The apperance of a homogeneous M-protein in a serum electrophoresis pattern provides evidence that a monoclonal population of cells producing an immunoglobulin with one class of heavy chain, and a single type of light chain, has reached a size of at least 5×10^9 cells. A classification of the diseases associated with the appearance of an M-protein is shown in Table 1.

Good estimates of the incidence of the various diseases in which M-proteins have been reported are not available because most of the reported series are based on serum samples obtained from sick patients, so that patients with plasma cell neoplasms are overrepresented. The application of serum electrophoresis as a screening test revealed an M-protein in 64 of 6,995 asymptomatic blood donors over the age of 25, an incidence of 0.9%. This sample represents about 70% of the adult population of four parishes in the district of Varmland, Sweden. The initial investigation demonstrated plasma cell myeloma in one and chronic lymphocytic leukemia in another [1]. An evaluation of the causes of death and the status of the surviving patients 11 years later showed that the original myeloma patient had died of the disease, the patient thought to have chronic lymphocytic leukemia had progressed and died of plasma cell myeloma, whereas another had

Table 1 Classification of Diseases Associated with an M-Protein

Controlled proliferation of an M-protein-producing clone
    Benign monoclonal gammopathy
    Chronic cold agglutinin syndrome
    Transient M-proteins
    Some patients with alpha heavy-chain disease

Uncontrolled proliferation (neoplasia) of an M-protein-producing clone
    Plasma cell myeloma
    Extramedullary plasmacytomas
    Plasma cell neoplasms associated with dermatological lesions
        Papular mucinosis
        Pyoderma gangrenosum
    Waldenström's macroglobulinemia
    Immunoglobulin-related amyloidosis
    Heavy-chain diseases: gamma, alpha, mu
    Chronic lymphocytic leukemia
    Diffuse lymphocytic lymphomas

died of a malignant lymphoma [2]. Thus, in this series of 64 asymptomatic patients with M-proteins evaluted for 11 years, a B-lymphocyte neoplasm was diagnosed in only 3 patients, and the remaining 61 are thought to have benign monoclonal gammopathy (BMG). The incidence of BMG increased with age to 2% in the eighth decade and 5.7% in the ninth decade. Based on the results of this Swedish screening study, the *prevalence* of BMG in asymptomatic adults is 61:6995 (0.872%), or 872 per 100,000 persons over the age of 25.

Most subjects with symptomatic disease did not have blood drawn for electrophoresis in the Swedish screening study. However, inquiry at local hospitals and interviews with doctors who knew the population revealed two additional patients with plasma cell myeloma [1]. Thus, at least 3 patients with myeloma were identified in 4 parishes with an adult population of about 10,000, which suggests that the *prevalence* of myeloma is about 30 per 100,000 persons over the age of 25.

These figures represent the numbers of patients found to have BMG or plasma cell myeloma at a certain time. Because patients with BMG survive for prolonged periods, whereas patients with plasma cell myeloma have a median survival of about 30 months, the numbers of patients with BMG tend to accumulate and greatly exceed the numbers of surviving myeloma patients at any point.

Investigators at the Mayo Clinic [3] have evaluated 241 BMG patients for 10 years or more. During this period 19% of the patients have progressed to develop myeloma, macroglobulinemia, amyloidosis, lymphoma, or chronic lymphocytic leukemia, and an additional 15% have shown a greater than 50% increase in the serum monoclonal protein, or the apperance of urine monoclonal protein. Thus, although BMG is associated with a good prognosis, about 2% of patients per year progress to develop a more aggressive form of the disease. BMG is now considered to be an early, stable stage in the development of plasma cell myeloma.

In the chronic cold agglutinin syndrome, the cold agglutinin is usually found to be a monoclonal macroglobulin, which may regress or remain constant for many years. Most cold agglutinins are of IgM class with kappa light chains [4]; however IgM/λ, IgA, and IgG cold agglutinins have been described [5-7]. Cold agglutinins have antibody activity against the Ii antigen system, and agglutinate the erythrocytes at temperature below that of the human body. Cold agglutinins have been shown to be cytotoxic to human B and T lymphocytes, neutrophils, and monocytes [8,9].

In addition, some patients with alpha heavy chain disease may have a "controlled proliferation," because the M-protein may disappear spontaneously or following treatment with antibiotics alone [10].

Uncontrolled (neoplastic) proliferation is distinguished by the fact that the clone continues to expand, resulting in an exponential increase in the serum M-protein concentration, the formation of tumors composed of lymphoplasmacytic cells, the appearance of lytic skeletal lesions, and other manifestations of the underlying disease.

The diagnosis associated with 3759 M-proteins detected in sera and urine submitted to an immunodiagnostic laboratory supervised by one of us (WP) is shown in Table 2. Plasma cell myeloma and macroglobulinemia accounted for more than half of these samples. Patients with other disorders (including those with BMG) formed the second largest group.

For the patients with neoplasms derived from B lymphocytes (plasma cell myeloma, macroglobulinemia, chronic lymphocytic leukemia, and diffuse lymphomas), there is good evidence that the M-proteins are produced by neoplastic plasma cells or plasmacytoid lymphocytes. Several morphological types of cells, each producing the same M-protein, or carrying the identical M-protein on the cell membrane, are found in patients with plasma cell myeloma and macroglobulinemia. The plasma cell and plasmacytoid lymphocytes that produce the M-protein predominate, but the monoclondal population of B lymphocytes that contain no M-protein in the cytoplasm but do synthesize an identical membrane-bound M-protein, can be demonstrated in the peripheral blood [11-13]. The M-protein decreases 50-90% in patients with plasma cell neoplasms and malignant lymphomas who respond to antineoplastic therapy [14].

Neoplasms derived from cells other than the B-lymphocyte lineage rarely have an associated M-protein. The M-protein is probably not produced by the neoplastic cell in most of these cases. Immunofluorescent studies of fresh tumor tissue from five patients

Table 2 Diagnosis in 3759 Patients with M-Proteins

Diagnosis	Patients Number	%
Plasma cell myeloma	1986	52.8
Macroglobulinemia	177	4.7
Primary amyloidosis	74	2.0
Lymphomas	194	5.2
Leukemias	87	2.3
Cancers	249	6.6
Collagen diseases	159	4.2
Other benign diseases (including BMG)	833	22.2
Total	3759	100.0

Source: From data collected by WP in an immunodiagnostic laboratory.

Table 3 Frequency of M-Proteins in Asymptomatic Adults and Patients with Various Diseases

Diagnosis	M-Proteins/Number of Sera Tested	%
Asymptomatic adults > 25 years [1]	64/6995	0.9
Cancer [17]	32/5043	0.6
Hodgkin's disease [14]	1/218	0.5
Nodular lymphomas [14]	4/292	1.4
Diffuse lymphomas and chronic lymphocytic leukemia [14]	45/641	7.0
Plasma cell myeloma [23]	253/255	99.2

with various types of cancer and serum M-protein showed that plasma cells surrounding and infiltrating the tumor were producing the same M-protein as was found in the serum; the cancer cells did not contain any immunoglobulin [15]. Since the frequency of M-proteins associated with cancer is no greater than the incidence in a group of blood donors (Table 3), it is also possible that this may represent the chance occurrence of a BMG in cancer patients. M-proteins have been reported most frequently in association with cancers of the gastrointestinal tract, breast, and lung [16]. It is not possible to state whether the incidence of M-proteins is greater in association with these cancers, because most other cancers, (e.g., skin cancer) have not been surveyed adequately by electrophoresis and immunoelectrophoresis. During a period of 4-43 months, the majority of M-proteins associated with cancer remain constant [17]. Regression of the M-protein following surgical removal of the cancer is exceedingly rare [15].

Although most neoplasms derived from cells other than the B-lymphocyte lineage do not appear to produce an M-protein, there is some evidence the M-proteins found in one child with acute leukemia [18] and another patient with chromic myelogenous leukemia [19] were produced by the leukemic cells.

The frequency of M-proteins in patients with various diseases is compared with that reported for asymptomatic adults over the age of 25 years in Table 3. M-proteins were not detected more commonly in patients with cancer, Hodgkin's disease, or nodular lymphomas than in asymptomatic adults, but have been found in 7% of lymphoma patients with a diffuse pattern or chronic lymphocytic leukemia. Virtually all plasma cell tumors produce an M-protein but some are unable to secrete the immunoglobulin [20], so that an M-protein is not found in the serum or urine of about 1% of these patients.

The frequency of immunoglobulin classes of M-protein in patients with plasma cell tumors and other disorders is shown in Table 4. It is of interest to compare the proportion of plasma cell tumors that produce various classes of M-proteins with the synthesis of each immunoglobulin by healthy individuals. Measurements of the synthesis of each immunoglobulin class by healthy individuals, expressing the result as the percentage of the total immunoglobulin produced per day, indicate that 52% is IgG, 37% IgA, 10% IgM, and 0.6% IgD [21]. It will be noted that there is a remarkable correlation between the frequency of plasma cell tumors producing each class of immunoglobulin and the proportion of normal immunoglobulin synthesized per day. This correlation is not exact, however, since 15% of the plasma cell tumors secrete only light chains. Some of these

Table 4 Classes of M-Proteins Associated with Various Disorders

Diagnosis	Number of Patients with M-Protein	IgG	IgA	IgM	IgD	IgE	Light-Chain Disease	Two or Three M-Peaks	Nonsecretor
Plasma cell neoplasms (myeloma and macroglobulinemia)	2163	55.9	20.1	8.6	1.3	0.05	12.5	1.2	0.4
Amyloidosis	74	29.7	18.9	5.4	1.4	—	44.6	—	—
Lymphomas and leukemias	281	33.8	6.8	49.1	0.4	0.7	—	—	—
Cancer	249	67.1	14.1	15.3	—	—	2.0	1.6	—
Collagen-vascular diseases	159	55.3	9.4	34.0	—	—	—	1.3	—
Benign monoclonal gammopathy	833	73.0	12.2	14.2	—	—	—	0.6	—
Total	3759								

Source: From data collected by WP in an immunodiagnostic laboratory.

tumor cells fail to synthesize heavy chains, produce abnormal heavy chains that cannot combine with light chains, and thus are not secreted, or have some other defect in the assembly of light and heavy chains. Despite the uncertainty about the class of heavy chain that neoplastic cells secreting only light chains may or may not produce the correlation between the production and frequency of the immunoglobulin class of myeloma is still reasonable. If the proportion of immunoglobulin class produced per day can be assumed to reflect the proportion of plasma cells making this immunoglobulin, this observation would support the view that plasma cells producing each class of immunoglobulin are equally susceptible to malignant transformation.

The pattern is distinctly different in patients with diffuse lymphomas and chronic lymphocytic leukemia: a higher proportion of these patients produce IgM and fewer produce IgG [17]. These findings are similar to those reported by Alexanian [14], who found that 14 of these patients produced IgG, 1 patient made IgA, 29 made IgM, and the immunoglobulin class was not identified in 5 patients. These observations are compatible with the view that the affected cell in patients with diffuse lymphomas and chronic lymphocytic leukemia is more primitive than the neoplastic cell in plasma cell neoplasms [22].

The pattern observed in patients with cancer and the connective tissue disease does not differ significantly from that of plasma cell tumors. In the group of patients with other benign conditions, however, the proportion producing IgG is significantly increased compared to the plasma cell neoplasms. This pattern conforms almost exactly with that reported by Kyle and Greipp [3] for 241 patients with BMG at the Mayo Clinic.

III. LABORATORY EVALUATION OF IMMUNOGLOBULIN-PRODUCING TUMORS

A. Serum Electrophoresis

Separation of proteins by electrical charge has been used for a long time as a screening procedure for the analysis of immunoglobulin abnormalities. The sensitivity of this technique depends on several variables such as the medium in which the proteins are separated, the buffer, the current, and the dye used for staining the membranes to mention only a few. When conventional cellulose acetate membrane is used, electropherogram will detect most serum M-proteins. The great majority of IgG, IgA, and IgM M-proteins will appear as narrow bands of gamma, beta, or rarely faster anodal mobility. In an early stage of the disease, or in remission, M-proteins may produce faint bands or these bands may not be visible. Because of their low concentration, IgD M-proteins and homogenous light chains may not be visible even in advanced disease. However, some modified techniques, such as plate isoelectric focusing [24], or use of nigrosin for staining the membrane, increase the rate of detection of discrete bands.

The process of measuring the proportion of M-protein in an electropherogram and calculating the concentration of M-protein as a percentage of the total serum protein has been used routinely in the majority of laboratories. The advantage of this procedure is its simplicity; however, several disadvantages limit its accuracy. Since the calculation involves two steps (i.e., estimation of the total serum protein and the electropherogram), the possibility of technical error is greater. Likewise, it is not always possible to differentiate accurately between the M-protein and the neighboring normal proteins. This is especially true in the case of IgA M-proteins, which are often broad-based, and in the case of small spikes such as IgD or free light chains, which are located in the area of normal gamma globulins. An additional error is introduced by the fact that even in the case of narrow well-defined spikes, there is always some proportion of normal immunoglobulins of identical electrophoretic mobility in the area occupied by the spike. Nevertheless, the fact that the electrophoretic techniques are readily available, easy to use, reproducible, and easy to interpret make this procedure one of the most popular tools in clinical immunology.

B. Immunoelectrophoresis

Immunoelectrophoresis (IEP) is used for a qualitative assessment of immunoglobulin abnormalities. As a rule IEP is used for identifying the class and type of M-protein(s), as well as for the analysis of the subclass and subtypes, detection of various parts of Ig molecules, and others. IEP may reveal the presence of M-proteins, even when they are not visible in the electropherograms.

The accuracy of IEP depends greatly on the availability of high-quality antisera. The concentration of antibody and the "purity" of antisera (i.e., the detection of specific antigens only) are crucial for the correct interpretation of IEP. Immunoelectrophoretic diagnosis depends also on the number of antisera that will be used in individual cases. For example, the use of anti-free kappa and anti-free lambda antisera significantly improved the detection and thus increased the incidence of free light chains in the serum, in addition to complete M-proteins [16]. Use of anti-Fab antisera may help in the detection of heavy chain disease protein. Since the precipitating properties of antibody-antigen complexes depend, among other factors, on the concentration of these components, appropriate dilution of sera with M-proteins may allow much easier identification of heavy and/or light chains in IEP. In a significant proportion of sera with IgM M-proteins, the light chains are not easily identifiable in IEP. Often in such cases concentration of IgM by the euglobulin precipitation technique will allow easy assessment of the light-chain type. In the case of unidentifiable light chains of IgA M-proteins, reduction and alkylation of the serum may allow the identification.

C. Quantitative Immunoglobulin Assays

Measurement of individual immunoglobulins is usually done with immunoquantitation plates, which are simply a supporting medium, usually agar or agarose, into which an appropriate antiserum is evenly incorporated. The fluid sample in which Ig is to be measured diffuses through the medium and a ring of precipitation is formed. The diameter of the precipitate ring is proportional to the logarithm of the antigen concentration in the sample [25]. In another method [26], the area of precipitation is proportionate to the antigen concentration. The accuracy of these techniques is far from perfect. The quality of antiserum incorporated into the medium (amount of Ab in milligrams per milliliter) influences precipitation and, thus, the visibility of precipitation rings. Time of incubation and the temperature play a role in the development of the precipitation ring in some but not all immunoplates. Other variables important for the accuracy of immunoquantitation are the quality of serum, the concentration, and some characteristics of the immunoglobulin to be measured. Immunoquantitation plates have limits of accuracy for both the lowest and the highest concentrations of immunoglobulins at which semilogarithmic linearity of the readings is preserved. Since the lowest concentration is usually 2-3 mg/100 ml (using ultra-low-level plates), this limitation is usually of no clinical significance. More important, however, is the upper limit of accuracy of immunoplates, which is about 4 g/dl for IgG and 2 g/dl for IgA. Above these values, accurate measurement cannot be obtained with immunoplates. In such cases many laboratories dilute the sera; however, the dilution per se introduces an additional error. Purified immunoglobulin in a buffer quantitated on immunoplates gives a different reading from that obtained when an identical quantity of the same immunoglobulin is dissolved in whole serum.

Immunoplates may give false high and false low readings. The former may be encountered when immunoglobulin is present in the serum in two or more forms of different molecular weights; for example, when 19S IgM and low-molecular-weight IgM such as 11S, 7S, or μ-heavy chain disease (HCD) protein are present. Likewise, the presence of 7S IgG together with lower-molecular-weight gamma-related globulins such as incomplete molecules, half-molecules, Fc fragment, or γ-HCD protein produces large precipitation rings and falsely high readings. The reason for this is faster diffusion of these low-molecular-weight molecules through the medium. When low-molecular-weight globulins are present in the serum along with "normal"-molecular-weight immunoglobulins, double precipitation rings are often observed on immunoplates. A dense precipitation ring is located centrally and a faint precipitation ring is observed peripherally. Such double precipitation rings should always raise the possibility of low-molecular-weight immunoglobulin-related fragments in the serum.

Low-molecular-weight proteins occasionally produce very weak precipitation reactions, and the ring is almost invisible. In such cases immersion of immunoplates into 7% acetic acid for 1 min will render the precipitation ring visible immediately.

False low readings may be observed when the serum is hyperviscous or when polymeric forms of immunoglobulin are present in the serum. There is some evidence that large amounts of immune complexes (IgM/IgG, IgG/IgG) may slow down diffusion of the involved immunoglobulins as well.

Another measurement technique used in some laboratories is the one-dimensional single electroimmunodiffusion, or rocket electrophoresis [27]. In this method, appropriate antiserum is incorporated into the medium, an antigen is applied into small wells, and then electrophoresis is performed. This technique detects small amounts of antigen.

Antisera to identify four subclasses of IgG and two subclasses of IgA have recently become available. Two kinds of antibodies—polyclonal and monoclonal—have been produced. Polyclonal antibodies have to be absorbed to render them specific and may be used for immunoprecipitation assays. Monoclonal antibodies may have restricted specificity and are occasionally poorly precipitating [28]. The usual method for quantification of subclasses is radial immunodiffusion. Enzyme-linked immunosorbent assays (ELISA), rocket immunoelectrophoresis, radioimmunoassays, and nephelometry have also been developed. Monoclonal Ig subclasses may also be detected by combined technique of isoelectric focusing followed by immunofixation [29].

D. Urine Proteins and Other Fluids

Examination of the urinary proteins may give an important clue to the presence of Bence Jones proteins (BJP) and whole M-proteins. It should be stressed that neither the heat-coagulation test nor dipsticks should be used for detection of Bence Jones proteinuria. The former may give both false-negative and false-positive results [30], while the latter does not detect BJP [31]. Urine investigation should include estimation of the total urinary protein excreted per 24 hr, subsequent concentration of urine, and then electrophoresis and immunoelectrophoresis of the concentrate. The degree of concentration depends on the total protein value. When the total protein is below 200 mg/24 hr, concentration of 500-fold or more is often necessary to detect small amounts of BJP. Urinary electropherogram easily shows abnormal bands corresponding to BJP or to other M-proteins. Final identification is usually made by IEP, by applying antisera against BJP of kappa and lambda types. The type of Bence Jones protein in the urine usually corresponds to the type of light chain of the serum M-protein. In rare cases, two Bence Jones proteins of different types have been found in the urine, whereas serum has contained either a single M-protein [32] or two M-proteins with the same type of light chain [33]. Exceedingly rare examples have been reported in which the light chains of serum M-proteins and urinary Bence Jones protein were different antigenic type [34].

Not infrequently whole M-proteins are found in urine as well as in serum. This information is important since it signifies increased glomerular permeability of the kidneys. In addition, two or three bands (spikes) of proteins related to BJP may be observed on urine electrophoresis. In such cases there are either monomers and polymers of BJP and/or there may be protelytic split of BJP into variable and constant parts with different electrophoretic mobilities. The electrophoretic pattern of urinary proteins may be helpful in the assessment of renal function in malignant diseases. Five different electrophoretic patterns can be identified. The preglomerular pattern occurs when there is an excess of a certain protein in the blood, which, by virtue of its low molecular weight, is filtered through the glomerular membrane and appears in the urine in quantities sufficient for visualization on electropherogram. The classic example of this is the Bence Jones protein that is spilled over to the urine after saturation of the tubular reabsorption mechanism. When the blood concentration of this protein is low and the tubular reabsorption mechanism is preserved, the amount of Bence Jones protein in the unconcentrated urine will be very small or undetectable. Appropriate concentration of the urine is therefore required. The second pattern is the glomerular pattern, which classically appears in the nephrotic syndrome. In such a case the glomerular membrane is damaged and a nonselective spilling over occurs. The electropherogram of the urine resembles that of the serum with albumin as a predominant (>25%) constituent of the total urinary protein. Examples of this pattern in malignancy are renal amyloidosis and immune complex glomerulitis, which may complicate lymphoma. The third, a tubular pattern, occurs when there is damage to the proximal tubule and inability to reabsorb proteins filtered through the glomerulus. Since the glomerulus filters proteins in certain relation to their molecular weight, the proteins that appear in the urine are mainly of low molecular weight. Electrophoresis usually shows low albumin (<25%) and high $alpha_{1-2}$-beta areas in which the majority of low-molecular-weight proteins are located. This pattern may appear in monocytic or myelomonocytic leukemia when damage to the proximal tubule occurs. Mixed patterns (glomerular-tubular) may occur in any of these conditions and in pyelonephritis. The last, postrenal, pattern usually appears in inflammatory conditions of the lower urinary tract.

Electrophoresis of concentrated urine may show several bands of cathodal postgamma mobility that have significant diagnostic importance, especially in leukemias. Although these bands represent nonimmunoglobulin, low-molecular-weight proteins, they are mentioned here since they present as spikes that resemble BJP. The most important in this group is lysozyme (LZM), an enzyme produced mainly by monocytes [35], and present in neutrophils as well. This enzyme is physiologically present in the urine in minute quantities, not exceeding 2 μg/ml. In monocytic and myelomonocytic leukemia, and occasionally in chronic myelogenous leukemia, the level of LZM in the plasma increases markedly and when the tubular mechanism of reabsorption becomes saturated, spilling over to the urine occurs. The excretion of LZM in the urine may reach values as high as 1347 μg/min [36]. When such urine is concentrated, LZM is visualized as a band of postgamma mobility. Bands of LZM may also be pronounced in abnormalities of tubular reabsorption mechanism such as the adult Fanconi's syndrome, in myeloma, cadmium poisoning, and others.

Another postgamma protein was described by Tishendorf et al. [37] in chronic myelogenous leukemia. This so-called cationic leukemia antigen (CLA) is excreted by cells of the neutrophil series, and appears in the urine as a postgamma protein of slightly less cathodic mobility than lysozyme.

M-proteins and BJP can be found in other fluids such as cerebrospinal fluid, saliva, synovial fluid, pleural, pericardial or peritoneal effusions, and others. In such instances M-proteins may diffuse into the fluid by exudation or transudation, or less commonly may be produced in situ by malignant immunoglobulin-producing cells. M components in the fluids mentioned above are usually identical to those detected in the serum and/or urine of the patient. In rare instances they are discordant [38]. In such cases M-proteins are most probably synthesized by a different clone of Ig-producing cells.

E. Tissue Biopsies

Lymphoplasmacytic tissue infiltrates are common in patients with plasma cell and other B-lymphocyte neoplasms. In addition, amorphous, acellular tissue deposits of amyloid are frequent, and rare instances of immunoglobulin deposits have been described. It is not within the scope of this chapter to discuss amyloidosis; the interested reader is referred to a few pertinent publications [39,40]. However, it should be stressed that immunochemical and physicochemical characteristics of amyloid have recently been characterized to a great extent; these show a relationship in the structure of amyloid to that of immunoglobulins, especially to the light chains in myeloma-related and "primary" amyloidosis [41], to calcitonin in medullary carcinoma of thyroid-related amyloid [42], or to γ-HCD protein in γ-HCD disease [5]. Conversion in vitro of glucagon and insulin into amyloid-like substance, detection of amyloid A, serum amyloid A, and amyloid λ-light-chain-related components in some "primary" and "secondary" types of amyloidosis [43-46] show clearly that the pathogenesis of amyloid infiltration is a complex phenomenon.

Much less frequent is the deposition of homogeneous immunoglobulin of class and type identical to the circulating M-protein in various organs and tissues. The mechanism of such deposition is poorly understood; however, in some cases; excessive production of immunoglobulin by plasma cells or lymphocytes and prominent spilling over to the extracellular space was implicated [47]. In others, tissue deposits appeared to rise as a result of blockage of the regional lymph nodes, back pressure with distention of lymphatic vessels, and "transduction" of lymph with M-protein into the extracellular space. Such deposits are usually detected as an incidental histological finding [47]; however, in a few cases they produced distinct clinical manifestations. Diarrhea and malabsorption syndromes have been reported due to IgM [48] or IgG [49] infiltration of the intestinal wall. Skin infiltration with IgG has been observed in lichen myxedematosus [50]. The diagnosis is usually made from immunofluorescent study of the involved organ. Another unusual type of infiltrate in which BJP with a strong tissue affinity was deposited in different tissues has recently been reported [51], and called systemic light-chain deposition disease.

F. Differentiation of Stable and Progressive Plasma Cell Neoplasia

The differentiation of BMG from malignant plasma cell myeloma may be difficult. Several markers have been developed to differentiate these conditions and yet none is absolutely reliable. Such markers as the presence of acid phosphatase or β-glucuronidase in the cells or of J-chain in the molecules of M-protein have only a little value. The presence of CALLA-positive cells [52] and pre-B cells [53] in the peripheral blood cannot differentiate with certainty progressive myeloma from more benign forms. However, the presence of CALLA-positive cells does seem to be associated with a more aggressive course of myeloma [52]. Likewise, high levels of serum deoxythymidine kinase are associated with shorter survival time [54]. More attention has been paid to the assessment of survival time [54]. More attention has been paid to the assessment of the thymidine labeling index (LI). The in vitro [^{3}H] thymidine labeling of bone marrow plasma cells is much higher in malignant myeloma than in BMG or "smoldering" myeloma. High LI (>3.0) is associated with a very short survival time [55]. Another prognostic test is estimation of the level of beta 2-microglobulin (B2m) in the serum [56-59]. When B2m level, which correletes with the myeloma cell turnover rate, exceeds 4.0 mg/L, the survival time is short. With levels > 6.0 mg/L, a low response rate to chemotherapy is observed [59].

In early stages, however, the lower level of B2m cannot differentiate between malignant myeloma and BMG [58]. Polyclonal immunoglobulins are suppressed in only 60-70% of cases [60]. Thus, in spite of the introductin of several markers, careful follow-up of patients with M-proteins remains the most important factor in the assessment of malignancy and its clinical course.

IV. PROGNOSTIC FACTORS AND CLINICAL STAGING

The principal variables that affect the survival of patients with myeloma are those associated with the myeloma cell mass, renal function, the myeloma growth rate, and certain intrinsic properties of myeloma cells in different patients.

Many clinical prognostic factors correlate strongly with the myeloma cell mass, which is estimated by dividing the total amount of M-protein produced in a patients per 24 hr by the amount of M-protein produced per myeloma cell during the same period [61]. By correlating the estimated myeloma cell mass with certain clinical manifestations, Durie and Salmon [62] devised a useful clinical staging system. They found that patients with marked anemia (hemoglobin < 85 g/L), hypercalcemia (serum calcium > 3.0 mmol/L), more than three osteolytic lesions, or high levels of M-proteins (IgG > 70 g/L, IgA > 50 g/L, or light-chain proteinuria > 12 g/24 hr) all had high myeloma cell numbers. Patients with one or more of these features were classified as having stage III disease. Patients with *all* of the following features had a low number of myeloma cells and were defined as having stage I disease: hemoglobin > 100 g/L, normal serum calcium levels, normal bones or a solitary lesion, low M-protein levels (IgG < 50 g/L; IgA < 30 g/L; light-chain priteinuria < 4.0 g/24 hr) and a serum albumin level > 30 g/L. Patients with intermediate values were classified as having stage II disease. These criteria enable clinicians to classify myeloma patients into groups with median survival expectations of 46 months for stage I, 32 months for stage II, and 23 months for stage III [63]. Serum beta 2-microglobulin levels, corrected [64] and uncorrected [65] for renal function, correlate strongly with estimates of myeloma cell mass by the Durie and Salmon clinical staging system.

Renal function is an important prognostic factor that is independent of the myeloma cell mass. Patients who present with an elevated serum creatinine level have a poor prognosis, regardless of whether the myeloma cell mass is low, intermediate, or high. For this reason, each Durie and Salmon stage is subclassified as A for those without and B for those with renal insufficiency.

The growth rate of a plasma cell tumor is a strong predictor of survival. The M-protein doubling time has been used as a measure of growth rate, and has been shown to correlate well with survival [66]. Patients with rapidly growing tumors (short M-protein doubling times) have the worst prognosis. The labeling index of marrow plasma cells provides another assessment of the growth fraction. Myeloma patients with a labeling index of less than 1% live significantly longer than those with values greater than this [67,68]. The pretreatment labeling index is an independent prognostic factor that can identify subgroups of patients with a poor prognosis within each stage. Serum beta 2-microglobulin (B2m) levels are influenced by the production rate (i.e., cell turnover) and the renal clearance of this small protein. In myeloma patients with normal renal function, serum B2m levels reflect largely the cell turnover rate, or growth rate of the myeloma cell population. Uncorrected serum B2m levels in myeloma patients correlate strongly with estimates of myeloma cell mass [65].

Several intrinsic properties of myeloma cells influence the survival of a patient. Perhaps the most important of these is the sensitivity of the myeloma cell to the drugs used for treatment; patients who respond to treatment will achieve median survivals up to four times longer than those whose disease is resistant and progresses. The type of light chain produced by the myeloma cell also influences survival. Patients producing lambda light chains have a significant shorter survival than those producing kappa [63,67,69,70]. The DNA content of myeloma cells before treatment has been shown to be a prognostic factor that is independent of stage [71]. Myeloma patients with aneuploid tumors were more likely to have advanced stage disease than those with diploid tumors. It was interesting that all patients with renal failure had aneuploid tumors. The ability of myeloma cells to form colonies in the culture system described by Takahashi et al. [72] signifies a grave prognosis. Myeloma colonies were detected in cultures of marrow cells from 14 of 71 myeloma patients at various stages of the disease. These colonies were most frequently obtained (10 cases) from patients who had entered the acute phase of the disease. Only 1 of the 14 patients survived beyond 1 year after the detection of myeloma colony formation.

Many staging systems have been devised for myeloma. The Durie Salmon system [62] has been described above. This staging system has difficulty in classifying myeloma patients into stage I and II groups with clearly different survival patterns, A simpler staging system, devised by Bataille et al. [73] groups patients into three stages on the basis of serum B2m and albumin levels. Stage I patients have levels of serum B2m < 6.0 mg/L and albumin > 30 g/l; for stage II levels of B are > 6.0 mg/L, but albumin is > 30 g/L; while stage III patients have serum albumin levels of < 30 g/L.

Diagnosis and Monitoring of Neoplasia

V. EVALUATING THE RESPONSE TO TREATMENT

The amount of M-protein produced by a plasma cell tumor reflects the tumor mass [74]. This makes it possible to evaluate the effectiveness of treatment by evaluating changes in the serum or urinary M-protein. For the majority of patients, a progressive increase or decrease in the serum M-protein concentration and the amount of M-protein excreted in the urine represent the best objective measurements for determining whether the disease is progressing or responding to treatment. Figure 1 illustrates the use of serum M-protein concentration in evaluating the response of a patient with myeloma. In this patient, the serum M-protein level continued to increase despite six courses of treatment with melphalan and prednisone. Therapy was therefore changed to intermittent courses of cyclophosphamide and prednisone. The serum M-protein concentration began to fall slowly and progressively after cyclophosphamide was introduced. With continued treatment, this improvement has been maintained.

We have measured the interval from the end of treatment until the M-protein returns to pretreatment values and divided this by the M-protein doubling time during relapse as an estimate of the myeloma cell-kill achieved by chemotherapy (Fig. 2). Using this method, we estimated the cell-kill achieved in 9 patients with IgG and IgA myeloma, in whom the M-protein was no longer visible on the electropherogram, on 12 occasions

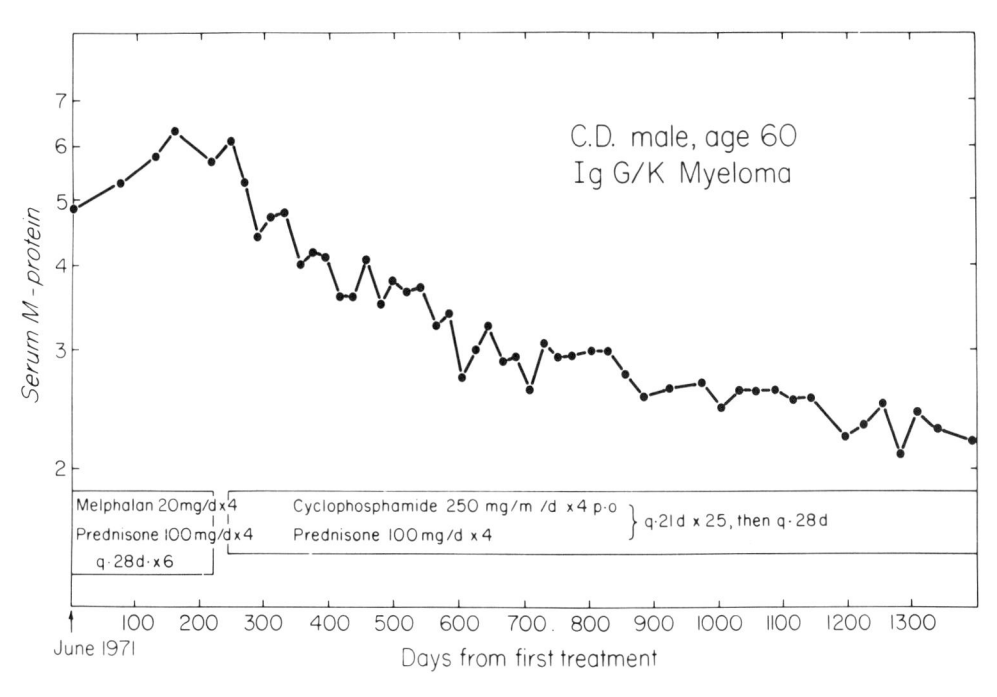

Figure 1 Serum M-protein as a guide to treatment. This patient did not respond to a combination of melphalan and prednisone, for the serum M-protein continued to rise while this treatment was given. The M-protein began to fall immediately after cyclophosphamide was started, and the patient's symptoms improved. The M-protein fell rapidly initially, and then more slowly, until a plateau was reached at about 2.5 g/dl. Treatment has continued, the M-protein remains at this plateau, and the patient remains well (December 1976).

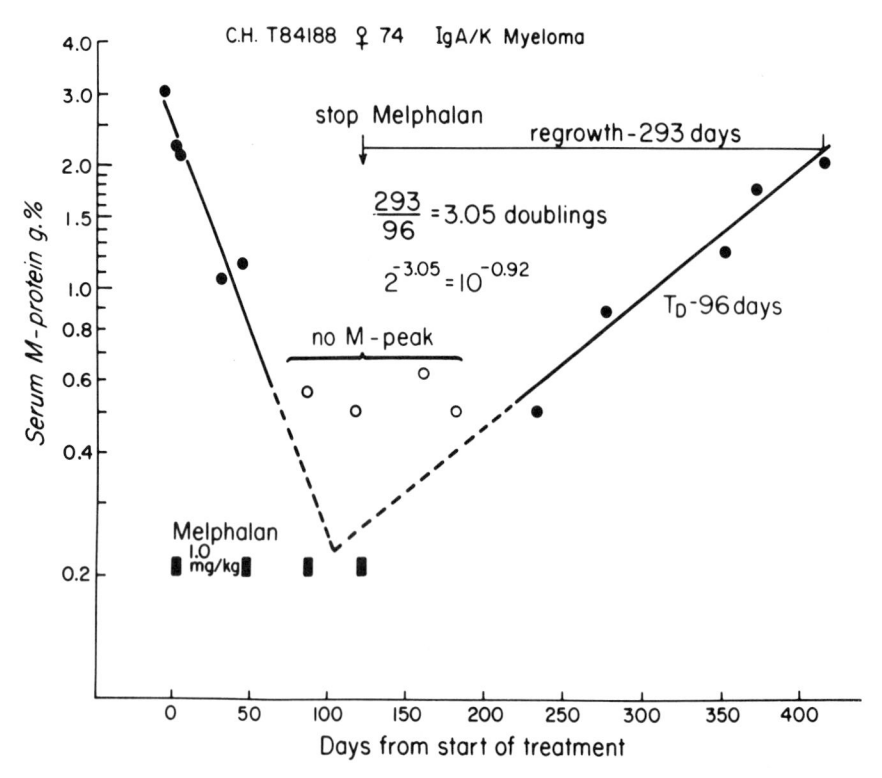

Figure 2 A method for estimating myeloma cell-kill following therapy. Therapy was stopped after the IgA M-protein disappeared following four courses of melphalan therapy. The M-protein reappeared after an interval and increased exponentially, with a doubling time of 96 days. The myeloma cell-kill is estimated by dividing the interval from the discontinuation of melphalan to complete relapse (when the M-protein has returned to the pretreatment value) by the M-protein doubling time.

following treatment. The cell-kill achieved following 9 courses of treatment was very close to 10^{-1} and showed no correlation with the subsequent survival of the patient. In contrast, the M-protein doubling time during relapse correlated strongly with survival. Those patients with short M-protein doubling times, indicating rapid growth of the relapsing myeloma, experienced the shortest survival times [66].

An adequate definition of response is required to guide treatment and assess prognosis. The definition should include patients with actual tumor regression and exclude those who show nonspecific improvement in symptoms or indirect manifestations of the disease. The two major definitions of myeloma response used by cooperative study groups are those developed by the Chronic Leukemia-Myeloma Task force [75] and the Southwest Oncology Group [76]. The major difference between these two sets of criteria is that the serum M-protein concentration must be converted to a "synthetic index" so that the effect of treatment on the production of M-protein (based on changes in the concentration-dependent fractional catabolic rate and blood volume) can be calculated for the Southwest Oncology Group criteria. Table 5 shows the criteria used by the National Cancer Institute of Canada Clinical Trials Group. These criteria were adapted from those of the Chronic Leukemia-Myeloma Task Force [75].

Table 5 NCI-C Response Criteria for Plasma Cell Myeloma

Response

    Serum M protein: a decrease to less than 50% of the baseline pretreatment concentration on two separate measurements taken at least 4 weeks apart.

    Urine M protein: a decrease to less than 10% of the baseline, pretreatment value on two separate measurements taken at least 4 weeks apart. If the baselin urine protein is less than 0.5 g/24 h, or if there is evidence of increasing renal failure, with rising serum creatinine levels, changes in 24 hr urine protein are not regarded as significant.

    "Stable response": the serum and/or urine protein level must fall below the response level described above, and remain stable for at least 4 months, with variations about the mean of less than + 15%.

Progression during initial treatment: any of the following

    Serum M protein: a progressive increase to at least 10 g/L above baseline.

    Urine M protein: a progressive increase in g/24 hr to at least 100% above baseline, with electrophoretic confirmation that the increase is due to light-chain proteinuria.

    The development of hypercalcemia.

    The progressive enlargement of plasmacytomas, or an increase in the number, or extent of lytic skeletal lesions. The collapse of osteoporotic vertebrae is not regarded as disease progression.

No change. Neither the criteria of response nor progression is satsified. Note that this group includes patients who have a minor fall in M protein level that fails to satisfy the criterial for response and also those in whom the M protein level does not change.

All criteria of response place a heavy emphasis on a major fall in the M-protein level. Alexanian and the Southwest Oncology Group [180] have reported that survival correlates with the level to which the M-protein synthesis index can be suppressed. This observation has been questioned by Palmer et al. [77], who found that for patients living for more than 3 months, the difference in the survival of responders and nonresponders is not significant for stage II patients and only modestly improved for stage III responders. Furthermore, in a recent trial of the value of maintenance therapy for myeloma patients [78], the National Cancer Institute of Canada treated 457 evaluable patients with melphalan and prednisone. Of these, 210 patients (46%) achieved a stable response, 47 patients (10%) showed only a minor fall or no change in the M-protein level, and 200 patients (44%) progressed, based on the criteria of response shown in Table 5. The median survival for the three groups were stable response, 48 months; no change, 49 months; progressors, 15 months. The fact that the patients who failed to achieve a response, but did not progress (the No Change group), survived just as well as the responder group clearly indicates that other important features of response are not included in the current definitions of response. One of these factors is the regrowth rate of the myeloma cell mass. As noted previously, the survival of responding myeloma patients does not correlate with the myeloma cell-kill achieved, but is strongly correlated with the M-protein doubling time during relapse [66]. Some measure of the "stability" of the disease, such as light chain isotype suppression [79,80], may help to identify myeloma patients in a stable plateau phase and provide improved guidance regarding the indications for additional treatment. The fact that some patients with myeloma can achieve a stable plateau phase and survive for long periods without a major change in the M-protein level suggests that we should regard only the progressor group as having experienced treatment failure and requiring a change in treatment. We should include the remaining patients in the stable plateau phase (responder group). Further experience in the use of serum B2m in monitoring the response to treatment should lead to improved response criteria.

REFERENCES

1. Axelsson U, Bachmann R, Hallen J. Frequency of pathological proteins (M components) in 6995 sera from an adult population. Acta Med Scand 1966;179:235-247.
2. Axelsson U. An eleven-year follow-up on 64 subjects with M-components. Acta Med Scand 1977;201:173-175.
3. Kyle RA, Greipp PR. Monoclonal gammopathies of undetermined significance. In: Wiernick PH, Canellos GP, Kyle RA, Schiffer CA, eds. Neoplastic diseases of the blood. New York/Edinburgh/London/Melbourne: Churchill Livingston, 1985:653-676.
4. Harboe M, Van Furth R, Schubothe H, Lind K, Evans RS. Exclusive occurrence of κ chains in isolated cold agglutinins. Scand J Haematol 1965;2:259-266.
5. Pruzanski W, Cowan DH, Parr DM. Clinical and immunochemical studies of IgM cold agglutinins with lambda light chains. Clin Immunol Immunopathol 1974;2:234-245.
6. Angevine CD, Andersen BR, Barnett WC. A cold agglutinin of the IgA class. J Immunol 1966;96:578-586.
7. Ambrus M, Bajtai G. A case of an IgG type cold agglutinin disease. Haematologica 1969;3:225-235.
8. Pruzanski W, Farid N, Keystone E, Armstrong M. The influence of homogeneous cold agglutinins on polymorphonuclear and mononuclear phagocytes. Clin Immunol Immunopathol 1978;4:277-285.
9. Pruzanski W, Farid N, Keystone E, Armstrong M, Greaves MF. The influence of homogeneous cold agglutinins on human B and T lymphocytes. Clin Immunol Immunopathol 1975;4:248-257.
10. Seligmann M. Immunochemical, clinical and pathological features of α-heavy chain disease. Arch Intern Med 1975;135:78-82.
11. Heller P, Yakulis V, Phoopalm N, Costea N, Cabana V, Nathan RD. Surface immunoglobulins on circulating lymphocytes in mouse and human plasmacytoma. Trans Assoc Am Physicians 1972;85:192-201.
12. Lindström FD, Hardy WR, Eberle BJ, Williams RC Jr. Multiple myeloma and benign monoclonal gammopathy: differentiation by immunofluorescences of lymphocytes. Ann Intern Med 1973;78:837-844.
13. Preud'homme JL, Seligmann M. In: Schwartz RD, ed. Progress in clinical immunology, vol. 2. New York: Grune & Stratton, 1974:121-174.
14. Alexanian R: Monoclonal gammopathy in lymphoma. Arch Intern Med 1975;135:62-66.
15. Williams RC Jr, Bailly RC, Howe RB. Studies of "benign" serum M-components. Am J Med Sci 1969;257:275-293.
16. Ameis A, Ko HS, Pruzanski W. M-components—A review of 1242 cases. Can Med Assoc J 1976;114:889-895.
17. Migliore PJ, Alexanian R. Monoclonal gammopathy in human neoplasia. Cancer 1968;21:1127-1131.

18. Cĕjka J, Bollinger RO, Schuit HRE, Lusher JM, Chang CH, Zuelzer WW. Macroglobulinemia in a child with acute leukemia. Blood 1974;43:191-199.
19. Shohet SB, Mohler WC. In vitro growth of peripheral blood cells from a patient with chronic myelogenous leukemia. Proc Am Assoc Cancer Res 1963;4:61.
20. Hurez D, Preud'homme JL, Seligmann M. Intracellular "monoclonal" immunoglobulin in non-secretory human myeloma. J Immunol 1970;104:263-264.
21. Waldmann TA, Strober W. In: Kallós P, Waksman BH, eds. Progress in allergy, Vol. 13. Basel: Karger 1969:1-110.
22. Warner NL, Potter M, Metcalf D. Multiple myeloma and related immunoglobulin-producing neoplasms, UICC Technical Report Series, vol. 13. Geneva: International Union Against Cancer, 1974:1-88.
23. Bergsagel DE, Griffith KM, Haut A, Stuckey WJ Jr. In: Haddow A, Weinhouse S, eds. Advances in cancer research, vol. 10. New York: Academic Press, 1967: 311-359.
24. Cornell FA, Plate isoelectric focusing technique for detection of M-components. Personal communication.
25. Fahey JL, McKelvey EM. Quantitative determination of serum immunoglobulins in antibody-agarplates. J Immunol 1965;94:84-90.
26. Mancini G, Carbonara AO, Heremans JF. Immunochemical quantitation of antigens by single radial immunodiffusion. Immunochemistry 1965;2:235-254.
27. Laurell CB. Antigen-antibody crossed electrophoresis. Anal Biochem 1965;10: 358-361.
28. Jefferis R, Reimir CB, Skvaril F. Evaluation of monoclonal antibodies having specificity for human IgG subclasses: results of an I.U.I.S./W.H.O. collaborative study. Immunol Lett 1985;10:223-252.
29. Mason DY, Erber B, Falini H, Stein K, Gatter KC. Immunoenzymatic labeling of haematological samples with monoclonal antibodies. In: Beverley PCL, ed. Monoclonal antibodies. Edinburgh: Churchill Livingstone, 1986:145-181.
30. Bernier GM, Putnam FW. Polymerism, polymorphism, and impurities in Bence Jones proteins. Biochim Biophys Acta 1965;86:295-308.
31. Smith JK. The significance of the "protein error" of indicators in the diagnosis of Bence Jones proteinuria. Acta Haematol 1963;30:144-152.
32. Dammacco F, Trizio D, Bonomo L. A case of IgA K-myelomatosis with two urinary Bence Jones proteins (BJK and BJL) and multiple chromosomal abnormalities. Acta Haematol 1941:309-320.
33. Maldonado JE, Silverstein MN, Williams RD Jr, Harrison EG Jr. Lymphosarcoma with Bence Jones proteinuria (type K) and γG serum M-component (type L): an asynchronous gammopathy. Abstracts of the simultaneous sessions. In: Proceedings of the 12th Congress International Society of Hematology, New York, 1968:43.
34. Engle RL Jr, Nachman RL. Two Bence Jones proteins of different immunological types in the same patient with multiple myeloma. Blood 1966;27:74-77.
35. Gordon S, Todd J, Cohn ZA. In vitro synthesis and secretion of lysozyme by mononuclear phagocytes. J Exp Med 1974;139:1228-1248.
36. Pruzanski W, Wilson DR. Renal handling of endogenous lysozyme in man. J Lab Clin Med 1977;90:61-67.
37. Tischendorf FW, Ledderose G, Wilmanns W, Tischendorf MM, Tischendorf GW. Cationic leukocyte antigen in urine of patients with chronic myelocytic leukemia. Nature 1973;245:379-380.
38. Dotten D, Pruzanski W. Multiple myeloma with discordant M-components in the serum and cerebrospinal fluid. Arch Intern Med 1981;141:1374-1376.
39. Glenner G. Amyloid deposits and amyloidosis. The β-fibrilloses. Parts 1 & 2. N Engl J Med 1980;302:1283-1292.

40. Editorial review. Chemical and clinical classification of amyloidosis in 1985. Scand J Immunol 1986;23:253-265.
41. Terry WD, Page DL, Kimura S, Isobe T, Osserman ER, Glenner GG. Structural identity of Bence Jones and amyloid fibril proteins in a patients with plasma cell dyscrasia and amyloidosis. J Clin Invest 1973;52:1276-1281.
42. Sletten K, Westermark P, Natvig JB. Characterization of amyloid fibril proteins from medullary carcinoma of the thyroid. J Exp Med 1976;143:993-998.
43. Husby G, Natvig JB. A serum component related to nonimmunoglobulin amyloid protein AS, a possible precursor of the fibrils. J Clin Invest 1974;53:1054-1061.
44. Husby G, Natvig JB. Sletten K. New third class of amyloid fibril protein. J Exp Med 1974;139:773-778.
45. Sletton K, Husby G. The complete aminoacid sequence of nonimmunoglobulin amyloid fibril protein AS in juvenile rheumatoid arthritis. Eur J Biochem 1974; 41:117-125.
46. Rosenthal CJ, Franklin EC. Variation with age and disease of an amyloid A protein-related serum component. J Clin Invest 1975;55:746-753.
47. Azar HA, Potter M. Multiple myeloma and related disorders, vol. 1. New York: Harper, 1973;4,8,11,136.
48. Pruzanski W, Warren RE, Goldie JH, Katz A. Malabsorption syndrome with infiltration of the intestinal wall by extracellular monoclonal macroglobulin. Am J Med 1973;54:811-818.
49. Prokipchuk EJ, Pruzanski W. Malabsorption syndrome with IgG(λ)M component: response to chemotherapy. Can Med Assoc J 1976;114:922-924.
50. Danby FW, Danby CWE, Pruzanski W. Papular mucinosis with IgG(κ) M-component. Can Med Assoc J 1976;114:920-922.
51. Randall RE, Williamson WC Jr, Mullinay F, Tung MY, Still WJS. Manifestations of systemic light chain deposition. Am J Med 1976;60:293-299.
52. Wearne AJ, Joshua DE, Brown RD Kronenberg H. Multiple myeloma. The relationship between CALLA (CD 10) positive lymphocytes in the peripheral blood and light chain isotype suppression. Br J Haematol 1987;67:39-44.
53. Pilarski LM, Mant MJ, Ruether BA. Pre-B cells in peripheral blood in multiple myeloma patients. Blood 1985;66:416-422.
54. Simonsson B, Kallander CFR, Brenning G, Killander A, Ahre A, Gronowitz JS. Evaluation of serum deoxythymidine kinase as a marker in multiple myeloma. Br J Haematol 1985;61:215-224.
55. Boccadoro M, Gavarotti P, Fossati G, et al. Low plasma cell ^{3}H thymidine incorporation in monoclonal gammopathy of undetermined significance (MGUS), smoldering myeloma and remission phase myeloma: a reliable indicator of patients not requiring therapy. Br J Haematol 1984;58:689-696.
56. Cuzick J, Cooper EH, MacLennon IC. The prognostic value of serum beta$_2$ microglobulin compared with other presentation features in myelomatosis. Br J Cancer 1985;52:1-6.
57. Bataille R, Durie BGM, Grenier J. Serum beta$_2$ microglobulin and survival duration in multiple myeloma: a simple reliable marker for staging. Br J Haematol 1983; 55:439-447.
58. Van Dobbenburgh OA, Rodenhuis S, Ockhuizen T et al. Serum beta$_2$-microglobulin: a real improvement in the management of multiple myeloma? Br J Haematol 1985;61:611-620.
59. Alexanian R, Barlogie B, Fritsche H. Beta 2 microglobulin in multiple myeloma. Am J Hematol 1985;20:345-351.
60. Pruzanski W, Gidon MS, Roy A. Suppression of polyclonal immunoglobulins in multiple myeloma: relationship to the staging and other manifestations at diagnosis. Clin Immunol Immunopathol 1980;17:280-286.

61. Salmon SE, Smith BA. Immunoglobulin synthesis and total body tumor cell number in IgG myeloma. J Clin Invest 1970;49:1114-1121.
62. Durie BGM, Salmon SE. A clinical staging system for multiple myeloma. Cancer 1975;36:842-854.
63. Bergsagel DE, Bailey AJ, Langley GR, MacDonald RN, White DF, Miller AB. The chemotherapy of plasma-cell myeloma and the incidence of acute leukemia. N Engl J Med 1979;301:743-748.
64. Scarfe JH, Anderson H, Palmer MK, Crowther D. Prognostic significance of pretreatment serum β_2-microglobulin levels in multiple myeloma. Eur J Cancer Clin Oncol 1983;19:1361-1364.
65. Bataille R, Grenier J, Sany J. Beta-2 microglobulin in myeloma: optimal use for staging, prognosis and pretreatment—a prospective study of 160 patients. Blood 1984;63:468-476.
66. Bergsagel DE. Assessment of the response of mouse and human myeloma to chemotherapy and radiotherapy. In: Drewinko B, Humphreys RM, eds. Growth kinetics and biochemical regulation of normal and malignant cells. Baltimore: Williams & Wilkins, 1977:705-717.
67. Durie GBM, Salmon SE, Moon TE. Pretreatment tumor mass cell kinetics and prognosis in multiple myeloma. Blood 1980;55:364-372.
68. Latreille J, Barlogie B, Johnston D, Drewinko B, Alexanian R. Ploidy and proliferative characteristics in monoclonal gammopathies. Blood 1982;59:43-51.
69. Shustik C, Bergsagel DE, Pruzanski W. κ and λ light chain disease. Survival rates and clinical manifestations. Blood 1976;48:41-51.
70. Alexanian R. Monoclonal gammopathy in lymphoma. Arch Intern Med 1975;135:62-66.
71. Bunn PA Jr, Krasnow S, Makuch RW, Schlam ML, Schechter GP. Flow cytometric analysis of DNA content of bone marrow cells in patients with plasma cell myeloma: clinical implications. Blood 1982;59:528-535.
72. Takahashi T, Lim B, Jamal N, et al. Colony growth and self-renewal of plasma cell precursors in multiple myeloma. J Clin Oncol 1985;3:1613-1623.
73. Bataille R, Durie GBM, Grenier J, Sany J. Prognostic factors and staging in multiple myeloma. A reappraisal. J Clin Oncol 1986;4:80-87.
74. Nathans D, Fahey JL, Potter M. The formation of myeloma protein by a mouse plasma cell tumor. J Exp Med 1958;108:121-130.
75. Chronic Leukemia-Myeloma Task Force. Guidelines for protocol studies. II. Plasma cell myeloma. Cancer Chemother Rep 1973;4:145-158.
76. Alexanian R, Bonnet J, Gehan E, et al. Combination chemotherapy for multiple myeloma. Cancer 1972;30:382-389.
77. Palmer M, Belch A, Brox L, Pollock E, Koch M. Are the current criteria for response useful in the management of multiple myeloma? J Clin Oncol 1987;5:1375-1377.
78. Belch A, Shelley W, Bergsagel DE, et al. A randomized trial of maintenance versus no maintenance melphalan and prednisone in responding multiple myeloma patients. Br J Cancer 1988;57:94-99.
79. Wearne A, Joshua DE, Kronenberg H. Light chain isotope associated suppression of surface immunoglobulin expression on peripheral blood lymphocytes in myeloma during plateau phase. Br J Haematol 1984;58:483-489.
80. Joshua DE, Wearne A, Kronenberg H. Indication for therapy in multiple myeloma: Should it be stage or stability. Lancet 1985;2:210.

33
Acute Phase Reactant Proteins

EDWARD H. COOPER

University of Leeds School of Medicine, Leeds, United Kingdom

Tissue injury and infection cause the normal homeostatic mechanisms in vertebrates to alter profoundly. The overall pattern of response is called the acute-phase response (APR). The APR starts shortly after injury and recruits a wide spectrum of cellular, hormonal, and biochemical changes (see [1,2] for review). This response occurs not only in wounding and bacterial infection but also in a wide variety of acute and chronic tissue injuries including cancer. In this Chapter we will focus on the nonspecific changes in plasma protein profiles that accompany the APR and the role of acute-phase reactant proteins as tumor markers in cancer. This topic was last reviewed in 1979 [2]; this chapter will concentrate on the clinical studies made during the past 10 years that reflect the interest in these proteins in contemporary clinical oncology.

The reaction involves the increased synthesis of certain proteins synthesized in the liver and the coincidental decrease in synthesis of other proteins. The proteins that increase synthesis during the APR are called acute-phase proteins (APRPs) and are defined arbitrarily by Kushner [1] as those proteins whose plasma concentrations rise by 25% or more. The general properties of the major APRPs are shown in Table 1. Most of the proteins such as C-reactive protein (CRP), alpha-1 acid glycoprotein (also named orosomucoid), alpha-1 protease inhibitor (also named antitrypsin), ceruloplasmin, and haptoglobin are well known and measured for a variety of reasons in general medicine. Serum amyloid A protein is a relatively new member of the APRPs, and has yet to enter the cadre of proteins for which there are well-established commercially available assays. More detailed consideration is required to bring it into focus.

I. SERUM AMYLOID A

Serum amyloid A (SAA) was first found in serum because it cross-reacted immunologically with antiserum raised against amyloid A protein purified from amyloid deposits [3]. SAA has been shown to have a molecular weight of approximately 14,000 daltons, although in serum it was shown to have an effective molecular weight of 85,000-200,000 daltons depending on the conditions [3,4]. This has been shown to be due to binding to

Table 1 Characteristics of the Main Acute Phase Reactant Proteins

Protein	Molecular Weight	Carbohydrate (%)	Serum Concentrations (g/1)
C reactive protein	110,000	0	<0.005
Serum amyloid A	180,000[a] 12,000	0	0.005-0.05
Alpha-1 acid glyprotein (orosomucoid)	41,000	42	0.55-1.4
Haptoglobin type 1-1 type 2.1 type 2-2	86,018 type 1-1	12	 1.0-2.3 0.9-3.8 1.9-3.5
Alpha-1 antitrypsin	54,000	12	1.9-3.5
Alpha-1 antichymotrypsin	68,000	27	0.3-0.6
Fibrinogen	340,000	3	2.0-4.5
Ceruloplasmin	132,000	11	0.15-0.60

[a]Combines with HDL portion of lipoproteins to give mol. wt. of 180,000.

high-density lipoproteins (HDL), of which SAA is considered to be an apolipoprotein [5]. SAA is synthesized in the liver [6] and is the structural unit of amyloid A fibrils [7].

SAA has been demonstrated to be an acute phase reactant, its serum concentration increasing greatly in response to tissue damage or inflammation. SAA has been reported to be a biochemical marker that differentiates between disseminated and local or regional neoplastic disease [8]. SAA levels have also been reported to be raised on a variety of chronic inflammatory diseases such as rheumatoid arthritis and systemic lupus erythematosus [9,10], inflammatory bowel disease, as well as in myelomatosis and benign monoclonal gammopathy [7].

A radial immunodiffusion assay for SAA has been developed by Chabners and Whicher [11] and we have used this method to examine SAA. Concentrations of serum SAA were found to be well correlated with CRP in inflammatory disease and cancer [12]. In general, the time course of the rise and fall of CRP and SAA following an acute injury is similar but the peak level of SAA is about 10 times higher than that of CRP.

After cholecystectomy, SAA levels rose from 18 ± 13 mg/L to about 5000 mg/L and fell with a half-life of 1.1 ± 0.3 days. It has been suggested that SAA levels can differentiate localized from disseminated malignant disease. However, in our studies the SAA levels showed considerable overlap between localized and metastatic cancer. In the context of clinical oncology, serum SAA levels would not seem to offer any advantages over CRP as an indicator of an inflammatory reaction.

II. GENERAL CONSIDERATIONS

APRPs can be measured by a variety of immunochemical methods. Single radial immunodiffusion is the simplest but requires 24-48 hr to obtain a result. The Laurell electroimmunoassay (rocket) is suitable for small batches of samples. Most routine laboratories today use nephelometry or turbidimetry, which has the advantages of speed. Rapid latex agglutination screening methods for CRP are available. When used with sera at two dilutions, these can give an indication of the intensity of an acute phase reaction within a few minutes.

When one is considering the APR in a a patient with cancer, the clinical context must always be kept in mind. The disturbance of the APRP profile following major curative surgery, such as an abdominoperineal resection, can take several weeks to return to baseline; the reaction reflects total tissue trauma and speed of healing. In any form of advanced cancer that is eventually fatal, the progressive decline of the patient's nutritional status will reflect on the levels of proteins such as albumin, transferrin, and prealbumin. Apart from the acute incidents of infection, mainly in immunosuppressed patients with hematological malignancies, a wide range of acute or chronic complications of cancer that include obstruction to organs, hemorrhage, and thrombosis may produce alterations in the steady state that can affect the levels of APRPs. It will nevertheless be seen that the APRP in cancer has sufficient consistency to provide useful information. Unlike the many tumor-associated antigens now available for monitoring cancer (Chapter 4), the APRPs provide information that reflects the strength of the signals from the host-tumor interaction and the response of the liver to these regulatory compounds. In an acute response, CRP, SAA, and ACT levels are the first to rise, followed by PI, Hp, Cp; conversely, CRP and SAA levels will be the first to fall toward normal when the response is terminated. In a chronic response, as in cancer or rheumatic disease, the APRPs are highly correlated so that the level of one of them (i.e., CRP or ACT) will reflect the likely level of the other APRPs. The CRP level in metastatic cancer will occasionally show a fluctuating level, suggesting that the rate-controlling signals are not being formed at a constant rate.

The acute-phase proteins are synthesized by the liver; each protein appears to be under separate control. The amount of protein synthesized is controlled by the local production of interleukins within the inflammatory site (see [13-15] for review).

Discussion of this complex topic is outside the scope of this chapter, even though it may be the basis of a future generation of clinically useful assays.

III. MICROHETEROGENEITY OF ACUTE-PHASE GLYCOPROTEINS

Many plasma glycoproteins exhibit heterogeneity of their carbohydrate side chain structures. This involves sugar changes within chains and the ratios of bi-, tri-, and tetra-antennary branching. The differences can be demonstrated by their relative affinity to lectins and made visible by lectin affinity crossed immunoelectrophoresis [16].

AGP has been most widely studied but similar heterogeneity is seen in ACT, Cp, and Hp. There are minor differences in the peak areas associated with disease. The most pronounced shift is generally seen in an acute inflammation when there is a tendency for the affinity to concanavalin A (Con A) to be reduced [17]. In patients with cancer, shifts in the lectin binding profiles occur in several acute-phase glycoproteins, but there is argument about their significance as discriminants [18-20]. Pregnancy and estrogen treatment of prostatic cancer cause marked alterations of the glycans of AGP, with a loss of affinity to Con A. The mechanism of the change is uncertain. In the rapid-onset APR it could be that the induction of glycosyl transferases is not synchronous with the increased rates of protein assembly. In the chronic response, selective loss from the circulation could influence the balance of glycan chains on the glycoproteins.

If the APRPs are evaluated as indicators of malignant disease, it becomes apparent that they lack the specificity required for the detection of early stage malignant disease. Being a nonspecific reaction, they cannot differentiate between malignant and acute or chronic inflammatory diseases. At first sight, this poor performance, coupled with the growing range of specific and general tumor markers available today (see Chapter 38), would suggest that measurements of acute phase reactants have no place in the management of cancer patients. Nevertheless, there are three main aspects of the acute phase reaction in cancer that, despite its lack of specificity, are the underlying reasons why clinical interest is sustained at a time when these proteins have been rejected as tumor markers in the classic terms of specificity and sensitivity.

First, the frequency of an acute-phase response in patients with many forms of solid tumors tends to increase with tumor burden. However, at each stage only a proportion of patients show an acute-phase response (Fig. 1). It is most likely that the inflammatory reaction around the tumor rather than the tumor itself produces the inflammatory response [20], but there are some exceptions. Second, if the acute-phase reaction is followed sequentially throughout the illness of a patients who will eventually die of metastatic cancer, a progressive rise of the APRPs, especially during the last 6 months of life occurs in almost all patients whose death is attributable to the cancer (Fig. 2). In these circumstances the acute-phase reaction behaves as a biochemical performance test, analogous to the Karnofsky scale. Third, in some malignant diseases such as leukemia or during intensive cancer therapy when there is considerable risk of septicemia a high and rising CRP can enhance the clinician's awareness that the patient has developed an infection.

The permutations of these basic phenomena of the acute-phase response in cancer have sustained clinicans' interest. They are best illustrated by considering the APR in cancers arising in various organ sites. Research on the biochemistry of the mediators of the acute-phase response and their action on the liver parenchymal cells may eventually provide new assays to dissect the components of the inflammatory response, but this is as yet still outside the realm of contemporary clinical biochemistry.

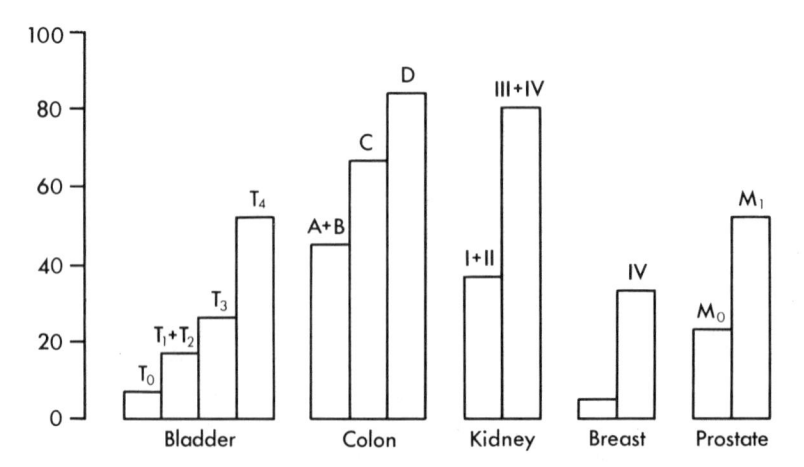

Figure 1 Percentage of tumors with raised APRPs according to tumor stage.

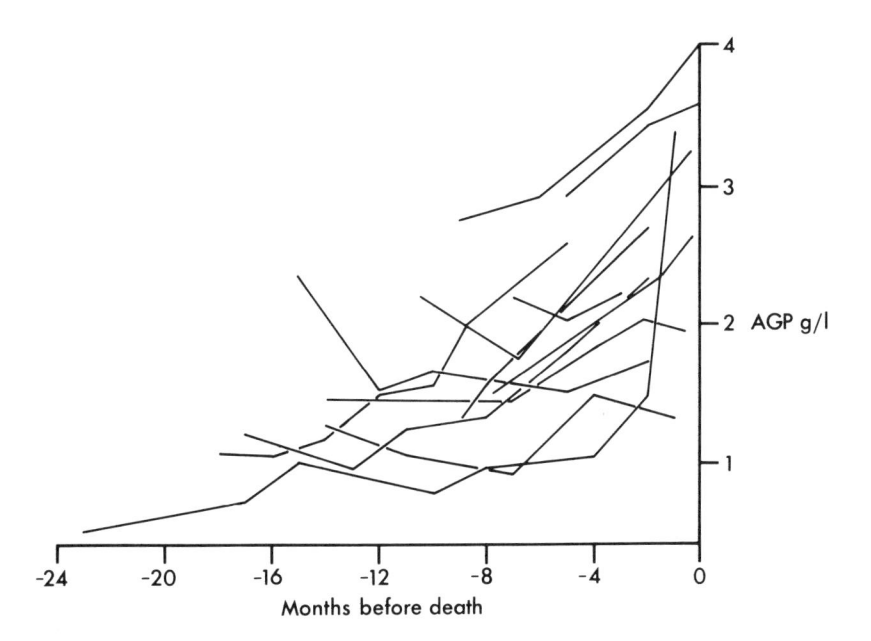

Figure 2 Levels of AGP in non-small-cell lung cancer during the last year of the patient's life, indicating the rise with growing tumor burden.

IV. HEAD AND NECK CANCER

Caldani et al. [21] have examined the ratio of serum alpha-1-acid glycoprotein (orosomucoid) to prealbumin (ORP) in patients with carcinoma of the head and neck prior to receiving any form of therapy. The level of this ORP ratio (mean ± SD) was higher in the patients than in controls: 5.29 ± 3.09, 2.63 ± 1.07. Lower ratios were found in patients with stages I and II disease compared to those with stages III and IV. The ratio appeared to be a strong prognostic factor for survival to 2 years; 51% of patients presenting with a ratio <6 were alive at 2 years compared to 24.5% with ratio >6. This was also noted in patients with stage III and IV disease, in whom the mean survival was 16 months and 7 months, as well as among the 89 patients who were treated with chemotherapy. In head and neck cancer the nutritional state of the patient was often poor due to mechanical factors interfering with eating. Since prealbumin is a well-recognized indicator of nutritional status, in this particular cancer the combination of negative reactant and a positive reactant belonging to the middle of the response range form a powerful prognostic factor. The AGP/prealbumin ratio has been evaluated previously in a variety of cancers by Hollinshead et al. [22], who clearly demonstrated its rise with increasing tumor burden.

V. GASTROINTESTINAL CANCER

The combination of APRPs and carcinoembryonic antigen CEA have been used to enhance the preoperative assessment of colorectal cancer [23] and gastric cancer [24]. The combination of CEA and AGP has been found to increase the detection rate of colorectal cancer, rising from 58% for CEA alone to 79%, without a significant fall in specificity. deMello et al. [25] subsequently made a detailed multivariate analysis of the predictive power of APRPs in gastrointestinal cancer. Preoperative CRP levels >50 mg/L in 100 cases of gastric cancer identified 50% of patients with technically inoperable tumors. Preoperative ACT levels could be used to classify patients with Dukes' B and C lesions into those with higher and lower risk of recurrence; a raised ACT enhanced the risk. They also described a discriminant function that included sex, ACT, CEA, gamma-glutamyl transferase (GGT), and phosphohexose isomerase levels that identified preoperatively 16 of 18 (89%) Dukes' D class patients and only misclassified 6 of 82 other patients with colorectal cancer. All the misclassified patients had Dukes' class C disease. It is possible to construct complex multivariate formulae, which include APRPs, for the differentiation of malignant from benign gastrointestinal disease [25] but the most successful method in our experience is to combine data from a questionnaire, CEA levels, and nonspecific markers that included CRP and GGT [26]. In another 100 patients it was later demonstrated that raised levels of AGP (>1.4 g/L) or CRP (>15 mg/L) were accurate preoperative predictors of fixation in colorectal cancer (specificity level 87 and 90%; sensitivity, 78 and 78%, respectively). A raised CEA level >50 ng/ml predicted 82% of the tumors with malignant fixation [27]. APRPs have been used in combination with CEA to predict recurrence [28] and to assess the biochemical status of patients with advanced disease. Milano et al. [29] demonstrated that the combination of transferrin, prealbumin, and albumin can be useful indices of patients' nutritional status. A striking fall in prealbumin levels occurs during the last 3 months of life in patients dying of metastatic colorectal cancer. However, none of these applications have made much impact on routine management. CEA is still the marker of choice for monitoring colorectal cancer.

Serum antiprotease levels (PI, ACT, and 2 macroglobulin), trypsin, and elastase have been measured in pancreatic cancer. Increased elastase and decreased alpha 2-macroglobulin levels were found in all patients, and the majority had coincidental increases in PI and ACT levels. Elastase and trypsin levels were highly correlated. Pancreatic cancer would seem to provide a complex situation in which part of the APRP response is that to be expected in cancer, and part is a modification of the antiprotease components due to the demand to react with excess protease production [30].

VI. LIVER CANCER

The presence of 1-PI Pi system Z allele may be a predisposing factor to the development of hepatocellular carcinoma (HCC). Deposits of PI are found in 55-73% of HCC [31,32]. PI with reduced sialylation of the glycan chains have been detected in HCC cell-derived PI. Immunohistochemical studies have also identified PI deposits in HCC in young women using oral contraceptives [31]. More recently it has been shown that HCC have a very high incidence of immunoreactive ACT. Although these changes are of interest as indicators of the protein biosynthetic activity in HCC cells, the serum levels of APRPs do not act as tumor markers in this form of cancer.

VII. LUNG CANCER

Serum APRP levels are often raised in advanced lung cancer, but cannot be relied on to differentiate between benigh and malignant diseases of the lung. Although Ganz et al.

[33] have reported that a fraction of AGP measured by radioimmunoassay (RIA) can differentiate lung cancer from benign lung cancer with a sensitivity of 89% and specificity of 84%, this has yet to be confirmed independently. At presentation APRPs can be used as an indicator of prognosis. Patients treated surgically are only a few percent of all those with lung cancers, nevertheless they have a real chance of prolonged survival. A raised haptoglobin [34] or ceruloplasmin [35] level prior to surgery has been reported as a sign of poor prognosis in patients with resectable tumors. By contrast, Muller [36], in a study of biochemical indices in 51 non-small-cell lung cancers treated surgically, found that only preoperative C_3 complement levels had any bearing on survival, and were related to stage, which is the strongest single determinant of survival. The changing attitudes toward more conservative surgical treatment of lung cancer could be a factor in these different results.

In patients with inoperable non-small-cell lung cancer, acute phase proteins have a prognostic significance. Caspers et al. [37] found that a discriminant function that included stage, performance status, and CRP levels produced an optimum stratification for survival probability in 90 patients with squamous cell lung cancer who were referred for radiotherapy. Such stratifications are helpful in clinical trials and when deciding whether to give radical or palliative treatment. In lung cancer, prognosis is generally so poor that the event being predicted (death) is often within 12 months. Under these circumstances APRPs become powerful predictors and their level is the biochemical analog of clinical performance status. In small-cell lung cancer, stage is a strong prognostic indicator at presentation. Serum enzyme levels including neuron-specific enolase, lactic dehydrogenase, and alkaline phosphatase are better prognostic indices than APRPs.

The microheterogeneity of AGP, as demonstrated by its affinity to concanavalin A lectin crossed immunoelectrophoresis (LCIE) has been shown to be different in lung cancer than benign inflammatory lung disease [19]. In a study of 54 patients with lung cancer and 16 with benign lung inflammation, they found that the concentration of ceruloplasmin combined with the heterogeneity of ceruloplasmin analyzed by crossed immunoelectrophoresis without lectin could discriminate from malignant lung disease with a sensitivity of 78% and specificity of 93% [20]. However, others, while demonstrating the microheterogeneity of AGP, did not find significant differences between the patterns in benign and malignant disease [18]. Differences in the LCIE precipitin profile of PI have been detected between serous effusions of benign and malignant origin. The most likely explanation is the presence of protease-antiprotease complexes caused by tumor excretion of proteases into the serous fluids that are cleared more slowly than from the blood. These complexes were absent in benign effusions [38].

The problem of tests that used to demonstrate and quantify the microheterogeneity of the APR glycoproteins is they are subject to technical and interpretive errors, for example, estimating areas under precipitin curves. Before we discard them as impractical, however, consideration should be given to whether applications of novel techniques such as lectin immunoassays might exploit these disease-associated changes in microheterogeneity.

VIII. UROGENITAL TUMORS

A. Carcinoma of the Kidney

Progess in the managment of carcinoma of the kidney has been slow. The disease presents late and there are few agents active against recurrent or metastatic disease. Metasta-

Table 2 Acute Phase Reactant Proteins in Kidney Carcinoma

Status	Number of Observations	ACT (g/1)	APG (g/1)	CRP[a] (mg/1)
Tumor free[b]	20	0.46 ± 0.17	0.73 ± 0.2	<10
				5-13
Preoperative[c]				
Group I	8	0.71 ± 0.13	1.3 ± 0.49	<10
				5-71
Group II	12	1.10 ± 0.4	2.0 ± 0.68	59
				5-153
Metastatic				
5-9[d]	9	0.89 ± 0.29	1.57 ± 0.54	32
				<10-70
3-4[d]	16	1.14 ± 0.52	2.03 ± 0.71	51
				<10-177
1-2[d]	20	1.37 ± 0.44	2.32 ± 0.70	88
				15-253

(mean ± SD)
[a]Median and range
[b]Following nephrectomy.
[c]Group I = Tumors confined to the kidney and its immediate vicinity; Group II = Tumors invading renal vein, local lymph nodes or associated with distant metastases.
[d]Months prior to death.

tic carcinoma of the kidney can run an aggressive course or enter an apparently static phase that lasts for many years. It is well known that erythrocyte sedimentation rate (ESR) is of prognostic significance in renal cancer. Jussela [39] and Hop and Van der Werf-Messing [40] have shown that ESR, sex, and renal vein involvement are powerful discriminants in predicting survival of patients with renal carcinoma. Richards et al. [41] in a study of 88 patients, found a progressive rise of ACT and AGP with increasing stage and with increasing tumor burden in patients with metastatic kidney cancer (Table 2). They concluded the APRPs at presentation give warning of extensive local or metastatic disease, but the system is insensitive to small tumor burdens. Babaian and Swanson [42] examined the serum haptoglobin levels in 116 patients with renal carcinoma. The haptoglobin level was normal in 87% of patients who were tumor free and raised in 58% with active disease. False-positve findings were rare. The main interest in these agents was following nephrectomy: a raised value had a positive predictive value (true positives equal/true positives + false positives) of 82%. Amyloid disease occurs in about 8% of kidney cancers, evidence that this is of the SAA type [43,44]; and a sustained APR with a raised SAA in a slow-growing tumor are some, but not all, of the factors involved in this form of secondary amyloidosis.

B. Prostatic Cancer

As in many forms of metastatic cancer, disseminated prostatic cancer will induce an acute-phase reaction that tends to mirror the tumor activity [45]. There has recently been interest in indices made of the ratio of an acute-phase reactant and a negative reactant plasma protein in prostatic cancer. Robey et al. [46] found significant differences between the AGP/prealbumin ratio of patients with no evidence of active disease after treatment and those with progressive disease. We were unable to confirm these results [47], the AGP levels being similar in the two groups, when patients receiving estrogen were excluded although the prealbumin and α2HS globulin levels were both significantly lower in the patients. Siddall et al. [48] observed in patients with metastatic prostatic cancer treated with a lutenizing hormone-releasing hormone (LHRH) agonist that the levels of AGP/prealbumin at presentation did not predict response to treatment, but after 6 months of treatment high ratios were an unfavorable prognostic indicator for long-term survival.

Prostatic cancer is often treated by hormone manipulation designed to lower the serum testosterone level; the main methods include castration, giving estrogens (stilbesterol, estramustine phophate), or LHRH agonists. Estrogens have a powerful effect on APRPs, causing an elevation in Cp and PI levels and a fall in Hp, AGP, and fibrinogen levels. Patients receiving estrogens do not show a normal APR to metastatic cancer. Castration and administration of LHRH agonists have no effect on APRPs.

C. Bladder Cancer

An acute-phase reaction is infrequently stimulated in patients with superficial bladder cancer. Chronic infection localized to the bladder does not induce an acute-phase response [49], but in cases that have invaded the muscle (T_3) or beyond the bladder wall (T_4), CRP level is an important prognostic factor. O'Quigley et al. [50], using a pretreatment CRP discriminant level of 12 mg/L, found the median survival in 61 T_3 cases to be >28 and 16 months, according to whether the patient had a CRP below or above this cut-off level, respectively, and in 41 T_4 cases the median survival was >28 and 5 months according to this discriminant level of CRP.

D. Ovarian Cancer

The search for reliable tumor markers for ovarian cancer has been underway for many years. CEA, placental alkaline phosphatase, and human placental lactogen were tried but generally found to be unreliable, which led to the interest in acute phase reactants as indicators of disease activity in patients following surgery. Muller et al. [51] identified haptoglobin as an indicator of tumor activity in ovarian cancer. Randle et al. [52] investigated serum seromucoid (mainly AGP) in ovarian cancer. Meerwaldt et al. [53], in a study of 93 patients lasting 2 years, found the combination of serum albumin, CRP, AGP, and phosphohexose isomerase to be useful adjuncts to clinical judgment in patients with advanced disease (Fig. 3).

Tamura et al. [54] described an immunosuppressive acidic protein (IAP) isolated from malignant ascitic fluids that is closely related to AGP. In ovarian cancer the mean ($\pm$SD) levels of IAP were 1085 $\pm$ 474 μg/ml, compared to levels in control of 298 $\pm$ 92 μg/ml and in those with benign tumors of 394 $\pm$ 93 μg/ml. Serial measurements of IAP were more helpful than CEA in following the progress of ovarian cancer. However, Morgensen et al. [55] found that AGP and IAP levels were highly correlated (r = 0.944). Although IAP is commercially available in Japan, it would appear to have no advantage over AGP.

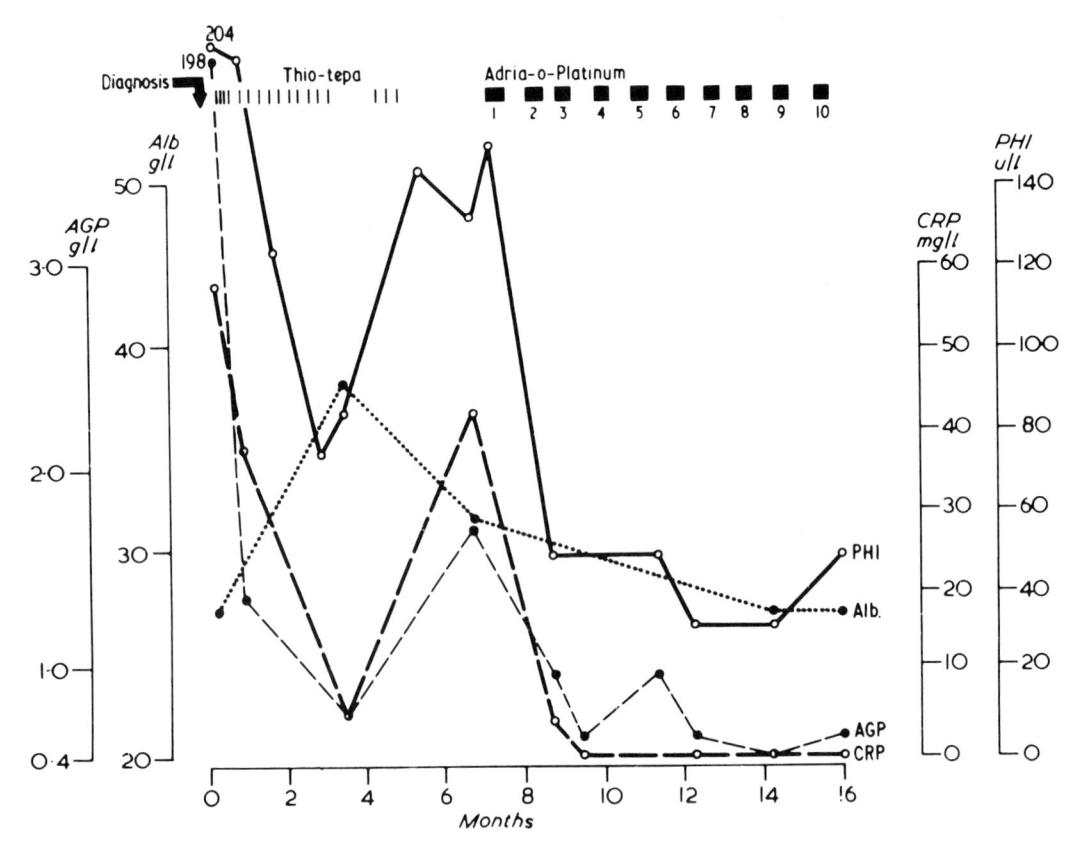

Figure 3 Findings in a 53-year-old patient, presenting with a stage IV adenocarcinoma, inoperable. Note the steep decrease in PHI, AGP, and CRP levels at first, followed by an increase when pleural effusions reappear and the stabilization of at normal levels after more aggressive chemotherapy.

CA-125 as a marker of ovarian cancer (see Chapter 21) has largely replaced the clinical need for APRPs to monitor patients at risk of recurrence.

IX. OTHER TUMORS

A. Breast Cancer

The use of APRPs to evaluate or monitor breast cancer has not been very successful. Several investigations involving over 700 patients have demonstrated that patients with stage IV disease tend to have raised APRPs levels, while such elevations are infrequent in patients with stage I-III disease [56,57]. Mortensen and Rudczynski [58] made a detailed study of CRP in breast cancer and failed to find evidence that its levels at the time of surgery or following treatment of the primary tumor had any significant prognostic value. The use of multiple markers has been advocated in an attempt to find a marker system for breast cancer. APRPs have been included in such schemes [57,59] but they were generally either too complicated or too insensitive to attract serious clinical attention. As with ovarian cancer, attention has now been switched to the new generation of breast cancer markers such as CA 15-3, and mucin-like cancer-associated antigen (MCA), which are undoubtedly more sensitive than APRPs for monitoring patients at risk of recurrence.

B. Lymphoid Tumors

During the 1970s several papers were published on the application of APRPs to the assessment and monitoring of Hodgkin's disease (HD) and non-Hodgkin's lymphoma (NHL). This research has been reviewed by Cooper and Stone [2]. Key points that emerged can be summarized briefly. The practical tests of disease activity were erythrocyte sedimentation rate (ESR), serum copper or ceruloplasmin levels, and other APRPs including AGP, CRP, and Hp. NHL and, to a lesser extent, HD may evolve into a chronic disease. Longitudinal studies over several years demonstrated the fluctuations of APRP activity with the waves of progression and response in extensive NHL; the individual APRPs were highly correlated [60]. In HD persistently raised APRPs are a sign of a bad prognosis [61].

During the past few years the research into APRPs in lymphoma has virtually ceased. Gobbi et al. [62], reporting on 85 patients with HD, found that AGP and Hp levels were elevated in patients with active disease compared to patients in remission. However, this relationship was restricted to active disease. The best discriminants between active and inactive disease were the levels of serum copper and iron combined with AGP. This is due in part to very well defined treatments aimed at curing HD and a better understanding of the subsets and management of patients with NHL. An example of the more dismissive conclusion is the study made by Margerison and Mann [63]. In a study of 29 children with HD, 22 with NHL, and 13 with nonmalignant lymphadenopathy, the levels of serum copper, ceruloplasmin, and ESR were correlated to disease activity. Raised levels occurred in patients with active advanced disease but were no more useful than the ESR.

C. C-Reactive Protein in Acute Leukemia

Microbial bacterial and fungal infection is a strong stimulus to C-reactive protein CRP production. High levels of CRP (>100 mg/L) are highly suggestive of infection in acute leukemia [64-66] but in more than one-third of documented infections the CRP remained below 100 mg/L [67], which led some authors to suggest 60 mg/L as a threshold. Chemotherapy, radiotherapy, and transfusion only cause a minor rise in CRP (<30 mg/L). The rise in CRP can precede a rising temperature, but more often the raised temperature is the first sign of infection. Once an infection is suspected, daily CRP measurements are helpful; the peak CRP level will be reached 24-48 hr after the onset of the fever and the efficacy of antimicrobial therapy can be judged by the rate of fall of the CRP. Despite these studies, CRP is still not widely used to investigate neutropenic pyrexia. Perhaps the main problem has been to make physicians appreciate that a single measurement of CRP gives limited information but that the trend in CRP is useful. The other problem is that the level of CRP may not be high on the first day of the fever; only daily measurements will accurately monitor the life-threatening episode.

The fever, bone pain, and high levels of CRP seen in patients with untreated acute leukaemai are usually not the result of infection but fall on treatment with cytostatics [68].

X. BIOLOGICAL EFFECTS OF APRPs IN CANCER

The precise role of the many large and subtle adjustments of the serum protein profile following acute injury or in chronic inflammation as seen in cancer and rheumatic diseases is still unclear. The general conclusion is that they will help to control local events at the site of inflammation. In cancer the cause of the reaction is unclear, but occasionally a relatively small focus of cancer may promote a marked response, as in kidney cancer, and the response stops when the primary tumor is excised. In animal models such as the transplantable VX2 carcinoma in rabbit, which produces large amounts of prostaglandin, marked stimulation of Cp and Hp occur to such an extent that the serum becomes blue within a week of transplantation [69]. Usually the APR response is associated with locally invasive or metastatic cancers, both of which are sites of proteolytic attack on the host cells at the growing edge of the cancer but are complicated by the products of cell death of tumor cells at a microscopic level and in the larger areas of necrosis found in large tumors as they outgrow their blood supply.

Evidence of how some of the individual APRPs may function has been derived in part from in vitro studies, but the way these functions are integrated is unclear as is whether they are modified in chronic reactions. The biological activity of some of the APRPs is well defined: PI and ACT are serine antiproteases and form a first line of defense against proteolytic enzymes. Hp is concerned with the capture and removal of hemoglobin (Hb) by the formation of Hp-Hb complexes. Fibrinogen is an essential component of the clotting cascade. Ceruloplasmin is a copper-transporting protein but its role in inflammation is more likely to result from its ability to act as a scavenger for free radicals. The role of AGP is still unknown. Being strongly acidic (Pi 2.4), it can bind basic compounds including some drugs and result in their blood levels being raised in advanced cancer [70]. There has been some evidence that it can suppress the action of certain lymphocytes and may be involved in the regulation of blood clotting. Perhaps the greatest uncertainty surrounds the actions of CRP and SAA, the proteins that document a first line of reaction and can show a response in excess of 100 times their normal concentration. SAA is reported to have an immunosuppressive function, while CRP selectively binds to a subset of human T cells, especially to the suppressive subset with IgG Fc receptors [71]. In vitro CRP can suppress the mediators of delayed hypersensitivity [72].

In summary, the acute-phase proteins, especially when measured sequentially, provide a guide to the behavior of metastatic cancer. They are more sensitive to change than albumin levels but less sensitive and less specific than corresponding measurements of an appropriate tumor-associated marker. They are indicators of risk: the risk of infection in leukopenic states, the risk of declining performance status in advanced disease, the risk of progression or recurrence in advanced tumors at presentation. However, when surgery adequately cures or controls the local tumor growth, and metastases or recurrence may not occur for several years, the levels of APRPs at presentation provide no guide as to what will happen to the patients several years later.

REFERENCES

1. Kushner I. The phenomenon of the acute phase response. Ann NY Acad Sci 1982; 389:39-48.
2. Cooper EH, Steon J. Acute phase reactant proteins in cancer. Adv Cancer Res 1979; 30:41-44.
3. Benson MD, Skinner M, Lian J, Cohen AS. "A" protein of amyloidosis. Isolation of a cross-reacting component from serum by affinity chromatography. Arthritis Rheum 1975;18:318-322.
4. Linke RP, Sipe JD, Pollock PS, Ignaczak TF, Glenner GG. Isolation of a low molecu-

lar weight serum component antigenically related to an amyloid fibril protein of unknown origin. Proc Natl Acad Sci USA 1975;72:1473-1476.
5. Benditt EP, Eriksen N. Amyloid protein SAA is associated with high density lipoprotein from human serum. Proc Natl Acad Sci USA 1977;74:4025-4028.
6. Benson MD, Kleiner E. Synthesis and secretion of serum amyloid protein A (SAA) by hepatocytes in mice treated with casein. J Immunol 1980;124:495-499.
7. Husby G, Lindström FD, Natvig JB, Dahlstrom V. Occurrence of amyloid-related serum proteins in patients with benign monoclonal gammopathy. Scand J Immunol 1977;6:659-663.
8. Rosenthal CJ, Sullivan IM. Serum amyloid A to monitor cancer dissemination. Ann Intern Med 1979;91:383-390.
9. Benson MD, Cohen AS. Serum amyloid A protein in amyloidosis, rheumatic and neoplastic diseases. Arthritis Rheum 1979;22:36-42.
10. Maury CPJ, Teppo AM, Wegelius O. Relationships between urinary sialylated saccharides, serum amyloid A and C-reactive protein in rheumatoid arthritis and systemic lupus erythematosus. Ann Rheum Dis 1982;412:268-271.
11. Chabners RE, Whicher JT. Qualitative radial immunodiffusion assay for serum amyloid A protein. J Immunol Methods 1983;59:95-103.
12. Raynes JG, Cooper EH. Comparison of serum amyloid A protein and C-reactive protein concentrations in cancer and non-malignant disease. J Clin Pathol 1983;36:798-803.
13. Kushner I, Mackiewicz A. Acute phase proteins as disease markers. Dis Markers 1987;5:1-11.
14. Dinarello CA. An update on human interleukin-1: from molecular biology to clinical relevance. J Clin Immunol 1985;5:287-297.
15. Whicher JT. Functions of the acute phase proteins in the inflammatory response. In: Arnaud P, Bienvenu J, Laurent P, eds. Marker proteins in inflammation, vol 2. New York: Walter de Gruyter, 1984:89-98.
16. Bøg-Hansen TC, Brogen CH. Identification of glycoproteins with one and with two or more binding sites to Con A by crossed immunoaffinoelectrophoresis. Scand J Immunol 1975;4:Suppl 2:135-140.
17. Raynes JG. Variations in the relative proportions of microheterogeneous forms of plasma glycoproteins in pregnancy and disease. Biomedicine 1982;36:77-86.
18. Bleasby AJ, Knowles JC, Cooke JN. Microheterogeneity of alpha 1-acid glycoprotein: lack of discrimination between benign and malignant inflammatory disease of the lung. Clin Chim Acta 1985;150:231-235.
19. Hansen JE, Larsen VA, Bøg-Hansen TC. The microheterogeneity of alpha-1-acid glycoprotein in inflammatory lung disease, cancer of the lung and normal health. Clin Chim Acta 1984;138:41-47.
20. Hansen JES, Iversen J, Lihme A, Bøg-Hansen TC. Acute phase reaction, heterogeneity and microheterogeneity of serum proteins as non-specific tumor markers in lung cancer. Cancer 1987;60:1630-1635.
21. Caldani C, Thyss A, Schneider M, Milano G, Buray L, Demard F. Orosomucoid: prealbumin ratio a marker of the host-tumour relationship in head and neck cancer. Eur J Cancer Clin Oncol 1988;24:653-658.
22. Hollinshead AC, Chuang CY, Cooper EH, Catalona WJ. Interrelationship of prealbumin and α1 acid glycoprotein in cancer patient sera. Cancer 1977;40:2993-2998.
23. Ward AM, Cooper EH, Turner R, Anderson JA, Neville AM. Acute phase reactant protein profiles: an aid to monitoring large bowel cancer by CEA and serum enzymes. Br J Cancer 1977;35:170-178.
24. Rashid SA, O'Quigley J, Axon ATR, Cooper EH. Plasma protein profiles and prognosis in gastric cancer. Br J Cancer 1982;45:390-394.

25. de Mello J, Struthers L, Turner R, Cooper EH, Giles GR, the Yorkshire Regional Gastrointestinal Cancer Research Group. Multivariate analyses as aids to diagnosis and assessment of prognosis in gastrointestinal cancer. Br J Cancer 1983;48:341-348.
26. Chisholm EM, Marshall RJ, Brown D, Cooper EH, Giles GR. The role of a questionnaire and four biochemical markers to detect cancer risk in a symptomatic population. Br J Cancer 1986;53:53-57.
27. Durdey P, Williams NS, Brown DA. Serum carcinoembryonic antigen and acute phase reactant proteins in the pre-operative detection of fixation of colorectal tumours. Br J Surg 1984;71:881-884.
28. Gray BN, Walker C. Monitoring of patients with carcinoma of the large intestine by use of acute phase proteins and carcinoembryonic antigen. Surg Gyn Obst 1983;156:770-780.
29. Milano G, Cooper EH, Goligher JC, Giles GR, Neville AM. Serum prealbumin, retinol-binding protein, transferrin, and albumin levels in patiens with large bowel cancer. J Natl Cancer Inst 1978;61:687-691.
30. Buamah PK, Skillen AW. Concentrations of protease and anti-protease in serum of patients with pancreatic cancer. Clin Chem 1985;31:876-877.
31. Palmer PE, Christopherson WM, Wolfe HJ. Alpha-1-antitrypsin, protein marker in oral contraceptive-associated hepatic tumours. Am J Clin Pathol 1977;68:736-739.
32. Thung S, Gruber M, Sarno E, Popper H. Distribution of fine antigens in hepatocellular carcinoma. Lab Invest 1979;41:101-105.
33. Ganz PA, Baras M, Ma PY, Elashoff RM. Monitoring the therapy of lung cancer with alpha-1-acid glycoprotein. Cancer Res 1984;44:5415-5421.
34. Bradwell AR, Burnett D, Newman CE, Ford HJ. Serum protein measurements of tumour mass and prognosis in carcinoma of the lung. In: Peeters H, ed. Protides of biological fluids, 27th Colloquium. Oxford: Pergamon Press, 1979:327-330.
35. Linder MC, Moor JR, Wright K. Ceruloplasmin assays in diagnosis and treatment of human lung, breast, and gastrointestinal cancers. J Natl Cancer Inst 1981;67:263-275.
36. Muller T, Marshall RJ, Cooper EH, Watson DA, Walker DA, Mearns AJ. The role of serum tumour markers to aid the selection of lung cancer patients for surgery and the assessment of prognosis. Eur J Cancer Clin Oncol 1985;21:1461-1466.
37. Caspers RJL, Pidcock NB, Cooper EH, van PUtten WLJ, Haije WG. The prognostic significance of acute phase proteins in patients with inoperable squamous cell carcinoma of the bronchus. Radiother Oncol 1984;2:107-113.
38. Bleasby AJ, Knowles JC, Cooper EH. Alpha-1-protease inhibitor in serous effusions: differences in lectin crossed immuno-affinoelectrophoresis precipitin profiles between benign and malignant disease. Int J Tissue React 1986;5:393-400.
39. Jussela H. Renal adenocarcinoma: clinical features and factors correlating to survival in 173 patients. Ann Chir Gynaecol Fenn 1978;67:3-43.
40. Hop WCH, Van der Werf-Messing BHP. Prognostic indexes for renal cell carcinoma. Eur J Cancer 1980;16:833-840.
41. Richards B, Robinson MRG, Pidcock NB, Cooper EH. Serum protein profiles in carcinoma of the kidney. Eur Urol 1982;8:32-36.
42. Babaian RJ, Swanson DA. Serum haptoglobin: a non-specific tumor marker for renal cell carcinoma. South Med J 1982;75:1345-1348.
43. Hind CR, Tennent GA, Evans DJ, Pepys MB. Demonstration of amyloid A (AA) protein and amyloid P component (AP) in deposits of systemic amyloidosis associated with renal adenocarcinoma. J Pathol 1983;139:159-166.
44. Pras M, Franklin EC, Shibolet S, Frangione B. Amyloidosis associated with renal cell carcinoma of the AA type. Am J Med 1982;73:426-428.
45. Trautner K, Cooper EH, Haworth S, Milford Ward A. An evaluation of serum pro-

tein profiles in the long-term surveillance of prostatic cancer. Scand J Urol Nephrol 1980;14:143-149.
46. Robey EL, Schellhammer PF, Wright GL Jr, el-Mahdi AM. Cancer serum index and prostatic acid phosphatase for detection of progressive prostatic cancer. J Urol 1985; 134:787-790.
47. Cooper EH, Hetherington JW, Siddall JK. Acute phase proteins in prostatic cancer. In: Peeters H, ed. Protides of the biological fluids, 31st colloquium. Oxford: Pergamon Press, 1986:629-632.
48. Siddall JK, Hetherington JW, Cooper EH, Newling DWW, Robinson MRG, Richards B, Denis L. Biochemical monitoring of carcinoma of prostate treated with an LH-RH Analogue (Zoladex). Br J Urol 1986;58:676-682.
49. Bastable JRG, Richards B, Haworth S, Cooper EH. Acute phase reactant proteins in the clinical managment of bladder cancer. Br J Urol 1979;51:283-289.
50. O'Quigley J, Haworth S, Cooper EH, Haije W, Van der Werf-Messing B, Richards B, Robinson MRG. Prognostic significance of serum proteins in invasive bladder cancer. A preliminary report of the EORTC Urological Group. Eur J Cancer 1981; 17:251-255.
51. Mueller WK, Handschumacher R, Wade WE. Serum haptoglobin in patients with ovarian malignancies. Obstet Gynecol 1971;38:427-435.
52. Randle GH, Good W, Cumberbatch KN. Serumucoid in serum in benign and malignant gynaecological disorders. J Obstet Gynaecol Br Emp 1974;81:479-482.
53. Meerwaldt JH, Haije WG, Cooper EH, Pidcock NB, van den Burg MEL. Biochemical aids in the monitoring of patients with ovarian cancer. Gynecol Oncol 1983;16: 209-218.
54. Tamura K, Shibata Y, Matsuda Y, Ishida N. Isolation and characterization of an immunosuppressive acidic protein from ascitic fluids of cancer patients. Cancer Res 1981;41:3244-3252.
55. Morgensen O, Mogensen B, Jakobsen A, Sell A. Tumor markers in ovarian cancer: a comparison between immunosuppressive acidic protein and alpha 1-acid glycoprotein (letter). Am J Obstet Gynecol 1986;154:1165-1166.
56. Thompson DK, Haddow JE, Smith DE, Ritchie RF. Elevated serum acute phase protein levels as predictors of disseminated breast cancer. Cancer 1983;51:2100-2105.
57. Cowen DM, Searle F, Milford Ward A, Benson EA, Smiddy FG, EAves G, Cooper EH. Multivariate biochemical indicators of breast cancer. An evaluation of their potential in routine practice. Eur J Cancer 1978;14:885-894.
58. Mortensen RF, Rudczynski AB. Prognostic significance of serum CRP levels and lymphoid cell infiltrates in human breast cancer. Oncology 1982;39:129-133.
59. Coombes RC, Powles TJ, Neville AM. Evaluation of biochemical markers in breast cancer. Proc R Soc Med 1977;70:843-845.
60. Child AJ, Spati B, Illingworth S, Barnard D, Corbett S, Simmons AV, Stone J, Worthy TS, Cooper EH. Serum beta 2 microglobulin and C-reactive protein in the monitoring of lymphomas: findings in a multicenter study and experience in selected patients. Cancer 1980;45:318-326.
61. Jayle MF, Ennuyer A, Chemama R. Intérêt clinique du dosage de l'haptoblobuline et due serumucoide dans la maladie de Hodgkin. Semin Hop Paris 1968;44:1861-1867.
62. Gobbi PG, Merlini G, Parrinello GA, Cavalli P, Ascari E. Serum alpha-1-acid glycoprotein, haptoglobin and C3 in Hodgkin's disease. A comparison with other acute phase indictors. Acta Haematol 1982;68:295-299.
63. Margerison AC, Mann JR. Serum copper, serum ceruloplasmin, and erythrocyte sedimentation rate measurements in children with Hodgkin's disease, non-Hodgkin's lymphoma, and non-malignant lymphadenopathy. Cancer 1985;55:1501-1506.

64. Mackie PH, Crockson RA, Stuart J. C-reactive protein for rapid diagnosis of infecttion in leukaemia. J Clin Pathol 1979;32:1253-1256.
65. Rose PE, Johnsdon SA, Meakin M, Mackie PH, Stuart J. Serial study of C-reactive protein during infection in leukaemia. J Clin Pathol 1981;34:263-266.
66. Schofield KP, Voulgari F, Gozzard DE, Leyland MJ, Beeching NJ, Stuart J. C-reactive protein concentration as a guide to antibiotic therapy in acute leukaemia. J Clin Pathol 1982;35:866-869.
67. Starke ID, de Beer FC, Donnelly JP, Catovsky D, Goldman JM, Galton DA, Pepys MB. Serum C-reactive protein levels in the management of infection in acute leukaemia. Eur J Cancer Clin Oncol 1984;20:319-325.
68. Timonen TT, Koistinen P. C-reactive protein for detection and follow-up of bacterial and fungal infections in severely neutropenic patients with acute leukaemia. Eur J Cancer Clin Oncol 1985;21:557-562.
69. Voelkel EF, Levin L, Alper CA, Tashjian AH Jr. Acute phase reactants ceruloplasmin and haptoglobin and their relationship to plasma prostaglandins in rabbits bearing the VS2 carcinoma. J Exp Med 1978;147:1078-1088.
70. Jackson PR, Tucker GT, Woods HF. Altered plasma drug binding in cancer: role of alpha 1-acid glycoprotein and albumin. Clin Pharmacol Ther 1982;32:295-302.
71. Mortensen RF, Osmond AP, Gewurz H. Effects of C reactive protein on the lymphoid system. 1. Binding to thymus-dependent thymocytes and alteration of their function. J Exp Med -975;141:821-832.
72. Mortensen RF, Brand D, Gewurz H. Effects of C reactive protein on lymphocyte functions. III-Inhibition of antigen-induced lymphocyte stimulation and lymphokine production. Cell Immunol 1977;28:59-67.

34
Fibronectin

BOLESLAW H. LIWNICZ and RAYMOND SAWAYA
University of Cincinnati Medical Center, Cincinnati, Ohio

Fibronectin (FN) is a major noncollagenous adhesion protein that allows for cell adherence to extracellular matrix or to the surface of other cells [1-4]. The FN-mediated adhesion may lead to a formation of simple structures such as tissue culture monolayers or complex forms such as animal organs. The adhesion may be permanent, securing the organ's architecure, or transient, facilitating nonspecific opsonization [5-8] and cell migration such as occurs during development [9,10], wound healing [8,11-14], or tumor formation [15,16]. Fibronectin plays a key role in embryogenesis. For example, anti-FN antibodies inhibit neuronal migration and synthetic peptides containing a key tetrapeptide sequence of FN cell-binding domain inhibit gastrulation [17-19]. There are three basic types of fibronectin-mediated adhesion: cell to cell, cell to extracellular matrix, and cross-linking of extracellular matrix. All participate in normal development and function of an organ, as well as in the growth of primary and metastatic neoplasms [1,20-22].

I. GENERAL CONSIDERATIONS

Fibronectins are a group of universal glycoproteins found in a wide variety of species from sea sponge to humans [23,24]. These proteins are remarkably well preserved and can be detected using the same antibodies in different mammals and in other, often philogenetically remote, species [23,25].

There are two forms of FN, plasma fibronectin (pFN) in the blood, and cellular fibronectin (cFN) on the cell surface [26]. The pFN is soluble in body fluids; the cFN is insoluble and occurs on the surface of a large variety of cells and attached to extracellular matrix. The pFN circulates in the form of a dimer; the cFN can be found as monomers, dimers, or multimers forming filamentous structures. Fibronectins, regardless of type, have similar biochemical structure based on 210-260 kD monomers containing 2-15% carbohydrates [1].

II. STRUCTURE AND FUNCTION OF FIBRONECTIN

A. Structure

Electron microscopic studies of fibronectin reveal a remarkably flexible macromolecule that varies in shape from an elongated, 160 nm long structure to a globular one, 20-30 nm in diameter [27-29]. The FN dimer is V-shaped with a 70 degree angle sustained by two apically positioned disulfide bonds; however, the structure of the dimer can be distorted when the molecule is bound to collagen [30] (Fig. 1).

Over 95% of the bovine pFN has been sequenced [31,32], while only smaller fragments of the human pFn were sequenced [33-36]. The complete primary structure of both human and rat fibronectin was deduced from cDNA [37,38]. The secondary structure shows three types of homologies (Fig. 2). Type I is composed of fingerlike loops, each 41-52 amino acids long, with two adjacent intraloop disulfide bonds [39,40]. Twelve type I homology regions were found in three different sites [39]. Two type II

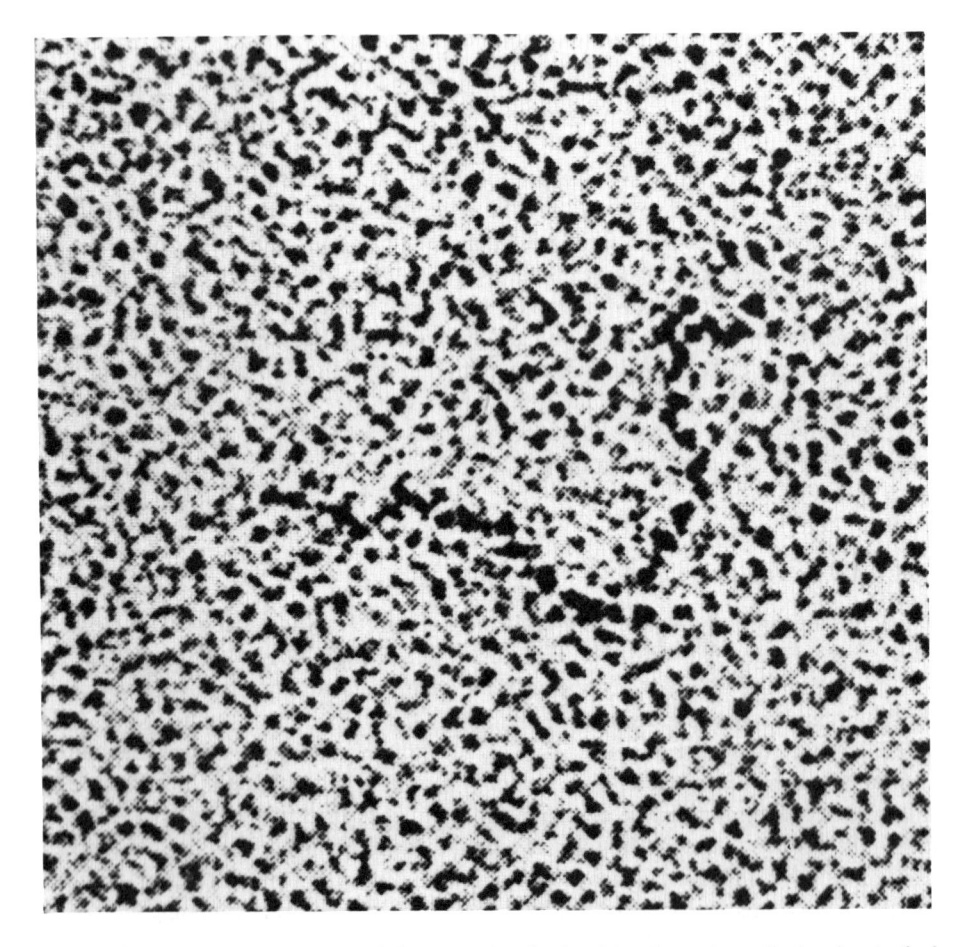

Figure 1 Electron micrograph of fibronectin obtained by the rotary shadowing technique. Note the characteristic V shape of the molecule (from ref. [30]).

Fibronectin

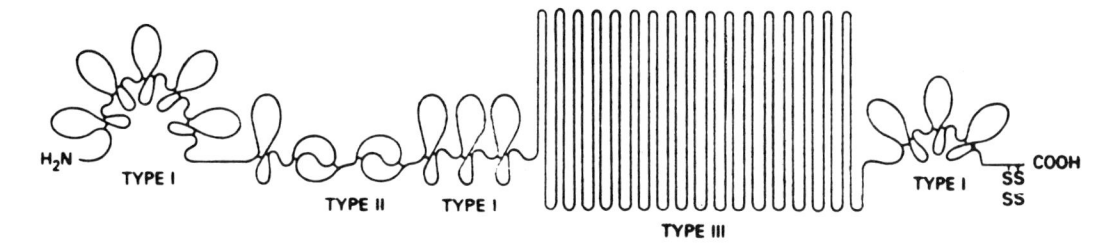

Figure 2 Schematic representation of the homologous repeat structure of the amino acid sequence of fibronectin. The sequence of a single plasma fibronectin polypeptide chain is shown as consisting of repeating series of three types of homologous units. Type I and type II homologous units are stabilized by intrachain disulfide bonds (from ref. [1]).

homologies occur in one site, each with 60 amino acid residues and 4 cysteine residues [40]. Type III homology occurs in one site composed of 16 regions, each 90 amino acids long [40]. The three types of homologies are distributed in five regions, some of which represent adhesion (functional) domains: amino terminal domain, collagen-binding domain, DNA-binding fragment, cell-binding region, and carboxyl terminal [1,2] (Fig. 3).

The 30 kD amino terminal domain is 259 amino acid residues long, grouped in 5 repeats of type I homology units. The domain has a strong affinity for fibrin and intermediate affinity for heparin [1]. The 40-45 kD collagen-binding domain has 339 amino acids in 4 type I homology units and 2 type II homology units. It has 24 half cysteines and 3 asparagine-linked carboxyl groups [37]. This domain binds collagen; however, the site of binding has been defined within a 29 kD region, and it is possible that within this region there are multiple collagen-binding sites [34,35]. The DNA-binding fragment is 35 kD with 268 amino acids grouped in 3 type III homology units. A carbohydrate moiety is attached to this fragment via asparagine [34-36]. This fragment also weakly binds heparin. The 34 kD cell-binding region contained within a 75 kD fragment is com-

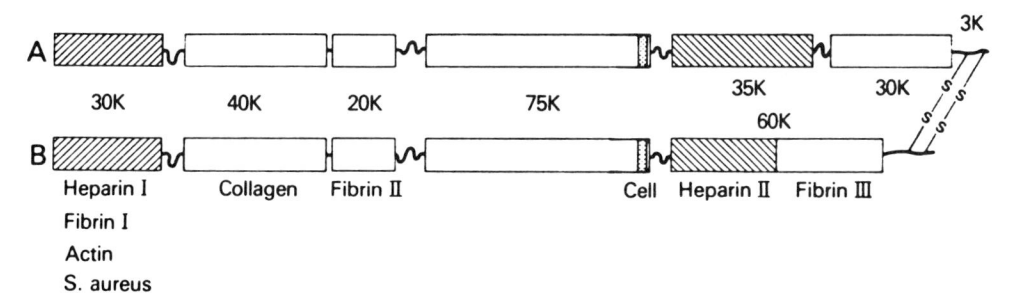

Figure 3 Functional domain structure of fibronectin. Fibronectin is a molecule composed of two or more disulfide-linked subunits. Each chain contains a similar linear sequence of modular domains, which can be separated by proteolytic cleavage. The apparent sizes of these domains in human plasma fibronectin are indicated by the numbers (K = kilodaltons). The ligands that they bind are indicated at the bottom. Note that there are two domains for binding to heparin and three for fibrin, which appear to differ in affinity (from ref. [25]).

posed of 300 amino acids in 3 type III homology regions. The cell-binding activity of this region is limited to amino acid residues 276-279 of bovine pFN within the third region of type III homology units [41,42]. There is a 95.8% homology between bovine and human third region of type III homology units [31,41,42]. The 30 kD carboxyl terminal is composed of 268 amino acids grouped in 3 type III homology units [31]. The carboxyl terminal has a high affinity for heparin [31]. The three domains with high affinity for heparin, for cells, and for DNA are within a 170 kD fragment composed of type III homologies; each domain is separated by a single type III homology region [31] (Fig. 2).

The fibronectin gene of chicken cFN is 48 kilobases long, of which 8 are exons. There are 48 exons, each 147 ± 37 bases, separated by introns 843 ± 647 bases long [1]. The fibronectin genes were studied in humans, cows, rats, and chickens [33,39,43-45]. The human fibronectin gene was assigned to chromosome 2 [22]. There are at least three different mRNAs generated from a single rat fibronectin gene [46], and there may be as many as 10 mRNAs for human FN [2,38,47,48]. Fibronectin genes are well preserved (e.g., in the carboxyl half of the polypeptide, only 54 of 700 amino acid residues are different when one compares rat and human FN [49]).

There are two known fibronectin cell surface receptors: a 110-160 kD glycoprotein and a smaller, 47 kD, glycoprotein. The fibronectin receptor isolated from MG-63 osteosarcoma is a 140 kD glycoprotein that, when run on polyacrylamide gel electrophoresis (PAGE), has two bands: 130 kD and 140 kD [50]. The receptor binds a lectin, wheat germ agglutinin, which has an affinity to N-acetyl glucosamine [51]. Similar proteins were found in FN receptors on normal rat kidney cells [51], BHK cells, NIH-3T3 cells, and mouse B16 melanoma cells [1]. Two monoclonal antibodies to avian FN, JG22, which binds to 150 kD and 125 kD proteins, and CSAT, which binds to 149 kD, 124 kD, and 115 kD proteins, inhibit cellular attachment to fibronectin and laminin substrates [52,53]. Microinjections of JG22 into the embryonic mesencephalon reduce the migration of neural crest cells at the site of injection [54]. In addition, JG22 was found to inhibit cell spread associated with serum spreading factor (vitronectin) and laminin [1]. The 47 kD glycoprotein was found on the surface of BHK cells [55]. This glycoprotein binds with wheat germ agglutinin [56].

B. Function

Fibronectin with its multiplicity of receptors can simultaneously bind to a solid surface and to a cell receptor [57-59]. This property allows for cell attachment to plastic surfaces of tissue culture flasks. Furthermore, FN facilitates cell spreading; thus it is an essential element of monolayer tissue culture initiation and maintenance [60]. When Latex microbeads are coated with FN, they attach to cultured cells. The cell attachment of beads occurs at a wide span of temperatures; however, only at about 37°C are the fibronectin-coated beads internalized by the cells [61]. Cell attachment is specifically associated with FN; if the same microbeads are coated with albumin, no attachment occurs [61]. The attachment of FN to beads can be considered an in vitro model of opsonization.

The cell attachment domain was found using competitive inhibition methods and small fragments were derived using enzyme digestion of FN or were obtained by artificial synthesis [4,62]. The cell attachment domain is a tetrapeptide Arg-Gly-Asp-Ser (RGDS). This tetrapeptide is a hydrophilic sequence and appears to be located in the beta-turn of

Fibronectin

the cell-binding region [63]. The affinity of the RGDS peptide based on competitive inhibition (Ki) of the binding of FN to fibroblast is 10^{-3}-10^{-4} M. Although the fragment has to be a tetrapeptide to remain active, valine, alanine or cysteine can all be substituted for serine. However, the remaining three amino acids cannot be altered; substitution of lysine for arginine, alamine for glycine, or asparagine for aspartate abolishes the cell attachment function of the tetrapeptide in each case [63]. The RGDS tetrapeptide competitively inhibits the FN-facilitated cell attachment and spreading. The tetrapeptide's competitive inhibition is not limited to FN, since the RGDS also inhibits the fibronigen-, laminin-, von Willebrand's factor-, and vitronectin-associated attachment, but it does not alter the collagen-cell attachment [64,65]. Although the primary structure of human complement components C3b and C3bi contains the peptide sequence Arg-Gly-Asp-Ser, the tetrapeptide does not competitively inhibit the binding of the C3b or C3bi to the surface of macrophages. However, substrate-bound synthetic peptide Gly-Arg-Gly-Asp-Ser-Pro (GRGDSP) can activate the C3b and C3bi receptors triggering phagocytosis [66].

Kinetic studies of FN attachment to fibroblasts led to a formulation of a two-pool model [57-59]. Pool I is short-term binding with a diffusion constant (Kd) of 3.6×10^{-8} M. Binding in this pool starts 2-10 min from the initiation of incubation, and after 15-20 min some bonds enter pool II. The binding in pool I is reversible [57]. When studied by immunofluorescence microscopy, the pattern of binding appears punctate or fibrillary. For all practical purposes, the binding in pool II is irreversible. It is most probably based on the formation of disulfide-bound multimers [89]. Studies using FN fragments have shown that only intact FN can bind to pool II and must be preceded by binding to pool I [59].

Fibronectin binding to a suspension of fibroblasts has shown that intact FN forms weaker bonds than its 75 kD fragment, with dissociation constants $Kd = 8 \times 10^{-7}$ M and 4×10^{-7} M, respectively [67]. However, smaller FN fragments form weaker bonds than intact FN. For example, inhibition constant (Ki) for an 11.5 kD fragment is 1-10×10^{-8} M [22,63]. For the GRGDS fragment, which contains the basic tetrapeptide, the Ki is 1×10^{-5} M [67]. These studies show that elements of FN fragments extending beyond the tetrapeptide potentiate the cell attachment [1]. Two hypotheses attempt to explain the potentiatin of cell attachment: (1) there is an additional cell attachment domain which is activated after the RGDS tetrapeptide is attached, and (2) the initial RGDS attachment triggers a conformational change in the surrounding FN molecule that increases its compatibility with the cell receptor [68] (Fig. 4). Regardless of the exact mechanism, the molecular bases for cell attachment are contained within a 70 kD FN fragment.

The properties of the 70 kD fragment were studied on fibroblasts, and the findings appear to be relevant to other cells. In hepatocytes, an 85 kD FN fragment containing GRGDS binds 30 times more strongly than the intact FN molecule [69]. The mechanism of binding of FN to platelets appears to differ from the binding to other cells. Although binding is mediated by the tetrapeptide domain, FN binds only to thrombin-activated platelets. Furthermore, binding of 120 kD fragment is three times weaker than that of a 11.5 kD fragment, which suggests that the FN receptor on platelets has a much lower requirement for regions outside the RGDS tetrapeptide to bind FN effectively [1].

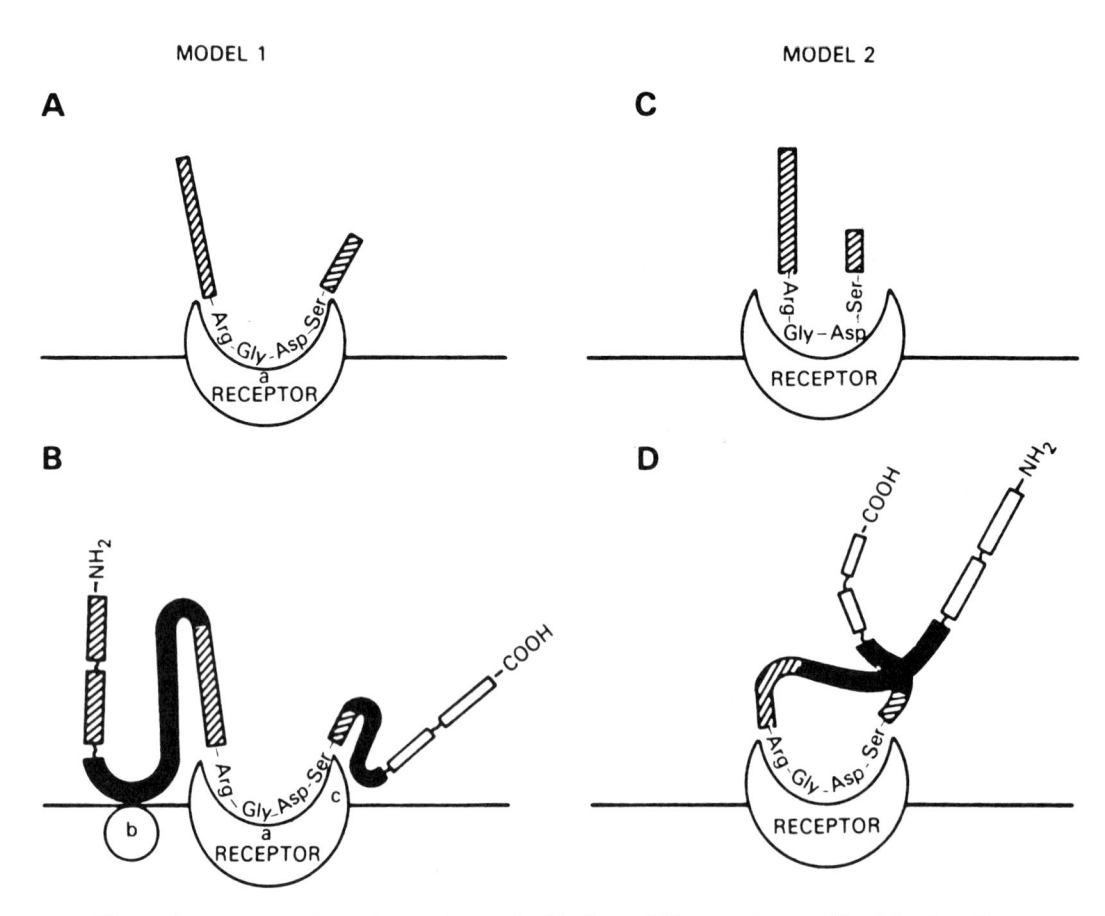

Figure 4 Two hypothetical models for the binding of fibronectin to a fibroblastic cell surface receptor. Arg-Gly-Asp-Ser is the minimum fibroblastic cell recognition signal for fibronectin. The 11,500 and 75,000-dalton cell-binding fragments are denoted by the hatched and solid areas, respectively. The remainder of a single fibronectin subunit is represented by the open boxes. (A,B) Model 1. In (A), the minimal recognition signal binds with only low affinity to a specific cell-surface receptor for fibronectin at the site indicated by a. For maximal interaction with its receptor, sections of the fibronectin molecule outside of the 11,500 dalton fragment shown must also contribute to the binding. In (B), two such possibilities are illustrated involving interactions in more distal portions of the 75 k fragment, Possibility b, which shows an interaction of fibronectin with another plasma membrane component (for example, a ganglioside), or with a second, distinct fibronectin receptor. Possibility c represents possible multiple interactions of fibronectin with a single receptor system. (C, D) Model 2. In (C), the minimal recognition signal binds weakly to a cell-surface receptor because of an inappropriate conformation. Interactions within the 75,000 dalton fragment, but outside of the 11,500 dalton fragment, are necessary in order for the recognition signal to acquire the optimal conformation for maximal binding (from ref. [1]).

III. FIBRONECTIN IN TUMORS

A. In Vitro Studies

Studies of FN in tissue cultures show that it plays an essential role in cell attachment and spreading [70,71]. Immunofluorescence studies and electron microscopy revealed a parallel orientation of the extracellular FN filaments and intracellular actin filaments. The bundles of actin filaments, known as tonofilaments or stress fibers, are responsible for the cellular shape and orientation, both of which are essential for the proper functioning of a cell (Fig. 5). The stress fibers do not extend extracellularily; thus, the essential information about the extracellular environment is transduced via FN connections. The interaction between FN and actin is reciprocal. On the one hand, cytochalasin, which disrupts actin filaments, also causes detachment of FN from the cell; on the other hand, addition of exogenous FN to tumor cells induces formation of stress fibers [72].

Decreased FN deposition was observed in the majority of the examined tumor cell lines [73,74]. When transformed fibroblasts were compared to their normal counterparts, the transformed cells showed features of neoplasia, such as uncontrolled proliferation and derangement of shape associated with the lack of tonofilaments. Analysis of the harvested tissue culture medium revealed low to no secretion of FN by the transformed cells. Subsequent addition of FN to the tissue culture medium was followed by the appearance of intracellular stress fibers, which is evidence that FN is decisive in shaping the cell morphology.

These results show that the neoplastic transformation of fibroblasts is associated with a defect in FN deposition; however, the transformed cells still retain functional FN receptors. Tonofilament formation can also be attained by the addition of biological modifiers, such as tissue polypeptide antigen 12-O-tetradecanoylphorbol-13-acetate (TPA) (phorbol esters), as shown in human erythroleukemia (HEL) cells [60,75]. Here, however, TPA appears to participate in the formation of FN receptors that are lacking in HEL cells and most probably on other cells originating from ascites and growing in suspension.

The defect of pericellular deposition of FN observed in the majority of neoplastic cell lines is still not understood. In hypothetical terms, there are the following possibilities: (1) a lack of FN secretion, (2) secretion of altered FN, and (3) increased pericellular proteolysis of the secreted FN. The first possibility seems unlikely since, as previously discussed, there is evidence that although secretion of FN by neoplastic cells is significantly diminished, it is present. Another possible cause of diminished pericellular FN deposition is an alteration of the secreted FN. Initial studies have shown that FN from transformed cell lines is indistinguishable from that of normal cell lines [76-78]. However, these studies were performed on a small number of cell lines that are not necessarily representative of the predominant trend [79]. In addition, point mutation in a key tetrapeptide of the FN cell attachment domain sufficient to decrease or abolish its affinity would not be detected by evaluating the products with polyclonal antibodies [80]. Finally, there is a significantly higher level of phosphorylation of serine residue in the RGDS tetrapeptide in transformed cells that can inactivate the cell attachment domain [76,81,82]. The enhanced proteolytic activity in the pericellular matrix has also been implicated as the cause of decreased cellular attachment of FN in transformed cells [83,84]. In addition, posttranslational defect of glycosylation can produce a FN molecule without carbohydrate moiety that may render it more susceptible to proteolytic digestion [49,68,85-87].

B. In Vivo Studies

Fibronectin plays a crucial role in the spread and metastasis of tumors. In the initial phase of metastasis, the cell detaches from the main tumor mass. Since FN anchors the cell to the extracellular matrix, the finding of an inverse correlation between the amount of pericellular FN and the metastatic potential of a tumor was to be expected [80]. In the next step, the tumor cell is bloodborne; therefore, it does not require FN attachment. Finally, after the cell arrives at the target organ, it penetrates the blood vessel wall and invades the surrounding parenchyma, for which the reestablishment of the cytoskeleton is essential. It begins with the attachment of FN to both the tumor cell surface and the basement membrane of the blood vessel. This attachment triggers the formation of tonofilaments, allowing for ameboid movement of the cell similar to diapedesis [88].

In vivo studies of FN in tumors evolved in two directions: biochemical studies of FN plasma levels, and immunohistochemical studies of FN antigens in the tissue. Initial immunohistochemical studies of tumor tissue showed a decrease of FN in tumors, which was most apparent in metastatic sites [18,45,89-119]. A general trend became obvious: the tumors that showed less FN were generally the more malignant, metastatic tumors. It was therefore suggested that a distinction between benign and malignant lesions can be made based on the relative amount of FN, with the malignant lesions not showing FN. Further studies showed, however, that this trend is not consistent; benign tumors are sometimes almost devoid of FN and malignant tumors are sometimes strongly positive for FN [80,94,120]. Furthermore, it was found that only carcinomas showed negative FN levels, whereas sarcomas, as well as benign soft tissue tumors, showed FN. This raised the possibility that FN immunohistochemistry can be used to differentiate carcinomas from sarcomas [121,122]. However, this differentiation is of limited practical value.

Early reports showed elevated plasma level of FN in patients and in experimental animals with tumors [123-127]. These findings were subsequently confirmed: higher plasma FN levels were shown in patients with carcinomas, sarcomas, and gliomas [128-134] and elevated urine FN levels were shown in patients with prostatic carcinoma [135] (Fig. 6). The cause of elevated plasma levels in tumor patients is still not understood. One plausible explanation is that the FN removed from the vicinity of tumor cells is released into plasma, causing elevation in plasma FN levels. Another is that the elevated plasma FN level results from a generalized immune response that triggers the FN-associated tumor cell opsonization [136]. It is also possible that both mechanisms play a role in affecting the levels of plasma FN in tumor patients. This phenomenon, although true, was shown by subsequent studies as not being consistent enough to be useful as a diagnostic marker. Some tumors are associated with lower, instead of higher, plasma FN levels and there are also considerable fluctuations in plasma FN levels among patients with the same tumor type [123,134,137,138]. In addition, other factors, such as inflammatory and infectious processes or obstructive hepatitis, can elevate plasma FN levels [139,140]. Conversely, trauma can lower the plasma FN level [141-144].

The findings of elevated plasma FN levels in patients with tumors and decreased FN production in tumor tissue were initially believed to represent potential methods for tumor detection and follow-up. It now appears that although clinical studies of FN have epidemiological significance (they show certain tendencies in tumor-burdened patients), no modalities are available that would allow for the diagnosis or follow-up of neoplasia based on serum level of FN or its immunohistochemical expression in tissue. Thus, there is a discrepancy between the significance of FN in tumor biology and its

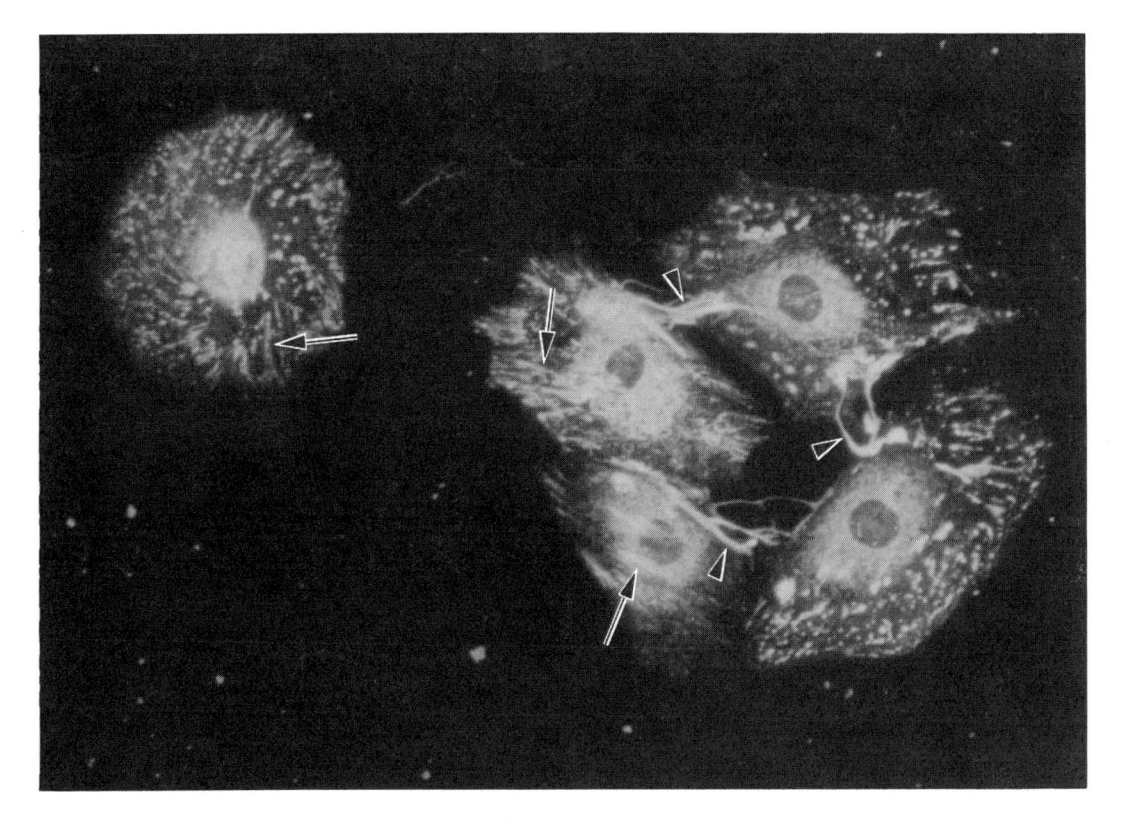

Figure 5 Fibronectin expression in tissue culture of human glioma (UC-11 MG). Note perinuclear distribution of intracellular FN, surface FN with parallel orientation reflecting the orientation of tonofilaments (arrows), and thick strands of extracellular FN (arrowheads) (indirect immunofluorescence using rabbit-antihuman FN antibody; original magnification 400).

clinical usefulness for tumor diagnosis. As discussed above, laboratory studies show alterations in the FN expression in both human and experimentally induced tumors. The qualitative and quantitative alterations in FN expression appear to correlate with the functional state of a tumor, as shown in metastasis or in FN-induced tumor differentiation, rather than with tumor type.

One of the factors limiting the practical application of diagnostic studies using FN is its ubiquitous character. FN is produced by the majority of cells; therefore, tumor detection based on FN polyclonal antibodies with a wide range of specificity is influenced by extraneous factors such as inflammation, trauma, or aging [145]. The problems with immunological diagnosis of tumors with FN as a tumor marker can be rectified by using monoclonal antibodies that detect FN isoforms generated by alternative splicing of mRNA precursors [37,46,146]. For example, monoclonal antibodies FN-3 and IST-9 detected a 10-fold higher level of FN ED segment in human SV-40 transformed fibroblasts than in nontransformed human fibroblasts [146]. In addition, the ED segment occurs only in cFN. Thus, the detection of FN using anti-ED monoclonal antibodies is

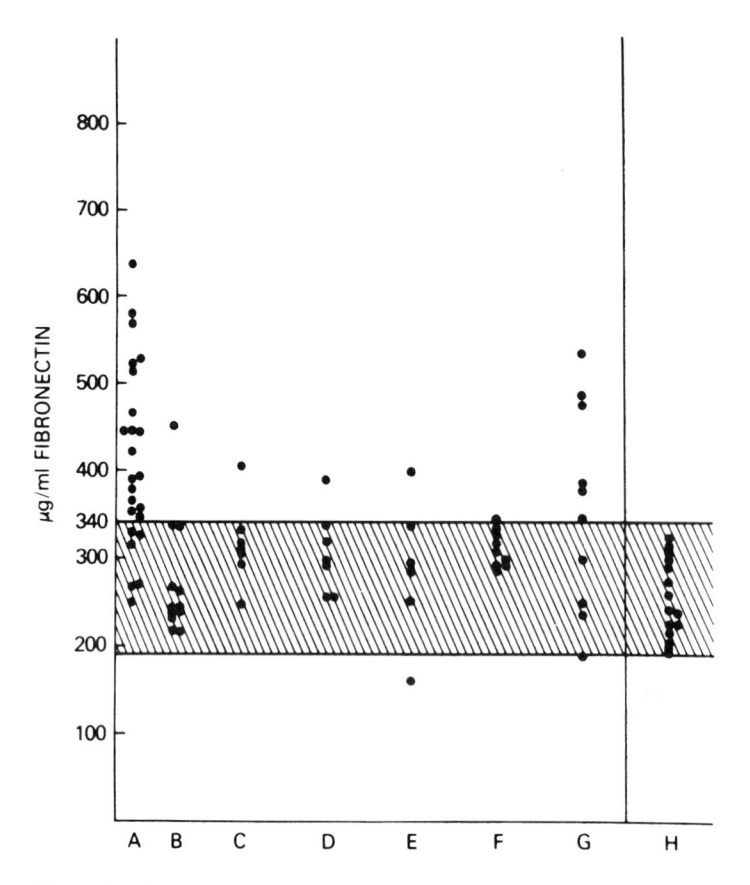

Figure 6 Plasma FN levels in patients with A, high-grade progressive astrocytoma; B, high-grade astrocytoma, stable course; C, low-grade astrocytoma; D, meningioma; E, spinal cord tumors; F, nonneoplastic CNS lesions; G, general neoplastic disease; and H, normal healthy volunteers (from ref. [131]).

not altered by the pFN level and is therefore suitable for plasma detection of tumor-derived cFN.

An alterantive approach also aimed at narrowing the specificity of anti-FN antibodies could be a simultaneous use of two antibodies to different epitopes on the tumor-released FN. Using a similar approach involving the simultaneous application of two antibodies—one against circulating tumor-associated FN and the other against tumor-released proteolytic enzyme—could be useful. Since there is evidence that FN is removed from the tumor vicinity by tumor-secreted proteolytic enzymes [147-149], a simultaneous elevation of both titers in plasma could be interpreted as increased tumor activity.

REFERENCES

1. Akiyama SK, Yamads SM. Fibronectin. Adv Enzymol 1987;59:1-57.
2. Hynes R. Molecular biology of fibronectin. Annu Rev Cell Biol 1985;1:67-90.
3. Oldberg A, Ruoslahti E. Interactions between chondroitin sulfate proteoglycan, fibronectin, and collagen. J Biol Chem 1982;257:4859-4863.

4. Pierschbacher MD, Hayman EG, Ruoslahti E. The cell attachment determinant in fibronectin. J Cell Biochem 1985;28:115-126.
5. Blumenstock FA, McKneally M, Saba TM. Quantification of immunoreactive opsonil alpha-2-surface binding (alpha-2-SB) glycoprotein in lung cancer patients. J Reticuloendothel Soc Abstr Suppl 1978;24:18a.
6. Blumenstock FA, Saba TM, Weber P, Laffin R. Biochemical and immunological characterization of human opsonil alpha-2-SB glycoprotein: its identity with cold-insoluble globulin. J Biol Chem 1978;253:4287-4291.
7. Bowersox JC, Richters A, Sorgente N. Altered characteristics of B16 melanoma cells induced chemically crosslinking fibronectin to cell surfaces. J Surg Oncol 1985; 29:11-14.
8. Kurkinen M, Veheri A, Roberts PJ, Stenman S. Sequential appearance of fibronectin and collagen in experimental granulation tissue. Lab Invest 1980;43:47-51.
9. Bronner-Fraser M. An antibody to a receptor for fibronectin and laminin perturbs cranial neural crest development. Dev Biol 1987;117:528-536.
10. Liotta LA, Rao CN, Barsky SH. Tumor invasion and the extracellular matrix. Lab Invest 1983;49:636-649.
11. Clark RA, DellaPelle P, Manseau E, Lanigan JM, Dvorak HF, Calvin RB. Blood vessel fibronectin increases in conjunction with endothelial cell proliferation and capillary ingrowth during wound healing. J Invest Dermatol 1982;79:269-276.
12. Clark RA, Lanigan JM, DellaPelle P, Manseau E, Dvorak HF, Calvin RB. Fibronectin and fibrin provide a provisional matrix for epidermal cell migration during wound reepithelialization. J Invest Dermatol 1982;79:264-269.
13. Fujikawa LS, Foster CS, Harrist TJ, Lanigan JM, Calvin RB. Fibronectin in healing rabbit corneal wounds. Lab Invest 1981;45:120-129.
14. Grinnell F, Billingham RE, Burgess L. Distribution of protein during wound healing in vivo. J Invest Dermatol 1981;76:81-189.
15. Gehlsen KR, Argraves WS, Pierschbacher MD, Ruoslahti E. Inhibition of in vitro tumor cell invasion by Arg-Gly-Asp-containing synthetic peptides. J Cell Biol 1988; 106:925-930.
16. Martin DE, Reece M, Reese AC. Effect of plasma fibronectin, macrophages, and glycosaminoglycans on tumor cell growth. Cancer Invest 1984;2:339-345.
17. Boucaut JC, Darribere T, Poole TJ, Aoyama M, Yamada KM, Thiery JP. Biologically active synthetic peptides as probes of embryonic development: a competitive peptide inhibitor of fibronectin function inhibits gastrulation in amphibian embryos and neural crest cell migration in avian embryos. J Cell Biol 1984;99:1822-1830.
18. Dunnington DJ, Kim U, Hughes CM, Monaghan P, Ormerod EJ, Rudland PS. Loss of myoepithelial cell characteristics in metastasizing rat mammary tumors reltive to their non-metastasizing counterparts. J Natl Cancer Inst 1984;72:455-466.
19. Thiery JP, Duband JL, Tucker GC. Cell migration in the vertebrate embryo: role of cell adhesion and tissue environment in pattern formation. Annu Rev Cell Biol 1985; 1:91-113.
20. Dvorak HF, Form DM, Manseau EJ, Smith BD. Pathogenesis of desmoplasia. I. Immunofluorescence identification and localization of some structural proteins of line 1 and line 10 guinea pig tumors and of healing wounds. J Natl Cancer Inst 1984; 73:1195-1205.
21. Mosher DF. Physiology of fibronectin. Annu Rev Med 1984;35:561-575.
22. Yamada KM, Akiyama SK, Hasegawa T, Hasegawa E, Humphries MJ, Kennedy DW, Nagata K, Urushihara H, Olden K, Chen W-T. Recent advances in research on fibronectin and other cell attachment proteins. J Cell Biochem 1985;28:79-97.
23. Akiyama SK, Johnson MD. Fibronectin in evolution: presence in invertebrates and isolation from *Microciona prolifera*. Comp Biochem Physiol 1983;76B:687-694.

24. Schwarzbauer JE, Paul JI, Hynes RO. On the origin of species of fibronectin. Proc Natl Acad Sci USA 1985;82:1424-1428.
25. Schoen RC, Bentley KL, Klebe RJ. Monoclonal antibody against human fibronectin which inhibits cell attachment. Hybridoma 1982;1:99-108.
26. Tsumagari J. Fibronectin in human hepatocellular carcinoma (HCC) and HCC cell lines. Acta Pathol Jpn 1987;37:413-423.
27. Erickson HP, Carrell N, McDonagh J. Fibronectin molecule visualized in electron microscopy: a long, thin, flexible strand. J Cell Biol 1981;91:673-678.
28. Erickson HP, Carrell NA. Fibronectin in extended and compact conformations. Electron microscopy and sedimentation analysis. J Biol Chem 1983;258:14539-14544.
29. Tooney NM, Mosesson MW, Amrani DL, Hainfeld JF, Wall JS. Solution and surface effects on plasma fibronectin structure. J Cell Biol 1983;97:1686-1692.
30. Engel J, Odermatt E, Engel A, Madri JA, Furthmayr H, Rohde H, Timpl R. Shapes, domain organizations and flexibility of laminin and fibronectin, two multifunctional proteins of the extracellular matrix. J Mol Biol 1981;150:97-120.
31. Skorstengaard K, Jensen MS, Petersen TE, Magnusson S. Purification and complete primary structures of heparin-, cell-, and DNA-binding domains of bovine plasma fibronectin. Eur J Biochem 1986;54:15-29.
32. Vibe-Pedersen K, Sahl P, Skorstengaard K, Petersen TE. Amino acid sequence of a peptide from bovine plasma fibronectin containing a free sulfhydryl group (cysteine). FEBS Lett 1982;142:27-30.
33. Garcia-Pardo A, Pearlstein E, Frangione B. Primary structure of human plasma fibronectin. The 29,000-Dalton NH_2-terminal domain. J Biol Chem 1983;258:12670-12674.
34. Gold LI, Garcia-Pardo A, Frangione B, Franklin EC, Pearlstein E. Subtilisin and cyanogen bromide cleavage products of fibronectin that retain gelatin-binding activity. Proc Natl Acad Sci USA 1979;76:4803-4807.
35. Pande H, Shiverly JE. NH_2-terminal sequences of DNA-, heparin-, and gelatin-binding tryptic fragments from human plasma fibronectin. Arch Biochem Biophys 1982;213:258-265.
36. Pande H, Calaycay J, Hawke D, Ben-Avram CM, Shively JE. Primary structure of glycosylated DNA-binding domain in human plasma fibronectin. J Biol Chem 1985;260:2301-2306.
37. Castellani P, Siri A, Rosellini C, Infusini E, Borsi L, Zari L. Transformed human cell release different fibronectin variants than do normal cells. J Cell Biol 1986;103:1671-1677.
38. Kornblihtt AR, Umezawa K, Vibe-Pedersen K, Baralle FE. Primary structure of human fibronectin: differential splicing may generate at least 10 polypeptides from a single gene. EMBO J 1985;4:1755-1759.
39. Petersen TE, Thogersen HC, Skorstengaard K, Vibe-Pedersen K, Sahl P, Sottrup-Jensen L, Magnusson S. Partial primary structure of bovine plasma fibronectin: Three types of internal homology. Proc Natl Acad Sci USA 1983;80:137-141.
40. Skorstengaard K, Thogersen HC, Petersen TE. Complete primary structure of the collagen-binding domain of bovine fibronectin. Eur J Biochem 1984;140:235-243.
41. Pierschbacher MD, Ruoslahti E, Sandelin J, Lind P, Peterson PA. The cell attachment domain of fibronectin. Determination of the primary structure. J Biol Chem 1982;257:9593-9597.
42. Pierschbacher MD, Hayman EG, Ruoslahti E. Synthetic peptide with cell attachment activity of fibronectin. Proc Natl Acad Sci USA 1983;80:1224-1227.
43. Kornblihtt AR, Vibe-Pedersen K, Baralle FE. Isolation and characterization of cDNA for human and bovine fibronectins. Proc Natl Acad Sci USA 1983;80:3218-3222.

44. Paul JI, Hynes RO. Multiple fibronectin subunits and their posttranslational modifications. J Biol Chem 1984;259:13477-13487.
45. Vibe-Pedersen K, Kornblihtt AR, Baralle FE. Expression of human alpha-globin/fibronectin gene hybrid generates two mRNAs by alternative splicing. EMBO J 1984;3:2511-2516.
46. Schwarzbauer JE. Three different fibronectin mRNAs arise by alternative splicing within the coding region. Cell 1983;35:421-431.
47. Kornblihtt AR, Vibe-Pedersen K, Baralle FE. Human fibronectin: cell specific alternative mRNA splicing generates polypeptide chains differing in the number of internal repeats. Nucleic Acid Res 1984;12:5853-5868.
48. Umezawa K. Kornblihtt AR, Baralle FE. Isolation and characterization of cDNA clones for human liver fibronectin. FEBS Lett 1985;186:31-34.
49. Bernard MP, Kolbe M, Weil D, Chu M-L. Human cellular fibronectin: Comparison of the carboxyl-terminal portion with rat identifies primary structural domains separated by hypervariable regions. Biochemistry 1985;24:2698-2704.
50. Akiyama SK, Yamada SS, Yamada KM. Characterization of 140-kD avian cell surface antigen as fibronectin-binding molecule. J Cell Biol 1986;102:442-448.
51. Pytela R, Pierschbacher MD, Ruoslahti E. Identification and isolation of 140 kD cell surface glycoprotein with properties expected of a fibronectin receptor. Cell 1985;40:191-198.
52. Decker C, Greggs R, Duggan K, Stubbs J, Horwitz A. Adhesive multiplicity in the interaction of embryonic fibroblasts and myoblasts with extracellular matrices. J Cell Biol 1984;99:1398-1404.
53. Horwitz A, Duggan K, Greggs R, Decker C, Buck C. The cell substrate attachment (CSAT) antigen has properties of a receptor for laminin and fibronectin. J Cell Biol 1985;101:2134-2144.
54. Bronner-Fraser M. Alterations in neural crest migration by a monoclonal antibody that affects cell adhesion. J Cell Biol 1985;101:610-617.
55. Aplin JD, Hughes RC, Jaffe CL, Sharon N. Reversible cross-linking of cellular components of adherent fibroblasts to fibronectin and lectin-coated substrata. Exp Cell Res 1981;134:488-494.
56. Oppenheimer-Marks N, Grinnell F. Inhibition of fibronectin receptor function by antibodies against baby hamster kidney cell wheat germ agglutinin receptors. J Cell Biol 1982;95:876-884.
57. McKeowan-Longo PJ, Mosher DF. Binding of plasma fibronectin to cell layers of human skin fibroblasts. J Cell Biol 1983;97:466-472.
58. McKeowan-Longo PJ, Mosher DF. Mechanism of formation of disulfide-bonded multimers of plasma fibronectin in cell layers of cultured human fibroblasts. J Biol Chem 1984;259:12210-12215.
59. McKeowan-Longo PJ, Mosher DF. Interaction of the 70,000-mol-wt amino-terminal fragment of fibronectin with the matrix-assembly receptor of fibroblasts. J Cell Biol 1985;100:364-374.
60. Jarvinen M, Ylanne J, Vartio T, Virtanen I. Tumor promoter and fibronectin induce actin stress fibers and focal adhesion sites in spreading human erythroleukemia (HEL) cells. Eur J Cell Biol 1987;44:238-246.
61. Grinnell FB, Geiger B. Interaction of fibronectin-coated beads with attached and spread fibroblasts. Exp Cell Res 1986;162:449-461.
62. McCarthy JB, Hagen ST, Furcht LT. Human fibronectin contains distinct adhesion- and motility-promoting domains for metastatic melanoma cells. J Cell Biol 1986;102:179-188.
63. Pierschbacher MD, Ruoslahti E. Cell attachment activity of fibronectin can be duplicated by small synthetic fragments of the molecule. Nature (London) 1984;309:30-33.

64. Hayman EG, Pierschbacher MD, Ruoslahti E. Detachment of cells from culture substrate by soluble fibronectin peptides. J Cell Biol 1985;100:1948-1954.
65. Silnutzer JE, Barnes DW. Effects of fibronectin-related peptides on cell spreading. In Vitro 1985;21:73-78.
66. Wright SD, Meyer BC. Fibronectin receptor of human macrophages recognizes the sequence Arg-Gly-Asp-Ser. J Exp Med 1985;162:762-767.
67. Akiyama SK, Hasegawa E, Hasegawa T, Yamada KM. The interaction of fibronectin fragments with fibroblastic cells. J Biol Chem 1985;260:13256-13260.
68. Bernard BA, Yamada KM, Olden K. Carbohydrates selectively protect a specific domain of fibronectin against proteases. J Biol Chem 1982;257:8549-8554.
69. Johansson S. Demonstration of high affinity fibronectin receptor on rat hepatocytes in suspension. J Biol Chem 1985;260:1557-1561.
70. Alitalo K, Kurkinen M, Vaheri A, Krieg T, Timpl R. Extracellular matrix components synthesized by human amniotic epithelial cells in culture. Cell 1980;19:1053-1062.
71. Pande G, Khar A. Differential expression of fibronectin in rat histiocytoma: possible role of fibronectin in tumor cell adhesion. Exp Cell Res 1988;174:41-48.
72. Kurkinen M, Wartiovaara J, Vaheri A. Cytochalasin B releases a major surface-associated glycoprotein, fibronectin, from cultured fibroblasts. Exp Cell Res 1978; 111:127-137.
73. Hayman EG, Engvall E, Ruoslahti E. Concomitant loss of cell surface fibronectin and laminin from transformed rat kidney cells. J Cell Biol 1981;88:352-357.
74. Neri A, Ruoslahti E, Nicolson GL. Distribution of fibronectin on clonal cell lines of a rat mammary adenocarcinoma growing in vitro and in vivo at primary and metastatic sites. Cancer Res 1981;41:5082-5095.
75. Hynes RO. Relationship between fibronectin and the cytoskeleton. Cell Surf Rev 1981;7:97-136.
76. Ali IU, Hunter T. Structural comparison of fibronectins from normal and transformed cells. J Biol Chem 1981;256:7671-7677.
77. Fukuda M, Hakomori S. Carbohydrate structure of galactoprotein A, a major transformation-sensitive glycoprotein released from hamster embryo fibroblasts. J Biol Chem 1979;254:5451-5457.
78. Wagner DD, Raymond I, Destree AT, Hynes RO. Similarities and differences between the fibronectins of normal and transformed hamster cells. J Biol Chem 1981; 256:11708-11715.
79. Borsi L, Allemanni G, Castellani P, Rosellini C, DiVinci A, Bardi L. Structural differences in the cell binding region of human fibronectin molecules isolated from cultured normal and tumor-derived human cells. FEBS Lett 1985;192:71-74.
80. Ruoslahti E. Fibronectin in cell adhesion and invasion. Cancer Metastasis Rev 1984; 3:43-51.
81. Ali IU. Phosphorylatin of fibronectin in quiescent and growing cell cultures. FEBS Lett 1983;151:45-47.
82. Ali IU, Rasheed S. Human normal and tumor cell lines: comparison of fibronectins and collagens. Cancer Invest 1987;5:17-24.
83. Nielson SE, Puck TT. Deposition of fibronectin in the course of reverse transformation of Chinese Hamster ovary cells by cyclic AMP. Proc Natl Acad Sci USA 1980; 77:985-989.
84. Sas DF, McCarthy JB, Furcht LT. Clearing and release of basement membrane proteins from substrates by metastatic tumor cell variants. Cancer Res 1986;46:3082-3089.
85. Armugham RG, Tanzer ML. Abnormal glycosylation of human cellular fibronectin in the presence of Swainsomine. J Biol Chem 1982;258:11883-11889.

86. Carnemolla B, Castellani P, Cutolo M, Boasi L, Zardi L. Lack of sialic acid in synovial fluid fibronectin. FEBS Lett 1984;171:285-288.
87. Zhu C-R B, Fisher SF, Pande H, Calaycay J, Shively JE, Laine RA. Human placental (fetal) fibronectin: increased glycosylation and higher protease resistance than plasma fibronectin. J Biol Chem 1984;259:3962-3970.
88. Morre DJ, Creck KE, Morre DM, Richardson CL. Glycosylation reactions and tumor establishment: modulation by vitamin A. Ann NY Acad Sci 1981;359: 367-388.
89. Alitalo K, Vaheri A. Pericellular matrix in malignant transofrmation. Adv Cancer Res 1982;37:111-158.
90. Alitalo K, Keski-Oja J, Hedman K, Vaheri A. Loss of different pericellular matrix components of rat cells transformed with a T-class ts mutant of Rous sarcoma virus. Virology 1982;119:347-357.
91. Asch BB, Kamat BR, Burstein NA. Interactions of normal dysplastic, and malignant mammary epithelial cells with fibronectin in vivo and in vitro. Cancer Res 1981;41:2115-2125.
92. Blumenstock F, Saba TM, Weber P, Cho E. Purification and biochemical characterization of macrophage stimulating alpha-2-globulin protein. J Reticuloendothel Soc 1976;19:157-172.
93. Bumol TF, Reisfeld RA. Biosynthesis and secretion of fibronectin in human melanoma cells. J Cell Biochem 1983;21:129-140.
94. Chen LB, Burridge K, Murray A, Walsh ML, Copple CD, Bushnell A, McDougall JK, Gallimore PH. Modulation of cell surface glycocalyx: Studies on large, external transformation-sensitive protein. Ann NY Acad Sci 1978;312:366-381.
95. Chen LB, Summerhayes I. Hsieh P, Gallimore PH. Possible role of fibronectin in malignancy. J Supramol Struct 1979;12:139-150.
96. Crouch EC, Stone KR, Bloch M, McDevitt RW. Heterogeneity in the production of collagens and fibronectin by morphologically distinct clones of a human tumor cell line: evidence for intratumoral diversity in matrix protein biosynthesis. Cancer Res 1987;47:6086-6092.
97. Der CJ, Stanbridge EJ. Lack of correlation between the decreased expression of cell surface LETS protein and tumorgenicity in human cell hybrids. Cell 1978; 15:1241-1251.
98. Furcht LT, Mosher DF, Wendelschafer-Crabb G, Woodbridge PA, Foidart J-M. Dexamethasone-induced accumulation of a fibronectin and collagen extracellular matrix in transformed human cells. Nature 1979;277:393-395.
99. Gahmberg CG, Kiehn D, Hakomori S. Changes in surface-labeled galactoprotein and in glycolipid concentration in cells transformed by temperature-sensitive polyoma virus mutant. Nature 1974;248:413-415.
100. Gilbert M-A, Noel P, Faucon M, Ceccatty MP. Comparative immunohistochemical localization of fibronectin and actin in human breast tumor cells in vivo and in vitro. I. Fibroadenoma. Virchows Arch 1982;B40:99-112.
101. Grimwood RE, Huff JC, Harbell JW, Clark RAF. Fibronectin in basal cell epithelioma: sources and significance. J Invest Dermatol 1984;82:145-149.
102. Hayman EG, Engvall E, Ruoslahti E. Butyrate restores fibronectin at cell surface of transformed cells. Exp Cell Res 1980;127:478-481.
103. Hayman EG, Oldberg A, Martin GR, Ruoslahti E. Co-distribution of heparan sulfate proteoglycan, laminin, and fibronectin in the extracellular matrix of normal rat kidney cells and their coordinate absence in transformed cells. J Cell Biol 1982; 94:28-35.
104. Hynes RO. Role of surface alterations in cell transformation: the importance of proteases and surface proteins. Cell 1974;1:147-156.

105. Hynes RO. Cell surface alterations in cell transformation. Biochim Biophys Acta 1976;458:73-107.
106. Jones PA, Laug WE, Gardner A, Nye CA, Fink LM, Benedict WF. In vitro correlates of transformation in C3H/10T-1/2 clone 8 mouse cells. Cancer Res 1976;36: 2863-2867.
107. Kahn P, Shin S-I. Cellular tumorgenicity in nude mice. Test of association among loss of cell-surface fibronectin, anchorage independence and tumor-forming ability. Cell Biol 1979;82:1-16.
108. Labat-Robert J, Birembant P, Adnet JJ, Mercantini F, Robert L. Loss of fibronectin in human breast cancer. Cell Biol Int Rep 1980;4:609-619.
109. Labat-Robert J, Birembaut P, Robert L. Adnett JJ. Modification of fibronectin distribution pattern in solid human tumors. Diag. Histopathol 1981;4:299-306.
110. Maizel JV. Polyacrylamide gel electrophoresis of viral proteins. Methods Virol 1971;5:179-246.
111. McKeever PE, Chronwall BM, Houff SA, Sever JL, Kornblith PL, Padgett BL, Walker DL, London WT. Glial and divergent cells in primary central nervous system tumor induced by JC virus isolated from human progressive multifocal leukoencephalopathy (PML). In: Polyomaviruses and human neurological diseases. New York: Alan R. Liss, 1983:239-251.
112. Nicolson GL. Metastatic tumor cell attachment and invasion utilizing vascular endothelial cell monolayers. J Histochem Cytochem 1982;30:214-220.
113. Pearlstein E, Gold LI, Garcia-Pardo A. Fibronectin: a review of its structure and biological activity. Mol Cell Biochem 1980;29:103-128.
114. Rajaraman R, Lonergan K. Fibronectin distribution pattern and three-dimensional growth behavior of mammalian cell under anchorage-independent conditions. Exp Cell Biol 1982;50:2-12.
115. Ruoslahti E, Englvall E, Hayman EG. Fibronectin: current concepts to its structure and functions. Collagen Res 1981;1:95-128.
116. Vaheri A, Ruoslahti E. Disappearance of a major cell-type specific surface glycoprotein antigen (SF) after transformation of fibroblasts by Rous sarcoma virus. Int J Cancer 1974;13:579-586.
117. Vaheri A, Ruoslahti E. Fibroblast surface antigen produced but not retained by virus-transformed human cells. J Exp Med 1975;142:530-535.
118. Vaheri A, Mosher DF. High molecular weight, cell-associated glycoprotein (fibronectin) lost in malignant transformation. Biochim Biophys Acta 1978;516:1-35.
119. Vaheri A, Alitalo K, Hedman K, Kurkinen M, Vartio T. Fibronectin and its loss in malignant transformation. Colloq Int Cent Nat Rech Sci 1979;287:249-254.
120. Smith HS, Riggs JL, Mosesson MW. Production of fibronectin by human epithelial cells in culture. Cancer Res 1979;39:4138-4144.
121. Scarpa S, Modesti A, Triche TY. Extracellular matrix synthesis by undifferentiated childhood tumor cell lines. Am J Pathol 1987;129:74-85.
122. Stenman S, Vaheri A. Fibronectin in human solid tumors. Int J Cancer 1981;27: 427-435.
123. Bruhn HD, Heimburger N. Factor VIII-related antigen and cold insoluble globulin in leukemia and carcinomas. Haemostasis 1976;5:189-192.
124. Mosher DF, Williams EM. Fibronectin concentration is increased in plasma of severely ill patients with disseminated intravascular coagulation. J Lab Clin Med 1978;91:729-733.
125. Puricelli L, de Kier BJ, Eiyan AM, Entin D, DeLusting ES. Levels of plasmatic fibronectin in mice bearing adenocarcinomas of different metastasizing ability. Cancer Lett 1985;29:189-195.

126. Siri A, Carnemolla B, Castellani P, Balza E, Raffanti S, Zardi L. Increased plasma fibronectin concentrations in tumor bearing mice. Cancer Lett 1983;21:117-123.
127. Zardi L, Lecconi C, Barbieri O, Carnemolla B, Picca M, Santi L. Concentration of fibronectin in plasma of tumor-bearing mice and synthesis by Ehrlich ascites tumor cell. Cancer Res 1979;39:3774-3779.
128. Choate JJ, Mosher DF. Fibronectin concentration in plasma of patients with breast cancer, colon cancer, and acute leukemia. Cancer 1983;51:1142-1147.
129. Hafter R, Klaubert W, Gollwitzer R, von Hugo R, Graeff H. Crosslinked fibrin derivatives and fibronectin in ascitic fluid from patients with ovarian cancer compared to ascitic fluid in liver cirrhosis. Thrombosis Res 1984;35:53-64.
130. Jones TR, Ruoslanti E, Schold SC, Bigner DD, Fibronectin and glial fibrillary acidic protein expression in normal human and anaplastic human gliomas. Cancer Res 1982;42:168-177.
131. Sawaya R, Cummins CJ, Smith BH, Kornbliht PL. Plasma fibronectin in patients with brain tumors. Neurosurgery 1985;16:161-165.
132. Scholmerich J, Volk BA, Kottgen E, Ehlers S, Gerok W. Differentialdiagnose des Aszites mittels einfacher laborparameter. Z Gastroent 1982;20:517-518.
133. Stathakis N, Fountanas A, Tsianoa E. Plasma fibronectin in normal subjects and in various disease states. J Clin Pathol 1981;34:504-508.
134. Todd HD, Coffee MS, Waalkes TP, Abeloff MD, Parsons RG. Serum levels of fibronectin and a fibronectin-like DNA-binding protein in patients with various diseases. J Natl Cancer Inst 1980;65:901-904.
135. Webb KS, Lin GH. Urinary fibronectin, potential as a biomarker in prostatic cancer. Invest Urol 1980;17:401-404.
136. Saba TM, Gregory TJ, Blumenstock FA. Circulating immunoactive and biossayable opsonic plasma fibronectin during experimental tumor growth. Br J Cancer 1980;41:956-965.
137. Pisano JC, Jackson JP, DiLuzio NR. Dimensions of humoral recognition factor depletion in carcinomatous patients. Cancer Res 1972;32:11-15.
138. Tatra G, Nasr F. Plasmafibronektin bei Tomoren. Des Weiblichen Genitaltraktes. Strahlentherapie 1983;159:606-609.
139. Lanser ME, Saba TM. Opsonic glycoprotein (plasma fibronectin) enhances bacterial phagocytosis by neutrophils. J Reticuloendothel Soc 1980;28:1a;
140. Matsuda M, Yamanaka T, Matsuda A. Distribution of fibronectin in plasma and in liver disease. Clin Chim Acta 1982;118:191-199.
141. Cho E, Saba TM. Effect of bilateral nephrectomy on plasma fibronectin level: its influence on acute fibronectin response to trauma and RE blockade. Adv Shock Res 1982;8:99-110.
142. Robbins AB, Doran JE, Reese AC. Effect of cholecystectomy on cold insoluble globulin. Arch Surg 1980;115:1207-1209.
143. Robbins AB, Doran JE, Reese AC. Cold insoluble globulin levels in operative trauma: serum depletion, wound sequestration, and biological activity: an experimental and clinical study. Am Surgeon 1980;46:663-672.
144. Saba TM, Scovill WA. Effect of surgical trauma on host defense. Surg Annu 1975; 7:71-102.
145. Labat-Robert J, Robert L. Modifications of fibronectin in age-related diseases: diabetes and cancer. Arch Gerontol Geriatr 1984;3:1-10.
146. Borsi L, Carnemolla B, Castellani P, Rosellini C, Vecchio D, Allemanni G, Chang SE, Taylor-Papadimitrion J, Pande H. Monoclonal antibodies in the analysis of fibronectin isoforms generated by alternative splicing of mRNA precursors in normal and transformed human cells. J Cell Biol 1987;104:595-600.
147. Liotta LA. Mechanism of cancer invasion and metastasis. In: DeVita VT, Hellman

S, Rosenberg SA, eds. Important advances in oncology 1985. Philadelphia: J.B. Lippincott, Co., 1985:28-41.
148. Nicolson GL. Transmembrane control of the receptors on normal and tumor cells. II. Surface changes associated with transformation and malignancy. Biochim Biophys Acta 1976;458:1-72.
149. Sloane BF, Dunn JR, Hann KV. Lysosomal cathepsin B: correlation with metastatic potential. Science 1981;212:1151-1153.

35
Immune Complexes and Malignancy

DANIEL R. VLOCK

Pittsburgh Cancer Institute, University of Pittsburgh School of Medicine, and Veterans Administration Medical Center, Pittsburgh, Pennsylvania

Immune complexes have been implicated in a wide variety of disease states. Their role in autoimmune diseases, glomerulonephritis, infectious diseases, and neoplasia has been under investigation for over 20 years. The role of immune complexes in malignancy is incompletely understood. There has been a large body of literature describing immune complexes in cancer. However, it is not clear if immune complexes in malignancy represent a response to the presence of cancer or are somehow involved with its pathogenesis.

This chapter will address the role of immune complexes in malignancy from a number of vantage points. What approaches are available to detect immune complexes and what are the limitations of the methods used? Do tumor-associated immune complexes produce systemic manifestations? Beyond their relatively uncommon paraneoplastic effects, it appears that immune complexes may be present in the vast majority of patients with cancer. If this is the case, is there a relationship between the level of immune complexes and the natural history of cancer? And, as a practical extension, is there a value in measuring immune complexes as a prognostic indicator?

Beyond an analysis of the relationship between immune complexes and clinical course, questions regarding the nature of the immune complexes present in the sera of patients with cancer are beginning to be addressed. While the makeup of tumor-associated immune complexes remains largely unknown, studies are beginning to characterize the antigens detected by host-derived antibodies. This chapter will summarize what is known about the detection, incidence, and systemic manifestations of immune complexes in malignancy and discuss their relationship to prognosis, pathophysiology, and potential therapeutic manipulations in patients with cancer.

I. DETECTION OF IMMUNE COMPLEXES

Technical difficulties encountered in the detection of immune complexes have hampered their evaluation. Tests for immune complexes should be sensitive, specific, and reproducible. Such tests should detect complement- and non-complement-binding immune complexes incorporating immunoglobulins of all classes. Tests should be able to differ-

Table 1 Neoplasia Associated with Glomerulopathies

Breast Carcinoma

Bronchogenic
   Squamous cell carcinoma
   Adenocarcinoma
   Small-cell carcinoma

Gastrointestinal
   Biliary
   Colonic
   Gastric
   Hepatoma
   Pancreatic
   Rectal

Genitourinary
   Hypernephroma
   Prostatic

Gynecological
   Ovarian
   Uterine
   Vulval
   Penile

Head and Neck
   Squamous cell carcinoma
   Thyroid

Hematological
   Lymphoma
   Leukemia
   Hodgkin's disease

Skin
   Basal cell carcinoma
   Melanoma
   Squamous cell carcinoma
   Condyloma acuminatum

Undifferentiated

entiate between free immunoglobulin and immunoglobulin attached to antigen. At present, no such test exists. Instead, one must contend with a large number of assays that are each capable of detecting only a subset of immune complexes present. In view of this, it is not surprising that numerous methods have been devised. A detailed description of these assays is beyond the scope of this paper, and has been reviewed elsewhere [1-3] However, a basic understanding of the methods used and a familiarity with a few of the more commonly used techniques are necessary.

There are two general approaches to the detection of immune complexes: demonstration of immune complex deposition in tissue and detection of circulating immune complexes. Immune complex deposition in tissue has had its greatest impact in the evaluation of renal diseases. The use of immunofluorescence and other immunohistochemical methods has identified granular or lumpy-bumpy deposits of immunoglobulin and complement on renal glomeruli. These lesions have been found in a wide variety of neoplasms (Table 1) and are similar to those produced in experimentally induced immune complex diseases [4]. In addition, if sufficient tissue is available, immune complexes can be eluted to allow subsequent analysis and identification of the antigen and antibody retrieved [5,6]. However, renal biopsy is not an acceptable method for routine immune complex detection and attention has turned to serological methods for the detection of *circulating* immune complexes.

As is suggested by their presence in serum, circulating immune complexes are infrequently deposited in tissue. This is due to a host of factors: the size and chemical composition of the antigen, the immunoglobulin class, and the ratio of antigen to antibody. Detection of circulating immune complexes may occur by either antigen-specific or antigen-non-specific methods. Use of antigen-specific methods has been unusual in medicine in general and particularly in oncology. These methods presume that the antigen is known. If this is the case, as with immune complexes containing viruses or drugs, detection can be accomplished by electromicroscopy [6], countercurrent immunoelectrophoresis [7,8], or anti-immunoglobulin antibody [9]. However, in most cases the antigen is not known, which diminishes the feasibility of antigen-specific approaches. In the majority of cases, one must therefore rely upon antigen-non-specific methods that use physiochemical and biological properties to differentiate immune complexes from free immunoglobulin. As previously mentioned, the number of methods available is extensive. Two methods deserve additional notice because they figure prominently in the ensuing discussion.

Assays using C1q, the first component of complement, are frequently used to detect circulating immune complexes. C1q binds to IgG_1, IgG_2, IgG_4 and IgM. Enhanced reactivity is found when these immunoglobulins are aggregated [10]. Various techniques involving precipitation [11] and C1q binding with polyethylene glycol precipitation [12] or as a solid phase assay with radiolabeled [13] or enzyme-linked immunoabsorbent assays (ELISA) [14] have been used. However, any assay utilizing C1q must take into account its limited specificity (it will not detect immune complexes composed of IgG_3, IgE, IgA) and its nonspecific binding with various polyanions such as DNA and heparin [15-17].

The Raji cell assay uses cultured lymphoblastoid B cells. These cells bear receptors with a high affinity for complement and immune complexes. However, unlike normal cells, they lack membrane-bound immunoglobulin. While Raji cells do have Fc receptors for IgG, they are of low affinity and do not interfere with the detection of immune complexes [18]. Therefore, in the assays used, any immune complexes bound to the cells reflect the amount of immune complexes in the test sample. Detection of immune complexes adhered to Raji cells can be accomplished utilizing immunofluorescence [19], radioimmunoassay [20], and ELISA [21]. A major disadvantage of the Raji cell assay is that it depends upon live cells that must be uniform and viable. The presence of a large percentage of nonviable cells can lead to nonspecific binding and an unacceptably high background.

The multiplicity of assays available has made it difficult to determine which of the assays currently available may have clinical utility. Herberman et al. [22] reported the results of an international study evaluating the 6 different assays in 30 patients with breast cancer, 30 patients with benign breast disease, and 30 normal controls. Although differences in levels of immune complexes were noted, none of the assays demonstrated any sufficient discriminatory ability. It was noted that the variations in the procedures used can have a marked effect on the outcome of assays.

Table 2 Ability of Various C1q Assays (% Positive) to Detect Immune Complexes in Patients with Cancer

Assay	Seminoma	Lymphoma	Nasopharyngeal carcinoma
C1q binding	14	70	44
Solid phase C1q RIA	0	30	11
C1q deviation test	0	10	11
C1q latex agglutination inhibition	0	0	33
C1q binding inhibition RIA	0	25	11

Source:: Modified from ref. [23].

The World Health Organization tested the sera of patients with idiopathic inflammatory disease, cancer, and tropical diseases for the presence of immune complexes utilizing a battery of 18 methods [23,24]. The lack of correlation between any of the techniques was striking. As illustrated in Table 2, the C1q technique used affected the sensitivity of the assay. As a result, the World Health Organization has recommended that at least three different assays based upon different mechanisms be used concurrently to detect immune complexes [24,25]. This was well illustrated in a study by Krapf et al. [25] in which three separate assays were used to detect immune complexes in patients with melanoma and breast cancer. As seen in Tabel 2, individual detection rates ranged from 30 to 48% for melanoma and 37 to 41% for breast carcinoma. However, once the results were combined, the detection rate for immune complexes rose to 85% for melanoma and 77% for breast carcinoma patients.

Newer techniques for the detection of tumor-associated immune complexes may obviate the need for multiple assays. Kelly et al. [26] have reported the use of zone electrophoresis as a sensitive, non-complement-dependent method to detect immune complexes. Encouraging results have recently been reported utilizing competitive immunoenzymatic assays with conglutinin and C1q [27]. Whether these new methods will produce more sensitive and specific results remains to be determined.

II. CLINICAL MANIFESTATIONS OF IMMUNE COMPLEXES

A. Renal Manifestations

One of the earliest suggestions that immune complexes were associated with malignancy was made by Lee et al. [28] in 1966, when the association between cancer and the nephrotic syndrome was noted. In that study, 11 patients with nephrotic syndrome developed a malignancy within 14 months of diagnosis. In 7 of the 11 patients, nephrotic syndrome preceded the diagnosis of cancer. In retrospectively reviewing all of their patients with nephrotic syndrome, they found the incidence of associated malignancy to be 11%. Since that initial report, others have confirmed this unexpectedly high association. Eagen and Lewis [29] reviewed the findings of 171 patients from the literature and their own experience. Of those patients, 114 (67%) had histological evidence of glomerulopathy. Electron microscopic examination of 19 kidney specimens was compatible with immune complex deposition. In 34 patients in whom a clinical course was described, the median survival was 12 months from the onset of nephrotic syndrome.

The median survival after diagnosis of malignancy was only 3 months, with cause of death secondary to progressive malignancy.

The best described group of patients with cancer and renal disease are those with Hodgkin's disease. Hodgkin's disease accounts for greater than 70% of all cases of hematological malignancy and the nephrotic syndrome reported [29,30]. Early studies by Plager and Stutzman [31] noted that the development of nephrotic syndrome could be related to clinical relapse of Hodgkin's disease. Most striking was that treatment of the underlying Hodgkin's disease resulted in a reduction of proteinuria. In three cases, relapse of Hodgkin's disease was paralleled by a recurrence of the nephrotic syndrome. Treatment delivered to areas apart from the kidneys (i.e., radiation to the mediastimum) could still lead to a remission of the nephrotic syndrome. These findings have been conformed by others [32,33].

The cause of the glomerular lesions remains to be elucidated. Sherman, et al. [33] noted that findings consistent with lipoid nephrosis and no evidence, by immunofluorescent staining, of immunoglobulin. Others have noted similar findings [34,35] and in one review of 27 renal biopsies by Moorthy et al. [36] only 3 cases showed findings other than lipoid nephrosis. However, other studies have pointed to the presence of immune deposits [37-39] and in the review by Eagen and Lewis [29] 26% of patients with Hodgkin's disease and nephrotic syndrome demonstrated electron dense deposits by electron microscopic examination.

Nephrotic syndrome associated with other hematological malignancies is very unusual, with fewer than 20 cases reported. Patients reported include those with non-Hodgkin's lymphoma [32,40-45], chronic lymphocytic leukemia [46-50], and Burkitt's lymphoma [32]. As seen in Table 3, the distribution of glomerulopathies is different in Hodgkin's disease, with a greater number of patients developing glomerulonephritis. The clinical course of these patients is not well described due to the paucity of cases. However, resolution of nephrotic syndrome has been reported with treatment of non-Hodgkin's lymphoma [41] and chronic lymphocytic leukemia [47].

In nonhematological malignancies, the association of nephrotic syndrome is less well described than in Hodgkin's disease [42,45,50-54]. However, the role of immune complex-mediated renal disease is more apparent. In the original 11 patients described by Lee et al. [28], all had nonhematological malignancies and 9 had membranous glomerulonephritis on light microscopic examination. The finding of glomerulonephritis in more than 75% of patients with carcinoma and nephrotic syndrome contrasts strikingly with patients with Hodgkin's disease, in whom the incidence is less than 7%. As with hematological malignancies, nephrotic syndrome can be temporally related to the onset of malignancy. In at least three reported cases, removal or treatment of the tumor resulted in resolution of the nephrotic syndrome [55,56].

Immunological evaluation of the glomerulopathies associated with malignancy was initially reported by Lewis et al. [57]. In that study immunoglobulins were eluted from the kidney of a patient with bronchogenic carcinoma and nephrotic syndrome. It was found that the eluted antibody reacted with an extract of the patient's tumor. These findings have been confirmed by Mendes-DaCosta [50], who in two patients with epidermoid lung carcinoma and nephrotic syndrome, was able to elute IgG antibody from the lungs and kidneys with reactivity against autologous and allogeneic lung carcinoma, oat cell carcinoma, and normal lung tissue. Ozawa et al. [58], in studies of three patients with clear cell renal carcinoma and one with an intrarenal malignancy, found immunoglobulin and complement on the glomeruli and tumor membranes of all patients. In

Table 3 Glomerulopathies Associated with Neoplasia and the Nephrotic Syndrome

Tumor	Patients (n)	Patients with renal Pathological changes (n)	Minimal Change (%)	Membranous (%)	Membrano-proliferative (%)	Amyloidosis (%)	Other[a]
Hodgkin's disease	63	52	41	6	0	29	6
Non-Hodgkin's lymphoma	7	6	29	29	14	0	14
Leukemia	9	9	11	11	0	0	78
Carcinoma	67	48	5	49	6	3	9
Embryonal tumors	12	9	8	8	0	0	58
Waldenstrom's macroglobulinemia and multiple myeloma	7	7	0	0	14	86	0

[a]Extracapillary glomerulonephritis, progressive sclerosing glomerulonephritis, proliferative glomerulonephritis, unclassified glomerulonephrosis (also includes renal dysplasia [2 cases], thrombotic microangiopathy [1 case].
Source: From ref. [29].

three patients, immunoglobulin isolated from tumor and sera was found to be reactive against their tumor as well as normal renal proximal tubules and jejunal mucosa. Costanza et al. [59] described a patient with colonic carcinoma and nephrotic syndrome in whom granular deposits of immunoglobulin and complement were found in the glomerular basement membrane. They noted similar deposits of carcinoembryonic antigen (CEA), which suggested that the renal lesions were caused by soluble immune complexes of CEA and anti-CEA.

While there has been extensive documentation of the pathological role of immune complexes in malignancy resulting in renal disease, only a small number of patients actually manifest nephrotic syndrome. It may be argued that the clinical manifestations of renal disease represent only the far end of the spectrum and the majority of patients do not show signs of immune complex deposition. Support for this comes from work by Sutherland et al. [60] in which the kidneys of 303 patients were examined postmortem. No patient had clinical evidence of renal disease before death by light microscopic examination. However, gamma globulin and complement were detected by immunofluorescence in 21 cases. The greatest number of subclinical immune complexes were found in patients with acute leukemia (57% of all positive cases). This contrasts with patients demonstrating clinically overt renal disease, of whom only 8.5% had acute leukemia. A similar postmortem study by Pascal et al. [61] reported an incidence of 55% of electron dense deposits in the glomeruli of cancer patients compared with 11% of normals. The difficulty with portmortem examinations is that background immunofluorescence increases make it difficult to determine true positives [60]. In a more recent study, Helin et al. [62] performed renal biopsies on 39 patients with cancer with no clinical evidence of renal disease. Of the 24 patients in whom enough tissue was available for electron microscopic evaluation, electron-dense deposits were noted in 46%.

B. Systemic Manifestations

Aside from renal manifestations associated with cancer and immune complexes, there have been very few reports describing other systemic manifestations. The association of cancer and rheumatological disorders has been well described [63,64] and would appear to be a likely area for examination. However, despite the relationship of malignancy with a wide range of musculoskeletal disorders, there have been few cases documenting immune complex-mediated disease. Bennett et al. [65] reported a patient who presented with polyarthritis, fever, confusion, positive antinuclear and rheumatoid factor, and a circulating factor that stimulated serotonin release from normal platelets. Exploratory laparotomy revealed adenocarcinoma of the ovary, which, after it was resected, led to normalization of the patient's symptoms and laboratory findings. Wallach [66] described two patients with locally advanced breast cancer who developed a lupus-like syndrome following local radiotherapy. Rheumatological symptoms and laboratory abnormalities resolved following therapy with steroids. It was postulated that the irradiation of malignant tissue might have initiated an immunological response. This contrasts with the role of radiotherapy in the treatment of autoimmune diseases.

A study of Awerbuch and Brooks [67] evaluated patients with malignant disease and associated hypertrophic osteoarthropathy or nonmetastatic polyarthritis. In patients with hypertrophic osteoarthropathy, there was no evidence of elevated levels of immune complexes systemically or in synovial fluid. However, in one patient with nonmetastatic polyarthritis, elevated circulating immune complexes and decreased levels of complement were noted in sera and synovial fluid, suggesting immune complex-mediated disease. A more recent article by Bradley and Pinals [68] described a patient with lung cancer and inflammatory arthritis. While elevated levels of circulating immune complexes were noted, synovial biopsy failed to reveal any evidence of immune complex deposition, which suggested that immune complexes may not be the cause of carcinoma-associated polyarthritis.

It is noteworthy that although immune complexes have been described in dermatological disorders [69], the additional association with malignancy has rarely been reported. A report by Miro et al. [70] described a patient who presented with arthralgia, myalgia, arthritis, skin rash, and polyneuritis who subsequently developed Hodgkin's disease. Immunofluorescent studies of intradermal vessels and temporal arteries revealed IgM, C_3, and fibrinogen, which suggested immune-complex-mediated injury.

Other paraneoplastic syndromes have been related to immune complexes. Havemann et al. [71] isolated immune complexes from seven patients with oat cell carcinoma. In four of seven patients, the immune complexes were noted to have similar activity to adrenocorticotropic hormone (ACTH) when tested by radioimmunoassay. Only one patient had Cushing's syndrome, and that patient had an unusually high level of immune complexes. A study by Reitan et al. [72] reported five cases of osteosclerotic myeloma with polyneuropathy. In two cases levels of immune complexes and rheumatoid factor were found to be elevated. While they were unable to document immune complex deposition in peripheral nerves, others [73] have reported it. Amlot et al. [74] studies 18 patients with Hodgkin's disease in whom symptomatic status correlated with circulating immune complexes. Elevated levels of C_3 were found in two of eight patients with "A" symptoms and nine of nine patients with "B" symptoms. They concluded that immune complexes can be correlated with "B" symptoms. Zimmerman et al. [75] reported a patient with gastric carcinoma and thrombotic thrombocytopenic purpura (TTP). Elevated levels of immune complexes paralleled the development of TTP, and antibodies isolated from the immune complexes were found to react with gastric carcinoma cells.

III. CIRCULATING IMMUNE COMPLEXES

With the advent of numerous assays to detect the presence of circulating immune complexes, there has been a virtual deluge of papers reporting the association between circulating immune complexes and malignancy (Table 4) [76-159]. A listing of the tumors in which elevated levels of circulating immune complexes have been noted is shown in Table 4. As can be seen, if only by the number of papers cited, there is mounting evidence that the presence of elevated levels of circulating immune complexes in cancer may represent the norm. This has implications regarding the relationship of circulating immune complexes to the stage of disease, prognosis, and perhaps the pathogenesis and therapy of malignancy.

One of the earliest papers reporting of elevated levels of circulating immune complexes in malignancy was in a study by Jerry et al. [76] Sera of 62 patients with various stages of malignant melanoma were examined for circulating immune complexes by the C1q binding assay. Elevated levels of circulating immune complexs were detected 74% of the time, with the incidence rising with extent of disease. Of note, 31 patients with stage I, II, and III disease were evaluated serially with weekly tests for 2 months. All of these patients demonstrated cyclic variations in immune complex levels. In three patients immune complex levels dropped following chemotherapy and reappeared several weeks

Table 4 Circulating Immune Complexes and Malignancy

Tumor	Reference
Bone	83,84
Brain	85,86
Breast	22,87-94
Gastrointestinal	95-104
Genitourinary	105,106
Gynecological	107-115
Head and neck	116
Hodgkin's disease	117-121
Leukemia/lymphoma/myeloma	81,82,122-128,130-133,174
Melanoma	76,134-139
Neuroblastoma	8,140,141
Pulmonary	143-148
Skin	149
General	1,15,24-27,77-80,150-158

later. Twenty-three patients receiving immunotherapy with bacillus Calmette-Guérin or levamisol showed continuous fluctuations while receiving therapy. The reason for these changes in detected levels of immune complexes is not clear. The authors postulated that it may be due to changes in the class of circulating immune complexes, immunoregulatory homeostatic mechanisms, or variations in circulating tumor antigen. However, despite these fluctuations, there was an overall increase in the level of circulating complexes with progression of disease.

Many subsequent studies have examined the overall incidence of circulating immune complexes in a wide variety of malignancies [77-80]. Teshima et al. [77] utilized the [^{125}I] C1q deviation test on the sera of 459 patients with various malignancies. Greater than 50% of sera showed a high inhibition of [^{125}I] C1q uptake. When a C1q binding assay was used, elevated immune complex levels were found in the sera of 83% of cancer patients. In the majority of cases, patients with residual or recurrent disease had higher levels of C1q binding than those with no evidence of disease. Rossen et al. [78], using a C1q binding assay, found elevated levels of circulating immune complexes in 83% of sera of patients with cancer. In addition, a higher level and incidence of circulating immune complexes were found in patients with residual disease compared with those with no evidence of disease. Conversely, Samoya et al. [79], using a precipitin test with monoclonal rheumatoid factor, found an incidence of only 29% of circulating immune complexes in patients with malignancy. Theofilopoulos et al. [80], using a Raji cell immunoassay, examind the sera of 517 patients with cancer. The overall incidence of immune complexes ranged from 16 to 52% depending upon the malignancy. A direct correlation was also noted with stage of disease and tumor burden.

Carpenter et al. [81,82] have shown a strong correlation between immune complex levels and prognosis in patients with leukemia. Sera from 186 patients with acute myeloid leukemia were evaluated. Overall, 70% of patients had elevated levels of circulating immune complexes. The duration of complete remission could be correlated with the presence or absence of immune complexes. Patients with elevated levels of circulating immune complexes at diagnosis achieved a complete remission 32% of the time, as opposed to 94% of patients without elevated levels of immune complexes. Median survival was less than 6 months in patients with immune complexes. Patients without immune complexes had a median survival of 14.5 months. Relationships were also found between the presence of immune complexes and the treatment time required to achieve a complete remission.

IV. POSSIBLE PATHOPHYSIOLOGICAL ROLE

From the preceding discussion it seems clear that elevated levels of immune complexes are present in a substantial number of patients with cancer, can be correlated with disease course, and have been implicated in paraneoplastic disorders. A major question is whether the immune complexes associated with malignancy represent a response to the presence of cancer or are somehow involved with immunosuppression and tumor promotion. Experimental evidence that immune complexes may be a mediator of immunosuppression or facilitate tumor growth exists in a number of model systems [157-183]. To date, there is no direct evidence implicating immune complexes as mediators of tumor growth or immunosuppression. However, concurrent with the recognition of immune complexes in malignancy, several studies have emerged that have suggested a role for immune complexes as potential immunosuppressive agents.

The ability of tumors to escape immunological destruction has been described by numerous researchers and is beyond the scope of this paper. Several possible mechanisms have been described including inappropriate or inadequate host immune response and modulation of immune detection [159-166]. A body of literature describes the ability of immune complexes to modify the host's immune response [2]. Several mechanisms have been put forth to explain antibody or immune complex-induced suppression of humoral and cellular immune responses. Humoral suppression could be the result of interaction with antigen and Fc receptors on B cells [165,166]; interaction with Fc receptors alone [167]; immune-complex-induced release of suppression factors [168]; effector cell blockade of antibody production [169]; interactions with T cells leading to activation of suppressor T cells [169,170] or suppression of T-cell factors [171,173]; inhibition of T- and B-cell cooperation [172]; or effects on macrophages [174-177].

Studies of the role of immune complexes as suppressors of cellular immune responses have focused on their potential role as blocking factor. Initial studies by Hellstrom and Hellstrom showed that a factor in serum could inhibit autologous cellular cytotoxicity [177]. Subsequent studies have shown that "blocking factor" is in the serum of patients with a variety of tumors, could be generated in vitro, can be abrogated by addition of excess tumor-specific antigen, and can be correlated with the size of tumor and prognosis [178,179]. In addition, Sjogren et al. [180] noted that blocking factor activity could be abrogated by separation of serum into low- and high-molecular-weight components, which suggests that blocking factor may be composed of antigen-antibody complexes. However, to date, there is no direct evidence that blocking factor is composed of tumor-associated antigen and antibody.

There have been relatively few studies evaluating the mechanisms by which immune complexes may mediate the host's response to cancer. Baldwin et al. [181] noted that sera from rats with aminoazo-dye-induced hepatomas inhibited the cellular cytotoxicity of effector cells in vitro. Sera from tumor-immune animals failed to block cytotoxicity.

However, papain-solubilized hepatoma-specific antigen inhibited cytotoxicty as well. The parallels between these results and the "blocking factor" described by Hellstrom et al. [178] are noteworthy. Jose and Seshadri [8] studied eight children with neuroblastoma and evaluated the relationship of circulating immune complexes by blocking activity. Sera of patients with progressive disease, and containing complexes of tumor antigen and antibody, demonstrated an ability to block lymphocyte cytotoxicity for autochthonous tumor cells. Sera from patients in remission demonstrated increased amounts of free tumor-associated antibody. In addition, it was found that the addition of shed tumor antigen produced blocking of lymphocytotoxicity.

Other investigators have reported the induction of suppressor cells by tumor antigen [182,182] or antigen-antibody complexes [184,185]. Circulating immune complex levels were noted to rise preceding nonspecific suppression of T-cell mitogenic responses [186]. Gershon et al. [187] reported that tumor antigen and antibody were able to activate suppressor T cells. Indeed, antigen and antibody were shown to enhance tumor growth in mice studied with several syngeneic tumors in a T-cell-dependent manner [188]. More recently, Ravikumar et al. [189] noted a relationship between immune complex formation and tumor growth in vitro. In addition, progressive tumor growth was related to a decrease in interleuken-1 and -2 production, leading the authors to postulate that circulating immune complexes may be involved in regulating interleukin production. Conversely, Hoon et al. [190] found that the formation of circulating immune complexes could not be clearly related to the ability of the tumor to metastasize in rats.

In our laboratory, we have examined the relationship of circulating immune complexes to the autologous response to cancer. Tumor-associated antigens of melanoma and other tumors have been detected by polyclonal xenogeneic and allogeneic antisera [190-196] and, more recently, monoclonal murine and human antibodies [197-203]. These heterologous tumor antibodies to human cancer have generally been found, when sufficiently studied, to react against normal cell constituents [203]. The advantage of studying host-derived antibodies is that because they are produced against physiologically relevant antigens, they offer a greater likelihood of detecting tumor-restricted antigens not previously identified by heteroantisera [204-207].

Antibody against autologous tumor cell surface antigens was first reported by Carey et al. [194,204] and Shiku et al. [205,206], who used sensitive microserological assays and cultured tumor cells. These authors demonstrated antibody against autologous cell surface antigens in one-third of melanoma patients studied. Antigens described in these studies were organized into three classes: those restricted to the tumor of a single individual (class I); those shared among tumors of a similar histotype or ontogeny and rarely, if ever, found in nonneoplastic cells (class II); those found in nonneoplastic cells as well as tumors (class III). However, the low incidence and titer of these antibodies have hampered further investigations and have raised questions regarding the relevance of autologous immune response. In melanoma, we have confirmed the work of others in identifying polyclonal autologous antibody in the sera of patients with specificity for individually restricted and common melanoma antigens. Six of 22 melanoma patients demonstrated significant titers of antibody reactivity in native sera against autologous melanoma cell cultures, of which 1 has been shown to detect an antigen of class I and another of class II [207].

We asked whether antibody might be detectable in the sera by dissociation of immune complexes formed by tumor antigens. Serum was acidified and ultrafiltered (AD&U) to

Table 5 Augmentation of Autologous Antibody Reactivity Following Immune Complex Dissociation

	Serum Antibody Titer	
Cell line	Native	Immune Complex Dissociated
Melanoma		
Y-Mel 81:370	0	0
Y-Mel 81:060	0	1:8
Y-Mel 81:090	0	1:64
Y-Mel 81:180	0	1:128
Y-Mel 81:170	0	1:256
Y-Mel 81:120	1:4	1:16
Y-Mel 78:010	1:64	1:128
Y-Mel 83:070	1:32	1:128
Y-Mel 82:550	0	1:128
Y-Mel 81:710	0	1:128
Y-Mel 84:420	1:32	1:262,044
Head and neck cancer		
UM-SCC 8	1:32	1:2048
UM-SCC 9	1:16	1:64
UM-SCC 17	0	1:1024
UM-SCC 21	1:4	1:128
UM-SCC 23	0	1:32
UM-SCC 30	0	1:4
UM-SCC 35	0	1:32
UM-SCC 36	0	1:8

remove substances of molecular weight <100,000, after the method described by Sjogren et al. [180] to evaluate the role of immune complexes in blocking factor. The acidified, ultrafiltered serum was tested in comparison to native untreated serum, or serum that had only been acidified but not ultrafiltered. Native sera from 12 patients were subjected to AD&U. Previously only five sera showed any IgG reactivity against autologous cultured melanoma cells. Following ultrafiltration, 11 of 12 sera showed reactivity of IgG against autologous cultured melanoma cells. These results are summarized in Table 5. Specificity analysis of four autologous melanoma systems have been performed and in each case it appears that a class II antigen is being detected.

Gupta and Morton [208] showed that the presence of tumor-associated antigen, antibody, and immune complexes could be correlated with prognosis in 20 patients with melanoma. We have demonstrated that the titer of autologous antibody can be correlated with stage of disease and clinical course [209]. This was initially pursued through serial studies of individual patients with melanoma. Serial studies offer a unique means by which the correlations between clinical course and autologous antibody reactivity may be evaluated. In four serial studies, autologous antibody reactivity was found to correlate with clinical course and disease progression. These serial autologous melanoma studies suggested that host antibody responses to cancer detect antigens that may be related to the clinical course and prognosis of malignancy. In a study of 43 stage I and II patients who were clinically free of disease, the presence of antibody directed against melanoma in either native or AD&U serum was found to be associated with eventual relapse (p =

0.0001). When results were subgrouped by stage of disease, Breslow depth, and hypopigmentation, antibody reactivity was still correlated with eventual relapse. In addition, the level of antibody titer could be correlated to the time of relapse. These findings suggest that tumor-reactive antibodies, the antigen against which they may be directed, or the immune complexes that are formed correlate with, and therefore may provide an explanation of, the natural history of melanoma.

These results have recently been expanded to include another malignancy: squamous cell carcinoma of the head and neck (SCCHN). Serological studies in patients with SCCHN have been limited. While some authors have documented changes in systemic immunoglobulin levels in patients with late-stage disease, the lack of reproducible results has made these observations difficult to interpret [210,211]. Unlike studies in other malignancies, such as melanoma [212-218], few efforts have been made to define tumor-associated antigens in SCCHN. Initial serological studies performed by Carey et al. [219] have shown the presence of autologous tumor-reactive antibody in patients with SCCHN and, in one patient, this reactivity correlated with clinical course. Analysis of the specificity of the antigen detected by the autologous antibody was limited, but it appeared to detect an antigen present on squamous cell carcinoma cells and not on normal tissue cells. Sera from 41 patients with SCCHN were evaluated for autologous antibody reactivity by protein A hemadsorption [219a]. Autologous antibody reactivity was noted in 24 of 41 systems tested with a median titer of 1:4. In the majority of cases, autologous antibody reactivity could only be detected in undiluted serum, which precluded further analysis. To determine if immune complex dissociation could augment autologous antibody reactivity in SCCHN, 12 serum samples from 8 patients were subjected to AD&U. Native serum samples from six of these patients had previously been shown to have reactivity against autologous SCCHN cells by protein A hemadsorption. After AD&U, all 12 sera from the 8 autologous systems demonstrated augmented IgG reactivity against SCCHN (Table 6).

As with melanoma, we have begun to examine whether autologous antibody reactivity could be correlated with clinical course [219b]. We have performed serial studies on six autologous systems. Sera were obtained from patients with SCCHN at several intervals during the course of their disease. As with melanoma, serial serological assays demonstrated correlations between tumor progression and clinical course. In four cases, rises in autologous antibody titers preceded the clinical diagnosis of recurrence by several months. Specificity analysis of the six autologous systems demonstrated a similar range of reactivity and were noted to detect a class II tumor.

The available data are insufficient to draw conclusions regarding the relationship of autologous antibody reactivity to clinical course. However, in four autologous melanoma and six SCCHN systems in which serial studies have been performed, correlations have been noted between antibody reactivity and clinical course. The exact nature of the antigen found in tumor-associated immune complexes remains to be determined. Previous work in ovarian and lung carcinoma has demonstrated that immune complexes, isolated from ascites or pleural effusions, may serve as a source of tumor-associated antibody and antigen [220-223]. To date, the antibodies isolated have lacked specificity or have not been adequately characterized. Recently, partial isolation and characterization of a melanoma-associated antigen by autologous antibody has been recognized. A 66,000 MW antigen was isolated from spent media and the immune complex dissociated ultrafiltrate [223]. Subsequent studies may allow a better understanding of the host-tumor interaction.

Table 6 Summary of Trials with *S. Aureus* and Protein A Immunoperfusion

Author	Patients (n)	Tumor	Method[a]	Number of Treatments	Dose	Interval	Response[b]	Toxicity
Basal [228]	1	Colon	SAC	20	35 grams	every 3 days	1 MR	Tumor pain
Bensinger [242]	18	Melanoma (7) Lung (1) Colon (2) Breast (3) Sarcoma (2) Thymoma (1) Renal (1) CML (1)	PAS	1-22	100-200 mg	—	4 MR	Tumor pain Hypoxemia Hypotension Tachycardia Fever, chills Nausea Diarrhea
Fer [240]	4	Breast	PACC	18	0.6-7.5 mg[c]	every 2-3 days	None	None
Kiprove [244]	1	Kaposi's sarcoma	PAS	3	200 mg	every 2 days	MR	Tumor pain
Korec [243]	11	Melanoma (2) Colon (5) Breast (1) Ovary (2) Prostate (1)	PAPB[d]	3-32	60 mg	1-2 ×/week	1 MR	Fever, chills Nausea Vomiting Bronchospasm
MacKintosh [238,239]	48[e]	Melanoma (7) Colon (9) Lung (9) Breast (7) Miscellaneous (16)	PASG	8+[f]	2-100 mg	2 ×/week	5 PR	Fever, chills Nausea Vomiting Diarrhea Bronchospasm Hypercalcemia Abnormal liver function tests

Messerschmidt [241]	16	Melanoma (1) Colon (4) Breast (6) Esophageal (2) Miscellaneous (3)	SAC PACC PAS	—	SAC PACC 100 mg PAS	—	None	Two treatment-related deaths with SAC, Fever Chills Hypotension Bronchospasm Neutropenia
Messerschmidt [246]	114	Breast (32) Kaposi's sarcoma (23) Colon (21) Miscellaneous (38)	PACC	1–48 200 mg	50 mg	3 ×/week	12 PR 9 MR 29 Stable	Fever, chills pain, nausea vomiting, rash Bronchospasm
Ray [245]	1	Colon	SAC	15	10% susp.	2 ×/week	MR	Tumor pain, Fever, chills
Terman [236]	4	Breast	PACC[g]	1–18	0.12–5 mg	1–17 days	3 PR 1 MR	Tumor pain Fever, chills Hypotension Tachycardia

[a]SAC, *S. Aureus* Cowan I; PAS, protein A silica bound; PACC, protein A collodion-charcoal bound; PAPB, protein A polyacrylamide beads; PASG, protein A sepharose gel.
[b]PR, partial response (>50%); MR, minimal response (>25%).
[c]Dose escalation.
[d]Perfusion column cooled to 15°C to reduce toxicity.
[e]Sixteen patients not evaluable.
[f]Only includes evaluable patients.
[g]Patients subsequently treated with cytosine arabinoside.

V. THERAPEUTIC TRIALS

If it is true that circulating immune complexes and tumor antigen may be involved in host immunosuppression, removal of these factors may allow restitution of host immune responsiveness. This has been the subject of several investigations. Numerous patients have been subjected to intensive plasmapheresis in the hopes of removing immune complexes or other immunosuppressive factors. It has been shown that plasmapheresis is capable of transiently reducing levels of circulating immune complexes [224]. Changes have also been noted in antibody-dependent cellular cytotoxicity and blocking factor [225,226]. The largest experience has been reported by Israel et al. [227]. Fifty-two patients with a variety of metastatic malignancies were subjected to plasma exchange two or three times a week for up to 12 weeks. Of the 45 evaluable patients, 14 (31%) achieved a partial response of 12 (26%) had stable disease. These results were not correlated with changes in circulating immune complexes and they await independent confirmation.

In the hope of improving these results, recent work has involved the use of a more "active" immunoadsorbent with plasma exchange perfusion over protein A. Protein A, a cell wall constituent of the Cowans I strain of *Staphylococcus aureus*, binds the Fc region of some immunoglobulins and immune complexes. If blocking factors are truly immune complexes, perfusion over protein A should remove them with subsequent antitumor effects. Initial work was performed by Bansal et al. [228], who observed tumoricidal effects in one patients with cancer of the colon after reinfusion of serum that has been perfused over immobilized staphylococcal protein A. A reduction in levels of IgG and circulating immune complexes was also noted.

Subsequent animals studies have documented regression and ablation of tumors [229-231]. Some of the more encouraging animal data have been reported in studies of cats spontaneously infected with feline leukemia virus (FeLV) who subsequently developed lymphosarcoma (LSA) [231,232]. Jones et al. [232] treated 16 cats with LSA using extracorporeal immunoadsorption over heat-killed *S. auerus* Cowans I. Nine of 16 cats achieved a complete remission of the LSA. Responses could be correlated with clearance on FeLV antigen, a drop in circulating immune complex levels, and the development of free antibody against FeLV. This work has been confimred by Day et al. [233] utilizing immunoadsorption over purified protein A. The immunoglobulins and complexes isolated by ex vivo immunoadsorption have been shown to contain FeLV envelope (gp70) and major core (p30) proteins, as well as IgG heavy and light chains [234]. In addition, it has been found that following ex vivo immunoadsorption over protein A, cytotoxic antibodies directed against gp70 occur and may be correlated with viral clearance and remission of LSA [233,235].

In humans, one of the initial studies utilizing plasma exchange in conjunction with perfusion over protein A was reported by Terman et al. [236,237]. Five patients with metastatic breast cancer were treated; four showed a reduction in tumor mass, leading the authors to postulate that perfusion of plasma over protein A may liberate cytotoxic tumor-associated antibodies. However, the concomitant use of cytosine arabinoside in all five patients has made interpretation of the results difficult. Subsequent studies have shown mixed results. MacKintosh et al. [238] reported that in 14 patients with metastatic cancer plasma perfusion over protein A resulted in 2 objective responses and 5 instances of disease stabilization. More recently, they have updated their experience to include 48 patients with a wide variety of malignancies [239]. Among 32 evaluable patients, 5 partial responses were observed. The authors noted that tumors involving skin, lymph nodes, and lung appeared to be the most response to therapy. Others have noted less encouraging results [240-244]. Fer et al. [240] attempted to confirm Terman's results in four patients with metastatic breast cancer without adding subsequent chemotherapy or surgery. No responses or toxicity were seen despite a subsequent dosage escalation. Five other trials [241-244] have used *S. aureus* or purified protein A to treat a total of 48 patients. Only seven minimal responses were seen.

Significant toxicity was noted in several of the trials. Young et al. [245] described the toxicities encountered in treatment of the five breast cancer patients reported by Terman [236,237]. In two patients marked cardiopulmonary toxicity was noted with hypotension, tachycardia, and a decrease in systemic and pulmonary vascular resistance. The authors reported that toxicity could be controlled by reducing the quantity of protein A and the perfusion rate. Messerschmidt et al. [241] noted significant toxicity in patients treated with immunoperfusion over *S. aureus* (SAC) or purified protein A bound to collodion-charcoal (PACC) or silica (PAS). Of the eight patients treated with SAC or PAS,, seven had marked to severe cardiopulmonary toxicity and two died of causes related to their treatment. In the nine patients treated with PACC in the manner reported by Terman et al. [236] no toxicity was noted. No responses were noted in the entire group of 16 patients.

The variability in response and toxicity to *S. aureus* and protein A immunoperfusion makes assessment of its efficacy difficult (Table 6). In most of the trials there was no uniform method of immunoperfusion or protein A immobilization. Terman has reported [236] that enterotoxins are present in some commercial preparations of protein A, with batch to batch variability. In addition, since Terman's preliminary trial [236] the steps used in the manufacture of protein A have been modified [240]. This may account for the failure of Fer et al. [240] to reproduce Terman's results.

More recently, Messerschmidt et al. [246] reported the first multi-institutional trial of protein A hemadsorption. Patients (n = 14) with a wide variety of tumors were treated with either offline or online with columns containing either 50 or 200 mg protein A. While no complete responses were noted, 12 partial (14%) and 9 minimal responses (10%) were documented. Toxicity with the protein A columns used in this study was far less than reported previously [245]; in the majority of patients treatment were not associated with any toxic side effects. This was attributed to greater purity of the protein A used and improvements in the silica binding to the column. Circulating immune complex levels were measured during the course of therapy but were not noted to change significantly.

The means by which protein A perfusion produces its antitumoral effects is not known. Smith et al. [247] found that commercial preparations of protein A are contaminated with *S. aureus* enterotoxin, which is a very potent inducer of interferon alpha and gamma. In cats with LSA, interferon titers have been noted to be elevated following ex vivo immunoperfusion [248,249]. This could account for the antitumoral effects reported. Bertram et al. [250] found that perfusion of serum over protein A columns leads to increased lymphocyte mitogenicity. Whether this was due to protein A or an enterotoxin contaminant was not discussed. An alternative mechanism was presented by Balint et al. [251], who found that serum perfused over protein A showed increased levels of C1q binding, implying that immune complexes were being created rather than removed. Characterization of these immune complexes found them to have a molecular weight ranging from 600,000 to 2,000,000 and be composed predominantly of IgG and protein A. Subsequent studies by Langone et al. [252] found that these complexes were able to generate anaphylatoxins C_{3a}, C_{4a}, and C_{5a}. Generation of anaphylatoxins by protein A was not found to be due to enterotoxin contamination. The authors postulated that the tumoricidal effects observed with protein A perfusion might be due to the creation of large immune complexes that generate high levels of anaphylatoxins.

At present it is difficult to know what place plasmapheresis and protein A immunoadsorption will have in cancer therapy. Until the mechanisms of action of this therapy is delineated, its potential role remains to be determined. However, well-documented responses have been seen and warrant additional studies.

V. CONCLUSION

Despite the limitations inherent in currently available methods to detect circulating immune complexes, the weight of evidence appears to favor the premise that elevated levels of circulating immune complexes are present in the majority of patients with cancer. It is likely that the immune complexes formed are small (possibly due to a large burden of tumor antigen) and that complexes in the sera of cancer patients are generally non-complement-binding. Such immune complexes would be difficult to detect with the individual assays currently available and may still be elusive when various different assay are used. The presence of small immune complexes would be less likely to mediate renal and other systematic manifestations, potentially explaining the low frequency with which these are noted. The level of circulating immune complexes may be a reflection of the underlying malignancy, and appears to be related to stage of disease, tumor burden, and prognosis.

What remains to be determined is the specific role immune complexes play in the pathophysiology of cancer. It is unclear if immune complexes are only an indicator of the presence of malignancy or are somehow involved in modifying the host's humoral and cellular immune responses. A large body of literature describes the role of immune complexes and antibody in modifying the immune response. However, direct proof of their immunoregulatory role in human malignancy is lacking. Despite the limitations of our current knowledge, continued evaluation of their role in cancer offers great promise. Unlike currently available monoclonal antibodies, host-derived immune complexes appear to be derived from a response to physiologically relevant antigens. As such, they offer a unique probe through which we may increase our understanding of the host's immune response and its relationship to disease status.

It may be that immune complexes play a pivotal role in the immune homeostasis of cancer. Manipulations that alter the balance of tumor antigen, antibody, and immune complexes may have therapeutic consequences. Removal of immune complexes by plasmapheresis or serum perfusion over protein A may allow increased cellular and humoral cytotoxicity. Removal of circulating tumor antigen or the creation of larger, complement-binding immune complexes (i.e., by binding to protein A) may allow greater tumoricidal effects. At present, one can only speculate that such manipulations will have any significant therapeutic effect. Once the nature of the antigen and antibody in the immune complexes is determined, it may be possible to understand how immune complexes affect the host's response to cancer.

ACKNOWLEDGMENTS

The author thanks Drs. Ronald B. Herberman, John M. Kirkwood, and Fred S. Kantor for advice and review of the manuscript, and Lynn Gambardella for her secretarial assistance. This work was supported by grant CA42922 from the National Cancer Institute. Dr. Vlock is the recipient of a VA Career Development Award.

REFERENCES

1. Theofilopoulos AN, Dixon FJ. Detection of immune complexes: techniques and implications. Hosp Prac 1980;May:107-121.
2. Theofilopoulos AN, Dixon FJ. The biology and detection of immune complexes. Adv Immunol 1979;28:89-220.
3. Williams RJ Jr. Immune complexes in clinical and experimental medicine. Cambridge: Harvard University Press, 1980.
4. Wilson CB, Dixon FJ. The renal response to immunological injury. In: Brenner BM, Rector FC, eds. The kidney. Philadelphia: W.B. Saunders, 1981:1237-1250.
5. Bartolotti SR. Quantitative elution studies in experimental immune complex and nephrotoxic nephritis. Clin Exp Immunol 1977;29:334-341.
6. Almeida JD, Waterson AP. Immune complexes in hepatitis. Lancet 1969;2:983-986.
7. Preheim L, Schiffman G, Komorowski R, Koethe S, Rytel JC. Immune complex nephritis associated with pneumococcal infection. Clin Res 1978;26:404A.
8. Jose DG, Seshadri R. Circulating immune complexes assay and role in blocking specific cellular activity. Int J Cancer 1974;13:824-833.
9. Notkins AL, Mahar S, Scheele C, Goffman J. Infectious virus-antibody complex in the blood of chronically infected mice. J Exp Med 1966;124:81-97.
10. Augener W, Grey HM, Cooper NR. The reaction of monomeric and aggregated immuglobulins with C1. Immunochemistry 1971;8:1011-1017.
11. Agnello V, Winchester RJ, Kunkel HG. Precipitin reactions of the C1q component of complement with aggregated gamma-globulin and immune complexes in gel diffusion. Immunology 1970;19:909-915.
12. Zubler RD, Nydegger U, Perin LA, Fehr K, McCormick J, Lambert PH, Miescher PA. Circulating and intra-articular immune complexes in patients with rheumatoid arthritis. Correlation of 125I-C1q binding activity with clinical and biological features of the disease. J Clin Invest 1976;57:1308-1319.
13. Hay FC, Nineham LJ, Roitt IM. Routine assay for the detection of immune complexes of known immunoglobulin class using solid phase C1q. Clin Exp Immunol 1976;24:396-405.
14. Ahlstedt S, Hanson LA, Wadsworth C. A C1q immunosorbent assay compared with thin-layer gel filtration for measuring IgG aggregates. Scand J Immunol 1976;5:293-298.
15. Rossen RD, Zubler RH, Day NK, Riesberg NK, Morgan AL, Gutterman JU, Hersh EM. Detection of immune complex-like materials in cancer patients' sera: a comparative study of results obtained with the C1q deviation and C1q binding tests. J Lab Clin 1978;91:191-204.
16. Cooper NR, Morrison DC. Binding and activation of the first component of human complement by the lipid region of lipopolysaccharides. J Immunol 1982;120:1862-1868.
17. Cooper NR, Jansen FL, Wevis RM Jr, Oldstone MB. Lysis of RNA tumor viruses by human serum: direct antibody-independent triggering of the classical complement pathway. J Exp Med 1976;144:970-984.

18. Theofilipoulos AN, Dixon FJ, Bokisch V. Binding of soluble immune complexes to human lymphoblastoid cells. I. Characterization of receptors for IgG, Fc and complement and description of the binding mechanism. J Exp Med 1976;144:970-984.
19. Theofilopoulos AN, Wilson AN, Bokisch VA, Dixon FJ. Binding of soluble immune complexes to human lymphoblastoid cells. II. Use of Raji cells to detect circulating immune complexes in animal and human sera. J Exp Med 1974;140:1230-1244.
20. Theofilopoulos AN, Wilson CB, Doxin FJ. The Raji cell radioimmune assay for detecting immune complexes in human sera. J Clin Invest 1976;57:169-182.
21. Raina SR, Kirkwood JM, Vlock DR. 1989. Unpublished data.
22. Herberman RB, Brodes M, Lambert PH, Luthra HS, Robins RA, Sizaret P, Theofilopoulos A. Report on international comparative evaluation of possible value of assays for immune complexes for diagnosis of human breast cancer. Int J Cancer 1981;27: 569-576.
23. Lambert PH, Dixon FJ, Zubler RH, Agnello V, Cambiaso C, Casali P, Clarke J, Cowdery JS, McDuffie FC, Hay FC, MacLennan ICM, Masson P, Muller-Eberhard HJ, Penttinen K, Smith M, Tappeiner G, Thoefilopoulos AN, Verroust P. A WHO collaborative study for the evaluation of eighteen methods for detecting immune complexes in serum. J Clin Lab Immunol 1978;1:1-15.
24. World Heath Organization. The role of immune complexes in disease. W.H.O. Tech Rep Seriex 1977;607.
25. Krapf F, Renger D, Schedel I, Fricke M, Kemper A, Deicher H. Circulating immune complexes in malignant diseases. Increased detection rate by simultaneous use of three assay methods. Cancer Immunol Immunother 1982;15:138-143.
26. Kelly RH, School MA, Harvey VS, Devenyi AG. Qualitative testing for circulating immune complexes by use of zone electrophoresis on agarose. Clin Chem 1980; 26:396-402.
27. Grassi R, Grazzini G, Sodini F, Ronchi MC, Nottolini R. Circulating immune complexes in the diagnosis of autoimmune and neoplastic disorders: new perspectives. Int J Biol Markers 1986;1:105-108.
28. Lee JL, Yamauchi H, Hopper J Jr. The association of cancer and the nephrotic syndrome. Ann Intern Med 1966;64:41-51.
29. Eagen JW, Lewis EJ. Glomerulopathies of neoplasia. Kidney Int 1977;11:297-306.
30. Gagliano RG, Costaniz JJ, Beathard GA, Sarles HE, Bell JD. The nephrotic syndrome associated with neoplasia: an unusual paraneoplastic syndrome report on a case and review of theliterature. Am J Med 1976;60:1026-1031.
31. Plager J, Stutzman L. Acute nephrotic syndrome as a manifestation of active Hodgkin's disease. Am J Med 1970;50:56-66.
32. Ghosh L, Muehrcke RC. The nephrotic syndrome: a prodrome to lymphoma. Ann Intern Med 1970;72:379-382.
33. Sherman RI, Susin M, Weksler ME, Becker EL. Lipoid nephrosis in Hodgkin's disease. Am J Med 1972;52:699-706.
34. Yum MN, Edwards JL, Klei S. Glomerular lesions in Hodgkin's disease. Arch Pathol 1975;99:645-649.
35. Routledge RC, Han IM, Jones PHM. Hodgkin's disease complicated by the nephrotic syndrome. Cancer 1976;38:1735-1740.
36. Moorthy AV, Zimmerman SW, Burkholder PM. Nephrotic syndrome in Hodgkin's disease. Evidence for pathogenic alternative to immune complex deposition. Am J Med 1976;61:471-477.
37. Froom DW, Franklin WA, Hano JE, Potter EV. Immune deposits in Hodgkin's disease with nephrotic syndrome. Arch Pathol 1972;94:547-553.
38. Sutherland JC, Markham RV, Mardiney MR. Subclinical immune complexes in the glomeruli of kidneys postmortem. Am J Med 1974;57:536-541.

39. Lewis MG, Loughridge LW, Phillips TM. Immunological studies in nephrotic syndrome associated with extrarenal malignant disease. Lancet 1971;2:134-135.
40. Sagel J, Muller J, Logan E. Lymphoma and nephrotic syndrome. S Afr Med J 1971; 45:79-80.
41. Muggia FM. Glomerulonephritis or nephrotic syndrome in malignant lymphoma, reticulum cell type. Lancet 1971;1:805-807.
42. Gluck MC, Gallo G, Lowenstein J, Baldwin DJ. Membranous glomerulonephritis: evaluation of clinical and pathologic features. Ann Intern Med 1973;78:1-12.
43. Hyman LR, Burkholder PM, Joo PA. Malignant lymphoma and nephrotic syndrome. A clinicopathologic analysis with light, immunofluorescence and electron microscopy of the renal lesions. J Pediatr 1973;82:207-212.
44. Rabkin R, Thatcher GN, Diamond LH, Eales L. The nephrotic syndrome, malignancy and immunosuppression. S Afr Med J 1973;47:605-606.
45. Gagliano RG, Costaniz JJ, Beathard GA, Sarles HE, Bell JD. The nephrotic syndrome associated with neoplasia: An unusual paraneoplastic syndrome. Report of a case and review of literature. Am J Med 1976;60:1026-1031.
46. Loughridge LW, Lewis MG. Nephrotic syndrome in malignant disease of non-renal origin. Lancet 1971;1:256-258.
47. Brudovsky HS, Samuels ML, Migliore PJ, Howe CD. Chronic lymphocytic leukemia, Hodgkin's disease, and the nephrotic syndrome. Arch Intern Med 1968;121:71-75.
48. Mandalenakis N, Mendoza N, Pirani CL, Pollack VE. Nodular glomerulonephritis: a clinical and pathological study based on renal biopsies. Medicine 1971;50:319-355.
49. Dathan JRE, Heyworth MF, MacIver AG. Nephrotic syndrome in chronic lymphocytic leukemia. Br Med J 1974;3:655-658.
50. Mendes-DaCosta CR, Dupont E, Hamers R, Hooghe R, Dupuis E, Potvliege R. Nephrotic syndrome in bronchogenic carcinoma: report of two cases with immunochemical studies. Clin Nephrol 1974;2:245-251.
51. Row PG, Cameron JS, Turner DR, Evans DG, White RHR, Ogg CS, Chantler C, Brown CB. Membranous nephropathy long-term follow-up and association with neoplasia. Q J Med 1975;44:207-239.
52. Higgins MR, Randall RE, Still WJS. Nephrotic syndrome with oat-cell carcinoma. Br Med J 1974;3:450-451.
53. Weintroub S, Stavorovsky M, Griffel B. Membranous glomerulonephritis. An initial symptoms of gastric carcinoma. Arch Surg 1975;110:833-858.
54. Lumeng J, Moran JE. Carotid body tumor associated with mild membranous glomerulonephritis. Ann Intern Med 1966;65:1266-1271.
55. Cantrell EG. Nephrotic syndrome cured by removal of gastric carcinoma. Br Med J 1969;2:739-740.
56. Couser WG, Wagonfield JB, Spargo BH, Lewis EJ. Glomerular deposition of tumor antigen in membranous nephropathy associated with colonic carcinoma. Am J Med 1974;57:962-970.
57. Lewis MG, Loughridge LW, Phillips TM. Immunological studies in nephrotic syndrome associated with extrarenal malignant disease. Lancet 1971;2:134-135.
58. Ozawa T, Pluss R, Lacher J, Boedecker E, Guggenheim D, Hammond W, McIntosh R. Endogenous immune complex nephropathy associated with malignancy. I. Studies on the nature and immunopathogenic significance of glomerular bound antigen and antibody, isolation and characterization of tumor specific antigen and antibody and circulating immune complexes. Q J Med 1975;44:523-541.
59. Costanza ME, Schwartz RS, Nathanson L. Carcinoembryonic antigen-antibody complexes in a patient with colonic carcinoma and nephrotic syndrome. N Engl J Med 1973;289:520-522.

60. Sutherland JR, Markham RV Jr, Mardiney MR Jr. Subclinical immune complexes in the glomeruli of kidneys postmortem. Am J Med 1974;57:536-541.
61. Pascal RR, Iannoclone PM, Rollwagen FM, Harding TA, Bennett SJ. Electron microscopy and immunofluorescence of glomerular immune complex deposits in cancer patients. Cancer Res 1976;36:43-47.
62. Helin H, Pasternack A, Penttinen K, Wager O. Glomerular electron-dense deposits and circulating immune complexes in patients with malignant tumors. Clin Nephrol 1980;14:23-30.
63. Caldwell DS, Musculoskeletal syndromes associated with malignancy. In: Kelly WN, Harris ED, Ruddy S, eds. Textbook of rheumatology. Philadelphia: W.B. Saunders, 1981:1658-1671.
64. Calabro JJ. Cancer and arthritis. Arthritis Rheum 1967;10:553-567.
65. Bennett RM, Ginsberg MH, Thomsen S. Carcinomatous polyarthritis. Arthritis Rheum 1976;19:953-958.
66. Wallach HW. Lupus-like syndrome associated with carcinoma of the breast. Arch Intern Med 1977;137:532-535.
67. Awerbuch MS, Brooks PM. Role of immune complexes in hypertrophic osteoarthropathy and non-metastatic polyarthritis. Ann Rheum Dis 1981;40:470-472.
68. Bradley JD, Pinals RS. Carcinoma polyarthritis: role of immune complexes in pathogenesis. J Rheumatol 1983;10:826-828.
69. Cormone RH, Asghar SS. Immunology and skin diseases. In: Turk J, ed. Current topics in immunology, vol 15. London: Edward Arnold, 1981.
70. Miro JM, Gatell JM, Llebaria C, Monjserrat E, SanMiguel JG. Hodgkin's disease associated with unusual immunological manifestations. Nouv Rev Fr Hematol 1982; 24:81-83.
71. Havemann K, Gropp C, Scheuer A, Scherfe T, Gramse M. ACTH-like activity in immune complexes of patients with oat-cell carcinoma of the lungs. Br J Cancer 1979; 39:43-50.
72. Reitan JB, Pape E, Fossa SD, Julsrud O, Slettnes ON, Solheim OP. Osteosclerotic myeloma and polyneuropathy. Acta Med Scand 1980;208:137-144.
73. Roussedy JD, Franck G, Grisar T, Rexnik T, Heynen G, Salmon J. Osteosclerotic myeloma with polyneuropathy and ectopic secretion of calcitonin. Eur J Cancer 1978;14:133-139.
74. Amlot PL, Slaney JM, Williams BD. Circulating immune complexes and symptoms in Hodgkin's disease. Lancet 1976;1:449-451.
75. Zimmerman SE, Smith FP, Phillips TM, Coffey RJ, Schein PS. Gastric carcinoma and thrombotic thrombocytopenic purpura: association with plasma immune complex concentrations. Br Med J 1982;284:1432-1434.
76. Jerry LM, Rowden G, Cano PO, Phillips TM, Deutsch GF, Capek A, Hartman D, Lewis MG. Immune complexes in human melanoma? A consequence of deranged immune regulation. Scand J Immunol 1976;5:845-859.
77. Teshima H, Wanebo H, Pinsky C, Day NK. Circulating immune complexes detected by 125I-C1q deviation test in sera of cancer patients. J Clin Invest 1977;59:1134-1142.
78. Rossen RD, Riesberg MA, Hersh EM, Gutterman JU. The C1q binding test for soluble immune complexes: clinical correlations obtained in patients with cancer. J Natl Cancer Inst 1977;58:1205-1215.
79. Samoyoa EA, McDuffie FC, Nelson AM, Go VLW, Luthr HS, Brumfield HW. Immunoglobulin complexes in sera of patients with malignancy. Int J Cancer 1977; 19:12-17.
80. Theofilopoulos AN, Andrews BS, Urist MM, Morton DL, Dixon FJ. The nature of immune complexes in human cancer sera. J Immunol 1977;119:657-663.

81. Carpenter NA, Fiere DM, Schull D, Lange GT, Lambert P. Circulating immune complexes and the prognosis of acute myeloid leukeima. N Engl J Med 1982 307:1174-1180.
82. Carpenter NA, Lange GT, Fiere DM, Fournie GJ. Clinical relevance of circulating immune complexes in human leukemia. Association in acute leukemia of the presence of immune complexes with unfavorable prognosis. J Clin Invest 1977;60:874-884.
83. Eiras AI, Robins RA, Baldwin RW, Byers VS. Circulating immune complexes in patients with bone tumors. Int J Cancer 1980;25:735-739.
84. Tsang KY, Singh I, Blakemore WG. Circulating immune complexes in human osteosarcoma. J Natl Cancer Inst 1979;62:743-748.
85. Detribolet N, Martin-Archard A, Louis JA. Circulating immune complexes in patients with gliomas. Acta Neurochir 1979;28(Suppl):473-474.
86. Martin-Achara A, Detribolet N, Louis JA, Zander E. Immune complexes associated with brain tumors: Correlation with prognosis. Surg Neurol 1980;13:161-163.
87. Baldwin RW, Hoffken K, Robins RA. Immune complexes in breast carcinoma. Rec Results Cancer Res 1979;67:85-87.
88. Gilead Z, Hatzubai A, Sulitzeanu D. Antigens in immune complexes from patients with breast cancer. Identification of autoantigens in immune complexes isolated from breast cancer effusions. Cancer Immunol Immunother 1982;13:153-157.
89. Gordon BR, Moroff S, Hurvitz AI, Matus RE, MacEwen EC, Good RA, Day NK. Circulating immune complexes in sera of dogs with benign and metastatic breast disease. Cancer Res 1980;40:3627-3631.
90. Hoffken K, Meredith ID, Robins RA, Baldwin RW, Davies CJ, Blamey RW. Circulating immune complexes in patients with breast cancer. Br Med J 1977;2:218-220.
91. Hoffken K, Meredith ID, Robins RA, Baldwin RW, Davis CJ, Blamey RW. Immune complexes and prognosis of human breast cancer. Lancet 1978;1:672-673.
92. Horvath M, Fekete B, Rahoty P. Investigation of circulating immune complexes in patients with breast cancer. Oncology 1982;39:20-22.
93. Krieger G, Kehl A, Wander HE, Solu AR, Rauschecke HF, Negal GA. Clinical significance of circulating immune complexes in metastatic breast cancer. Int J Cancer 1983;31:207-211.
94. Papsidero LD, Nemoto T, Snyderman MC, Chu TM. Immune complexes in breast cancer as directed by C1q binding. Cancer 1979;44:1636-1640.
95. Costanza ME, Pinn V, Schwartz RS, Nathanson L. Carcinoembryonic antigen-antibody complexes in a patient with colonic carcinoma and nephrotic syndrome. N Engl J Med 1974;289:520-522.
96. Hobbiss J, Cooper KM, Moore M, Gowland E, Schofield PF. Limitations of immune complex measurements in colorectal disease. J Surg 1983;70:473-477.
97. Kapsopoulou-Dominos K, Anderer FA. Circulating carcinoembryonic antigen immune complexes in sera of patients with carcinomata of the gastrointestinal tract. Clin Exp Immunol 1979;35:190-195.
98. Mavligit GM, Stuckey S. Colorectal carcinoma evidence for circulating CEA-Anti-CEA complexes. Cancer 1983;146-149.
99. Mihas AA, Volanakis JE, Schrohenloher RE, Mihas TA. Circulating immune complexes in carcinoma of the pancreas. Anticancer Res 1981;1:155-158.
100. Ristow SS, Rossen RD, Frud DS, McKhann CF. Circulating immune complexes in colon cancer patient's sera. Cancer 1979;43:1320-1327.
101. Staab HJ, Anderer FA, Stumpf E, Fischer R. Are circulating CEA immune complexes a prognostic marker in patients with carcinoma of the gastrointestinal tract. Br. J. Cancer 1980;42:26-33.

102. Sanz ML, Oehling A, Lera JM, Hernandez JL, Fernandez J. Quantification of circulating immune complexes in cancer of the gastrointestinal tract. Allergol Immunopathol 1982;10:373-380.
103. Steele G, Lahey S, Rudrilk M, Ross D, Deasy J, Zamcheck N, Osteen R, Wilson R. Circulating immune complexes in patients with colorectal cancer. Am J Surg 1983; 145:549-553.
104. Vellacott KD, Baldwin RW, Balfour TW, Hardcastle JD. Circulating immune complexes in patients with benign and malignant colorectal tumors. Br J Surg 1981; 68:402-404.
105. Babain RJ, Watson DA, Jones JM. Immune complexes in urine and serum of patients with bladder cancer. J Urol 1984;131:463-466.
106. Pesce AJ, Phillips TM, Ooi BS, Evans A, Shank RA, Lewis MG. Immune complexes in transitional cell carcinoma. J Urol 1980;123:486-488.
107. Barber HRK, Dorsett B. Immunologic aspects of gynecologic cancer. Cancer 1981; 48:472-483.
108. Begent RHJ, Chester KA, Walker LC, Tucker DF. Circulating immune complexes as markers of response to chemotherapy in malignant teratomas and gestational trophoblastic tumors. Br J Cancer 1982;45:217-222.
109. Berkowitz RJ, Lahey SJ, Rudrick ML, Rayner AA, Goldstein DP, Steele G. Circulating immune complexes in patients with molar pregnancy. Obstet Gynecol 1983;61:165-168.
110. Clayton LA, Gall SA, Dawson JR, Creasman WT. Immune complexes in ovarian cancer. Gyncol Oncol 1982;13:203-212.
111. Clarke AG, Vasey DP, Symonds EM, Faratian B, McLaughlin JJ, Price MR, Baldwin RW. Levels of circulating immune complexes in patients with ovarian cancer. Br J Obstet Gynecol 1982;89:231-237.
112. Dodd JK, Hicks LJ, Tyler JPA, Crandon AJ, Hudson CN. Circulating IgG specific immune complexes as a potential tumor marker in gynecologic malignancies. Gynecology 1983;16:232-239.
113. Mooney NA, Townsend PA, Wiltshaw E, Evans DG, Shanti-Raju K, Poulton TA. An assessment of sequential measurements of immune complex levels in ovarian cancer with respect to clinical progress. Gynecol Oncol 1982;15:207-213.
114. Poulton TA, Crowther ME, Hay FC, Nineham LJ. Immune complexes in ovarian cancer. Lancet 1978;2:72-73.
115. Seth P, Balachandran N, Malaviva AN, Kumar R. Circulating immune complexes in carcinoma of uterine cervix. Clin Exp Immunol 1979;38:77-82.
116. Maxim PE, Veltri RW, Sprinkle PM, Pusateri RJ. Soluble immune complexes in sera from head and neck cancer patients: a preliminary report. Otolaryngology 1978;86:428-432.
117. Akolkar PN, Advani SH, Gothoskar BP. Circulating immune complexes in Hodgkin's disease. Neoplasma 1983;30:73-79.
118. Amlot PL, Slaney JM, Williams BD. Circulating immune complexes and symptoms in Hodgkin's disease. Lancet 1986;1:449-451.
119. Brandeis WE, Tan C, Wang Y, Good RA, Day NK. Circulating immune complexes, complement and complement component levels in childhood Hodgkin's disease. Clin Exp Immunol 1980;39:551-561.
120. Long JC, Dvorak AM, Quay SC, Stamatos C, Chi S. Reaction of immune complexes with Hodgkin's disease tissue cultures: radioimmune assays and immunoferritin electron microscopy. J Natl Cancer Inst 1979;62:787-797.
121. Long JC, Hall CL, Brown CA, Stamatos C, Weitzman SA, Carey K. Binding of soluble immune complexes in serum of patients with Hodgkin's disease to tissue cultures derived from the tumor. N Engl J Med 1977;297:295-299.

122. Balestrieri G, Tincanio R, Cattaneo R, Rossi G, Verzura P, Marins G, Calculli G, Marinone G. Circulating immune complexes in human acute leukemia. Br J Hematol 1981;49:269-274.
123. Berini M, Galleto G, Comoglio PM. Immune complexes and circulating antibodies against autologous leukemic cells in patients with acute leukemia. Immunology 1982;45:569-575.
124. Claque RB, Kumar S, Hann IM, Jones PHM, Hold PJL. Relevance of circulating immune complexes in childhood acute lymphoblastic leukemia. Int J Cancer 1978; 22:227-230.
125. Doan R, Ollier-Hoartman MP, Schwartzenberg L, Hartman L. Study of circulating immune complexes in three hematologic diseases: acute meyloid leukemia, acute lymphatic leukemia, and hematosarcoma. Cancer Immunol Imminother 1980;8: 161-166.
126. Faldt R, Ankerst J. Possibly specific immune complexes in sera of patients with untreated acute myelogenous leukemia. Int J Cancer 1980;26:309-314.
127. Gergely J, Phi NC, Fust G, Medgyesi GA, Puskas E. Immune complexes in human myeloma and rat immunocytoma. Ric Clin Lab 1980;10:27-34.
128. Gunven P, Klein G, Rosen A, Mukojima T. EBV-associated immune complexes and recurrent Burkitt's lymphoma. IARC Sci Publ 1978;24:875-882.
129. Heier HE, Landaas TD, Marton PE. Circulating immune complexes and prognosis in human malignant lymphoma: a prospective study. Int J Cancer 1979;23:292-298.
130. Hollan SR, Fust G, Mod A, Puskas E, Nguyen CP. Studies of the composition and elimination of circulating immune complexes in leukemia. Haematology (Budap) 1980;13:263-282.
131. Hubbard RA, Aggio MC, Lozzio BB, Wust CJ. Correlation of circulating immune complexes and disease status in patients with leukemia. Clin Exp Immunol 1981; 43:46-53.
132. Minden P, Odom LF, Tubergen DG, Hardtke MA, Sharpton TR, Rose B, Zlotnick A, Carr RI. Immune complexes in children with leukemia: relationship to disease in patients receiving BCG therapy. Cancer 1980;45:460-468.
133. Mod A, Carpenter NA, Fust G, Lambert PH, Miescher P, Hollan S. Serial measurement of circulating immune complexes in the sera of patients with leukemia. Clin Lab Immunol 1980;4:15-20.
134. Brozini P, Delguidice G, Natali S, Serino G, Nembri P, Manca F, Zanussi C. Immune complexes in malignant skin melanoma. Quantitation by three methods, purification and characterization of their non-tumor related constituents. Boll Inst Sierother Milan 1980;59:366-373.
135. D'Amelio R, Cooke B, Hobbs JR. Circulating immune complexes in human malignant melanoma. Tumori 1982;68:469-472.
136. Gupta RK, Theofilopoulos AN, Dixon FJ, Morton DL. Circulating immune complexes as a possible cause for anticomplementary activity in humans with malignant melanoma. Cancer Immunol 1979;6:211-219.
137. Kristensen E, Brandslund I, Nielsen H, Suehag SE. Prognostic value of assays for circulating immune complexes and natural cytotoxicity in malignant skin melanoma (stages I and II). Cancer Immunol Immunother 1980;9:31-36.
138. Ruell P, Murray E, McCarthy WH, Hersey P. Evaluation of assays to detect immune complexes as an immunodiagnostic tool in patients with melanoma. Oncoden Biol Med 1982;3:1-12.
139. The TH, VanDerGiesseni M, Huiges HA, Koops HST, VanWingerden I. Immune complexes in peripheral blood polymorphonuclear leukocytes of malignant melanoma patients. Clin Exp Immunol 1978;32:387-391.

140. Brandeis WE, Helson L, Wang Y, Good RA, Day NK. Circulating immune complexes in sera of children with neuroblastoma. J Clin Invest 1978;62:1201-1209.
141. Jose DG, Seshadri R. Circulating immune complexes in human neuroblastoma: direct assay and role in blocking specific cellular immunity. Int J Cancer 1974;13:824-838
142. Dent PB, Louis JA, McCulloch PB, Dunnett CW, Cerottini JC. Correlation of elevated C1q binding activity and carcinoembryonic antigen levels with clinical features and prognosis in bronchogenic carcinoma. Cancer 1980;45:130-136.
143. Fekete B, Fust G, Pal A, Ansyal I, Jakab A. Clinical correlates of circulating immune complex levels in advanced lung cancer. A discrimination analysis. Oncology 1983;40:106-110.
144. Fust G, Fekete B, Angyal I, Jakab A. Evaluation of different methods for detecting circulating immune complexes. Studiesin patients with lung cancer. J Immunol Methods 1981;46:259-276.
145. Gropp C, Hauerman K, Scherfe T, Ax W. Incidence of circulating immune complexes in patients with lung cancer and their effect on antibody-dependent cytotoxicity. Oncology 1980;37:71-76.
146. Guy K, DiMario U, Irvine WJ, Hunter AM, Hadley A, Horne NW. Circulating immune complexes and autoantibodies in lung cancer. Br J Cancer 1981;43:276-283.
147. Lowe J, Segal-Eiras A, 'Iles PB, Baldwin RW. Circulating immune complexes in patients with lung cancer. Thorax 1981;36:56-59.
148. Tincani A, Balestrieri G, Moraschini AT, Fadini A, Tassi G, Rugarli C. Circulating immune complexes in patients with lung neoplasms. Ric Clin Lab 1980;10:397-403.
149. Delire M. Detection and possible pathogenic role of circulating immune complexes in skin tumors of children. Dermatologia 1980;161:90-97.
150. Gauci L, Carain J, Serrou B. Immune complexes in the context of the immune response in cancer patients. In: Serron B, Rosenfeld C, eds. Immune complexes and plasma exchange in cancer patients. New York: Elsevier/North Holland, Biomedical Press, 1980.
151. Helin H, Pasternack A, Hakala T, Penttinen K, Wager O. Glomerular electron-dense deposits and circulating immune complexes in patients with malignant tumors. Clin Nephrol 1980;14:23-30.
152. Inman RD, Day NK. Immunologic and clinical aspects of immune complex disease. Am J Med 1981;70:1097-1106.
153. Jennette JC. Consistent fluctuations in quantities of circulating immune complexes during progressive and regressive phases of tumor growth. Am J Pathol 1980;100:403-410.
154. Nydegger YE, Davis JS. Soluble immune complexes in human disease. CRC Crit Rev Clin Lab Sci 1980;11:123-170.
155. Poskitt PKF, Poskitt TR. The L1210 assay for immune complexes: application in cancer patients and correlation with disease prognosis. Int J Cancer 1979;24:560-566.
156. Rossen RD, Reisberg MD, Singer D, Suki WN, Duffy J, Hersh FM, Schloeder FX, Hill LL, Eknoyan G. The effect of age on the character of immune complex disease: a comparison of the incidence and relative size of materials reactive with C1q in sera of patients with glomerulonephritis and cancer. Cancer Res 1979;58:65-79.
157. Theofilopoulos AN, Dixon FJ. Immune complexes in human disease. Am J Pathol 1980;100:530-583.

158. Yoshida R, Zanadzki ZA. Circulating immune complexes in patients with neoplastic disorders. Oncology 1980;37:152-156.
159. Baldwin RW, Robins RA. Factors interfering with immunological rejection of tumors. Br Med Bull 1976;32:118-123.
160. Perry LL, Greene MI. T cell subset interactions in the regulation of syngeneic tumor immunity. Fed Prox 1981;40:39-44.
161. Kumar RK, Penny R. Escape of tumors from immunological destruction. Pathology 1982;14:173-179.
162. Roth JA. Tumor induced immunosuppression. Surg Gynecol Obstet 1983;156:233-240.
163. Nepom GT, Hellstrom I, Hellstrom KE. Suppressor mechanisms in tumor immunity. Experientia 1983;139:235-242.
164. Lewis MG, Phillips TM, Rowden G. Beneficial and detrimental effects of humoral immunity in malignancy. Pathobiol Annu 1978;8:217-239.
165. Oberbarnscheidt J, Kolsch E. Direct blockade of antigen-reactive B lymphocytes by immune complexes. An "off" signal for precursors of IgM-producing cells providing by the linkage of antigen- and Fc-receptors. Immunology 1978;35:151-157.
166. Stockfinger B, Lemmel E. Fc receptor dependency of antibody mediated feedback regulation: On the mechanism of inhibition. Cell Immunol 1978;40:395-403.
167. Ryan JC, Henkart PA. Inhibition of B lymphocyte activation by interaction with Fc receptors. Immunol Commun 1976;5:455-468.
168. Masuda T, Miyama M, Kuribayashi K, Yodol J, Takabayashi A, Kyoizumi S. Immunological properties of Fc receptor on lymphocytes. 5. Suppressive regulation of humoral immune response by Fc receptor bearing B lymphocytes. Cell Immunol 1978;39:238-249.
169. Schrader JW, Nodal GJV. Effector cell blockade. A new mechanism of immune hyporeactivity induced by multivalent antigens. J Exp Med 1974;139:1582-1598.
170. Moretta L, Webb SR, Gross CE, Lydyard PM, Cooper MD. Functional analysis of two human T cell subpopulations: help and suppression of B-cell responses by T cells bearing receptors for IgM or IgG. J Exp Med 1977;146:184-200.
171. Moretta L, Mingari MC, Moretta A, Cooper MD. Human T lymphocytes subpopulations: studies of the mechanism by which T cells bearing Fc recptors for IgG suppress T-dependent B cell differentiation induced by pokeweed mitogen. J Immunol 1979;122:984-992.
172. Hoffman MK, Kappler JW, First JA, Oettgen HF. Regulation of the immune response. V. Antibody-mediated inhibition of T and B cell cooperation in the in vitro response to red cell antigens. Eur J Immunol 1974;4:282-286.
173. Fridman WH, Golstein P. Immunoglobulin-binding factor present on and produced by thymus-processed lymphocytes (T cells). Cell Immunol 1980;11:442-455.
174. Ptak W, Zembala M, Gershon RK. Intermediary role of macrophages in the passage of suppressor signals between T-cell subsets. J Exp Med 1979;148:424-434.
175. Taylor RB, Basten A. Suppressor cells in humoral immunity and tolerance. Br Med Bull 1976;32:152-157.
176. Rao VS, Groszicki RL, Mitchell MS. Specific in vivo inhibition of macrophage receptors for cytophilliac antibody by soluble immune complexes. Cancer Res 1979;39:174-182.
177. Hellstrom I, Hellstrom KE. Studies on cellular immunity and its serum-mediated inhibition in moloney-virus-induced mouse sarcomas. Int J Cancer 1977;4:587-600.

178. Hellstrom I, Sjogren HO, Warner GA. Blocking of cell-mediated tumour immunity by sera from patients with growing neoplasms. Int J Cancer 1971;7:226-237.
179. Hellstrom KE, Hellstrom I, Nepom JJ. Specific blocking factors—are they important? Biochem Biophys Acta 1977;473:121-148.
180. Sjogren HO, Hellstrom I, Bansal SC. Suggestive evidence that the "blocking antibodies" of tumor-bearing individuals may be antigen-antibody complex. Proc Natl Acad Sci USA 1971;68:1372-1375.
181. Baldwin RW, Price MR, Robins RA. Inhibition of hepatoma-immune lymph node cell cytotoxicity by tumor-bearer serum and solubilized hepatoma antigen. Int J Cancer 1973;11:527-535.
182. Greene MT, Benacerraf B. Studies on hapten specific T cell immunity and suppression. Immunol Rev 1980;50:103-186.
183. Morgan EL, Tempelis CH. The requirement for the Fc portion of antibody in antigen-antibody complex-mediated suppression. J Immunol 1978;120:1669-1671.
184. Morgan EL, Tempelis CH. The role of antigen-antibody complexes in mediating immunologic unresponsiveness in the chicken. J Immunol 1977;119:1293-1298.
185. Roderick ML, Steele G Jr, Ross DS. Serial circulating immune complex levels and mitogen responses during progressive tumor growth in WF rats. J Natl Cancer Inst 1983;70:1113-1118.
186. Gershon RK, Mokyr MB, Mitchell MS. Activation of suppressor T cells by tumor cells and specific antibody. Nature 1974;250:594-596.
187. Kirkwood JM, Gershon RK. A role of suppressor T-cells in immunological enhancement of tumor growth. Prog Exp Tumor Res 1974;19:257-264.
188. Ravikumar T, Steele G, Rodrick M. Effects of tumor growth on interleukins and circulating immune complexes: mechanisms of immune unresponsiveness. Cancer 1984;53:1373-1378.
189. Hoon DBS, Ziola B, Carlsen S, Warrington R, Ramshaw I. Circulating immune complexes and immunoglobulin M-CLAB rheumatoid factor in rats bearing mammary adenocarcinomas which vary in ability to metastasize. Cancer Res 1983;43:114-119.
190. Chee DI, Boddie IA, Roth JA. Production of melanoma-associated antigens by a defined malignant cell strain grown in chemically defined medium. Cancer Res 1976;36:1503-1509.
191. Roth JA, Slocum HK, Pellegrino A, Holmes EC, Reisfeld R. Purification of soluble human melanoma-associated antigens. Cancer Res 1976;36:2360-2364.
192. Bystryn JC, Smalley JR. Identification and solubilization of iodinated cell-surface human melanoma associated antigens. Int J Cancer 1977;20:165-172.
193. McCabe RP, Ferrone S, Pellegrino MA. Purification and immunologic evaluation of human melanoma-associated antigens. Int J Cancer 1977;20:165-172.
194. Carey TE, Lloyd KO, Takahashi T, Travassos LR, Old LJ. Cell surface antigen of human malignant melanoma: solubilization and partial characterization. Proc Natl Acad Sci USA 1979;76:2898-2902.
195. Galloway DR, McCable RP. Pellegrino A, Ferrone S, Reisfeld RA. Tumor-associated antigens in spent medium of human melanoma cells: immunochemical characterization with xenoantisera. J Immunol 1981;126:62-66.
196. Heaney-Kieras J, Brystryn JC. Identification and purification of 75K dalton cell-surface human melanoma associated antigen. Cancer Res 1982;42:2310-2316.
197. Koprowski HA, Steplewski A, Herlyn D, Herlyn M. Study of antibodies against human melanoma produced by somatic cell hybrids. Proc Natl Acad Sci USA 1978;75:3405-3409.

198. Dippold WG, Lloyd KO, Li LTC, Ikeda H, Oettgen HF, Old LJ. Cell surface antigens of human malignant melanoma: definition of six antigenic systemic with mouse monoclonal antibodies. Proc Natl Acad Sci 1980;77:2183-2187.
199. Woodbury RG, Brown JP, Yeh MY, Hellstrom I, Hellstrom K. Identification of a cell surface protein, p97 in human melanomas and certain other neoplasms. Proc Natl Acad Sci USA 1980;77:2183-2187.
200. Mitchell KF, Fuhrer JP, Steplewski Z, Koprowski H. Biochemical characterization of human melanoma cell surfaces: dissection with monoclonal antibodies. Proc Natl Acad Sci USA 1980;77:7287-7291.
201. Harper JR, Bumol TF, Reisfeld RA. Characterization of monoclonal antibody 155.8 and partial characterization of its proteoglycan antigen of human melanoma cells. J Immunol 1984;132:2096-2104.
202. Bumol TF, Reisfeld RA. Unique glycoprotein-proteoglycan complex defined by monoclonal antibody on human melanoma cells. Proc Natl Acad Sci USA 1979; 79:1245-1249.
203. Tai T, Paulson JC, Cahan LD, Irie RF. Ganglioside GM2 as a human tumor antigen (OFA-1-1). Proc Natl Acad Sci USA 1983;80:5382-5396.
204. Carey TE, Takahashi T, Resnick L, Oettgen HF. Cell surface antigens of human malignant melanoma. I. Mixed hemadsorption assays for humoral immunity to cultured autologous melanoma cells. Proc Natl Acad Sci USA 1976;73:3278-3282.
205. Shiku H, Takahashi T, Oettgen GF, Old LJ. Cell surface antigens of human malignant melanoma. II. Serological typing with immune adherence assays and definition of two new surface antigens. J Exp Med 1978;144:873-881.
206. Shiku H, Takahashi T, Resnick L, Oettgen HF, Old LJ. Cell surface antigens of human malignant melanoma. III. Recognition of auto-antibodies with unusual characteristics. J Exp Med 1977;145:784-789.
207. Kirkwood JM, Vlock DR. Autologous antibody to melanoma is augmented following acid dissociation and ultrafiltration of serum. Cancer Res 1984;44:4177-4182.
208. Gupta RK, Morton DL. Dynamics of tumor-associated antigen, antibody and immune complexes circulating in melanoma patients. Proc Am Assoc Cancer Res 1985;26:1120 (abstr).
209. Vlock DR, Kirkwood JM. Serial studies of autologous antibody reactivity to melanoma: relationship to circulating immune complexes and clinical course. J Clin Invest 1985;76:849-854.
210. Katz AE, Nysather JO, Harker LA. Major immunoglobin ratios in carcinoma of the head and neck. Ann Otol Rhinol Laryngol 1978;87:412-416.
211. Brown AM, Lally ET, Frankel A, Harvick R, Davis LW, Rominger CJ. The association of the IgA levels of serum and whole saliva with the progression of head and neck cancer. Cancer 1975;35:1154-1162.
212. Carey TE, Takahashi T, Resnick L, Oettgen HF. Cell surface antigens of human malignant melanoma. 1. Mixed hemadsorption assays for humoral immunity to cultured autologous melanoma cells. Proc Natl Acad Sci USA 1976;73:3278-3282.
213. Galloway DR, McCabe RP, Pellegrino MA, Ferrone S, Reisfeld RA. Tumor-associated antigens in spent medium of human melanoma cells: immunochemical characterization with xenoantisera. J Immunol 1981;126:62-66.
214. Koprowski H, Steplewski Z, Herlyn D, Herlyn M. Study of antibodies against human melanoma produced by somatic cell hybrids. Proc Natl Acad Sci USA 1978;75:3405-3409.
215. Dippold WG, Lloyd KO, Li LTC, Ikeda H, Oettgen HF, Old LJ. Cell surface antigens of human malignant melanoma: definition of six antigenic systems with mouse monoclonal antibodies. Proc Natl Acad Sci USA 1980;77:2183-2187.

216. Tai T, Paulson JC, Cahan LD, Irie RF. Ganglioside GM2 as a human tumor antigen (OFA-I-1). Proc Natl Acad Sci USA 1983;80:5382-5396.
217. Mattes MJ, Cordon-Cardo C, Lewis JL, Old LJ, Lloyd KO. Cell surface antigens of human ovarian and endometrial carcinoma defined by mouse monoclonal antibodies. Proc Natl Acad Sci USA 1984;81:568-572.
218. Lloyd KO. In: Herberman RB, ed. Human tumor antigens: detection and characterization with monoclonal antibodies. Basic and clinical tumor immunology. Boston: Martinus Nijhoff, 1983:159-214.
219. Carey TE, Kimmel KA, Schwartz DR, Richter DE, Baker SR, Krause CJ. Antibodies to human squamous cell carcinoma. Otolaryngol Head Neck Surg 1983; 91:482-491.
219a. Vlock DR, Scalise D, Schwartz DR, Richter DE, Krause CJ, Baker SR, Carey TE. Incidence of serum antibody reactivity to autologous head and neck cancer cell lines and augmentation of antibody reactivity following acid dissociation and ultrafiltration. Cancer Res 1989;49:1361-1365.
219b. Vlock DR, Arnold B, Humpierres J, Schwartz DR, Carey TE. Serial studies of autologous antibody reactivity to squamous cell carcinoma of the head and neck. Proc Am Assoc Cancer Res 1989;30:1415 (abstr).
220. Dorsett BH, Iochim HL, Stolback L, Barber HRK. Isolation of tumor-specific antibodies from effusions of ovarian carcinomas. Int J Cancer 1975;16:779-786.
221. Cronin WJ, Dorsett BH, Iochim HL. Isolation of lung carcinoma antibodies from human complexes and production of heterologous antisera. Cancer Res 1982;42: 292-300.
222. Lutz PA, Dawson JR. Activity of antibodies recovered from immune complexes of ovarian cancer patients. Cancer Immunol Immunother 1984;17:180-189.
223. Oian H, Feng J, Fu T. The study of antibodies and antigens dissociated from the immune complexes extracted from ovarian carcinoma ascitic fluid. Gynecol Oncol 1985;20:100-108.
224. Israel L, Edelstein R, Mannoni P, Radio E, Greenspan EM. Plasmapheresis in patients with disseminated cancer: clinical results and correlation with changes in serum protein. Cancer 1977;40:3146-3154.
225. Samak R, Edelstein R, Israel L, Boglicki D, Samak M. Repeated plasma exchange in patients with advanced cancer: biological and immunological findings. In: Serrou B, Rosenfield C, eds. Human cancer immunology. New York: Elsevier/North Holland, 1981:221-242.
226. Hersey P, Edwards A, Adams F, Milton GW, Nelson DS. Antibody dependent cell-mediated cytotoxicity against melanoma cells induced by plasmapheresis. Lancet 1976;1:825-827.
227. Israel L, Edelstein R, Samak R. Clinical results of multiple plasmapheresis in patients with advanced cancer. In: Serrou B, Rosenfield C, eds. Human cancer immunology. New York: Elsevier/North Holland, 1981:309-328.
228. Bansal SC, Bansal BR, Thomas HL. Ex vivo removal of serum IgG in a patient with colon carcinoma. Some biochemical, immunological and histological observations. Cancer 1978;42:1-18.
229. Messerschmidt, GL, Bowles C, Alsaker R. Prognostic indicators of tumor response to *Staphylococcus aureus* Cowan strain I plasma perfusion. J Natl Cancer Inst 1983;71:535-538.
230. Terman DS, Yamamoto T, Mattioli M. Extensive necrosis of spontaneous canine mammary adenocarcinoma after extracorporeal perfusion over staphylococcus auereus cowans. I. Description of acute tumoricidal response: morphologic, histologic, immunohistochemical immunologic and serologic findings. J Immunol 1980; 124:795-805.

231. Jones FR, Yoshida LH, Lagiges WC, Kenny MA. Treatment of feline leukemia and reversal of FeLV by ex vivo removal of IgG: a preliminary report. Cancer 1980;46:675-684.
232. Jones FR, Grant CK, Snyder HW Jr. Lymphosarcoma and persistent feline leuemia virus in pet cats: a system to study responses during extracorporeal treatments. J Biol Response Mod 1984;3:286-292.
233. Day NK, Engelman RW, Liu TW, Trang L, Good RA. Remission of lymphoma leukemia in cats following ex vivo immunosorption therapy using *Staphylococcus aureus* protein. J Biol Response Mod 1984;3:278-285.
234. Snyder HW Jr, Jones FR, Day NK, Hardy WD Jr. Isolation and characterization of circulating feline leukemia virus-immune complexes from plasma of persistently infected pet cats removed by ex vivo immunosorption. J Immunol 1982;128:2726-2730.
235. Liu WT, Engleman RW, Trang LQ, Hau K, Good RA, Day NK. Appearance of cytotoxic antibody to viral GP70 on feline lymphoma cells during ex vivo immunoadsorption therapy: quantitation, characterization and association with remission of disease and disappearance of viremia. Proc Natl Acad Sci USA 1984;81:3516-3520.
236. Terman DS, Young JB, Shearer WT, Ayus C, Lehane D, Mattioli C, Espada R, Howell J, Yamamoto T, Zaleski HI, Miller L, Frommer P, Henry JF, Tillquist R, Cook G, Daskal Y. Preliminary observations of the effects of breast adenocarcinoma of plasma perfused over immobilized protein A. N Engl J Med 1981;305:1195-1200.
237. Terman DS. Staphyloccal protein A in neoplastic disease. J Biol Response Mod 1984;3:316-324.
238. MacKintosh FR, Bennett K, Schiff S, Shields J, Hall SW. Treatment of advanced malignancy with plasma perfused over staphylococcal protein A. West J Med 1983;139:36-40.
239. MacKintosh FR, Bennett K, Hall SW. Clinical response to protein A infusions and in vitro correlations. J Biol Response Mod 1984;3:336-340.
240. Fer MF, Beman J, Stevenson JC, Maluish A, Moratz C, Delawier T, Foon K, Herberman RB, Oldham RK, Terman DS, Young JB, Daskal Y. A trial of autologous plasma perfusion over protein A in patients with breast cancer. J Biol Response Mod 1984;3:352-358.
241. Messerschmidt GL, Bowles CA, Henry DH, Deisseroth AB. Clinical trials with staphylococcal aureus and protein A in the treatment of malignant disease. J Biol Response Mod 1984;3:325-329.
242. Bensinger WI, Buckner CD, Clift RA, Thomas ED. Clinical trials with protein A. J Biol Response Mod 1984;3:347-351.
243. Korec S, Smith FP, Schein PS, Phillips TM. Clinical experiences with extracorporeal immunoperfusion of plasma from cancer patients. J Biol Response Mod 1984;3:330-335.
244. Kiprov DD, Lippert R, Jones FR, Lagios MD, Balkint JP, Cohen RJ. Extracorporeal perfusion of plasma over immobilized protein A in a patient with Kaposi's sarcoma and acquired immunodeficiency. J Biol Response Mod 1984;3:341-346.
245. Ray PK, Idiculla A, Mark R, Rhoads JE Jr, Thomas H, Bassett JG, Cooper DR. Extracorporeal immunoadsorption of plasma from metastatic colon carcinoma patient by protein A-containing nonviable staphylococcus aureus. Clinical, biochemical, serologic and histologic evaluation of the patient's response. Cancer 1982;49:1800-1809.
246. Messerschmidt GL, Henry DH, Snyder HW Jr, Bertrram J, Mittelman A, Ainsworth S, Fiore J, Viola MV, Louie J, Ambinder E. Protein A immunoadsorption in the treatment of malignant disease. J Clin Oncol 1988;6;203-212.

247. Young JB, Ayus JC, Miller LK, Divinr GW, Frommer JP, Miller RR, Terman DS. Cardiopulmonary toxicity in patients with breast carcinoma during plasma perfusion over immobilized protein A. Pathophysiology of reaction and attenuating methods. Am J Med 1983;75:278-288.
248. Smith EM, Johnson HM, Blalock JE. Staphylococcus aureus protein A induces the production of interferon-alpha in human lymphocytes and interferon-alpha/beta in mouse spleen cells. J Immunol 1983;130:773-776.
249. Yamamoto JK, Good RA, Johnson HM, Engelman R, Tyler R, Machida D, Tran X, Day NK. Augmentation of serum interferon titers prior to remission of leukemia in cats with ex vivo immunoadsorption. Fed Proc 1983;42:838 (abstr).
250. Bertram JH, Bouquiren D, Hengst JCD, Grunberg SM, Mitchell MJ. On the mechanism of action of the staph protein A column. Proc Am Assoc Cancer Res 1984; 25:919 (abstr).
251. Balint J, Ikeda Y, Langone JJ. Tumoricidal response following perfusion over immobilized protein A: identification of immunoglobulin oligomers in serum after perfusion and their partial characterization. Cancer Res 1984;44:734-743.
252. Langone JJ, Das C, Bennett D, Terman DS. Generation of human C3a, C4a and C5a anaphylatoxins by protein A of staphylococcus aureus and immobilized protein A reagents used in serotherapy of cancer. J Immunol 1984;133:1057-1063.

36
Pancarcinoma T and Tn Epitopes: Autoimmunogens and Diagnostic Markers that Reveal Incipient Carcinomas and Help Establish Prognosis

GEORG F. SPRINGER* and PARIMAL R. DESAI*

Evanston Hospital and Northwestern University, Evanston, Illinois

WILFRED WISE†, SHEILA C. CARLSTEDT*, and HERTA TEGTMEYER*

Evanston Hospital, Evanston, Illinois

RHONA STEIN

Center for Molecular Medicine and Immunology, Newark, New Jersey

EDWARD F. SCANLON

Northwestern University, Evanston, Illinois

The earliest detection and subsequent surgical removal of incipient clinical carcinomas (CAs) at the tumor in situ or $T_1 N_0 M_0$ (T_1: lung CA ≤ 3 cm with no evidence of invasion more proximal than lobar bronchus; breast CA ≤ 2 cm in greatest dimension) stage is today's overriding goal in cancer therapy. Yet most CAs are ≥ 8 years old when first detected by clinicopathological methods, and the outcome of curative surgery is uncertain. Once a CA is more advanced, a generally fatal outcome (exceptions: uterine cervix, some skin CAs) cannot be avoided.

We report here on the nature of Thomsen-Friedenreich (T) and Tn antigens (Ags) and their application in the detection and monitoring of CA, primarily by measurement of the patient's *auto*immune responses to T, and secondarily by immunohistochemistry (IHC), where they permit detection of even single CA cells in otherwise normal-appearing lymph nodes [1], and, most importantly, their CA cell surface density aids in prognostication [2-7].

The CA-associated T Ag is unique among cancer Ags, since it induces measurable autoimmune responses that occur in a CA patient from the onset of disease onwards [cf. 2,8,9]. These tests detected 85.4% of 41 $TisN_0M_0$ and $T_1 N_0 M_0$, and so far cancers in 62% of 34 patients suspected of having CA, but who continued to have negative biopsy

---

Present affiliations:
*University of Health Sciences, Chicago Medical School, and Veterans Administration Medical Center, North Chicago, Illinois
† St. Francis Hospital and Loyola Univ. Med. School, Evanston, Illinois

or x-ray results, months to years before the CA was finally diagnosed by standard means. Tn antigen is also autoimmunogenic [69].

T and Tn are generally occluded Ags [10-13], except in some immunoprivileged sites [cf. 14,15, section I.B]. Accordingly, all humans have anti-T and -Tn antibodies (Abs) in their serum (induced largely by their intestinal flora) [16].

We first noted that live breast (adeno) CA cells specifically adsorb anti-T and -Tn Abs [17,18] and in the presence of complement human anti-T killed approximately 95% of T Ag-rich murine mammary CA cells in vitro, provided that the T was unmasked [17]. T and Tn were not demonstrable by Abs in benign breast glandular tissue, except as traces, nor in the healthy postnatal tissues tested [2,18].

I. CHARACTERISTICS OF PANCA MARKERS, T AND Tn (EXCLUSIVE OF IHC)

A. The Nature and Specificities of T and Tn

T and Tn were discovered as causes of the polyagglutination of human red blood cells (RBC) by autochthonous and all human antisera [10-13]. On RBCs, T is transitory and results from the action of microbial sialidases [10,12, cf. 2], while RBC Tn [13] is persistent and due to a somatic pleiotropic mutation [19-22,54].

Neither T nor Tn was thought to be associated with malignant tumors [10,13,16,22, cf. 2], until we and later others found them both in breast CAs [17,18,23-26], where they appear *uncovered*, probably incompletely synthesized precursors of the normal human blood group MN Ags [2,12] and most *O*-glycosidically linked heterosaccharides [27-29]. T and Tn were soon established as hallmarks of the most common deadly CAs [2] (see also section V.A), which amount to ~75% of all cancers [30]. The authenticity of T in humans is supported by humoral and cellular anti-T responses in CA patients [23,25, cf. 2,31,77].

T and Tn glycoproteins (GPs) have been prepared intact from human breast CA DU 4475 [32], and homologues closely similar in qualitative and quantitative carbohydrate and amino acid composition were obtained by partial degradation of MN GPs isolated from normal human group O RBC [23,32]. Tn Ag has recently been obtained from Tn RBCs (Desai PR, Tegtmeyer H, Springer GF, in preparation). The chemical and some physicochemical properties of T and Tn Ags and epitopes (EPs) have been established using glycosidases [32,34,35], glycosyltransferase (T synthesis) [19-21] and proteases [33,36], and by standard amino acid and carbohydrate analyses [32,33].

The T hapten is D-galactopyranosyl $\beta(1 \rightarrow 3)$ *N*-acetyl-D-galactosamine (Gal$\beta 1 \rightarrow 3$ –GalNAc) [19-21,27,28,32-35,37-41], and Tn EP is GalNAcα-*O*-Ser/Thr [34,35,42,43] which was shown first on desialylated O,MM and O,NN Ags; *O*-α-GalNAc was uncovered by this desialylation and demonstrated by precipitin and hapten inhibition studies combined with quantitative colorimetry and ninhydrin degradation. This subterminal GalNAc amounted to ~10% of M and N's total GalNAc [37,38,cf. 34,35].

In guinea pig L-10 CA, T and Tn, that elicited anti-T- and anti-Tn-specific Abs in rabbits, were found exclusively as glycolipids [44,46]. Here, the haptens are presumably linked β- to ceramide [45-48]. Asialoganglioside M$_1$ (AS-G$_{M1}$) had some T activity [44,46]. Glycolipids are usually not as accessible in vivo to Abs as blood group GPs [45]. AS-G$_{M1}$ does occur in reactive form in some leukemias, in nonmalignant hematological disorders, and in healthy tissues [47,48].

T is transformed to Tn by $\beta 1 \rightarrow 3$-galactosidase [32,34,35,37,42]. Immunoreactive sialyl-Tn is frequently expressed in CAs [49,50], but in contrast to Tn it also occurs in healthy structures such as cervix uteri mucin [51,52], colorectal mucus [53], and healthy RBCs [34,35,37,38,54] and is therefore only conditionally CA-associated.

The only nonCA-immunoaccessible tissues found to be associated with T and/or Tn are other malignant tumors: T lymphomas, leukemias, and thymomas [2,9,31,55-57,82].

B. Occurrence of T and Tn Ags and EPs in Malignant Tumors

T and/or Tn were initially detected in all primary breast CA tissues (18 T, 14 Tn) but not in the six benign breast glands tested [18].

Later it was shown that Tn on hematopoietic cells (Tn syndrome) may precede leukemias and thus be a prognostic indicator of malignancy [cf. 2,56,57]. The pathogenesis of the Tn syndrome precludes the appearance of T [19-21].

All types of CA, which are universally of epithelial origin, possess T and Tn [2,4-7,18, 26,31,58-64]. In most CA types, T and Tn are present on 80-90% of the specimens [2,5,18,23,25,31,63-67]. Within a given CA only certain cells of the same type [5,7] or particular cells of various types of epithelia may have T and/or Tn. T and Tn may also be found at the cell membrane in some lymphomas, Hodgkin's disease, and in leukemias, but not in soft tissue sarcomas and central nervous system (CNS) tumors, nor in nine *benign* tumors studied (neuroma, 1; parotid mixed, 2; thyroid adenoma, 1; cervix adenoma, 2; tendon sheath tumor, 1; ovarian cysts, 2) [cf. 2,31]. It is of interest that although the CNS tumors reacted negatively, the brain [70] and enteric ganglion plexuses [61] have gangliosides with immunoreactive T-specific EPs.

II. ANTI-T AND –Tn MONOSPECIFIC POLYCLONAL ABS AND MONOCLONAL ABS AND LECTINS

Authentication of the specificity of anti-T and –Tn Abs and lectins was done by a wide range of immunochemical procedures [12,17,27,28,32,33,37-41,54,62,63,71-73]. In most of these assays, chemically and physically well-defined Ags, EP clusters, haptens, and cross-reacting control substances were used [27,28,32,37,39,40,74,75]. Glycosidases [27,28,32,34,35,39,40,43,76] and glycosyltransferases [19-21] were used to produce some of the Ags in vitro.

A. Anti-T and -Tn Abs, Their Measurements, and Specific Inhibition

1. Hemagglutination, Its Specific Inhibition, and Hemolysis

Standard hemagglutination and hemagglutination inhibition (HI) tests were performed and interpreted as described earlier [23,72]; T and Tn agglutination tests were performed between 17°C and 21°C to obtain maximal specificity [10-12,16]. Anti-T agglutinin levels remain remarkably constant from about 3 years of age through adulthood [10,11, 77]. Specific hemolytic anti-T cytotoxicity determination of sera from patients with anaplastic gastrointestinal (GI) CA has been described admirably elsewhere [73].

Breast CA detection by rapid anti-T agglutinin assays using an autoanalyzer (variation coefficient, 2.57%) was achieved, no increase in sensitivity was found [77; cf.8,23,78,79]. We read anti-T titer scores (at least two persons independently) by microscope. The readings of the same sample showed average standard deviation of ±0.58 for same-day readings by different individuals and of ±0.73 for readings performed by the same person on different days [9].

2. Solid-Phase Immunoassays

These quantitative assays developed by us more than doubled the sensitivity of CA detection by anti-T measurement over hemagglutination without loss of specificity.

Immunofluorescent. Solid-phase immunoassay for anti-T (SPIA-T) measures serum anti-T IgM (~60% of total serum anti-T Igs and 7-14% of total IgM), and relates it to total IgM [16,23,80,81]. Among CA patients, total IgM concentration was usually within the wide normal range (50-410 mg/dl). However, in most CA patients anti-T IgM, but not other anti-T Ig levels, differed substantially from the norm, including those in patients without cancer. Anti-T IgM levels were depressed in most preoperative CA patients [cf. 2,78,80,81]; but in ~5% of CA patients with no indication of liver disease, in which anti-T levels are elevated [23], anti-T IgM was and remained at a highly elevated plateau (as high as ~55 mg/dl [78,80]), indicating an active IgM immune response characteristic of carbohydrate Ags [83].

The difference in anti-T IgM levels between patients with CA and control subjects was accentuated when the total IgM concentration in the same specimen was taken into account. A person's mg/dl anti-T IgM relative to total IgM, expressed as the value Q_M, was calculated by the formula:

$$Q_M = \frac{(\text{Anti-T IgM})^2}{\text{Total IgM}} \times 100$$

The normal range of Q_M was from ≥ 100 to ≤ 360 (see III.C.2. [80,81, cf. 2]). The average percentage variation in Q_M values of the same samples determined on different days was $< \pm 3\%$.

Enzymatic. We developed, based on our direct-binding enzyme immunoassay (EIA) [32], simple enzyme-linked immunosorbent assays (ELISAs) to quantitate anti-T, anti-Tn, and total-IgM levels in patient sera by using RBC and CA-T and CA-Tn solid phases (Desai PR, Springer GF, to be published).

B. Anti-T and -Tn Abs and Lectins

1. Polyclonal, Monospecific Human and Animal Anti-CA-T and -Tn Abs

Because T and Tn are concealed from the immune system in nonCA tissues [10,11,13] and sialyl-Tn appears readily accessible only in some tissues, all humans have anti-T, -Tn and usually very low anti-sialyl-Tn Ab levels [84; Desai PR, Tegtmeyer H, Springer GF, in preparation].

Polyclonal human anti-T and anti-Tn Abs from blood group $A_1 B$ donors were affinity-purified using O MN RBC [cf. 71]; specificities are ascertained as described below for monoclonal antibodies (MoAbs). Rabbit hyperimmune polyclonal anti-T Abs were induced with enzymatically maximally desialylated O RBCs (T RBCs) [6,61,62].

AS-ovine submaxillary mucin (AS-OSM) (<2% Neu5Ac) was up to >100-fold stronger as an inhibitor in 56 of 58 human sera (collected at random) than OSM in both HI with Tn RBCs and in ELISA [32]. Thus, a great majority of anti-Tn Abs are directed against Tn EPs. Inhibition by O RBC-derived Tn paralleled that by AS-OSM in all instances.

2. Anti-T and -Tn MoAbs Induced with T- and Tn-Specific Ags

Anti-T, -Tn, and -Sialyl-Tn MoAbs Elicited with O T RBC, O Tn RBC, and O RBC-Derived Tn-Specific Ags and OSM. Rodent anti-T MoAbs were first elicited with T

RBCs as Ag; these were predominantly IgM but there were some IgGs [85-87]. MoAbs were recognized as true anti-T if they were specifically and completely inhibited by authentic T Ag, by its glycopeptides, with T EP clusters at the amino terminal part of the AS-glycophorin A, and by the *free* T hapten [32,35,36,39]. (This excluded MoAb H.24 [cf. 8]). MoAbs that had lost most of their specific activities within ~1 year after publication of their properties are not listed.

We evoked with desialylated O RBC a rat anti-T ascites HT-8 (IgM$_\kappa$) (formerly Ca 3114), that cross reacted moderately with Tn but not with A_1, A_2, B, I, H(O) RBC. A MoAb with anti-T-like specificity, HH8 (IgM), was induced in mice with Gal-A^b glycolipid, prepared by deglycosylation of A^a type 3 chain glycolipid. In distinction to usual anti-T, HH8 was inhibited almost equally well by both anomers of Galβ1 → 3GalNAc [29, cf. 35].

Anti-Tn MoAb HTn-1 (IgM$_\kappa$) (formerly Ca 2921) was produced in rats [86,87] with O T RBCs (known to have Tn EPs [37,38,54]). HTn-1 was strongly inhibited by O RBC Tn Ag, AS-OSM, GalNAc, and me-αGalNAc [32] but not by OSM, A_1 or A_2 Ag.

Murine IgM$_\kappa$ MoAbs with high anti-Tn activity were selected among over 50 anti-Tn active hybridomas elicited with either Tn Ag derived chemically from normal O RBCs (MoAbs BaGS-1 to -3) or with Ca.L.'s O Tn RBC (MoAbs BaGS-4 to -7) as immunogens [32].

Anti-Tn ascites fluids were prepared from the BaGS MoAbs. The fluids were absorbed with O MN RBC [71] and the supernatant centrifuged for 18 hr, at 4°C, 6,000 x g; the ascites were collected and stored under standard conditions. The ascites fluids were 25-100 times as active as corresponding culture supernatants. All anti-Tn fluids were specifically adsorbed onto Tn RBC and CA cells, and were inhibited by CA-Tn and RBC-Tn Ags. None of the culture supernatants or ascites fluids of these MoAbs had anti-A_1, -A_2, -B, -H(O), -I, or -MN activities [9,32]. The anti-T titer of ascites anti-Tn was <0.1% of the anti-Tn titer; in the supernates of anti-Tn culture fluids, anti-T activity was 2% of that of anti-Tn [9]. The most active hapten toward these anti-Tn and anti-sialyl-Tn human polyclonal, monospecific, and rodent MoAbs (supernatants and ascites) was me-αGalNAc [32]. BaGS-6 was the only MoAb specifically inhibited by Neu5Ac (Desai PR et al., in preparation).

A highly active anti-Tn MoAb (FBT 3, IgM) was generated [56,88] with Tn RBCs. FBT 3 reacted specifically with 71% of 147 CAs and with 1 of 13 lymphomas. Anti-sialyl-Tn murine MoAbs TKH1 and TKH2 were induced with OSM as Ag (TKH2 reacted slightly with O RBC) [cf. 84].

3. Anti-T, -Tn, and -Sialyl-Tn MoAbs Unwittingly Evoked with Human CA Tissues

Because of the panCA nature of T and Tn Ags [2,25], it is not surprising that anti-CA-T and anti-CA-Tn MoAbs were generated unintentionally with CA tissue. The murine MoAb RS1-114 (IgM$_\kappa$), evoked with human lung adenoCA line 549, is the first anti-CA-T MoAb; it has been extensively and well characterized [58].

MoAb B72.3 was evoked with human breast CA metastasis [89]. Its combining site encompasses the Tn EP [8,32]. Many diagnostic studies have been performed with this MoAb [e.g., 89,90], now recognized as an anti-sialyl-Tn [50], although the extent of agglutinability of authentic Tn RBCs (Ca.L. and Ca.C.) by B72.3 did not change after exhaustive desialylation, and Neu5Ac did not inhibit it (Springer GF, et al., to be submitted). MoAb MLS 102 (anti-sialyl-Tn) was induced with LS 180 colon CA in mice [91]. Hirohashi et al. produced with human lung CA LU-65, IgM MoAbs NCC-LU-35 and -81 and defined them as anti-Tn [92]; however, they also reacted well with sialyl-

Tn [50] and, to a lesser extent, with blood group A. The anti-sialyl-Tn Abs were inhibited by the same carbohydrates in the same sequence as anti-Tn.

4. Anti-T and -Tn Lectins from Higher Plants and a Mollusk

Most investigations of the role of the interrelated T/anti-CA-T and Tn/anti-CA-Tn systems have focused on the former using peanut agglutinin (PNA) lectin [cf. 93], which has high affinity for the T disaccharide but also reacts well with unrelated structures, such as GalN, α-Glc, and D-Fuc [93], and with healthy breast gland tissue [18,67,94,95].

Human colorectal CAs and corresponding "normal" colorectal mucosae distant from the CA were obtained at surgery and compared concomitantly using labeled PNA by IHC and polyacrylamide gel electrophoresis (PAGE, after tissue homogenization) [65]. Both assays showed that only CAs had PNA-positive GPs.

In contrast to PNA, human anti-T Abs are highly susceptible to Neu5Ac charge interference [96], and have an absolute requirement for terminal β-D-Gal residues, preferably linked $1 \rightarrow 3$ to GalNAc. Two more anti-T plant lectins from (i) *Artocarpus intergrifolia* [132] and (ii) *Amaranthus caudatus* [133], both more specific than PNA, were recently defined.

Salvia sclarea [97] and *Marrubium candidissimum* [98] lectins can be made strictly anti-Tn-specific if they are somewhat diluted before use. *Vicia villosa* lectin, particularly its B_4 fraction, has, in addition to anti-Tn, significant anti-Cad specificity [68] and cross-reacts with T (Springer GF, Tollefsen SE, Tegtmeyer H, to be published). The lectin from the snail *Helix pomatia* reacts strongly with GalNAcα-O- in proper spatial conformation and is a powerful anti-A, -Tn and -Forssman reagent [cf. 2; see, however, section V.B].

III. AUTOIMMUNE RESPONSES TO CA-ASSOCIATED T AND Tn, AND THEIR ROLE IN DETECTION AND MONITORING OF THE DYSCRASIA

T is exceptional among CA-associated EPs. The patient's cellular and humoral autoimmune responses against CA Ags T and Tn are used for CA detection [2,69]. This permits sensitive and specific detection of CAs at their first clinical appearance (Tables 1-3) and frequently even before [23,25,59,78; see section III.D]. All tables list the results of only the initial test.

A. Preparation of Sterile, Apyrogenic, and Human-Leukocyte-Antigen-Free T and Tn Ags Suitable for In Vivo Studies in Humans

For diagnostic (and therapeutic) studies MN GP [33], is converted to T with sialidase [12,23] and Tn is prepared from T with $\beta 1 \rightarrow 3$ galactosidase [cf. 32]. The procedures include three and four preparative steps, respectively, each of which by itself kills acquired immunodeficiency syndrome (AIDS) viruses, hepatitis viruses, and any other known virus (Centers for Disease Control, Atlanta, GA, 1986). All lots are pretested by laboratories approved by the Food and Drug Administration (FDA) for lack of pyrogens (*U.S. Pharmacopoeia*, 21st ed., 1988).

B. Cell-Mediated Immune Response (CMIR) to CA-T and CA-Tn

1. In Vivo CMIR to T: Delayed-Type Skin Hypersensitivity Reaction to T Ag

When possible the initial T test for delayed-type skin hypersensitivity reaction to T Ag (DTHR-T) was performed before other diagnostic procedures. Patients with earlier or

Table 1 Lung and Breast CA Detection by Intradermal Delayed-Type Hypersensitivity Response to Erythrocyte-Derived T Ag (DTHR-T) at *Initial* Patient Visit

Category	DTHR-T +/ Total Tested	% Positive
Lung		
CA		
Adeno		
Bronchial		
Stage I $(T_1 N_0 M_0)$[a]	11/11	100.0
Stage I $(T_2 N_0 M_0)$	18/20	90.0
Stage II	1/1	(100.0)[b]
Stage III	40/46	87.0
Total	70/78[c]	89.7
Bronchioloalveolar		
Stage I $(T_1 N_0 M_0)$	2/3	(66.7)[b]
Stage II	2/2	(100.0)[b]
Stage III	3/3	(100.0)[b]
Total	7/8	(87.5)[b]
Squamous-cell		
Stage I $(T_1 N_0 M_0)$	2/2	(100.0)[b]
Stage I $(T_2 N_0 M_0)$	5/6	(83.3)[b]
Stage II	1/2	(50.0)[b]
Stage III	12/15	80.0
Total	20/25	80.0
Squamous-adeno		
Stage III	3/3	(100.0)[b]
Giant-cell anaplastic		
Stage III	8/9	(88.9)[b]
Small-cell		
Stage III	19/21	90.5
Grand total	**127/144**	**88.2**
Other pleuropulmonary cancers[d]	2/4	(50.0)[b]
Benign diseases		
Hamartoma (6), carcinoid (3)[e]	1/9	(11.1)[b]
Other diseases[f]	5/55	9.1
Total	6/64	9.4
Breast		
AdenoCA		
Stage 0 $(Tis N_0 M_0)$	18/23	78.3
Stage I	48/67	71.6
Stages II and III[g]	79/87	90.8
Stage IV	21/22	95.5
Total	166/199	83.4
Benign diseases[h]	19/198	9.6
Healthy controls	1/127	< 1

[a]Staging is picTNM throughout.
[b]Cohorts of <10 persons.
[c]2/4 CA tissues from the negatively reacting patients had T Ag but were hypoergic.
[d]Positive: 1/2 mesothelioma patients and 1/2 with malignant carcinoid.
[e]Positive: 1/6 hamartomas.
[f]See text.
[g]Cancer tissues from 5 of the negatively reacting patients were available: 3 had T, 2 lacked it, but all had Tn.
[h]All but 2 of 19 patients with positive DTHR-T had severe hyperplastic premalignant disease [101,134].

Table 2 Detection of CAs Other than Those of Lung and Breast by DTHR-T at *Initial* Patient Visit

Category	DTHR-T +/ Total Tested	% Positive
Upper aerodigestive tract		
CA		
Adeno[a]	5/6	(83.3)[b]
Squamous-cell[c]	9/13	69.2
Mucoepidermoid	2/2	(100.0)[b]
Thyroid		
CA		
Papillary	1/5	(20.0)[b]
Thymus		
Malignant thymoma	1/1	(100.0)[b]
Pancreas[d]		
AdenoCA	23/26	88.5
Benign disease		
Pancreatitis	0/14	0
Gastrointestinal tract		
CA		
Stomach[e]	4/5	(80.0)[b]
Colon, adeno	13/15	86.7
Carcinoid, metastatic	0/2	(0)[b]
Benign diseases		
Colon, ulcerative colitis; diverticulitis, adenoma	0/5	(0)
Genitourinary tract		
CA		
Transitional-cell, bladder[f]	23/27	85.2
Adeno and squamous-cell[g]	9/11	81.8
Malignant melanoma[h]	9/12	75.0
Leukemia/lymphoma[i]	3/12	25.0
Central nervous system[j]	0/9	(0)[b]
Sarcomas[k]	0/5	(0)[b]
Benign diseases not related to lung, breast, pancreas, colon, genital tract		
Tumors[l]	0/18	0
Other diseases[m]	0/27	0

[a]Positive: 4/4 esophagus, 1/2 submaxillary glands.
[b]Cohorts of <10 persons.
[c]Positive: 3/4 tongue and floor of mouth; 1/2 soft palate ([+] $TisN_0M_0$, [−] $T_2N_0M_0$); 2/2 adenoids; 1/1 vocal cord; 1/1 neck; 1/3 esophagus.
[d]See [102].
[e]Positive: 3/4 adeno; 1/1 squamous CA.
[f]1/1 $TisN_0M_0$ positive.
[g]Positive: 2/3 hypernephromas; 3/3 endometrial, adeno; 1/2 ovary, adeno; 1/1 prostate, adeno; 1/1 seminoma; 1/1 penis, squamous.
[h]All stages.
[i]Positive: 0/1 B lymphoma; 1/5 lymphoma/lymphocytic leukemia; 1/3 mixed lymphoma (see text, section III B); 1/3 Hodgkin's disease, the only patient with active disease.
[j]Five astrocytoma (stages II-IV); 1 glioblastoma; 3 malignant chordoma.
[k]Sarcomas: 1 skin-fibro sarcoma; 1 osteosarcoma; 1 leiomyosarcoma; 2 liposarcomas.
[l]Patients: 4 pituitary adenomas; 3 meningiomas; 4 parotid (2 mixed, 1 Warthins, 1 venous); 2 thyroid adenomas; 3 ovarian cyst, simple or dermoid; 1 extrauterine serous; 1 prostate adenoma.
[m]Only major disorders listed: 6 cardiovascular disease, high blood pressure; 4 diabetes mellitus; 3 alcoholism with liver disease; 2 myasthenia gravis; and 1 each, chronic ear, nose, throat infection; sarcoidosis; Parkinsonism; achalasia (esophagogastric); infected bunion; Wilson's disease; tracheal stenosis; post-Guillain-Barré; hypothyroidism; lymphoblastic granulomatosis; leukoplakia vulvae; psoriasis.

Table 3 CA Detection with Solid-Phase Immunoassay by Quantitating Serum Anti-IgM and Total IgM at Initial Visit (Values Expressed as $Q_M = [(\text{anti-T IgM})^2/\text{Total IgM}] \times 100^a$)

Category	Number of Subjects Tested	Depressed ($\leqslant 100$)	Normal (100 to 360)	Elevated ($\geqslant 360$)	% Positive
Lung					
CA					
AdenoCA					
Stage I ($T_1 N_0 M_0$)	11	10	0	1	100
Stage I ($T_2 N_0 M_0$)	10	8	1	1	90
Stage II	2	2	0	0	(100)[b]
Stage III	28	23	4	1	85.7
Total	51	43	5	3	90.2
Squamous cell					
Stage I ($T_1 N_0 M_0$)	2	2	0	0	(100)[b]
Stage I ($T_2 N_0 M_0$)	6	5	0	1	(100)[b]
Stage II	2	0	0	2	(100)[b]
Stage III	9	5	0	4	(100)[b]
Total	19	12	0	7	100
Small-cell					
Stage III	20	17	2	1	90
Grand total	**90**	**72**	**7**	**11**	**92.2**
Benign diseases[c]	32	2	30	0	6.3
Breast					
AdenoCA (all types)					
Stage 0 ($TisN_0 M_0$)	8	7	1	0	(87.5)[b]
Stage I	35	28	7	0	80.0
Stages II and III	45	34	8	3	82.2
Stage IV	4	4	0	0	(100)[b]
Total	92	73	16	3	82.6
Benign diseases[d]	39	6	33	0	15.4
Upper aerodigestive tract:					
Squamous-cell CA[e]	4	4	0	0	(100)[b]
Pancreas					
AdenoCA[f]	19	18	1	0	94.7
Pancreatitis	5	0	5	0	(0)[b]
Genitourinary system[g]					
CA					
Transitional-cell, bladder[h]	11	9	2	0	81.8
Non-CA diseases not listed under breast, lung, or pancreas	9	0	9	0	(0)[b]
Healthy controls	67	5	62	0	7.5

[a] For further description of Q_M, see text.
[b] Cohorts of <10 persons.
[c] 1/2 patients with carcinoid and 1/5 with COPD had a depressed Q_M (the latter still positive after 22 months).
[d] 4/6 positively reacting patients had premalignant hyperplastic disease.
[e] One patient each with cancer of the tongue, soft palate ($TisN_0 M_0$), pharynx, esophagus.
[f] One patient (not listed) with duodenal adenocarcinoma invading the pancreas was also Q_M-positive.
[g] One seminoma patient with negative Q_M.
[h] One of one $TisN_0 M_0$ urinary bladder cancer patient showed positive results.

additional cancers were excluded, as were pregnant women and CA patients < 3 weeks after surgery or chemotherapy and < 3 months after radiation therapy. For DTHR-T determination, 1 mg T Ag (1% w/v) in buffered saline (BS) was injected intradermally (ID). The control was BS containing 1 mg MN Ag. Also, all subjects received concomitantly ID one standard dose of common Ags to ascertain their normergy [31].

Injections were made 4-6 cm apart into the upper outer arm. Induration and erythema were measured at ~24 and ~48 hr [99] and standardized in collaboration with Drs. H. Oettgen and S. Krown (Memorial-Sloan Kettering, NY). After reactions to MN Ag were subtracted from those to T Ag, an induration of ⩾4 mm diameter was considered positive [cf. 31].*

We consecutively skin-tested 951 patients and controls. Recorded are only patients who underwent subsequent biopsy and/or surgery for their lesions. The two largest patient groups tested together with healthy controls are listed in Table 1.† All others are listed in Table 2.

Adenocarcinoma (AdenoCA) predominated among the 144 lung CA patients (59.7%). DTHR-T was positive in 89.7% of 78 patients with bronchial adenoCA and in 90.5% of 21 patients with small-cell lung CA (Table 1). The sensitivity of DTHR-T for the 16 $T_1 N_0 M_0$ lung CA patients was 93.8% and for all lung CA patients it was 88.2%. A positive DTHR-T was found in 83.4-88.5% of breast, pancreatic, and colon CAs and also among patients with CAs of other organs (Tables 1 and 2). Among patients with head-neck squamous CA, 9 of 13 (69.2%) reacted positively while only 1 of 5 with papillary thyroid CA showed positive results.

The DTHR-T results overall are most encouraging, because of their high sensitivity (except in thyroid CA) and also their high specificity: >99% for 127 healthy, 100% for 45 persons with benign diseases unrelated to breast, lung, pancreas, colon, and genitourinary tract (Table 2), and ⩾90% for persons with benign diseases of the organs from which the CAs arose (Tables 1 and 2).

Patients with incipient CAs generally had the highest incidence of positive DTHR-Ts (lung, adeno and squamous, soft palate, urinary bladder CA) (Tables 1 and 2). This finding may emphasize an important role of the cellular over the humoral immune responses in most CAs, because the preexisting anti-T Ab levels decreased concomitantly to levels considered as CA positive. This dovetails with reports of others [109,134].

A positive DTHR is thought to mirror precisely local defenses against the primary lesion in a patient, as first pointed out in 1911 by von Pirquet [100] and accepted since. The ability to mount a delayed-type hypersensitivity response correlates positively with a moderate increase in T-helper lymphocytes versus cytotoxic/suppressor cells [109]. DTH-mediated immune answers are thought by many to be of considerably higher anticancer efficiency than humoral responses [cf. 109].

Skin biopsies of positive DTHR-Ts and positive DTHR-PPDs in CA patients regularly showed, besides other profound differences, a relative increase in T-helper cells in the DTHR-T by itself and compared to the DTHR-PPD of the same patients [111]. Recent

---

*When skin testing was performed on the outer thigh in persons walking about freely; the reactions ensuing were uninterpretable in CA and healthy persons 24 and 48 hr after injection [25,31].
†Included are only those persons on whom, to the best of our knowledge after a most careful study, we had a complete history of relevant diseases in accord with rigid criteria for test subject evaluation. The same diligence and tedium were applied predominantly by one of us (W.W. M.D., Int. Med.) with the help of S.C., to all patients and control subjects reported on here. No effort was spared to retrieve missing, possibly important details.

studies suggest that application of autologous DTHR-mediating lymphocytes may shift from a continuing humoral response to a possibly more efficient cell-mediated antitumor response [109].

DTHR-T was negative in 58 of 64 benign lung disease patients (Table 1): 14 of 14 with lobar, lobular, acute (both bacterial and viral) pneumonias; granulomatous, fungal diseases; cysts; effusive tuberculosis; silicosis; chronic interstitial fibrosis; emphysema and benign lung tumors [2,59,78]; 37 of 40 patients with chronic obstructive pulmonary disease (COPD), severe chronic indurating pneumonias, and/or abscess; 2 of 4 with sarcoidosis; and 5 of 6 with hamartomas. The positive reactions of one patient each with pneumonia and sarcoidosis were negative in < 1 year [2].

A positive DTHR-T (as opposed to humoral anti-T responses) in CA patients generally persisted for several years after attempted curative surgery, even if there was no clinical recurrence. On the average, the DTHR-T turned negative within 26 months after surgery [nine stage I: seven breast and two lung adenoCA; four stage II ductal breast CA patients]. Patients were tested at 0.5 year intervals for ≥6½ years. The DTHR-T remained negative in all these patients who are well and without any clinical sign of cancer [cf. 59].

These results do not dispense one from long-term monitoring; thus, a woman who had undergone a mastectomy in April, 1981, for stage I ductal breast CA had a positive DTHR-T in August 1981. The reaction was negative from mid-1983 until February, 1985. In October, 1985, both DTHR-T and anti-T Ab scores became positive and remained positive through May, 1988, when, despite negative results of frequent medical clinical check-ups, extensive lung metastases were first detected.

We tested 198 patients with standard clinicopathological and/or x-ray diagnosis of benign breast disease, of whom 90.4% reacted negatively; 17 of the 19 patients with false-positive results had premalignant disease (Table 1, footnote h). Five more patients had initially "false-positive" DTHR-T. However, after our findings, reinterpretation of the original slides or a second biopsy shortly thereafter showed that all had breast CA; in these cases, the DTHR-T detected CA first [2,8].

Table 2 lists DTHR-T results on smaller groups of CA patients and corresponding nonCA controls: 87.5% of the eight patients with adeno- or mucoepidermoid CA of the upper aerodigestive tract showed positive results. In patients with pancreatic disease, studied predominantly by Dr. R. Goodale, 88.5% of 26 patients with pancreatic CA were DTHR-T positive [102] while simultaneous carcinoembryonic antigen (CEA) (Roche) determinations detected 50.0%. None of 14 patients with pancreatitis were DTHR-T positive (Table 2), but CEA levels (>3.5 ng/ml) indicated CA in 4 of 9 (44.4%). In one patient with negative results of DTHR-T and solid-phase immunoassay for anti-T (SPIA-T) (not listed), the surgeon insisted on the diagnosis of pancreatic CA, until surgery revealed only a malrotated bowel.

Table 2 shows that none of the patients with intracranial or intraspinal malignancies nor any of those with various soft tissue sarcomas reacted positively [2,31,59,103]. Reactions of patients with benign diseases are also listed in Table 2, footnotes l and m. Six anergic CA patients (four lung, two breast) are not included.

2. In Vitro CMIR to T and Tn

In vitro CMIR to CA-associated T Ag measured with the leukocyte migration inhibition (LMI) assay was much less sensitive but almost as specific as the DTHR-T. MN Ag was inactive [31,103,104]. A "superoxide" leukocyte adherence inhibition assay for T Ag yielded superior results [105]: 11 of 13 (85%) breast CA patients tested positive, as did

83% of 23 lung adenoCA patients with *early* operable disease. This finding of the highest positive rate in the earliest CA cases correlates with those of the DTHR-T (Tables 1 and 2) and SPIA-T studies (Table 3). LAI was positive in 3 of 19 patients (16%) with benign pulmonary disease but in 0 of 34 healthy controls. Slightly less sensitive LAI-T and -Tn results in CA patients were reported by Thomson et al. [69; Thomson DMP, personal communication), MN Ag was inactive.

C. Humoral Anti-CA-T and -Tn Responses

The specific ability of preexisting human anti-T Abs to kill CA cells effectively in the presence of complement in vitro has been described [17].

1. Hemagglutination and Hemolysis Tests

Hemagglutination tests first focused on the determination of anti-T Ab levels in sera of breast disease patients compared to healthy persons [7,18,23,25]; they now encompass patients with most types of CA and control populations [2,8,23,25,59,73,78,106-108]. The normal range of anti-T scores was between 20 and 25 in serum pools of healthy humans of the midwestern United States [2,23,59]. Among 287 consecutive preoperative patients with breast, respiratory, or gastrointestinal tract CAs, 36.6% had severely depressed anti-T hemagglutinin scores (<12), while 4.2% had highly elevated anti-T scores ($\geqslant 44$). Anti-T levels were depressed in 9.4% of 309 patients with benign diseases and in 8.5% of 200 putatively healthy persons ($p < 0.001$ for CA patients versus nonCA groups) [23,59,78].

Curative surgery in the first 42 breast and 3 lung CA patients resulted in a substantial increase (including overshoot) of anti-T scores ($>25\%$) 1-5 months later in 73.8% of the breast CA and in all three (100%) lung CA patients; only 1 of 42 patients with a biopsy showing hyperplastic benign breast disease and none of 22 with non-CA-related surgery had any significant change in anti-T levels ($p < 0.001$) [2,23]. The rebound implicates CA, rather than a genetic defect, as the cause of depressed anti-T. In CA patients, anti-ABO Abs did not change along with anti-T [77] and anti-sheep RBC hemolysin levels remained unaltered while anti-T hemolysin levels decreased in parallel with the increase of the CA burden [73]. Thus, the anti-T depression in CA patients is specific. A renewed decrease of anti-T, in the few patients investigated, preceded clinically recognized CA recurrence by 2 months to 3 years [2,59; to be published]. Determination of anti-T Ab levels at ~6 month intervals seems suitable in monitoring stable postsurgical CA patients [2,59].

Thatcher et al.'s studies of 55 patients with untreated, histologically proved, disseminated melanoma [106] showed significantly depressed serum anti-T agglutinin levels in the patients compared with 60 healthy persons ($p < 0.0005$). Significant increase in anti-T Abs followed one infusion of *Corynebacterium parvum* in 14 melanoma patients indicating immunotherapeutic relevance of *C. parvum* but not of bacillus Calmette-Guérin (BCG) injected in nine other melanoma patients. We reported earlier strong expression of T and Tn EPs in *C. parvum* and their absence in BCG [79]. Pretreatment anti-T titers were statistically significantly higher among the melanoma patients who responded to chemoimmunotherapy (n = 21); 50% survival time was ~ 120 weeks compared to 20 weeks for the nonresponders (n = 34); the anti-T titers therefore had prognostic meaning while tumor burden was irrelevant [106].

Bray et al. [73] showed that all 40 normal human sera tested had anti-T Abs that specifically lysed O T RBCs in the presence of complement. Eighty-three percent of 18 postoperative (>3 months) patients with massive metastatic gastrointestinal CA had depressed lytic serum anti-T levels correlating directly with disease burden, whereas 11 patients with nondemonstrable or minimal disease had no decrease in anti-T hemolysin ($p < 0.001$). Anti-T cytotoxin usually correlated well with anti-T agglutinins.

Vos and Brain [107] found that mean anti-T scores of 103 and 54 Indian women (in South Africa) with breast and cervical CA, respectively, were statistically significantly lower than in 200 healthy controls. The impetus to begin CA studies was the epidemic of esophageal CA for over 40 years among black men in rural areas of Transkei (South Africa). The mean anti-T score for 95 men with esophageal CA was 4.2, compared with 8.0 for 323 age-matched healthy black men [107].

Among >500 children in geographic areas with and without high risk for esophageal CA, anti-T was highly elevated exclusively in the geographic areas with a high risk of esophagus CA for men. Vos et al. concluded that the T Ag itself may be decisive in the development of CAs [108], either via intracellular growth defects or by alteration of plasma membranes that facilitate cell entry by viral or other agents.

Tn expression in a patient's myelomonocytic leukemia blood cells, as studied by IHC, increased in parallel with the duration and severity of the disease, indicating that a hematopoietic subclone expressing Tn became dominant with time; Tn detection by IHC seemed more sensitive than hemagglutination [56]. It is interesting that the patient described by Ness et al. [57] had the Tn syndrome prior to myelomonocytic leukemia.

2. Solid-Phase Serum Immunoassays

The SPIA-T more than doubled the sensitivity of CA detection over the hemagglutination assay [80,81]. The Q_M values of 368 healthy persons and preoperative patients were measured; 216 of these had CA; ~85% of the subjects studied also underwent the DTHR-T test. The general usefulness of the SPIA-T is attested to by the high statistical significance of its extraordinary ability to detect a high proportion of clinicopathologically very early CAs.

The Q_M assay detected CA in 92.2% of 90 lung CA patients (Table 3). Among the lung *adeno*CA patients, 100% of the 11 with $T_1N_0M_0$ and 90.2% of all 51 had positive Q_M (Table 3). The sensitivity was thus highest in the earliest stages, which is in agreement with DTHR-T data (Tables 1 and 2).

It is remarkable that 100% of 19 patients with squamous cell lung CA of all stages reacted positively (Table 3), while in the DTHR-T only 80.0% of 25 did so (Table 1) and that squamous CA patients had by far the highest number (36.8%) of abnormally high Q_M values. The dichotomies in squamous cell CAs between CMIR and humoral immune reactivity, and between depressed and elevated Q_M, may be of import in understanding aspects of CA pathogenesis. Of 20 patients with small-cell lung CA, 90% (all stage III) had positive Q_M (Table 3).

The disease spectrum of 32 patients with benign lung disorders was similar to that listed for DTHR-T specificity 93.7% (see section III.B.1; Table 3, footnote c).

The SPIA-T was somewhat less sensitive in detection of breast CAs (82.6% overall) than of other CAs, for which overall sensitivity was 91.9% (Table 3).

SPIA-T detected different types of breast CA equally well. Positive Q_M was most frequent among $TisN_0M_0$ (87.5%) stage IV excepted. Three breast CA patients stages

II and III had a strongly elevated Q_M (normal liver function tests); the remaining 73 had depressed Q_M. Of the 39 patients with benign breast disease, 6 reacted positively (see Table 3, footnote d).

Nearly 95% of 19 patients with pancreatic CA were Q_M-positive; the 1 patient with negative results was borderline (Q_M 100.2) [81]. All five patients with pancreatitis reacted negatively. The nine patients without CA were all Q_M-negative (Table 3). Of 67 presumably healthy persons, 5 had positive Q_M (Table 3); this figure was <1% for 127 persons tested with DTHR-T (Table 1).

D. The T/Anti-T System

1. Detection of Earliest Clinical and Preclinical Stages of CA

The unique sensitivity and specificity of our CA detection assays are evident from all three tables. Especially striking are the results for $TisN_0M_0$ and $T_1N_0M_0$ CA patients (Table 2, footnotes c and f; Table 3, footnotes e and h).

2. Detection of CA in Patients Before Clinicopathological Confirmation

We evaluated longitudinally 34 patients, 32 with breast and lung disease, who showed positive results repeatedly in our assay(s), although for many months and frequently years they had consistently negative biopsy and/or radiographic results and were thought to be free of CA. As of May 1989, 21 of these 34 (61.8%) patients have developed biopsy-verified CA within 3 months to 10 years at the site of the original lesions [cf. 2,78]. These cancers were detected by anti-T Ab agglutination and/or DTHR-T [8,71,72]. It is of preeminent importance that our anti-T Ab and DTHR-T assays have high predictive value (sensitivity) in the diagnosis of CA at a stage when all standard tests are still negative. The situation may be similar in lymphomas/leukemias: a patient with repeated histological diagnosis of benign pseudolymphoma showed consistently positive results in our tests and had T Ag in the lymphoma biopsy tissues; after 8 years, the patient was finally confirmed by histological study to have advanced mixed-cell malignant lymphoma [9; Springer GF, Wise W, and Carlstedt SC, in preparation].

E. Statistical Aspects

Detailed, highly significant statistical data on smaller cohorts of similar patient groups were reported earlier [2]. Because our more recent DTHR-T and SPIA-T results are consistent with the earlier data given, we present here the results of only the Pearson χ-square statistic.

For the DTHR-T data (Tables 1 and 2), all p values of comparisons between patients with CA and those with benign diseases of the same organ were in the 0.00-0.000 range except for thyroid CA. The p values for CA versus healthy (except for thyroid CA) were < 0.000 throughout.

The χ-square statistic, performed on all Q_M data (Table 3), usually gave p values of 0.0000 for Q_Ms of patients compared with those of patients with corresponding benign diseases and healthy persons. Comparison of the few lung CA patients with elevated Q_M to the cases with benign lung diseases gave a p value of 0.008.

IV. IN VIVO LOCATION AND IMAGING OF CA-SPECIFIC T AND Tn EPS

A. PNA Extract

PNA was first used for immunoscintigraphy of CAs because of its high affinity (but imperfect specificity) for T EPs. Results and conclusions in most studies were similar; an arbitrary selection is presented here.

The preferential affinity of radioiodinated PNA for T was first shown in in vivo radio-immunodetection of a T Ag-expressing murine lymphoma and TA3 CA; high tumor-to-background ratio and rapid blood clearance made background subtraction unnecessary. Scintigraphic images were obtained with [^{131}I] PNA in a colon CA and in two of the eight breast CA patients with large lung and liver metastases [137]. The immunogenicity of PNA was not discussed.

B. Murine Anti-sialyl-Tn MoAb

Extensive studies of the distribution of radiolabeled anti-sialyl-Tn MoAb B72.3 gave positive scans in 14 of 27 (51.9%) CA patients [90]. To detect a majority of lesions, their minimum diameter had to be ⩾4 cm. Selective tumor location by [^{131}I] B72.3, given before surgery, was proved by direct comparison of surgically removed tumor and normal tissues in 17 of 20 (85%) CA cases; 5.7% of 210 histologically normal tissues reacted positively [90].

V. T AND Tn: CA MARKERS AND PREDICTORS OF AGGRESSIVENESS

A. IHC and Immunoadsorption Studies

After our observations of T in breast CA [18,25], Laurent et al. [26], also using immunoadsorption of human anti-T serum, showed a high prevalence of T Ag in primary breast and other CA cultures. Howard and Taylor [7,112] introduced the T Ag as a sensitive and specific IHC marker to diagnose breast CA in surgical specimens, using standard tissue sections from 24 clear cases of CA and 27 histologically defined benign lesions. Staining was by the indirect immunoperoxidase procedure, using either A_1B persons' serum or affinity-purified human anti-T as primary Ab. All CA tissues stained strongly with anti-T; highly differentiated CA cells had more T Ag than the dedifferentiated CAs, which is in agreement with Ab absorption data [2,18,25,59]. Breast epithelium proper was negative in benign fibrocystic lesions and fibroadenomata; in a few benign lesions, some myoepithelial cells stained positively. Thus, IHC differentiation between adjoining malignant and benign tissues was achieved [112]. It is relevant to remember that Ab Fc receptors do not react with breast CA cells (negative in 97 of 97 cases) [113].

This pioneering IHC work was followed by studies with PNA; only a few of the many valuable publications using it are discussed. The great merit of IHC techniques in studies of T and Tn Ags in CAs of the breast [3,7,26,66,67,89,94,95,112], urinary bladder [5,6,63,115-117], colon [60,62,64,65,118], prostate [4], kidney [120], ovary [135], and leukemias [55,56] is evident on comparison with ordinary histopathology. The results were applied successfully in CA prognosis. The more recent inclusion of T's immediate precursor, Tn, has substantially improved the accuracy of the predictions of CA's aggressiveness in regard to lymph node invasion, distant metastasis, and time of death [3,5,63,66,88].

A careful analysis of PNA binding sites in breast CA tissues and an attempt to correlate these sites and their T density with tumor type, histological grading, staging, and endocrinological receptor status was carried out by Stegner's group [95] on 120 invasive breast CAs. In 117 of 120 CAs positive cytoplasmic reactions predominated. Three malignant tissues failed to stain. Histological grade of primary CA did not correlate with any immunohistochemical features [95].

Others reported positive PNA binding results in 17 of 22 breast CAs, showing T Ag diffusely in the cytoplasm; luminal staining was strictly confined to highly differentiated CAs. The five PNA-negative CAs were histologically the most anaplastic [121], thus permitting the speculation that the majority of these had high Tn density [cf. 2,3]. Of 22 benign breast lesions used in the same study [121], 19 stained along the luminal cytoplasmic membrane only and sometimes in intraluminal secretions.

PNA was especially helpful in differentiating early intraductal breast CA from benign hyperplastic lesions in three-dimensional studies; CAs had a discontinuous porous glandular structure, in contrast to a continuous network in epitheliosis and papilloma [67]. Intraductal CA showed supranuclear, cell membrane, parenchymal, and intracytoplasmic luminal staining. These important studies may be increased in sensitivity and specificity for CA diagnosis with Abs such as our anti-T or -Tn MoAbs [32,58,87].

In normal colon epithelium, PNA was found to bind to the supranuclear stalk portions of epithelial cells, indicating no more than nascent Gal termini in the Golgi cisternae [118]. In $TisN_0M_0$ CA, PNA was found in the apical cytoplasm of columnar cells. A positive PNA reaction in the apex of goblet cells was not necessarily an indication of malignancy [118]. In two recent thorough studies for T EP expression in normal, premalignant, and carcinomatous colonic tissues using a MoAb, polyclonal rabbit hyperimmune Abs, and PNA [62,122], PNA had the highest sensitivity and, as expected, the lowest specificity in CA detection. Sensitivity of the MoAb was 76%, and specificity, 100%. The rabbit Ab was intermediate in both. Degree of malignancy and extent of staining with any anti-T reagent did not correlate. Sensitivities for PNA and polyclonal rabbit anti-T were 72% with paraffin-embedded colon CA; they increased to >90% with frozen sections [61].

Specific detection of CA with anti-T and -Tn MoAbs was first reported in 1984 and 1985 [cf. 2,66]. Sections of two ductal and three lobular breast CAs were used. The anti-T MoAb stained all five CA specimens well whereas the anti-Tn MoAb stained the two ductal but only one of the three lobular CAs [66]. The first reported CA-induced anti-T MoAb is RS1-114 [58]; immunogen was human lung CA line A549 cells. This MoAb reacted well with 75% of surgical CA specimens and also with cultured CA cells [32,58]. Overall it reacted similarly to RBC anti-T MoAbs [32]. Normal tissues did not react, with two minor exceptions.

We found that five affinity-purified human anti-T Ab pools were fully absorbed (>88->95%) [71] by five different highly purified colon CA cell suspensions obtained from surgical specimens (>90% viability) used in a National Cancer Institute (NCI) phase II study [119]. Anti-T MoAb HH8 [29] was absorbed to >95% by all five colon CA cell suspensions; MoAb RS1-114 was completely absorbed by two of five but not at all by the other three, and MoAb HT-8 was not absorbed by any of them. This points to the need for MoAb cocktails, because of rigidity of the MoAb combining sites, and to the possibility that not all spatial, structural, and ionic charge arrangements of the T and Tn hapten groupings, which may be linked either a-O to Ser/Thr or β to ceramide or be located terminally on large carbohydrate complexes, are recognized by all anti-T (or anti-Tn) MoAbs.

Anti-T MoAb HT-8 was negative with colon CA goblet cells (Dr. M. I. Filipe, Guy's Hospital, London, 1985, personal communication). We confirmed this observation and also found that lung and colon goblet cells of fetuses >110 days old immunostained with anti-Tn MoAb BaGS-2 as the primary Ab but not with anti-T MoAb HT-8 [123]. The mucosal colon cells of these fetuses gave a membranous and coarsely granular cytoplasmatic staining with both anti-T and -Tn MoAbs [123]. These findings agree with the careful, detailed work of Cooper and Reuter [118] on T in colon adenomas, polyps, and on $TisN_0M_0$ colon CA and columnar cells (see above).

Normal heterogeneous human colon mucins (immunoprivileged) have numerous carbohydrates with terminal Tn and sialyl-Tn EPs but not the T EP [53]. However, a colon galactosyltransferase that endows AS-OSM with T activity was reported by Corfield et al. [124]. We showed T in two of two colon adenoCA lines with human polyclonal, monospecific sera: HT-29 (abundant T) and LS-174 (sparse T) [31]. Stein et al. reported the same findings with these lines using anti-T MoAb RS1-114 [58].

Cohorts with urinary bladder lesions were studied with PNA, polyclonal rabbit hyperimmune anti-T, and anti-T MoAbs (evoked with human RBC T GP). The latter were superior in diagnosis and prognostication of 73 cases [6, cf. 5]. Normal urothelial tissue did not react with anti-T and anti-Tn [5,6,63]; T Ag expression in bladder CAs and histological tumor grades correlated significantly ($p < 0.05$), and 68% of patients with strong T Ag expression [cf. 115] experienced recurrence ($p < 0.01$). Blasco et al. [116] found similar relations for urothelial CA and T Ag expression. T Ag may have a more ominous meaning in urinary bladder CAs than in breast [66] and other CAs, such as those from the colon [122], where it indicates a relatively benign course [cf. 7,112]. This observation warns against generalization as to the pathogenic import of CA-associated EP T among primary CAs of different organs; however, no exceptions have been reported for the dire outlook for CAs with high Tn density (see section B, below).

The first studies of T on CA [18] also revealed that breast CA cells express reactive Tn Ag whereas ordinary benign and healthy breast glandular tissues and all other healthy and nonCA diseased tissues do not [cf. 2]. Tn Ag and its pathogenetic, clinical, and diagnostic significance in CA have been less extensively investigated [2,3,8,63,66,88,125, cf. 5] than T [cf. 32]. The recently described murine anti-Tn MoAb FBT 3 specifically and intensely stained cell membranes, cytoplasm, and perinuclear regions of CAs [88]. Background staining was insignificant.

In each of three patients with breast CA, primary CA and local and distant recurrent metastases removed >30 months (three in one patient 30-74 months) later showed both T and Tn. The patient with the three metastases studied developed a second primary CA in the contralateral breast 5 years after the first; the T/Tn ratios in both primary CAs and in the metastases were closely similar. Thus, T and Tn are at least predominantly nonmodulating Ags, a tenet supported by studies on in vitro subcultures of flow-cytometrically separated T EP-bearing urinary bladder CA cells of line 647^{V_1}: strong and weak T was maintained unchanged for >30 weeks in culture. T expression was largely cell-cycle independent, as measured with PNA [117].

B. T and Tn as Predictors of Aggressiveness

The density of T Ag expression on the CA cell membrane permits one to predict the aggressiveness of some major primary CAs [2,3-7,59,60,64,66,112,121,122]. The most reliable predictions may be made if density of Tn is determined simultaneously. With polyclonal monospecific anti-T and anti-Tn Ab pools from A_1B donors, we and others found in most surgical specimens a predominance of *Tn* EPs in undifferentiated, aggressive tumors and a higher density of T EPs in well-differentiated CAs [3,7,18,31,63]. Thus, more Tn than T was found in 27 of 30 (90.0%) anaplastic breast CAs but only in 2 of 25 (8.0%) well-differentiated ones [66].

Human breast-CA-derived epithelial cultures showed both Tn and T when stained (immunoperoxidase) with human polyclonal monospecific anti-Tn and -T as primary Abs. Strong Tn reactivity was found in the highly malignant DU 4475 breast CA line [66].

Of 23 benign breast lesions, 19 (82.6%) had no T or Tn as determined with human Abs. The four positive tissues had atypical lobular and/or ductal hyperplasia; two of these tissues were from patients who had Tn and invasive CA in the contralateral breast [66].

We studied only two animal cancers, TA3-St and Eb (both murine) and their sublines TA3-Ha and ESb. There was a strong correlation between invasiveness and absolute density of T and Tn receptors. Live cells, especially from the highly invasive, metastatic subline Ha, had Tn $\gg$ T on their outer cell membrane than the poorly invasive parent line [66].

Noteworthy in relation to our subject are findings of Fenlon et al. [126] and Leathem and Brooks [127], who observed a strong positive correlation, using IHC, between the aggressiveness of primary human breast CAs and their affinity for HPA. Reactivity with HPA accurately predicted or retrospectively proved ($p < 0.01$) among primary breast CAs of 305 patients the extent of axillary lymph node invasion, early or late first CA recurrence, and patient survival time. The nature of the innate specificity of the large HPA combining groove, aside from its avid reactivity with appropriately spaced GalNAcα-*O*- as in blood group A and Forssman Ags, remains obscure, however, despite investigations for over a decade [128] [see II.B.4]. Leathem and Brooks concluded that "... *HPA appears to recognize an as yet undefined biological marker associated with both high metastatic potential and aggressive tumour behaviour*..." and that "... *regardless of histopathological* features, the metastatic potential of an individual primary breast cancer might be assessed at the time of operation..." (italics added) by IHC using only HPA and disregarding other cytological criteria such as histological and nuclear grade; this procedure cannot be done during surgery [127].

There are caveats to these exciting investigations: Leathem and Brooks [127] exempt postmenopausal women from these findings, whereas Fenlon et al. [126] do not. Included in the positive findings are subjects with blood group A, which is present in breast epithelia of blood group A persons [129] but frequently decreases and may disappear with increasing CA malignancy [cf. 3] (see sections II.B.1 and 4). Tn, which is predominant in patients with poorly differentiated breast CA, in the breast CA line DU 4475 [32], and in murine mammary CA TA3-Ha [40], reacts avidly with HPA as do A and Forssman Ag [cf. 128].

Human breast, lung, a few gastrointestinal CAs, and murine cancer sublines studied in which Tn predominated over T on the cell membranes are more aggressive than those that have predominantly T. These observations [66] permit the hypothesis that the "as yet undefined biological marker... with aggressive tumour behaviour" of Leathem and Brooks, detected by HPA, encompasses the Tn EP [3]; this is now supported by IHC studies on human breast CA [Springer GF et al., in preparation].

In conclusion, there is strong evidence that Tn and T are *functional* markers (rather than just sensitive and specific CA indicators detectable by IHC). T and Tn are involved in CA pathogenesis [2-7,63,64,106-108]. In addition to the attributes described in this review, they are adhesion molecules involved in the attachment of CA cells to healthy cells [130,136]. Tn and T EPs partake in fundamental interactions of incipient CA with their microenvironment demonstrated by the correlation of T and Tn with the scope of aggressiveness as well as by strong cellular and humoral autoimmune responses by the patient to even incipient CA [28,78]. T and Tn aid in CA monitoring and in and in stratification for adjuvant treatment [cf. 2,8,66,127]. T is being used therapeutically with positive results since 1974 in patients with breast CA stages IV (metastatic and recurrent) and III [9,23,131].

NOTE ADDED IN PROOF

Recently, Ghazizadeh et al. [136], using MoAb HT-8, found 67% of 30 ovarian CAs to react strongly T positive; among the 10 most anaplastic (grade 3) CAs, 80% reacted positively; of the 10 benign ovarian tumors, 3 stained weakly positive. Again, a pool of anti-T MoAbs, likely would have shown higher sensitivity (see p. 602).

PicTNM: postsurgical, pathological staging at beginning of immunology testing [114].

ACKNOWLEDGMENTS

We are grateful to Dr. Theodora Cohen (Dept. Statistics, Northwestern University) who performed most statistical evaluations since 1984 [cf. 2] and all donors of rare substances: Dr. B. M. Longenecker for anti-T MoAb H.24; Dr. J. Schlom for MoAb B72.3; Drs. M. Hanna and R. McCabe for purified colon CA cell suspensions; Dr. H. Clausen for anti-T MoAb HH8. Our work was supported by U.S. National Cancer Institute Grant Nos. CA 22540 and CA 19083, and The Elsa U. Pardee Foundation. G.F.S. is Julia S. Michels Investigator in Surgical Oncology.

REFERENCES

1. Seitz RC, Fischer K, Stegner HE, Poschmann A. Detection of metastatic breast carcinoma cells by immunofluorescent demonstration of Thomsen-Friedenreich antigen. Cancer 1984;54:830-836.
2. Springer GF. T and Tn, general carcinoma autoantigens. Science 1984;224:1198-1206.
3. Springer GF. Tn epitope (N-acetyl-D-galactosamineα-O-serine/threonine) density in primary breast carcinoma: a functional predictor of aggressiveness. Mol Immunol 1989;26:1-5.
4. Ghazizadeh M, Kagawa S, Izumi K, Kurokawa K. Immunohistochemical localization of T antigen-like substance in benign hyperplasia and adenocarcinoma of the prostate. J Urol 1984;132:1127-1130.
5. Weinstein RS, Schwartz D, Coon JS. Blood group antigens and ploidy as prognostic factors in urinary bladder carcinoma. In: Fenoglio-Preiser CM, Weinstein RS, Kaufman N, eds. New concepts in neoplasia as applied to diagnostic pathology (Intl Acad Pathology Monograph, vol. 27). Baltimore: Williams & Wilkins, 1986:225-241.
6. Ohoka H, Shinomiya H, Yokoyama M, Ochi K, Takeuchi M, Utsumi S. Thomsen-Friedenreich antigen in bladder tumors as detected by specific antibody: a possible marker of recurrence. Urol Res 1985;13:47-50.
7. Howard DR, Taylor CR. An antitumor antibody in normal human serum: reaction of anti-T with breast carcinoma cells. Oncology 1980;37:142-148.

8. Springer GF, Desai PR, Robinson MK, Tegtmeyer H, Scanlon EF. The fundamental and diagnostic role of T and Tn antigens in breast carcinoma at the earliest histologic stage and throughout. In: Dao TL, Brodie A, Ip C, eds. Tumor markers and their significance in the management of breast cancer (Progress in Clinical and Biological Research, vol. 204). NewYork: Alan R. Liss, 1986:47-70.
9. Springer GF, Scanlon EF, Desai PR. Pancarcinoma T and Tn receptors, members of a sialic acid-containing glycoprotein superfamily, in human tumor pathogenesis, in early CA (immuno)detection, prognosis and treatment. Cancer Metastasis Rev 1990; in press.
10. Friedenreich V. The Thomsen hemagglutination phenomenon. Copenhagen: Levin and Munksgaard, 1930.
11. Lind PE, McArthur NR. The distribution of "T" agglutinins in human sera. Aust J Exp Biol Med Sci 1947;25:247-250.
12. Springer GF, Ansell NJ. Inactivation of human erythrocyte agglutinogens M and N by influenza viruses and receptor-destroying enzyme. Proc Natl Acad Sci USA 1958; 44:182-189.
13. Moreau R, Dausset J, Bernard J, Moullec J. Anémie hémolytique acquise avec polyagglutinabilitédes hématies par un nouveau facteur présent dans le sérum humain normal [anti-Tn]. Bull Soc Méd Hôp (Paris) 1957;73:569-587.
14. Barker CP, Billingham RE. Immunologically privileged sites. In: Kunkel HG, Dixon FJ, eds. Advances in immunology, vol. 25. New York: Academic Press, 1977;1-54.
15. Pierce-Cretel A, Pamblanco M, Strecker G, Montreuil J, Spik G. Heterogeneity of the glycans O-glycosidically linked to the hinge region of secretory immunoglobulins from human milk. Eur J Biochem 1983;114:169-178.
16. Springer GF, Tegtmeyer H. Origin of anti-Thomsen-Friedenreich (T) and Tn agglutinins in man and in White Leghorn chicks. Br J Haematol 1981;47:453-460.
17. Springer GF, Desai PR. Relation of human blood-groups MN to cancer cell surface antigens and to receptors for oncogenic viruses. Ann Clin Lab Sci 1974;4:294-298.
18. Springer GF, Desai PR, Banatwala I. Blood group MN antigens and precursors in normal and malignant human breast glandular tissues. J Natl Cancer Inst 1975;54: 335-339.
19. Desai PR, Springer GF. Biosynthesis of blood group T-, N-, and M-specific immunodeterminants on human erythrocyte antigens. J Immunogenet 1979;6:403-417.
20. Springer GF, Desai PR, Schachter H, Narasimhan S. Enzymatic synthesis of human blood group M-, N- and T-specific structures. Naturwissenschaften 1976;63:488-489.
21. Cartron JP, Andreu G, Cartron J, Bird GWG, Salmon C, Gerbal A. Demonstration of T-transferase deficiency in Tn-polyagglutinable blood samples. Eur J Biochem 1978; 92:111-119.
22. Sturgeon P, Luner SJ, McQuiston DT. Permanent mixed-field polyagglutinability (PMFP). Vox Sang 1973;25:498-512.
23. Springer GF, Desai PR, Scanlon EF. Blood group MN precursors as human breast carcinoma-associated antigens and "naturally" occurring human cytotoxins against them. Cancer 1976;37:169-176.
24. Anglin JH Jr, Lerner MP, Nordquist RE. Blood group-like activity released by human mammary carcinoma cells in culture. Nature (London) 1977;269:254-255.
25. Springer GF, Desai PR. Cross-reacting carcinoma-associated antigens with blood group and precursor specificities. Transplant Proc 1977;9:1105-1111.
26. Laurent JC, Noël P, Faucon M. Expression of a cryptic cell surface antigen in primary cell cultures from human breast cancer. Biomedicine 1978;29:260-261.
27. Lloyd KO, Kabat EA. Immunochemical studies on blood groups. XLI. Proposed structures for the carbohydrate portions of blood group A, B, H, Lewis[a] and Lewis[b] substances. Proc Natl Acad Sci USA 1968;61:1470-1477.

28. Donald ASR, Creeth JM, Morgan WTJ, Watkins WM. The peptide moiety of human blood group active glycoproteins associated with the ABO and Lewis groups. Biochem J 1969;115:125-127.
29. Clausen H, Stroud M, Parker J, Springer GF, Hakomori S-I. Monoclonal antibodies directed to the blood group A associated structure, Galactosyl-A: specificity and relation to the Thomsen-Friedenreich antigen. Mol Immunol 1988;25:199-204.
30. Silverberg E, Boring CC, Squires TS. Cancer statistics, 1990. CA − Cancer J Clinicians 1990;40:9-26.
31. Springer GF, Murthy MS, Desai PR, Scanlon EF. Breast cancer patient's cell-mediated immune response to Thomsen-Friedenreich (T) antigen. Cancer 1980;45:2949-2954.
32. Springer GF, Chandrasekaran EV, Desai PR, Tegtmeyer H. Blood group Tn-active macromolecules from human carcinomas and erythrocytes: characterization of and specific reactivity with mono- and poly-clonal anti-Tn antibodies induced by various immunogens. Carbohydr Res 1988;178:271-292.
33. Springer GF, Nagai Y, Tegtmeyer H. Isolation and properties of human blood-group MN and meconium-Vg antigens. Biochemistry 1966;5:3254-3272.
34. Springer GF, Desai PR. Common precursors of human blood group MN specificities. Biochem Biophys Res Commun 1974;61:470-475.
35. Springer GF, Desai PR. Human blood-group MN and precursor specificities: structural and biological aspects. Carbohydr Res 1975;40:183-192.
36. Springer GF, Yang HJ. Isolation and partial characterization of blood group M- and N-specific glycopeptides and oligosaccharides from human erythrocytes. Immunochemistry 1977;14:497-502.
37. Springer GF, Huprikar SV. On the biochemical and genetic basis of the human blood-group MN specificities. Haematologia 1972;6:81-92.
38. Springer GF, Tegtmeyer H, Huprikar SV. Anti-N reagents in elucidation of the genetical basis of human blood-group MN specificities. Vox Sang 1972;22:325-343.
39. Codington JF, Yamazaki T, van den Eijnden DH, Evans NA, Jeanloz RW. Unequivocal evidence for a β-D-configuration of the galactose residue in the disaccharide chain of epiglycanin, the major glycoprotein of the TA3-Ha tumor cell. FEBS Lett 1979;99:70-72.
40. Bhavanandan VP, Codington JF. Selective release of the disaccharide 2-acetamido-2-deoxy-3-O-(β-D-galactopyranosyl)-D-galactose from epiglycanin by endo-N-acetyl-α-D-galactosaminidase. Carbohydr Res 1983;118:81-89.
41. Kim Z, Uhlenbruck G. Untersuchungen über T-Antigen und T-Agglutinin. Z Immun Exp Klin Immunol 1966;130:88-99.
42. Dahr W, Uhlenbruck G, Gunson HH, Van Der Hart M. Molecular basis of Tn-polyagglutinability. Vox Sang 1975;29:36-50.
43. Desai PR, Springer GF, Yang HJ. Aspects of human blood group MN and their precursors (Abstract). In: Proc XIV Congr Intl Soc Blood Transfusion, Finland, 1975;67.
44. Desai PR, Springer GF. Carcinoma glycolipids possess immunogenic blood group Tn-, T-, N- and M-activities. In: Chester MA, Heinegård D, Lundblad A, Svennson S, eds. Glycoconjugates, Proc 7th Intl Conf, Lund, Sweden: Rahms, 1983:431-432.
45. Rauvala H, Finne J. Structural similarity of the terminal carbohydrate sequences of glycoproteins and glycolipids. FEBS Lett 1979;97:1-8.
46. Springer GF, Cantrell JL, Desai PR, Tegtmeyer H. Human blood groups M-, N-, T-, and Tn-specific substances in lipidic extracts of line 10 hepatocarcinoma of Strain-2 guinea pigs. Clin Immunol Immunopathol 1982;22:9-19.
47. Nakahara K, Ohashi T, Oda T, et al. Asialo G_{M1} as a cell-surface marker detected in acute lymphoblastic leukemia. N Engl J Med 1980;302:674-677.

48. Solomon FR, Higgins TJ. A monoclonal antibody with reactivity to asialo G_{M1} and murine natural killer cells. Mol Immunol 1987;24:57-65.
49. Kurosaka A, Nakajima H, Funakoshi I, Matsuyama M, Nagayo T, Yamashina I. Structures of the major oligosaccharides from a human rectal adenocarcinoma glycoprotein. J Biol Chem 1983;256:11594-11598.
50. Kjeldsen T, Clausen H, Hirohashi S, Ogawa T, Iijima H, Hakomori S-I. Preparation and characterization of monoclonal antibodies directed to the tumor-associated O-linked Sialosyl-2 → 6 α-N-acetylgalactosaminyl (Sialosyl-Tn) epitope. Cancer Res 1988;48:2214-2220.
51. Yurewicz EC, Moghissi KS. Purification of human midcycle cervical mucin and characterization of its oligosaccharides with respect to size, composition, and microheterogeneity. J Biol Chem 1981;256:11895-11904.
52. Yurewicz EC, Matsuura F, Moghissi KS. Structural studies of sialylated oligosaccharides of human midcycle cervical mucin. J Biol Chem 1987;262:4733-4739.
53. Podolsky DK. Oligosaccharide structures of isolated human colonic mucin species. J Biol Chem 1985;260:15510-15515.
54. Springer GF, Desai PR. Tn epitopes, immunoreactive with ordinary anti-Tn antibodies, on normal, desialylated human erythrocytes and on Thomsen-Friedenreich antigen isolated therefrom. Mol Immunol 1985;22:1303-1310.
55. Berdinskikh MS, Pavlyuchenkova RP, Kiseleva AS, Zotikov EA, Kosyakov PN. Identification of Thomsen's antigen in leukocytes of leukemic patients. Bull Exp Biol Med 1987;103:334-336.
56. Roxby DJ, Morley AA, Burpee M. Detection of the Tn antigen in leukaemia using monoclonal anti-Tn antibody and immunohistochemistry. Br J Haematol 1987;67:153-156.
57. Ness PM, Garratty G, Morel PA, Perkins HA. Tn polyagglutination preceding acute leukemia. Blood 1979;54:30-34.
58. Stein R, Chen S, Grossman W, Goldenberg DM. A human lung carcinoma monoclonal antibody specific for the Thomsen-Friedenreich antigen. Cancer Res 1989;49:32-37.
59. Springer GF, Murthy SM, Desai PR, Fry WA, Tegtmeyer H, Scanlon EF. Patients' immune response to breast and lung carcinoma-associated Thomsen-Friedenreich specificity. Klin Wochenschr 1982;60:121-131.
60. Altavilla G, Lanza G Jr, Rossi S, Cavazzini L. Morphologic changes, mucin secretion, carcinoembryonic antigen (CEA) and peanut lectin reactivity in colonic mucosa of patients at high risk for colorectal cancer. Tumori 1984;70:539-548.
61. Ørntoft TF, Mors NPO, Eriksen G, Jacobsen NO, Poulsen HS. Comparative immunoperoxidase demonstration of T-antigens in human colorectal carcinomas and morphologically abnormal mucosa. Cancer Res 1985;45:447-452.
62. Yuan M, Itzkowitz SH, Boland CR, et al. Comparison of T-antigen expression in normal, premalignant, and malignant human colonic tissue using lectin and antibody immunohistochemistry. Cancer Res 1986;46:4841-4847.
63. Nishiyama T, Matsumoto Y, Watanabe H, Fujiwara M, Sato S. Detection of Tn antigen with *Vicia villosa* agglutinin in urinary bladder cancer: its relevance to the patient's clinical course. J Natl Cancer Inst 1987;78:1113-1118.
64. Boland CR, Montgomery CK, Kim YS. Alterations in human colonic mucin occurring with cellular differentiation and malignant transformation. Proc Natl Acad Sci USA 1982;79:2051-2055.
65. Kellokumpu I, Kellokumpu S, Andersson LC. Identification of glycoproteins expressing tumour-associated PNA-binding sites in colorectal carcinomas by SDS-GEL electrophoresis and PNA-labelling. Br J Cancer 1987;55:361-365.
66. Springer GF, Taylor CR, Howard DR, et al. Tn, a carcinoma-associated antigen, reacts with anti-Tn of normal human sera. Cancer 1985;55:561-569.

67. Ohuchi N, Nose M, Abe R, Kyogoku M. Lectin-binding pattern of breast carcinoma: significance on structural atypism. Tohoku J Exp Med 1984;143:491-499.
68. Bailly P, Tollefsen SE, Cartron JP. Glycolipid specificity of the B_4 lectin from *Vicia villosa* seeds. Glycoconjugate J 1985;2:401-408.
69. Thomson DMP, Springer GF, Desai PR, Scanzano R, Gubersky M, Shenouda G. Comparison by leukocyte adherence inhibition of human immune response to cancer-associated immunogens, Thomsen-Friedenreich (T) and Tn, myelin basic protein, and organ-specific cancer neoantigens. Clin Immunol Immunopathol 1988;49:231-241.
70. Friedenreich V, Andersen TT. On the existence, outside human blood, of substances convertible into receptors T. Acta Pathol Microbiol Scand 1929;VI:236-251.
71. Springer GF, Horton RE. Blood-group isoantibody stimulation in man by feeding blood group-active bacteria. J Clin Invest 1969;48:1280-1291.
72. Springer GF. a) Enzyme-treated erythrocytes; b) Inhibition of hemagglutination by antibodies and antibody-like reagents in semiquantitative tube tests. In: Williams CA, Chase MW, eds. Methods in immunology and immunochemistry, vol. IV. New York: Academic Press, 1977;14-19:67-72.
73. Bray J, Maclean GD, Dusel FJ, McPherson TA. Decreased levels of circulating lytic anti-T in the serum of patients with metastatic gastrointestinal cancer: a correlation with disease burden. Clin Exp Immunol 1982;47:176-182.
74. Kabat EA. In: Kabat EA, Mayers MM, eds. Experimental immunochemistry, 2nd ed. Springfield: Charles C Thomas, 1964:361-475.
75. Codington JF, Sanford BH, Jeanloz RW. Glycoprotein coat of the TA3 cell. Isolation and partial characterization of a sialic acid containing glycoprotein fraction. Biochemistry 1972;11:2559-2564.
76. Walker-Nasir E, Codington JF, Jahnke MR, Fuller TC, Jeanloz RW. Isolation and partial characterization of surface components of cell line MDA-MB-231 derived from a human metastatic breast carcinoma. J Natl Cancer Inst 1982;69:371-380.
77. Schneider AW, Fischer K, Stegner HE, Poschmann A. Automatic determination of anti-T antibodies in patients with breast carcinoma and controls. Tumor Diagn Ther 1986;7:78-84.
78. Springer GF, Desai PR, Tegtmeyer H, et al. Responses of patients to their carcinoma-associated T and Tn-specific antigens. In: Cimino F, Birkmayer GD, Klavins JV, Pimentel E, Salvatore F, eds. Human tumor markers, Proc 3rd Intl Conf Berlin-New York: Walter de Gruyter 1987:315-332.
79. Springer GF, Desai PR, Murthy MS, Tegtmeyer H, Scanlon EF. Human carcinoma-associated precursor antigens of the blood group MN system and the host's immune responses to them. In: Ishizaka K, Kallós P, Waksman BH, de Weck AL, eds. Progress in allergy, vol. 26. Basel, Switzerland: S Karger 1979:42-96.
80. Springer GF, Desai PR. Detection of lung- and breast carcinoma by quantitating serum anti-T IgM levels with a sensitive, solid-phase immunoassay. Naturwissenschaften 1982;69:346-348.
81. Desai PR, Springer GF. Carcinoma detection by quantitation and interrelation of serum anti-T IgM and total IgM. Protides Biol Fluids Proc 1984;31:421-424.
82. Bird GWG, Wingham J, Pippard MJ, et al. Erythrocyte membrane modification in malignant diseases of myeloid and lymphoreticular tissues. Tn-polyagglutination in acute myelocytic leukaemia. Br J Haematol 1976;33:289-294.
83. Heidelberger M. Lectures in immunochemistry. New York: Academic Press, 1956:120-131.
84. Kjeldsen T, Hakomori S, Springer GF, Desai PR, Harris T, Clausen H. Coexpression of sialosyl-Tn (NeuAcα2 → 6GalNAcα1 → O-Ser/Thr) and Tn (GalNAcα1 → O-Ser/Thr) blood group antigens on Tn erythrocytes. Vox Sang 1989;57:81-89.

85. Rahman AFR, Longenecker BM. A monoclonal antibody specific for the Thomsen-Friedenreich cryptic T antigen. J Immunol 1982;129:2021-2024.
86. Metcalfe SM, Svvennsen RJ, Springer GF, Tegtmeyer H. Monoclonal antibodies to tumour associated Thomsen-Friedenreich (T) and Tn antigens. J Immunol Methods 1983;62:M8.
87. Metcalfe S, Springer GF, Svvennsen RJ, Tegtmeyer H. Monoclonal antibodies specific for human Thomsen-Friedenreich (T) and Tn blood group precursor antigens. Protides Biol Fluids Proc 1984;32:765-768.
88. Roxby DJ, Skinner JM, Morley AA, Weeks S, Burpee M. Expression of a Tn-like epitope by carcinoma cells. Br J Cancer 1987;56:734-737.
89. Nuti M, Teramoto YA, Mariani-Costantini R, Horan Hand P, Colcher D, Schlom J. A monoclonal antibody (B72.3) defines patterns of distribution of a novel tumor-associated antigen in human mammary carcinoma cell populations. Int J Cancer 1982;29:539-545.
90. Colcher D, Esteban JM, Carrasquillo JA, et al. Quantitative analyses of selective radiolabeled monoclonal antibody localization in metastatic lesions of colorectal cancer patients. Cancer Res 1987;47:1185-1189.
91. Kurosaka A, Fukui S, Kitagawa H, et al. Mucin-carbohydrate directed monoclonal antibody. FEBS Lett 1987;215:137-139.
92. Hirohashi S, Clausen H, Yamada T, Shimosato Y, Hakomori S-I. Blood group A cross-reacting epitope defined by monoclonal antibodies NCC-LU-35 and -81 expressed in cancer of blood group O or B individuals: its identification as Tn antigen. Proc Natl Acad Sci USA 1985;82:7039-7043.
93. Pereira MEA, Kabat EA, Lotan R, Sharon N. Immunochemical studies on the specificity of the peanut (*Arachis hypogaea*) agglutinin. Carbohydr Res 1976;51:107-118.
94. Newman RA, Klein PJ, Rudland PS. Binding of peanut lectin to breast epithelium, human carcinomas, and a cultured rat mammary stem cell: use of the lectin as a marker of mammary differentiation. J Natl Cancer Inst 1979;63:1339-1346.
95. Böcker W, Klaubert A, Bahnsen J, et al. Peanut lectin histochemistry of 120 mammary carcinomas and its relation to tumor type, grading, staging, and receptor status. Virchows Arch [Pathol Anat] 1984;403:149-161.
96. Springer GF, Desai PR. Extent of desialation of blood group MM, NN, and MN antigens required for reactivity with human anti-T antibody and *Arachis hypogaea* lectin. J Biol Chem 1982;257:2744-2746.
97. Bird GWG, Wingham J. Haemagglutinins from *Salvia*. Vox Sang 1974;26:163-166.
98. Bird GWG, Wingham J. Anti-Tn from *Marrubium candidissimum*. Blood Transfusion Immunohaematol 1981;25:347-348.
99. Hollinshead AC, Stewart THM, Herberman RB. Delayed-hypersensitivity reactions to soluble membrane antigens of human malignant lung cells. J Natl Cancer Inst 1974;52:327-338.
100. von Pirquet CE. Allergy. Arch Inter Med 1911;I:259-288, and 383-436.
101. Dupont DL, Page WD. Risk factors for breast cancer in women with proliferative breast disease. N Engl J Med 1985;312:146-151.
102. Goodale RL, Springer GF, Shearen JG, Desai PR, Tegtmeyer H. Delayed-type cutaneous hypersensitivity to Thomsen-Friedenreich (T) antigen in patients with pancreatic cancer. J Surg Res 1983;35:293-297.
103. Springer GF, Desai PR, Fry WA, Goodale RL, Shearen JG, Scanlon EF. Patient's immune response to CA-associated T antigen. In: Moloy PJ, Nicolson GL, eds. Cellular oncology, new approaches in biology, diagnosis, and treatment (Cancer Research Monographs, vol. 1). New York: Praeger Press, 1983:99-130.

104. Bernhard MI, Wanebo HJ, Helm J, Pace RC, Kaiser DL. Leukocyte migration inhibition responses to MCF-7, murine mammary tumor virus, and Thomsen-Friedenreich antigen in a series of cancer patients. Cancer Res 1983;43:1932-1937.
105. Ichinose Y, Yagawa K, Kaku M, Hara N, Ohta M. Immune reactivity to Thomsen-Friedenreich antigen in patients with lung cancer detected by superoxide assay-leukocyte adherence inhibition test. Cancer Res 1985;45:4473-4477.
106. Thatcher N, Hashmi K, Chang J, Swindell R, Crowther D. Anti-T antibody in malignant melanoma patients. Cancer 1980;46:1378-1382.
107. Vos GH, Brain P. Heterophile antibodies, immunoglobulin levels, and the evaluation of anti-T activity in cancer patients and controls. S Afr Med J 1981;60:133-136.
108. Vos GH, Rose EF, Vos D. Antibody responses to T-activated red cells in children from high- and low-risk areas of cancer of the oesophagus in Transkei. S Afr Med J 1981;59:56-60.
109. Bretscher PA. In vitro analysis of the cellular interactions between unprimed lymphocytes responsible for determining the class of response an antigen induces: specific T cells switch a cell-mediated response to a humoral response. J Immunol 1983;131:1103-1107.
110. Zabel PL, Noujaim AA, Shysh A, Bray J. Radioiodinated peanut lectin: a potential radiopharmaceutical for detection of tumours expressing the T-antigen. Proc 3rd World Congress Nuclear Med and Biol, Paris, 1982;3670-3673.
111. Obara T, Springer GF, Klein-Szanto A, Tegtmeyer H. Immunohistological differences in man between delayed-type skin hypersensitivity reactions (DTHR) to Thomsen-Friedenreich (T) antigen and to PPD [Abstract]. Proc Am Assoc Cancer Res 1986;27:15.
112. Howard DR, Taylor CR. A method for distinguishing benign from malignant breast lesions utilizing antibody present in normal human sera. Cancer 1979;43:2279-2287.
113. Turner DT, Connolly CE, Isaacson P, Turnbull AR. Receptors for Fc and complement in human breast carcinoma. Clin Oncol 1978;4:87-92.
114. Beahrs OH, Henson DE, Hutter RP, Myers MH, eds. Manual for staging of cancer, 3rd ed. Philadelphia: JB Lippincott, 1988.
115. Yokoyama M, Ohoka H, Oda H, Oda T, Utsumi S, Takeuchi M. Thomsen-Friedenreich antigen in bladder cancer tissues detected by monoclonal antibody. Acta Urol Jpn 1988;34:255-258.
116. Blasco E, Torrado J, Belloso L, Arocena F, Gutierrez-Hoyos A, Cuadrado E. T-antigen. A prognostic indicator of high recurrence index in transitional carcinoma of the bladder. Cancer 1988;61:1091-1095.
117. Coon JS, Watkins JR, Pauli BU, Weinstein RS. Flow cytometric analysis of heterogeneity in blood group-related antigen expression in a human urinary bladder carcinoma cell line, 647V^1. Cancer Res 1985;45:3014-3021.
118. Cooper HS, Reuter VE. Peanut lectin-binding sites in polyps of the colon and rectum. Adenomas, hyperplastic polyps, and adenomas with in situ carcinoma. Lab Invest 1983;49:655-661.
119. Hoover HC Jr, Surdyke MG, Dangel RB, Peters LC, Hanna MG Jr. Prospectively randomized trial of adjuvant active-specific immunotherapy for human colorectal cancer. Cancer 1985;55:1236-1243.
120. Ghazizadeh M, Kagawa S, Kurokawa K. Immunohistochemical studies of human renal cell carcinomas for ABO(H) blood group antigens, T antigen-like substance and carcinoembryonic antigen. J Urol 1985;133:762-766.
121. Howard DR, Ferguson P, Batsakis JG. Carcinoma-associated cytostructural antigenic alterations: detection by lectin binding. Cancer 1981;47:2872-2877.

122. Itzkowitz SH, Yuan M, Montgomery CK, et al. Expression of Tn, sialosyl-Tn, and T antigens in human colon cancer. Cancer Res 1989;49:197-204.
123. Barr N, Taylor CR, Young T, Springer GF. Are pan carcinoma T and Tn differentiation antigens? Cancer 1989;64:834-841.
124. Corfield AP, Roe R, Wagner SA, Clamp JR, Williamson RCN. A combined biochemical-histological approach to carbohydrate changes in colonic mucus glycoproteins during malignant transformation. Br J Cancer 1986;54:198.
125. Desai PR, Tegtmeyer H, Chandrasekaran EV, Springer GF. Autoantigenic adhesion molecule Tn in pathogenesis of carcinomas and some leukemias. J Tumor Marker Oncol 1987;2:233-244.
126. Fenlon S, Ellis IO, Bell J, Todd JH, Elston CW, Blamey RW. *Helix pomatia* and *Ulex europaeus* lectin binding in human breast carcinoma. J Pathol 1987;152:169-176.
127. Leathem AJ, Brooks SA. Predictive value of lectin binding on breast-cancer recurrence and survival. Lancet 1987;i:1054-1056.
128. Baker DA, Sugii S, Kabat EA, Ratcliffe RM, Hermentin P, Lemieux RU. Immunochemical studies on the combining sites of Forssman hapten reactive hemagglutinins from *Dolichos biflorus, Helix pomatia*, and *Wistaria floribunda*. Biochemistry 1983;22:2741-2750.
129. Springer GF, Rose CS, György P. Blood group mucoids: their distribution and growth promoting properties for *Lactobacillus bifidus var penn*. J Lab Clin Med 1954;43:532-542.
130. Springer GF, Cheingsong-Popov R, Schirrmacher V, Desai PR, Tegtmeyer H. Proposed molecular basis of murine tumor cell-hepatocyte interaction. J Biol Chem 1983;258:5702-5706.
131. Springer GF, Desai PR, Scanlon EF. T and Tn pancarcinoma autoantigens in tumor pathogenesis, very early cancer detection, follow-up and treatment, Proc 15th Intl Cancer Congress, Hamburg, 1990: in press.
132. Sastry MVK, Banarjee P, Patanjali SR, Swamy MJ, Swarnalatha GV, Surolia A. Analysis of saccharide binding to *Artocarpus integrifolia* lectin reveal specific recognition of T-antigen (β-D-Gal(1 $\rightarrow$ 3)D-GalNAc). J Biol Chem 1986;261:11726-11733.
133. Rinderle SJ, Goldstein IJ, Matta KL, Ratcliffe RM. Isolation and characterization of amaranthin, a lectin present in the seeds of *Amaranthus caudatus*, that recognizes the T- (or cryptic T)-antigen. J Biol Chem 1989;264:16123-16131.
134. Black MM, Zachrau RE. Stepwide mammary carcinogenesis: immunological considerations. In: Zander J, Baltzer J, eds. Early breast cancer. Berlin: Springer-Verlag, 1985:64-72.
135. Ghazizadeh M, Oguro T, Sasaki Y, Aihara K, Araki T, Springer GF. Immunohistochemical and ultrastructural localization of T antigen in ovarian tumors. Am J Clin Pathol 1990;93:315-321.
136. Schlepper-Schäfer J, Springer GF. Carcinoma autoantigens T and Tn and their cleavage products interact with Gal/GalNAc-specific receptors on rat Kupffer cells and hepatocytes. Biochim Biophys Acta 1989;89:266-272.
137. Holt S, Wilkinson A, Suresh MR, et al. Radiolabelled peanut lectin for the scintigraphic detection of cancer. Cancer Letters 1984;25:55-60.

37
Serum Isoenzymes in Cancer Diagnosis and Management

DONALD W. MERCER

Montefiore Hospital and Pittsburgh Cancer Institute, Pittsburgh, Pennsylvania

For decades human cancer has been associated with abnormal production of a variety of circulatory enzymes. During this time enzymologists have searched for a unique enzyme "marker" of malignancy. Although a wide variety of enzymes have been assayed in connection with the management of cancer patients, no single enzyme has proven to be a sensitive and specific indicator of malignancy [1-3].

An alternative approach to overcoming some of the problems associated with enzymes as tumor markers is the measurement and use of isoenzymes. The term *isoenzyme* is usually used as an operational term to describe multiple forms of enzyme proteins with the same catalytic activity but with different molecular forms that can be separated by suitable methods such as electrophoresis or column chromatography. In some instances the reason for multiplicity is known, such as those arising from genetically determined differences in primary structure. In other cases, the reason for the multiplicity is not known and posttranslational modifiction is usually assumed. For convenience in this chapter, all examples of multiple enzyme forms, regardless of the nature of multiplicity, will be referred to as isoenzymes.

Isoenzyme determinations in the past were usually performed by electrophoresis, which takes advantage of the differences in electrical charge associated with isoenzymes. Subsequent identification of the electrophoretically separated isoenzymes on supporting media such as polyacrylamide and agarose is based on the measurement of their catalytic activity as determined by histochemical staining and densitometric scanning.

Although this technique is characterized by great resolving power for the qualitative identification of isoenzymes, precise quantitation of isoenzymes, especially at low levels of activity, is limited for several reasons. Problems such as asymmetrical distribution of isoenzymes along the axis of scanning, poor diffusion of substrate into the supporting media, and loss of staining dye from the supporting media are a few of the reasons for unreliable quantitative results when electrophoretic techniques are used. Failure to detect the presence of an isoenzyme marker by electrophoresis, especially in patients with early-stage malignancies in whom the activity levels are often low, may therefore be the fault of the technique.

Table 1 Serum isoenzymes in Cancer Diagnosis and Management

Serum Isoenzymes	Cancer Type and Site
Group I (established markers, approved for routine use)	
Postatic acid phosphatase (PAP)	Adenocarcinoma of prostate
Group 2 (promising markers, for investigational use only)	
Placental alkaline phosphatase (PLAP)	Germ cell tumors of testis
BB-creatine kinase (BB-CK)	SCLC, adenocarcinoma of breast
Neuron-specific enolase (NSE)	SCLC, neuroblastoma
Lactate dehydrogenase 1 (LDH-1)	Germ cell tumors of testis
Salivary amylase (S-AMS)	Adenocarcinoma of ovary and lung
Galactosyltransferase-II (GT II)	Adenocarcinoma of pancreas
5'-Nucleotide phosphodiesterase-V (NPD V)	Primary hepatocellular carcinoma, lever metastasis
Variant beta hexosaminidase (Var Hex-β)	Liver metastasis
Thymidine kinase 1 (TK-1)	Leukemias, lymphomas

Improvements in analytical tools for isoenzyme analysis, utilizing hybridoma-monoclonal antibody techniques, have recently been made. Antigenic specificity among isoenzymes of the same type enzyme is utilized to develop highly sensitive and specific immunoassays. This immunological approach, which often uses monoclonal antibodies as immunochemical reagents, appears to have revived and stimulated interest in isoenzymes as tumor markers.

With several assays now available commercially and other assays almost ready for routine clinical use, a re-evaluation of isoenzymes as tumor markers is in order. This chapter will examine the most promising isoenzymes in cancer diagnosis and management from the viewpoint of the following topics: history and biochemistry, methods of measurement, and clinical utility. Table 1 lists 10 isoenzymes and their associated cancer type and site that in the author's judgment, show potential as tumor markers. This table is organized into two groups of isoenzymes: those established and approved for routine clinical use and those with clinical promise that are for investigational use only.

I. PROSTATIC ACID PHOSPHATASE

Prostatic acid phosphatase (PAP) is one of several isoenzymes of acid phosphatase (ACP, EC:3.1.3.2). Virtually all tissues contain acid phosphatase of some type and most of it is located within the lysosomes. Acid phosphatase, a glycoprotein with molecular weight of 100 Kd, hydrolyzes esters of orthophosphoric acid at an acid pH. The highest amount of acid phosphatase activity has been found in the cells and secretions of the prostate gland.

Measurement of serum acid phosphatase levels has been of clinical interest since the Gutmans [4] in 1938 first reported increased serum ACP activity in patients with metastasizing carcinoma of the prostate. Subsequent studies have diminished the effective use of total serum acid phosphatase levels as a sensitive and specific indicator of prostatic carcinoma because of the numerous reports that ACP activity is also increased in sera of patients with Gaucher's disease [5], multiple myeloma [6], leukemia [7], Paget's disease [8], kidney disease [8], rectal carcinoid tumor [9], thrombocytopenia [10], and liver disease [11].

In the 1950s, attempts were made to assay ACP in a manner that would be specific for the diagnosis of prostatic cancer. These techniques centered around the use of selective substrates such as beta-glycerophosphate, thymolphthalein phosphate, and alpha-naphthyl phosphate or selective inhibitors such as L-tartrate. The clinical usefulness of selective substrates as inhibitors was criticized because of the number of false-positive results observed in patients with nonprostatic conditions.

Attempts by Lam [12] in the 1960s to enhance the diagnostic specificity of ACP focused on the use of electrophoretic techniques to separate ACP isoenzymes. Prostatic tissue was the only tissue to exhibit large amounts of ACP-2 or PAP and electrophoretic patterns from patients with carcinoma of prostate were found to reflect those of prostatic tissue [12]. Although the electrophoretic method for ACP was more specific and preferable to direct assay with selective substrates, the technique of gel electrophoresis involving histochemical staining and densitometric scanning was time-consuming, semiquantitative, and insensitive to low PAP activities.

The recent availability of specific PAP antibodies [13] that react only with prostatic acid phosphatase has led to the development of more precise and specific immunological methods. A variety of PAP immunoassays, such as counterimmunoelectrophoresis (CIE), radioimmunoassay (RIA), and enzyme immunoassay (EIA) are currently available. However, even though PAP immunoassays are extremely precise and specific, the sensitivity of the assay for detection of prostatic cancer in early-stage malignancy is still very low. Table 2 summarizes representative data found in clinically staged prostatic cancer patients. The number of prostatic cancer patients with a positive result correlates with progression of disease.

However, in the early, more treatable stages A and B of prostate cancer, low detection rates (30% average) are found. Heller [18] has recently compiled data in clinically staged prostatic patients with similar results from 22 different investigators. Thus, while an elevated PAP level is highly sensitive for late-stage disease, detection levels observed for early-stage disease are too low for use in screening asymptomatic individuals.

Nevertheless, in patients with a confirmed diagnosis of prostatic malignancy and elevated pretreatment PAP levels, past experience has shown that the monitoring of therapy with serial PAP testing plays a significant role. In such cases, PAP levels may fall or return to normal following successful treatment, while persistent or increasing levels indicate residual or recurrent disease. Figure 1 illustrates the return of serum PAP to normal levels after orchiectomy in a patient with late-stage prostatic carcinoma.

Although PAP is quite tissue-specific, it is not tumor-specific. Thus, PAP levels may be elevated in nonmalignant conditions of the prostate, such as benign prostatic hyperplasia (BPH). About 10% of patients with BPH have been found to have slightly elevated PAP levels and serial studies of PAP are recommended to help differentiate between malignant and benign conditions. Also prostate manipulative procedures such as prostatic massage, rectal examination, or biopsy have been found to give rise to elevated PAP levels. However, these increases in PAP usually last only a few days.

Development of a more sensitive immunological assay for PAP would appear not to enhance the clinical usefulness of PAP measurements since existing RIA and EIA procedures are already sensitive enough to detect normal range levels of PAP in serum. Im-

Table 2 Serum PAP Levels in Clinically Staged Prostatic Cancer

Reference	Assay	Upper Limit of Normal	Incidence of Elevated Levels; and (%)			
			Stage A	Stage B	Stage C	Stage D
Foti [14]	RIA	+4 SD[a]	8/24 (33)	26/33 (78)	22/31 (71)	23/25 (92)
Abbott Laboratories [15]	EIA	2.2 ng/ml	6/21 (29)	9/22 (41)	6/11 (55)	18/25 (72)
Davies [16]	RIA	+2 SD[a]	8/19 (44)	4/14 (29)	5/10 (50)	27/34 (79)
Bruce [17]	RIA	+2 SD[a]	4/32 (13)	10/38 (26)	2/14 (14)	11/14 (76)

[a]in healthy individuals
SD, standard deviation.

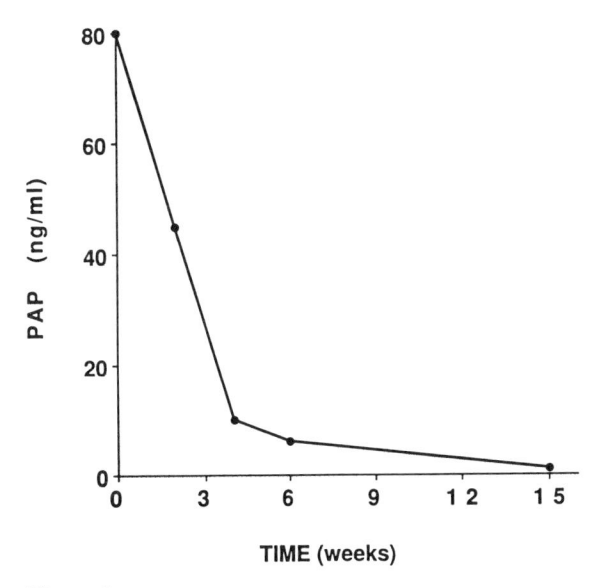

Figure 1 The change in serum PAP levels following bilateral orchiectomy in a 68-year-old white man with probable metastases.

provements in the detection of early-stage prostatic carcinoma will most likely come with the use of markers in combination.

Simultaneous measurement of other antigens such as prostatic-specific antigen (PSA) and PAP has recently been shown by Kuriyama et al. [19] to be of additive value. At this time the multiple marker approach appears to be a logical one, especially in view of the lack of a single ideal tumor marker for prostatic cancer. For a more detailed discussion and review of the multiple marker approach in prostate cancer (see Chapter 4).

II. PLACENTAL ALKALINE PHOSPHATASE

Placental alkaline phosphatase (PLAP) is one of several isoenzymes of alkaline phosphatase [ALP, EC:3.1.3.1]. Alkaline phosphatase, with a molecular weight of 160 Kd, hydrolyzes at an alkaline pH a wide variety of esters of phosphoric acid. Liver, bone, and intestinal ALPs are more commonly measured serum isoenzymes and their measurements are often used to identify the tissue origin of an unexpected elevated total serum alkaline phosphatase.

Placental isoenzyme is normally only present in serum during pregnancy, but there has been interest in the placental isoenzyme as a tumor marker ever since Fishman discovered a placental-like alkaline phosphatase (Regan isoenzyme) in the tissue of a patient with lung carcinoma. Another placental-like isoenzyme associated with cancer is the Nagao isoenzye. Both Regan and Nagao isoenzymes have been reported in the sera of patients with a wide variety of cancers [21]. A comprehensive review describing the role of multiple forms of ALP in cancer detection and monitoring has recently been compiled by Moss [22].

In spite of the many reports describing the potential use of PLAP as a tumor marker, current routine use in the clinical laboratory is limited primarily due to limitations in assays based on PLAP's unique property of heat resistance (56°C at 30 min). Many investigators have found the heat treatment procedure difficult to control and unreliable, especially when attempting to measure low activity levels of PLAP.

However, with the development and use of monoclonal antibodies as immunochemical reagents, greatly improved PLAP assays are now available. A review by Tucker [23] draws attention to these advances and Table 3 summarizes representative data from recent studies. The highest prevalence of elevated PLAP-like isoenzyme levels [80%] has been demonstrated in testicular tumors of the seminoma type, whereas monoclonal antibody (H17E2) was used in a solid-phase immunoassay. A low incidence of elevated levels (10%) was observed in nonseminomatous germ cell tumors. Although no large-scale study comparing stage of disease and level of PLAP has been reported with this assay, several case reports of longitudinal studies have shown PLAP levels to correlate closely with tumor burden [24]. Studies evaluating the specificity of PLAP in sera of patients with a variety of benign conditions have not yet been reported with the H17E2 monoclonal antibody tests. However, significant elevations of PLAP have been demonstrated in about 29% of smokers.

Another monoclonal antibody, H317 (25), has been used to detect PLAP in sera and this assay has been evaluated in sera of patients with a variety of ovarian tumors (31% incidence of elevated rates). In contrast to testicular cancer, ovarian PLAP does not appear to be related to stage of disease or tumor burden. Moreover, this assay does not give rise to false-positive elevations in smokers, but elevated levels have been reported in patients (20%) with nonmalignant disease of the ovary.

The presence of PLAP in breast and lung cancer has been studied with another newly developed monoclonal antibody assay [26]. In patient sera, an incidence of 5.2% and 11.2% elevated levels was reported for breast and lung cancer, respectively. Excellent correlation was observed in this study between PLAP activity and histopathological staging of tumor tissue.

Thus, in view of the low incidence of PLAP in patients with cancer other than testicular cancer, the use of PLAP in screening asymptomatic individuals is not recommended. Nevertheless, as new immunoassays become available, PLAP will probably be useful in monitoring response to treatment, especially in patients exhibiting pretreatment PLAP in abnormal amounts.

Table 3 Serum PLAP Levels in Various Types of Malignancies

Reference	Monoclonal Antibody	Assay	Type of Malignancy	Incidence[a] (Elevated Levels):(%)
Tucker [23,24]	H17E2	EIA	Testicualr (seminoma)	14/16 (88)
			Testicular (nonseminoma)	1/10 (10)
McDicken [25]	H317	EIA	Ovarian (all types)	28/89 (31)
Van de Voorde [26]	EG	EIA	Breast (all types)	15/266 (5)
			Lung (all types)	15/133 (11)

[a]Patients with early- and late-stage disease combined.

Reports have also shown that combination testing involving PLAP with other markers can be helpful. For example, the combined use of PLAP and CA-125 appears promising in helping to differentiate between a benign and malignant condition of the ovary [27]. A sensitive but not sufficiently specific marker, CA-125, is used in combination with a specific but insensitive ovarian marker, PLAP. See Chapter 4 for additional discussions of the utility of marker combinations.

III. CREATINE KINASE BB

Creatine kinase BB (CK-BB) is one of several isoenzymes of the energy transfer enzyme, creatine kinase [CK, EC:2.7.3.2]. CK is present in the form of a dimer of two molecules, each having a molecular weight of 40 Kd. The monomers are designated CK-M (for "muscle") and CK-B (for "brain"), and they hybridize to form three isoenzymes CK-BB, CK-MB, and CK-MM. Determination of serum CK-MB isoenzyme is widely used as a diagnostic test for detection of myocardial infarction.

Sensitive and quantitative methods such as column chromatography [28] were developed in the early 1970s to measure CK-MB. These same methods were also found to be excellent for the measurement of CK-BB. Beginning in the late 1970s and early 1980s, reports of elevated CK-BB isoenzyme levels in sera of cancer patients, especially those with metastatic disease, began to appear. Patients with prostatic carcinoma [29, 30], breast cancer [31], lung cancer [32], colon cancer [33], and various other malignancies were reported to have increased amounts of serum CK-BB. Table 4 illustrates the incidence of elevated levels in patients with a variety of malignancies. Comprehensive reviews regarding CK-BB activity levels in sera of cancer patients have been compiled by King [36] and Griffiths [37]. Although the incidence of elevated values among different laboratories vary considerably, these variations are probably related to the use of different antisera and different sample handling procedures. CK-BB is very sensitive to inactivation and special precautions with regard to sample handling are needed to minimize loss of activity prior to assay [38].

Few correlation studies with progression and remission of cancer have been carried out, but in two studies of a longitudinal nature, Thompson [31] and Carney [32] demonstrated good correlation between clinical response and CK-BB. However, CK-BB isoen-

Table 4 Serum BB-CK Levels in Various Types of Malignancies

Reference	Assay	Upper Limit of Normal	Malignancy	Incidence[a] (Elevated Levels): (%)
Mercer [33]	Column	1.5 U/L	Colon	7/12 (58)
Mercer [34]	Column	2.5 U/L	Breast	30/47 (64)
Thompson [31]	RIA	3.0 ng/ml	Breast	27/34 (79)
Feld [29]	RIA	6.2 µg/L	Prostate	11/46 (24)
Silverman [30]	RIA	10 µg/L	Prostate	15/17 (89)
Usui [35]	EIA	1.0 ng/ml	Lung (SCLC)	28/40 (79)

[a]All patients with late-stage disease.

zyme is not restricted to malignant disease since it has been detected in various nonmalignant disorders such as benign prostatic hypertrophy, brain trauma, and muscle disease. Benign conditions usually result in low and transient levels of CK-BB, which can be readily identified when serial studies and time course patterns are analyzed.

Recent reports [39] have suggested that a variant form of CK isoenzyme, macro-CK type 2 (M-CK2) might also serve as a tumor marker. This variant form, which is electrophoretically distinguishable from CK-BB, is thought to be of mitochondrial origin. Although M-CK2 has been shown to be a useful marker for monitoring disseminated malignancy, poor sensitivity has been demonstrated in early-stage malignancies [33].

In general both CK-BB and M-CK2 do not rate very highly as tumor markers since they appear to be insensitive to early-stage disease and very nonspecific. However, they do appear reliable enough to support a diagnosis of cancer and to monitor therapy. CK-BB especially deserves continuing study with more reliable methods that eliminate the possibility of CK-BB inactivation during the assay. Recent reports that CK-BB is transformed to CK-BB' in vivo [40] offer a possible explanation for the low levels of activity often found in early-stage cancer patients. Recent studies in my laboratory [34] have demonstrated markedly increased incidence levels (40%) in patients with active breast cancer when both CK-BB and CK-BB' activities were measured.

IV. NEURON-SPECIFIC ENOLASE

Neuron-specific enolase (NSE) is one of several isoenzymes of the enzyme enolase [EC: 4.2.1.11]. Enolase, which is present in all cells, is a glycolytic enzyme that catalyzes the conversion of 2-phosphoglycerate to phosphoenolypyruvate. Enolase isoenzymes are dimers of 100 Kd, formed from combinations of three subunits: alpha, beta, and gamma. The dimer of gamma, which was first isolated from brain tissue in 1972 [41], is found in elevated amounts (10-100-fold) in neural and neuroendocrine tissues [42].

Conventional biochemical assays are inadequate to detect nanogram levels of NSE found in serum. However, two sensitive immunochemical assays have been recently developed that provide investigators the opportunity to assess the role of NSE as a serum tumor marker.

To date, serum NSE measurements by immunoassay have been performed primarily in patients with neuroblastoma and small-cell carcinoma of the lung (SCLC). High incidence levels (approaching 90%) have been found in children with neuroblastoma [43]. As shown in Table 5, an excellent correlation exists between the incidence of NSE elevation

Table 5 Serum NSE Levels in SCLC

Reference	Assay	Upper Limit of Normal (ng/ml)	Incidence (Elevated Levels) and (%) Early	Incidence (Elevated Levels) and (%) Late
Esscher [43]	RIA	25.0	74/103 (72)	101/103 (98)
Cooper [44]	RIA	13.5	26/38 (68)	34/39 (87)
Savoia [45]	EIA	21.9		24/26 (54)
Carney [45]	RIA	12.0	15/38 (39)	49/56 (87)

and the extent of disease in patients with SCLC. Also of significance is the high incidence level (60% mean) observed in early-stage disease. To date, no other tumor marker in use today has exhibited incidence levels this high for early-stage disease.

Sequential measurement studies of NSE in patients (n = 23) receiving chemotherapy for SCLC also revealed good correlation between disease state and NSE concentration [48]. Thus, NSE appears to be an excellent indicator of tumor burden and to correlate closely with response to therapy, especially in children with neuroblastoma and in adults with SCLC.

With regard to specificity, NSE determinations in sera of patients with other types of lung cancer and nonlung malignancies have shown elevated levels, although these have been low in magnitude and incidence [49]. Also evaluated in a separate study was the presence of NSE in nonmalignant disease. Normal amounts of NSE were observed in a study of 138 patients [44]. Thus, NSE appears to be a relatively specific neuroendocrine marker, especially with respect to SCLC.

V. LACTATE DEHYDROGENASE ISOENZYME 1

Lactate dehydrogenase (LDH, EC:1.1.1.27) is a ubiquitous cytoplasmic enzyme that converts pyruvate to lactate in malignant tumors. LDH has a molecular weight of 140 Kd and its isoenzymes exist as tetramers composed of various combinations of two different subunits: H and M. Five isoenzymes are possible and LDH-5 is the prominent form in most malignant tissues.

Recently Liu and others [50-53], using an electrophretic LDH isoenzyme technique, found elevated amounts of LDH-1 in serum of patients with germ cell tumors of the testis. Abnormal levels of LDH-1 activity in 81% of patients with stage III germ cell malignancy of the testis were observed in both seminoma and nonseminoma cell types. The incidence rate in early-stage germ cell tumors was not evaluated in this study. However, this report did demonstrate increases in LDH-1 with progression of disease, normal LDH-1 levels in patients with no clinical evidence of disease, and decreases in LDH-1 levels reflecting the patients' response to therapy.

Although abnormal serum levels of LDH-1 are often associated with several nonmalignant conditions such as myocardial necrosis, renal infarction, and hemolytic anemia, the use of serial determinations should help to identify these benign conditions. For example, testicular tumors would give rise to increasingly elevated LDH-1 patterns whereas conditions such as acute myocardial infarction would produce a transient spike of high LDH-1 levels.

It therefore appears that LDH-1 measurements in patients with testicular cancer may be of clinical utility, especially in cases of seminoma cell type tumors, in which alpha fetoprotein (AFP) and human chorionic gonadotropin (HCG) levels are often normal. Also of important consideration in future studies of early-stage testicular cancer will be the use of a sensitive and quantitative immunoassay that is capable of distinguishing between normal and slightly elevated LDH-1.

VI. SALIVARY AMYLASE

Amylase [AMS, EC:3.2.1.1.] is a lysosomal enzyme that converts large polysaccharides into smaller components. AMS, with a molecular weight of 40-50 Kd, is found in rich amounts in the pancreas and salivary glands. Two major isoenzymes of AMS, P and S, are found predominantly in the pancreas and salivary gland, respectively. Unexplained elevations of total AMS levels are frequently encountered in sera of patients and measurements of P and S isoenzymes are utilized to assist in the identification of the tissue of origin.

The isoenzymes of amylase can be quantified by electrophoresis, column chromatography, selective inhibition, or immunoassay. A commercial inhibition kit (Pharmacia) that uses wheat germ protein as an inhibitor of S-AMS has been in use for several years [54]. Immunological assays utilizing the production of monoclonal antibodies prepared from highly purified S-amylase [55] will be available soon and these new assays will undoubtedly stimulate more widespread use of S-AMS as a tumor marker.

Hyperamylasemia has been previously reported most frequently in malignancies of the lung and ovary [56,57]. The type of isoenzyme found in sera of cancer patients with various tumors such as lung and ovary is usually the S-type, but unfortunately the incidence is very low (about 10%). However, in several cases in which pretreatment levels of S-AMS were elevated, subsequent sequential monitoring of S-AMS appeared to provide clinically significant data [58].

The presence of a variant, more acidic form of S-AMS has recently been demonstrated in cyst fluids from patients with serous ovarian tumors [59]. Work is currently in progress to evaluate the presence of the variant amylase in sera of patients with ovarian cancer. However, variant amylase may be a slightly modified form of S-AMS produced and found only within tumor cyst fluid.

VII. GALACTOSYLTRANSFERASE II

Galactosyltransferase II (GT II) is one of two isoenzymes of galactosyltransferase [GT, EC:2.4.1.22]. GT II is an enzyme with a molecular weight of 76 Kd that catalyzes the transfer of galactose from uridine diphosphate (UDP)-galactose to acceptors such as oligosaccharides of various glycoproteins. It is usually associated with cell membranes. The concept of GT as a tumor marker originated when it was detected in sera of patients with various types of cancer [60]. In 1975, a cancer-associated isoenzyme of GT, GT II, was found by Podolsky and Weiser to be a better marker than total GT [61]. Podolsky extended his results, confirming the presence of GT II in a wide range of cancer sera [62] and finding that it was an excellent marker (sensitivity 67%, specificity 98%) to differentiate between benign and malignant pancreatic disease.

Podolsky's initial studies utilized the complex technique of discontinuous polyacrylamide gel electrophoresis followed by radiochemical assay of gel slices. However, in more recent studies using different separation techniques such as ion-exchange chromatography [63] and isoelectric focusing [64], no evidence of a cancer-associated GT II was demonstrated.

Several investigators have recently attempted to improve the assay by purifying GT II and preparing monoclonal antibodies directed against GT II [65-67]. Uemura's group [67] appears to have prepared successfully a monoclonal antibody named 3872 with GT II specificity. An immunoassay for serum GT II level was developed and preliminary results show the new assay to be highly sensitive for ovarian (87%, n = 38) and pancreatic (77%, n = 13) cancer [68]. Continued use of highly sensitive and specific immunoassays is needed to establish GT II as a reliable tumor marker.

VIII. 5' NUCLEOTIDE PHOSPHODIESTERASE V

5'-Nucleotide phosphodiesterase [NPD: EC:3.1.4.1.] is an exonuclease that hydrolyzes DNA to 5' nucleotides. It is widely distributed, with the highest activities occurring in kidney, pancreas, liver, and colon. Tsou and co-workers have studied extensively the isoenzyme distribution of NPD in tissues and serum using electrophoretic separation and histochemical staining on polyacrylamide gels [69]. In sera of healthy donors, four isoenzymes could be detected and in sera of patients with primary hepatocellular carcinoma an additional fast-moving electrophoric fraction (NPD-V) was observed [70]. NPD-V was also detected (87%, n = 15) by Tsou [71] in lung cancer patients with liver metastasis. In another study by Tsou [72], NPD-V serum levels correctly predicted liver metastasis in eight of nine breast cancer patients.

Specificity studies are limited but evidence of elevated NPD-V levels in smokers (35%) and in patients with cirrhosis have been reported [69,71]. Although these reports by Tsou suggest that NPD-V may be useful in predicting liver metastases, confirmatory studies by other investigators are needed, along with a new and highly sensitive assay, to evaluate fully the potential of NPD-V as an early-stage marker of liver metastases.

IX. VARIANT BETA HEXOSAMINIDASE

Beta hexosaminidase [EC: 3.2.1.30] exists in most tissues as a lysosomal glycosidase with a molecular weight of about 150 Kd. The enzyme has two major isoenzymes, a thermolabile form (Hex A) with a pI near 5.0 and a thermostable form (Hex B) with a more neutral pI near 7.5. The presence of a variant thermolabile beta hexosaminidase with pI slightly less than Hex B has been found in metastatic tumor tissues of human liver and in sera of patients with liver metastasis [73,74]. Recent studies by Plucinsky have demonstrated the presence of variant beta hexosamidase in the sera of 36% (n = 36) of colorectal cancer patients [75] and in 80% (n = 108) of a large and mixed group of cancer patients [76]. In the latter report, the incidence rate for variant isoenzyme increased to nearly 100% in patients with advanced metastatic disease. Although control studies were limited, variant isoenzyme was detected in several patients with nonmalignant disease of the gastrointestinal tract.

To date, all studies by the Plucinsky group have utilized the technique of analytical isoelectric focusing, which is at best semiquantitative. Thus, a more sensitive and reliable quantitative method is needed to explore the possible use of variant isoenzyme, especially as a sensitive early-stage marker and as a marker for monitoring therapy. Clinical studies performed by other investigators are also needed to confirm the above results.

X. THYMIDINE KINASE 1

Thymidine kinase 1 is one of two isoenzymes of thymidine kinase [TK, EC:2.7.1.21]. TK plays an important role in DNA synthesis by catalyzing the phosphorylation of deoxythymidine (dT) to deoxythymidine monophosphate (dTMP). TK-1, a cytoplasmic isoenzyme, is present in high concentrations in dividing cells and absent in resting cells. TK-2, a mitochondrial isenzyme, is not related to cell growth and its levels remain relatively constant during the cell cycle.

Elevated amounts of TK-1 in serum have been demonstrated in malignancies associated with high rates of proliferation. Reports of elevations in hematological malignan-

Table 6 Author's Ratings of 10 Serum Isoenzymes as Tumor Markers

	Ideal Marker	PAP	PLAP	CK-BB	NSE	LD-1	AMS-S	GT-II	5-NPD-V	Variant Beta-HEX	TK-1
Screening (asymptomatic individuals)	5	0	0	0	0	0	0	0	0	0	0
Incidence of elevation (early-stage cancer)	5	1	0	1	3	NE	NE	NE	NE	NE	0
Incidence of elevation (late-stage cancer)	5	4	2	3	4	4	2	4	4	3	2
Normal levels in benign disease	5	4	4	2	4	2	4	2	2	2	2
Monitoring response to therapy and tumor recurrence	5	4	4	4	4	4	NE	NE	NE	NE	3
Quantitative and sensitive assay	5	5	5	2	5	2	2	2	2	2	4
Confirmatory reports	5	5	5	3	5	4	3	2	2	2	3
Total point score	35	23	20	15	25	16	11	10	10	9	14

NE, not evaluated; Rating scale: 5 highest possible score; for example, 100% sensitivity; 0 lowest possible score; for example, 0% sensitivity.

cies [77], brain [78], small-cell lung [79], and breast carcinomas [80] have recently appeared. Reports of elevated TK-1 levels in several nonmalignant conditions have also appeared [80,81]. Many additional studies are needed before TK-1's potential as a tumor marker can be fully evaluated.

A sensitive radioenzyme assay was recently introduced for the quantitation of TK-1. This assay measures the amount of radioactive substrate (ATP) incorporated into product (dTMP) during a 4 hr incubation. Availability of this assay and the development of immunoassays should help to stimulate more widespread clinical testing of TK-1.

XI. SUMMARY

The 10 isoenzyme markers discussed here represent those that in the author's judgment show promise as effective tumor markers. The relative usefulness of these isoenzymes as tumor markers is summarized in Table 6. Each isoenzyme is evaluated by a rating system, with a scale of 0-5 points in each of seven categories. The hypothetical ideal tumor marker received 5 points in all seven categories for a total score of 35. Unfortunately, less than perfect scores ranging from 9 to 26 were found for the 10 isoenzymes evaluated here.

The five best isoenzymes were neuron-specific enolase (26 points), prostatic acid phosphatase (23 points), placental alkaline phosphatase (20 points), thymidine kinase 1 (16 points), and lactate dehydrogenase 1 (16 points). In general, low isoenzyme scores can be attributed to the problems exhibited by all tumor markers: insensitivity to early-stage malignancies and false-positive elevations in nonmalignant diseases. Nevertheless, each of the 10 isoenzymes described here has potential clinical usefulness to support a diagnosis of cancer and/or to assist in the monitoring of therapy.

REFERENCES

1. Schwartz MK. Enzyme tests in cancer. In: Statland BE, Per Winkel, eds. Clinics in laboratory medicine. Philadelphia: W.B. Saunders, 1982;2:479-491.
2. Stefanini M, Enzymes, isoenzymes, and enzyme variants in the diagnosis of cancer. Cancer 1985;55:1931-1936.
3. Schwartz MK. Enzymes in cancer. Clin Chem 1983;19:10-22.
4. Gutman AB, Gutman EB. An acid phosphatase occurring in the serum of patients with metastasizing carcinoma of the prostate gland. J Clin Invest 1938;17:473-478.
5. Li CY, Chuda RA, Lam K, et al. Acid phosphatase in human plasma. J Lab Clin Med 1973;82:446-460.
6. Frenkel EP, Tourtelbotte CD. Elevated serum acid phosphatase associated with multiple myeloma. Arch Intern Med 1962;110;345-349.
7. Li CY, Yam LT, Lam KW. Acid phosphatase isoenzymes in human leukocytes in normal and pathologic conditions. J Histochem Cytochem 1970;18:473-481.
8. Schoenfeld MR. Serum acid phosphatase as an index of kidney disease. J Urol 1963; 90:373-379.
9. Davidson ED, McDougas WS. Elevated serum acid phosphatase levels with rectal carcinoid tumor. Gastroenterology 1976;70:114-116.
10. Oski FA, Naiman JL, Diamond LK. Use of the plasma acid phosphatase value in the differentiation of thrombocytopenic states. N Engl J Med 1963;268:1423-1431.
11. Schoenfeld MR. Serum acid phosphatase: a measure of the hepatic injury. Clin Res 1962;11:398.

12. Lam KW, Li O, Li CY, et al. Biochemical properties of human prostatic acid phosphatase. Clin Chem 1973;19:483-487.
13. Shulman S, Mamood L, Gondar MJ, et al. The detection of prostatic acid phosphatase by antibody reactions in gel diffusions. J Immunol 1964;93:474-480.
14. Foti AG, Cooper JF, Herschman H, et al. Detection of prostatic cancer by solid-phase radioimmunassay of serum prostatic acid phosphatase. N Engl J Med 1977;397:1357-1361.
15. Abbott Laboratories, Diagnostic Division, North Chicago, IL. Data on file. September, 1985.
16. Davies SN, Gochman N. Evaluation of a monoclonal antibody-based immunoradiometric assay for prostatic acid phosphatase. Am J Clin Pathol 1983;79:1-4.
17. Bruce AW, Mahan DE, Sullivan LD, et al. The significance of prostatic acid phosphatase in adenocarcinoma of the prostate. J Urol 1981;125:357-360.
18. Heller JE. Prostatic acid phosphatase: its current clinical status. J Urol 1987;137:1091-1103.
19. Kuriyama M, Wang MC, Lee CL, et al. Multiple marker evaluation in human prostate cancer with the use of tissue-specific antigens. J Natl Cancer Inst 1982;68:99-105.
20. Cohen P, Dix D. Routine screening for prostatic cancer by assay of serum acid phosphatase: a modest proposal. Clin Chem 1984;30:171.
21. Fishman WH. Clinical and biological significance of an isoenzyme tumor marker—PLAP. Clin Biochem 1987;20:387-392.
22. Moss DW. The role of multiple forms of alkaline phosphatase in cancer detection and monitoring. In: Cimino F, Birkmayer GD, Klavins JV, et al. Human tumor markers. Berlin: Walter de Gruyter, 1987:829-843.
23. Tucker DF, Ward AM. Monoclonal antibody assays for placental alkaline phosphatase. In: Kupchik HZ, ed. Cancer diagnosis in vitro using monoclonal antibodies. New York: Marcel Dekker, 1988;39:267-287.
24. Tucker DF, Oliver RTD, Travers P, et al. Serum marker potential of placental alkaline phosphatase-like activity in testicular germ cell tumors evaluated by H17E2 monoclonal antibody assay. Br J Cancer 1985;51:631-639.
25. McDicken IW, McLaughlin PJ, Tromans PM, et al. Detection of placental-type alkaline phosphatase in ovarian cancer. Br J Cancer 1985;52:59-64.
26. Voorde A, Groote GD, Waele PD, et al. Screening of sera and tumor extracts of cancer patients using a monoclonal antibody diverted against human placental alkaline phosphatase. Eur J Cancer Clin Oncol 1985;21:65-71.
27. Eerdekens MW, Nouwen EJ, Pollet DE, et al. Placentyl alkaline phosphatase and cancer antigen 125 in sera of patients with benign and malignant diseases. Clin Chem 1985;31:687-690.
28. Mercer DW. Separation of tissue and serum creatine kinase isoenzymes by ion-exchange column chromatography. Clin Chem 1974;20:36-40.
29. Feld RD, Van Steirteghem AC, Zweig MH, et al. The presence of creatine kinase BB isoenzyme in patients with prostatic cancer. Clin Chim Acta 1980;100:267-273.
30. Silverman LM, Dermer GB, Zweig MH, et al. Creatine kinase BB: a new tumor-associated marker. Clin Chem 1979;25:1432-1435.
31. Thompson RJ, Rubery ED, Hilary MJ. Radioimmunoassay of serum creatine kinase-BB as a tumor marker in breast cancer. Lancet 1980;2:673-675.
32. Carney DN, Zweig MH, Ihde DC, et al. Elevated serum creatine kinase BB levels in patients with small cell lung cancer. Cancer Res 1984;44:5399-5407.
33. Mercer DW, Talamo TS. Multiple markers of malignancy in sera of patients with colorectal carcinoma: preliminary clinical studies. Clin Chem 1985;31:1824-1828.

34. Mercer DW, Barry G, Losos F, et al. Multiple markers of malignancy in sera of patients with breast carcinoma. Clin Chem 1988;34:1297.
35. Usui A, Fujita K, Imaizumi M, et al. Determination of creatine kinase isoenzymes in sera and tissues of patients with various lung carcinoma. Clin Chim Acta 1987; 164:47-53.
36. King JS. Cancer-related changes in creatine kinase isoenzymes and variants. In: Cimino F, Birkmayer GD, Klavins JV, et al., eds. Human tumor markers. Berlin: Walter de Gruyter, 1987:857-868.
37. Griffiths JC. Creatine kinase isoenzyme 1. In: Statland BE, Per Winkel, MD, eds. Clinics in laboratory medicine. 1982;2:493-506.
38. Abbott LB, Lott JA. Reactivation of serum creatine kinase isoenzymes BB in patients with malignancies. Clin Chem 1984;30:1861-1863.
39. Koren IH, Freedman M, Miller D, et al. Macro-creatine kinase 2: a possible marker of gastrointestinal cancer? Surgery 1983;94:631-634.
40. Chastain SL, Ketchum CH, Grizzie WE. Stability and electrophoretic characteristics of creatine kinase BB extracted from human brain and intestine. Clin Chem 1988; 34:489-492.
41. Moore BW. Chemistry and biology of two proteins, S-100 and 14-3-2, specific to the nervous system. In: Pfeiffer CC, Smythias JR, eds. International review of neurobiology. New York: Academic Press, 1972;15:215-235.
42. Hullin DA, Brown K, Kynoch PAM, et al. Purification, radioimmunoassay, and distribution of human brain 14-3-2 protein (nervous system-specific enolase) in human tissue. Biochem Biophys Acta 1980;628:98-108.
43. Ishiguro Y, Kato K, Ito T, et al. Nervous system-specific enolase in serum as a marker for neuroblastoma. Pediatrics 1983;72:696-700.
44. Esscher T, Steinholtz L, Bergh J, et al. Neuron specific enolase: a useful diagnostic serum marker for small cell carcinoma of the lung. Thorax 1985;40:85-90.
45. Cooper EH, Splinter TAW, Brown DA, et al. Evaluation of a radioimmunoassay for neuron specific enolase in small-cell lung cancer. Br J Cancer 1985;52:333-398.
46. Savoia M, Schwartz D, Smith C, et al. NSE as a tumor marker. Clin Chem 1987; 33:928.
47. Carney DN, Marangos PJ, Ihde DC, et al. Serum neuron-specific enolase: a marker for disease extent and response to therapy of small-cell lung cancer. Lancet 1982; 1:583-585.
48. Splinter TAW, Cooper EH, Kho GS, et al. Neuron-specific enolase as a guide to the treatment of small cell lung cancer. Eur J Cancer Clin Oncol 1987;34:191-196.
49. Gerbitz KD, Summer J, Schumacher I, et al. Enolase isoenzymes as tumor markers. J Clin Chem Clin Biochem 1986;24:1009-1016.
50. Liu F, Fritsche HA, Trujillo JM, et al. Serum LD-1 in patients with advanced testicular cancer. Am J Clin Pathol 1982;78:178-183.
51. Von Eyben FE, Skude G. Lactate dehydrogenase (S-LDH) and its isoenzyme S-LDH-1, in serum are markers of testicular germ cell tumors. Clin Chem 1984;30:340-341.
52. Vladutiu AD. Serum lactate dehydrogenase isoenzymes 1 (LDH 1) in a patient with seminoma. Clin Chem 1983;29:1552-1553.
53. Lippert M, Papadopoulos N, Javadpour N. Role of lactate dehydrogenase isoenzymes in the testicular cancer. Urology 1981;18:50-53.
54. O'Donnell MD, Fitzgerald O, McGeeney KF. Differential serum amylase determination by use of an inhibitor, and design of a routine procedure. Clin Chem 1977;23: 560-566.
55. Mifflin TE, Benjamin DC, Bruns DE. Rapid quantitative, specific measurements of pancreatic amylase in serum with use of a monoclonal antibody. Clin Chem 1985; 31:1283-1288.

56. Amman RW, Berk JE, Fridhandler L, et al. Hyperamylasemia with carcinoma of the lung. Ann Intern Med 1973;78:521-525.
57. Hodes ME, Sisk CJ, Karn RC, et al. An amylase-producing serous cystadenocarcinoma of the ovary. Oncology 1985;42:242-247.
58. Yagi C, Mijata J, Hanoi J, et al. Hyperamylasemia associated with endometrial carcinoma of the ovary: case report and immunohistochemical study. Gynecol Oncol 1986;25:250-255.
59. Von Kley H, Cramer S, Bruns DE. Serous ovarian neoplastic amylase (SONA). A potentially useful marker for serous ovarian tumors. Cancer 1981;48:1444-1449.
60. Weiser MM, Wilson JR. Serum levels of glycosyltransferase and related glycoproteins as indicators of cancer; biological and clinical implications. Crit Rev Clin Lab Sci 1981;14:189-239.
61. Weiser MM, Podolsky DK, Isselbacher KJ. Cancer-associated isoenzyme of serum galactosyltransferase. Proc Natl Acad Sci 1976;73:1319-1322.
62. Podolsky DK, McPhee MS, Alpert E, et al. Galactosyltransferase isoenzyme II in the detection of pancreatic cancer: comparison with radiologic, endoscopic and serologic tests. N Engl J Med 1981;304:1313-1318.
63. Kim YD, Weber GF, Tomita JT. Galactosyltransferase variant in pleural effusion. Clin Chem 1982;28:1133-1136.
64. Davey RA, Rozelle MH, Cahill EJ, et al. Serum glactosyltransferase isoenzymes as markers for solid tumors in humans. Eur J Cancer Clin Oncol 1984;20:75-79.
65. Podolsky DK, Weiser MM. Purification of galactosyltransferase "isoenzymes" I and II. J Biol Chem 1979;254:3983-3990.
66. Ram BP, Munjan DD. Isolation and characterization of cancer-associated galactosyltransferase isoenzyme. Clin Chem 1984;30:1656-1663.
67. Uemura M, Winant RC, Sikie BI, et al. Characterization and immunoassay of human tumor-associated galactosyltransferase isoenzyme II. Cancer Res 1988;48:5325-5334.
68. Uemura M, Winant RC, Brandt AE. Immunoassay of serum galactosyltransferase isoenzyme II in cancer patients and control subjects. Cancer Res 1988;48:5335-5341.
69. Tsou KC, Lo KW. 5-nucleotide phosphodiesterase and liver cancer. Methods Cancer Res 1982;19:273-300.
70. Tsou KC, McCoy MG, Ledir S. 5'-Nucleotide phosphodiesterase isoenzyme pattern in the sera of human hepatoma patients. Cancer Res 1973;33:2215-2217.
71. Tsou KC, McCoy MG, Enterline HT, et al. 5'-nucleotide phosphodiesterase isoenzyme and hepatic cancer in NCI-Mayo Clinic panel sera. J Natl Cancer Inst 1973;51:2005-2006.
72. Tsou KC, Lo KW, Rosato EF. Evaluation of 5'-nucleotide phosphodiesterase isoenzyme-V as a predictor for liver metastasis in breast cancer patients. Cancer 1982;50;191-196.
73. Alhadeff JA, Holzinger RT. Presence of an atypical thermolabile species of beta-hexosaminidase β in metastatic-tumor tissue of human liver. Biochem J 1982;201:95-99.
74. Alhadeff JA, Prorok JJ, Dura PA. Atypical β-hexosaminidase in sera of cancer patients with liver metastases. Cancer Res 1984;44:5422-5426.
75. Plucinsky MC, Prorok JJ, Alhadeff JA. β-hexosaminidase from colon and sera of Dukes-classified colorectal cancer patients: activity levels, isoenzyme patterns, and kinetic properties. J Natl Cancer Inst 1986;77:57-62.
76. Plucinsky MC, Prorok JJ, Alhadeff JA. Variant serum β-heoxsaminidase as a biochemical marker of malignancy. Cancer 1986;58:1484-1487.

77. Vertongen F, Fondu P, VonDen Henle B, et al. Thymidine kinase and thymidine phosphorylase activities in various types of leukemia and lymphoma. Tumor Biol 1984;5:303-311.
78. Granowitz JS, Kolland CFR, Hazberg H, et al. Deoxythymidine-kinase in cerebrospinal fluid: a new potential "marker" for brain tumor. Acta Neurochir 1984;73:1-12.
79. Granowitz JS, Steinholtz L, Kallander CFR, et al. Serum deoxythymidine kinase in small-cell carcinoma of the lung. Cancer 1986;58:111-118.
80. McKenna PG, O'Neill KL, Abram WP, et al. Thymidine kinase activities in mononuclear leukocytes and serum from breast cancer patients. Br J Cancer 1988;57:619-622.
81. Hazberg H, Granowitz JS, Killander A, et al. Serum thymidine kinase in vitamin B12 deficiency. Scand J Haematol 1984;32:41-45.
82. Kallander CFR, Granowitz JS, Olding-Stenkvist E. Rapid diagnosis of varicella-zoster virus infection by detection of viral deoxythymidine kinase in serum and vesicle fluid. J Clin Microbiol 1983;17:280-287.

38
Mucin Glycoproteins as Tumor Markers

WILLIAM F. FELLER

Georgetown University School of Medicine, Washington, D.C.

JERRY G. HENSLEE, ROBERT J. KINDERS, GEORGE L. MANDERINO, JOSEPH T. TOMITA, and HARRY G. RITTENHOUSE*

Abbott Laboratories, North Chicago, Illinois

Historically, the term *mucin* has been used to designate biological polyanions on cells, which are stained by various dyes, including Alcian blue [1]. At the molecular level, mucins have been considered a mixture of mucoproteins and mucopolysaccharides [2,3]. More recent studies have characterized mucins on human cells and in fluids as high-molecular-weight glycoproteins, greater than 200 kDa in the dissociated form, and often greater than 1000 kDa aggregates; usually comprised of more than 40-50% carbohydrate; containing oligosaccharide groups predominantly in the O-glycosidic linkage with either serine or threonine; and having a peptide core rich in serine, threonine, glycine, alanine, and proline [3,4]. Exceptions have been found to these generalizations, for instance, the presence of both N-linked and O-linked oligosaccharide on a mucin containing the CA 19-9 epitope [5]. Mucins may also have different properties in plasma or tumor cells due to degradation or altered oligosaccharide synthesis. For example, the mucin molecule bearing the CA 125 epitope isolated from ovarian cancer cell lines [6] or ascites fluid [7] can contain less than 25% carbohydrate. The recent cloning and expression of the gene for the core peptide of mucin from a breast tumor cell line, the PUM locus [8], should lead to a better understanding of this complex glycoprotein family.

Many of the more clinically useful tumor markers identified over the last 10 years are mucins. In view of the extensive effort over many decades to find clinically useful tumor-associated antigens, it may seem somewhat surprising that mucin antigen assays were not developed earlier. Animal mucins share many biochemical properties with human mucins [2,9] and have been identified in animal tumors [10-12]. In a few cases tumor-associated monoclonal antibodies (MABs) to human mucins cross-react with animal mucins, for example, B72.3 reactivity with ovine submaxillary mucin [13]. Early assays may have measured human mucins indirectly, in part in the seromucoid fraction of blood, or the total sialic acid content in serum. An early example of a high-molecular-mass glycoprotein antigen (CSAp) detected on a human colon carcinoma by hamster antibodies was characterized as having the properties of a mucin [14].

Present affiliation: Specialty Laboratories, Inc., Santa Monica, California

The emergence of mucin tumor markers is largely due to the development of MABs. Tumor and cell line materials are often used to immunize mice and a small number of highly specific MABs were eventually produced. These MABs recognized tumor cells but not their normal cell counterparts or normal blood cells. MABs selected in this manner have been extensively studied by immunochemical and serological analysis. Detailed studies usually demonstrated a lack of absolute tumor specificity, but certain antigens detected by these MABs appeared to be tumor-associated for particular organs. An unusual feature of some of the early serum assays using these MABs was that the same antibody could be used both as a capture antibody on a solid phase and as a labeled probe, indicating the multiple epitope nature of the mucin antigens. Later work revealed that repeated sequences in both the oligosaccharide and peptide portions of these mucin antigens accounted for the multiplicity of epitopes.

CA 19-9, one of the first mucin tumor markers to be characterized, was shown to be a sialylated Lewis[a] (Le[a]) structure present on glycolipids of certain tumors [15,16]. The CA 19-9 immunoassay was initially presumed to measure serum glycolipid and hence was called a monosialoganglioside assay [17]. Later work demonstrated that the antigen measured in serum was actually a high-molecular-weight glycoprotein and that the CA 19-9 epitope is expressed on both glycolipids and glycoproteins [18].

Many MABs with tumor specificity have subsequently been shown to recognize mucin antigens and several immunoassays have been developed using MABs to detect mucin antigens in sera of cancer patients. Further studies have shown that the detection of circulating mucin often correlates with tumor burden or tumor recurrence in adenocarcinomas, especially in advanced disease. Increased levels of antigen measured by the assay may reflect higher concentration of mucin molecules and/or increased synthesis of the specific MAB-reactive epitope. In this chapter we describe the utility and present status, as well as potential future developments, of this most interesting glycoprotein family. For more detailed discussions of the biochemistry and immunology of these molecules, the reader is referred to reviews by Feizi [4,19] and Hakomori [20] and their colleagues.

I. IMMUNOLOGY OF MUCINS

Many classic tumor markers, such as carcinoembryonic antigen (CEA), alpha fetoprotein (AFP), and prostatic alkaline phosphatase (PAP) are N-linked glycoproteins. Clinically useful antibodies to these markers recognize peptide epitopes. It is not surprising that antibodies to N-linked carbohydrates are not clinically useful in detecting these antigens due to the limited variability in N-linked carbohydrate structures and their presence on normal serum glycoproteins. It is intriguing that many of the newly identified tumor-associated antigens of interest in clinical oncology are carbohydrate epitopes on glycolipids and O-linked glycoproteins [19-22]. Many differentiation antigens present on cell surfaces are glycolipids or O-linked glycoproteins. Since differentiation antigens undergo both qualitative and quantitative alterations in various stages of differentiation and oncogenesis [23-26], and since the O-glycosidic linkage is not commonly found on normal serum glycoproteins, the importance of O-linked carbohydrate epitopes should not be surprising.

Unlike the classic tumor markers, elevated serum levels of which are due to tumor cell proliferation and the expression of the genes encoding these polypeptides, several possible mechanisms result in altered carbohydrate antigens in mucins. Mechanisms that alter carbohydrate structure, density, or organization on mucins can result in the generation of tumor markers even without enhanced production of the peptide core. A glycoprotein may contain many antigenic epitopes, each consisting of up to seven monosaccharides and/or amino acids in each epitope [27]. Even the addition of a single sugar residue can alter a carbohydrate's antigenic activity [15]. Since subtle changes in a complex carbohydrate structure can lead to antigenic differences in only some of the carbohydrate chains, MABs that detect these differences are very useful. In contrast, polyclonal antibodies are not as specific unless affinity-purified since they also detect other carbohydrate epitopes, notably blood group substances, present on normal as well as cancer cells.

The biosynthesis of O-linked carbohydrate determinants occurs by the sequential addition of single sugars to a growing precursor chain, each addition being made from a nucleotide and catalyzed by a specific glycosyltransferase [21,22]. Depending on the relative activities of glycosyltransferases, the availability of their donor and acceptor substrates, the proportion of various oligosaccharide structures, and other subcellular variables, the antigenicity of glycoproteins and glycolipids will vary. Conversely, incomplete biosynthesis or the actions of glycosidases may reveal cryptic antigens that can disappear or reappear at various stages of differentiation or cell transformation. Although an antigen may be formed by each new carbohydrate substitution, the antigenic activity of the underlying structure might be masked [28-30]. For example, the decreased expression of blood group (ABH) antigens in tumors of some cancer patients is due to either incomplete synthesis [30] or masking by sialic acid [31]. Another example of this phenomenon is the interchangeability of the I/i antigen. Erythrocytes of neonates are enriched for the i antigen. During the first year of life, the i antigen diminishes and is replaced by the I antigen, which persists through adulthood. The i antigen consists of an unbranched chain of three or four repeating N-acetyl lactosamines [32] that is converted to the I antigen when an N-acetyl lactosamine branch is attached to an internal galactose residue [33,34]. The i antigen cannot be detected on monocytes, but is revealed after treatment of the cells with neuraminidase [35].

Conversely, the reappearance of the i antigen in certain cancers of adults [36-38] is evidence that aberrant glycosylation in cancer cells can result from re-expression of carbohydrate synthesis characteristic of a certain stage of fetal development. Other well-defined examples of rapid alterations in glycoprotein and glycolipid antigenicity during normal differentiation include modification of type 2 chains responsible for the appearance of X-hapten (Le^x), SSEA-1 (stage-specific mouse embryonic antigen) [38,39], and transition of globo-series antigen expression through P(k), P, and Forssman antigens [40].

Despite our increasing knowledge of aberrant glycosylation in tumor cells, little is known about the biological significance of these changes. One can speculate that these subtle changes affect metastatic potential, cell recognition or adherence, and that the aberrations correlate with degree of malignancy. Many of the newly developed MABs against mucins associated with human carcinomas are found to be directed to carbohydrate sequences expressed by some normal epithelia but more strongly expressed on glycoproteins and/or glycolipids of carcinomas. A good example is the sialylated blood group Le^a defined by the CA 19-9 MAB [15]. This epitope is a normal component of certain tissues of some individuals [15,41,42] and is expressed strongly on glycolipids and glycoproteins extracted from adenocarcinomas of the colon and pancreas but not from normal colon [15,23]. It has been detected in normal pancreatic glycolipids [42] and normal seminal fluid glycoprotiens [43]. It has been determined that both the sialic acid and the fucose on sialylated lacto-N-fucopentaose II are essential components of the CA 19-9 epitope [15]. Other studies suggest that another glycolipid structure, paragloboside, may be present as a tumor-associated component of serum mucins [7,44].

Another family of mucin molecules of interest in cancer immunology is the human milk epithelial antigens recognized by various antibodies generated against human milk fat globule (HMFG), breast carcinoma cell lines, and fractions of metastatic breast cell membranes [45-50]. Both polyclonal and monoclonal antibodies have been described. Some HMFG antigenic epitopes are expressed on lactating human mammary cells, breast carcinoma, ovarian carcinoma, and other carcinomas, but not tumors from mesenchymal tissues [45,51]. Similarly to the colorectal tumor-associated antigen CA 19-9, the HMFG antigens are carried on mucins in the serum [47]. Unlike CA 19-9, some HMFG antigens are associated only with glycoproteins in tumors and are not present on glycolipids [47]. The 115D8 MAB, which is also generated from mice immunized with HMFG, reacts with a mucin termed MAM-6 [52]. This antigen has been found in many carcinomas as well as a few normal tissues including lung epithelium [52,53]. Another MAB, F36/22, generated against a breast tumor cell line [54] reacts with a mucin expressed on adenocarcinomas of the breast and ovaries.

It should be noted that none of the above antibodies are organ-specific. These antibodies recognize epitopes on mucins commonly found in a variety of epithelial tissues and adenocarcinomas. A MAB generated to deglycosylated HMFG immunogen [55], the SM-3 MAB, is a notable exception since it shows very high specificity for cancerous but not normal breast tissue. This MAB does not bind to native mucin in milk, indicating the presence of mucin core peptide in an incompletely glycosylated form in breast cancers.

The OC 125 MAB generated against a human serous cystadenocarcinoma cell line OCVA 433 [56] reacts with the following: nonmucinous ovarian epithelial tumors [56,57]; human milk [58]; the central airway and normal lung tissue [59]; human seminal fluid [6]; normal cervical mucus [60]; coelomic epithelium during embryonic development [61]; and in fetal pleura, pericardium, peritoneum, mullerian duct, and amnion [57]. The antigen is not detected in normal ovary in the fetus or adults [56,57,60]. Despite its presence in human milk, the CA 125 determinant appears to be quite distinct from the HMFG family of antigens both chemically and clinically. Although the antigen is a high-molecular-weight glycoprotein, it does not appear to be a typical mucin [6]. The nature of the CA 125 epitope is not completely understood, as discussed in the next section.

Monoclonal antibody B72.3 was generated against a membrane-enriched fraction from a mammary carcinoma metastasis [62]. It reacts strongly to approximately 50% of breast carcinomas [63], to a majority of colon and ovarian carcinomas [64], and minimally to normal adult liver, spleen, heart, breast, uterus, and other tissues [64]. The antigen detected by B72.3, designated TAG-72, is a high-molecular-weight glycoprotein [65].

Substantial work has been done on glycolipids to study aberrant glycosylation [66] since they can be purified to homogeneity and their structures elucidated. Mucins, in contrast, are extremely heterogeneous and difficult to purify to homogeneity. Elucidation of their carbohydrate structures is more difficult than that of glycolipids, and automated methods used to delineate protein or DNA epitopes are not applicable. MABs reactive with carbohydrate epitopes on mucins have allowed more rapid characterization of these complex glycoproteins.

A key problem in characterizing MABs directed against mucins is to establish whether the epitope is carbohydrate, peptide, or a combination. Examples of each case are discussed in the following section.

Abolition of immunoreactivity of mucins following treatment with proteases does not necessarily prove antipeptide specificity of the antibody in question. Proteolysis may result in destruction of conformation or reduced density of carbohydrate epitopes, thus reducing antibody binding. Conservation of antigenic activity following heat treatment (boiling for up to 1 hr) is usually indicative of a carbohydrate epitope, yet there are examples of peptide determinants capable of surviving boiling, including B-galactosidase-binding protein [67] and CEA [68]. Susceptibility to mild periodate oxidation is another hallmark of carbohydrate epitopes. However, not all carbohydrate is destroyed by mild oxidation [69,70] and residual carbohydrate chains can confuse the interpretation of the results. Exo- and endoglycosidases may be used, but such methods do not always yield the complete digestion of carbohydrate epitopes.

Inhibitors of N-glycosylation, such as tunicamycin, have been used in mucin characterization demonstrating the O-linked nature of the carbohydrate. Chemical deglycosylation may be more effective in destroying carbohydrate epitopes but complete deglycosylation is not readily achieved [71,72].

II. BIOCHEMISTRY OF TUMOR-ASSOCIATED MUCIN ANTIGENS

Mucin tumor markers have been shown to be high-molecular-weight glycoproteins (Table 1), often in excess of 1×10^6 Da under nonreducing, nondenaturing conditions [7,18, 41,70,73-80]. When analyzed in the presence of reducing and denaturing agents, these antigens are often observed as lower-molecular-weight species. For example, the CA 125 epitope has been found on glycoproteins ranging from approximately 200 to 600 kDa [6]. The majority of CA 125 activity [6] was found in the 200 kDa molecular mass material after being heated in 6 M urea followed by molecular size exclusion chromatography in the presence of sodium dodecyl sulfate (SDS) and urea [6]. In the presence of SDS and dithrothreitol (DTT), CA 19-9 activity was detected in molecular sizes ranging from 200 to greater than 2000 kDa, but after treatment with the harsher dissociating conditions of DTT and 8 M urea, all CA 19-9 activity was recovered in a 210 kDa molecular mass species [5]. The sources of CA 19-9 and CA 125 antigens showing predominance of activity in 200 kDa species were tumor cell lines. SDS-polyacrylamide gel electrophoresis (PAGE) analysis of DF3 antigen purified from malignant effusions demonstrated two distinct molecular species of 350 and 450 kDa [74]. CA 549 antigen, purified from the T-417 cell line, was found in two species of different molecular weights, 400 and 500 kDa, when subjected to SDS-PAGE under reducing conditions [78]. Affinity-purified CA1 antigen appeared as two components, 350 and 390 kDa, in SDS-PAGE [77].

TAG-72, identified by MAB B72.3, was initially extracted from LS174T xenografts under gentle conditions to minimize shearing of the antigen and then subjected to further purification [65]. On SDS-PAGE or molecular size exclusion chromatography, in the presence of 6 M guanidine HCl, TAG-72 antigen was found only as a very-high-molecular-weight species ($>1 \times 10^6$ Da). The lower-molecular-mass TAG-72 antigen observed in earlier studies was apparently due to shearing of the native antigen. Different sources of the mucin tumor markers studied (e.g., xenograft tissue, secreted mucins from cell lines, patient sera, or tumor tissue) may be responsible for the different biochemical properties reported for these mucins [6,81].

The mucin core protein from milk fat globule membrane containing the HMFG-1 and HMFG-2 MAB-defined epitopes was obtained by subjecting the milk mucin to hydrogen

Table 1 Clinically Useful Mucin Tumor Markers

Antigen	Source[a]	MW of Mucin (kDa)	AB[b]:Epitope	Other Properties
HMFG [49,51,55,96] PAS-0 [45]	HMFG membrane	400	HMFG-1:peptide; HMFG-2:peptide	50% CHO; gene contains 60 bp repeats
Deglycosylated HMFG mucin [55,96]	HMFG membrane	68	SM-3:peptide	Epitope also encoded on mRNA from MCF-7 cells
NCRC-11[c] antigen [102,167]	Breast carcinoma	400	NCRC-11:CHO	Also binds HMFG-1, HMFG-2, and 115D8 MABs
DF3 antigen [97,74]	Breast carcinoma membrane	330–450	DF3:sialic acid dependent with possible peptide involvement	Gene contains 60 bp repeats
MAM-6 [90,180]	HMFG membrane	400	115D8 MAB	May contain N-linked CHO; gene contains 120 bp repeats
DCA [84,86]	Malignant effusion from breast or ovarian adenocarcinomas	400	F36/22:sialic acid independent with CHO and possible peptide involvement	Contains no blood group activity; 1.42 g/ml CsCl
CTA [131]	Ascites from ovarian cancer patients	High MW glycoprotein	85/34:i/I	MAB reactive to O-linkages; N-linkages tested were unreactive
CA 549 antigen [78]	T-417 cell line	900	BC4E549:sialic acid independent, periodate insensitive	Binds WGA, PNA, and RCA but not Con A
EMA [50]	Skimmed human milk	50–1000	Goat, rabbit: heat and protease stable	Gal and GlcNAc predominant CHO; binds PNA, WGA, but not ConA
CA 19-9 antigen [5,15]	SW 1116 cell line	200–800, 900 and 2000	19-9:SiLe[a]	85% CHO; may contain N-linked CHO; 35% Ser, Thr, Pro
DU-PAN-2 antigen [73,82]	HPAF cell line, ascites from pancreatic cancer patient	1000	DU-PAN-2:sialic acid dependent and possible peptide involvement	80% CHO; 60% Ser, Thr, Pro, Gly, Ala; MAB not reactive to glycolipid

CA 50 antigen [183]	COLO 205 cell line	High MW glycoprotein	C50:SiLe^a, Si-lactotetraose	Antigen expression independent of Le gene
Span-1 antigen [93,94]	SW 1990 cell line	660	Span-1:sialic acid dependent and abolished by periodate	1.40-1.45 g/ml CsCl; Span-1 MAB competes with 19-9 MAB for epitope but Span-1 antigen present in Le^{a,b} tissue
CA 125 antigen [6]	OVCA 433 cell line	200-400 and 1000	OC125:peptide and possible CHO involvement	24% CHO; may contain N-linked CHO; 1.36-1.46 g/ml CsCl
MOv1 and MOv2 antigens [80;101]	Mucinous ovarian cystadenocarcinoma	20,000	MOv1:CHO; MOv2:Le^a	MOv1 antigen is 1.45 g/ml CsCl and not present on glycolipid; MOv2 antigen is 1.52 g/ml CsCl on mucin and glycolipid
COTA [79,155,156]	Colon tumor, mucinous ovarian adenocarcinoma	3,000-15,000	Goat: heat-stable; SP21 MAB: CHO- and sialic-acid dependent	Distinct from CEA, CSAp, and 19-9
TAG-72 [13,65]	LS174T cell line	1000	B72.3:Si2-6GalNAcl-O-Ser, Thr with possible peptide involvement	Contains Le^a, Le^b, H type 2, SiLe^x, and SiLe^a; B72.3 not reactive to glycolipid
CSAp [75,85]	GW-39 cell line	4000	Goat: destroyed by DTT; Mu-1: conformationally dependent; Mu-2, Mu-4: sialic acid dependent; Mu-3	Contains Le^a and Le^b activities
CMA [89,157,158]	Human colonic mucosa and colonic tumors	15,000	Rabbit: contains a colon-specific determinant	68-86% CHO; 48% Ser, Thr, Pro, Gly, Ala
CA antigen [77,99]	Hep2 and urothelium	670	CA1: sialic-acid-dependent and possible peptide involvement	50% Ser, Thr, Pro, Gly, Ala

[a] Indicates cell lines/tissues used to characterize MABs immunologically and biochemically.
[b] AB is murine MAB or, where noted, polyclonal antibody from goat or rabbit.
[c] Antigen designation indicates a MAB-defined antigen.

fluoride treatment [55]. The native mucin molecule is larger than 400 kDa, while the dominant polypeptide species after complete deglycosylation is 68 kDa, which demonstrates the very significant contribution of carbohydrate to the mucin's molecular mass.

Enzymes and lectins have been used to characterize mucin antigen. Chondroitinase ABC, hyaluronidase, and/or chondro-4-sulfatase digestions did not affect relevant MAB binding to TAG-72 [65], DF3 [74], and DU-PAN-2 [82] antigens, showing that these determinants are not carried on proteoglycan type oligosaccharides. These properties differentiate them from tumor antigens such as the high-molecular-weight melanoma antigen [83]. Lectin binding studies on DF3 [74], CA 549 [78], PAS-O [45], and F36/22 [54,84] antigens showed that each preferentially bound to lectins having galactose, N-acetylglucosamine, and/or galactose β-(1 $\rightarrow$ 3) N-acetylgalactosamine specificities, whereas binding to concanavalin A (ConA) lectin, which has N-linked mannose binding specificity, was minimal. These lectin studies reflect the great predominance of O-linkages over N-linkages or the absence of N-linkages in the antigens, which is a key characteristic of mucins.

Mucins also commonly contain blood group activities, for example, TAG-72 [65], CSAp [85], and CA 19-9 antigens [18]. Blood group substance activities, however, were not detected on the DU-PAN-2 antigen [73] nor on the F36/22 antigen [86].

DU-PAN-2 [73,81], TAG-72 [65], CA 19-9 [18,76], MOv1 [80], and F36/22 [84] were all shown to have buoyant densities of approximately 1.45 g/ml in cesium chloride, which is characteristic of mucins [87]. The degree of antigen purity was observed to influence the buoyant density obtained for the CA 125 antigen from the OVCA 433 cell line [6]. Prior to immunoaffinity purification, the CA 125 antigen had a buoyant density of 1.42 g/ml whereas a density of 1.36 g/ml was obtained after immunoaffinity purification. The CA 125 antigen, therefore, has a lower buoyant density than is typically described for mucins. A possible exception is the CA 125 preparation from human milk [58], which had a buoyant density of 1.41 g/ml.

CA 125 antigen isolated from the OVCA 433 cell line [6] had other characteristics not typical of mucins. It was found to contain 24% carbohydrate by mass, whereas other mucinlike antigens contain a greater carbohydrate content typical of mucins: 77.4% and 80.8% for DU-PAN-2 [73], 68% for a mucin glycoprotein from human milk described as having both CA 19-9 and CA 125 activities [58], at least 50% for milk mucin or PAS-O containing HMFG-1 activity [45,55], 85% for CA 19-9 containing glycoprotein [5], and 70% for ASGP-1, the major cell surface mucin of a rat mammary adenocarcinoma cell line [88].

The carbohydrate compositions for several tumor-associated mucins have been reported. Galactose and N-acetylglucosamine were the predominant sugar residues in these mucins making up more than 50% of the total carbohydrate of DU-PAN-2 [73], CA 125 [6], CA 19-9 [5], and PAS-O antigens [45,55]. In the CMA (colonic mucoprotein antigen) mucin, a component of colonic mucosa [89], a somewhat lower galactose and N-acetylglucosamine content (45%) was found. The detection of mannose in the CA 125 [6] and CA 19-9 [5] antigens suggests the presence of some N-glycosidically linked structures in certain mucins. Another mucin, MAM-6, has also been reported to contain N-linked structures [90]. Definitive conclusions regarding the presence of N-linked oligosaccharides on mucins are difficult because of the possibility of trace contamination of mucin preparations by N-linked glycoproteins.

The O-glycosidic linkage of carbohydrate to protein is between the hydroxyl groups of serine or threonine and the reducing ends of N-acetylgalactosamine [41]. The high content of O-linked carbohydrate structures in mucins parallels a relatively high content of serine and threonine amino acid residues. Serine and threonine together form 26% of the total amino acid residues in DU-PAN-2 [73,91], 28.5% in CA 19-9 [5], 20.6% in CA1 [77], 18.5% in HMFG-1 purified milk mucin [55], 33.6% in ASGP-1 [88], and 23.0% for tumor CMA [89]. It is not presently known how many of the serine and/or threonine residues are linked to carbohydrate. Other amino acid residues that are also abundant in mucin glycoproteins include proline, alanine, and glycine. The amino acid profile of the CA 19-9 antigen is different from typical mucins due to the absence of cysteine and the relatively high amount of acidic amino acids and their amide derivatives [5].

As mentioned earlier, one of the first objectives in characterizing a MAB-defined epitope on mucin molecules is to establish whether the epitope is carbohydrate or peptide in nature. Some epitopes may demonstrate characteristics of both. The CA 19-9 MAB-defined epitope structure is the blood-group-related sialylated Le^a carbohydrate [17]. Neuraminidase or fucosidase treatment of CA 19-9 results in complete loss of antigen activity, whereas extensive pronase treatment results in CA 19-9 activity residing in much smaller glycopeptides [5].

A related MAB, CA 50, binds to sialylated Le^a but exhibits broader specificity than the CA 19-9 MAB [92]. If the fucose residue in the sialylated Le^a structure is absent, the CA 50 MAB still binds whereas the CA 19-9 MAB requires the fucose residue in the epitope. Consequently, CA 50 expression can occur in Le^{a-b-} individuals, in contrast to the CA 19-9 antigen that cannot. Span-1, another MAB related to CA 19-9, has also been described [93]. The epitopes recognized by CA 19-9 and Span-1 MABs are probably structurally related because these MABs compete strongly with each other in antigen-binding experiments and both require sialic acid [94]. Studies have demonstrated a broader specificity for Span-1 than for CA 19-9, since Span-1 MAB specifically binds to tissue from Le^{a-b-} individuals [93].

Certain lectins were found to compete with the HMFG-1 and HMFG-2 MABs in binding milk mucin antigen, which suggested carbohydrate epitopes [51]. Subsequent studies, however, have provided strong evidence that both epitopes are at least partially peptide [55,95,96]. Due to the high carbohydrate density of mucins, lectins can bind multiple sites. These lectins may bind MAB-defined carbohydrate epitopes or sterically prevent MAB binding to polypeptide epitopes on the mucin core. Both HMFG MABs bind strongly to deglycosylated milk mucin, the mucin core protein [55], and precipitated protein from in vivo translated mRNA [95]. It is interesting that the SM-3 MAB, generated against deglycosylated milk mucin (and thus an amino-acid-defined epitope), appears to be tumor-associated [55]. In related work, the milk core protein was found to be encoded by 60 base pair (bp) tandem repeats, which may explain the observed polymorphism of milk mucin [96]. cDNA encoding the DF3 epitope also has recently been isolated, indicating peptide involvement in the epitope [97]. A 60 bp tandem repeat has been identified, which agrees with polymorphism observed with the DF3-containing mucin. It has been reported, however, that the DF3 epitope requires sialic acid [98], indicating carbohydrate involvement in the integrity of the epitope. Neuraminidase-sensitivity of the DF3 epitope has been confirmed by others (Caminiti, W. et al., Abbott Laboratories, 1988, unpublished observations). Repeated sequences may be a common theme of mucins, since another mucin gene encoding the MAM-6 mucin has been reported to contain a repeated 120 bp sequence [90].

The antigenic activity of CA 125 is dependent on the conformational integrity of the mucin's peptide core. The CA 125 epitope is completly destroyed by protease treatment [6]. In contrast, exoglycosidase treatment increases antigen activity, which may be due to peptide-dependent binding sites. Reduction and alkylation of the mucin also reduced CA 125 MAB binding whereas periodate oxidation had no effect [6]. Although peptide is known to be involved in the CA 125 epitope, the role of carbohydrate in the conformational integrity of the structure of this epitope remains undetermined.

Several MAB-defined tumor-associated epitopes on mucins demonstrate both carbohydrate and protein characteristics. Proteolytic treatment of mucins decreased antigenic activities defined by the MABs B72.3 [65], DU-PAN-2 [73], and CA1 [77,99], suggesting that the conformational integrity of these epitopes is dependent on peptide. In each case, however, neuraminidase treatment of the mucin also significantly decreased antigenic activity, demonstrating that sialic acid is either a component of these epitopes or required for conformational integrity. Consistent with peptide involvement in the epitope, glycolipids containing B72.3, DU-PAN-2, or CA1 epitopes have not been detected.

Immunochemical studies of the B72.3 MAB have recently revealed the carbohydrate structure of its epitope [13]. Inhibition experiments with defined oligosaccharide conjugates and MABs with known specificities showed that B72.3 recognized Neu Acα2 $\rightarrow$ 6GalNAcα1 $\rightarrow$ O-Ser/Thr [13]. B72.3 therefore, is a good example of an epitope involving both carbohydrate and peptide.

In some cases neuraminidase treatment of suspected carbohydrate epitope had no effect on antigenic activity; MOv2 [80], CA 549 [78], and W5 [100] antigens were all significantly reduced in activity by periodate oxidation but not by neuraminidase treatment. Thus, these carbohydrate epitopes either do not contain sialic acid or sialic acid is present but inaccessible to cleavage by the enzyme. The MOv2 MAB was recently shown to recognize the Le^a-active hapten, which does not contain sialic acid [101]. Neuraminidase treatment of the F36/22 antigen also failed to perturb MAB binding. This epitope may involve carbohydrate because extensive protease treatment of the F36/22 antigen produced monovalent glycoproteins carrying single F36/22 MAB-combining sites [86]. W1 and W9 [100], NCRC-11 [102], COTA [79], and Mu-2 and Mu-4 [85] are other MAB-defined tumor-associated structures found on mucins that may involve carbohydrate.

Studies have been performed to identify characteristics shared by tumor-associated epitopes found on mucins. In one study in which coexpression of DU-PAN-2, CA 125, TAG-72, and MOv2 was investigated, only DU-PAN-2 and CA 19-9 were coexpressed on the same mucin molecule [81]. Since MABs to none of the above epitopes crossreact, it was confirmed that distinct epitopes were recognized by these MABs. In another study, MABs reactive against CA 19-9 and CA 125 were shown not to bind to ductal carcinoma antigen, which contains the F36/22 MAB-defined epitope [86], demonstrating that the three epitopes are immunologically distinct and not coexpressed on the ductal tumor antigen. It should be emphasized that mucins from some sources, for example, cancer patients' ascites fluids, may coexpress epitopes not similarly coexpressed on mucins from other sources. In a third study, mucin purified by NCRC-11 MAB immunoaffinity chromatography was shown to be bound by HMFG-1 and HMFG-2 MABs in addition to the NCRC-11 MAB, indicating that all three antigens are present on this mucin. In addition, results from competitive inhibition assays suggest that the NCRC-11 MAB-defined epitope is topographically linked to the HMFG-1 and HMFG-2-defined epitopes [102].

III. CLINICAL APPLICATION OF MUCIN MONOCLONAL ANTIBODIES

The clinical use of tumor markers is organized into four areas: monitoring patients with epithelial-type cancer after initial treatment; early cancer detection (i.e., screening of asymptomatic individuals); a diagnostic aid in symptomatic patients suspected of having cancer; and histological studies to determine organ source of metastatic disease in cases where standard pathological analysis cannot determine the primary organ source. The most useful clinical aspect of mucin markers at this time is monitoring malignant disease after primary treatment. Although none of the mucin markers detected by monoclonal antibody assays are specific for a given cancer, the markers are generally grouped according to their primary clinical utility for a specific organ (Table 2). Recent clinical studies covering the use of mucin markers have focused on the following major epithelial cancers: pancreas, breast, ovary, lung, and colon.

Initial clinical studies with immunoassays have been aimed at establishing their sensitivity using the sera of patients with metastatic or advanced disease. Sensitivities of 70-80% in patients with advanced disease have been often achieved. Specificity of mucin immunoassays can be determined from studies of seropositivity of normal healthy controls or studies of patients with benign disease. In general, a useful immunoassay should have less than 2% seropositivity for healthy controls and less than 8-10% for benign disease of the same organ. Some mucin immunoassays use cutoff values to yield less than 1% positivity for healthy controls and less than 5% seropositivity for patients with benign disease.

A. Pancreatic Cancer

Four mucin antigens have demonstrated clinical utility in the diagnosis of pancreatic cancer: CA 19-9, CA 50, DU-PAN-2, and Span-1. Since most pancreatic cancers are generally diagnosed late in the clinical course of disease, clinical studies rely on sera from patients with biopsy-proven pancreatic cancer.

CA 19-9, also termed gastrointestinal cancer-associated antigen (GICA), is found on glycolipid and mucin components of tumor cell membranes and on mucin molecules in sera of patients with pancreatic and gastrointestinal cancer. Several studies [103-109] utilizing a variety of immunoassays have shown that 70-80% of patients with biopsy-verified pancreatic cancer have abnormally elevated CA 19-9 activity in their sera. Additional studies have shown that 50-70% of patients with either hepatocellular [103,110] or advanced colorectal cancer [105,111] have elevated levels of CA 19-9 activity in their sera. Patients with the Le^{a-b-} phenotype are genetically incapable of expressing the CA 19-9 antigen and are never seropositive in the CA 19-9 assay [112,113]. Lewis-negative individuals represent approximately 7% of the total U.S. population [114]. In Japan this number is much higher, approximating 20% [114].

Two clinical studies on CA 50, a mucin marker closely related to CA 19-9, have shown that 71% and 83% of patients with pancreatic cancer had elevated serum levels of CA 50. With the use of both the CA 50 and CA 19-9 immunoassays, 95% of patients with pancreatic cancer tested seropositive with one or both antigens [115].

DU-PAN-2 is another mucin marker with elevated levels in the sera of patients with pancreatic cancer. It is antigenically distinct from CA 19-9 and its appearance in sera is not restricted to the Lewis genotype since Le^{a-b-} patients secrete high levels of the marker

Table 2 Serological/Tissue Assays for Mucins

Assay Antibody	Isotype	References
Breast		
115D8	IgG2b	52, 159
DF3	IgG1	121
CA 15-3 (115D8/DF3)		122-128, 160
67D11		161, 162
F36/22	IgG3	54
CTA (F36/22, M85/34)	IgG3/IgM	130
W1		190
M29/M38	IgG1	129
M26/M38	IgM/IgG1	129
MSA (3E1.2)	IgM	117, 163, 181, 182
CA 549		118, 119
HMFG-1	IgG1	47, 120
HMFG-2	IgG1	47, 120
CSLEX-1[a]	IgM	164-166
HME	Rabbit	48
NCRC-11	IgM	167
A-80	IgM	168
MA6/MA9		169
SM3	IgG1	55
Ovarian		
CA 125	IgG1	132-142
MOv2/MOv8	IgM/IgG2a	170
OM-1	IgM	171
Pancreatic		
CA 19-9	IgG1	103-109
CA 50	IgM	115, 172
DU-PAN-2	IgM	173, 174
Span-1	IgM	93, 116
Colon		
CA 195		152, 153
CA 72 (B72.3)	IgG1	151
CSAp	Goat	14, 75
1D3	IgG1	175
Lung		
FH6	IgM	176, 177
CSLEA1	IgG	143
CSLEX1	IgM	178, 143
Miscellaneous		
CA1	IgM	99

[a] Also pancreas.

[107]. A recent clinical study of both cancer tissue and sera from pancreatic cancer patients revealed that 20 of 22 (91%) of the patients expressed the antigen in cancer tissue, including Le^{a-b-} patients. Elevated levels of DU-PAN-2 were found in the sera of 14 of 21 (67%) of patients with pancreatic cancer including all of the Le^{a-b-} patients. These serological results were compared to those obtained by the CA 19-9 assay, in which 16 of 21 (73%) were positive. When the two markers were used together, 95% of all the patients were seropositive [107].

Another mucin marker, Span-1, detected with a MAB generated against the human pancreatic cell line SW 1990, has been found to be elevated in the sera of patients with pancreatic cancer [116]. Like the other pancreatic mucin markers, the Span-1 epitope is neuraminidase-sensitive. Similar to DU-PAN-2 and CA 50, the Span-1 MAB reacts with pancreatic cancer tissues from patients with the Le^{a-b-} phenotype. In a recent clinical study [93], 62 of 67 (93%) patients with pancreatic cancer were Span-1 seropositive. No normal subjects had Span-1 levels above the cutoff value. In a small population of patients with stage I pancreatic cancer, the Span-1 marker was elevated in two of four (50%) of patients. In patients with stages II, III, and IV cancer, all sera were Span-1 seropositive. In a comparative clinical study [93] with Span-1, CA 19-9, and DU-PAN-2 assays, it was shown that Span-1 had the highest overall sensitivity at 93%. Ca 19-9 was the next most sensitive at 85% and DU-PAN-2 gave the lowest sensitivity at 38%. The Span-1 marker was also elevated in patients with hepatocellular cancer (59%), gastric cancer (23%), and colorectal cancer (13%). It was noted that the sera from 5% of patients with chronic pancreatitis gave false-positive results compared to 15% in the CA 19-9 assay. In summary, four mucin tumor markers defined by different MABs, CA 19-9, CA 50, DU-PAN-2, and Span-1, appear to be useful diagnostic aids in patients suspected of having pancreatic cancer. Sensitivities approaching 95% can be achieved either with Span-1 alone or by various combinations of the other mucin markers.

B. Breast Cancer

The utility of mucin tumor markers to monitor breast cancer was first indicated by Ceriani et al. [48]. Subsequent reports have confirmed these findings. The clinical results of mucin marker assays are described in Table 3. Three of the assays, HMFG-2, CA 549, and MSA, use a single MAB as capture and probe antibody in solid-phase immunoassays. The other assays, CA 15-3, M29/M38, and CTA, use different MABs as catcher and probe.

The MSA assay [117] uses an IgM MAB raised against human breast carcinoma cells in a competitive enzyme immunoassay (EIA) format. A recent study showed that elevated levels of the MSA antigen were detected in the sera of 93% of patients with metastatic breast cancer [117]. Elevated levels of MSA were detected in 75% of stage I patients and in 89% of stage II patients. None of the nonsmoking healthy controls were MSA seropositive. Of interest, 70% of patients with ovarian cancer, 60% of patients with colon cancer, and 71% of patients with lung cancer were also positive by the MSA assay. A comparative study of CEA and MSA levels in this group of breast cancer patients showed that MSA was a more sensitive marker of disease than CEA.

Studies with CA 549 showed that 32 of 35 (91%) of women with metastatic breast cancer had elevated serum levels of this marker [118]. None of the healthy women was CA 549 seropositive. This mucin tumor marker was used to monitor response to treat-

Table 3 Primary Indications for Serological Assays for Mucins

Immunoassay	Reference	Monoclonal Antibody	Immunogen	Control (% Range)	Advanced Cancer (% Range)
Breast cancer					
CA 15-3	122-128, 160, 179	115D8 DF3	HMFG; cancer cell membranes	1-1.3	63-94
HMFG-2	47, 120	HMFG-2	HMFG	3-16	53-72
M29/M38	129	M29/M38 M26/M38	Purified mucin from milk/ ascites cell lines	0-1	67-76
MSA	117, 163, 181, 182	3E 1.2	Breast cancer tissue	0-3	82-93
CTA	130	M85/34, F36/22		0	72
CA 549	118, 119	CA 549	Breast cancer cell line	0	91
Pancreatic cancer					
CA 19-9	103-109	CA 19-9	Colon cancer cell line	0-1	70-79
CA 50	115, 172, 183	CA 50	Colorectal cell line	0-1	71-83
DU-PAN-2	173, 174	DU-PAN-2	Pancreatic cancer cell line	0	67-77
Span-1	92, 116	Span-1	Pancreatic cancer cell line	0	93
Ovarian cancer					
CA 125	132-142 184-188	CA 125	Ovarian cancer cell line	0.2-1	63-97
Colon cancer					
CA 195	152-154	CA 195		0-5	56-70
TAG 72	151	B72.3	Membranes; breast cancer metastasis	1	56

ment in women with metastatic breast cancer [119]. In women with disease progression, 16 of 18)88%) showed at least a 25% increase in serum CA 549 levels. In women with disease regression, 9 of 12 (75%) showed a decrease in CA 549 levels. It was concluded that CA 549 may be useful in monitoring response to treatment of women with metastatic breast cancer.

The HMFG-2 assay has been evaluated as a marker for both breast and ovarian cancer. In a study of 239 individuals [120], 3% of healthy controls and 4% of women with benign breast disease had elevated serum HMFG-2 levels, whereas 50% of women with preoperative stage I or II breast cancers, 47% of patients with breast cancer in apparent remission, and 73% of patients with metastatic breast cancers had elevated levels. Twenty-seven percent of patients with benign diseases of other origins had elevated HMFG-2 levels.

The CA 15-3 assay uses the MAB 115D8 [52] as the catcher and the MAB DF3 [121] as the probe. In a recent study, the clinical results of the CA 15-3 assay and CEA levels were compared in women with metastatic breast cancer [122]. CA 15-3 values over 25 U/ml were seen in 16% of women without axillary metastases. In contrast, 54% of women with axillary lymph node involvement and 91% of women with metastatic breast cancer were CA 15-3 positive. This study also showed that 82% of all patients tested with early metastasis were CA 15-3 positive, compared to 54% of women with CEA values greater than 10 ng/ml. Other clinical studies using the CA 15-3 assay showed similar results [123-128]: 70-80% of women with metastatic breast cancer were CA 15-3 seropositive and CA 15-3 was a more sensitive marker for metastatic breast cancer than CEA.

A recent study of CA 15-3 and CEA in serial plasma samples of 53 women with metastatic breast cancer showed that antigen levels must increase or decrease by greater than 25% to correlate with disease progression or regression, respectively [127]. Variations in antigen levels of less than 25% were considered to be indicative of disease stability. About 75% of women had increasing CA 15-3 serum levels as disease progressed, compared to 58% of women with increasing CEA levels.

The M29/M38 assay is based on a double antibody immunoassay. In a study of women with metastatic breast cancer, 93% had elevated serum levels by the M29/M38 assay compared with 83% for the CA 15-3 assay on the same panel [129]. When a combination of M29/38 and M26/38 assays was used with cutoff values to exclude all healthy controls, a 76% sensitivity was still achieved.

The cryptic tumor antigen (CTA) assay is a new immunoassay for detecting a breast cancer tumor-associated mucin [130]. It utilizes two MABs, M85/34 [131], which is reactive with the i/I oligosaccharide antigen sequence expressed on O-glycosidic linkages, and the other, F36/22 [54], directed to the ductal carcinoma antigen epitope. Neuraminidase treatment of serum specimens in this assay reveals otherwise cryptic epitopes. It was reported that 14 of 23 (61%) of women with progressive breast disease had rising CTA levels [130]. This assay appears to be as sensitive as the CA 15-3 assay, with improved specificity.

Over 20 breast-cancer-associated antimucin MABs have been developed and used in serological assays for breast cancer patients during the past 5 years. Many have achieved up to 80-90% sensitivity in detecting elevated levels of mucins in the sera of women with metastatic breast cancer. Many of the reported serological studies, however, have emphasized sensitivity at the expense of specificity. Several studies have reported false-positive rates of 3-6% for healthy controls and 10-20% false-positive results in sera from women with benign breast disease [54,117,121].

C. Ovarian Cancer

CA 125 antigen was found to be expressed in the tumors of more than 80% of nonmucinous human ovarian carcinomas [57] and several studies have shown that the measurement of this mucin is particularly useful in monitoring response to treatment of ovarian cancer. Serum levels of CA 125 have also been found to be useful in predicting the malignant status of pelvic masses before exploratory surgery [132-133]. Only 1% of healthy individuals have serum CA 125 levels greater than 35 U/ml, whereas more than 80% of ovarian cancer patients have antigen levels above this cutoff [134]. Elevated CA 125 serum levels have also been found in the majority of patients with carcinoma of the endometrium, fallopian tubes, endocervix [135], and pancreas [134]. A significant number (20-30%) of patients with lung, breast, and colorectal cancer also have elevated levels of CA 125 in their serum [134].

In monitoring disease, a very good correlation between CA 125 levels and disease progression, stability, or remission has been found. In the initial report on the use of CA 125 for monitoring epithelial ovarian cancer [134], changes in serum antigen level correlated with disease status in 93% of cases. It was shown that as tumor progressed, CA 125 increases preceded clinical detection of disease recurrence by 1-14 months, with a median of 3 months, in approximtely 80% of reported cases [136]. Persistent rising CA 125 values have been consistently associated with progressive disease.

Elevation of serum CA 125 levels (>35 U/ml) at the time of a second-look surveillance predicts persistence of disease with high specificity. The usefulness of the assay, however, is limited as a general predictor of recurrence because only 60% of patients with recurrent ovarian cancer have CA 125 serum levels greater than 35 U/ml [137,138]. One study has shown that CA 125 levels in peritoneal lavage fluid is a more sensitive predictor of residual disease than are serum levels. Antigen levels above 35 U/ml in peritoneal lavage fluid accurately predicted disease status in 80% of patients surveyed [139].

In patients with a pelvic mass, two studies have shown the predictive value of determining serum CA 125 levels before exploratory surgery [132-133]. One of these studies showed that 76 of 77 patients with benign pelvic masses had serum CA 125 levels <65 U/ml, whereas 13 of 14 patients with gynecological malignancies had antigen levels >65 U/ml.

CA 125 has also been evaluated for screening and detection since CA 125 is elevated in patients with stage I and II malignant ovarian disease. A major study [140] of 1020 women attending a gynecological clinic in Boston showed that CA 125 levels were elevated (>65 U/ml) in the sera of 1.8% of healthy women. If pregnant women were excluded, only 1% of healthy controls showed positive results. A second study [141] of 915 nuns with a median age of 55 revealed that only 0.5% of healthy women over 50 were CA 125 positive (>65 U/ml). Based on actual ovarian cancer incidence of 20 cases/100,000 nonsymptomatic women, this screening assay would yield 500 false-positive and only 20 true-positive cases per 100,000 women screened, that is, 1 true positive for 25 false positives. Attempts to reduce this false-positive load in a screening situation are being investigated by using secondary or confirmatory tumor assays such as B72.3 or CA 15-3. Other studies are being performed to increase detection of localized disease (stages I and II) using additional markers. A study [142] of 11 women with stages I and II ovarian cancer using CA 125, HMFG-2, and placental alkaline phosphatase concurrently increased the sensitivity of cancer detection from 18% with one assay to 64% when all three assays were used.

D. Lung Cancer

Markers associated with lung cancer have been identified and some clinical data have recently been collected. As with the CA 19-9 marker, these antigens have been both glycolipids and mucins. Paragloboside/Le^x antigen serum concentrations were elevated in 60% of patients with lung cancer and 5% of healthy controls [44]. SiLe^x as detected by CSLEX1 MAB was elevated in the sera of 34% of lung cancer patients and 3% of normal persons, compared to SiLe^a, which was elevated in 50% and 0% in lung cancer patients and normal persons, respectively [143]. In another study using the CSLEX1 MAB serum assay, the SiLe^x sensitivity as a marker for lung cancer increased with advancing stages. In addition, antigen levels decreased with disease regression and increased with progression of disease [144]. FH6 MAB also has SiLe^x specificity, but differs from the specificity for CSLEX1. FH6 MAB is specific for extended SiLe^x carbohydrate structures, whereas CSLEX1 MABs binds all SiLe^x structures. An RIA utilizing FH6 MAB detected five of seven lung cancer patients with elevated antigen levels, whereas levels in all 17 normals were at or below the cutoff [145]. The serum antigen detected by FH6 MAB was characterized as a mucin type glycoprotein [146], demonstrating the presence of the SiLe^x structure on mucins. Another MAB, 43-9F, binds to soluble mucins of human squamous lung carcinoma cells and recognizes an epitope that contains both the Le^a and internal X-determinants [147]. This MAB reacted to difucosyl Le^a-X determinant on 16 of 22 squamous lung carcinoma tissue specimens, but 0 of 16 small cell carcinoma tissue specimens [148]. Other MABs recognizing carbohydrate epitopes, including Le^x, have been reported and these MABs demonstrate specificity for small-cell lung cancer [149]. However, at this time a serum mucin marker for lung cancer has not been identified that has the clinical utility of mucin markers in other cancers (e.g., breast and pancreas).

Mucin markers in sputum may provide assays for early detection of lung cancer or identification of very-high-risk individuals. Sputum cytological studies, which rely on cell morphology and detect cell surface structures, have been used in the management of lung cancer [150]. Mucins and their tumor-associated structures, which are present on cell surfaces, are also in the soluble fraction of sputum and may serve as lung cancer markers including TAG-72 activity, which has been measured in the sputum of lung cancer patients (Plate,C. Abbott Laboratories, 1988, unpublished results).

E. Colorectal Cancer

The B72.3 MAB has been used to measure the serum levels of TAG-72 antigens in patients with colorectal cancer [151]. The cutoff was set to exclude 99% of healthy controls. Elevated levels of TAG-72 were found in 15 of 26 (58%) of rectal cancer patients and in 14 of 25 (56%) of patients with colon cancer. Although the test showed high specificity, the low sensitivity may limit its clinical utility.

A newer mucin MAB, CA 195, has been shown to recognize Le^a as well as sialylated Le^a (CA 19-9 epitope) [152,153]. The CA 195 assay detected 29 of 80 (36%) patients with colon cancer, primarily those with Duke's C lesions. When one combined CEA and CA 195 assay results, 59% of patients with Duke's C colon cancer could be detected, compared to 33% with CEA alone. A recent serological study comparing levels of CEA and CA 195 [154] using 108 colon cancer patients found similar trends: increased sensitivity when the results of the two assays are combined. It is becoming apparent that combinations of marker assays may increase the sensitivity and specificity of early cancer detection.

IV. SENSITIVITY AND SPECIFICITY OF MUCIN TUMOR MARKERS

The precise quantity of mucin glycoproteins in the serum of normal individuals and cancer patients is not currently known. Since the epitopes recognized by antimucin MABs are often oligosaccharides, which are repeated in variable amounts on different molecules, the actual weight equivalent of the measured antigen may only be approximated. The units used in the assays are usually arbitrary, based on reference standards. From the known sensitivity of RIA and EIA and the amount of serum or plasma required to perform the assays, it is reasonable to assume that many mucins occur in at least the ng/ml range. For example, the 99% confidence level for CA 19-9 is 37 U/ml, above which levels in serum from 1% of normal healthy subjects are elevated. A CA 19-9 unit corresponds to 0.59 ng/ml of antigen as defined by the manufacturer, indicating that CA 19-9 levels up to approximately 22 ng/ml are present in healthy individuals. These concentrations are well within the sensitivity range for EIA/RIA technology.

Technical sensitivity is not a problem for these antigens, but the normal background levels present difficulty, as is also the case for traditional glycoprotein tumor markers, such as CEA, AFP, and PAP. Clinical specificity depends, of course, on the specificity and affinity of antibodies in the configured assay. Unlike the situation with traditional markers, the selection of antibodies to different epitopes on a mucin can have a very significant effect on an immunoassay's clinical sensitivity and specificity for cancer. For traditional tumor markers, the antibodies are directed to the peptide region of the molecule and the assays measure a direct gene product. Mucin immunoassays, on the contrary, often use antibodies directed totally, or in part, to the carbohydrate portion of the molecule. The display of the specific carbohydrate on different mucins in serum may be quite heterogeneous. The selection of antibodies will determine which portion of a heterogeneous family of mucins in serum will be detected, hence the resulting clinical correlation.

Characterization of mucin antigens has recently demonstrated that certain purified mucin preparations coexpress several specific epitopes including CA 19-9, CA 125, and SiLe^x [7]. One would expect mucin assays to differ in specificity and sensitivity but also to share common features. In general, mucin levels are elevated in the serum of patients with adenocarcinomas, particularly breast, pancreas, ovary, lung, and colon/rectum. Greater than 50% of patients with advanced cancer usually show elevated levels, whereas earlier stages of cancer are not often detected. A wide range of sensitivity has been reported for some of the assays, probably due to different cutoff values and different populations of cancer patients. Although one organ site may be more often identified with a particular assay, mucin tumor markers are not organ-specific. For example, CA 15-3, CA 549, and CTA assays primarily detect breast cancer but elevated serum values are also observed in patients with other adenocarcinomas. It is likely that future mucin assays will be designed as general markers for adenocarcinomas.

Among the currently available mucin tumor markers, CA 125 appears to be the most specific and sensitive. Greater than 70% of patients with advanced ovarian cancer have elevated CA 125 levels. With the notable exception of patients with severe liver disease and endometriosis, patients with benign diseases do not usually have elevated CA 125 levels. Although CA 125 levels are not often elevated in early asymptomatic ovarian cancer, recent studies have indicated the potential screening value of this tumor marker.

V. SUMMARY

Extensive biochemical studies have shown that mucin tumor antigens have a range of molecular sizes from 200 to greater than 1000 kDa. The molecular size of mucin antigens can be dramatically affected by the source and method of purification. Mucin antigens vary from 24 to 80% in carbohydrate content and their density is usually greater than 1.40 g/ml. Galactose and N-acetyl glucosamine are the predominant sugar residues in many mucins, whereas mannose is usually present in low levels or absent. The amino acids serine, threonine, alanine, glycine, and proline are abundant in mucins. An O-glycosidic linkage between the carbohydrate and protein of mucins is the most common linkage encountered. The gene encoding the core peptide for at least one mucin tumor marker, HMFG, has been identified, sequenced, and expressed. These findings may lead to a better understanding of the multiepitope nature of mucin tumor markers.

The advent of hybridoma technology has yielded several monoclonal antibodies that have been used to identify the presence of tumor-associated mucins in the sera of cancer patients. Elevated levels of mucin antigens have been found in the serum of most patients with advanced adenocarcinomas. Many studies have shown that tumor-associated markers are useful in monitoring patients following cancer treatment. Clinically useful immunoassays have been developed for monitoring patients with ovarian, breast, and pancreatic adenocarcinomas. Although individual mucin tumor markers show limited utility in detecting early adenocarcinoma, recent studies using multiple mucin markers have suggested that early detection, at sensitivities greater than 50%, can be achieved.

VI. RECENT DEVELOPMENTS

Since this review article was written, numerous studies involving the role of mucins as tumor markers have been reported. The following is an update of the most recent studies.

Levels of one of the most extensively studied mucin markers, CA 19-9, were not sufficiently elevated in serum to be used to screen cancers of the esophagus [191,192], head and neck [193], or pancreas [194]. CA 19-9 does appear to be useful in detecting recurrence of pancreatic cancer following pancreatectomy, since serum CA 19-9 levels were elevated in 11 of 14 postsurgical patients 1-7 months prior to clinically apparent recurrence [195]. In contrast, CEA was a poorer predictor of recurrence in this study. In another study, 70% of patients with pancreatic cancer had elevated serum CA 19-9 levels, whereas only 33% had elevated serum CEA levels and discordance between the two markers was found in 44% of the patients [196].

In patients with colorectal cancer [197,198] and gastric cancer [199] CA 19-9 levels were identical in peripheral and draining venous blood, whereas CEA levels were higher in the draining blood. This and other observations suggested different mechanisms for affecting blood levels of CA 19-9 and CEA in cancer patients. Serum levels of CA 19-9 were increased in patients during administration of recombinant interferon-γ, implying that this treatment might augment in vivo localization of anti-CA 19-9 in patients expressing this antigen [200].

Some recent studies have measured CA 19-9 levels in body fluids other than sera or plasma in an attempt to improve sensitivity and clinical utility. Although serum CA 19-9 levels were elevated in the minority of patients with gastric cancer, CA 19-9 levels in gastric juice obtained from patients with gastric cancer and acute gastritis during endoscopy were significantly higher than found in normal subjects [201]. This study suggested a possible role for CA 19-9 measurement in gastric juice for detecting earlier malignancy. CA 19-9 levels were also found to be elevated in cervical mucus from patients with cervical adenocarcinoma and it was concluded that CA 19-9 in conjunction with CEA and CA 125 in cervical mucus might prove useful in detecting cervical adenocarcinoma [202]. When CA 19-9, TPA, and CEA levels were compared in the urine of patients with transitional cell bladder cancer, CA 19-9 was the best marker, with a sensitivity of 71.6% and specificity of 91.6% [203].

CA 50, a mucin marker closely related to CA 19-9, also has been further investigated. In a study involving 266 patients with colorectal cancer, 66% of those who experienced relapse had elevated serum CA 50 levels [204]. This study also showed that a rise in CA 50 levels after resection of patients with Dukes' level A-C disease was indicative of recurrence up to 9 months prior to clinical evidence of relapse. CA 50 was found to be a more sensitive marker than CA 19-9 for squamous cell carcinoma of the esophagus (41% versus 13%, respectively) [191], and for squamous cell carcinoma of the head and neck (27% versus 13%, respectively) [193]. When used in combination with CA 50, CEA sensitivity was increased from 41 to 58% in patients with esophageal cancer [191] and from 18 to 43% in patients with squamous cell carcinoma of the head and neck [193]. In a study in which five tumor markers were evaluated in bronchial lavage fluid, the simultaneous determination of CA 50, CEA, and neuron-specific enolase was found to be useful in the diagnosis of lung malignancy [205]. In another study, the monoclonal antibody CA 50 was used in three different formats—an inhibition assay, an immunoradiometric assay (IRMA), and a dissociation-enhanced lanthanide fluoro-immunoassay (DELFIA)—to measure plasma CA 50 levels in patients with pancreatic cancer [206]. It was interesting that the three assays displayed different specificities. The inhibition test that could detect both single and multiepitope antigens gave slightly better differential diagnosis of cancer from chronic pancreatitis, despite better technical performances by the other two tests. A recent review summarizes the effects of various factors including gender, pregnancy, menstruation, addiction to alcohol, smoking, and benign disease on the specificity of several markers including CA 19-9 and CA 50 [207].

CA 19-9 and CEA values were compared with serum CA 72.4 values in patients with gastrointestinal cancers [208]. CA 72.4 is a second-generation assay for detecting the TAG-72 antigen using the B72.3 and CC49 MABs. In sera of patients with pancreatic cancer, 22% of patients had elevated CA 72.4 values compared to 82% with elevated CA 19-9 values [208]. In serum specimens from patients with colorectal cancer, 32% showed elevated CA 72.4 levels whereas 58% had elevated CEA levels. Combining the two markers did not improve sensitivity. CA 72.4 appears to have the most promise in gastric cancer. It was elevated in the sera of 59% of gastric cancer patients whereas CA 19-9 and CEA were elevated in 52% and 25% of the cases, respectively. Combining CA 72.4 and CA 19-9 improved sensitivity to 70%. In an immunohistochemical study of TAG-72 antigen using sections of colonic tissue from patients with ulcerative colitis, TAG-72 expression in quiescent disease generally increased with duration of disease. In contrast, there was no correlation between TAG-72 expression and duration of disease in patients with active disease [209]. It was also concluded in this study that increased TAG-72 expression correlated with severity of dysplasia, suggesting a correlation of this marker with dysplasia-to-carcinoma progression. The frequent expression of TAG-72 in inflamed colonic mucosa, however, may limit the utility of TAG-72 for the detection or management of colon cancer.

Recent clinical studies of mucin markers in breast cancer have been reported for CA 15-3 [210-212], MSA [213-216], and CA 549 [217,218] as well as for the more recently described mucin markers MCA [212,219-223] and BCM [224,225]. As in other studies, serum CA 15-3 levels were elevated in the majority of patients with advanced breast disease (67% of stage IV) but in the minority (7%) of those with stage I disease [210]. Serum CA 15-3 values correlated with the clinical course of 26 of 27 patients in this study. In a study of 500 patients with nonmammary epithelial carcinomas, CA 15-3 levels were found to be most frequently elevated in patients with ovarian (46%), respiratory (26%), and liver (30%) carcinomas [211]. CA 15-3 levels increased with progressive stages of epithelial cancer and were especially elevated when distant metastases were present. This was particularly evident in ovarian cancer; 70% of patients with metastases had elevated serum CA 15-3 values. Combining CA 15-3 and CA 125 results led to slightly increased sensitivity for detecting ovarian cancer, suggesting that CA 15-3 might have a role here [211].

Elevated MSA levels were also found in the sera of the majority of patients with advanced breast cancer (73% and 87% for stages III and IV, respectively) and, as with CA 15-3, there was good correlation between MSA values and the clinical course of the disease [215]. In another study 76% of stages I and II and 86% of stages III and IV breast cancer sera had elevated MSA values [216]. These results suggested that MSA might be useful not only to monitor breast cancer but also to detect early disease. Elevations of MSA values in 18% of patients with benign breast diseases [216], however, might limit the utility of MSA in patients with early breast cancer.

Serum values of MSA, CEA, and beta-2 microglobulin were compared in 186 cases of breast cancer [214]. MSA values were elevated in 79% of stage I/II patients whereas 25% and 12% of these patients had elevated levels of beta-2 microglobulin and CEA, respectively. In 25 patients monitored over 3-9 months, MSA levels correlated much better with clinical course (92%) than did beta-2 microglobulin (28%) or CEA (36%) [214]. When these markers were combined it was found that beta-2 microglobulin enhanced MSA sensitivity, especially in early stages of breast cancer. MSA remains a very promising marker, yet further studies in more laboratories are needed before firm conclusions can be drawn.

As with CA 15-3 and MSA, serum levels of CA 549 were elevated in the majority of patients with late-stage breast cancer and this marker also tracked the clinical course of this disease [217,218]. CA 549 values were especially elevated in breast cancer patients with metastasis to liver and bone [217], whereas patients in remission had normal CA 549 levels. Patients with ovarian cancer (38%) had the highest CA 549 levels of those with nonmammary epithelial carcinomas.

BCM (breast cancer mucin) is detected by three assays: an enzyme immunoassay, CTA-EIA [130,131]; a radioimmunoassay, BCM-RIA [224]; and an automated microparticle immunofluorescence assay, BCM-IMx [225]. These assays detected elevated BCM levels in the majority of breast cancer patients with advanced disease and fewer patients with early breast cancer. BCM levels also correleated well with the clinical course of breast cancer. Simultaneous analysis of BCM and CA 15-3 values revealed a correlation of 0.86 and 0.73 for serum specimens obtained from healthy donors and breast cancer patients, respectively [224].

The assay for mucin-like cancer-associated antigen, MCA, uses MAB b-12 [223]. MCA levels were elevated in 20% of stage I and 37.7% of stage III breast cancer sera [221]. MCA levels were elevated in 80% of breast cancer patients with active metastases but decreased as patients responded to therapy [222]. In a study of 191 patients with breast cancer (122 with progressive disease and 79 with no evidence of disease) there was good correlation (0.84) between MCA and CA 15-3 [226]. MCA was found, however, to be less sensitive and less specific than CA 15-3. Combining MCA and CA 15-3 resulted in higher sensitivity but lower specificity.

CA 15-3 was compared with CEA [227-230], MCA [231], or both CEA and MCA [212, 219, 220] in patients with breast cancer. CA 15-3 was significantly better than CEA for detecting recurrent and/or metastatic disease [228-230]. The simultaneous use of CA 15-3 and CEA improved CA 15-3's sensitivity from 73% to 86% [229] and, in another study, from 65% to 77% [230]. Sensitivity was generally improved by the combined use of all three markers for advanced breast cancer: for visceral metastases, sensitivities were 19.44%, 58.33%, 66.66%, and 72.77% for CEA, MCA, CA 15.3, and all three combined, respectively [220]. Elevated MCA levels correctly predicted recurrence in 43% of cases compared to 36% for CA 15.3 and 33% for CEA [219]. The use of CEA in combination with CA 15-3 to improve the sensitivity for detecting breast cancer recurrence has not been firmly established.

Observations from clinical studies on various breast cancer mucin markers have shown clinical interrelationships that are now being substantiated by biochemical and immunochemical studies. Although levels of mucin markers are elevated in the majority of patients with advanced breast cancer, elevated levels are present in only a minority of patients with early breast cancer. This lack of sensitivity for early disease has limited the clinical application of these markers in managing breast cancer.

In addition to immunoassays used to measure mucin antigens in biological fluids, immunohistochemical studies have contributed to our understanding of the expression of breast cancer-associated mucin antigens. The mucin core protein epitope defined by MAB SM-3, for example, is highly expressed in primary carcinomas of lung, colon, and ovary in addition to breast, but levels are significantly reduced in corresponding normal tissues [232]. The A-80 MAB, which was raised against mucin purified from the human colon adenocarcinoma cell line LS174T, stained strongly 100% of 38 breast carcinomas whereas fibroadenomas were negatively (n = 10) or weakly (n = 2) stained [233]. Strong A-80 reactivity was also observed in other epithelial malignancies [233,234]. Two recent review articles on mucin antigens in breast cancer discuss both serum and tissue applications [235,236].

Evidence for the value of monitoring ovarian cancer with CA 125 continues to accrue [237-240]. Other studies have examined the predictive value of preoperative CA 125 serum levels for differentiating benign from malignant pelvic masses [241,242]. The preoperative predictive value of clinical examination and ultrasound in postmenopausal ovarian cancer patients was significantly enhanced when CA 125 values were added [241]. CA 125 combined with CA 15-3 significantly improved sensitivity and specificity to differentiate benign from malignant masses preoperatively [242]. A panel composed of CA 125, CA 15-3, and TAG-72 detected 93% whereas CA 125 alone detected 85% of ovarian carcinomas [243].

CA 125 has been investigated for early detection of ovarian cancer [244,245] as well as for second-look surgery [246]. Sera collected within 18 months preceding diagnosis of ovarian cancer had CA 125 levels greater than 30 U/ml in 50% of the cases and 33% had levels in excess of 65 U/ml [245]. This is in contrast to CA 125 values in the sera of healthy controls: only 7% and 0.9% were greater than 30 or 65 U/ml, respectively. In a phase I study of 1082 women, 1% over 50 years of age and at risk for ovarian cancer had CA 125 levels above 65 U/ml [244]. This is consistent with an earlier study that determined that 1% positive controls (no ovarian malignancy) would yield 500 false-positive cases for only 20 true-positive cases for every 100,000 women screened. All patients with ovarian cancer with complete or partial clinical remission with CA 125 values above 35 U/ml had confirmed residual tumor at second-look surgery [246]. CA 125's sensitivity for ovarian cancer was increased from 75 to 90% when combined with MCA, CA 19-9, and CEA but specificity was lowered from 93 to 73% [247]. CA 125 also has been investigated as a marker for uterine sarcoma: three of four patients with stage I and five of seven with advanced and recurrent disease had elevated preoperative levels [248].

An immunoassay has been described for a new ovarian cancer-associated mucin antigen, CA 54/61 [249]. This assay utilizes two MABs, MA54 and MA61, which recognize different carbohydrate epitopes on the antigen. CA 54/61 levels were elevated in the sera of 64% of mucinous histological type ovarian cancers. Since CA 125 levels are usually not elevated in the sera of mucinous type ovarian cancers, further studies are needed to determine whether this marker might complement CA 125 in managing patients with ovarian cancer.

The serum levels of three different carbohydrate epitopes related to Le^a were measured in patients with benign diseases [250]. MAB FH-7 recognizes $2 \rightarrow 6$ sialylated Le^a, levels of which were elevated in 33% of the sera. The $2 \rightarrow 3$, $2 \rightarrow 6$ disialylated Lc$_4$ structure, which is recognized by the FH-9 MAB, was present in elevated amounts in 27% of the sera. In contrast, levels of CA 19-9, which is $2 \rightarrow 3$ sialylated Le^a, was elevated in only 14% of the benign sera. CA 19-9 levels were elevated in 36%, $2 \rightarrow 6$ sialylated Le^a in 25%, and disialyl Lc$_4$ in 17% of a serum panel from patients with various cancers [250]. Biochemical analysis indicated that these antigens were mucin-like glycoproteins.

A radioimmunoassay for extended sialylated Le^x using MAB FH-6 was used to measure mucin antigen in pleural effusions from patients with lung adenocarcinoma [251]. Sixty-four percent of patients with lung adenocarcinoma, 8% with nonadenocarcinomatous lung cancer, and 0% with pulmonary benign disease showed elevated levels. Elevations of mucin marker levels in ascites or pleural fluids from cancer patients have been frequently observed.

Two recently described mucin antigens are NCC-CO-450 [252] and LISA 1-6 [253]. Levels of NCC-CO-450 mucin antigen defined by the corresponding MAB were found elevated in sera in 3% of normal controls, 56% of those with colon carcinomas, and 40% of patients with gastric carcinomas by RIA [252]. MABs LISA 101 and KL-6 were used in an immunoassay detecting elevated LISA 1-6 levels in 6.5% of patients with benign lung diseases, 7.1% of patients with pancreatitis, 63% of patients with lung adenocarcinoma, and 92% of patients with pancreatic carcinoma [253].

Several conclusions can be drawn from these very recent clinical studies on mucin markers. Many of these studies have confirmed the clinical usefulness of mucin markers for detecting metastatic disease in patients with ovarian, pancreatic, and breast cancers. There appears to be increased interest in using a combination of mucin markers on the same patient sample. Also, the use of body fluids other than plasma or sera (e.g., urine, bronchial secretion, and breast duct fluids) to detect abnormal levels of mucin markers has attracted greater clinical interest.

Numerous biochemical and immunochemical characterization studies have been performed on mucins containing cancer-associated epitopes. These studies have involved epitope characterization and more general analyses of these mucin proteins. The OC 125 epitope as analyzed on human amniotic mucins may involve both sugar residues of the nonreducing end and protein in maintaining epitope structure [254]. The theory of involvement of sugar residues was based on the diminution of CA 125 immunoreactivity by either mild periodate or exoglycosidase treatment [58,254]. This is in contrast to an earlier study [6], which concluded that the CA 125 epitope was primarily or exclusively peptide in nature. The precise role of carbohydrate in the CA 125 antigen is not completely clear. Experiments using synthetic peptides indicated that the DF3 epitope consisted of peptide although carbohydrate was involved in maintaining DF3 immunoreactivity as shown by decreased immunoreactivity on neuraminidase treatment of the DF3 antigen [255].

Protease treatment destroyed the 1D3 epitope [256] and the CAR-5 epitope defined by MAB BD-5 [257], which suggests that these epitopes consist of protein. The MCA-b-12 epitope [223] and the MAB 660 epitope [258] may also be peptides. On the other hand, the epitopes on CA 54/61 antigen defined by MABs MA54 and MA61 showed characteristics of carbohydrates [249], as did the mucin epitopes defined by MAB 3E1.2 [259], MAB YH206 [260], MAB AR-3 [261], MABs Span-1 and Ypan-1 [262], and MAB MLS 102, which recognized a cluster of NeuAcα2 → 6GalNAc [263]. MAB MSW 113 recognized sialylated Le^a yet was not identical to the CA 19-9 MAB also reactive to sialylated Le^a [264]. MAB Onc-M26 was shown to bind to the sialylated Le^x structure [265]. The M1 mucin was shown to contain at least seven epitopes; in addition to the Le^a and Le^b blood group antigens, there were at least five different peptide epitopes on this mucin [266].

Biochemical characteristics of mucins including high-molecular weight, 1.45 g/ml buoyant density in cesium chloride, high content of threonine, serine, and proline, and other mucin-defining properties were found for the following cancer-associated mucin antigens: DF3 [255], 1D3 [256], CAR-5 [257], MAB 660 defined antigen [258], MSA [259], YH206 [260], CAR-3 [261], MAM-6 [267], EMA [268], Span-1/Ypan-1 [262], TAG-72 [269], MAB defined sulfomucin [270], gp580 [271], and Ca antigen [272]. High-molecular-weight proteins isolated from tumor derived cell lines were also shown to have biochemical properties characteristic of mucins [273-279]. Structures also have been determined for O-linked oligosaccharides of human milk mucin [280] and human meconium mucin [281,282].

Eleven MABs against different sialomucins, including MAM-6, were used to analyze MAM-6 [267]. These antibodies reacted to sialomucins generated from common precursor molecules produced by various cell lines. In one study it was concluded that MAM-6 and HMFG were identical mucins and that MAM-6 constitutes a family of sialomucins. MSA and DF3 antigens were also shown to be related [259]. These investigations have further revealed the relationships between several of the breast cancer mucin markers described during the last several years. In other studies, heterogeneity of epitope expression on the breast cancer mucins PAS-O [283] and gp580 [271] on a cell-by-cell basis have been observed with several tumor-derived cell lines.

Relationships between other cancer-associated mucins were also observed. Span-1 and Ypan-1 epitopes were found on the same mucin molecule [262]. MAB B72.3 was utilized in the immunopurification of TAG-72, which was then used to generate nine distinct second-generation MABs [284]. In this group, MAB CC49 was subsequently used to immunopurify TAG-72 for composition studies [269] and is used in the CA-72.4 assay. Colonic mucins in inflammatory bowel diseases were shown to express aberrant epitopes defined by MABs 19-9, B72.3, DU-PAN-2, and CSLEX1 [285]. The CAR-3 [261] and CAR-5 [257] epitopes were found to be expressed on the same mucin molecules as Le^a, Le^b, and CA 19-9, whereas the YH206 MAB-defined epitope was not coexpressed with either of two mucin markers tested, CA 19-9 or DU-PAN-2 [260]. Also, the mucin-like glycoprotein gp580, which binds MAB GP21:56, did not demonstrate other immunoreactivities tested, including TAG-72 [271].

Although most of the progress in tumor-associated mucin markers has been made using monoclonal antibodies, polyclonal antibodies have been raised against purified mucins. Studies with polyclonal antibody against LS174T mucin [273,274], SW1990 mucin [278], and mucin purified from the pleural fluid of lung adenocarcinoma [286] indicated that the protein and not the carbohydrate components of these mucins were the principal antigenic determinants recognized. Mucins from the LS174T and SW1990 cell lines were found to coexpress protein epitopes [278]. Polyclonal antibody to LS174T mucin and HT29 mucin stained mucosa from human normal colon and colon cancer [273,274] and human normal colon [277], respectively.

Several investigations on the biosynthesis of tumor-associated mucins have been reported [287-291]. MABs immunoreactive to MAM-6 were used to immunoprecipitate proteins during their synthesis [287]. These proteins were shown to be 220, 200, and 500 kDa precursors to a 450 kDa mature MAM-6 and 380, 350, and 700 kDa precursors to 650 kDa mature MAM-6. Likewise, immunoprecipitation with MABs identified mucin precursors synthesized by MCF7 cells as 170, 185, 260, and 275 kDa proteins [288]. These mucin precursors were found to contain N-linked oligosaccharides that may be lost in the native glycoprotein [287,288]. Biosynthesis involving O-glycosylation of mucins by 13762 rat mammary adenocarcinoma cells has been delineated [289-291]. Incomplete glycosylation of mucins by tumor cells was proposed to be the result of late oligosaccharide initiation due to changes in enzyme localization in tumor versus normal cells [289,291].

The molecular biology of mucins is an area about which we are learning more. The same 60-base pair tandem repeat sequence was reported for encoding the core protein of DF3 antigen [255] and polymorphic epithelial mucin [292], although the sequence initially reported for DF3 did contain some differences [293]. Synthetic peptides based on this sequence were used for defining peptide epitopes of several MABs [255, 292,294]. Although different, the epitopes defined by MABs HMFG-2 and SM-3 were mapped to the same 14 amino acid segment in the 20 amino acid repeating unit, while the HMFG-1 epitope appeared to be clearly distinct [292]. The epitopes recognized by MABs BC1, BC2, and BC3 appeared to be in a different region of the repeating unit, which consisted of a 9 amino acid region [294], 8 of which were in the 10 amino acid synthetic peptide that may contain the DF3 epitope [255]. Synthetic peptides based on the repeating unit of mucin core protein are thus proving to be powerful tools in establishing the amino acid composition and the relationships between protein epitopes on mucins. As the genes encoding other mucin molecules are sequenced, the use of synthetic peptides based on these sequences will undoubtedly be more frequently used to probe peptide epitopes of mucins. Synthetic peptides can be used as immunogens for generating MABs with finely defined immunoreactivities to the mucin polypeptide or to generate monospecific polyclonal antibodies. Polyclonal antibody against mucin preparations shown to be primarily reactive to mucin peptide [273,274,277,286] may be made more specific for a defined epitope by immunoaffinity purification using synthetic peptide as the immunoabsorbent. Detailed mapping of mucin protein epitopes as an approach to producing tumor-specific antibodies has been suggested elsewhere [292].

Mucins present in normal and malignant breast tissue [295] and in normal and malignant lung tissue [296] were shown to be products of the polymorphic gene locus [PUM].

These and other studies continue to show that the variation in the number of tandem repeats at the mucin gene level contributes to the electrophoretic heterogeneity of mucin-type glycoproteins by encoding mucin polypeptides of varying sizes. cDNAs have been cloned encoding the protein backbone for human small intestinal mucin [297]. A 69 base pair tandem repeat sequence was delineated that had little homology with human breast mucin and porcine submaxillary gland mucin [298]. The porcine mucin is encoded by genes composed of a 243 base pair tandem repeat sequence [298]. Very recently, the sequence of a pancreatic mucin cDNA was shown to contain the same 60-base pair tandem repeat of breast mucin [299].

The number of MABs reactive to tumor-associated mucins continues to expand. Unfortunately, the use by investigators of an inconsistent system of letters and numbers to name MABs and mucins has caused considerable confusion. Molecular biological, biochemical, and immunological analysis of many tumor-associated mucins has revealed many similarities in both antigens and epitopes. As these reagents become better characterized, it should be possible to consolidate and rename them in an orderly fashion, perhaps based on the epitope rather than the MAB.

ACKNOWLEDGMENT

The authors thank Ms. C. Dostalek for her patience and valuable assistance in the preparation of the manuscript.

REFERENCES

1. Mowry RW. The special value of methods that color both acidic and vicinal hydroxyl groups in the histochemical study of mucins. Ann NY Acad Sci 1963;106: 402-423.
2. Spiro RG. Glycoproteins. Adv Protein Chem 1973;27:349-467.
3. Gottschalk A, ed. Glycoproteins, their composition, structure and function, 2nd ed. Amsterdam: Elsevier, 1972.
4. Feizi T, Childs RA. Carbohydrates as antigenic determinants of glycoproteins. Biochem J 1987;245:1-11.
5. Klug TL, LeDonne NC, Greber TF, Zurawski VR Jr. Purification and composition of a novel gastrointestinal tumor-associated glycoprotein expressing sialylated lacto-N-fucopentaose II (CA 19-9). Cancer Res 1988;48:1505-1511.
6. Davis HM, Zurawski VR Jr, Bast RC Jr, Klug TL. Characterization of the CA 125 antigen associated with human epithelial ovarian carcinomas. Cancer Res 1986;46: 6143-6148.
7. Venegas M, Shaw S, Lovell L, et al. Purification of ascitic fluid glycoproteins containing certain tumor-associated antigen markers. Glycoconjugate J 1989;6;511-524.
8. Swallow DM, Gendler S, Griffiths B, et al. The human tumor-associated epithelial mucins are coded by an expressed hypervariable gene locus PUM. Nature 1987;328: 82-84.
9. Eckhardt AE, Timpte CS, Abernethy JL, et al. Structural properties of porcine submaxillary gland apomucin. J Biol Chem 1987;262:11339-11344.
10. Codington JF, Sanford BH, Jeanloz RW. Glycoprotein coat of the TA3 cell: isolation and partial characterization of a sialic acid-containing glycoprotein fraction. Biochemistry 1972;11:2559-2564.
11. Carraway KL, Spielman J. Structural and functional aspects of tumor cell sialomucins. Mol Cell Biochem 1986;72:109-120.

12. Hill HD Jr, Schwyzer M, Steinman HM, Hill RL. Ovine submaxillary mucin. Primary structure and peptide substrates of UDP-N-acetylgalactosamine mucin transferase. J Biol Chem 1977;252:3799-3804.
13. Kjeldsen T, Clausen H, Hirohashi S, et al. Preparation and characterization of monoclonal antibodies directed to the tumor-associated O-linked sialosyl-2 → 6α-N-Acetylgalactosaminyl (sialosyl-Tn) epitope. Cancer Res 1988;48:2214-2220.
14. Pant KD, Shochat D, Nelson MO, Goldenberg DM. Colon-specific antigen-p (CSAp). I: Initial clinical evaluation as a marker for colorectal cancer. Cancer 1982; 50:919-926.
15. Magnani JL, Nilsson B, Brockhaus M, et al. A monoclonal antibody defined antigen associated with gastrointestinal cancer is a ganglioside containing sialylated lacto-N-fucopentaose II. J Biol Chem 1982;257:14365-14369.
16. Arends JW, Verstijnen C, Bosman FT, et al. Distribution of monoclonal antibody-defined monosialoganglioside in normal and cancerous human tissues: an immunoperoxidase study. Hybridoma 1983;2:219-229.
17. Magnani JL, Brockhaus M, Smith DF, et al. A monosialoganglioside is a monoclonal antibody defined antigen of colon cancer. Science 1981;212:55-56.
18. Magnani JL, Steplewski Z, Koprowski H, Ginsburg V. Identification of the gastrointestinal and pancreatic cancer-associated antigen detected by monoclonal antibody 19-9 in the sera of patients as a mucin. Cancer Res 1983;43:5489-5492.
19. Feizi T. Demonstration by monoclonal antibodies that carbohydrate structures of glycoproteins and glycolipids are onco-developmental antigens. Nature 1985;314: 53-57.
20. Hakomori SI. Tumor-associated carbohydrate antigens. Annu Rev Immunol 1984;2: 103-126.
21. Rittenhouse HG, Manderino GL, Hass GM. Mucin-type glycoproteins as tumor markers. Lab Med 1985;16:556-560.
22. Hakomori S. Aberrant glycosylation in cancer cell membranes as focused on glycolipids: overview and perspectives. Cancer Res 1985;45:2405-2414.
23. Feizi T. Monoclonal antibodies reveal saccharide structures of glycoproteins and glycolipids as differentiation and tumor-associated antigens. Biochem Soc Trans 1984;12:545-549.
24. Milstein C, Lennox E. The use of monoclonal antibody techniques in the study of developing cell surfaces. Curr Top Dev Biol 1980;14:1-32.
25. Hakomori S. Glycospingolipids in cellular interaction, differentiation and oncogenesis. Ann Rev Biochem 1981;50:733-764.
26. Feizi T. Carbohydrate differentiation antigens. Trends Biochem Sci 1981;6:333-335.
27. Kabat EA, ed. Structural concepts in immunology and immunochemistry, 2nd ed. New York: Holt, Rinehart and Winston, 1976:143-197.
28. Feizi T. The blood group Ii system; a carbohydrate antigen system defined by naturally monoclonal or oligoclonal autoantibodies of man. Immunol Commun 1981;10:127-156.
29. Tang PW, Scudder P, Mehmet H, et al. Sulfate groups are involved in the antigenicity of keratan sulfate and mask i antigen expression on their poly N-acetyllactosamine backbones. An immunochemical and chromatographic study of keratan sulfate oligosaccharides after desulphation or nitrosation. Eur J Biochem 1986;160: 537-545.
30. Schoentag R, Primus FJ, Kuhns W. ABH and lewis blood group expression in colorectal carcinoma. Cancer Res 1987;47:1695-1700.
31. Kassulke J, Stutman O, Yunis E. Blood group isoantigens in leukemic cells: reversibility of isoantigenic changes by neuraminidase. J Natl Cancer Inst 1971;46:1201-1208.

32. Nieman H, Watanabe K, Hakomori S, et al. Blood group i and I activities of "Lacto-N-norhexaosylceramide" and its analogues: the structural requirements for i-specificities. Biochem Biophys Res Commun 1978;81:1286-1293.
33. Watanabe K, Hakomori S, Childs RA, Feizi T. Characterization of a blood group I-active ganglioside. Structural requirements for I and i specificities. J Biol Chem 1979;254:3221-3228.
34. Feizi T, Childs RA, Watanabe K, Hakomori S. Three types of blood group I specificity among monoclonal anti-I autoantibodies revealed by analogues of a branched erythrocyte glycolipid. J Exp Med 1979;149:975-980.
35. Thorpe SJ, Feizi T. Species differences in the expression of carbohydrate differention antigens on mammalian blood cells revealed by immunofluorescence with monoclonal antibodies. Biosci Rep 1984;4:673-685.
36. Gilbert ER, Crookston MC. Agglutinability of red cells by anti-i in patients with thalassemia major and other haematological disorders. Nature 1964;201:1138.
37. Kannagi R, Levery SB, Hakomori S. Sequential change of carbohydrate antigen associated with differentiation of murine leukemia cells; Ii antigenic conversion and shifting of glycolipid synthesis. Proc Natl Acad Sci (USA) 1983;80:2844-2848.
38. Gooi HC, Feizi T, Kapadia A, et al. Stage-specific embryonic antigen involves $\alpha 1 \to 3$ fucosylated type 2 blood group chains. Nature (London) 1981;292:156-158.
39. Hakomori S, Nudelman E, Levery S, et al. The hapten structure of a developmentally regulated glycolipid antigen (SSEA-1) isolated from human erythrocytes and adenocarcinoma: a preliminary note. Biochem Biophys Res Commun 1981;100:1578-1586.
40. Willison KR, Karol RA, Suzuki A, et al. Neutral glycolipid antigens as developmental markers of mouse teratocarcinoma and early embryos: an immunologic and chemical analysis. J Immunol 1982;129:603-609.
41. Feizi T, Gooi HC, Childs RA, et al. Tumor-associated and differentiation antigens on the carbohydrate moieties of mucin-type glycoproteins. Biochem Soc Trans 1984;12:591-596.
42. Hansson GC, Karlsson KA, Larson G, et al. Mouse monoclonal antibodies against human cancer cell lines with specificities for blood group and related antigens. J Biol Chem 1983;258:4091-4097.
43. Uhlenbruck G, Van Meensel-Maene U, Hanisch FG, Dienst C. Unexpected occurrence of the Ca 19-9 tumor marker in normal human seminal plasma. Hoppe-Seylers Z Physiol Chem 1984;365:615-617.
44. Myoga A, Taki T, Arai K, et al. Detection of patients with cancer by monoclonal antibody directed to lactoneotetraosylceramide (paragloboside). Cancer Res 1988;48:1512-1516.
45. Shimizu M, Yamauchi K. Isolation and characterization of mucin-like glycoproteins in human milk fat globule membranes. J Biochem (Tokyo) 1982;91:515-524.
46. Koldovsky U, Wargalla U, Hilkens J, et al. Reactions of monoclonal antibodies against human milk fat globule membranes with embryonal tissue. Protides Biol Fluids 1984;31:1021-1024.
47. Burchell J, Wang D, Taylor-Papadimitriou J. Detection of the tumor-associated antigens recognized by the monoclonal antibodies HMFG-1 and 2 in serum from patients with breast cancer. Int J Cancer 1984;34:763-768.
48. Ceriani RL, Sasaki M, Sussman H, et al. Circulating human mammary epithelial antigens in breast cancer. Proc Natl Acad Sci (USA) 1982;79:5420-5424.
49. Taylor-Papadimitriou J, Peterson JA, Arklie J, et al. Monoclonal antibodies to epithelium-specific components of the human milk fat globule membrane: production and reaction with cells in culture. Int J Cancer 1981;28:17-21.
50. Ormerod MG, Steele K, Westwood JH, Mazzini MN. Epithelial membrane antigen: partial purification, assay and properties. Br J Cancer 1983;48:533-541.

51. Burchell J, Durbin H, Taylor-Papadimitriou J. Complexity of expression of antigenic determinants, recognized by monoclonal antibodies HMFG-1 and HMFG-2 in normal and malignant human mammary epithelial cells. J Immunol 1983;131: 508-513.
52. Hilkens J, Hilgers J, Buijs F. Monoclonal antibodies against milk fat globule membranes useful in carcinoma research. Protides Biol Fluids 1984;31:521-524.
53. Wagennaar SJ, Hilgers J, Schmitz-DuMoulin F. Patterns of expression of some new antigens of human bronchial carcinomas. Protides Biol Fluids 1985;45:379-385.
54. Papsidero LD, Nemoto T, Croghan GA, Chu TM. Expression of ductal carcinoma antigen in breast cancer sera as defined using monoclonal antibody F36/22. Cancer Res 1984;44:4653-4657.
55. Burchell J, Gendler S, Taylor-Papadimitriou, et al. Development and characterization of breast cancer reactive monoclonal antibodies directed to the core protein of human milk mucin. Cancer Res 1987;47:5476-5482.
56. Bast RC Jr, Feeney M, Lazarus H, et al. Reactivity of a monoclonal antibody with human ovarian carcinoma. J Clin Invest 1981;68:1331-1337.
57. Kabawat SE, Bast RC Jr, Welch WR, et al. Immunopathologic characterization of a monoclonal antibody that recognizes common surface antigens of human ovarian tumors of serous, endometrioid and clear cell types. Am J Clin Pathol 1983;79: 98-104.
58. Hanisch F-G, Uhlenbruck G, Dienst C, et al. CA 125 and CA 19-9: two cancer-associated sialylsaccharide antigens on a mucus glycoprotein from human milk. Eur J Biochem 1985;149:323-330.
59. Nouwen EJ, Pollet DE, Eerdekens MW, et al. Immunohistochemical localization of placental alkaline phosphatase, carcinoembryonic antigen and cancer antigen CA 125 in normal and neoplastic lung. Cancer Res 1986;46:866-876.
60. deBruijn HWA, vanBeeck-Calkoen-Carpay-T, Jager S, et al. The tumor marker CA 125 is a common constituent of normal cervical mucus. Am J Obstet Gynecol 1986; 154:1088-1091.
61. Kabawat SE, Bast RC Jr, Bahan AK, et al. Tissue distribution of a coelomic-epithelium-related antigen recognized by the monoclonal antibody OC125. Int J Gynecol Pathol 1983;2:275-285.
62. Colcher D, Horan Hand P, Nuti M, Schlom JA. A spectrum of monoclonal antibodies reactive with human mammary tumor cells. Proc Natl Acad Sci (USA) 1981; 78:3199-3203.
63. Nuti M, Teramoto YA, Meriani-Costantini R, et al. A monoclonal antibody (B72.3) defines patterns of distribution of a novel tumor-associated antigen in human mammary carcinoma cell populations. Int J Cancer 1982;29:539-545.
64. Stramignoni D, Bowen R, Atkinson B, Schlom J. Differential reactivity of monoclonal antibodies with human colon carcinomas and adenomas. Int J Cancer 1983; 31:543-552.
65. Johnson VG, Schlom J, Paterson AJ, et al. Analysis of a human tumor-associated glycoprotein (TAG-72) identified by monoclonal antibody B72.3. Cancer Res 1986; 46:850-857.
66. Hakomori S. Blood group glycolipid antigens and their modifications as cancer antigens. Am J Clin Pathol 1984;82:635-648.
67. Carding SR, Thorpe SJ, Childs RA, Feizi T. Production and characterization of monoclonal antibodies to B-galactoside binding lectin of bovine heart muscle. Biochem J 1984;220:253-260.
68. Shively JE, Beatty JD. CEA-related antigens: molecular biology and clinical significance. CRC Crit Rev Oncol Hematol 1985;2:355-399.

69. Woodward MP, Young WW Jr, Bloodgood RA. Detection of monoclonal antibodies specific for carbohydrate epitopes using periodate oxidation. J Immunol Methods 1985;78:143-153.
70. Basbaum CB, Chow A, Macher BA, et al. Tracheal carbohydrate antigens identified by monoclonal antibodies. Arch Biochem Biophys 1986;249:363-373.
71. Edge AS, Faltynek CR, Hof L, et al. Deglycosylation of glycoproteins by trifuoromethane sulfonic acid. Anal Biochem 1981;118:131-137.
72. Tang PW, Feizi T. Neoglycolipid micro-immunassays applied to the oligosaccharides of human milk galactosyltransferase detect blood-group related antigens on both O- and N-linked chains. Carbohydr Res 1987;161:133-143.
73. Lan MS, Khorrami A, Kaufman B, Metzgar RS. Molecular characterization of a mucin-type antigen associated with human pancreatic cancer. J Biol Chem 1987;262:12863-12870.
74. Sekine H, Ohno T, Kufe DW. Purification and characterization of a high molecular weight glycoprotein detectable in human milk and breast carcinomas. J Immunol 1985;135:3610-3615.
75. Shochat D, Pant KD, Goldenberg DM. Colon-specific antigen-p (CSAp). II: Further characterization in colorectal and pancreatic cancer. Cancer 1982;50:927-931.
76. Kalthoff H, Kreiker C, Schmiegel W-H, et al. Characterization of CA19-9 bearing mucins as physiological exocrine pancreatic secretion products. Cancer Res 1986;46:3605-3607.
77. Bramwell ME, Bhavanandan VP, Wiseman G, Harris H. Structure and function of the Ca antigen. Br J Cancer 1983;48:177-183.
78. Kress BC, Dollar LA. Koda JE, Gaur PK. Purification and characterization of CA-549, a new circulating breast cancer marker. Abstract. J Cell Biochem Suppl 1987;11D:160.
79. Pant KD, Zamora PO, Rhodes BA, et al. Characterization of a common antigen of colorectal and mucinous ovarian tumors, COTA. Tumor Biol 1984;5:243-254.
80. Miotti S, Aguanno S, Canvari J, et al. Biochemical analysis of human ovarian cancer-associated antigens defined by murine monoclonal antibodies. Cancer Res 1985;45:826-832.
81. Lan MS, Bast Jr RC, Colnaghi MI, et al. Co-expression of human cancer-associated epitopes on mucin molecules. Int J Cancer 1987;39:68-72.
82. Lan MS, Finn OJ, Fernsten PD, Metzgar RS. Isolation and properties of a human pancreatic adenocarcinoma-associated antigen, DU-PAN-2. Cancer Res 1985;45:305-310.
83. Morgan AC, Crane MM, Rossen RD. Measurement of a monoclonal antibody-defined, melanoma-associated antigen in human sera: correlation of circulating antigen levels with tumor burden. J Natl Cancer Inst 1984;72:243-249.
84. Papsidero LD, Croghan GA, Johnson EA, Chu TM. Immunoaffinity isolation of ductal carcinoma antigen using monoclonal antibody F36/22. Mol Immunol 1984;21:955-960.
85. Nocera M, Shochat D, Primus FJ, et al. Representation of epitopes on colon-specific antigen-p defined by monoclonal antibodies. J Natl Cancer Inst 1987;79:943-948.
86. Papsidero LD, Johnson EA. Physiochemical purification and immunological characteristics of ductal carcinoma antigen. Int J Cancer 1986;37:697-703.
87. Carlstedt I, Lindgren H, Sheehan JK, et al. Isolation and characterization of human cervical-mucus glycoproteins. Biochem J 1983;211:13-22.
88. Sherblom AP, Buck RL, Carraway KL. Purification of the major sialoglycoproteins of 13762 MAT-B1 and MAT-C1 rat ascites mammary adenocarcinoma cells by density gradient centrifugation in cesium chloride and guanidine hydrochloride. J Biol Chem 1980;255:783-790.

Mucin Glycoproteins

89. Gold DV. Colonic mucoprotein antigen (CMA). In: Herberman, ed. Compendium of assays for immunodiagnosis of human cancer. New York: Elsevier North Holland, 1979:231-236.
90. Hilkens J, Lightenberg M, Buijs F, Hageman PH. Biosynthesis and molecular organization of epithelial sialomucins. Abstract. J Cell Biochem Suppl 1988;12E:117.
91. Lan MS, Hollingsworth MA, Metzgar RS. Properties of DU-PAN-2 Epitope. Abstract. J Cell Biochem Suppl 1987;11D:160.
92. Lindholm L, Holmgren J. Svennerholm L, et al. Monoclonal antibodies against gastrointestinal tumor-associated antigens isolated as monosialogangliosides. Int Arch Allergy Appl Immunol 1983;71:178-181.
93. Chung YS, Ho JJU, Kim YS, et al. The detection of human pancreatic cancer associated antigen in the serum of cancer patients. Cancer 1987;60:1636-1643.
94. Ho JJL, Chung Y, Fujimoto Y, et al. Mucin-like antigens in a human pancreatic cancer cell line: identified by murine monoclonal antibodies, Span-1 and YPan-1. Cancer Res 1988;48:3924-3931.
95. Gendler SJ, Burchell JM, Duhig T, et al. Cloning of partial cDNA encoding differentiation and tumor associated mucin glycoproteins expressed by human mammary epithelium. Proc Natl Acad Sci (USA) 1987;84:6060-6064.
96. Gendler SJ, Taylor-Papadimitriou J, Duhig T, Burchell J. A highly immunogenic region of the polymorphic gene coding for the human mammary mucin is made up of 60 base pair tandem repeats. Abstract. J Cell Biochem Suppl 1988;12E:128.
97. Abe M, Siddiqui J, Hayes D, Kufe D. Isolation and sequencing of a cDNA coding for the human DF3 breast carcinoma-associated antigen. Abstract. J Cell Biochem Suppl 1988;12E:127.
98. Abe M, Kufe D. Identification of a family of high molecular weight tumor associated glycoproteins. J Immunol 1987;139:257-261.
99. Ashall F, Bramwell ME, Harris H. A new marker for human cancer cells: 1. The Ca antigen and the Ca 1 antibody. Lancet 1982;2:1-10.
100. Linsley PS, Ochs V, Horn D, et al. Heritable variation in expression of multiple tumor associated epitopes on a high molecular weight mucin-like antigen. Cancer Res 1986;46:6380-6386.
101. Leoni T, Magnani JL, Miotti S, et al. The antitumor monoclonal antibody MOv2 recognizes the Lewis A hapten. Hybridoma 1988;7:129-139.
102. Price MR, Edwards S, Owainati A, et al. Multiple epitopes on a human breast-carcinoma-associated antigen. Int J Cancer 1985;36:567-574.
103. Del Villano BC, Brennan S, Brock P, et al. Radioimmunometric assay for a monoclonal antibody-defined tumor marker, CA 19-9. Clin Chem 1983;29:552.
104. Ritts RE Jr, Del Villano BC, Go VLW, et al. Initial clinical evaluation of an immunoradiometric assay for CA 19-9 using NCI serum bank. Int J Cancer 1984;33:339-345.
105. Herlyn M, Sears HF, Steplewski Z, and Koprowski H. Monoclonal antibody detection of circulating tumor-associated antigen. I. Presence of antigen in sera of patients with colorectal, gastric, and pancreatic carcinoma. J Clin Immunobiol 1982;2:135-140.
106. Satake K, Kanazawa G, Kho I, et al. Clinical evaluation of carbohydrate antigen 19-9 and carcinoembryonic antigen in patients with pancreatic carcinoma. J Surg Oncol 1985;29:15-21.
107. Takasaki H, Uchida E, Tempera MA, et al. Correlative study on expression of CA 19-9 and DU-PAN-2 in tumor tissue and in serum of pancreatic cancer patients. Cancer Res 1988;48:1435-1438.

108. Fujii Y, Albers GHR, Carre-Llopis A, Escribano MJ. The diagnostic value of the foetoacinar pancreatic (FAP) protein in cancer of the pancreas; a comparative study with CA 19-9. Br J Cancer 1987;56:495-500.
109. Ritts RE, Nagorney DM, Jacobsen DJ, Zurawski VR. A prospective study of CA 19-9 and CEA in abdominal surgery patients. Abstract. Antigens of human pancreatic adenocarinomas: Their role in diagnosis and therapy. Sponsored by Organ Systems Coordinating Center, 1987.
110. Kew MC, Berger EL, Koprowski H. Carbohydrate antigen 19-9 as a serum marker of hepatocellular carcinoma: comparison with alpha foetoprotein. Br J Cancer 1987;56:86-88.
111. Herlyn M, Sears HF, Verrill H, Koprowski H. Increased sensitivity in detecting tumor-associated antigens in sera of patients with colorectal carcinoma. J Immunol Methods 1984;75:15-21.
112. Brockhaus M, Wysocka M, Magnani JL, et al. Normal salivary mucin contains the gastrointestinal cancer associated antigen detected by monoclonal antibody 19-9 in the serum mucin of patients. Vox Sang 1985;48:34-38.
113. Bara J, Zabaleta EH, Mollicone R, et al. Distribution of GICA in normal gastrointestinal and endocervical mucosae and in mucinus ovarian cysts using antibody NS 19-9. Am J Clin Pathol 1986;85:152-159.
114. Mourant AE. Distribution of human blood groups and other polymorphisms. London: Oxford University Press, 1976;574.
115. Harmenberg U, Wahren B, Wiechel K. Tumor markers carbohydrate antigens CA 19-9 and CA-50 and carcinoembryonic antigen in pancreatic cancer and benign diseases of the pancreatobiliary tract. Cancer Res 1988;48:1985-1988.
116. Chung YS, Tanaka H, Yamamoto T, et al. Clinical evaluation of pancreatic cancer associated antigen Span-1. Abstract. Dig Dis Sci 1986;31:1127.
117. Stacker SA, Sacks NPM, Golder J, et al. Evaluation of MSA as a serum marker in breast cancer: a comparison with CEA. Br J Cancer 1988;57:298-303.
118. Demers LM, Harvey HA, Lipton A, Gaur PK. The evaluation of a new breast cancer tumor marker, CA 549, in patients with breast cancer. Abstract. Breast Cancer Res Treat 1987;10:141.
119. Bray KR, Suchocki, Gaur PK. Correlation of serum levels of CA 549 to disease progression and remission in breast cancer patients. Abstract. FASEB J 1988;2:6813.
120. Dhokia B, Pectasides D, Self C, et al. A low pH enzyme linked immunoassay using two monoclonal antibodies for the serological detection and monitoring of breast cancer. Br J Cancer 1986;54:885-889.
121. Hayes DF, Sekine H, Ohno T, et al. Use of a murine monoclonal antibody for detection of circulating plasma DF3 antigen levels in breast cancer patients. J Clin Invest 1985;75:1671-1678.
122. Pons-Anicet DMF, Krebs BP, Mira R, Namer M. Value of CA 15-3 in the follow up of breast cancer patients. Br J Cancer 1987;55:567-569.
123. Fujino N, Haga Y, Sakamoto K, et al. Clinical evaluation of an immunoradiometric assay for CA15-3 antigen associated with human mammary carcinomas comparison with carcinoembryonic antigen. Jpn J Clin Oncol 1986;16:335-346.
124. Namer M, Krebs BP, Hery M, et al. CEA versus CA 15-3 as markers for metastasized breast cancer. Abstract. Proc Ann Meet Am Soc Clin Oncol 1986;5:24.
125. Schmidt-Rhode P, Sturm G, Schulz KD, et al. Preliminary findings with a new tumor marker (CA 15-3) in breast cancer. In: Wust G, ed. Tumor markers. Darmstadt, Germany: Steinkopff Verlag, 1987;186-197.
126. Hayes DF, Zurawski VR, Kufe DW. Comparison of circulating CA 15-3 and carcinoembryonic antigen in patients with breast cancer. J Clin Oncol 1986;4:1542-1550.

127. Tondini C, Hayes DF, Gelman R, et al. Comparison of CA 15-3 and CEA in monitoring the clinical course of patients with metastatic breast cancer. Cancer Res 1988;48:4107-4112.
128. Kiang DT, Greenberg LJ, Kennedy BJ. Comparison of plasma CA 15-3 and CEA changes in advanced breast cancer. Proc Am Assoc Cancer Res 1988;29:681.
129. Linsley PS, Brown JP, Magnani JL, Horn D. Monoclonal antibodies reactive with mucin glycoproteins found in sera from breast cancer patients. Cancer Res 1988; 48:2138-2148.
130. Kinders RJ, Slota J, Moore BS, et al. Abbott CTA: a new serum assay for the assessment of breast cancer patients. Abstract. Breast Cancer Res Treat 1987;10: 46.
131. Anderson B, Slota J, Kundu S, et al. Characterization of monoclonal antibodies to paragloboside and sialosyl-PG, and an improved chromatogram binding assay for rapidly identifying antibodies to tumor antigens. J Cell Biochem Supplement 1987;11D:157.
132. Einhorn N, Bast RC, Knapp RC, et al. Preoperative evaluation of serum CA 125 levels in patients with primary epithelial ovarian cancer. Obstet Gynecol 1986; 67:414-416.
133. Malkasian GD, Knapp RC, Lavin PT, et al. Preoperative evaluation of serum CA 125 levels in premenopausal and postmenopausal patients with pelvic mass–discrimination of benign from malignant disease. Amer J Obstet Gynecol 1988;159: 341-346.
134. Bast RC Jr, Klug TL, St. John E, et al. A radioimmunassay using a monoclonal antibody to monitor the course of epithelial ovarian cancer. N Engl J Med 1983; 309:883-887.
135. Niloff JM, Klug TL, Schaetzl E, et al. Elevation of serum CA 125 in carcinomas of the fallopian tube, endometrium, and endocervix. Am J Obstet Gyencol 1984; 148:1057-1058.
136. Knapp RC, Lavin PT, Schaetzl E, et al. Elevation of CA 125 prior to recurrence of ovarian cancer. Proc Soc Gynecol Oncol 1985;16:30.
137. Niloff JM, Bast RC Jr, Schaetzl EM, et al. Predictive value of CA 125 antigen levels at second look procedures in ovarian cancer. Am J Obstet Gynecol 1985;151: 981-986.
138. Bereck JS, Knapp RC, Malkasian GD, et al. CA 125 serum levels correlate with second look operations among ovarian cancer patients. Obstet Gynecol 1986;67: 685-689.
139. Bast RC Jr. CA 125 levels in peritoneal fluid in benign and malignant disease. Personal communication.
140. Niloff M, Knapp RC, Schaetzl E, et al. CA 125 antigen levels in obstetric and gynecologic patients. Obstet Gynecol 1985;64:703-707.
141. Zurawski VR, Broderick SF, Pickens P, et al. Serum CA 125 in a group of non-hospitalized women. Relevance for the early detection of ovarian cancer. Obstet Gynecol 1987;69:606-611.
142. Ward BG, Cruickshank DJ, Tucker DF, Love S. Independent expression in serum of three tumor associated antigens: CA 125, placental alkaline phosphatase and HMFG2 in ovarian carcinoma. Br J Obstet Gynecol 1987;94:696-698.
143. Chia D, Terasaki PI, Suyama N, et al. Use of monoclonal antibodies to sialylated Lewis^x and sialylated Lewis^a for serological tests of cancer. Cancer Res 1985;45: 435-437.
144. Kawahara M, Chia D, Terasaki PT, Roumanas A, et al. Detection of sialylated Lewis^x antigen in cancer sera using a sandwich radioimmunoassay. Int J Cancer 1985;36:421-425.

145. Fukushi Y, Kannagi R, Hakomori S, et al. Location and distribution of difucoganglioside (VI3NeuAcV3III3Fuc2nLc6) in normal and tumor tissues defined by its monoclonal antibody FH6. Cancer Res 1985;45:3711-3717.
146. Kannagi R, Fukushi Y, Tachikawa T, et al. Quantitative and qualitative characterization of human cancer-associated serum glycoprotein antigens expressing fucosyl or sialyl-fucosyl type 2 chain polylactosamine. Cancer Res 1986;46:2619-2626.
147. Martensson S, Due C, Pahlsson P, et al. A carbohydrate epitope associated with human squamous lung cancer. Cancer Res 1988;48:2125-2131.
148. Pettijohn DE, Stranahan PL, Due C, et al. Glycoproteins distinguishing non-small cell from small cell human lung carcinoma recognized by monoclonal antibody 43-9F. Cancer Res 1987;47:1161-1169.
149. Fargion S, Carney D, Mulshine J, et al. Heterogeneity of cell surface antigen expression of human small cell lung cancer detected by monoclonal antibodies. Cancer Res 1986;46:2633-2638.
150. Benbassat J, Regev A, Slater PE. Predictive value of sputum cytology. Thorax 1987;42:165-172.
151. Klug TL, Sattler MA, Colcher D, Schlom J. Monoclonal antibody immunoradiometric assay for an antigenic determinant (CA72) on a novel pancarcinoma antigen (TAG-72). Int J Cancer 1986;38:661-669.
152. Gaur PK, Lofaro LR, Bray KR, et al. Incidence of elevated serum levels of CA195 in colon cancer according to disease stage. Abstract. Proc Am Assoc Cancer Res 1988;29:1541.
153. Gupta MK, Arciaga R, Bukowski RM, Gaur PK. Monoclonal antibody defined antigens Ca 195 and CA 19-9 in colon cancer: comparison with CEA. Abstract. J Cell Bilchem Suppl 1988;12E:137.
154. Schwartz MK, Schwartz D, Smith C, Nisselbaum J. Comparison of CEA and CA 195. Abstract. Proc Am Assoc Cancer Res 1988;29:698.
155. Pant KD, Stewart TA, Berry CDA, Rhodes BA. Production of monoclonal antibody SP-21 to colon-ovarian tumor antigen, COTA. Hybridoma 1986;5:129-135.
156. Pant KD, Shah VO, O'Rourke AT, et al. Further characterization of COTA and a comparison of SP-21 (anti COTA) to other anti-colorectal antibodies. J Tumor Marker Oncol 1988;3:1-14.
157. Gold DV, Miller F. Comparison of human colonic mucoprotein antigen from normal and neoplastic mucosa. Cancer Res 1978;38:3204-3211.
158. Gold DV, Miller F. A mucoprotein with colon-specific determinants. Tissue Antigens 1978;11:362-371.
159. Hilkens J, Bonfrer JM, Kroezen V, et al. Comparison of circulating MAM 6 and CEA levels and correlation with the estrogen receptor in patients with breast cancer. Int J Cancer 1987;39:431-435.
160. Tobias R, Rothwell C, Wagner J, et al. Development and evaluation of a radioimmunoassay for the detection of a monoclonal antibody defined breast tumor associated antigen 115D8/DF3. Abstract. J Am Assoc Clin Chem 1985;31:986.
161. Kantor JA, Feller WF, Hilkens J, Hilgers J. Use of monoclonal antibodies to detect circulating epithelial membrane antigens in breast cancer patients. J Tumor Marker Oncol 1987;1:225-229.
162. Feller WF, Kanto J, Hilkens J, Hilgers J. Circulating differentiation antigens in the serum of patients with breast cancer and benign epithelial proliferation. Abstract Proceedings International Breast Cancer Research Conference, London, UK. Abst 4-08:126, 1985.
163. Stacker SA, Sacks NPM, Thompson CH, et al. A serum test for the diagnosis and monitoring of the progress of breast cancer. In: Ceriani RL, ed. Immunological approaches to the diagnosis and therapy of breast cancer. New York: Plenum Press, 1987;217.

164. Kawahara M, Terasaki PI, Chia D, et al. Use of four monoclonal antibodies to detect tumor markers. Cancer 1986;58:2008-2012.
165. Chia D, Terasaki PI, Wiseman C, et al. The use of multiple tumor markers to monitor cancer patients. Proceedings UCLA Symposium: Altered Glycosylation in Tumor Cells. J Cell Biochem Suppl 1987;11D:
166. Chia D, Terasaki PI, Wiseman C, et al. The use of multiple tumor markers to monitor cancer patients. In: Reading CL, Hakomori S, Marcus, eds. UCLA symposia on molecular and cellular biology new series, Vol 79. Altered glycosylation in tumor cells. New York: Alan R. Liss, Inc., 1988;295-300.
167. Price MR, Crocker G, Edwards S, et al. Identification of a monoclonal antibody defined breast carcinoma antigen in body fluids. Eur J Cancer Clin Oncol 1987;23: 1169-1176.
168. Gould VE, Gooch GT, Rittenhouse HR, Manderino GL. Selective expression of a novel mucin-type glycoprotein in human tumors: immunohistochemical demonstration with MAb A-80. Hum Pathol 1988;19:623-627.
169. Schecter R, Major P. Double antibody RIA in conjunction with immunoblotting for the specific detection of a serum breast tumor marker. Abstract. J Cell Biochem Suppl 1987;11D:122.
170. Diotti A, Santoro O, Mantovani L, et al. A double determinant radioimmunoassay MOv2/MOv8 for monitoring ovarian carcinomas: definition of the methodology. Int J Biol Markers 1987;2:161-168.
171. DeKretser TA, Thorne HJ, Picone D, Jose DG. Biochemical characterization of the monoclonal antibody defined ovarian carcinoma associated antigen SGA. Int J Cancer 1986;37:705-712.
172. Holmgren J, Lindholm L, Persson B, et al. Detection by monoclonal antibody of carbohydrate antigen CA-50 in serum of patients with carcinoma. Br Med J 1984; 288:1479-1482.
173. Sawabu N, Toya D, Takemori Y. Measurement of a pancreatic cancer associated antigen (DU-PAN-2) detected by a monoclonal antibody in sera of patients with digestive cancers. Int J Cancer 1986;37:693-696.
174. Metzgar RS, Rodriguez N, Finn O, et al. Detection of a pancreatic cancer associated antigen (Du-PAN-2 antigen) in serum and ascites of patients with adenocarcinoma. Proc Natl Acad Sci USA 1984;81:5242-5246.
175. Bhattacharya-Chatterjee M, Petrelli NJ, Herrera L, et al. Expression of a high molecular weight mucin-type glycoprotein in colon cancer sera as defined with monoclonal antibody 1D3. Abstract. Proc Am Assoc Cancer Res 1987;28:741.
176. Kannagi R, Hakomori S, Imura H. Cancer-associated serum glycoprotein antigens expressing Type 1 and Type 2 chain polylactosamines. Abstract. J Cell Biochem Suppl 1987;11D:138.
177. Kannagi R, Fukushi Y, Hakomori S. Cancer-associated mucin detected by monoclonal anti-carbohydrate antibodies. Gann 1986;13:812-815.
178. Hirota M, Fukushima K, Terasaki PI, et al. Sialosylated Lewis x in the sera of cancer patients detected by a cell binding inhibition assay. Cancer Res 1985;45: 1901-1905.
179. Colomer R, Sole LA, Navarro M, et al. CA 15-3; early results of a new breast cancer marker. Anticancer Res 1986;6:683-684.
180. Hilkens J, Kroezen V, Bonfrer JMG, et al. MAM-6 antigen, a new serum marker for breast cancer monitoring. Cancer Res 1986;46:2582-2587.
181. Stacker SA, Thompson CH, Lichtenstein M, et al. Detection of breast cancer using the monoclonal antibody 3E1.2. In: Ceriani RL, ed. Proceedings, international workshop on monoclonal antibodies and breast cancer. Boston: Martinus Nijhoff, 1985:233.

182. Sacks NPM, Stacker SA, Thompson CH, et al. Comparison of mammary serum antigen (MSA) and CA15-3 levels in the serum of patients with breast cancer. Br J Cancer 1987;56:820-824.
183. Haglund C, Kuusela P, Jalanko H, Roberts PJ. Serum CA 50 as a tumor marker in pancreatic cancer: a comparison with CA 19-9. Int J Cancer 1987;39:477-481.
184. Canney PA, Moore M, Wilkinson PM, James RD. Ovarian cancer antigen CA 125: a prospective clinical assessment of its role as a tumor marker. Br J Cancer 1984;50:765-769.
185. Crombach G, Zippel HH, Wurz H. Clinical significance of cancer antigen 125 (CA 125) in ovarian cancer. Cancer Detect Prev 1983;6:623.
186. Bast RC Jr, Knapp RC. The emerging role of monoclonal antibodies in the clinical management of epithelial ovarian carcinoma. In: DeVita VT, Hellman S, Rosenberg SA, eds. Important advances in oncology, 1987. Philadelphia: J B Lippincott 1987;39-53.
187. Kreienberg R, Melchart F. CA 125—A new radioimmunoassay for monitoring patients with epithelial ovarian carcinoma. Cancer Detect Prev 1983;6:619.
188. Lien LC, Hu XF, Liu WS. A monoclonal antibody RIA for an antigenic determinant (Ca 125) in ovarian cancer patients. Chin J Obstet Gynecol 1985;20:257-260.
189. Little LE, Lofaro LR, Gaur PK. Role of the Lewis A epitope on the circulating tumor associated antigen CA 195. Abstract. FASEB J 1988;2:6813.
190. Linsley PS, Ochs V, Laska S, et al. Elevated levels of a high molecular weight antigen detected by antibody W1 in sera from breast cancer patients. Cancer Res 1986;46:5444-5450.
191. Munck-Wikland E, Kuylenstierna R, Wahren B, et al. Tumor markers carcinoembryonic antigen, CA 50 and CA 19-9 and squamous cell carcinoma of the esophagus. Cancer 1988;62:2281-2286.
192. McKnight A, Mannell A, Shperling I. The role of carbohydrate antigen 19-9 as a tumour marker of oesophageal cancer. Br J Cancer 1989;60:249-251.
193. Gustafsson H, Franzen L, Grankvist K, et al. Glycoprotein tumour markers in head and neck neoplasms—a consecutive study of CA-50, CA 19-9, and CEA. J Cancer Res Clin Oncol 1988;114:394-398.
194. Frebourg T, Bercoff E, Manchon N, et al. The evaluation of CA 19-9 antigen level in the early detection of pancreatic cancer. Cancer 1988;62:2287-2290.
195. Glenn J, Steinberg WM, Kurtzman SH, et al. Evaluation of the utility of radioimmunoassay for serum CA 19-9 levels in patients before and after treatment of carcinoma of the pancreas. J Clin Oncol 1988;6:462-468.
196. Pleskow DK, Berger HJ, Gyves J, et al. Evaluation of a serologic marker, CA19-9, in the diagnosis of pancreatic cancer. Ann Intern Med 1989;110:704-709.
197. Tabuchi Y, Deguchi H, Saitoh Y. Carcinoembryonic antigen and carbohydrate antigen 19-9 levels of peripheral and draining venous blood in colorectal cancer patients. Cancer 1988;62:1605-1613.
198. Talbot RW, Nagorney DM, Pemberton JH, et al. Comparison of portal and peripheral blood levels of carcinoembryonic antigen, CA 19-9, and CA 125 tumor-associated antigens in patients with colorectal and pancreatic cancer. Cancer Res 1989;49:542-543.
199. Tabuchi Y, Deguchi H, Saitoh Y. Histopathological and immunohistochemical studies on the elevation mechanism of cancer associated antigens, CEA and CA19-9, in gastric cancer patients. Nippon Geka Gakkai Zasshi 1988;89:1181-1191.
200. O'Connell MJ, Ritts RA, Moertel CG, et al. Recombinant interferon-γ lacks activity against metastatic colorectal cancer but increases serum levels of CA 19-9. Cancer 1989;63:1998-2004.
201. Farinati F, Nitti D, Cardin F, et al. CA 19-9 determination in gastric juice: role in

identifying gastric cancer and high risk patients. Eur J Cancer Clin Oncol 1988;24: 923-927.
202. Fujii S, Konishi I, Nanbu Y, et al. Analysis of the levels of CA125, carcinoembryonic antigen, and CA19-9 in the cervical mucus for a detection of cervical adenocarcinoma. Cancer 1988;62:541-547.
203. Tizzani A, Cassetta G, Cicigoi A, et al. Tumor markers (CEA, TPA, and CA 19-9) in urine of bladder cancer patients. Int J Biol Markers 1987;2:121-124.
204. Persson BE, Stakle E, Pahlman L, et al. A clinical study of CA-50 as a tumor marker for monitoring of colorectal cancer. Med Oncol Tumor Pharmacother 1988; 5:165-167.
205. Macchia V, Mariano A, Caralacanti M, et al. Tumor markers and lung cancer: correlation between serum and bronchial secretion levels of CEA, TPA, CA-50, NSE, and ferritin. Int J Biol Markers 1987;2:151-156.
206. Masson P, Palsson B, Andren-Sandberg A. CA-50 in patients with pancreatic disease—an evaluation of three different laboratory techniques. Scan J Clin Lab Invest 1988;48:751-755.
207. Touitou Y, Bogdan A. Tumor markers in non-malignant diseases. Eur J Cancer Clin Oncol 1988;24:1083-1091.
208. Heptner G, Domschke S, Domschke W. Comparison of CA 72.4 with CA 19-9 and carcinoembryonic antigen in the serodiagnostics of gastrointestinal malignancies. Scand J. Gastroenterol 1989;24:745-750.
209. Thor A, Itzkowitz SH, Schlom J, et al. Tumor-associated glycoprotein (TAG-72) expression in ulcerative colitis. Int J Cancer 1989;43:810-815.
210. Kallioniemi O-P, Oksa H, Aaran R-K, et al. Serum CA 15-3 assay in the diagnosis and follow-up of breast cancer. Br J Cancer 1988;58:213-215.
211. Colomer R, Ruibal A, Genolla J, Salvador L. Circulating CA 15-3 antigen levels in non-mammary malignancies. Br J Cancer 1989;59:283-286.
212. Eskelinen M, Tikanoja S, Collan Y. Use of tumor markers CA 15-3, MCA, and CEA in breast cancer diagnostics. J Tumor Marker Oncol 1989;4:39-44.
213. Tjandra JJ, Busmanis I, Russell IS, et al. The association of mammary serum antigen (MSA) with the histopathological findings in localized breast cancer. Br J Cancer 1988;58:815-817.
214. Tjandra JJ, McLaughlin PJ, Russell IS, et al. Comparison of mammary serum antigen (MSA) with β_2-microglobulin (β_2 M) and carcinoembryonic antigen (CEA) assays in patients with breast cancer. Eur J Cancer Clin Oncol 1988;24:1633-1640.
215. Tjandra JJ, Russell IS, Collins JP, et al. Application of mammary serum antigen assay in the management of breast cancer: a preliminary report. Br J Surg 1988; 75:811-817.
216. Stacker SA, Thompson CH, Sacks NP, et al. Detection of mammary serum antigen in sera from breast cancer patients using monoclonal antibody 3E1.2. Cancer Res 1988;48:7060-7066.
217. Bhargava AK, Nemoto T, Ferng S, et al. Serum levels of cancer-associated antigen CA-549 in patients with advanced breast cancer. J Tumor Marker Oncol 1989;4: 283-293.
218. Beveridge RA, Chan DW, Bruzek D, et al. A new biomarker in monitoring breast cancer: CA 549. J Clin Oncol 1988;6:1815-1821.
219. Bieglmayer C, Szepesi T, Neunteufel W. Follow-up of metastatic breast cancer patients with a mucin-like carcinoma-associated antigen: comparison to CA 15-3 and carcinoembryonic antigen. Cancer Lett 1988;42:199-206.
220. Rasoul-Rockenschaub S, Zielinski CC, Kubista E, et al. Diagnostic value of mucin-like carcinoma-associated antigen (MCA) in breast cancer. Eur J Cancer Clin Oncol 1989;25:1067-1072.

221. Bombardieri E, Gion M, Mione R, et al. A mucinous-like carcinoma-associated antigen (MCA) in the tissue and blood of patients with primary breast cancer. Cancer 1989;63:490-495.
222. Cooper EH, Forbes MA, Hancock AK, et al. An evaluation of mucin-like carcinoma associated antigen (MCA) in breast cancer. Br J Cancer 1989;59:797-800.
223. Stahli C, Caravatti M, Aeschbacher M, et al. Mucin-like carcinoma-associated antigen defined by three monoclonal antibodies against different epitopes. Cancer Res 1988;48:6799-6802.
224. Chu TM, Constantine R, Nemoto T. Serum level of cryptic tumor antigens in breast cancer patients as determined by two monoclonal antibodies (M85/F36) and its comparison with CA 15-3. J Clin Lab Anal 1989;3:267-272.
225. Konrath JG, Manderino GL, Przywara LW. IMx BCM, an automated microparticle enzyme immunoassay for breast cancer mucin. Abstract. Proceedings of 41st National Meeting of Am Assoc Clin Chem, 1989:69.
226. Steger GG, Mader R, Derfler K, et al. Mucin-like cancer-associated antigen (MCA) compared with CA 15-3 in advanced breast cancer. Klin Wochenschr 1989;67:813-817.
227. Eskelinen M, Tikanaja S, Valkamo E, et al. Cancer associated antigen 15-3 in the diagnostics of breast tumors. Scand J Clin Lab Invest 1988;48:653-658.
228. Barak V, Carlin D, Sulker A, et al. CA 15-3 serum levels in breast cancer and other malignancies—correlation with clinical course. Isr J Med Sci 1988;24:623-627.
229. Omar YT, Behbehani AE, Al Naqeeb N, et al. Carcinoembryonic antigen and breast carcinoma antigen (CA 15-3) in preoperative staging and postoperative monitoring of patients with carcinoma of the breast. Int J Biol Markers 1988;3:165-171.
230. Martoni A, Ercolino L, Bellanova B, et al. CA 15-3 and CEA plasma level monitoring in patients with breast cancer. Int J Biol Markers 1988;3:154-158.
231. Gozdz SS, Kowalska MM, Sluszniak JT, et al. Pretreatment concentrations of breast carcinoma antigen (CA 15.3) and mucin-like carcinoma-associated antigen in patients with carcinoma of the breast. Tumor Biol 1989;10:103-108.
232. Girling A, Bartkova J, Burchell J, et al. A core protein epitope of the polymorphic epithelial mucin detected by the monoclonal antibody SM-3 is selectively exposed in a range of primary carcinomas. Int J Cancer 1989;43:1072-1076.
233. Gould VE, Shin SS, Manderino GL, et al. Selective expression of a novel mucin-type glycoprotein in human tumors: immunohistochemical demonstration with Mab A-80. Hum Pathol 1988;19:623-627.
234. Jansson D, Gould VE, Gooch GT, et al. Immunohistochemical analysis of colon carcinomas applying exocrine and neuroendocrine markers. APMIS 1988;96:1129-1139.
235. Price MR. High molecular weight epithelial mucins as markers in breast cancer. Eur J Cancer Clin Oncol 1988;24:1799-1804.
236. Burchell J, Taylor-Papadimitriou J. Antibodies to human milk fat globule molecules. Cancer Invest 1989;7:53-61.
237. van der Burg ME, van Putten WL, Lammer FB. Evaluation of the clinical relevance of CA 125 in the management of ovarian cancer. Abstract. Proceedings of the Fourth European Conference on Clinical Oncology and Cancer Nursing, 1987:212.
238. Hoffman L, Muller-Hagen S, Schafer E, Klapdor R. CA 125 monitoring in the evaluation of the course and prognosis of metastasizing ovarian cancer. Abstract, Proceedings of the Fourth European Conference on Clinical Oncology and Clinical Nursing, 1987:213.
239. Quarante M, Abbate I, Correale M, et al. Role of CA 125 as tumor marker in monitoring of epithelial ovarian carcinoma. Abstract, Proceedings of the Fourth European Conference on Clinical Oncology and Clinical Nursing, 1987:224.
240. Jager W, Adam R, Wildt L, Lang N. Serum CA 125 as a guideline for the timing

of a second look operation and a second line treatment in ovarian cancer. Arch Gynecol Obstet 1988;243:91-99.
241. Finkler NJ, Benacerref B, Lavin PT, et al. Comparison of serum CA 125, clinical impression, and ultrasound in the preoperative evaluation of ovarian masses. Obstet Gynecol 1988;72:659-664.
242. Yedema C, Massuger L, Hilgers J, et al. Preoperative discrimination between benign and malignant ovarian tumors using a combination of CA 125 and CA 15-3 serum assays. Int J Cancer Suppl 1988;3:6-17.
243. Einhorn N, Knapp RC, Bast RC Jr, et al. The CA 125 assay used in conjunction with CA 15-3 and TAG-72 assays for discrimination between malignant and nonmalignant diseases of the ovary. Abstract. Proceedings of the Fourth European Conference on Clinical Oncology and Cancer Nursing, 1987:205.
244. Einhorn N, Bast RC, Eklund E, et al. CA 125 in early detection of ovarian cancer. Abstract. Proceedings First Meeting of the International Gynecologic Cancer Society, 1987:19.
245. Zurawski VR, Orjaseter H, Andersen A, Jellum E. Elevated serum CA 125 levels prior to diagnosis of ovarian neoplasia: relevance for early detection of ovarian cancer. Int J Cancer 1988;42:677-680.
246. Mogensen O, Mogensen B, Jakobsen A, Sell A. Measurement of the ovarian cancer-associated antigen CA 125 prior to second look operation. Eur J Cancer Clin Oncol 1988;24:1835-1837.
247. Koelbl H, Schieder K, Neunteufel W, Bieglmayer C. A comparative study of mucin-like carcinoma-associated antigen (MCA), CA-125, CA-19-9, and CEA in patients with ovarian cancer. Neoplasia 1989;36:473-478.
248. Patsner B, Mann WJ. Use of serum CA-125 in monitoring patients with uterine sarcoma. Cancer 1988;62:1355-1358.
249. Nozawa S, Yajima M, Kojima K, et al. Tumor-associated mucin-type glycoprotein (CA54/61) defined by two monoclonal antibodies (MA54 and MA61) in ovarian cancers. Cancer Res 1989;49:493-498.
250. Kannagi R, Kitahara A, Itai S, et al. Quantitative and qualitative characterization of human cancer-associated serum glycoprotein antigens expressing epitopes consisting of sialyl or sialyl-fucosyl type 1 chain. Cancer Res 1988;48:3856-3863.
251. Iguchi H, Hara N, Miyazaki K, et al. Elevation of sialyl stage-specific mouse embryonic antigen levels in pleural effusion in patients with adenocarcinoma of the lung. Cancer 1989;63:1327-1330.
252. Sakurai Y, Hirohashi S, Shimosato Y, et al. Selection of monoclonal antibody reactive with a high-molecular weight glycoprotein circulating in the body fluid of gastrointestinal cancer patients. Cancer Res 1988;48:4053-4058.
253. Kohno N, Kyoizumi S, Tanabe M, et al. Detection of a circulating tumor-associated antigen with a murine monoclonal antibody, LISA 101, selected by reversed indirect enzyme-linked immunosorbent assay. Cancer Res 1989;49:3412-3419.
254. Hanisch F-G, Uhlenbruck G, Peter-Katalinic J, Egge H. Structural studies on oncofetal carbohydrate antigens (CA 19-9, CA 50, and CA 125) carried on O-linked sialyl-oligosaccharides on human amniotic mucins. Carbohydr Res 1988;178:29-47.
255. Abe M, Kufe D. Structural analysis of the DF3 human breast carcinoma-associated protein. Cancer Res 1989;49:2834-2839.
256. Gangopadhyay A, Chatterjee SK, Bhattacharya M, Barlow JJ. Characterization of the antigen recognized by monoclonal antibody 1D3. Cancer Biochem Biophys 1988;10:107-115.
257. Prat M, Medico E, Garrino C, Comoglio PM. Biochemical and immunological properties of the human carcinoma antigen CAR-5 defined by the monoclonal antibody BD-5. Int J Cancer 1989;44:67-74.

258. Decaens C, Gautier R, Bara J, et al. A new mucin-associated oncofetal antigen, a marker of early carcinogenesis in rat colon. Cancer Res 1988;48:1571-1577.
259. Stacker SA, Tjandra JJ, Xing P-X, et al. Purification and biochemical characterisation of a novel breast carcinoma associated mucin-like glycoprotein defined by antibody 3E1.2. Br J Cancer 1989;59:544-553.
260. Hinoda Y, Imai K, Ban T, et al. Immunochemical characterization of adenocarcinoma-associated antigen YH206. Int J Cancer 1988;42:653-658.
261. Prat M, Medico E, Rossino P, et al. Biochemical and immunological properties of the human carcinoma-associated CAR-3 epitope defined by the monoclonal antibody AR-3. Cancer Res 1989;49:1415-1421.
262. Ho JJ, Chung Y, Fujimoto Y, et al. Mucin-like antigens in a human pancreatic cancer cell line identified by murine monoclonal antibodies SPan-1 and YPan-1. Cancer Res 1988;48-3924-3931.
263. Kurosaka A, Kitagawa H, Fukui S, et al. A monoclonal antibody that recognizes a cluster of a disaccharide, NeuAcα2 → 6GalNAc, in mucin-type glycoproteins. J Biol Chem 1988;263:8724-8726.
264. Kitagawa H, Nakada H, Numata Y, et al. Immunoaffinity isolation of a sialy-Le^a oligosaccharide from human milk. J Biochem Tokyo 1988;104:591-594.
265. Magnani JL, Linsley PS, Brown JP, Horn D. Monoclonal antibody Onc-M26, which detects elevated levels of mucin in sera from breast cancer patients, binds a sugar sequence in sialylated lacto-N-fucopentaose III. Abstract. IX International Symposium on Glycoconjugates, 1987.
266. Bara J, Gautier R, Le Pendu J, Oriol R. Immunochemical characterization of mucins. Biochem J 1988;254:185-193.
267. Hilkens J, Buijs F, Ligtenberg M. Complexity of MAM-6, an epithelial sialomucin associated with carcinomas. Cancer Res 1989;49:786-793.
268. Hendrick JC, Collette J, Claes S, Franchimont P. Epithelial membrane antigen (EMA) distribution in various biological fluids. Eur J Cancer Clin Oncol 1988;24:1589-1594.
269. Sheer DG, Schlom J, Cooper HL. Purification and composition of the human tumor-associated glycoprotein (TAG-72) defined by monoclonal antibodies CC49 and B72.3. Cancer Res 1988;48:6811-6818.
270. Yamori T, Ota DM, Cleary KR, et al. Monoclonal antibody against human colonic sulfomucin: immunochemical detection of its binding sites in colonic mucosa, colorectal primary carcinoma, and metastases. Cancer Res 1989;49:887-894.
271. North SM, Steck PA, Spohn WH, Nicolson GL. Development and characterization of a syngeneic monoclonal antibody to a rat mammary tumor metastasis-associated mucin-like cell-surface antigen (gp580). Int J Cancer 1988;42:607-614.
272. Bhavanandan VP. Malignancy-related cell surface mucin-type glycoproteins. Indian J Biochem Biophys 1988;25:36-42.
273. Byrd JC, Nardelli J, Siddiqui B, Kim YS. Isolation and characterization of colon cancer mucin from xenografts of LS174T cells. Cancer Res 1988;48:6678-6685.
274. Byrd JC, Lamport DT, Siddiqui B, et al. Deglycosylation of mucin from LS174T colon cancer cells by hydrogen fluoride treatment. Biochem J 1989;261:617-625.
275. Siddiqui B, Byrd JC, Fearney FJ, Kim YS. Comparison of metabolically labeled mucins of LS174T human colon cancer cells in tissue culture and xenograft. Tumor Biol 1989;10:83-94.
276. Shimamoto C, Deshmukh GD, Rigot WL, Boland CR. Analysis of cancer-associated colonic mucin by ion-exchange chromatography: evidence for a mucin species of lower molecular charge and weight in cancer. Biochim Biophys Acta 1989;991:284-295.

277. Maoret J-J, Font J, Augeron C, et al. A mucus-secreting human colonic cancer cell line: purification and partial characterization of the secreted mucins. Biochem J 1989;258:793-799.
278. Nardelli J, Byrd JC, Ho J, et al. Pancreatic cancer mucin from xenograft of SW-1990 cells: isolation, characterization, and comparison to colon cancer mucin. Pancreas 1988;3:631-641.
279. Moosic JP, Mosley AM, Huang K, et al. Characterization of two monoclonal antibodies that recognize high molecular weight colon antigens. Cancer Lett 1989;44:127-136.
280. Hanisch F-G, Uhlenbruck G, Peter-Katalinic J, et al. Structures of neutral O-linked polylactosaminoglycans on human skim milk mucins. J Biol Chem 1989;264:872-883.
281. Hounsell EF, Lawson AM, Feeney J, et al. Identification of a novel oligosaccharide backbone structure with a galactose residue monosubstituted at C-6 in human foetal gastrointestinal mucins. Biochem J 1988;256:397-401.
282. Capon C, Leroy Y, Wieruszeski J.-M., et al. Structures of O-glycosidically linked oligosaccharides isolated from human meconium glycoproteins. Eur J Biochem 1989;182:139-152.
283. Moss L, Greenwalt D, Cullen B, et al. Cell-to-cell heterogeneity in the expression of carbohydrate-based epitopes of a mucin-type glycoprtein on the surface of human mammary carcinoma cells. J Cell Physiol 1988;137:310-320.
284. Muraro R, Kuroki M, Wunderlich D, et al. Generation and characterization of B72.3 second generation monoclonal antibodies reactive with the tumor-associated glycoprotein 72 antigen. Cancer Res 1988;48:4588-4596.
285. Haviland AE, Borowitz MJ, Lan MS, et al. Aberrant expression of monoclonal antibody-defined colonic mucosal antigens in inflammatory bowel disease. Gastroenterology 1988;95;1302-1311.
286. Imokawa M, Okabe H, Ochi Y, Kajita Y. Clinical significance of mucin-like high molecular weight glycoprotein originated from lung cancer as tumor marker. Clin Chim Acta 1989;180:45-58.
287. Hilkens J, Buijs J. Biosynthesis of MAM-6, an epithelial sialomucin. J Biol Chem 1988;263:4215-4222.
288. Linsley PS, Kallestad JC, Horn D. Biosynthesis of high molecular weight breast carcinoma associated mucin glycoproteins. J Biol Chem 1988;263:8390-8397.
289. Hull SR, Carraway KL. Mechanism of expression of Thomsen-Friedenreigh (T) antigen at the cell surface of a mammary adenocarcinoma. FASEB J 1988;2:2380-2384.
290. Spielman J, Hull SR, Sheng Z, et al. Biosynthesis of a tumor cell surface sialomucin. J Biol Chem 1988;263:9621-9629.
291. Carraway KL, Hull SR. O-glycosylation pathway for mucin-type glycoproteins. Bioessays 1989;10:117-121.
292. Gendler S, Taylor-Papdimitriou J, Duhig T, et al. A highly immunogenic region of a human polymorphic epithelial mucin expressed by carcinomas is made up of tandem repeats. J Biol Chem 1988;263:12820-12823.
293. Siddiqui J, Abe M, Hayes D, et al. Isolation and sequencing of a cDNA coding for the human DF3 breast carcinoma-associated antigen. Proc Natl Acad Sci USA 1988;85:2320-2323.
294. Xing PX, Tjandra JJ, Reynolds K, et al. Reactivity of antihuman milk fat globule antibodies with synthetic peptides. J Immunol 1989;142:3503-3509.
295. Griffiths B, Gordon A, Burchell J, et al. The breast tumor-associated epithelial mucins and the peanut lectin binding urinary mucins are coded by a single highly polymorphic gene locus 'PUM.' Dis Markers 1988 6;185-194.

296. Griffiths B, Bobrow LG, Happerfield L, Swallow DM. Expression of the hypervariable PUM locus in normal and malignant lung: the tumour-associated epitopes are present but masked in normal tissue. Dis Markers 1988;6:195-202.
297. Gum JR, Byrd JC, Hicks JW, et al. Molecular cloning of human intestinal mucin cDNAs: sequence analysis and evidence for genetic polymorphism. J Biol Chem 1989;264:6480-6487.
298. Timpte CS, Eckhardt AE, Abernethy JL, Hill RL. Porcine submaxillary gland apomucin contains tandemly repeated, identical sequences of 81 residues. J Biol Chem 1988;263:1081-1088.
299. Lan MS, Batra SK, Qi W-N, et al. Cloning and sequencing of a human pancreatic tumor mucin cDNA. J Biol Chem 1990; in press.

39
Placental Proteins as Tumor Markers

GLENN D. BRAUNSTEIN

*Cedars-Sinai Medical Center–UCLA School of Medicine,
Los Angeles, California*

A large variety of proteins are synthesized by the human placenta. Many of the proteins are identical to those produced by normal organs in nonpregnant adults, while others are relatively unique to the placenta and have been termed "placental-specific" proteins [1]. Among the latter are placental isoenzymes including placental alkaline phosphatase and placental cystine aminopeptidase, hormones such as human chorionic gonadotropin (HCG), human placental lactogen (HPL), and a growth hormone variant [2], as well as proteins with no well-defined biological activity such as pregnancy-specific beta-1 glycoprotein (SP-1). Although for many years the "placental-specific" proteins were considered to be made only by placental tissue, the development of methodology that allows measurement of minute quantities of these proteins in tissues and bodily fluids has resulted in the finding that low levels of the "placental-specific" proteins are present in normal nonpregnant individuals. Thus, no placental protein can be considered to be completely unique to the placenta. The major difference between the placenta and other tissues in the nonpregnant adult is the quantity of the protein produced.

For many decades it has been known that some cancers may synthesize placental proteins. In the case of gestational and nongestational choriocarcinoma, the production of placental proteins such as HCG was expected since these tumors are of trophoblastic origin. The production of placental proteins by these tumors was, therefore, considered to be eutopic. In contrast, the production of placental proteins by nontrophoblast-derived tumors was designated ectopic production. However, it is now clear that the difference between eutopic and ectopic production is quantitative and not qualitative.

For a protein to serve as a tumor marker, it should be easily and precisely measurable in blood or urine, present in all patients with the tumor at all stages of the disease, be absent from the bodily fluids of patients without the cancer (including those with nonneoplastic diseases that include cancer in the differential diagnosis), and should correlate quantitatively with the tumor burden, especially while the patient is receiving therapy, so that it may serve as an objective parameter to monitor the effects of the therapy. In the case of trophoblastic neoplasms, HCG and, to a slightly lesser extent, SP-1 come close to satisfying these criteria. However, in the majority of patients with nontrophoblastic cancers, none of these criteria are adequately satisfied for any of the "placental-specific" proteins described thus far.

As noted above, a myriad of "placental-specific" proteins have been described and many have been preliminarily investigated as potential tumor markers. In most cases, insufficient information is available to examine their sensitivity, specificity, and utility for monitoring changes in tumor burden to allow adequate evaluation. Sufficient information on HCG, HPL, and SP-1 is available to examine their potential as tumor markers.

I. HUMAN CHORIONIC GONADOTROPIN

A. Introduction

HCG is a glycoprotein hormone composed of two noncovalently linked subunits designated alpha and beta. The alpha subunit contains 99 amino acids and is virtually identical to the alpha subunits of the other glycoprotein hormones, luteinizing hormone (LH), follicle-stimulating hormone, and thyroid-stimulating hormone. The beta subunit contains 145 amino acids that differ considerably from the composition of the beta subunits of follicle-stimulating hormone and thyroid-stimulating hormone. There is a great deal of homology between the first 115 amino acids of the beta subunit of HCG and LH, but HCG contains an additional 24 carboxy terminal amino acids. This carboxy terminal protein also contains O-linked oligosaccharides that are not present on LH. The differences in the beta subunits between HCG and the other glycoprotein hormones confers the biological and immunological specificity on the hormone [3].

HCG is synthesized and secreted primarily in the syncytiotrophoblast of the placenta. During normal pregnancy, the hormone may be detected as early as 6-9 days following conception and rises in a logarithmic fashion during the first trimester, peaking in maternal serum and urine 8-12 weeks after the last menstrual period. The levels then fall, reaching a nadir at approximately 17 weeks followed, by a plateau throughout the rest of pregnancy [4]. In early pregnancy the free beta subunit of HCG is also secreted and the maternal serum levels of the beta subunit parallel those of intact HCG [5]. The free alpha subunit of HCG is also secreted throughout pregnancy, but in contrast to HCG and its free beta subunit, the levels increase steadily throughout gestation [4]. A hyperglycosylated form of alpha subunit (big alpha subunit) has also been noted in placental extracts and sera from pregnant women [6]. In addition, a low-molecular-weight fragment of the beta subunit of HCG is present in the urine of pregnant women. This fragment, designated the beta-core fragment, accounts for approximately one-third to almost 100% of the immunoreactive HCG moieties in the urine at various times during pregnancy [7]. The half-life of disappearance of HCG from the circulation following normal delivery or therapeutic abortion is biexponential, with a rapid component of approximately 6 hr and a slower component of 24-36 hr [8]. The relatively long half-life of the hormone in the circulation is due to the carbohydrate portion of the molecule, especially sialic acid, since removal of sialic acid rapidly shortens the disappearance time of the hormone [9].

The primary biological activity of HCG is luteotrophic, although the hormone also possesses a small amount of follicle-stimulating hormone activities [10]. The major physiological function of HCG during pregnancy is to maintain the secretion of progesterone by the corpus luteum following conception to allow implantation of the blastocyst to take place. In addition, HCG may be important for male fetal sexual differentiation, through stimulation of fetal Leydig cell production of testosterone. A number of other functions of HCG have been postulated but not clearly demonstrated [11].

An HCG-like substance has been found in nonpregnant individuals. A large number of different organs contain immunoreactive HCG, with the highest concentration being found in the pituitary [12]. Immunohistochemical techniques have also localized the HCG-like substance in the human pituitary and recent immunoelectron microscopic studies have shown the presence of a previously undescribed gonadotroph that contains the HCG-like substance [13]. In vitro biosynthesis of HCG has been demonstrated in human fibroblast cell lines and by human fetal kidney. The HCG-like substance in nonpregnant humans also circulates in the blood with average concentrations of less than 100 pg/ml (500 mIU/ml) in adult men and premenopausal women and levels up to several hundred pg/ml in postmenopausal women. At the present time, there is no known physiological function of the HCG or the HCG-like substance in nonpregnant humans [12].

Numerous methods for measuring HCG have been described since its discovery in 1927. Bioassays utilizing mice, rats, frogs, and rabbits were used initially for pregnancy diagnosis and for the detection of HCG present with gestational and nongestational trophoblastic disease. During the 1960s, agglutination inhibition assays and radioimmunoassays were developed utilizing polyclonal anti-HCG antibodies. These assays were hampered by the cross-reaction between HCG and LH. In 1972, a radioimmunoassay was developed that utilized antiserum against the purified beta subunit of HCG [14]. This "beta-HCG" radioimmunoassay was relatively specific for HCG and its beta subunit and showed little cross-reaction with physiological concentrations of LH. Subsequently, highly specific radioimmunoassays and immunometric assays were developed that used antisera directed against the HCG-specific carboxy terminal portion of the HCG-beta subunit, and, most recently, two site "sandwich" immunometric assays using combinations of monoclonal antibodies. This latter technology has resulted in extremely sensitive and specific methods for measuring HCG and free-HCG subunits. Several of the currently available commercial qualitative pregnancy tests are capable of detecting 25 mIU/ml HCG in serum or urine without cross-reaction with LH [15,16].

B. HCG as a Tumor Marker

1. Gestational Trophoblastic Disease

Gestational trophoblastic disease includes hydatidiform mole, invasive mole, and choriocarcinoma. In each of these conditions, HCG is an excellent tumor marker since it is present in all patients with active disease and changes in its serum or urine concentrations accurately reflect changes in tumor burden.

Quantitative HCG measurements are useful for the diagnosis of gestational trophoblastic disease. Serum concentrations above 200 IU/ml or 24 hr urine concentrations greater than 300,000 IU 100 or more days after the last menstrual period are strongly suggestive of trophoblastic disease, although an occasional woman bearing multiple fetuses may have HCG levels of this magnitude [17,18]. The accuracy of a single serum measurement of HCG for diagnostic purposes is enhanced with the addition of ultrasound examination of the uterus. A serum HCG concentration greater than 83 IU/ml associated with absent fetal heart motion on ultrasound has an approximately 90% sensitivity in diagnosing the presence of hydatidiform mole [19]. Serial HCG measurements are also useful in the diagnosis of gestational trophoblastic disease, since such patients generally have a steady increase in HCG concentrations as the disease progresses while pregnant patients will experience a fall in serum and urine HCG levels following the normal peak at 8-12 weeks after the last menstrual period.

In patients with a noninvasive hydatidiform mole, surgical evacuation results in the rapid initial disappearance of HCG from the serum of urine during the first 3-4 weeks, followed by a more gradual disappearance of the hormone [20]. HCG may be detected in the urine for 8 weeks in approximately half of the patients following removal of the molar tissue and is detectable in some patients up to 40 weeks following removal [21]. In the serum, HCG is commonly measurable for 60-80 days following molar evacuation [22]. In an individual patient, the length of time for HCG to reach an undetectable level depends on the initial concentration of HCG in the serum and the possible presence of viable trophoblast tissue remaining in the uterus following curettage. Chemotherapy is generally withheld unless the serum concentrations of HCG plateau for 2 or more weeks or increase after evacuation. After HCG levels have remained undetectable for 3 consecutive weeks, the patient's disease is considered to be in remission. Serum HCG concentrations should be measured monthly for 6 months and then at 2 month intervals for the next 6 months [23]. If HCG levels have remained undetectable during that period of time, the patient has less than 0.1% risk of a recurrence of the trophoblastic tumor [24]. Pregnancy is generally intradicted during the year following the time when the HCG levels reached the normal range.

The concentration of HCG in bodily fluids provides prognostic information in patients with malignant trophoblastic disease. Urine HCG concentrations greater than a 100,000 IU/24 hr or serum concentrations greater than 40 IU/ml in the presence of metastatic disease are a poor prognostic sign [23,25]. Additional poor prognostic factors include a duration of more than 4 months between the antecedent pregnancy and the onset of symptoms or the start of therapy, the presence of liver or brain metastases, unsuccessful prior therapy, and the presence of high concentrations of the free alpha subunit of HCG [23,25-27]. Elevated concentrations of the free beta subunit of HCG may be seen in patients with all types of gestational trophoblastic disease, but the relative concentration of free beta to intact HCG is highest in those patients with choriocarcinoma and reflects a greater degree of immaturity of the trophoblast in choriocarcinoma than in hydatiform mole [28]. In patient harboring cerebral metastasis, the cerebrospinal fluid concentrations of HCG are generally elevated. During pregnancy the serum CSF/HCG ratio is 60:1 or greater. In patients with brain metastases from choriocarcinoma, the ratio is less than 60:1 and generally less than 45:1 [29,30].

The concentration of HCG in the urine or serum roughly reflects the tumor burden [31], and there is a correlation between the initial HCG level and the amount of chemotherapy that an individual patient will require to achieve remission [32]. During therapy the changes in HCG concentrations correlate quite well with the change in tumor burden as assessed by radiological or radionucleotide measurements of metastatic deposits [31]. It should be noted, however, that when chemotherapy is started there may be a transient rise in HCG levels, presumably due to tissue necrosis and release of preformed hormone into the circulation.

The response to chemotherapy in patients with malignant gestational trophoblastic disease is quite good. Patients who do not have any of the poor prognostic signs have shown a 95-100% cure rate in various series with single-agent chemotherapy. Single-agent therapy is less effective in those patients with poor prognostic findings, curing only 20-40%. However, combination chemotherapy is effective in 70-80% of such high-risk patients [23,25,33-35]. Monitoring HCG levels following the achievement of a remission should follow the schedule used to evaluate patients who have achieved remission following evacuation of a benign hydatidiform mole.

2. Germ Cell Tumors of the Testis

Germ cell tumors of the testicle account for 95% of testicular cancers and are broadly classified into two major groups: seminoma accounting for 33-50% of tumors, and nonseminomatous germ cell tumors. The latter have been subclassified into several histological patterns including embryonal carcinoma, teratocarcinoma, adult teratoma, and choriocarcinoma.

Shortly after the discovery of HCG in the urine of pregnant women, Zondek described the presence of HCG in the urine of a patient with a metastatic testicular tumor [36]. Since that time numerous studies have indicated that HCG is present in the serum and urine of many patients with testicular tumors.

As shown in Table 1 [9,37-53], approximately 11% of patients with a pure seminoma have immunoreactive HCG in their circulation. Immunohistochemical studies have demonstrated that most often the HCG originates in trophoblastic giant cells present in the tumor. Approximately 45% of patients with nonseminomatous tumors have immunoreactive HCG in their serum (Table 1). The cellular sources of HCG in these patients include the syncytiotrophoblastic cells and trophoblastic giant cells [54]. The fact that less than half of the patients with testicular tumors have HCG in their serum reduces the utility of HCG measurements as a screening tool for testicular cancer in patients with scrotal masses. In addition, most patients presenting with testicular tumors have neoplasms confined to their testes and spermatic cord (stage I), and in this group the HCG levels are lowest [51].

Measurements of HCG and alpha fetoprotein (AFP) have been found to be useful in reducing staging errors following orchiectomy [43]. Following removal of an HCG-producing testicular tumor, HCG disappears from the circulation with a half-life of approximately 2 days [55]. To use HCG measurements effectively as a tumor marker after orchiectomy, serial levels must be measured. If the HCG does not disappear from the circulation with the anticipated half-life or if the levels rise, residual tumor tissue is pres-

Table 1 Immunoreactive HCG in Sera of Patients with Germ Cell Tumors of the Testis

Tumor Type	n	% Positive
Seminoma	1027	10.6
Nonseminoma	1041	45.4
Embryonal carcinoma	82	51.2
Teratocarcinoma	20	7
Choriocarcinoma	12	100

Source: Data from [9,37-52].

ent. However, two caveats must be kept in mind. First, HCG must be measured by a very specific assay that does not cross-react with LH. Some patients will develop primary hypogonadism following orchiectomy, radiation therapy, or chemotherapy and this will result in an elevation in LH levels that may be detected in some nonspecific commercial HCG assays [56]. Differentiation between HCG and LH may be made through the use of more specific assays, such as the newer two-site immunometric enzyme assays. Alternatively, the patient may be given several injections of testosterone and the HCG serum concentration measured subsequently. The testosterone will decrease an elevated LH level if hypogonadism is present, but will not alter the HCG concentration of residual HCG-producing testicular tumor tissue remains [57]. A second caveat in utilizing HCG measurements to detect residual tumor is that in some patients transient elevation of HCG may be seen following a course of chemotherapy; this is presumably due to tumor necrosis with release of intracellular HCG [58].

The major utility of HCG measurements is in monitoring the effects of therapy. As noted above, serial HCG (and AFP) measurements are useful in determining whether orchiectomy has cured clinical stage I disease in a marker-positive neoplasm [43]. Serial measurements of both HCG and AFP are useful for monitoring the effects of therapy in patients with metastatic disease. The HCG concentrations roughly correlate with tumor burden [59,60]. Persistently rising levels of HCG (or AFP) indicate progressive disease. In most instances, a decline in tumor marker level indicates tumor regression. However, in approximately 10% of patients there is a discordance between the marker levels that decline and tumor growth [51]. This is most often seen in patients with tumors containing cellular elements. It is assumed that the chemotherapy and/or radiotherapy destroys the marker-producing cells while other tumor elements continue to grow [61,62]. In some patients with complex tumors, the disappearance of the tumor marker despite the persistence of a mass reflects the conversion of the residual tumor to a benign teratoma.

An area of controversy is the question of whether the presence of HCG is of prognostic importance. Early studies using relatively insensitive bioassays or nonspecific immunoassays suggested that the presence of HCG in the blood or urine of a patient with a testicular neoplasm indicated a poor prognosis [59,63-67]. It is likely that the high concentrations of HCG that were required to be detected in those assay systems reflected a larger tumor burden, which is the more important prognostic indicator. When specific and sensitive HCG assays are used and the confounding effects of tumor mass are controlled for, the presence of HCG in the serum of patients with testicular tumors does not appear to indicate an adverse prognosis [58]. Indeed, with the high rate of remissions in patients with metastatic germ cell tumors using modern chemotherapy regimens, any prognostic information that the presence of HCG in the serum may have had in the past may be clinically irrelevant today.

An area in which the prognostic significance of HCG may be relevant concerns the approximately 11% of patients with pure seminomas who have immunoreactive HCG in their circulation. Early reports using sensitive assays suggested that the presence of HCG in such patients was a poor prognostic sign and indicated that some patients have unsuspected nonseminomatous germ cell tumor in association with the seminoma [68,69]. Certainly, if both HCG and AFP or if AFP alone is present in what appears to be a pure seminoma, nonseminomatous germ cell tumor elements are present in the primary tumor or metastases and the patient should be treated for nonseminomatous germ cell tumor

Table 2 Immunoreactive HCG in Sera of Patients with Cancer

Tumor or Site	n	% Positive
Islet cell	104	39.4
Gynecological	1020	28.9
Ovary	633	28.6
Endometrium	348	16.7
Cervix	976	33.8
Vulva/vagina	37	27.0
Carcinoid	41	26.8
Gastrointestinal	2165	18.0
Oropharynx	298	14.8
Esophagus	124	17.7
Gastric	232	20.7
Small intestine	24	16.7
Colon/rectum	693	9.8
Hepatic	281	22.1
Biliary	26	30.7
Pancreatic	200	20.0
Lung	1365	17.4
Breast	3031	16.8
Melanoma	244	13.9
Genitourinary	658	11.8
Renal	119	6.7
Bladder	176	23.3
Prostate	363	8.0
Sarcoma	136	11.8
Hematopoietic	544	6.1
Lymphoma	339	5.3
Miscellaneous	576	10.4
Total	11213	18.0

Source: Data from [5,9,37,73-136].

rather than a seminoma. However, otherwise pure seminomas can produce HCG if they have scattered trophoblast giant cells. HCG does not appear to have an adverse prognostic implication in patients with low-stage seminomas [52,70,71]. However, if the patient has widespread metastases beyond the regional lymph nodes, the presence of HCG does appear to affect the prognosis adversely [70].

3. Nontrophoblastic Neoplasms

Before the development of sensitive and specific assays for HCG, the production of HCG by nontrophoblastic malignancies appeared to be an unusual event. Case reports described this phenomenon in patients with lung cancer, hepatocellular neoplasms, adrenal cortical carcinoma, melanoma, renal cell carcinoma, and breast carcinoma [72]. The clinical manifestations of the HCG in these patients included isosexual precocious pu-

Table 3 Immunoreactive HCG in Sera of Control Patients Without Cancer

Tumor or Site	n	% Positive
Premenopausal females	430	0.7
Postmenopausal females	242	0
Normal males	438	0.2
Castrated males	4	0
Blood donors	165	2.4
Benign disease	1963	3.6
Gynecological	528	6.6
Gastrointestinal	318	6.0
Breast	254	4.3
Lung	174	2.9
Genitourinary	84	0
Neurologic	10	0
Hypothyroid	10	0
Unspecified	650	0
Other normal	172	0
Miscellaneous	691	2.7
Total	4125	2.4

Source: Data from [5,9,37,73-136].

berty in boys with hepatoblastomas and gynecomastia in adult males, especially those with bronchogenic carcinoma. The development of the sensitive and relatively specific beta-HCG radioimmunoassay in 1972 allowed investigators to examine the utility of using HCG as a tumor marker for nontrophoblastic tumors. The combined data from a large number of published series indicate that approximately 18% of patients with nontrophoblastic cancers have immunoreactive HCG in their sera (Table 2) [5,9,37,73-136].

Although initial studies were quite enthusiastic about the potential of HCG as a tumor marker for nontrophoblastic malignancies, the experience during the last 15 years has indicated that there are many problems with using HCG as a serum tumor marker. The first problem is the finding that HCG is not specific for malignancy. In many of the series that evaluated immunoreactive HCG in the sera of cancer patients, sera from nor-

Table 4 Immunohistochemical Detection of HCG in Benign Tissues

Tissue	n	% Positive
Gastric	12	75.0
Endometrium	29	13.8
Ovary	70	8.6
Breast	54	3.7
Colon/rectum	43	0
Prostate	8	0
Total	216	9.7

Source: Data from [142-170].

mal male subjects, female subjects, blood donors, and patients with benign diseases were also evaluated. As shown in Table 3, 2-3% of these control patients have been found to have immunoreactive HCG in their sera [5,9,37,73-136]. Since an HCG-like substance has been found in the blood of normal nonpregnant individuals (generally at levels below 0.5 ng/ml or 2.5 mIU/ml) [11], and is identifiable by immunohistochemical techniques in benign disease tissues (Table 4), it is not surprising that an occasional normal individual or a patient with a benign disorder may have a level of immunoreactive HCG above 1-2 ng/ml (5-10 mIU/ml), which is the lower limit of sensitivity in many of the assays used to examine the sera of patients for HCG. Indeed, the majority of positive levels in patients with benign disorders fall within the 1-3 ng/ml (5-15 mIU/ml) range. The serum levels of immunoreactive HCG in the majority of patients with cancer who have the hormone in their circulation are also in the low range with 42% falling between 1 and 2 ng/ml (5-10 mIU/ml), 29% between 2 and 3 ng/ml (10-15 mIU/ml), 9% between 3 and 4 ng/L (15-20 mIU/ml), 8% between 4 and 5 ng/ml (20-25 mIU/ml), and 11% above 5 ng/ml (25 mU/ml) [74,84,88,109,115,116,134]. Raising the "normal range" of serum HCG levels to 5 ng/ml (25 mIU/ml) would virtually eliminate the positive results in patients without neoplasms, but would reduce the frequency of positive results in patients with cancer to approximately 2%. Thus, measurements of HCG in the serum are of no utility in screening patients for the presence of cancer. If one requires 100% specificity (i.e., 5 ng/ml), the sensitivity of the test is too low to be of use. If 1 ng/ml is used as the cutoff point, the number of false-positive results in a population will exceed the number of true-positive results, rendering the test meaningless.

Another problem with using HCG as a tumor marker is the generally poor correlation with the stage of the disease. It was anticipated that since the concentration of immunoreactive HCG in the sera of patients with trophoblastic disease correlated with the tumor burden, a similar correlation would be found in patients with nontrophoblastic neoplasms that produced the hormone. Although some investigators have noted a correlation between the stage of the tumor and the frequency of the presence of HCG in the blood [108,118,133], most of the studies have failed to demonstrate such a correlation [73,79, 88,98,103,109,116,118,137]. As an example, Table 5 shows the combined data of five published series that examined the relationship between stage of disease in patients with breast cancer and the presence or absence of immunoreactive HCG in the serum [76,79, 80,84,119]. As will be noted, approximately 13% with locally invasive breast cancer had immunoreactive HCG in their circulation prior to surgery, but 30% were found to have HCG postoperatively, while only 29% of individuals who had metastatic breast cancer had HCG in their circulation. Of particular note is the finding that a greater number

Table 5 Relationship of Disease Stage and Presence of HCG in Serum of Patients with Carcinoma of the Breast

	n	% Positive
Local: preoperative	325	13
Local: postoperative	40	30
Disease free	111	44
Metastatic	301	29

Source: Data from [76,79,80,84,119].

Table 6 Relationship of Disease and Stage and Presence of HCG in Serum of Patients with Cervical Carcinoma

Stage	n	% Positive
I	55	14.5
II	117	41.9
III	223	54.7
IV	30	56.7

Source: Data from [103,108,118,137].

of patients who were considered to be free of breast cancer at the time of evaluation were found to have immunoreactive HCG in their blood. A somewhat better correlation has been found in patients with cervical carcinoma (Table 6) [103,108,118,137], although this finding has not been uniform.

Serial measurements of HCG in patients with nontrophoblastic malignancies generally have been of little utility in monitoring changes in tumor growth. Although individual patients described in case reports and some series have shown a good correlation between changes in tumor burden during or following therapy, and changes in serum HCG levels, most studies have shown that there is generally as much discordance between tumor growth and changes in HCG levels as there is concordance [73,79,80,82,84,88,89,98,106, 108,114,118,119,138,140,142]. There are many explanations for this finding. As noted above, most of the patients with HCG-producing neoplasms have HCG concentrations that are at the limits of assay sensitivity. In most assays, the precision of the measurements is least at the lower limit of sensitivity and many of the commercial assays in use are subject to various artifacts that can alter measurements. These artifacts include cross-reaction with LH, serum protein abnormalities, the presence of heterophilic antibodies, and other nonspecific factors. In addition, many studies have shown spontaneous fluctuations in HCG concentration in patients with neoplasms not related to change in tumor burden or therapy [73,81,84,89,95,98,111,116,119,132]. Another exaplanation for the discordance has come from the immunohistochemical studies of tumor tissue stained for HCG. As shown in Table 7, approximately 25% of tumor tissues contain neoplastic cells that stain positively for HCG [142-170]. Each of the studies has shown a great deal of heterogeneity in the tumors, with some tumors containing only isolated cells or clusters of cells that stain positively for HCG, while other tumors contain sheets of cells that are HCG positive. Thus, as has been demonstrated for patients with complex germ cell tumors of the testes, it is likely that chemotherapy or radiation therapy of some cancers may destroy the HCG-producing clone of cells but not the cells that do not produce the hormone.

Most studies have indicated that the presence of HCG in the serum of a patient with nontrophoblastic cancer has little or no prognostic value. One study did suggest that in patients with breast carcinoma, those with little or no HCG had a better response to chemotherapy, with a longer disease-free interval prior to relapse, than those with higher HCG concentrations [84]. Other studies on patients with breast carcinoma as well as evaluations of patients with hepatocellular, ovarian, or gastric carcinoma have failed to show any difference in survival of patients between those who had immunoreactive HCG detected in their circulation and those who did not [73,143,152,159].

Table 7 Immunohistochemical Detection of HCG in Tumor Tissue

Tumor	n	% Positive
Colon/rectum	101	52
Lung	493	34
Pancreas	13	31
Esophagus	39	28
Breast	594	24
Bladder	29	21
Ovary	252	19
Cervix	123	18
Gastric	280	17
Endometrium	124	14
Prostate	51	8
Miscellaenous	105	15
Total	2183	25

Source: Data from [142-170].

For the majority of patients with suspected or proven nontrophoblastic cancer, serum HCG measurements are not useful as a tumor marker. They are neither sensitive nor specific, they correlate poorly with the stage of the disease, show a poor correlation with the clinical course of the patient, and provide little or no prognostic information. However, before HCG is discarded as a viable tumor marker, it must be emphasized that the studies upon which these conclusions have been based have only examined serum using methods that measure intact HCG or free HCG-beta subunit. A number of molecular species of immunoreactive HCG have been found in cancer patients that may not be measured by the assays used in the studies detailed above (Table 8).

Perhaps the most exciting moiety currently being studied is the beta-core fragment of HCG. This substance is a fragment of the beta subunit and is devoid of the carboxy terminal region as well as much of the carbohydrate. It has not been found in the serum,

Table 8 Molecular Species of Immunoreactive HCG Detected in Patients with Cancer

Isolated alpha or beta subunits
Abnormal glycosylated forms
    Hyperglycosylated alpha and beta subunits
    Hypoglycosylated (asialo-HCG)
    Unique sugar chains in choriocarcinoma HCG
Fragments
    Beta core
    HCG beta without carboxyterminal peptide
    Asialo-HCG-beta-carboxyterminal peptide

but is present in the urine of pregnant patients and patients with cancer [171]. In three studies 56% (range, 25-78%) of patients with cancer were found to have the beta-core fragment in their urine [110,136,172]. The beta core fragment has been found in 6% of the urines from nonpregnant, healthy women, indicating that the presence of this fragment is not specific for cancer [136]. Nevertheless, early studies have shown a correlation between the level of the beta core fragment and stage of disease in women with gynecological malignancy [136]. There is no information available regarding any prognostic significance of the presence of this fragment nor whether serial measurements of the fragment will be useful as an objective parameter for monitoring the effects of therapy.

II. HUMAN PLACENTAL LACTOGEN

A. Introduction

HPL is a protein produced by the syncytiotrophoblast of the placenta during pregnancy. It has a molecular weight of 22,279 daltons and belongs to the prolactin-growth hormone gene family [173]. A macromolecular form of HPL is present in placenta and sera from pregnant women [174]. HPL can be detected by immunohistochemical techniques in the developing trophoblast of the early embryo, but most assays do not consistently detect HPL in material serum prior to the eighth week of pregnancy [175]. From that time until just prior to term, the concentrations in maternal serum progressively rise. HPL has anti-insulin and lipolytic activities and is one of the hormones responsible for the insulin resistance that occurs during late prenancy [173].

B. HPL as a Tumor Marker

1. Gestational Trophoblastic Disease

In patients with gestational trophoblastic disease, HPL levels are low in contrast to the HCG concentrations which are generally high, especially in hydatidiform mole [176-178]. Thus, HPL measurements have not been useful as a tumor marker for gestational trophoblastic disease. It has been suggested that HPL measurements may help in the differentiation between trophoblastic disease, in which HCG concentrations are high and HPL levels are low, and twin or other multiple pregnancies, in which both HCG and HPL levels are elevated and reflect placental mass [178].

2. Germ Cell Tumors of the Testis

HPL may also be produced by nonseminomatous germ cell neoplasms of the testes, especially choriocarcinoma [176,177]. However, the serum concentrations of HPL are quite low relative to the concentrations of HCG and the hormone is present in the serum of only a minority of patients with germ cell tumors [179,180]. HPL measurements are not useful for monitoring the effects of therapy of germ cell tumors because of the low concentrations of the hormone in the majority of patients [179-181].

3. Nontrophoblastic Tumors

There is a considerable amount of variation in the studies that have evaluated serum HPL concentrations in patients with nontrophoblastic malignancies. A composite of these studies is presented in Table 9 [82,88,90,113,116,118,120,181,129,132,141,177,182]. Although overall it would appear tht 14.3% of the patients studied have immunoreactive

Table 9 Immunoreactive HPL in Sera of Patients with Cancer and Control Patients

Tumor or site	n	% Positive
Patients with Cancer		
Gynecological	344	30.8
Ovary	219	38.4
Cervix	125	17.6
Gastrointestinal	274	13.5
Esophagus	16	0
Gastric	46	26.1
Colon/rectum	174	9.2
Hepatic	24	29.2
Biliary	13	15.4
Leukemia	13	7.7
Lung	64	4.7
Prostate	57	0
Miscellaneous	60	15
Total	1438	14.3
Controls		
Normal volunteers	455	0
Benign disease	180	0
Breast	57	0
Gastrointestinal	54	0
Genitourinary	19	0
Other	50	0
Total	630	0

Source: Data from [82,88,90,113,116,118,120,129,132,141,177,186].

HPL in their sera, caution is required in evaluating the results since there are marked discrepancies between studies. For instance, an early study by Weintraub and Rosen found that only 5.5% of patients with a variety of cancers were positive for HPL at a level greater than 1 ng/ml [177]. In contrast, a survey performed by Das and co-workers with an assay of similar sensitivity found HPL in 51% of the sera of patients with malignancy [120]. The reason for the discrepancy between studies is unclear. None of the studies found immunoreactive HPL in normal control patients (Table 9).

Of interest is the finding of immunohistochemical techniques that almost half of the cancer tissues studied contained cells that stained positively for HPL (Table 10) [143,165,184,185]. As has been found with HCG, tumors demonstrated much heterogeneity with only a small portion of the tumor cells staining positively for HPL. Two studies examined benign breast lesions for the presence of HPL and found that 13% had HPL-positive cells, indicating that HPL is not a specific tissue marker for nontrophoblastic malignancy [184,185].

The presence of HPL in the serum or tumor tissue of patients with cancer does not correlate with the presence of other placental proteins [88,165]. There are few data on the prognostic significance of the finding of HPL in patients with cancer. Horne and co-

Table 10 Immunohistochemical Detection of HPL in Tumor and Benign Tissues

Site	n	% Positive
Tumor		
Gastrointestinal	19	78.9
Breast	163	57.1
Uterus	23	52.2
Lung	99	20.2
Cervix	12	16.7
Ovary	20	10.0
Miscellaneous	8	62.5
Total	344	46.2
Control: benign breast disease	211	13.3

Source: Data from [143,165,184,185].

workers considered that patients with breast carcinoma whose tumor stained positively for HPL had a worse prognosis than those whose tumors were negative for HPL [143]. This report has been confirmed. HPL elevations in patients may be transient [132] and the levels of HPL in patients with nontrophoblastic tumors are generally at the limits of sensitivity of the assays. Several investigators have noted discordance between changes in HPL and changes in tumor burden during therapy and, therefore, serial HPL measurements are of little utility [82,139,141].

III. PREGNANCY-SPECIFIC BETA-1 GLYCOPROTEIN

A. Introduction

SP-1 is a glycoprotein (32.6% carbohydrate content) with a molecular weight of 42,300 daltons, and is produced by syncytiotrophoblast of the placenta [186]. It is detected in maternal serum approximately 10 days after conception and rises rapidly during the first trimester and more slowly throughout the remainder of gestation [175]. The maternal serum concentrations correlate with trophoblastic mass [175]. The SP-1 gene has close structural homology with the carcinoembryonic antigen (CEA) gene, indicating that SP-1 belongs to the CEA family [187]. The function of SP-1 during pregnancy is unknown and no biological activity has been proven. An immunologically similar material has been shown to be produced by human fibroblasts and glial cells and to be present in granulocytes [188-190].

1. Gestational Trophoblastic Disease

SP-1 has been found in the serum of virtually all patients harboring hydatidiform mole and about 70% of patients with choriocarcinoma [191,192]. HCG and SP-1 concentrations are correlated in patients with choriocarcinoma and during therapy the levels of both markers tend to change in parallel [193]. As a general rule, if SP-1 is present in the serum, HCG will be present, since the concentrations of the latter are higher in the majority of patients [192]. The differences between the rate of changes of HCG and SP-1 during therapy of gestational trophoblastic disease may be accounted for by differences in the half-life of disappearance of both glycoproteins, since SP-1 disappears with a half-life of 36-40 hr, which is slightly longer than the average for HCG [186]. In addition, both HCG and SP-1 have a great deal of microheterogeneiety in terms of their carbohydrate content and such differences may lead to alterations in the half-life of disappearance [186]. In patients with hydatidiform mole, the higher the level of SP-1 the greater the risk of the patient developing choriocarcinoma [194].

2. Germ Cell Tumors of the Testis

SP-1 is also produced by the syncytiotrophoblastic cells and trophoblast giant cells found in nonseminomatous germ cell tumors. The data from three series indicate that approximately 16% of patients with such neoplasms have immunoreactive SP-1 in their circulation [53,179,191]. The low frequency of positivity of SP-1 in comparison to HCG and alpha fetoprotein makes SP-1 measurements in patients with germ cell neoplasms less useful. In adddition, there has been some discordance between HCG and SP-1 measurements during therapy, with the changes SP-1 concentration correlating less well with the clinical findings than the HCG changes [179]. Thus, SP-1 does not appear to be a useful tumor marker for germ cell tumors of the testes.

3. Nontrophoblastic Tumors

SP-1 is produced by nontrophoblastic tumors both in vivo and in vitro. The combined data from the literature suggest that approximately one-quarter of patients with nontrophoblastic malignancies will have SP-1 in their blood (Table 11) [121,132,185,191, 193,195-205]. However, there has been much variation between reports, with a range from 2.4 to 71.4% [198,202,203]. Some of the discrepancy may be due to the use of different antibodies against SP-1. SP-1 has been shown to have some heterogeneity in regards to carbohydrate content, and also exists in two forms that differ in molecular size and electrophretic mobility [206].

There has been similar variation in the studies that have examined tumor tissue for the presence of SP-1 producing cells (Table 12) [143,165,184,185,199,201]. For instance, although overall about 34.3% of the breast cancer specimens that have been examined have been positive for SP-1, the range in different series extends from 0 to 76% [143, 184]. In addition, four studies have shown SP-1-staining cells in 7% of benign breast disease tissue, with a range in the studies from 3 to 32% [184,185,201,207].

Since SP-1 is produced by fibroblast and glial cells and has been found in granulocytes, it is not surprising that an SP-1-like substance has been identified by multiple investigators in the serum of nonpregnant normal adults [208-210]. Approximately 6% of normal control patients and patients with benign disorders have been found to have immunoreactive SP-1 in their circulation (Table 13) [121,132,185,191,193,195-205]. Thus, as has been found with HCG and HPL, the presence of SP-1 in the serum is not indicative of cancer.

Several investigators have examined the prognostic significance of SP-1 in patients with breast cancer and have suggested that the presence of SP-1 in the tissue adversely effects the prognosis and correlates with presence of metastatic disease [143,202,207]. The prognostic significance of SP-1 in patients with other types of tumors has not been evaluated.

Table 11 Immunoreactive SP-1 in Sera of Patients with Cancer

Tumor or site	n	% Positive
Myeloma-Waldenström's gammaglobulinemia	303	69.0
Leukemia	66	63.6
Sarcoma	20	45.0
Lung	69	29.0
Lymphoma	144	22.2
Gastrointestinal	330	16.1
Gastric	20	30.0
Colon/rectal	284	14.1
Hepatic	21	19.0
Biliary	2	50.0
Pancreatic	3	66.7
Ovary	74	13.5
Breast	714	9.4
Melanoma	41	0
Miscellaneous	68	35.3
Total	1829	25.5

Source: Data from [121,136,185,191,193,195–205].

Table 12 Immunohistochemical Detection of SP-1 in Tumor and Benign Tissues

Site	n	% Positive
Tumor		
Lymphoid	2	100
Lung	99	67.7
Genitourinary	3	66.7
Gastrointestinal	19	52.6
Uterus	23	43.5
Breast	367	34.3
Cervix	18	33.3
Ovary	20	10.0
Bone	3	0
Total	554	40.4
Control: benign breast disease	324	6.8

Source: Data from [143,165,184,185,199,201].

Table 13 Immunoreactive SP-1 Sera of Patients Without Cancer

Control	n	% Positive
Normal	818	2.7
Benign disease	498	9.6
Gastrointestinal	65	43.0[a]
Hematological	100	20.0
Breast	153	9.1
Infectious	64	3.1
Rheumatological	74	2.7
Allergic	42	0
Total	1316	5.6

[a]Primarily inflammatory bowel disease.
Source: Data from [121,132,185,191,193,195-205].

IV. SUMMARY AND CONCLUSIONS

Among the three placental proteins discussed, HCG is the only clinically useful tumor marker, and the value of HCG measurements is restricted to patients with gestational and nongestational trophoblastic disease. In patients with gestational trophoblastic disease, HCG levels may serve as an adjunct for the diagnosis, provide prognostic information, and be an objective parameter to evaluate the effects of therapy. Little or no additional information is obtained from HPL or SP-1 measurements. In patients with germ cell neoplasms of the testis, HCG measurements add useful information for clinical staging and monitoring of therapy, although discordance between tumor growth and HCG levels can be found in patients whose tumors contain several different elements. Therefore, AFP measurements must be made as well in these patients to monitor disease activity. Neither HPL nor SP-1 measurements are useful in these patients. None of the placental proteins are useful for screening, as prognostic indicators, or for evaluating the effects of therapy in groups of patients with nontrophoblastic neoplasms. In some patients with nontrophoblastic malignancies, each of the markers may accurately reflect changes in tumor burden during therapy. However, the problems with specificity and sensitivity of the tests and the fact that the majority of patients whose tumors produce the hormone have circulating concentrations that are at the limits of detection of the assays decrease the utility of these measurements and render them cost-ineffective for routine patient care.

REFERENCES

1. Bohn H. Systematic identification of specific oncoplacental proteins. In: Fishman WH, ed. Oncodevelopmental markers. Biologic, diagnostic, and monitoring aspects. New York: Academic press, 1983:69-86.
2. Frankenne F, Pirens G, Horward-Poncelet M, Demoulin A, Scippo ML, Hennen G. Determination of the physiology of maternal growth hormones during pregnancy and partial characterization of the placental GH variant. J Clin Endocrinol Metab 1988;66:1171-1180.

3. Birkin S, Canfield RE. Chemistry and immunochemistry of human chorionic gonadotropin. In: Segal SJ, ed. Chorionic gonadotropin. New York: Plenum Press, 1980:65-88.
4. Braunstein GD, Rasor J, Adler D, Danzer H, Wade ME. Serum human chorionic gonadotropin levels throughout normal pregnancy. Ann J Obstet Gynecol 1976; 126:678-681.
5. Ozturk M, Bellet D, Manil L, Hennen G, Frydman R, Wands J. Physiological studies of human chorionic gonadotropin (hCG), αhCG and βhCG as measured by specific monoclonal immunoradiometric assays. Endocrinology 1987;120:559-558.
6. Blithe DL, Nisula RC. Variations in the oligosaccharides on free and combined α-subunits of human choriogonadotropin in pregnancy. Endocrinology 1985;117: 2218-2228.
7. Kato Y, Braunstein GD. Beta-core fragment is a major form of immunoreactive urinary chorionic gonadotropin in human pregnancy. J Clin Endocrinol Metab 1988; 66:1197-1201.
8. Reyes FI, Winter JSD, Faiman C. Postpartum disappearance of chorionic gonadotropin from the maternal and neonatal circulation. Am J Obstet Gynecol 1985; 153:486-489.
9. Vaitukaitis JL, Ross GT, Braunstein GD, Rayford PL. Gonadotropins and their subunits: basic and clinical studies. Rec Progr Horm Res 1976;32:289-331.
10. Nisula BC, Taliadouros GS, Carayon P. Primary and secondary biologic activities intrinsic to the human chorionic gonadotropin molecule. In: Segal SJ, ed. Chorionic gonadotropin. New York: Plenum Press, 1980:17-36.
11. Braunstein GD, Production of human chorionic gonadotropin by nontrophoblastic tumors and tissues. In: Tomoda Y, Mitzutani S, Narita O, Klopper A, eds. Placental and endometrial proteins: basic and clinical aspects. The Netherlands, VNU Science Press BV, 1988:493-502.
12. Braunstein GD. Physiologic function of human chorionic gonadotropin during pregnancy. In: Mochizui M, Hussa R, eds. Excerpta medical international congress series of the satellite symposium on placental proteins. The Netherlands: Elsevier Science Publishers, 1988:33-41.
13. Odell WD, Hammond E, Griffin J. A chorionic gonadotropin secreting human pituitary cell. Abstracts of the 70th Annual Meeting of the Endocrine Society, June 8-11, 1988; abstract 49.
14. Vaitukaitis JL, Braunstein GD, Ross GT. A radioimmunoassay which specifically measures human chorionic gonadotropin in the presence of human luteinizing hormone. Am J Obstet Gynecol 1972;113:751-758.
15. Braunstein GD, Kelley L, Farber S, Sigall ER, Wade ME. Two rapid, sensitive, and specific immunoenzymatic assays of human choriogonadotropin in urine evaluated. Clin Chem 1986;32:1413.
16. Braunstein GD, Kelley L, Farber S, Wade ME. Two rapid, sensitive, and specific immunoenzymatic assays of choriogonadotropin in serum compared. Clin Chem 1987;33:1947.
17. Delfs E. Quantitative chorionic gonadotrophin. Prognostic value in hydatidiform mole and chorionepithelioma. Obstet Gynecol 1957;9:1-24.
18. Braunstein GD, Karow WG, Gentry WC, Rasor J, Wade ME. First-trimester chorionic gonadotropin measurements as an aid in the diagnosis of early pregnancy disorders. Am J Obstet Gynecol 1978;131:25-32.
19. Romero R, Horgan JG, Kohorn EI, Kadar N, Taylor KJ, Hobbins JC. New criteria for the diagnosis of gestational trophoblastic disease. Obstet Gynecol 1985;66:553-558.
20. Morrow CP, Kletzky OA, Disaia PJ, Townsend DE, Mishell DR, Nakamura RM.

Clinical and laboratory correlates of molar pregnancy and trophoblastic disease. Am J Obstet Gynecol 1977;128:424-430.
21. Bagshawe KD. Human chorionic gonadotropin (HCG)-measurement. Radioimmunoassay. In: Bearson SA, Yalow RS, eds. Methods in investigative and diagnostic endocrinology, part III, peptide hormones. Amsterdam: North-Holland, 1973:756-763.
22. Yuen BH, Cannon W. Molar pregnancy in British Columbia: estimated incidences and post evaluations regression patterns of the beta subunit of human chorionic gonadotropin. Am J Obstet Gynecol 1981;139:316-319.
23. Ross GT, Goldstein DP, Hertz R, Lipsett MB, Odell WD. Sequential use of methotrexate and actinomycin D in the treatment of metastatic choriocarcinoma and related trophoblastic diseases in women. Am J Obstet Gynecol 1965;93:223-229.
24. Bagshawe KD. Clinical applications of hCG. Adv Exp Med Biol 1984;76:313-324.
25. Hammond CB, Weed JC Jr, Currie JL. The role of operation in the current therapy of gestational trophoblastic disease. Am J Obstet Gynecol 1980;136:844-858.
26. Vaitukaitis JL, Ebersole ER. Evidence for altered synthesis of human chorionic gonadotropin in gestational trophoblastic tumors. J Clin Endocrinol Metab 1976; 42:1048-1055.
27. Dawood MY, Saxena BB, Landesman R. Human chorionic gonadotropin and its subunits in hydatidiform mole and choriocarcinoma. Obstet Gynecol 1977;50: 172-181.
28. Fan C, Goto S, Furuhashi Y, Tomoda Y. Radioimmunoassay of the serum free β-subunit of human chorionic gonadotropin in trophoblastic disease. J Clin Endocrinol Metab 1987;64:313-318.
29. Bagshawe KD, Harland S. Immunodiagnosis in monitoring of gonadotrophin-producing metastases in the central nervous system. Cancer 1976;38:112-118.
30. Soma H, Kikuchi K, Takayama M, et al. Concentrations of SPI and beta-hCG in serum and cerebrospinal fluid and concentrations of hCG in urine in patients with trophoblastic tumour. Arch Gynecol 1981;230:321-327.
31. Bagshawe KD. Choriocarcinoma. The clinical biology of the trophoblast and its tumours. Baltimore: Williams & Wilkins, 1969.
32. Hammond DB, Hertz R, Ross GT, Lipsett MB, Odell WD. Primary chemotherapy for nonmetastatic gestational trophoblastic neoplasms. Am J Obstet Gynecol 1967; 98:71-78.
33. Hammond DB, Parker RT. Diagnosis and treatment of trophoblastic disease. A report for the Southeastern regional center. Obstet Gynecol 1970;35:132-143.
34. Goldstein DP. Preferable management of gestational trophoblastic disease: Benign and malignant. In: Reid DR, Christian CD, eds. Controversy in obstetrics and gynecology II. Philadelphia: WB Saunders, 1974:219-234.
35. Jones WB, Lewis JL Jr. Treatment of gestational trophoblastic disease. Am J Obstet Gynecol 1974;120:14-20.
36. Zondek B. Versuch einer biologischen (hormonalen) diagnostik beim malingnen hudentumor. Chirurg 1930;2:1072-1073.
37. Braunstein GD, Vaitukaitis JL, Carbone PP, Ross GT. Ectopic production of human chorionic gonadotropin by neoplasms. Ann Intern Med 1973;78:39-45.
38. Waldmann TA, McIntire KR. The use of a radioimmunoassay for alpha-fetoprotein in the diagnosis of malignancy. Cancer 1974;34:1510-1515.
39. Cochran JS, Walsh PC, Porter JC, Nicholson TC, Madden JD, Peters PD. The endocrinology of human chorionic gonadotropin-secreting testicular tumors: new methods in diagnosis. J Urol 1975;114:549-555.
40. Lange PH, McIntire KR, Waldmann TA, Hakala TR, Fraley EE. Serum alpha-feto-

protein and human chorionic gonadotropin in the diagnosis and management of non-seminomatous germ-cell testicular cancer. N Engl J Med 1976;295:1237-1240.
41. Newland ES, Dent JJ, Kardana A, Searle F, Bagshawe KD. Serum α1-fetoprotein and HCG in patients with testicular tumours. Lancet 1976;2:744-745.
42. Edsmyr F, Wahren B, Silfversward C. Usefulness of immunology and hormonal markers in the treatment of testis tumor. Int J Radiat Oncol Biol Phys 1976;1:279-284.
43. Scardino PT, Cox HD, Waldmann TA, McIntire KR, Mittemeyer B, Javadpour N. The value of serum tumor markers in the staging and prognosis of germ cell tumors of the testis. J Urol 1977;118:994-999.
44. Schultz H, Sell A, Norgaard-Pedersen B, Arends J. Serum alpha-fetoprotein and human chorionic gonadotropin as markers for the effect of postoperative radiation therapy and/or chemotherapy in testicular cancer. Cancer 1978;42:2182-2186.
45. Javadpour N, McIntire KR, Waldmann TA. Human chorionic gonadotropin (HCG) and alpha-fetoprotein (AFP) in sera and tumor cells of patients with testicular seminoma. Cancer 1978;42:2768-2772.
46. Thompsom DK, Haddow JE. Serial monitoring of serum alpha-fetoprotein and chorionic gonadotropin in males with germ cell tumors. Cancer 1979;43:1820-1829.
47. Anderson T, Waldmann TA, Javadpour N, Glatstein E. Testicular germ-cell neoplasms: recent advances in diagnosis and therapy. Ann Intern Med 1979;90:373-385.
48. Scardino PT, Skinner DG. Germ-cell tumors of the testis: improved results in a prospective study using combined modality therapy and biochemical tumor markers. Surgery 1979;86:86-93.
49. Narayana AS, Loening S, Weimer G, Culp DA. Serum markers in testicular tumors. J Urol 1979;121:51-53.
50. Ward AM. Markers in germ cell tumours: the current state of the art, AFP, βhCG and AHL kinetics. In: Anderson CK, Jones WG, Ward AM, eds. Germ cell tumors. London: Taylor and Francis, 1981:207-215.
51. Norgaard-Pedersen B, Schultz H, Arends J, et al. Biochemical markers for testicular germ-cell tumors in relation to histology and stage: Some experience from the Danish Testicular Cancer (DATECA) Study from 1976 through 1981. Ann NY Acad Sci 1983;417:390-399.
52. Mirimanoff RO, Shipley WU, Dosoretz DE, Meyer JE. Pure seminoma of the testis: The results of radiation therapy in patients with elevated human chorionic gonadotropin titers. J Urol 1985;134:1124-1126.
53. DeBruijn HWA, Sleifjer DT, Koops HS, Suurmeijer AJH, Marrink J, Ockhuizen T. Significance of human chorionic gonadotropin, alpha-fetoprotein, and pregnancy-specific beta-1-glycoprotein in the detection of tumor relapse and partial remission in 126 patients with nonseminomatous cancer. Cancer 1985;55:829-835.
54. Kurman RJ, Scardino PT, McIntire KR, Waldmann TA, Javadpour N. Cellular localization of alpha-fetoprotein and human chorionic gonadotropin in germ cell tumors of the testis using an indirect immunoperoxidase technique. A new approach to classification utilizing tumor markers. Cancer 1977;40:2136-2151.
55. Willemse PH, Sleijfer DT, Schraffordt KH, et al. Tumor markers in patients with non-seminomatous germ cell tumors of the testis. Oncodev Biol Med 1981;2:117-128.
56. Javadpour N, Soares T. False-positive and false-negative alpha-fetoprotein and human chorionic gonadotropin assays in testicular cancer: a double-blind study. Cancer 1981;48:2279-2281.
57. Catalona WJ, Vaitukaitis JL, Fair WR. Falsely positive specific human chorionic gonadotropin assays in patients with testicular tumors: conversion to negative with testosterone administration. J Urol 1979;122:126-128.

58. Lang PH. Markers for germ cell tumors of the testis. In: Fishman WH, ed. Oncodevelopmental markers. Biologic diagnostic, and monitoring aspects. New York: Academic Press, 1983:241-257.
59. Twombly GH, Temple HM, Dean AL. Clinical value of the Aschheim-Zondek test in the diagnosis of testicular tumours. JAMA 1942;118:106-111.
60. Patton JF, Seitzman DM, Zone RA. Diagnosis and treatment of testicular tumors. Am J Surg 1960;99:525-532.
61. Braunstein GD, McIntire KR, Waldmann TA. Discordance of human chorionic gonadotropin and alpha-fetoprotein in testicular teratocarcinomas. Cancer 1973; 31:1065-1068.
62. Perlin E, Engeler JE Jr, Edson M, Karp D, McIntire KR, Waldmann TA. The value of serial measurements of both human chorionic gonadotropin and alpha-fetoprotein for monitoring germinal cell tumors. Cancer 1976;37:215-219.
63. Boctor SM, Kurohara SS, Badib AO, Murphy GP. Current results from therapy of testicular tumors. Cancer 1969;24:870-875.
64. Gangai MP. The A-Z tier as an aid in treatment of testicular tumors. J Urol 1965; 94:589-591.
65. Notter G, Ranudd NE. Treatment of malignant testicular tumors—a report on 355 patients. Acta Radiol 1964;2:273-301.
66. Wilson JM, Woodhead DM. Prognostic and therapeutic implications of urinary gonadotropin levels in the management of testicular neoplasia. J Urol 1972;108: 754-756.
67. Keogh B, Hreshchyshyn MM, Moore RH, Merrin CE, Murphy G. Urinary gonadotropins in management and prognosis of testicular tumor. Urology 1975;4:496-503.
68. Butcher DN, Gregory WM, Gunter PA, Masters JR, Parkinson MC. The biological and clinical significance of HCG-containing cells in seminoma. Br J Cancer 1985; 51:473-478.
69. Morgan DA, Caillaud JM, Bellet D, Eschwege F. Gonadotrophin-producing seminoma: a distinct category of germ cell neoplasm. Clin Radiol 1982;33:149-153.
70. Lange PH, Nochomovitz LE, Rosai J, et al. Serum alpha-fetoprotein and human chorionic gonadotropin in patients with seminoma. J Urol 1980;124:472-478.
71. Swartz Da, Johnson DE, Hussey DH. Should an elevated human chorionic gonadotropin titer alter therapy for seminoma. J Urol 1984;131:63-65.
72. Braunstein GD. hCG expression in trophoblastic and nontrophoblastic tumors. In: Fishman WH, ed. Oncodevelopmental markers. Biologic, diagnostic, and monitoring aspects. New York: Academic Press, 1983:351-371.
73. Braunstein GD, Vogel CL, Vaitukaitis JL, Ross GT. Ectopic production of human chorionic gonadotropin in Ugandan patients with hepatocellular carcinoma. Cancer 1973;32:223-226.
74. Goldstein DP, Kosasa TS, Skarim AT. The clinical application of a specific radioimmunoassay for human chorionic gonadotropin in trophoblastic and nontrophoblastic tumors. Surg Gynecol Obstet 1974;138:747-751.
75. Sheth NA, Saruiya JN, Ranadive KJ, Sheth AR. Ectopic production of human chorionic gonadotrophin by human breast tumours. Br J Cancer 1974;30:566-570.
76. Tormey DC, Waalkes TP, Ahmann D, et al. Biological markers in breast carcinoma. Cancer 1975;35:1095-1100.
77. Fishman W, Raam S, Stolbach LL. Markers for ovarian cancer: Regan isoenzyme and other glycoproteins. Semin Oncol 1975;2:211-216.
78. Fishman WH, Inglis NR, Vaitukaitis J, Stolbach LL. Regan isoenzyme and human chorionic gonadotropin in ovarian cancer. Natl Cancer Inst Monogr 1974;42:63-73.
79. Franchimont P, Zangerle PF, Nogarede J, et al. Simultaneous assays of cancer-associ-

ated antigens in various neoplastic disorders. Cancer 1976;38:2287-2295.
80. Stolbach L, Inglis N, Lin C, et al. Measurement of regan isoenzyme hCG, CEA, and histaminase in the serum and effusion fluids of patients with carcinoma of the breast, ovary, or lung. In: Fishman WH, Sell S, eds. Onco-developmental gene expression. New York: Academic Press, 1976:433-434.
81. Gailani S, Chu TM, Nussbaum A, Ostrander M, Christoff N. Human chorionic gonadotrophins (hCG) in non-trophoblastic neoplasms. Assessment of abnormalities of hCG and CEA in bronchogenic and digestive neoplasms. Cancer 1976;38:1684-1688.
82. Samaan NA, Smith JP, Rutledge FN, Schultz PN. The significance of measurement of human placental lactogen, human chorionic gonadotropin, and carcinoembryonic antigen in patients with ovarian carcinoma. Am J Obstet Gynecol 1976;126:186-189.
83. Hagen C, Gilby ED, McNeilly AS, Olgaard K, Bondy PK, Ree LH. Comparison of circulating glycoprotein hormones and their subunits in patients with oat cell carcinoma of the lung and uraemic patients on chronic dialysis. Acta Endocrinol 1976; 83:26-35.
84. Tormey DC, Waalkes TP, Simon RM. Biological markers in breast cancer. II. Clinical correlations with human chorionic gonadotrophin. Cancer 1977;39:2391-2396.
85. Coombes RC, Powles TJ, Gazet JC, et al. A biochemical approach to the staging of human breast cancer. Cancer 1977;40:937-999.
86. Franchimont R, Zangerle PF, Hendrick JC, Reuter A, Colin C. Simultaneous assays of cancer associated antigens in benign and malignant breast diseases. Cancer 1977; 39:2806-2812.
87. Franchimont R, Hendrick JC, Reuter A, Zangerle PF. Ectopic production of HCG, its subunits and of casein. Proceedings of th V International Congress of Endocrinology, Hamburg, July 19-24, 1976.
88. Sheth NA, Suraiya JN, Sheth AR, Ranadive KJ, Jussawalla DJ. Ectopic production of human placental lactogen by human breast tumors. Cancer 1977;39:1693-1699.
89. Stone M, Bagshawe KD, Kardana A, Searle F, Dent J. Beta-human chorionic gonadotrophin and carcino-embryonic antigen in the management of ovarian carcinoma. Br J Obstet Gynecol 1977;84:375-379.
90. Broder LE, Weinbraub BD, Rosen SW, Cohen MH, Tejada F. Placental proteins and their subunits as tumor markers in prostatic carcinoma. Cancer 1977;40:211-216.
91. Kahn CR, Rosen SW, Weintraub BD, Fajans SS, Gorden P. Ectopic production of chorionic gonadotropin and its subunits by islet-cell tumors. N Engl J Med 1977; 297:565-569.
92. Sufrin G, Mirand EA, Moore RH, Chu TM, Murphy GP. Hormones in renal cancer. J Urol 1977;117:433-438.
93. Williams RR, McIntire KR, Waldmann TA, et al. Tumor-associated antigen levels (carcinoembryonic antigen, human chorionic gonadotropin, and alpha-fetoprotein) antedating the diagnosis of cancer in the Framingham study. J Natl Cancer Inst 1977;58:1547-1551.
94. Waalkes TP, Gehrke CW, Tormey DC, et al. Biologic markers in breast carcinoma. Cancer 1978;41:1871-1882.
95. Cowen DM, Searle F, Ward AM, et al. Multivariate biochemical indicators of breast cancer: an evaluation of their potential in routine practice. Eur J Cancer 1978; 14:885-893.
96. Dash RJ, Dutta TK, Purohit OP, Jayakumar RV, Gupta BD. Prevalence of ectopic hCG production in non-endocrine malignancy. Ind J Cancer 1978;15:23-27.
97. Dosogne-Guerin M, Stolarczyk A, Brokowski A. Prospective study of the α and β subunits of human chorionic gonadotrophin in the blood of patients with various benign and malignant conditions. Eur J Cancer 1978;14:525-532.

98. Rutanen EM, Seppala M. The hCG-beta subunit radioimmunoassay in nontrophoblastic gynecologic tumors. Cancer 1978;41:692-669.
99. Hattori M, Fukase M, Yoshimi H, Matsukura S, Imura H. Ectopic production of human chorionic gonadotropin in malignant tumors. Cancer 1978;42:2328-2333.
100. Tubbs RR, Velasco ME, Benjamin SP. Immunocytochemical identification of human chorionic gonadotropin. Comparative study of diaminobenzidine and 3-Amino, 9-ethylcarbazole, a nonhazardous chromogen. Arch Pathol Lab Med 1979;103:534-536.
101. Bender RA, Weintraub BD, Rosen SW. Prospective evaluation of two tumor-associated proteins in pancreatic adenocarcinoma. Cancer 1979;43:591-595.
102. Cove DH, Woods KL, Smith SCH, et al. Tumour markers in breast cancer. Br J Cancer 1979;40:710-718.
103. Dash RJ, Dutta TK, Gupta SK, Purohit OP, Jayakumar RV, Gupta BD. Ectopic human chorionic gonadotropin in carcinoma cervix. Ind J Med Res 1979;70:478-482.
104. Menon M, Stefani SS. Evaluation of human chorionic gonadotropin-β levels in prostatic carcinoma. Urol Int 1980;35:291-293.
105. Blackman MR, Weintraub BD, Rosen SW, Kourides IA, Steinwascher K, Gail MH. Human placental and pituitary glycoprotein hormones and their subunits as tumor markers: a quantitative assessment. J Natl Cancer Inst 1980;65:81-93.
106. Carenza L, DiGregorio R, Mocci C, Moro M, Pala A. Ectopic human chorionic gonadotropin: gynecological tumors and nonmalignant conditions. Gynecol Oncol 1980;10:32-38.
107. Lindstedt G, Lundberg PA, Hedman LA. Circulating choriogondotropin β subunit in a patient with primary amenorrhea and embryonal ovarian carcinoma. Clin Chim Acta 1980;104:195-200.
108. Donaldson ES, VanNagell JR, Pursell S, et al. Multiple biochemical markers in patients with gynecologic malignancies. Cancer 1980;45:948-953.
109. Gropp C, Havemann K, Scheuer A. Ectopic hormones in lung cancer patients at diagnosis and during therapy. Cancer 1980;46:347-354.
110. Papaetrou PD, Sakarelou NP, Braouzi H, Fessas PH. Ectopic production of human chorionic gonadotropin (hCG) by neoplasms: the value of measurements of immunoreactive hCG in the urine as a screening procedure. Cancer 1980;45:2583-2592.
111. Kirschner MA, Lippman A, Berkowitz R, Mayrer E, Drejka M. Estrogen production as a tumor marker in patients with gonadotropin-producing neoplasms. Cancer Res 1981;41:1447-1450.
112. Oberg K, Wide L. hCG and hCG subunits as tumour markers in patients with endocrine pancreatic tumours and carcinoids. Acta Endocrinol 1981;98:256-260.
113. Shinde SR, Adil MA, Sheth AR, Koppikar MG, Sheth NA. Ectopic human placental lactogen and β-human chorionic gonadotropin in gastric fluid of patients with malignant and non-malignant conditions of the stomach. Oncology 1981;38:277-280.
114. Tsalacopoulos G, Bloch B. Ectopic production of the beta subunit of human chorionic gonadotrophin by malignant ovarian neoplasms. S Afr Med J 1982;62:487-488.
115. Ayala AR, Saad A, Vazquez X, Ramirez-Wiella G, Perches RD. Human chorionic gonadotropin immunoreactivity in serum of patients with malignant neoplasms. Am J Reprod Immunol 1983;3:149-151.
116. Monteiro JCMP, Barker G, Ferguson KM, Wiltshaw E, Neville AM. Ectopic production of human chorionic gonadotrophin (hCG) and human placental lactogen (hPL) by ovarian carcinoma. Eur J Cancer Clin Oncol 1983;19:173-178.

117. Prinz RA, Bermes EW, Kimmel JR, Marangos PJ. Serum markers for pancreatic islet cell and intestinal carcinoid tumors: a comparison of neuron-specific enolase, B-human chorionic gonadotropin and pancreatic polypeptide. Surgery 1983; 94:1019-1023.
118. Das S, Mukherjee K, Bhattacharya S, Chowdhury JR. Ectopic production of placental hormones (human chorionic gonadotropin and human placental lactogen) in carcinoma of the uterine cervix. Cancer 1983;51:1854-1857.
119. Monteiro JCMP, Ferguson KM, McKinna JA, Greening WP, Neville AM. Ectopic production of human chorionic gonadotrophin-like material by breast cancer. Cancer 1984;53:957-962.
120. Das S, Bhattacharya S, Mukherjee K, Roychowdhury J. Placental hormones as possible markers in gynaecological cancer. Eur J Gynaecol Oncol 1985;6:101-104.
121. Crowther ME, Grudzinskas JG, Poulton TA, Gordon YB. Trophoblastic proteins in ovarian carcinoma. Obstet Gynecol 1979;53:59-61.
122. Demura H, Jibiki K, Demura R, Shizume K. Immunological and biological studies on ectopic production of human chorionic gonadotropin by neoplasms. Cancer Treat Rep 1979;63:1215.
123. Wahren B, Lidbrink E, Wallgren A, Eneroth P, Zajicek J. Carcinoembryonic antigen and other tumor markers in tissue and serum or plasma of patients with primary mammary carcinoma. Cancer 1978;42:1870-1878.
124. Mercer DW, Talamo TS. Multiple markers of malignancy in sera of patients with colorectal carcinoma: preliminary clinical studies. Clin Chem 1985;31:1824-1828.
125. Cauchi MN, Koh SH, Lim D, Hay DL. Oncofetal antigens in cancer of the cervix and ovary. Br J Cancer 1981;44:403-409.
126. Wurz H, Geiger W, Grah H, Hoffman M. Simultaneous assays of SP_1 (PSβG), SP_3 (α_2PAG), CEA, AFP and β-HCG in the serum of patients with breast cancer and other nontrophoblastic malignancies. In: Lehmann FG, ed. Carcino-embryonic proteins. Amsterdam: Elsevier/North Holland Biomedical Press, 1979:487-490.
127. teVelde ER, deKroon RA, duBoueuff JA, et al. Studies on serum levels of carcinoembryonic antigen, alpha-fetoprotein, human chorionic gonadotrophin and alpha-2-pregnancy associated glycoprotein in patients with invasive cervical cancer. In: Lehmann FG, ed. Carcino-embryonic proteins. Amsterdam: Elsevier/North Holland Biomedical Press, 1979:185-190.
128. Reddy MN, Rochman H, Hunter RL, Fang VS, DeMeester T. Carcinoembryonic antigen, k-casein and β-human chorionic gonadotropin the staging of lung cancer. In: Lehmann FG, ed. Carcino-embryonic proteins. Amsterdam: Elsevier/North Holland Biomedical Press, 1979:173-174.
129. Morrow JS, Levin DL, Kyle RA, Rosen SW. Placental proteins as tumor markers in human monoclonal gammopathies: results in 109 patients. Tumor Biol 1985; 6:221-232.
130. Caffier H, Brandau H. Serum tumor markers in metastatic breast cancer and course of disease. Cancer Detect Prev 1983;6:451-457.
131. Scheuer A, Grun R, Lehmann F-G. Peptide hormones in liver cirrhosis and hepatocellular carcinoma. Oncodev Biol Med 1981;2:1-10.
132. Szymendera JJ, Kaminska JA, Nowacki MP, Szawlowski A, Gadek A. The serum levels of human α-fetoprotein, AFP, choriogonadotropin, hCG, placental lactogen, hPL, and pregnancy-specific β_1-glycoprotein, SP_1, are of no clinical significance in colorectal carcinoma. Eur J Cancer Clin Oncol 1981;17:1047-1052.
133. Canal P, Bugat R, Soula G, Combes PF. Determination de la βHCG circulante ectopicque dans les tumeurs solides. Pathol Biol 1981;29:150-154.
134. Ruibal A, Monne J, Richart C, Gomez-Sanchez D. La subunidad beta de la hor-

mona gonadotrofica corionica como marcadore de las neoplasias digestivas. Rev Esp Oncol 1979;26:13-18.
135. Behera D, Malik SK, Sharma BR, Dash RJ. Circulating hormones in lung cancer. Ind J Med Res 1984;79:636-640.
136. Cole LA, Wong Y, Elliott M, et al. Urinary human chorionic gonadotropin free β-beta and β-core fragment: a new marker of gynecological cancer. Cancer Res 1988; 48:1356-1360.
137. Gupta SK, Buckshee K, Talwar GP. Ectopic secretion of human chorionic gonadotropin by gynaecological tumours. Ind J Med Res 1984;80:189-195.
138. Sussman HH, Weintraub BD, Rosen SW. Relationship of ectopic placental alkaline phosphatase to ectopic chorionic gonadotropin and placental lactogen. Discordance of three "markers" for cancer. Cancer 1973;33:820-823.
139. Muggia FM, Rosen SW, Weintraub BD, Hansen HH. Ectopic placental proteins in nontrophoblastic tumors. Cancer 1975;36:1327-1337.
140. Rutanen E-M, Stenman U-H, Ranta T, Nieminen U, Seppal M. Discordant expression of carcinoembryonic antigen and human chorionic gonadotropin in gynecologic malignancy. In: Peeters H, ed. Protides of the biological fluids. Oxford: Pergamon Press, 1976:401-404.
141. Stanhope CR, Smith JP, Britton JC, Crosley PK. Serial determination of marker substances in ovarian cancer. Gynecol Oncol 1979;8:284-287.
142. McManus LM, Naughton MA, Martinex-Hernandez A. Human chorionic gonadotropin in human neoplastic cells. Cancer Res 1976;36:3476-3481.
143. Horne CHW, Reid IN, Milne GD. Prognostic significance of inappropriate production of pregnancy proteins by breast cancers. Lancet 1976;2:279-282.
144. Hustin J. Immunohistochemical demonstration of several tumour markers in neoplastic and preneoplastic states of the uterine mucosa. Gynecol Obstet Invest 1978;9:3-15.
145. Buckley CH, Fox H. An immunohistochemical study of the significance of hCG secretion by large bowel adenocarcinomata. J Clin Pathol 1979;32:368-372.
146. Bellet D, Arrang JM, Contesso G, Caillaud JM, Bohuon C. Localization of the β subunit of human chorionic gonadotrophin on various tumors. Eur J Cancer 1980;16:433-439.
147. Nishiyama T, Stolbach LL, Rule AH, DeLellis RA, Inglis NR, Fishman WH. Expression of oncodevelopmental markers (Regan isozyme, β-hCG, CEA) in tumor tissues and uninvolved bronchial mucosa. An immunohistochemical study. Acta Histochem Cytochem 1980;13:245-253.
148. Wilson TS, McDowell EM, McIntire KR, Trump BF. Elaboration of human chorionic gonadotropin by lung tumor. Arch Pathol Lab Med 1981;105:169-173.
149. Kodama T, Kameya T, Hirota T, et al. Production of alpha-fetoprotein, normal serum proteins, and human chorionic gonadotropin in stomach cancer: histologic and immunohistochemical analyses of 35 cases. Cancer 1981;48:1647-1655.
150. Ito H, Tahara E. Human chorionic gonadotropin in human gastric carcinoma. A retrospective immunohistochemical study. Acta Pathol Jpn 1983;33:287-296.
151. Bychkov V, Rothman M, Bardawil WA. Immunocytochemical localization of carcinoembryonic antigen (CEA), alpha-fetoprotein (AFP), and human chorionic gonadotropin (HCG) in cervical neoplasia. Am J Clin Path 1983;79:414-420.
152. Mohabeer J, Buckley CH, Fox H. An immunohistochemical study of the incidence and significance of human chorionic gonadotrophin synthesis by epithelial ovarian neoplasms. Gynecol Oncol 1983;16:78-84.
153. Wittekind C, Wachner R, VonKleist S. Immunohistological distribution pattern of CEA, SP-1 and HCG in mammary carcinoma and their axillary lymph node metastases. Verh Dtsch Ges Pathol 1984;68:97-99.

154. Wachner R, Wittekind C, VonKleist S. Immunohistological localization of β-HCG in breast carcinoma. Eur J Cancer Clin Oncol 1984;20:679-684.
155. Wachner R, Wittekind C, VonKleist S. Localisation of CEA, β-HCT, SP1, and keratin in the tissue of lung carcinomas. An immunohistochemical study. Virchows Arch 1984;402:415-423.
156. Purnell DM, Heatfield BM, Trump BF. Immunocytochemical evaluation of human prostatic carcinomas for carcinoembryonic antigen, nonspecific cross-reacting antigen, β-chorionic gonadotrophin, and prostate-specific antigen. Cancer Res 1984; 44:285-292.
157. Casper S, VanNagell JR, Powell DF, et al. Immunohistochemical localization of tumor markers in epithelial ovarian cancer. Am J Obstet Gynecol 1984;149:154-158.
158. Kimura N, Ghandur-Mnaymen L. An immunohistochemical study of keratin, carcinoembryonic antigen, human chorionic gonadotropin and alpha-fetoprotein in lung cancer. Tohoku J Exp Med 1985;145:23-38.
159. Lee AK, Rosen PO, DeLellis RA, et al. Tumor marker expression of breast carcinomas and relationship to prognosis. An immunohistochemical study. Am J Clin Pathol 1985;84:687-696.
160. Okano T. Immunohistochemical and clinical study of hCG-β producing tumors. Jpn J Clin Oncol 1979;9:215-224.
161. Burg-Kurland CL, Purnell DM, Combs JW, Hillman EA, Harris CC, Trump BF. Immunocytochemical evaluation of human esophageal neoplasms and preneoplstic lesions for β-chorionic gonadotropin, placental lactogen, α-fetoprotein, carcinoembryonic antigen, and nonspecific cross-reacting antigen. Cancer Res 1986;46:2936-2943.
162. Yamase HT, Wurzel RS, Nieh PT, Gondos B. Immunohistochemical demonstration of human chorionic gonadotropin in tumors of the urinary bladder. Ann Clin Lab Sci 1985;15:414-417.
163. Nadji M, Tabei SZ, Castro A, Morales AR. Immunohistochemical demonstration of human chorionic gonadotropin in mammary carcinoma. Cancer Treat Rep 1979; 63:1183.
164. Manabe T, Adachi M, Hirao K. Human chorionic gonadotropin in normal, inflammatory, and carcinomatous gastric tissue. Gastroenterology 1985;89:1319-1325.
165. Harach HR, Skinner M, Gibbs AR. Biological markers in human lung carcinoma: an immunopathological study of six antigens. Thorax 1983;38:937-941.
166. Kasurinen J, Syrjanen KJ. Stainability of the peptide hormones in gastrointestinal apudomas as demonstrated by immunoperoxidase kits. Scand J Gastroenterol 1984;19:167-172.
167. Hayata T, Crum CP, Fenoglio CM, Richart RM. The simultaneous expression of human chorionic gonadotropin and carcinoembryonic antigen in the female genital tract. diagn Gynecol Obstet 1981;3:309-314.
168. Tsai D-Z. HCG production in uterine cervical cancers and characterization of ectopic HCG-β. Keio J Med 1985;34:91-104.
169. Campo E, Palacin A, Benasco C, Quesada E, Cardesa A. Human chorionic gonadotropin in colorectal carcinoma. An immunohistochemical study. Cancer 1987;59: 1611-1616.
170. Kuida CA, Braunstein GD, Shintaku P, Said JW. Human chorionic gonadotropin expression in lung, breast, and renal carcinoma. Arch Pathol Lab Med 1987;112: 282-285.
171. Kato Y, Kelley L, Braunstein GD. Beta-core fragment of human chorionic gonadotropin. In: Tomoda Y, Mizutani S, Narita O, Klopper A, eds. Placental and endometrial proteins: basic and clinical aspects. The Netherlands: VNU Science Press BV, 1988:87-90.

172. Fukutani K, Libby JM, Panko WB, Scardino PT. Human chorionic gonadotropin detected in urinary concentrations from patients with malignant tumors of the testis, prostate, bladder, ureter and kidney. J Urol 1983;129:74-77.
173. Talamantes F, Ogren L. The placenta as an endocrine organ: polypeptides. In: Knobil E, Neill J, et al., eds. The physiology of reproduction. New York: Raven Press, 1988:2093-2144.
174. Calvert IS, Weintraub BD, Panton L, Gaffney L, Rosen SW. Eutopic and ectopic macromolecular human placental lactogen. Am J Obstet Gynecol 1985;152:103-111.
175. Braunstein GD, Rasor JL, Engvall E, Wade ME. Interrelationships of human chorionic gonadotropin, human placental lactogen, and pregnancy-specific-β_1-glycoprotein throughout normal human gestation. Am J Obstet Gynecol 1980;138:1205-1213.
176. Samaan N, Yen CC, Friesen H, Pearson OH. Serum placental lactogen levels during pregnancy and in trophoblastic disease. J Clin Endocrinol 1966;26:1303-1308.
177. Weintraub BD, Rosen SW. Ectopic production of human chorionic somatomammotropin by nontrophoblastic cancers. J Clin Endocrinol 1971;32:94-101.
178. Selenkow HA, Saxena BN, Dana CL, Emerson K Jr. Measurement and pathophysiologic significance of human placental lactogen. Proceedingsof an international symposium, the foeto-placental unit. Excerpta Med Int Congr Ser 1968;183:340-362.
179. Braunstein GD, Thompson R, Princler GL, McIntire KR. Trophoblastic proteins as tumor markers in nonseminomatous germ cell tumors. Cancer 1986;57:1842-1845.
180. Szymendera JJ, Zborzil J, Sikorowa L, Kaminska JA, Gadek A. Value of five tumor markers (AFP, CEA, HCG, HPL, and SP$_1$) in diagnosis and staging of testicular germ cell tumors. Oncology 1981;38:222-229.
181. Szymendera JJ, Zborzil J, Sikorowa L, Lenko J, Kaminska JA, Gadek A. Evaluation of five tumor markers (AFP, CEA, HCG, HPL and SP$_1$) in monitoring therapy and follow-up of patients with testicular germ cell tumors. Oncology 1983;40:261-267.
182. Sheth NA, Adil MA, Shinde SR, Sheth AR. Paraendocrine behavior of tumours of the gastrointestinal tract with reference to human placental lactogen. Br J Cancer 1980;42:610-612.
183. Monteiro JCMP, Biswas S, Al-Awgati MA, Greening WP, McKinna JA, Neville AM. Serum levels of human placental lactogen and pregnancy-specific β_1-glycoprotein in breast cancer. Br J Cancer 1982;46:279-282.
184. Eiermann W, Brutting G, Prechtel K. Detection of pregnancy specific proteins β_1-(SP1) glycoprotein and human placental lactogen in benign breast disease. In: Lehmann FG, ed. Carcino-embryonic proteins, Vol II. Elsevier/North Holland Biomedical Press, 1979:477-480.
185. Prechtel K, Eiermann W, Groh M, Brutting G, Hogel B. Ein beitrag zur bewertung des nachweises von schwangerschaftsspezifischen hormonen (human placental lactogen and beta-1-glycoprotein) in brustdrüsenexzidaten bei mastopathie und mammacarcinom. Pathologe 1983;4:12-16.
186. Rosen SW. New placental proteins: chemistry, physiology and clinical use. Placenta 1986;7:575-594.
187. Watanabe S, Chou JY. Human pregnancy-specific β_1-glycoprotein: a new member of the carcinoembryonic antigen gene family. Biochem Biophys Res Commun 1988;152:762-768.
188. Rosen SW, Kaminska J, Calvert IS, Aaronson SA. Human fibroblasts produce "pregnancy-specific" beta-1 glycoprotein in vitro. Am J Obstet Gynecol 1979;134:734-738.

189. Hiekinheimo M, Paasivuo R, Wahlstrom T. Cells from normal brain and gliomas synthesize pregnancy-specific β_1-glycoprotein-like material in vitro. Br J Cancer 1981;43:654-658.
190. Heikinheimo M, Gahmberg CG, Bohn H, Andersson LC. Oncoplacental protein SP_1 —a constitutive and inducible late differentiation marker of the human meylomonocytic lineage. Blood 1987;70:1279-1283.
191. Tatarinov YS. The diagnostic value of circulating trophoblast-specific β_1-glycoprotein (TSG) in cancer patients. Br J Cancer 1980;41:821-824.
192. Lee JN, Salem HT, Al-Ani A-A TM, et al. Circulating concentrations of specific placental proteins (human chorionic gonadotropin, pregnancy-specific beta-1 glycoprotein, and placental protein 5) in untreated gestational trophoblastic tumors. Am J Obstet Gynecol 1981;139:702-704.
193. Tatarinov YS. Trophoblast-specific beta$_1$-glycoprotein as a marker for pregnancy and malignancies. Gynecol Obstet Invest 1978;9:65-97.
194. Tsakok FHM, Koh S, Chua SE, et al. Prognostic significance of the new placental proteins in trophoblastic disease. Br J Obstet Gynecol 1983;90:483-486.
195. Searle F, Leake BA, Bagsahawe KD, Dent J. Serum-SP_1-pregnancy-specific-β-glycoprotein in choriocarcinoma and other neoplastic disease. Lancet 1978;1:579-590.
196. Yoshioka N, Takahashi K, Oshimo T, Watanabe T. Incidence of pregnancy-specific beta-1-glycoprotein in sera of malignant tumor patients. Toh J Exp Med 1977;122:383-385.
197. Würz H. Serum concentrations of SP_1 (prenancy-specific-β_1-glycoprotein) in healthy, nonpregnant individuals, and in patients with nontrophoblastic malignant neoplasms. Arch Gynecol 1979;227:1-6.
198. Engvall E, Yonemoto RH. Is SP_1 (pregnancy specific β_1 glycoprotein) elevated in cancer patients? Int J Cancer 1979;23:759-761.
199. Horne CHW, Pugh-Humphreys RGP, Bremner RD. Practical and theoretical considerations in the detection and measurement of immunoreactive pregnancy-specific β_1-glycoprotein (SP_1) in tumours. In: Lehmann F-G, ed. Carcino-embryonic proteins, Vol I. Amsterdam: Elsevier/North Holland Biomedical Press, 1979:301-311.
200. Rosen SW, Gail MH, Tormey DC. Use of circulating pregnancy-specific β_1 glycoprotein as a marker in carcinoma of the breast of women. J Natl Cancer Inst 1982;69:1067-1071.
201. Sorensen S, Andersen J, Norgaard T. Pregnancy-specific β_1-glycoprotein (SP_1) in serum and tissue from patients with benign and malignant breast tumours. Br J Cancer 1984;49:663-667.
202. Fagnart OC, Cambiaso CL, Lejeune MD, Noel G, Maisin H, Masson PL. Prognostic value of concentration of pregnancy-specific β_1-glycoprotein (SP1) in serum of patients with breast cancer. Br J Cancer 1985;36:541-544.
203. Fagnart OC, Tasiaux N, Masson PL. Pregnancy-specific β_1-glycoprptein in serum in monoclonal gammopathies: relationship with serum β_2-microglobulin, and cellular origin. Clin Chem 1986;32:2150-2154.
204. Marty F, Huber PR, Ludwig C. SP1-really a tumor marker for lymphoid malignancy? Clin Chem 1986;32:2123-2124.
205. Fagnart OC, Mareschal JC, Cambiasco CL, Masson PL. Particle-counting immunoassay (PACIA) of pregnancy-specific β_1-glycoprotein, a possible marker of various malignancies and Crohn's ileitis. Clin Chem 1985;31:397-401.
206. Sorensen S. Pregnancy-"specific" β_1-glycoprotein (SP_1): purification, characterization, quantification and clinical application of malignancies (a review). Tumour Biol 1984;5:275-301.

207. Kuhajda FP, Bohn H, Mendelsohn G. Pregnancy-specific beta-1 glycoprotein (SP-1) in breast carcinoma. Cancer 1984;54:1392-1396.
208. Sorensen S, Trentemoller S. Non-specific serum/plasma response in radioimmunoassay of pregnancy-specific β_1-glycoprotein (SP_1) in healthy, non-pregnant subjects. Clin Chim Acta 1984;140:101-108.
209. Mueller UW, Jones WR. Identification of an SP-1-like protein in non-pregnancy serum: isolation using a monoclonal antibody. J Reprod Immunol 1985;8:111-120.
210. Heikinheimo M, Wahlstrom T, Lehto V-P, Stenman U-H. Evidence for the presence of oncoplacental SP_1-like-protein in normal nonpregnant serum. J Clin Endocrinol Metab 1985;61:188-191.

Index

Acid phosphate, 5, 614
Acute phase reactant proteins (APRPs), 521
  biological effects, 531–532
  increase with tumor burden, 524
  in bladder cancer, 529
  in breast cancer, 530
  in kidney cancer, 527–528
  in lung cancer, 526–527
  in liver cancer, 526
  in lymphomas, 531
  in ovarian cancer, 529–530
  in prostate cancer, 528–529
Acute phase glycoproteins
  concanavalin A (Con A) affinity, 523
  microheterogeneity, 523
Adenocarcinoma, 385, 387, 389
Adenosine deaminase binding protein, 473
Adenosine deaminase, 365
Adoptive immunotherapy, 64
Alkaline phosphatase, 109, 363, 426
Aliesterase, 364
Alpha-1 acid glycoprotein, 40, 43, 522, 525, 527-529
Alpha-1 alkaline phosphatase isoenzymes, 414

Alpha-Fetoprotein (AFP)
  immunoassay of, 84, 89–90, 103,136
  immunotherapy with anti-AFP, 416–418
  in head and neck cancer, 361
  in uterine tumors, 391
  in testicular tumors, 397–398, 677–678
  in hepatocellular carcinoma, 406
  sub-fractions, 410
Alpha-lactalbumin, 225
Amyloidosis, 503, 510
Analytical variance, 36
Anaplastic carcinoma, 437
Autocrine growth factor, 457
Avidin, 131

Banking sera and tissues, 55–66
Bayes' equation and formula, 5
Bence Jones protein, 508, 509
Beta human chorionic gonadotropin, 43, 48, 49, 102, 103, 397–398
Beta protein, 362
Beta-2 microglobulin, 363, 511
Beta-galactosidase, 129
Biological response modifiers (BRM), 58, 179

Biological variance, 36
Biotin, 131
    -labeled DNA probes, 214, 215
Bladder cancer, 485–495
    acute phase reactant proteins, 529
    blood-group related antigens, 488
    cytogenetics, 492
    growth factors, 492
    intermediate filaments, 494
    monoclonal antibodies to, 486
    oncogenes, 492
    P-glycoprotein, 494
    tumor-associated antigens, 485
Blood group antigens, 292
Bone tumors, 423–429
    benign, 423
    Ewing's sarcoma, 424
    markers, 423–429
    metastatic tumors, 428
    multiple myeloma, 423
    osteosarcomas, 425
    parathyroid tumors, 427
Breast cancer, 223–237
    acute phase reactant proteins, 530
    antibreast antibiotics, 224
    antigen catabolism, 233
    antigen metabolism, 233
    antigen release, 233
    immunoassays, 225
    milk fat globule, 226
    monoclonal antibodies to, 69–74, 229–231
    mucin glycoproteins, 641
    multiple markers, 44–46
Burkitt's lymphoma, 256

C-reactive protein (CRP), 40, 521, 522, 526, 528, 531
C1q binding assay, 557
CA 15-3
    breast tumor-associated, 327, 643, 651
    combination testing, 45
    immunoassay of, 230
CA 19-9
    combination testing, 42, 47
    gastrointestinal tumor-associated, 291–293, 641, 649
CA 50
    combination testing, 41–42
    gastrointestinal tumor-associated, 293, 412, 641, 650
CA 72-4, 48, 66, 647, 650
CA 125
    combination testing, 46–47
    immunoassay of, 84
    ovarian tumor-associated, 323, 329, 331, 389, 645, 652
CA 195
    combination testing, 40–41
    gastrointestinal tumor-associated, 292, 641, 649
CA 549
    breast tumor-associated, 643, 650
    combination testing, 45
CA M26, 45, 69
CA M29, 45, 69, 643
Calcitonin, 42–43, 365, 437, 454–455, 461
Carcinoembryonic antigen (CEA)
    combination testing, 37, 39, 43–44
    immunoassay of, 84, 88–90, 98–100
    in bone cancer, 425
    in central nervous system tumors, 318
    in head and neck cancer, 359
    in hepatocellular cancer, 412
    in lung cancer, 461
    in pancreatic tumors, 289
    in thyroid cancer, 440
    in uterine tumors, 388
Casein, 225
Central nervous system (CNS) tumors, 317–320
    circulating antigens, 317
        D2 glycoprotein, 318
        glial fibrillary acidic protein, 318
        myelin basic protein, 318
        neuron specific enolase, 318
        S-100 protein, 318
    circulating antibodies, 319

Index

[Central nervous system tumors]
  immune complexes, 319
  localization studies, 319
  multiple markers, 318–319
Cerebrospinal fluid (CSF) markers, 318, 463
Cervical cancer, 385–389, 391
Chemiluminescence, 107
Cholangiocellular carcinoma, 403
Chromogranin A (CgA), 454, 461
Chromosome changes
  in renal cancer, 476
Chrondrotin sulfate proteoglycan, 304–308
Cirrhosis, 403–405
Cluster analysis, 75, 172–173
Colorectal cancer
  acute phase reactant proteins, 525–526
  flow cytometry, 175–176
  mucin glycoproteins, 647
  multiple markers, 39–41
Combination testing, 9, 78–79, 387, 389, 399, 463
  parallel, 10
  series, 10–11
Cox model, 32
Creatine kinase
  BB isoenzyme, 43–45, 619
  macro CK-2, 40
Cytogenetics, 492

Death rate, 32
Digoxin assay, 163–168
Discriminant analysis, 32, 76–77, 79
DNA hybridization techniques, 206
DNA ploidy, 425
DNA polymerase chain reaction (PCR), 221
Double-determinant assays, 225
Du-Pan-2, 41–42, 292, 641

Ectopic hormones, 454
Efficiency
  in laboratory testing, 3
Electrochemical detection, 107–109
  for IgG, 113–116
Electrochemical interferences, 120
Electrophoresis, 506–507
Endometrial cancer, 385–389, 391
Endometriosis, 389
Enzyme tracers
  for luminescence detection, 95–97
Enzyme-linked immunoassays
  electrochemical detection, 107–123
  in breast cancer, 76–79
  luminescence detection, 95–104
  novel detection systems, 125–140
    competitive binding, 127–134
    noncompetitive binding, 135–140
  strategies for, 83–91
Epidermal growth factor (EGF), 319, 457
Epithelial membrane antigens
  in breast cancer, 225
Epitope mapping, 73
Epstein-Barr virus (EBV), 368
Estrogen receptor protein, 84
Ewing's sarcoma, 424
Experimental design, 13–14, 23

False-positive rate, 5
Ferritin, 47, 136, 318, 361, 412
Fetoacinal protein (FAP), 293
Fiber optic immunosensors, 145–170
  digoxin assay, 163
  distal sensing, 147
  evanescent wave sensing, 147–151
  ferritin assay, 158
Fibronectin (Fn), 537–546
  cellular, 537
  function, 540
  in tumors, 543
  plasma, 537
  plasma levels, 545
  structure, 528
Flow cytometry, 172–182
  drug resistance assessments, 176–177
  in neoplasms
    colorectal, 175
    hematological, 174–175

[Flow cytometry]
  urological, 175
  oncogene and oncogene product detection, 179
  tumor proliferation assessment, 177
Flow injection analysis/electrochemical detection, 120
Fluorescence, 107
Fluorescence polarization immunoassay, 90
Fluorescent probes, 155–157
Follicular carcinoma, 433

Galactosyl transferase isoenzyme II, 622
Galactosyl transferase, 328
Gangliosides
  GD2 and GD3, 309–311
Gastrointestinal cancer
  acute phase reactant proteins, 525–526
  flow cytometry, 175–176
  mucin glycoproteins, 641–643, 647
  multiple markers, 39–41, 41–42
  tumor associated antigens, 289–294
Gene analysis, 205
Germ cell tumors, 397, 677–679, 684, 687
Gestational trophoblastic disease, 675–677, 684, 686
Glial fibrillary acidic protein (DFAP), 318
Glioma-associated membrane antigens, 317
Gliomas, 317
Glycoproteins
  in head and neck tumors, 362
  mucin, 631–656
Glycosyltransferase, 225
Group based multivariate analysis, 31
Group based reference limit, 30
Growth factors, 492
Gynecological neoplasms
  flow cytometry, 174–175
  ovarian, 323–332, 645
  uterine, 385–392

Hazard rate, 32
Head and neck carcinoma
  acute phase reactant proteins, 525
  adjuvant therapy, 358
  diagnostic aid, 357
  enzymes, 363
  hormones, 366
  immune parameters, 366
  lipids, 367
  metabolic byproducts, 366
  multiple markers, 373–374
  neck mass evaluation, 358
  oncofetal antigens, 359
  proteins, 361
  SCC antigen, 370–373
  tumor associated antigens, 370
  viral markers, 368
Heparinized plasma, 58
Hepatic dysfunction, 471
Hepatitis B virus (HBV), 403
Hepatocellular carcinoma (HCC)
  acute phase reactant proteins, 526
  alpha-fetoprotein monitoring, 413
  dietary and environmental factors, 406
  immunotherapy, 415
  incidence and causes, 403
  liver cirrhosis, 405
  tumor markers, 406–415
Herpes simplex virus (HSV), 369, 389
Heterogeneous immunoassays, 83, 95, 107–108
Hodgkin's disease (HD), 243
Homeostatic autoregressive model, 34, 35
Homogeneous immunoassay, 90, 108, 125–140
  competitive binding, 127–134
Hook effects, 86
Horseradish peroxidase, 97
Human chorionic gonadotropin (HCG), 43, 48, 49, 102, 103, 397–398, 673–689

Index

Human milk fat globule (HMFG), 69, 223, 225–226, 229, 328
Human papilloma virus (HPV), 370, 388
Human placental lactogen (HPL), 686–688
Hydroxyproline, 427–428
Hypercalcemia, 470, 471

Immune complexes, 562–564
  circulating, 562–564
  detection of, 555–558
  pathophysiological role, 564–567
  renal manifestations, 558–561
  systemic manifestations, 561–562
  therapeutic trials, 570–572
Immunoassays
  development and evaluation, 69–81
  double determinant, 225
  electrochemical detection, 107–123
  enhanced detection systems, 95–104
  heterogeneous, 83, 95, 107–108
  IMx, automated, 87–91
  latex microparticle capture assay, 86
  luminescence detection, 95–104
  novel homogeneous systems
    apoenzyme reactivation, 130
    associated enzyme sensitive, 139
    biotin enzyme avidin, 131–133
    cloned enzyme donor (CEDIA), 133–134
    enzyme channeling, 131
    enzyme enhancement, 138
    enzyme inhibitory, 136
    enzyme modulator-mediated (EMMIA), 130
    enzyme multiple immunometric technique (EMIT), 127–129
    hybrid antibody, 135
    proximal linkage, 138
  substrate-labeled fluorescein (SLFIA), 129
  sandwich, 225
Immunocytology, 247
Immunoelectrophoresis, 507
Immunofluorescence, 247
Immunoglobulins, 424, 501
  assays, 507
  M-protein diseases, 501
  prognostic factors, 511
  serum electrophoresis, 506
  tissue biopsies, 510
  treatment response, 513
  urine proteins, 508–509
Immunohistochemistry
  lung tumors, 459
  ovarian tumors, 332
  prostatic tumors, 352
Immunolocalization
  central nervous system tumors, 319
  hepatocellular, 416
  ovarian, 331
  prostate, 348–349
  renal cell, 477
  uterine, 386
Immunological monitoring, 65
Immunometric assay, 145–146, 152–155, 158–164
Immunophenotyping
  practical panels, 257–258
  prognostic significance, 256
Immunoprofiles
  beta cell malignancies, 255
  beta-lineage, ALL, 253
  T-cell malignancies, 254
  T-lineage, ALL, 252
Immunoradioscinitigraphy, 473
Immunosensors
  digoxin assay, 163
  evanescent wave, 147
  ferritin assay, 158
Insulin-like growth factor-1, 457
Interferon, 66
Interlaboratory comparisons, 55
Interleukin-2 receptor, 66
Interleukin-1, 470
Intermediate filaments, 494

Intraindividual normal ranges, 51
Isoenzymes, 414, 613–625
    5″ nucleotide phosphodiesterase V, 623
    creatine kinase BB, 619–620
    galactosyltransferase II, 611–623
    in hepatocellular cancer, 414
    lactate dehydrogenase isoenzyme I, 261
    neuron-specific enolase, 620–621
    placental alkaline phosphatase, 617–619
    prostatic acid phosphatase, 614–617
    Regan, 391, 414
    salivary amylase, 621–622
    thymidine kinase I, 623–624
    variant beta hexoseaminidase, 623

Laboratory tests
    in cancer patients, 27
Lactic dehydrogenase, 43–44
Latex microparticle assay, 86
Leukemia (acute nonlymphocytic, ANLL), 265
    antigenic heterogeneity, 277
    asynchronous expression, 278
    genetic probes, 279–280
    mixed lineage, 279
    monoclonal antibodies, 266, 274, 280
    useful antigens, 265–273
Lewis antigens, 488
Linear discriminants, 20
Lipid associated aialic acid, 43, 44, 328
Liver cancer, 403–418, 526
Logistic discrimination, 20
Logistic regression analysis, 32
Luminescence detection, 95
Lung cancer
    acute phase reactant proteins, 526
    chromosome and gene deletions, 456
    growth factors and receptors, 457
    monoclonal antibodies, 458

mucin glycoproteins, 646
multiple markers, 42–44
proto-oncogene amplification, 455
serum and body fluid markers, 460–463
tissue markers associated with, 454–455
types of, 453
Lymphocytic leukemia (ALL) and lymphomas, 243–264
Lymphoid proliferation associated antigens, 178
Lymphokine-activated killer (LAK) cells, 65
Lysozyme, 509

M-protein, 501, 503, 504
Macroglobulinemia, 502–503
Mammary serum antigen (MSA), 45
Markov chain, 33
Medical diagnosis
    statistical methods, 13–22
Medullary thyroid carcinoma (MTC), 442
Medulloblastomas, 317
Melanoma
    antimelanoma immunoconjugates, 298
    GD3 antigen, 309–311
    incidence, 297
    p97 antigen, 299–303
Menigiomas, 317
Milk fatty globule membrane, 69
Monitoring with markers, 22–24
Monoclonal antibodies
    to acute nonlymphocytic leukemia
    MAb CD2, 272, 278
    MAb CD7, 272, 278
    MAb CD10, 272, 278
    MAb CD11b, 270, 277, 278
    MAb CD13, 269, 274–276
    MAb CD14, 270, 276, 278
    MAb CD15, 269–270, 276, 278
    MAb CD33, 269, 274–276
    MAb CD34, 270, 277, 278
    MAb CD56, 272, 279
    MAb HLA-DR, 269, 273, 278

Index

[Monoclonal antibodies]
  to bladder cancer, 486–487
  to breast cancer
    MAb 115D8, 229
    MAb 3E1.2, 230
    MAb AB13, 230
    MAb B72-3, 230
    MAb DF3, 229
    MAb F36/22, 230
    MAb GP-15, 231
    MAb HMFG-1, 229
    MAb HMFG–2, 229
    MAb M26, 70
    MAb M29, 70
    MAb M38, 70
    MAb Mc1, 229
    MAb Mc3, 231
    MAb Mc5, 229
    MAb Mc8, 321
  to kidney cancer, 474–475
  to lung cancer
    MAb 624H12, 460
    MAb 70304, 460
  to lymphocytic leukemias and lymphomas, 244–245
  to melanoma
    MAb 9.2.27, 305
    MAb 48.7, 305
    MAb MG-22, 310
    MAb p97, 300
  to ovarian cancer
    MAb OC 125, 324
    MAb OVTL3, 332
    PAb NB/70K, 324
  to pancreatic cancer
    MAb 19-9, 290–292
    MAb 50, 293
    MAb 195, 292
    MAb Du-Pan-2, 293
  to prostatic cancer
    MAb 7E11-C5, 351
    MAb 83-21, 351
    MAb KR-P8, 350–351
    MAb PD-41, 351
    MAb TURP 27, 351
mRNA detection
  in ANLL, 280
Mucin glycoproteins
  biochemistry, 635–640
  immunology, 632–635
  monoclonal antibodies
    to breast cancer, 69–70, 643
    to colorectal cancer, 647
    to lung cancer, 646
    to ovarian cancer, 645
    to pancreatic cancer, 641–643
  recent developments, 649
  sensitivity and specificity, 647–648
Multiple marker assays
  clinical utility, 39–51
  discrimination rules, 19–22
  in breast cancer, 44–46
  in colorectal cancer, 39–41
  in leukemias and lymphomas, 260
  in lung cancer, 42–44
  in ovarian cancer, 46–48
  in pancreatic cancer, 41–42
  in prostate cancer, 49–50
  in testicular cancer, 48–49
Multiple myeloma, 423
Multiprime assay, 213
Multivariate analysis, 27
  group-based, 31–32
Myelin basic protein (MBP), 318
Myelodysplasias, 174
Myeloma
  prognostic factors, 511
  clinical staging, 513–514

Negative nodes, 28
Negative predictive value, 17–19, 21
Neopterin, 66
Neuroblastoma, 190, 317
Neuroendocrine lung tumor, 454
Neurofilaments, 318
Neuron-specific enolase, 43–44, 318, 365, 442, 454, 461, 620
Nick translation, 212
Non-competitive assay, 84, 108, 126, 135
Non-Hodgkin's lymphomas (NHL), 243
NPGP complex, 227, 229, 231–234

Nucleotide phosphodiesterase V (5'), 623

Oncogene families, 191
  ATP-binding domain, 193
  DNA-binding domain, 194
  kinase domain, 191–193
  hormone-binding domain, 194
Oncogene products, 179–182, 189
Oncogenes
  abl, 190
  c-jun, 456
  c-myb, 456
  c-raf-1, 456
  Her-2/neu, 190
  in bladder cancer, 492
  in lung, 455–456
  in renal, 477
  in thyroid, 442–443
  L-myc, 190
  myc, 456
  N-myc, 190
  P-53, 456
  ras, 456
Oncolipids, 367
Osteocalcin, 427
Osteosarcomas, 425
Ovarian cancer
  acute phase reactant proteins, 529–530
  circulating tumor markers, 323–325
  differential diagnosis, 328–329
  immunohistochemistry, 332
  radioimmunoscintigraphy, 331–332
  residual and recurrent disease, 329–330
  screening, 325–328

P-glycoprotein, 494
Pancreas cancer-associated antigen (PCAA), 42, 289
Pancreas-specific antigen (PaA), 42
Pancreatic cancer
  tumor associated antigens
    CA 19-9, 290–292
    CA 50, 293
    CA 195, 292
    CEA, 289
    Du-Pan-2, 293
    fetoacinar protein (FAP), 293
    pancreatic oncofetal antigen, 289
Parallel testing, 10
Parathyroid adenomas, 427
Pelvic masses, 329
Peptides, 192
  conserved region, 192
  synthetic, 190
Phosphohexose isomerase, 365
Placental alkaline phosphatase, 46–49, 328, 364, 399
Placental proteins
  5 (PP-5), 391
  human chorionic gonadotropin, 673–689
    in germ cell tumors of testis, 677–679
    in gestational trophoblastic disease, 675–677
    in nontrophoblastic neoplasms, 679–684
  human placental lactogen, 684–686
    in germ cell tumors of testis, 684
    in gestational trophoblastic disease, 684
    in nontrophoblastic neoplasms, 684–686
  pregnancy-specific beta-1 glycoproteins, 391, 686–689
    in germ cell tumors of testis, 687
    in gestational trophoblastic disease, 686–687
    in nontrophoblastic tumors, 687
Plasma cell neoplasms, 502
Plasma cell myeloma, 502
Plasmaphoresis, 570
Polyepitopic antigens, 224
  breast epithelial cells, 224
Polymerase chain reaction (PCR), 219
Polystyrene beads, 86

Positive nodes, 28
Positive predictive value, 17–19, 21
Predictive value
  computer programs, 11
  effect of prevalence, 8–9
  in screening, 3
  PAP model, 5–9
Progesterone receptor protein, 84
Proliferating cell nuclear antigen, 178
Prostate specific antigen, 84, 339–347
  combination testing, 49–50
  immunoassay, 340–342
  molecular characteristics, 339
  stability, 342
Prostatic acid phosphatase, 5, 84, 339, 347–349
  biological function, 349
  combination testing, 49–50
  radioimmunodetection, 348
Prostatic cancer
  acute phase reactive proteins, 528–529
  biological function, 344–345
  differential diagnosis, bladder from prostate, 346
  immunohistopathology, 345–346
  monitoring, 343
  monoclonal antibodies
    7E11-C5, 351
    83.21, 351
    KR-P8, 350–351
    PD-41, 351
    Turp-27, 352
  multiple markers, 49–50, 349–350
  prostate-antigen binding globulin, 344
  prostate-specific antigen, 339–347
  radioimmunodetection, 348–349
  staging, 342
Protein A immunoperfusion, 568–572
Proto-oncogenes, 205

Radioimmunoassay of
  CA 15-3, 230, 327, 643, 651
  CA 19-9, 291–293, 641, 649
  CA 50, 293, 412, 641, 650
  CA 72-4, 647, 650
  CA 125, 84, 323, 329, 331, 389, 645, 652
  CA 195, 292, 641, 649
  CA 549, 643, 650
  Du-Pan–2, 292, 641
  PSA, 341
  SCC, 386
Raji cell assay, 557
Receiver operating characteristic curves (ROC), 17–19, 21, 407
Recursive partitioning, 21
Reference values, 27
Regan isoenzyme, 391, 414
Renal cell carcinoma, 468
  acute phase reactant proteins, 527–528
  antigenic determinants, 472–473
  chromosome changes, 476
  hypercalcemia, 471
  immunoradioscinitigraphy, 473–476
  paraneoplastic peptides, 469
  pyrexia and anemia, 470
Restriction enxyme, 207
Restriction fragment-length polymorphisms (RFLPs), 477

Salivary amylase, 621–622
Sandwich assays, 84–85, 225
Screening, 3, 22
Secretor status, 491
Seminoprotein-1 (SP-1), 397
Sensitivity, 3, 17–19, 21
Sensor
  digoxin, 163–168
  ferritin, 153–155, 158–164
Series testing, 11
Serine proteases, 345
Serotonin, 455
Serum amyloid A (SAA), 521–522

Serum bank, 55
  examples of utilization, 64–66
  fiscal aspects, 63–64
  quality control, 62–63
Sialic acid, 43, 44, 427
Sialylated Lewis x antigen (SLXA), 40, 43, 45, 47
Sialytransferase, 225
Southern blot, 209
Specificity, 3, 17–19, 21
Squamous cell carcinoma-associated antigen (SCC), 370
Squamous cell carcinoma, 385–387, 389
Statistical methods, 13–25
  experimental design, 13–14, 23
  for medical diagnosis, 13–22
Steroid hormone receptors, 192, 194, 472
Stochastic process model, 33
Synaptophysin, 454

T and Tn antigens
  anti T and Tn antibodies, 590
  autoimmune responses, 592
  immunoassays, 599–600
  immunoscinitigraphy, 601
  occurrence in tumors, 589
T-cell clonality, 66
Testicular cancer, 397
  clinical value, 399
  markers AFP and HCG, 397–399
Thymodine kinase 1, 623
Thyroglobulin (TG), 432
Thyroid hormone receptor, 191
Thyroid tumors, 431
  calcitonin as markers, 437–440
  CEA as marker, 440–441
  cell-bound markers, 441–442
  DNA analysis, 445

  immune competence, 443–445
  oncogenes, 442
  thyroglobulin as marker, 432–437
Time series, 32
Tissue bank, 55
  examples of utilization, 64–66
  fiscal aspects, 63–64
  quality control, 62–63
Tissue polypeptide antigen (TPA), 43–44, 47, 318, 363, 412
Total variance, 36
Transcription, regulatory genes, 192
Transferrin, 457
Transferrin receptor, 178
Transition probability, 33
Transitional cell carcinoma, 488
Tumor localization, 319
Tumor proliferation, 177
Tumor-infiltrating lymphocytes, 61

Urine proteins, 508
Uterine tumor markers
  CA 125, 389
  CEA, 388
  HCG, 389
  squamous cell carcinoma (SCC) antigen, 386–387, 391–392
  TA-4, 386
  viral antigens, 388
Uterine tumors
  cervical cancer, 385, 386, 387–389, 391
  endometrial cancer, 385, 386–387, 389, 391
  squamous cell carcinoma, 385–387, 389
  squamous cell carcinoma (SCC) antigen, 386–387, 391–392